Statistics in Action

UNDERSTANDING A WORLD OF DATA

Ann E. Watkins
Richard L. Scheaffer
George W. Cobb

 Key Curriculum Press
Innovators in Mathematics Education

Project Editor	Cindy Clements
Editors	Anna Werner, Mary Jo Cittadino
Project Administrator	Michael Hyett
Editorial Assistants	Kristin Burke, Carrye De Mers, Heather Dever, Sigi Nacson, Amy Vidali
Mathematics Reviewer	Mary Parker, Austin Community College, Austin, Texas
AP Teacher Reviewers	Angelo DeMattia, Columbia High School, Maplewood, New Jersey Beth Fox-McManus, formerly of Alan C. Pope High School, Marietta, Georgia Dan Johnson, Silver Creek High School, San Jose, California
Multicultural Reviewers	Gil Cuevas, University of Miami, Coral Gables, Florida Genevieve Lau, Skyline College, San Bruno, California Beatrice Lumpkin, Malcolm X College (retired), Chicago, Illinois
Accuracy Checkers	Dudley Brooks, Monica Johnston
Editorial Production Manager	Deborah Cogan
Production Editor	Jacqueline Gamble
Production Director	Diana Jean Parks
Production Coordinator	Charice Silverman
Text Designer	Kathy Cunningham
Compositor	TSI Graphics
Art Editor	Laura Murray
Illustrator	Wendy Wray
Technical Artist	Matt Perry
Technical Art Consultant	Brett Garrett
Photo Researcher	Margee Robinson
Art and Design Coordinators	Caroline Ayres, Kavitha Becker
Cover Designer	Greg Dundis
Cover Photo Credits	*top:* Strauss-Curtis/Corbis; *center left:* Hulton Deutsch Collection/Corbis; *center right:* Tom Nebbia/Corbis; *bottom:* Bettman/Corbis
Prepress	TSI Graphics
Printer	Von Hoffmann Press
Executive Editor	Casey FitzSimons
Publisher	Steven Rasmussen

Key Curriculum Press
1150 65th Street
Emeryville, CA 94608
editorial@keypress.com
http://www.keypress.com

Printed in the United States of America
10 9 8 7 6 5 4 3 2 1   08 07 06 05 04 03
ISBN: 1-931914-27-3

About the Authors

Ann E. Watkins is Professor of Mathematics at California State University, Northridge. She received her Ph.D. in education from the University of California, Los Angeles. She is a former president of the Mathematical Association of America and a Fellow of the American Statistical Association. Dr. Watkins has served as co-editor of the *College Mathematics Journal,* as a member of the Board of Editors of the *American Mathematical Monthly,* and as Chair of the Advanced Placement Statistics Development Committee. She was selected as the 1994–1995 CSUN Outstanding Professor and won the 1997 CSUN Award for the Advancement of Teaching Effectiveness. Before moving to CSUN in 1990, she taught for the Los Angeles Unified School District and at Pierce College in Los Angeles. In addition to numerous journal articles, she is the co-author of books based on work produced by the Activity-Based Statistics Project (co-authored by Watkins and Scheaffer and now published by Key College Publishing), the Quantitative Literacy Project, and the Core-Plus Mathematics Project.

Richard L. Scheaffer is Professor Emeritus of Statistics at the University of Florida. He received his Ph.D. in statistics from Florida State University. He then joined the faculty of the University of Florida and served as chairman of the Department of Statistics for 12 years. Dr. Scheaffer's research interests are in the areas of sampling and applied probability, especially in their applications to industrial processes. He has published numerous papers and is co-author of four college-level textbooks. In recent years, much of his effort has been directed toward statistics education throughout the school and college curriculum. He was one of the developers of the Quantitative Literacy Project in the United States. As a developer on this project, Dr. Scheaffer helped form the basis of the data analysis emphasis in mathematics curriculum standards recommended by the National Council of Teachers of Mathematics. Dr. Scheaffer also directed the task force that developed the Advanced Placement Statistics Program and served as its first Chief Faculty Consultant. He continues to work on educational projects at the elementary, secondary, and college levels. Dr. Scheaffer is Fellow and past president of the American Statistical Association, from whom he received a Founders Award.

George W. Cobb is the Robert L. Rooke Professor of Statistics at Mt. Holyoke College, where he served a three-year term as Dean of Studies. He received his Ph.D. in statistics from Harvard University. In addition to his fundamental contributions to the emerging science of confectionery ballistics (the statistics of firing gummy bears from a launcher), he is an expert in statistics education with a significant publication record in this field. He chaired the joint committee on undergraduate statistics of the Mathematical Association of America and the American Statistical Association. He also led the STATS project of the Mathematical Association of America, which helped professors of mathematics learn to teach statistics. He is the author of *Introduction to Design and Analysis of Experiments,* published by Key College Publishing. Dr. Cobb served on the National Research Council's Committee on Applied and Theoretical Statistics. Over the past two decades, he has frequently served as an expert witness in lawsuits involving alleged employment discrimination. Dr. Cobb is a Fellow of the American Statistical Association.

Contents

A Note to Students from the Authors

Whether you talk about income, prices of goods and services, sports, health, politics, or the weather, data enter the conversation. In fact, in this age of information technology, data come at you at such a rapid rate that you can catch only a glimpse of the masses of numbers. The only way to cope intelligently with this quantitative world and make informed decisions is to gain an understanding of the basic concepts of statistics and practice what you have learned with real data.

What's in This Book

This book is designed for an introductory statistics course—either an introductory college course or its high school equivalent, Advanced Placement Statistics—and includes all of the standard topics for that course. Beginning in Chapter 1 with a court case about age discrimination, you will be immersed in real problems that can be solved only with statistical methods. You will learn to

- explore, summarize, and display data
- design surveys and experiments
- use probability to understand random behavior
- make inferences about populations by looking at samples from those populations
- make inferences about the effect of treatments from designed experiments

How This Book Is Different

Statistical work is more active than it was a generation ago. Computers and graphing calculators have automated the graphical exploration of data, and in the process have made the practice of statistics a more visual enterprise. Statistical techniques are also changing as simulations allow statisticians (and you) to shift the emphasis from following recipes for calculations to paying more attention to statistical concepts. Your instructor has selected this book for you because he or she

- wants you to learn this modern, data-analytic approach to statistics
- encourages you to be an active participant in the classroom
- wants you to see real data (If you have only pretend data, you can only pretend to analyze it.)
- believes that statistical analyses must be tailored to the data
- uses graphing calculators or statistical software for data analysis and for simulations

Throughout this textbook you will see many graphical displays, lots of real data, activities that introduce each major topic, computer printouts, questions for you to discuss with your class, and practice problems so you can be sure you understand the basics before you move on. These features grow out of the vigorous changes that have been reshaping the practice of statistics and the teaching of statistics over the last quarter century.

The most basic question to ask about any data set is, "Where did the data come from?" Good data for statistical analysis must come from a good plan for data collection. Thus, *Statistics in Action* gives an honest and thorough treatment to the design and analysis of both experiments and surveys.

What You Should Know Before You Start

You will be using this book in an introductory statistics course; thus, you aren't required to know anything yet about statistics. You may find that your perseverance in trying to understand what you read will contribute more to your success in statistics than your skill with algebra. However, basic topics from algebra, such as the equation for a line, slope, exponential equations, and the idea of a logarithm, will come up throughout the book. Be prepared to review those as you go along, if the need arises.

Acknowledgments

This book is a product of what we have learned from the statisticians and teachers who have been actively involved in helping the introductory statistics course evolve into one that emphasizes activity-based learning of statistical concepts while reflecting modern statistical practice. This book is written in the spirit of the recommendations from the MAA's STATS project and Focus Group on Statistics, the ASA's Quantitative Literacy projects, and the College Board's AP Statistics course. We hope that it adequately reflects the wisdom and experience of those with whom we have worked and who have inspired and taught us.

It has been an awesome experience to work with the Key Curriculum staff and field-test teachers, who always put the interests of students and teachers first. Their commitment to excellence has motivated us to do better than we ever could have done on our own. Steve, Casey, Mary Jo, Anna, Dudley, Bill, and the rest of the staff have been professional and astute throughout. Our deepest gratitude goes to Cindy Clements, our editor, who has been a joy to work with. (Not all authors say that— and mean it—about their editors.) Cindy was an outstanding statistics and calculus teacher before coming to Key. She brings an extraordinary intellectual curiosity and talent for teaching and for statistics to her current position. Her organizational skills, experience in the classroom, and insight have improved every chapter of this text.

Ann Watkins
Dick Scheaffer
George Cobb

Statistics
in Action

UNDERSTANDING A WORLD OF DATA

CHAPTER 1

A CASE STUDY
OF STATISTICS IN ACTION

Were older workers discriminated against during a company's downsizing? When an older worker felt he was unfairly laid off, his lawyers called on a statistician to help evaluate the claim.

Robert Martin turned 55 in 1991. Earlier in that same year, the Westvaco Corporation, which makes paper products, decided to downsize. They laid off several members of their engineering department, where Robert Martin worked, and he was one of those who lost their jobs. Later that year, he sued Westvaco, claiming he had been laid off because of his age. A major piece of Martin's case was based on a statistical analysis of the ages of the employees at Westvaco.

In the two sections of this chapter, you will get a chance to try your hand at two very different kinds of statistical work, called exploration and inference. **Exploration** is an informal and open-ended examination of data for patterns. Your goal will be to uncover and summarize patterns in data from Westvaco that bear on the *Martin* case. You will try to formulate and answer basic questions like, "Were those who were laid off older on average than those who weren't laid off?" You can use any tools—graphs, averages, and so on—that you think might be useful. **Inference,** which you'll get to in the second section, is quite different from exploration in that it follows strict rules, and its focus is to judge whether the patterns you found are the sort you would expect. You'll use inference to decide whether the patterns you find in the Westvaco data are the sort you would expect from a company that does not discriminate on the basis of age or whether further investigation is needed into possible age discrimination.

The purpose of this first chapter is to give you a head start with the ideas of statistical thinking, before you get involved with the details of learning the methods. It is very easy to get caught in the trap of doing rather than understanding, of asking how rather than why. You can't *do* statistics unless you learn the methods, but you must not get so caught up in the details of the methods that you lose sight of what they mean. Doing and thinking, method and meaning, will compete for your attention throughout the course.

In this chapter you will learn the basic ideas of

- exploring data—uncovering and summarizing patterns
- making inferences from data—deciding whether or not an observed feature of the data could reasonably be attributed to chance

These will remain key components of statistical ideas that you'll develop and study throughout this book.

1.1 Discrimination in the Workplace: Data Exploration

At the beginning of 1991, Robert Martin was one of 50 people working in the engineering department of Westvaco's envelope division. That spring, Westvaco's management went through five rounds of planning for a reduction in their workforce. In Round 1, they decided to eliminate 11 positions, and in Round 2, 9 more. By the time the layoffs ended, after all five rounds, only 22 of the 50 workers had kept their jobs, and the average age in the department had fallen from 48 to 46.

Display 1.1 shows the data provided by Westvaco to Martin's lawyers.[1] Each row corresponds to one worker, and each column corresponds to a characteristic of the workers: job title, whether hourly or salaried, the date of birth, the date of hire, and age as of the first of January 1991 (shortly before the layoffs). The next-to-last column (RIF) tells how the worker fared in the downsizing: a 1 means chosen for layoff in Round 1 of planning for the reduction in force; a 2 means chosen in Round 2, and so on for Round 3, 4, or 5; and a 0 means "not chosen for layoff."

Variables provide information about cases.

The subjects (or objects) of statistical examination are often called **cases.** Here, in the rows, the cases are individual Westvaco employees. Their characteristics, in the columns, are **variables.** If you pick a row and read across, you get information about a single case. (For example, Robert Martin, in Row 44, was salaried, was born in September 1937, was hired in October 1967, was chosen for layoff in Round 2, and was 54 on January 1, 1991.) Although reading across may seem the natural way to read the table, in statistics you will often find it useful to pick a column and read down. This gives you information about a single variable as you range through all the cases. For example, pick *Age,* read down the column, and notice the *variability* in the ages. It is variability like this—the fact that individuals differ—that can make it a challenge to see patterns in data and to figure out what they mean.

Variability is what statistics is all about.

Just imagine: If there had been no variability—if all the workers had been of just two ages, say, 30 and 50, and Westvaco had laid off all the 50-year-olds and kept all the 30-year-olds—the conclusion would be obvious and there'd be no need for statistics. But real life is more subtle than that. The ages of the laid-off workers varied, as did the ages of the workers retained. *Statistical methods were designed to cope with such variability.* In fact, you might define statistics as the science of learning from data in the presence of variability.

A distribution is a record of variability.

Although the bare fact that the ages vary is easy to see in the data table, the pattern of the ages is not so easy to see in a column of numbers. This pattern—what the values are and how often each occurs—is called their **distribution.** In order to see that pattern, a graph is better. The **dot plot** in Display 1.2 shows the distribution of the ages of the 14 hourly employees who worked in the engineering department just before the layoffs began.

---

[1]The statistical analysis in the lawsuit used all 50 employees in the engineering department of the envelope division, with separate analyses for exempt (salaried) and nonexempt (hourly) workers.

Row	Job Title	Pay	Birth Mo	Birth Yr	Hire Mo	Hire Yr	RIF	Age 1/1/91
1	Engineering Clerk	H	9	66	7	89	0	25
2	Engineering Tech II	H	4	53	8	78	0	38
3	Engineering Tech II	H	10	35	7	65	0	56
4	Secretary to Engin Manag	H	2	43	9	66	0	48
5	Engineering Tech II	H	8	38	9	74	1	53
6	Engineering Tech II	H	8	36	3	60	1	55
7	Engineering Tech II	H	1	32	2	63	1	59
8	Parts Crib Attendant	H	11	69	10	89	1	22
9	Engineering Tech II	H	5	36	4	77	2	55
10	Engineering Tech II	H	8	27	12	51	2	64
11	Technical Secretary	H	5	36	11	73	2	55
12	Engineering Tech II	H	2	36	4	62	3	55
13	Engineering Tech II	H	9	58	11	76	4	33
14	Engineering Tech II	H	7	56	5	77	4	35
15	Customer Serv Engineer	S	4	30	9	66	0	61
16	Customer Serv Engr Assoc	S	2	62	5	88	0	29
17	Design Engineer	S	12	43	9	67	0	48
18	Design Engineer	S	3	37	6	74	0	54
19	Design Engineer	S	3	36	2	78	0	55
20	Design Engineer	S	1	31	3	67	0	60
21	Engineering Assistant	S	6	60	7	86	0	31
22	Engineering Associate	S	2	57	4	85	0	34
23	Engineering Manager	S	2	32	11	63	0	59
24	Machine Designer	S	9	59	3	90	0	32
25	Packaging Engineer	S	3	38	11	83	0	53
26	Prod Spec—Printing	S	12	44	11	74	0	47
27	Proj Eng—Elec	S	9	43	4	71	0	48
28	Project Engineer	S	7	49	9	73	0	42
29	Project Engineer	S	8	43	4	64	0	48
30	Project Engineer	S	6	34	8	81	0	57
31	Supv Engineering Serv	S	4	54	6	72	0	37
32	Supv Machine Shop	S	11	37	3	64	0	54
33	Chemist	S	8	22	4	54	1	69
34	Design Engineer	S	9	38	12	87	1	53
35	Engineering Associate	S	2	61	9	85	1	30
36	Machine Designer	S	2	39	4	85	1	52
37	Machine Parts Cont—Supv	S	10	28	8	53	1	63
38	Prod Specialist	S	9	27	10	43	1	64
39	Project Engineer	S	7	25	9	59	1	66
40	Chemist	S	12	30	10	52	2	61
41	Design Engineer	S	4	60	5	89	2	31
42	Electrical Engineer	S	11	49	3	86	2	42
43	Machine Designer	S	3	35	12	68	2	56
44	Machine Parts Cont Coor	S	9	37	10	67	2	54
45	VH Prod Specialist	S	5	35	9	55	2	56
46	Printing Coordinator	S	2	41	1	62	3	50
47	Prod Dev Engineer	S	6	59	11	85	3	32
48	Prod Specialist	S	7	32	1	55	4	59
49	VH Prod Specialist	S	3	42	4	62	4	49
50	Engineering Associate	S	8	68	5	89	5	23

Display 1.1 The data in *Martin v. Westvaco.*

Source: *Martin v. Envelope Division of Westvaco Corp.*, CA No. 92-03121-MAP, 850 Fed. Supp. 83 (1994).

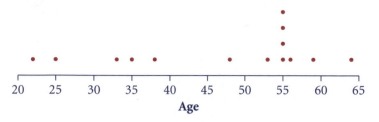

Display 1.2 Ages of the hourly workers. (Each dot is a worker; the ages are shown by the position of the dots along the scale below them.)

Display 1.2 provides some useful information about the variability in the ages but by itself doesn't tell anything about possible age discrimination in the layoffs. For that, you need something like Display 1.3 to distinguish between those hourly workers who lost their jobs and those who didn't.

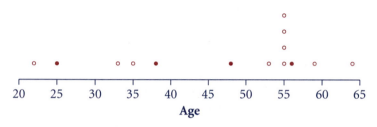

Display 1.3 Hourly workers: Ages of those laid off (open circles) and those retained (solid dots).

Discussion: Exploring the *Martin v. Westvaco* Data

D1. Suppose you were on a jury in the *Martin v. Westvaco* case. How would you use the information in Display 1.1 to decide if Westvaco tended to lay off older workers (for whatever reason)?

D2. Does the dot plot in Display 1.3 show a clear-cut case of age discrimination in layoffs of hourly workers at Westvaco, a possible case of age discrimination, or no discrimination?

D3. The dot plots in Display 1.4 show the ages of the hourly workers laid off and retained by round. For example, in the top dot plot, the open circles show the ages of the hourly workers laid off in Round 1 and the solid dots show the ages of the hourly workers whose jobs survived Round 1. (There is no plot for Round 5 because no hourly workers were chosen for layoff in that round.) Compare the round-by-round information you get in Display 1.4 with the summary for all rounds in Display 1.3. Which display provides stronger support for Martin's claim that Westvaco discriminated against older workers?

D4. The dot plot in Display 1.5 is like Display 1.3 except that it gives data for the salaried workers. Compare the plots for the hourly and salaried workers. Which provides stronger evidence in support of a claim of age discrimination?

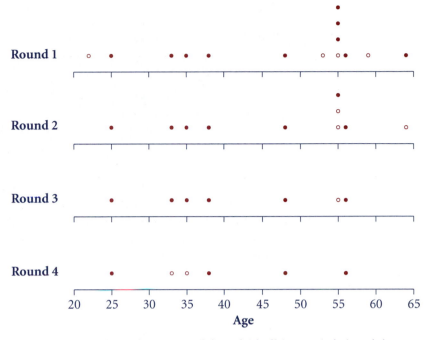

Display 1.4 Hourly workers: Ages of those laid off (open circles) and those retained (solid dots) in each round.

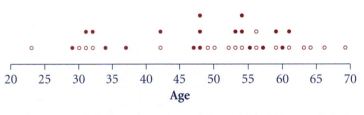

Display 1.5 Salaried workers: Ages of those laid off (open circles) and those retained (solid dots).

D5. Display 1.6 is the counterpart of Display 1.4. It shows, for the salaried workers, the ages of those laid off and those retained in each of the five rounds. Compare the pattern for the salaried workers with the pattern for the hourly workers in Display 1.4.

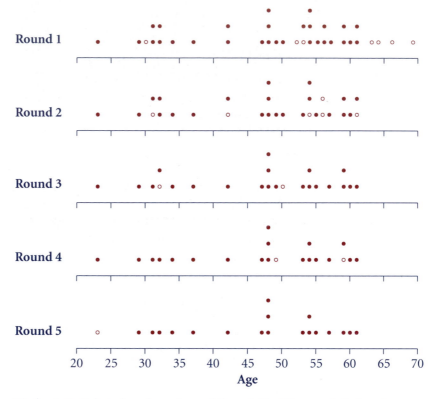

Display 1.6 Salaried workers: Ages of those laid off (open circles) and those retained (solid dots) in each round.

D6. The summary table shown here classifies salaried workers using two yes/no questions: Under 40? and Laid off? (In employment law, 40 is a special age because only those 40 or older belong to what is called the "protected class," the group covered by the law against age discrimination.)

		Laid Off?			
		Yes	No	Total	% Yes
Under 40?	Yes	4	5	9	44.4
	No	14	13	27	51.9
	Total	18	18	36	50.0

a. Does the pattern in this table support Martin's claim of age discrimination? Why or why not?

b. Construct a similar table for salaried workers, but this time use 50 instead of 40 to divide the ages. (Your two age groups will be those under 50 and those 50 or older.) Does the evidence in this new table provide stronger or weaker support for Martin's case? Explain.

c. How do you account for the different messages from the two tables? Both provide evidence; how do you judge the evidence from the two tables taken together?

D7. Whenever you think you have a message from data, you should be careful not to jump to conclusions. The patterns in the Westvaco data might be "real"—they reflect age discrimination on the part of management. On the other hand, the patterns might be the result of chance—management wasn't discriminating on the basis of age but simply by chance happened to lay off a larger percentage of older workers. What's your opinion about the Westvaco data: Do the patterns seem "real"—too strong to be explained by chance?

D8. You may feel as if the analysis so far ignores important facts like worker qualifications. That's true. However, the first step is to decide if, based on the data in Display 1.1, older workers were more likely to be laid off. If not, Martin's case fails. If so, it is then up to Westvaco to justify its actions. List several specific considerations that might justify the fact that older workers were more likely to be laid off.

■ Practice

P1. Construct a dot plot, similar to the one in Display 1.3, comparing the ages of hourly workers who lost their jobs at some point during the first three rounds to the ages of hourly workers who still had their jobs at the end of Round 3. How do the ages differ?

P2. In D6, you looked at a pair of summary tables for salaried workers.

a. Construct similar tables for the hourly workers. Which of your tables (using age 40 or age 50) provides stronger evidence of age discrimination?

b. In what ways is the pattern in these tables similar to the patterns in D6 for salaried workers?

Summary 1.1: Data Exploration

Data exploration, or exploratory analysis, is a purposeful investigation to find patterns in data, using such tools as tables and graphs to display those patterns, and statistical concepts such as distributions and averages to summarize them.

- Table displays, with cases in rows and variables in columns, help you look at how variables differ from case to case.

- The distribution of a variable tells you the set of values that the variable takes on, together with how often each value occurs.

- Dot plots, which show the values of a variable along a number line, provide you with a visual display of the distribution of a variable.

- Plots of distributions give you a sense of how large or small the values are, which values occur most often, how spread out the values are, and whether there are any values that appear to be unusually large or small.

Remember that statistics involves coping with variability, but you have to understand that variability before you can use it intelligently to draw conclusions. All the features of data exploration that you have investigated here will be important when you move on to the inference phase of a statistical investigation.

■ Exercises

E1. Explore whether hourly workers at Westvaco were more likely than salaried workers to lose their jobs. Start by constructing a table to summarize the relevant data. What do you conclude?

E2. Twenty-two workers kept their jobs. Explore whether the age distributions are similar for the hourly and salaried workers who kept their jobs.

  a. Show the two age distributions in a pair of dot plots on the same scale. How do these distributions differ?

  b. Do the dot plots support a claim that Westvaco was more inclined to keep older workers if they were salaried rather than hourly?

E3. It may seem natural to think that cases will always be individuals, but it is also possible to have cases that are groups of individuals.

  a. Use Display 1.1 to create a data table whose five cases are Round 1 through Round 5 and whose three variables are the *total number* of employees laid off in that round, the *number* of employees laid off in that round who were 40 or older, and the *percentage* laid off in that round who were 40 or older.

  b. Describe any patterns you find in the table and what you think they might mean.

E4. "Last hired, first fired" is shorthand for "When you have to downsize, start by laying off the newest person, then the person hired next before that, and work back in reverse order of seniority." (The person who's been there longest will be the last to be laid off.) Examine the Westvaco data: How was seniority related to the decisions about layoffs in the engineering department at Westvaco? What explanation(s) can you suggest for the patterns you find?

E5. Many tables in the newspaper and elsewhere are arranged with cases as rows and variables as columns. Pick one of these displays and tell what the cases and variables are.

  - A table of major league baseball standings

  - A list of the day's activity on the New York Stock Exchange

  - The nutritional summary on a food package

1.2   Discrimination in the Workplace: Inference

Overall, the exploratory work on the Westvaco data set in Section 1.1 shows that older workers were more likely than younger ones to be laid off, and they were laid off earlier. One of the main arguments in the court case, along the lines set out in D7, was about what those patterns mean:

- Can we infer from them that Westvaco has some explaining to do?

- Or are the patterns of the sort that might happen even if there was no discrimination?

A comprehensive analysis of *Martin v. Westvaco* will have to wait for its reappearance among the case studies of Chapter 12, when you'll know more of the concepts and tools of statistics. For now, though, you can get a pretty good idea of how the analysis goes by working with only a subset of the data.

The ages of the ten hourly workers involved in Round 2 of the layoffs, arranged from youngest to oldest, were 25, 33, 35, 38, 48, 55, 55, 55, 56, and 64. The three who were laid off were age 55, 55, and 64. Display 1.7 shows the same data in a dot plot.

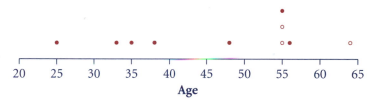

Display 1.7   Hourly workers: Ages of those laid off (open circles) and those retained (solid dots) in Round 2.

Use a summary statistic to "condense" the data.

To simplify the statistical analysis to come, it will help to "condense" the data into a single number, called a **summary statistic.** One possible summary statistic is the average, or mean, age of the three who lost their jobs:

$$\frac{55 + 55 + 64}{3} = 58 \text{ years}$$

What to make of the data requires balancing two points of view. On one hand, the pattern in the data is pretty striking. Of the five people under age 50, all kept their jobs. Of the five who were 55 or older, only two kept their jobs. On the other hand, the number of people involved is pretty small: just three out of ten. Should you take seriously a pattern involving so few people? Listen to two imaginary people taking sides in an argument that was at the center of the statistical part of the *Martin* case.

Martin: Look at the pattern in the data. All three of the workers laid off were much older than the average age of all workers. That's evidence of age discrimination.

Westvaco: Not so fast! You're looking at only ten people total, and only three positions were eliminated. Just one small change and the picture would be entirely different. For example, suppose it had been the 25-year-old instead of the 64-year-old who was laid off. Switch the 25 and the 64 and you get a totally different set of averages:

Actual data: 25   33   35   38   48   **55**   **55**   55   56   **64**

Altered data: **25**   33   35   38   48   **55**   **55**   55   56   64

See! Just one small change and the average age of the three who were laid off is *lower* than the average age of the others.

	Average Age	
	Laid Off	Retained
Actual data	58.0	41.4
Altered data	45.0	47.0

Martin: Not so fast, yourself! Of all the possible changes, you picked the one that is most favorable to your side. If you'd switched one of the 55-year-olds who got laid off with the 55-year-old who kept his or her job, the averages wouldn't change at all. Why not compare what actually happened with *all* the possibilities that might have happened?

Westvaco: What do you mean?

Martin: Start with the ten workers, treat them all alike, and pick three at random. Do this over and over, to see what typically happens, and compare the actual data with these results. Then we'll find out how likely it is that their average age would be 58 or more.

Discussion: Picking Workers at Random

The dialog describes one age-neutral method for choosing which workers to lay off: Pick the three completely at random, with all sets of three having the same chance to be chosen.

D9. If you pick three of the ten ages at random, do you think you are likely to get an average age of 58 or more?

D10. If the probability of getting an average age of 58 or more turns out to be small, does this favor Martin or Westvaco?

Activity 1.1 shows you how to estimate the probability that if you choose three workers at random, just by chance you will get an average age of 58 years or more. To do that, you use **simulation,** a procedure in which you set up a model of a chance process (drawing three ages out of a box) that copies—or simulates— a real situation (selecting three employees at random to lay off).

Simulation requires a chance model.

Activity 1.1 By Chance or by Design?

What you'll need: paper or 3 × 5 cards, a box or other container

Let's test the process suggested by Martin's advocate.

1. *Create a model of a chance process.* Write each of the ten ages on identical pieces of paper or 3 × 5 cards, and put the ten cards in a box. Mix them thoroughly, draw out three (the ones to be laid off), and record the ages.

2. *Compute a summary statistic.* Compute the average of the three numbers in your sample to one decimal place.

3. *Repeat the process.* Repeat steps 1 and 2 nine more times.

4. *Display the distribution.* Pool your results with the rest of your class and display the summary statistics in a dot plot.

5. *Estimate the probability.* Calculate the number of times your class got an average age of 58 years or more. Estimate the probability that simply by chance the average age of those chosen would be 58 years or more.

6. *Interpret your results.* What do you conclude from the magnitude of your class's probability estimate?

Your simulation was completely age-neutral: All sets of three workers had exactly the same chance of being selected for layoff, regardless of age. The simulation tells you what sorts of results are reasonable to expect from that sort of age-blind process.

Here are the first 4 of 1000 repetitions from such a simulation. (The ages in red are those selected for layoff.) The average ages of these groups—42.7, 48.0, 42.7, and 37.0—are highlighted by the blue dots in the distribution in Display 1.8.

										Average age
25	33	35	38	**48**	55	55	**55**	56	64	42.7
25	**33**	35	38	48	55	55	**55**	**56**	64	48.0
25	33	**35**	**38**	48	55	55	**55**	56	64	42.7
25	33	35	**38**	**48**	55	55	55	56	64	37.0

Mean Age

Display 1.8   Results of 1000 repetitions: The distribution of the average age of the three chosen for layoff by chance alone. (Each dot represents 5 points.)

A distribution records variability in a chance process.

Display 1.8 is a dot plot that shows the distribution of average ages for 1000 repetitions of the process you used in Activity 1.1. This is the second important use of distributions to show variability: In the last section, you used distributions to show the set of values of one variable, *age*. Here, you have a distribution that shows the variability in a process, as you go from one repetition to the next.

The simulation tells what kind of data to expect if workers are selected at random for layoff.

Out of 1000 repetitions, only 46, or almost 5%, gave an average age of 58 or older. So it is not at all likely that just by chance you'd pick workers as old as the three Westvaco picked. Did the company discriminate? There's no way to tell from the numbers alone. However, if your simulations had told you that an average of 58 or larger is easy to get by chance alone, then the data would provide no evidence of discrimination. If, on the other hand, it turns out to be very unlikely to get a value this large simply by chance, as in this case, statistical logic says to conclude that the pattern is more than simple coincidence. It is then up to the company to explain why their decision-making process led to such a large average age for those laid off.

Here we take a closer look at the logic.

It might help your understanding of how this logic applies to *Martin v. Westvaco* if we imagine a realistic argument between the advocates for each side.

Martin: Look at the pattern in the data: All three of the workers laid off were much older than average.

Westvaco: So what? You could get a result like that just by chance. If chance alone can account for the pattern, there's no reason to look for any other explanation.

Martin: Of course you *could* get this result by chance. The question is whether it's easy or hard to do. If it's easy to get an average as large as 58 by drawing at random, I'll agree that we can't rule out chance as one possible explanation. But if an average that large is really hard to get from random draws, we agree that chance alone can't account for the pattern. Right?

Westvaco: Right.

Martin: Here are the results of my simulations. If you look at the three hourly workers laid off in Round 2, the probability of getting an average age of 58 or more by chance alone is only 5%. And if you do the same computations for the entire engineering department, the probability is a lot less, about 1%. What do you say to that?

Westvaco: Well . . . I'll agree that it's really hard to get patterns that extreme simply by chance, but that by itself still doesn't *prove* discrimination.

Martin: No, but I think it leaves you with some explaining to do!

In the actual case, Martin and Westvaco reached a settlement out of court before the case went to trial.

The logic you've just seen is basic to all statistical inference. But it's not easy to understand. In fact, it took mathematicians centuries to come up with it! It wasn't until the 1920s that a brilliant British biological scientist and mathematician, R. A. Fisher, realized that results of agricultural experiments should be analyzed carefully to see if observed differences could be attributed to chance alone or to treatment. Calculus, in contrast, was first understood in 1665! Precisely because it *is* so important, you will see the logic of using randomization as a basis for statistical inference over and over again throughout this book. You'll have lots of time to practice with it.

Key Steps in a Simulation

Once again, a simulation goes like this:

1. *Model.* Set up a model of a process where chance is the only cause of being selected.

   In Activity 1.1 related to the *Martin v. Westvaco* case, the model for an age-neutral chance process was to draw three numbers *at random* from the set of ten ages. You put the ages of the workers who could be laid off in a box and selected three by random draw.

2. *Repetition.* Gather data by repeating the process.

   In the activity, you repeated the process of drawing ages many times.

3. *Distribution.* Display the distribution of outcomes using a summary statistic, and determine how likely the actual result would be.

   In the activity, you used the average age to summarize the results, although other summaries also could be used. For each repetition, you computed the average age of those laid off and added that average to your dot plot. Repeated simulation showed that the chance of an average age of 58 or older was only about .05.

4. *Conclusion.* If the probability is small (and the definition of "small" will vary depending on the situation), conclude that chance may not be the only cause. If the probability isn't small, conclude that you can reasonably attribute the result to chance alone.

   In the *Martin v. Westvaco* case, it turned out that the probability was small enough (1 chance out of 20) that Westvaco had some explaining to do. However, it wasn't small enough to serve as evidence of discrimination in a court case (which requires a probability of .025 or smaller).

Discussion: The Logic of Inference

D11. Why must we estimate the probability of getting an average age of 58 *or more* rather than the probability of getting an average age of 58?

D12. **How unlikely is "too unlikely"?** The probability you estimated in Activity 1.1 is in fact exactly equal to .05. In a typical court case, a probability of .025 or less is required to serve as evidence of discrimination.

   a. Did the Round 2 layoffs of hourly workers in the *Martin* case meet the court requirement?

   b. What if the probability in the *Martin* case had been .01 instead of .05? Or .10 instead of .05? How would that have changed your conclusions?

D13. **A trustworthy friend?** A friend wants to bet with you on the outcome of a coin toss. The coin looks fair, but you decide to do a little checking. You flip the coin and it lands heads. You flip again: also heads. A third flip: heads. Flip: heads. Flip: heads. You continue to flip, and the coin lands heads 19 times in 20 tosses.

   a. Explain why the evidence—19 heads in 20 tosses—makes it hard to believe the coin is fair.

   b. Design and carry out a simulation to estimate how unusual this result would be if the coin were fair.

■ Practice

P3. Suppose that three workers were laid off from a set of ten whose ages were the same as in the *Martin* case. However, this time, the ages of those laid off were 48, 55, and 55.

25   33   35   38   **48**   55   **55**   **55**   56   64

   a. Use the dot plot in Display 1.8 to estimate the probability of getting an average age as large as that of those laid off here.

   b. What would your conclusion be if Westvaco had laid off workers of these three ages?

P4. At the end of Round 3, only six hourly workers were left. Their ages were 25, 33, 35, 38, 48, and 56. In Round 4, the 33- and 35-year-olds were chosen for layoff. Think about how you would repeat Activity 1.1 using the data from Round 4.

   a. What is the average age of the workers actually laid off?

   b. Describe a simulation for finding the distribution of the average age of two workers laid off at random. Use your calculator or a statistical program to do the simulation. Repeat your simulation 20 times.

   c. What is your estimate of the probability of getting an average age of 34 or more if two workers are picked at random for layoff in Round 4? Does this one part of the data (Round 4, hourly) provide evidence in Martin's favor?

Summary 1.2: Statistical Inference

Inference is a statistical procedure that involves deciding whether an event can reasonably be attributed to chance or whether you should look for—and perhaps investigate—some other explanation. In the *Martin* case, you used inference to determine whether the relatively high average age of the laid-off employees could reasonably be due to chance.

Simulation is a useful device for inference.

- First you set up a *model* of a process in which chance is the only factor influencing the outcomes.

- The next stage is *repetition*—you gather data by repeating the process in order to determine the likelihood of different outcomes.

- Then you plot the *distribution* of outcomes, using a summary statistic in order to determine how likely the actual result would be.

- Finally, you reach—or infer—a *conclusion,* evaluating the likelihood of getting your actual data in light of the chance-generated data.

If the probability of getting your actual data is small, conclude that chance may not be the only cause for getting the data. If the probability isn't small, conclude that you can reasonably attribute the result to chance alone. In the *Martin* case, the probability was about .05, which was considered small enough to warrant asking for an explanation from Westvaco but not small enough to present in court as clear evidence of discrimination.

■ Exercises

E6. Revisit the idea of the simulation in Activity 1.1, this time for all 14 hourly workers and using a different summary statistic. Because the age class protected by law is those 40 or older, use as your summary statistic the number of hourly workers laid off who were 40 or older. The ages shown here are of the hourly workers, with those laid off in red. Note that, out of 10 hourly workers laid off by Westvaco, 7 were in the protected class.

22 25 **33 35** 38 48 **53**

55 55 55 55 56 **59 64**

a. Write the 14 ages on 14 slips of paper and draw 10 at random for layoff. How many of the 10 were in the protected class of 40 or older? Repeat your simulation a total of 20 times using your calculator or statistical software, and make a dot plot of the number who were laid off in each repetition.

b. What is your estimate of the probability that, just by chance, seven or more of the ten hourly workers who were laid off would be in the protected class?

c. Do you conclude that seven out of ten could reasonably be due to chance alone, or should Westvaco be asked for an explanation?

d. Discuss the advantages and disadvantages of using actual ages rather than just the information about whether a person is 40 or older.

E7. Ten hourly workers were left after Round 1. Their ages were

25 33 35 38 48 **55 55 55** 56 **64**

The ages of the four workers laid off in Rounds 2 and 3 are in red type. They have an average age of 57.25. In the questions, consider the combined layoffs in Rounds 2 and 3.

a. Describe how to simulate the chance of getting an average age of 57.25 or more using the methods of Activity 1.1 and your calculator or statistical program.

b. Repeat your simulation a total of ten times and make a dot plot of the average age of the four hourly workers laid off in Rounds 2 and 3. What is your conclusion?

E8.   ***Which summary statistic?*** Activity 1.1 asked you to use the average age to summarize the set of three ages of the workers chosen for layoff. Here are some other possible choices.

- Sum of the ages of the three who were laid off

- Average age of those laid off minus average age of those retained

- Number of employees 55 or older who were laid off

- Age of the youngest worker who was laid off

- Age of the oldest worker who was laid off

- Middle (median) of the ages of the three who were laid off

a. Are any of these possible summary statistics equivalent to the average age?

b. Which of these possible summary statistics would it be reasonable to use?

E9.   ***Snow in July?*** You have spent some time in Oz. You think the date is July 4 back in Kansas, but you can't be sure because days may not have the same length in Oz as on Earth. A friendly tornado puts you and your dog Toto down in Kansas. However, you see snow in the air (data). Which of the following is the inference you should make?

- If this is Kansas, it is very unlikely to be snowing on July 4. Therefore, this probably isn't Kansas.

- If it is July 4, it is very unlikely to be snowing in Kansas. Therefore, this probably isn't July 4.

- If it is snowing in Kansas on July 4, it is time to go back to Oz.

- If this is Kansas and it is July 4, it probably isn't really snowing.

E10. For some situations, instead of using simulation, it is possible to find exact probabilities by counting equally likely outcomes. Suppose only two out of the 10 hourly workers had been laid off in Round 2 and that those two workers were 55 and 64, with an average age of 59.5 years. It is straightforward, though tedious, to list all possible pairs of workers who might have been chosen. Here's the beginning of a systematic listing. The first nine outcomes all include the 25-year-old and one other. The next eight outcomes all include the 33-year-old and one other but not the 25-year-old because the pair {25, 33} was already counted.

Count	Pair Chosen (red = laid off)	Average Age
1	**25 33** 35 38 48 55 55 55 56 64	29.0
2	**25** 33 **35** 38 48 55 55 55 56 64	30.0
3	**25** 33 35 **38** 48 55 55 55 56 64	31.5
·		
·		
·		
9	**25** 33 35 38 48 55 55 55 56 **64**	44.5
10	25 **33 35** 38 48 55 55 55 56 64	34.0
11	25 **33** 35 **38** 48 55 55 55 56 64	35.5
. . .		

a. How many possible pairs are there? (Don't list them all!)

b. How many pairs give an average age of 59.5 years or older? (Do list them.)

c. If the pair is chosen completely at random, then all possibilities are equally likely, and the probability of getting an average age of 59.5 or older equals the number of possibilities with an average of 59.5 or more divided by the total number of possibilities. What is the probability for this situation?

d. Is the evidence of age discrimination strong or weak?

Statistics in Action: What Next?

In this chapter, you have explored the data from the *Martin* case, looking for evidence you consider relevant. After that, you saw how to use statistical reasoning to test the strength of the evidence: Are the patterns in the data solid enough to support a conclusion of age discrimination, or are they the sort that you would expect to occur even if there was no discrimination?

The next two chapters deal with data exploration. Chapter 2 looks at distributions, much as you did in Section 1.1. What graphs and summaries are useful for finding and showing patterns? Chapter 3 does the same sort of thing for exploring relationships between pairs of variables. For example, how are employee age and seniority at Westvaco related?

Chapter 4 deals with the role of random samples in surveys and random assignment of treatments to subjects in experiments. With exploration, what you see is all you get. Often, though, you want more—you want to generalize to an entire population. When the Gallup Organization interviews a sample of 1500 carefully chosen people, the goal is not just to learn about those 1500 people; it's to learn about the whole adult population in the country. If you pause and think, it's not obvious that you can learn much about 200 million people just from talking to 1500, but the surprising fact is that you can—if the 1500 were selected at random.

Chapters 5, 6, and 7 deal with sampling distributions and probability. Instead of using a simulation, you will learn to use probability theory to find the characteristics of distributions like the one you generated in Activity 1.1.

Chapters 8 through 11 deal with inference—how to extend the logic you used in Section 1.2 to answer a great variety of questions. When the news media report the results of a poll and give the margin of error, what does that mean and where does that number come from? Do patients with AIDS-related complex (ARC) do better when given the drug AZT alone or AZT combined with acyclovir? Does living in a city reduce the ability of your lungs to get rid of harmful particles in the air you breathe? Can special exercises help a baby learn to walk sooner?

Finally, Chapter 12 is a review presented as a set of case studies. This last chapter tells the stories of several investigations, integrating all three elements of statistical thinking—collecting data, exploration, and inference. You'll have a chance to apply what you've learned to study the salaries of professional athletes, data from experiments to compare treatments for producing bigger and better flowers, and then end with a final look at the *Martin* case.

■ Review Exercises

E11. People with asthma often use an inhaler to help open up the lungs and breathing passages. A drug company has come up with a new compound to put in the inhaler that, they believe, will open up the lungs of the user even more than the standard compound tends to do. The new compound B and the standard compound A are each tested on five volunteers with asthma, with the ten volunteers being randomly split into the two treatment groups. The measurements are the increase in lung capacity (in liters) one hour after the use of the inhaler.

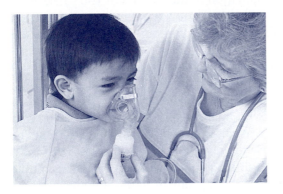

Compound A	Compound B
1.03	1.11
0.45	1.01
0.32	0.44
0.64	1.41
1.29	1.04

a. By simply studying the data in the table, do you think compound B does better than compound A in increasing lung capacity?

b. Construct a dot plot, plotting the A's and B's with different symbols. Does it now appear that B tends to give larger measurements than A?

c. Find the average increase for compound A and for compound B. When you compare these means, does it look to you as if B is better than A at opening up the lungs?

E12. If you are studying the effects of poverty and plan to construct a data set whose cases are villages in Bolivia, what would be some meaningful variables?

E13. Refer to the scenario of E11. If there really is no difference between compound A and compound B, then the apparent differences between the two data sets are due to chance alone. (Recall that the treatments were randomly assigned to the volunteers.) Your task, now, is to see if the observed difference in the means of each treatment group could be attributed to chance alone.

a. Place the ten data values on separate slips of paper and mix them in a bag. Select five at random to play the role of the A treatment group; the other five play the role of the B treatment group. This time you will use as your summary statistic the difference between the means of each treatment group. Calculate this difference for your samples.

b. Repeat the procedure from part a until you have at least 20 simulated differences between means for the two groups. (You may use a calculator or computer to simulate the choices for participants for the two groups.) Plot the simulated differences on a dot plot.

c. Compute the difference between the means for the actual data. Mark this difference on the dot plot. How many simulated differences exceed this one? In light of this (small) simulation, do you think the actual difference could be due to chance alone? Explain your reasoning.

E14. In this exercise, you will follow the same steps as in E10 to find the probability of getting an average age of 58 or more when drawing three hourly workers at random in Round 2. The number of ways to pick three different workers from ten to lay off is

$$_{10}C_3 = \binom{10}{3} = 120$$

a. List the ways that give an average age of 58 or more.

b. Compute the probability of getting an average age of 58 or more when three workers are selected for layoff at random.

c. How does this number compare with the results of your class simulation in Activity 1.1? Why do the two probabilities differ (if they do)?

E15. How would your reasoning and conclusions change if the five oldest workers among the entire group of ten hourly workers in Round 2 were all age 55 (so that the ages of the ten were 25, 33, 35, 38, 48, 55, 55, 55, 55, 55) and the three chosen for layoff were all 55? Is the evidence of age discrimination stronger or weaker than in the actual case?

CHAPTER 2
EXPLORING DISTRIBUTIONS

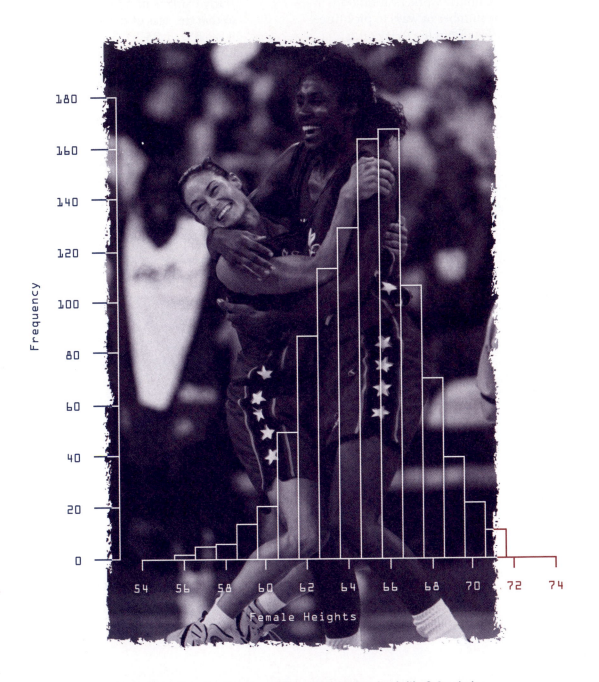

What does the distribution of female heights look like? Statistics gives you the tools to visualize and describe large sets of data.

"Raw" data—a long list of values—is hard to make sense of. Suppose, for example, that you are applying to the University of Michigan at Ann Arbor and wonder how your SAT I score of 1190 compares with those of the students who attend that university. If all you have is raw data—a list of the SAT I scores of the 22,000 students at the University of Michigan—it would take a lot of time and effort to make sense of the numbers.

Suppose instead that you read the summary in their college guide, which says "the middle 50% of the scores were between 1170 and 1340, with half the scores above 1210 and half below." Now you know that although your 1190 is in the bottom half of the scores, it is not far from the center value of 1210 and higher than the bottom quarter.

Notice that the summary of the scores gives you two different kinds of information: the **center** 1210 and the **spread** from 1170 to 1340, for the middle 50%. Often that's all you need, especially if the **shape** of the distribution is one of a few standard shapes you'll learn about in this chapter.

These three features—shape, center, and spread—can sometimes take you a surprisingly long way in data analysis. For example, in Chapter 1 you did a simulation to answer the question "If you choose 3 people at random from a set of 10 people and compute the average age of the ones you choose, how likely is it that you get an average of 58 years or more?" But generally you don't need to do all this work! Using shape, center, and spread, it is possible to get an answer without doing a simulation. This remarkable fact first began to come to light in the late 1600s and helped make statistical inference possible in the 20th century before the age of computers. In the next several chapters, you'll learn how to make good use of these facts.

In this chapter, you begin your systematic study of distributions by learning how to

- make and interpret different kinds of plots
- describe the shapes of distributions
- choose and compute a central or typical value
- choose and compute a useful measure of spread (variability)
- work with the normal, or bell-shaped, curve

2.1 The Shapes of Things: Visualizing Distributions

Summaries simplify. In fact, summaries can sometimes *over*simplify, which means that it is important to know when to use summaries and which summaries to use. Often the right choice depends on the shape of your distribution. To help you build your visual intuition about how shape and summaries are related, this first section of the chapter introduces various shapes and asks you to estimate some summary values visually. (Later sections will tell you how to compute summary values numerically.)

Activity 2.1 introduces one of the most important common shapes and one of the common ways this shape is produced. What happens when different people measure the same distance or the same feature of very similar objects? In the next activity, you'll measure a tennis ball with a ruler, but the results you'll get reflect what happens even if you use very precise instruments under carefully controlled conditions. For example, a 10-gram platinum weight is used for calibration of scales all across the United States. When scientists at the National Institute of Standards and Technology use an analytical balance for its weekly weighing, they face a similar challenge because of variability.

Activity 2.1 Measuring Diameters

What you'll need: a tennis ball and a ruler with a centimeter scale

1. With your partner, plan a method for measuring the diameter of the tennis ball with the centimeter ruler.

2. Using your method, make two measurements of the diameter of your tennis ball to the nearest millimeter.

3. Combine your data with that of the rest of the class to form a dot plot. Speculate first, though, about the shape you expect for the distribution.

4. *Shape.* What is the approximate shape of the plot? Are there clusters and gaps or unusual values (outliers) in the data?

5. *Center and spread.* Choose two numbers that seem reasonable for completing this sentence: "Our typical diameter measurement is about —?—, give or take about —?—." (There is more than one reasonable set of choices.)

6. Discuss some possible reasons for the variability in the measurements. How could the variability be reduced? Can the variability be eliminated entirely? (We will return to these issues in Chapter 4.)

Distributions come in a variety of shapes, but four of the most common basic shapes are illustrated in the rest of this section.

The uniform distribution is rectangular.

Uniform or Rectangular Distribution

Is there any reason to believe that more babies are born in one month than in another? Or should the number of births be fairly uniform across the year? Display 2.1 shows the U.S. births and deaths (in thousands) for 1997. Display 2.2 shows a plot of the birth data, along with a smooth approximation of the distribution.

Month	Births (in thousands)	Deaths (in thousands)
1	305	218
2	289	191
3	313	198
4	342	189
5	311	195
6	324	182
7	345	192
8	341	178
9	353	176
10	329	193
11	304	189
12	324	192

Display 2.1 Births and deaths in the United States, 1997.
Source: Centers for Disease Control and Prevention.

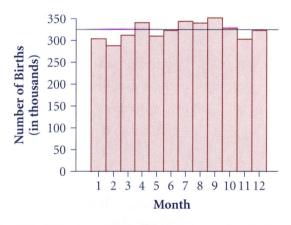

Display 2.2 Births per month, 1997. An example of a (roughly) uniform distribution.

The plot shows that there is actually little change from month to month; that is, we see a roughly **uniform distribution** of births across the months. You can use the smooth approximation as the basis for a short verbal summary: "The distribution of births is roughly uniform over the months January through December, with about 325,000 births per month."

Computers and many calculators generate random numbers between 0 and 1 that have a uniform distribution. Display 2.3 shows a dot plot of 1000 random numbers generated by Minitab statistical software. The flat line across the top is a smoothed version of the plot. For this smooth approximation, the percentage of outcomes in any interval, such as [0.2, 0.4], is given by the percentage of the total area that lies above the interval. Because 20% of the total area lies above the interval [0.2, 0.4], the smooth approximation tells us that 20% of the random numbers fell between 0.2 and 0.4. (You'll learn more about this kind of graph in the next section.)

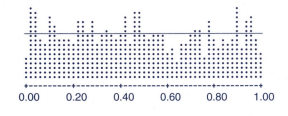

| 0.00 | 0.20 | 0.40 | 0.60 | 0.80 | 1.00 |

Display 2.3   Dot plot of 1000 random numbers from a uniform distribution showing a smooth approximation. Each dot represents 2 points.

Discussion: Uniform Distribution

D1. Think of other scenarios that you would expect to give rise to uniform distributions

  a. over the days of the week

  b. over the digits 0, 1, 2, . . . , 9

D2. Think of scenarios that you would expect to give rise to very nonuniform distributions

  a. over the months of the year

  b. over the days of the month

  c. over the digits 0, 1, 2, . . . , 9

  d. over the days of the week

■ Practice

P1. Plot the number of deaths per month given in Display 2.1. Do they appear to be uniformly distributed over the months? Use your plot as the basis for a verbal summary of the way deaths are distributed over the months of the year.

P2. Display 2.3 shows 1000 numbers randomly selected from a uniform distribution on the interval [0, 1]. Now imagine a uniform distribution on [0, 2].

  a. What value divides the plot in half, with half the numbers below that value, half above?

  b. What values divide the area into quarters?

c. What values enclose the middle 50% of the data?

d. What percentage of the values lie between 0.4 and 0.7?

e. What values enclose the middle 95% of the data?

Normal Distribution

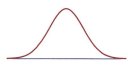

The normal distribution is bell-shaped.

The measurements of the diameter of a tennis ball taken by your class probably were not uniform. More likely, they piled up around some central value with a few being far away on the low side and a few being far away on the high side. This common bell shape has an idealized version—the **normal distribution,** which is especially important in statistics.

Pennies minted in the United States are supposed to weigh 3.110 grams, but a tolerance of 0.130 grams is allowed in either direction. Display 2.4 shows a plot of the weights of 100 pennies.

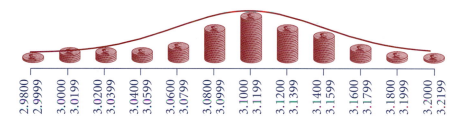

Display 2.4   Weights of pennies.

Source: W. J. Youden, *Experimentation and Measurement* (National Science Teachers Association, 1985), p. 108.

The smooth curve superimposed on the graph of the pennies is an example of a normal curve. No real-world examples match the curve perfectly, but many plots of data are approximately normal. The idealized normal shape is perfectly symmetric—the right side is a mirror image of the left side, as shown in Display 2.5. There is a single peak, or **mode,** at the line of symmetry, and the curve drops off smoothly on both sides, flattening toward the *x*-axis but never quite reaching it, stretching infinitely far in both directions. On either side of the mode, at about 60% of the height of the highest point of the curve, are points of inflection, where the curve changes from concave down to concave up.

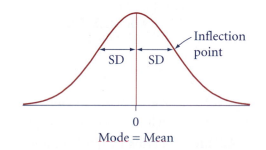

Display 2.5   A normal curve, showing the line of symmetry and points of inflection.

The center and spread for a normal distribution are the mean and standard deviation.

To estimate the center and spread for a normal distribution, start with the line of symmetry. The point where it cuts the *x*-axis is the **mean** (or average). This value is where the area under the curve would balance if you cut it out of cardboard and held a finger under it. For all normal distributions, the mode and mean are equal. To measure spread, estimate the horizontal distance from the line of symmetry to either point of inflection. This distance is called the **standard deviation,** or *SD* for short.

Example: Averages of Random Samples

Display 2.6 shows the distribution of average ages computed from 1000 sets of 5 workers chosen at random from the 10 hourly workers in Round 2 of the Westvaco case, discussed in Chapter 1. Notice that apart from the bumpiness, the shape is roughly normal. Estimate the mean and standard deviation.

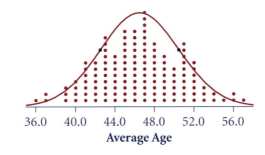

Display 2.6 Distribution of average age for groups of five workers drawn at random. Each dot represents about 8 points.

Solution

The curve shown in the display has center at 46.5 and inflection points at 42.5 and 50.5. Thus, the estimated mean is 46.5, and the estimated standard deviation is 4. A typical random sample of 5 workers has an average age of 46.5, give or take 4 years or so. ∎

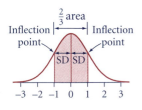

It is difficult to locate inflection points, especially when curves are drawn by hand, so a more reliable way to estimate the standard deviation is to use areas. For a normal curve, roughly $\frac{2}{3}$ of the total area under the curve is between the vertical lines through the two inflection points. In other words, the interval that stretches for one standard deviation on either side of the mean accounts for roughly $\frac{2}{3}$ of the area. For the distribution in Display 2.6, roughly $\frac{2}{3}$ of the dots are in the interval 46.5 ± 4 or $[42.5, 50.5]$.

Activity 2.1 and the last two examples together illustrate the three most common ways that normal distributions arise in practice:

- through variation in measurements (diameters of tennis balls)
- through natural variation in populations (weights of pennies)
- through variation in averages of random samples (average ages)

All three scenarios are quite common, which makes the normal distribution especially important in statistics.

Discussion: Normal Distribution

D3. Determine these summaries visually.

    a. Estimate the center and spread for the penny weight data in Display 2.4, and use your estimates to write a summary sentence.

    b. Estimate the mean and standard deviation for your class data from Activity 2.1.

■ Practice

P3. Sketch a normal distribution with mean 0 and standard deviation 1. This distribution is called a **standard normal distribution.**

P4. For each of the normal distributions in Display 2.7, estimate the mean and standard deviation visually, and use your estimates to write a verbal summary of the form "a typical SAT score is roughly (mean), give or take (*SD*) or so." Then check to see that this interval contains roughly $\frac{2}{3}$ of the total area under the curve.

    a. SAT verbal scores

    b. ACT scores

    c. heights of women attending college

    d. single-season batting averages for professional baseball players in the decade of the 1910s

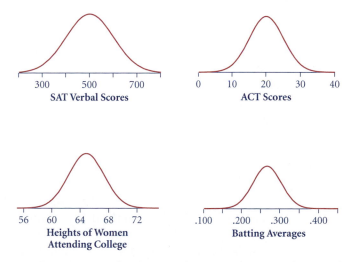

Display 2.7   Four distributions that are approximately normal.

Skewed Distribution

Skewed left

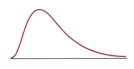

Skewed right

Both the uniform (rectangular) and normal distributions are symmetric. That is, if you smooth out minor bumps, the right side of the plot is a mirror image of the left side. Not all distributions are symmetric, however. Many common distributions show bunching at one end and a long tail stretching out in the other direction. These distributions are called **skewed.** The direction of the tail tells whether the distribution is **skewed right** (tail stretches right toward the high values) or **skewed left** (tail stretches left toward the low values).

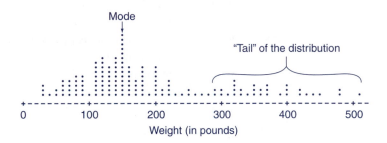

Display 2.8   Weights of bears in pounds.

Source: MINITAB data set from *MINITAB Handbook,* 3rd ed.

The dot plot of Display 2.8 shows the weights in pounds of 143 wild bears. It is skewed right (toward the higher values) because the tail of the distribution stretches out in that direction. In everyday conversation, you might describe the two parts of the distribution as "normal" and "abnormal." Usually, bears weigh between about 50 and 250 pounds (this part of the distribution even looks approximately normal), but if someone shouts "Abnormal bear loose!" you had better run for cover because that unusual bear is likely to be big! The "unusualness" is all in one direction.

Often the bunching in a skewed distribution happens because values "bump up against a wall"—either a minimum that values can't go below, like 0 for measurements and counts, or a maximum that values can't go above, like 100 for percentages. For example, the distribution in Display 2.9 shows the grade-point averages of college students (mostly first-year students and sophomores) taking an introductory statistics course at the University of Florida during the spring of 1999. It is skewed left (toward the smaller values). The maximum grade-point average is 4.0, for all A's, so the distribution is bunched at the high end because of this wall. The skew is to the left: An unusual GPA would be one that is low compared to most GPAs for students in the class.

Display 2.9   Grade-point averages of 61 statistics students. Each dot represents 2 points.

The center and spread for skewed distributions are the median and quartiles.

For skewed distributions, the center and spread are not as clear-cut as they are for normal distributions. Because there is no line of symmetry, the idea of center is ambiguous. Moreover, because the left and right halves of a skewed distribution don't match, "distance to the point of inflection" is ambiguous, too. To get around this problem, people often report the **quartiles,** three numbers that divide the values into fourths. This lets you describe a distribution as in the introduction to the chapter: "The middle 50% of the SAT scores were between 1170 and 1340, with half above 1210 and half below."

To estimate these values from a dot plot, first draw a vertical line at the value that divides the dots into two halves. This value, called the **median,** is the measure of center. To measure spread, repeat the halving process with each half of the data: Draw a vertical line that cuts each half into two pieces with equal numbers of dots on either side. These values are the **lower quartile** and **upper quartile.** They enclose the middle 50% of the values.

Example

Divide the bears' weights in Display 2.10 into four equal parts, and estimate the median and quartiles. Write a short summary of this distribution.

Solution

There are 143 dots in Display 2.10, so there are about 71 or 72 dots in each half and 35 or 36 in each quarter. The value that divides the dots in half is about 155. The values that divide the two halves in half are roughly 115 and 250. Thus, the middle 50% of the bear weights are between about 115 and 250 pounds, with half above about 155 and half below. ■

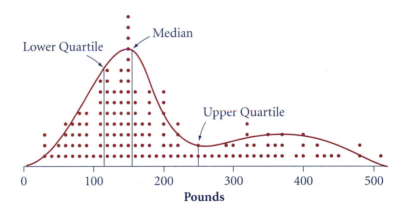

Display 2.10   Estimating center and spread for the weights of bears.

Discussion: Skewed Distribution

D4. Decide whether each distribution below will be skewed. Is there a wall that leads to bunching near it and a long tail away from it? If so, describe this wall.

a. Sizes of islands in the Caribbean

b. Average per capita incomes for the nations of the United Nations

c. Lengths of pant legs cut and sewn to be 32″ long

    d. The times for 300 university students of introductory psychology to complete a one-hour timed exam

    e. The lengths of reigns of Japanese emperors

D5. Make up a scenario (name the cases and variables) whose distribution you would expect to be skewed right because of a wall. What is responsible for the wall?

D6. Make up a scenario whose distribution you would expect to be skewed left because of a wall. What is responsible for the wall?

D7. Which would you expect to be the more common direction of skew, right or left? Why?

■ Practice

P5. Match each plot in Display 2.11 with its median and quartiles, that is, the set of values that divide the area into fourths.

    a. 0.15, 0.50, 0.85

    b. 0.50, 0.71, 0.87

    c. 0.63, 0.79, 0.91

    d. 0.35, 0.5, 0.65

    e. 0.25, 0.50, 0.75

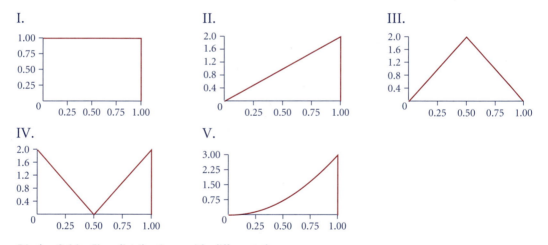

Display 2.11   Five distributions with different shapes.

P6. The U.S. Environmental Protection Agency's *National Priorities List Fact Book* tells the number of hazardous waste sites for each of the U.S. states and territories. For 1992, the numbers ranged from 0 to 102, the middle 50% of the values were between 6 and 22, half were above 10, and half below. Sketch what the distribution might look like.

Source: *World Almanac and Book of Facts 1994*, p. 173.

P7. Estimate the median and quartiles for the distribution of GPAs in Display 2.9. Then write a verbal summary of the same form as in the example.

Bimodal Distribution

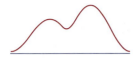

A bimodal distribution
has two peaks.

Many distributions, including the normal, and many skewed distributions as well, have only one peak (**unimodal**), but some have two (**bimodal**) or even more. When your distribution has two or more obvious peaks or **modes,** it is worth asking whether your cases represent two or more groups. For example, Display 2.12 shows the life expectancies for females from countries on two continents—Europe and Africa.

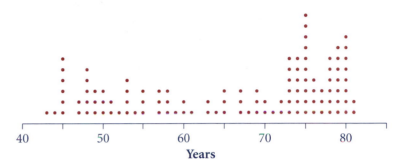

Display 2.12 Life expectancy of females by country on two continents.
Source: Population Reference Bureau, World Population Data Sheet, 1996.

Europe and Africa are quite different in their socioeconomic conditions, and the life expectancies reflect those conditions. If you make separate plots for the two continents, the two peaks become essentially one peak in each plot, as shown in Display 2.13. And, yes, Europe is a mixture as well: east and west with means about 75 and 79, respectively.

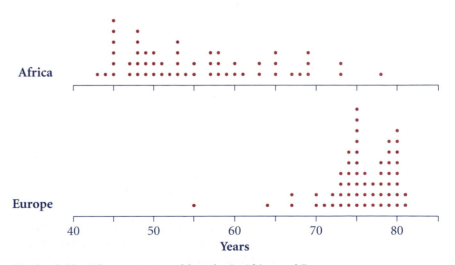

Display 2.13 Life expectancy of females in Africa and Europe.

Although it makes sense to talk about the center of the distribution of life expectancies for Europe, or of those for Africa, notice that it doesn't really make sense to talk about "the" center of the distribution for both continents together. Instead you could tell the locations of the two peaks. But finding the reason for the two modes and separating the cases into two distributions, tells even more.

Other Features: Outliers, Gaps, and Clusters

An unusual value, or **outlier,** is a value that stands apart from the bulk of the data. Outliers always deserve special attention. Sometimes they are mistakes—a typing mistake, a measuring mistake—sometimes they are atypical for other reasons—a really big bear, a faulty lab procedure—and sometimes they are the key to an important discovery.

In the late 1800s, John William Strutt, third Baron Rayleigh (English, 1842–1919), was studying the density of nitrogen using samples from the air outside his laboratory (from which known impurities were removed) and samples produced by a chemical procedure in the lab. He saw a pattern in the results that you can observe in the plot of his data in Display 2.14.

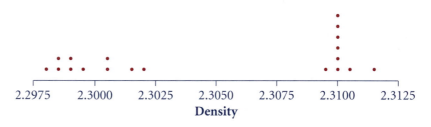

Display 2.14 Lord Rayleigh's densities of nitrogen.

Source: *Proceedings of the Royal Society 55* (1894).

Lord Rayleigh saw two clusters separated by a gap. (There is no formal definition of a **gap** or a **cluster,** so you will have to use your best judgment about them. For example, some people call a single outlier a cluster of one; others don't. You could also argue that the value at the extreme right is an outlier, perhaps because of a faulty measurement.)

When Rayleigh checked the clusters, it turned out that the 10 values to the left had all come from the chemically produced samples and the 9 to the right had all come from the atmospheric samples. What did this great scientist conclude? The air samples on the right might be denser because of something in them besides nitrogen. This hypothesis led him to discover inert gases like radon in the atmosphere.

Summary 2.1: Visualizing Distributions

Distributions have different shapes, and different shapes call for different summaries.

- If your distribution is uniform (rectangular), it's often enough simply to tell the range of the set of values and the approximate frequency with which each occurs.

- If your distribution is normal (bell-shaped), you can give a good summary with the mean and the standard deviation. The mean lies at the center of the distribution, and the standard deviation is the horizontal distance from the center to the points of inflection, where the curvature changes. To estimate it, find the distance on either side of the mean that encloses about two-thirds of the cases.

- If your distribution is skewed, you can give the values (quartiles) that divide the distribution into fourths.

- If your distribution is bimodal, it isn't useful to report a single center. One reasonable summary is to locate the two peaks. However, it is even more useful if you can find another variable that divides your set of cases into two groups centered at the two peaks.

Later in the chapter, you will study the various measures of center and spread in more detail and learn how to compute them.

■ Exercises

E1. Sketch the shape you would expect each distribution to have.

  a. Age of each person who died last year in the United States

  b. Age of each person who got his or her first driver's license in your state last year

  c. SAT scores for all students in your state taking the test this year

  d. Selling prices of all cars sold by General Motors this year

E2. Describe each distribution below as bimodal, skewed right, skewed left, approximately normal, or roughly uniform.

  a. The incomes of the world's 100 richest people

  b. The birth rates of Africa and Europe

  c. The heights of soccer players on the last U.S. Woman's World Cup team

  d. The last two digits of telephone numbers in the town where you live

  e. The length of time students used to complete a chapter test, out of a 50-minute class period

E3. Sketch these distributions:

  a. A uniform distribution that shows the sort of data you would get from rolling a fair die 6000 times

  b. A roughly normal distribution with mean 15 and standard deviation 5

  c. A distribution that is skewed left, with half its values above 20, half below, and that has the middle 50% of its values between 10 and 25

  d. A distribution that is skewed right, with the middle 50% of its values between 100 and 1000 and with half the values above 200 and half below

E4. The plot in Display 2.15 shows the last digit of the social security numbers of the students in a statistics class. Describe this distribution.

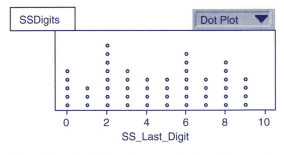
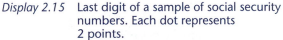

Display 2.15 Last digit of a sample of social security numbers. Each dot represents 2 points.

E5. The dot plot in Display 2.16 gives the ages of the officers who attained the rank of colonel in the Royal Netherlands Air Force.

  a. What are the cases? Describe the variables.

  b. Describe this distribution in terms of shape, center, and spread.

c. What kind of wall might there be that causes this shape? Generate as many possibilities as you can.

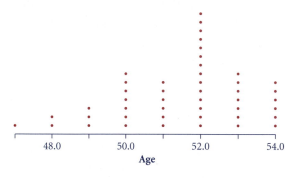

Display 2.16 Ages of colonels. Each dot represents 2 points.

Source: Data and Story Library at Carnegie Mellon University, *http://lib.stat.cmu.edu/DASL.*

E6. The dot plot in Display 2.17 shows the distribution of the number of inches of rainfall in Los Angeles for the seasons 1877–78 through 2001–2002.

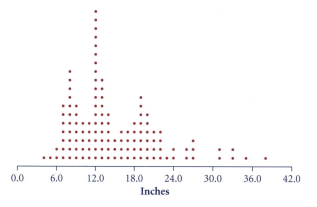

Display 2.17 Los Angeles rainfall.

Source: *Los Angeles Times.*

a. What are the cases? Describe the variables.

b. Describe this distribution in terms of shape, center, and spread.

c. What kind of wall might there be that causes this shape? Generate as many possibilities as you can.

E7. The distribution in Display 2.18 shows measurements of the strength in pounds of 22s yarn (22s refers to a standard unit for measuring yarn strength). What is the basic shape of this distribution? What feature makes it uncharacteristic of that shape?

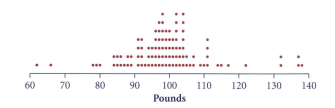

Display 2.18 Strength of yarn.

Source: Data and Story Library at Carnegie Mellon University, *http://lib.stat.cmu.edu/DASL.*

E8. Although a uniform distribution gives a reasonably "smooth" approximation to the actual distribution of births over months (Display 2.2), you can "blow up" the graph to see departures from the uniform pattern, as in Display 2.19. Do these deviations from the uniform shape form their own pattern, or do they appear haphazard? If you think there's a pattern, describe it.

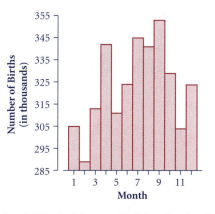

Display 2.19 A "blow up" of the distribution of births over months, showing departures from the uniform pattern.

E9. Draw a graph similar to that in Display 2.19 for the data on deaths in the United States in Display 2.1, and summarize what you find.

E10–11. *Nielsen ratings.* Every week many newspapers publish the Nielsen report of the numbers of people who watch prime-time network television shows.

Display 2.20 gives the estimated number of viewers who watched each television program from start to finish. This week was special because it ended the season and featured the very last new episode of *Seinfeld*.

Program	Network	Viewers (millions)
1 Seinfeld	NBC	76.26
2 Seinfeld Clips	NBC	58.53
3 ER	NBC	47.78
4 Touched by an Angel	CBS	20.47
5 The X-Files	FOX	18.76
.	.	.
.	.	.
.	.	.
50 48 Hours	CBS	10.18
51 Dr. Quinn Medicine Woman	CBS	10.15
52 Beverly Hills, 90210	FOX	10.15
.	.	.
.	.	.
.	.	.
101 Malcolm and Eddie (Tue.)	UPN	2.32

Display 2.20   Nielsen estimates of television show viewers.

Source: *Los Angeles Times,* May 20, 1998.

E10. The dot plot in Display 2.21 shows the distribution of the Nielsen ratings.

a. In the Nielsen data, what are the cases? Describe the variables.

b. Describe the basic shape of the distribution in Display 2.21. Note any outliers and any gaps or clusters in the distribution.

c. Find the median number of people who watched a prime-time television show. Is there a lot of spread (variability) in the numbers of viewers? The middle half of the ratings are between what two values?

d. What can you say about how the number of people watching the last episode of *Seinfeld* compared to the number who watch a typical television show?

e. The dot plot in Display 2.22 shows the Nielsen estimates of viewers for an ordinary week for which there was nothing special, such as the last *Seinfeld* episode. Compare the shape, center, and spread of this distribution with the one in Display 2.21.

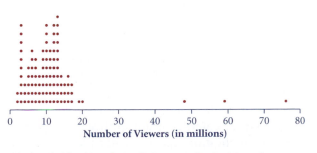

Display 2.21   Number of viewers of television shows in millions, per Nielsen ratings.

Display 2.22   Dot plot of Nielsen ratings for a less unusual week.

Source: *www.foxnews.com.*

E11. The dot plots in Display 2.23 can be used to compare the distributions of the ratings for the six networks.

a. Describe the basic shape of the distribution for each network. Note any outliers and any gaps or clusters in the distribution.

b. Compare the center and spread of the ratings for FOX and for NBC. For which of the six networks are the ratings centered highest? Lowest?

c. Which network has the most variability in its ratings? The least variability?

d. From looking at the plots, rank the six networks according to the popularity of their shows.

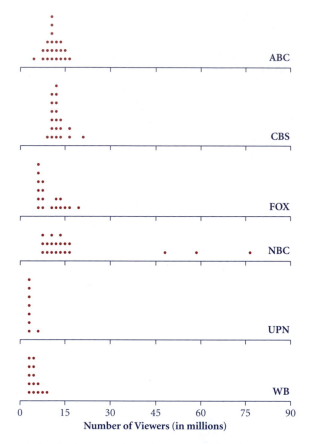

Display 2.23 Dot plots of Nielsen ratings of television shows by network.

2.2 Graphical Displays for Distributions

Plots should present the essentials quickly and clearly.

As you saw in the last section, the best way to summarize a distribution often depends on its shape. To see the shape, you need a suitable graph. In this section, you'll learn how to make and interpret three kinds of plots for quantitative variables.

Pet cats typically live about 12 years, but some have been known to live for 28 years. Is that typical of domesticated predators? What about domesticated nonpredators, like cows and guinea pigs? Or wild mammals? The rhinoceros, a nonpredator, lives an average of 15 years, with a maximum of about 45 years. On the other hand, the grizzly bear, a wild predator, lives an average of 25 years, with a maximum of about 50 years. Do meat-eaters typically outlive vegetarians in the wild? Often you can find answers to questions like these in a plot of the data.

Mammal	Gestation Period (days)	Average Life Span (years)	Maximum Life Span (years)	Speed (mph)	Wild (1 = yes; 0 = no)	Predator (1 = yes; 0 = no)
Baboon	187	20	45	*	1	0
Grizzly bear	225	25	50	30	1	1
Beaver	105	5	50	*	1	0
Bison	285	15	40	*	1	0
Camel	406	12	50	*	1	0
Cat	63	12	28	30	0	1
Cheetah	*	*	14	70	1	1
Chimpanzee	230	20	53	*	1	0
Chipmunk	31	6	8	*	1	0
Cow	284	15	30	*	0	0
Deer	201	8	20	30	1	0
Dog	61	12	20	39	0	1
Donkey	365	12	47	40	0	0
Elephant	660	35	70	25	1	0
Elk	250	15	27	45	1	0
Fox	52	7	14	42	1	1
Giraffe	425	10	34	32	1	0
Goat	151	8	18	*	0	0
Gorilla	258	20	54	*	1	0
Guinea pig	68	4	8	*	0	0
Hippopotamus	238	41	54	20	1	0
Horse	330	20	50	48	0	0
Kangaroo	36	7	24	40	1	0
Leopard	98	12	23	*	1	1
Lion	100	15	30	50	1	1
Monkey	166	15	37	*	1	0
Moose	240	12	27	*	1	0
Mouse	21	3	4	*	1	0
Opossum	13	1	5	*	1	1
Pig	112	10	27	11	0	0
Puma	90	12	20	*	1	1
Rabbit	31	5	13	35	0	0
Rhinoceros	450	15	45	*	1	0
Sea lion	350	12	30	*	1	1
Sheep	154	12	20	*	0	0
Squirrel	44	10	23	12	1	0
Tiger	105	16	26	*	1	1
Wolf	63	5	13	*	1	1
Zebra	365	15	50	40	1	0

Display 2.24 Facts on mammals.

Source: *World Almanac and Book of Facts 2001*, p. 237.

Cases and Variables, Quantitative and Categorical

Many of the examples in this section are based on the data about mammals in Display 2.24. For wild mammals, longevity is taken from records kept on mammals in captivity, and maximum longevity is the largest longevity on record. The column Wild is coded 1 if the mammal is wild and 0 if it is domestic. The column Predator is coded 1 if the mammal preys on other animals for food and 0 if it does not. The asterisks (*) mark missing values.

In Display 2.24, each row (each mammal) is a case. In general, the cases in a data set are the individual people, cities, mammals, or other items being studied. Measurements and other properties of the cases are organized into columns, one column for each variable. Thus, average longevity and speed are variables, and, for example, 30 mph is the value of the variable *speed* for the case *grizzly bear*. *Speed* is a **quantitative variable** because the speeds are numbers that can be compared in a meaningful way. *Wild* is a **categorical variable,** as is *predator*—although the values 0 and 1 are numbers, the numbers are actually substitutes for the categories "no" and "yes."

More About Dot Plots

Dot plots show individual cases as dots.

You've already seen dot plots beginning in Chapter 1. As the name suggests, dot plots show individual cases as dots (or other plotting symbols such as **x**). When reading a dot plot, keep in mind that different statistical software packages make dot plots in different ways. Sometimes one dot represents two or more cases, and sometimes values have been rounded. With a small data set, different rounding rules can give different shapes.

Display 2.25 shows a dot plot of the speeds of the mammals.

Display 2.25 Dot plots of the speeds of mammals.

When are dot plots most useful?

As you saw in Section 2.1, a dot plot shows shape, center, and spread. They tend to work best when

- you have a relatively small number of values to plot
- you want to see individual values, at least approximately
- you want to see the shape of the distribution
- you have one group or a small number of groups you want to compare

Discussion: More About Dot Plots

D8. Classify each variable in Display 2.24 as quantitative or categorical.

D9. Consider the mammals' speeds in Display 2.24.

a. Count the number of mammals that have speeds ending in a 0 or a 5.

b. How many would you expect to end in a 0 or a 5 just by chance?

c. What are some possible explanations for the fact that your answers in parts a and b are so different?

■ Practice

P8. In the listing of the Westvaco data in Chapter 1 on page 5, which variables are quantitative? Which are categorical?

P9. Decide on a reasonable scale, and make a dot plot of the gestation periods of the mammals listed in Display 2.24. Describe the shape, center, and spread from this dot plot. Write a sentence using shape, center, and spread to summarize the distribution of gestation periods for the mammals. What kinds of mammals have longer gestation periods?

Histograms

Histograms show groups of cases as rectangles or bars.

A dot plot shows individual cases as dots. A **histogram** shows groups of cases as rectangles or bars. In fact, you can think of a histogram as a dot plot with bars drawn around the dots and the dots erased. This makes the height of the bar a visual substitute for the number of dots. The plot in Display 2.26 is a histogram of the mammal speeds. Like the dot plot of a distribution, a histogram shows shape, center, and spread. The vertical axis gives the number of cases (called **frequency** or count) that are represented by each bar. For example, four mammals have speeds of 30 to 35 miles per hour.

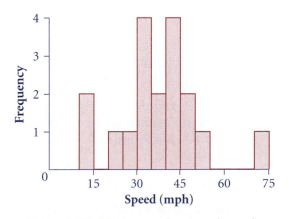

Display 2.26   Histogram of mammal speeds.

Borderline values go in the box on the right.

Most statistical software places a value that falls at the dividing line between two bars into the bar on the right. For example, in Display 2.26, the bar going from 30 to 35 would contain values such that $30 \le speed < 35$.

Changing the width of the bars in your histogram can sometimes change your impression of the shape of the distribution. For example, the histogram of the speeds of mammals in Display 2.27 has fewer and wider bars than the histogram in Display 2.26 and shows a more symmetric, bell-shaped distribution. Now there appears to be one peak rather than two. If there are few values in the data set, it is difficult to identify peaks. In this situation, it is better to use a plot that identifies individual values, like a dot plot or a stemplot.

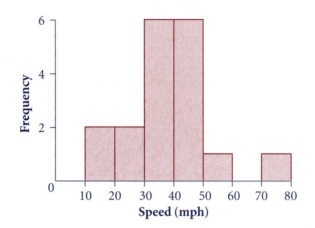

Display 2.27   Speeds of mammals with a wider-bar histogram.

There is no "right answer" to the question of which bar width is best, just as there is no rule that tells a photographer when to use a zoom lens for a close-up. Different versions of a picture bring out different features; the job of a data analyst is to find a version that shows important features of the data.

When are histograms most useful?

Histograms work best when

- you have a large number of values to plot
- you don't need to see individual values exactly
- you want to see the general shape of the distribution
- you have only one distribution or a small number of distributions you want to compare
- you can use a calculator or computer to draw the plots for you

Relative frequency histograms show proportions instead of counts.

A histogram shows frequencies on the vertical axis. To make a histogram into a **relative frequency histogram,** divide the frequency for each bar by the total number of values in the data set, and show these relative frequencies on the vertical axis.

Example

Display 2.28 shows the relative frequency distribution of life expectancies for 250 countries around the world. What proportion of the countries have life expectancies of 64 years or more?

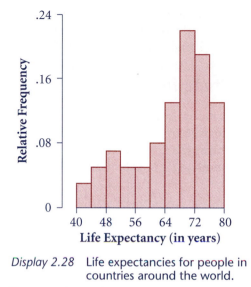

Display 2.28   Life expectancies for people in countries around the world.

Source: Population Reference Bureau.

Solution

Locate the interval of values of 64 or more on the *x*-axis. What proportion of the total area is taken up by the bars over that interval? A rough visual estimate is about $\frac{2}{3}$ of the area: Roughly $\frac{2}{3}$ of the countries have life expectancies of at least 64 years. Now suppose you want a more precise estimate. The proportion of countries with life expectancies of 64 years or greater is the sum of the heights of the four bars of the histogram to the right of 64, or about $0.13 + 0.22 + 0.19 + 0.13 = 0.67$.   ■

Discussion: Histograms

D10. Describe the center and spread of the distribution of mammal speeds based first on the histogram in Display 2.26, then based on the histogram in Display 2.27. How much difference does the bar width make for this data set?

D11. In what sense does a histogram with narrow bars as in Display 2.26 give you more information than a histogram with wider bars as in Display 2.27? In light of your answer, why don't we make all histograms with very narrow bars?

D12. Does using relative frequencies change the shape of a histogram? What information is lost or gained when presenting a relative frequency histogram rather than a frequency histogram?

■ Practice

P10. Using a calculator or computer, make histograms of the average longevities and the maximum longevities of the mammals. Describe how the distributions differ in terms of shape, center, and spread. Why do these differences occur?

P11. Convert your histograms of the average longevities and the maximum longevities of the mammals to relative frequency histograms. Do the shapes of the histograms change?

P12. In the histogram for life expectancies (Display 2.28), which will be larger, the mean (balance point) or the median (value that divides the area into a right half and a left half)? Explain your reasoning.

Stemplots

Both the dot plot and the histogram show the shape, center, and spread of a distribution of data, but neither retains the exact values. The plot in Display 2.29 shows the key features of the distribution and preserves all of the original numbers. It is a **stem-and-leaf plot** or **stemplot** of the mammal speeds.

```
1 │ 1 2
2 │ 0 5
3 │ 0 0 0 2 5 9
4 │ 0 0 0 2 5 8
5 │ 0
6 │
7 │ 0
```

3│9 represents 39 miles per hour

Display 2.29   Stemplot of mammal speeds.

A stemplot shows cases as digits.

The numbers on the left, called the **stems,** are the tens digits of the speeds. The numbers on the right, called the **leaves,** are the ones digits of the speeds. The leaf for the speed of 39 is printed in bold. If you turn your book 90° counterclockwise, you see what looks something like a dot plot or histogram, and you can see the shape, center, and spread of the distribution, just as you can from those plots.

The stemplot in Display 2.30 displays the same information but with **split stems:** Each stem from the original plot has become two stems. If the ones digit is 0, 1, 2, 3, or 4, it is placed on the first line for that stem. If the ones digit is 5, 6, 7, 8, or 9, it is placed on the second line for that stem.

```
1 │ 1 2
. │
2 │ 0
. │ 5
3 │ 0 0 0 2
. │ 5 9
4 │ 0 0 0 2
. │ 5 8
5 │ 0
. │
6 │
. │
7 │ 0
```

3│9 represents 39 miles per hour

Display 2.30   Stemplot of mammal speeds, using split stems.

Spreading out the stems in this way is similar to changing the width of the bars in a histogram. The goal here, as always, is to find a plot that conveys the essential pattern of the distribution as clearly as possible.

You have compared two data distributions by constructing dot plots on the same scale (see Display 2.13, for example). Another way to compare two distributions is to construct a back-to-back stemplot. Such a plot for the speeds of predators and nonpredators is shown in Display 2.31. The predators tend to have the faster speeds, or at least there are no slow predators!

```
    Predator  |     |  Nonpredator
              |   1 |  1 2
              |   . |
              |   2 |  0
              |   . |  5
        0 0   |   3 |  0 2
          9   |   . |  5
          2   |   4 |  0 0 0
              |   . |  5 8
          0   |   5 |
              |   . |
              |   6 |
              |   . |
          0   |   7 |
```

3 | 9 represents 39 miles per hour

Display 2.31   Back-to-back stemplot of mammal
speeds for predators and nonpredators.

Usually, only two digits are plotted on a stemplot, one digit for the stem and one digit for the leaf. If the values contain more than two digits, the values may either be truncated (the extra digits simply cut off) or rounded. For example, if the speeds had been given to the nearest tenth of a mile, 32.6 miles per hour could either be truncated to 32 miles per hour or rounded to 33 miles per hour.

As with the other plots, the rules for making stemplots are flexible. Do what seems to work best to help your reader see the important features of the data.

The stemplot of mammal speeds in Display 2.32 was made by statistical software. Although it looks a bit different from the handmade plot in Display 2.31, it is essentially the same. In the first two lines, N = 18 means that 18 cases were plotted; N* = 21 means that there were 21 cases in the original data set for which speeds were missing; and Leaf Unit = 1.0 means that the ones digits were graphed as the leaves. The numbers in the left column keep track of the number of cases, counting in from the extremes. The 2 on the left in the first line means that there are 2 cases on that stem. If you skip down three lines, the 4 on the left means that there are a total of 4 cases on the first 4 stems.

```
Stem-and-leaf of Speeds    N = 18
Leaf Unit = 1.0            N* = 21
         2    1 12
         2    1
         3    2 0
         4    2 5
         8    3 0002
        (2)   3 59
         8    4 0002
         4    4 58
         2    5 0
         1    5
         1    6
         1    6
         1    7 0
```

Display 2.32 Stem-and-leaf plot of mammal speeds
made by statistical software.

When are stem-and-leaf plots most useful?

Stemplots are useful when

- you are plotting a single quantitative variable

- you have a relatively small number of values to plot

- you would like to see individual values exactly, or, when the values contain more than two digits, you would like to see approximate individual values

- you want to see the shape of the distribution clearly

- you have two (or sometimes more) groups you want to compare

Discussion: Stemplots

D13. Describe the shape, center, and spread of the distribution of mammal speeds from the stemplot in Display 2.30 or Display 2.32. Compare your answer to that of D10.

D14. What information is given by the numbers in the leftmost column of the bottom half of the plot in Display 2.32?

D15. Discuss how you might construct a stemplot of the data on gestation periods for the mammals given in Display 2.24. Note that some of these values are three-digit numbers, so you will have to decide on a rule for stems and leaves.

■ Practice

P13. Make a back-to-back stemplot of the average longevities and maximum longevities from Display 2.24. Compare the two distributions.

P14. Examine Display 2.31 and describe how the speeds of predators and nonpredators seem to differ in terms of shape, center, and spread. Explain why you should expect these differences.

Activity 2.2   Do Units of Measurement Affect Your Estimates?

In this experiment, you will see if you and your class estimate lengths better in feet or in meters.

1. Your instructor will randomly split the class into two groups.

2. If you are in the first group, you will estimate the length of your classroom in feet. If you are in the second group, you will estimate the length of the room in meters. Do this by looking at the length of the room; no pacing the length of the room allowed.

3. Find an appropriate and meaningful way to plot the two data sets so that you can compare them.

4. Do the students in your class tend to estimate more accurately in feet or in meters? What is the basis for your decision?

5. Why split the class randomly into two groups instead of simply letting the left half of the room estimate in feet and the right half in meters?

Bar Graphs for Categorical Data

Bar graphs show frequencies for categorical data as heights of bars.

You now have three different types of plots to use with quantitative variables. What about categorical variables? How can you plot the outcomes? You could make a dot plot, or you could make what looks like a histogram but is called a **bar graph.** There is one bar for each category, and the height of the bar tells the frequency. (Remember that a bar graph has categories on the horizontal axis, whereas a histogram has measurements—values from a quantitative variable.)

The bar graph in Display 2.33 shows the frequency of mammals in the table that fall into the categories of wild and domestic. (Note that the bars are separated so that there is no suggestion that the variable can take on the value of, say, 1.5.)

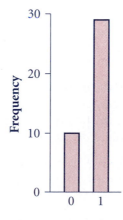

Display 2.33   Bar graph showing frequency of domestic (0) and wild (1) mammals.

Display 2.34 shows the proportion of the female labor force aged 25 and older in the United States that falls into various educational categories. The coding used in the plot is as follows:

1. none–8th grade
2. 9th grade–11th grade
3. high school graduate
4. some college, no degree
5. associate degree

6. bachelor's degree
7. master's degree
8. professional degree
9. doctorate degree

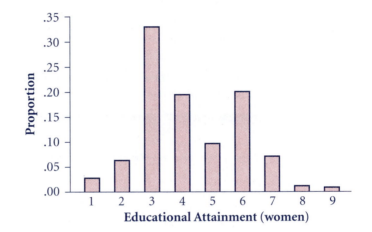

Display 2.34 The female labor force 25 years and older by educational attainment.

Source: U.S. Census Bureau, March 1999 Current Population Survey, *www.census.gov/population/www/socdemo/educ-attn.*

The variable on the horizontal axis reflects the amount of formal education received. Even though it is labeled with numerical values here, attained education, as defined above, is best thought of as a categorical variable rather than a measurement. This bar graph, then, shows the relative frequencies for a categorical variable.

Discussion: Bar Graphs

D16. In the bar graph of Display 2.33, would it matter if the order of the bars were reversed? In the bar graph of Display 2.34, would it matter if the order of the first two bars in the graph were reversed? Comment on how we might define two different types of categorical variables.

D17. Examine the grouped bar graph in Display 2.35.

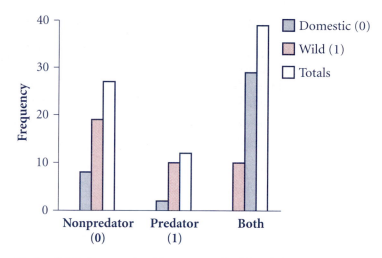

Display 2.35 Bar graph of frequency of wild and domestic mammals by predator status.

a. Describe what the height of each bar represents.

b. How can you tell from this bar graph whether a predator from our list is more likely to be wild or domestic?

c. How can you tell from this bar graph whether a nonpredator or a predator is more likely to be wild?

■ Practice

P15. Display 2.36 for the male labor force is the counterpart of Display 2.34. What are the cases, and what is the variable? Describe the distribution you see here. How does the distribution for female education compare to the distribution for male education? Why is it better to look at relative frequency bar graphs rather than frequency bar graphs to make this comparison?

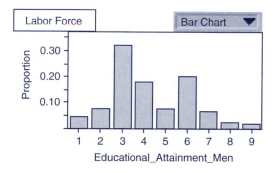

Display 2.36 The male labor force 25 years and older by educational attainment.

Source: U.S. Census Bureau, March 1999 Current Population Survey, *www.census.gov/population/www/socdemo/educ-attn.*

P16. From the data in Display 2.23, make a bar graph showing the number of prime-time shows for each network.

Summary 2.2: Graphical Displays of Data

When a variable is quantitative, you can use dot plots, stemplots (or stem-and-leaf plots), and histograms to display the distribution of values. From each, you can see shape, center, and spread. However, the amount of detail varies, and you should choose a plot that fits both your data set and your reason for analyzing it.

- Stemplots can retain the actual data values.
- Dot plots show approximations to the data values.
- Histograms show only intervals of values, losing the actual data values, and are most appropriate for large data sets.

A bar graph shows the distribution of a categorical variable.

When you look at a plot, you should attempt to answer these four questions:

- Where did this set of data come from?
- What are the cases and the variables?
- What is the shape, center, and spread of this distribution? Does the distribution have any unusual characteristics such as clusters, gaps, or outliers?
- What are possible interpretations or explanations of the patterns you see in the distribution?

■ Exercises

E12. Suppose you collect this information for each student in your class: age, hair color, number of siblings, gender, miles he or she lives from school. What are the cases? What are the variables? Classify each variable as quantitative or categorical.

E13. The dot plot in Display 2.37 shows the distribution of the ages of the pennies in a sample collected by a statistics class.

a. Where did this set of data come from? What are the cases and the variables?

b. What are the shape, center, and spread of this distribution?

c. Does the distribution have any unusual characteristics? What are possible interpretations or explanations of the patterns you see in the distribution? That is, why does the distribution have the shape it does?

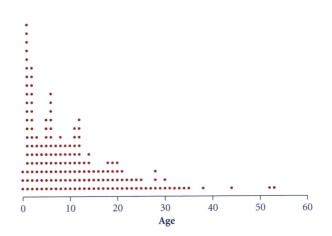

Display 2.37   Age of pennies. Each dot represents 4 points.

E14. How do you expect the distributions of average life expectancies to compare for wild and domesticated mammals?

a. Write your prediction in a sentence or two. Cover shape, center, and spread.

b. Use the data in Display 2.24 to make a back-to-back stemplot to compare average life expectancies.

c. Write a short summary comparing the two distributions.

E15. The graphs in Display 2.38 below appeared in a story on the "changing course of fast food." What kinds of graphs are these? Study the graphs, and then write a story that might have been in the paper.

E16. Using your knowledge of the variables and what you think the shape of the distribution might look like, match each of the variables in the list below with the appropriate histogram in Display 2.39.

   I. Scores on a fairly easy examination in statistics

  II. Heights of a group of mothers and their 12-year-old daughters

 III. Numbers of medals won by medal-winning countries in the 2000 Summer Olympics

 IV. Weights of grown chickens in a barnyard

E17. Using the technology available to you, make histograms of the average longevity and maximum longevity data (Display 2.24) using bar widths of 4, 8, and 16 years. Comment on the main features of the shapes of these plots, and determine which bar width appears to display these features best.

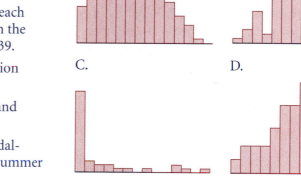

A.   B.

C.   D.

Display 2.39   Four histograms with different shapes.

Number of Fast-Food Restaurants in the United States

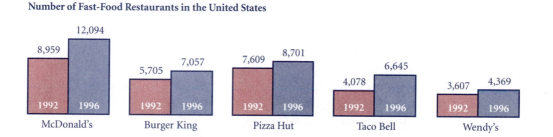

Change in Average Revenue per U.S. Restaurant Open at Least One Year

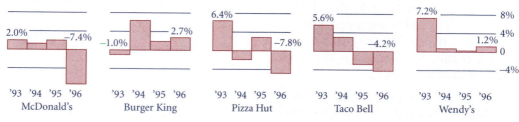

Display 2.38   Fast food restaurants.

Source: *USA Today,* June 6, 1997.

E18. The histogram in Display 2.40 shows the distribution of average ages for 1000 random samples of size 3 chosen from the set of 10 hourly workers involved in the second round of layoffs at Westvaco.

   a. Estimate the mean and standard deviation.

   b. Very roughly, what percentage of the 1000 averages would you estimate are within one standard deviation of the mean? Within two standard deviations? Three standard deviations?

   c. For this set of 1000 repetitions, about how many samples had an average age of 58 or more? What percentage of 1000 is this?

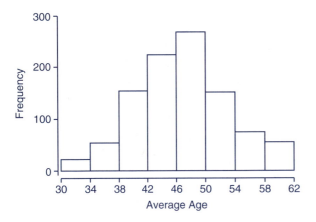

Display 2.40   Average ages for 1000 random samples.

E19. The histogram in Display 2.41 shows the distribution of SAT I math scores for 1999–2000.

   a. Estimate the mean and standard deviation.

   b. Roughly what percentage of the SAT I math scores would you estimate are within one standard deviation of the mean? Within two standard deviations? Three standard deviations?

   c. For SAT I verbal scores, the shape was similar, but the mean was 9 points lower and the standard deviation was 2 points smaller. Draw a smooth curve to show the distribution of SAT I verbal scores.

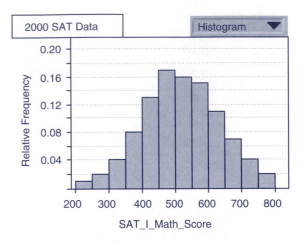

Display 2.41   Relative frequency histogram of SAT I math scores, 1999–2000.

Source: College Board Online,
www.collegeboard.org/sat/cbsenior/yr2000/nat/natsdm00.

E20. Display 2.42 shows the distribution of the heights of U.S. males between the ages of 18 and 24. The heights are rounded to the nearest inch.

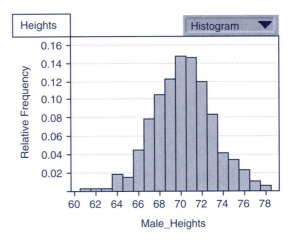

Display 2.42   Heights of males, 18 to 24 years old.

Source: *Statistical Abstract of the United States,* 1991.

   a. Draw a smooth curve to approximate the histogram.

   b. Estimate the mean and standard deviation.

c. Estimate the proportion of men aged 18 to 24 who are 74 inches tall or less.

d. Estimate the proportion of heights that fall below 68 inches.

e. Explain why, in the histogram of Display 2.42, you can find proportions either by adding the heights of the bars or by adding the areas of the bars. Is this true of every histogram?

f. Why should you say that the distribution of heights is "approximately" normal rather than simply saying it is normally distributed?

E21. The plots in Display 2.43 show a form of back-to-back histogram called a population pyramid. Describe how the population distribution of the United States differs from the population distribution of Mexico.

E22. Look through newspapers and magazines to find an example of a graph that is either misleading or difficult to interpret. Redraw the graph to make it clear.

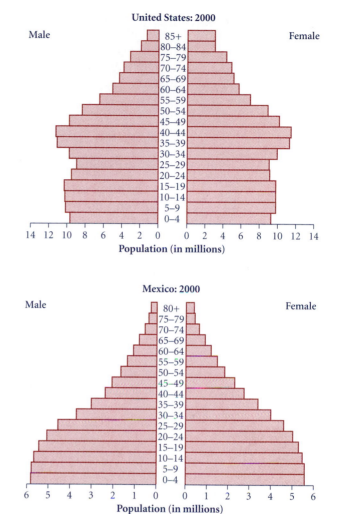

Display 2.43 Population pyramids for the United States and for Mexico for 2000.

Source: U.S. Census Bureau, International Data Base, *www.census.gov/ipc/www/idbpyr.*

2.3 Measures of Center and Spread

So far you have relied on visual methods for estimating summary numbers to measure center and spread. In this section, you will learn how to compute exact values of those same summaries directly from the data.

Measures of Center

The two most commonly used **measures of center** are the mean and the median.

The mean is the balance point.

> The **mean,** $\bar{x}$, is the same number that you called the "average" in your mathematics classes. To compute it, add all the values of x, and divide by the number of values, n:
>
> $$\bar{x} = \frac{\sum x}{n}$$
>
> (The symbol $\sum$, for sum, means to add up all of the values of x.)

The mean is the balance point of a distribution. To estimate the mean visually on a dot plot or histogram, find where you would have to place a finger below the horizontal axis in order to balance the distribution as if it were a tray of blocks. (See Display 2.44.) If a distribution is approximately normal, it balances at the line of symmetry, so the mean is on the horizontal axis directly below the highest point of the bell curve.

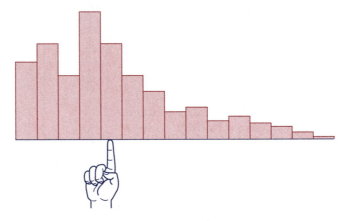

Display 2.44   The mean is the balance point of a distribution.

The median is the halfway point.

The **median** is the value that divides the data into halves as shown in Display 2.45. To find it, list all of the values in order, and select the middle one, or the average of the two middle ones. If there are n values, you can find the median at, or surrounding, position $\frac{n+1}{2}$.

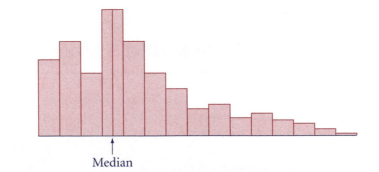

Median

Display 2.45   The median divides the distribution into two equal areas.

Example

The ages of the hourly workers involved in Round 2 of the layoffs at Westvaco were 25, 33, 35, 38, 48, 55, 55*, 55*, 56, and 64* (* means laid off in Round 2). The two dot plots in Display 2.46 show the distributions before and after the second round. What was the effect of Round 2 on the mean age? On the median age?

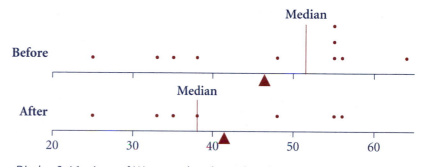

Display 2.46   Ages of Westvaco hourly workers before and after Round 2, showing the means and medians.

Solution

Means:

Before: The sum of the 10 ages is 464, so the mean age is $\frac{464}{10}$ or 46.4 years.

After: There are 7 ages, and their sum is 290, so the mean is $\frac{290}{7}$ or 41.4 years.

The layoffs reduced the mean age by 5 years.

Medians:

Before: Because there are 10 observations, $n = 10$, so $\frac{(n + 1)}{2} = \frac{(10 + 1)}{2} = 5.5$, and the median is halfway between the fifth ordered value, 48, and the sixth, 55. So the median is $\frac{(48 + 55)}{2}$ or 51.5 years.

After: There are 7 ages, so $\frac{(n + 1)}{2} = \frac{(7 + 1)}{2} = 4$. The median is the fourth ordered value, or 38 years.

The layoffs reduced the median age by 13.5 years.   ■

Discussion: Measures of Center

D18. Find the mean and median for each ordered list, and contrast their behavior.

    a. 1 2 3   b. 1 2 6

    c. 1 2 9   d. 1 2 297

D19. As you saw in D18, typically the mean is more affected than the median by an outlier.

    a. Use the fact that the median is the halfway point and the mean is the balance point to explain why this is true.

b. For the distributions of mammal speeds in Display 2.31, the means are 43.5 mph for predators and 31.5 for nonpredators. The medians are 40.5 and 33.5. What is it about the distributions that causes the means to be farther apart than the medians?

c. What is it about the shapes of the plots in Display 2.46 that explains why the means change so much less than the medians?

■ Practice

P17. Find the mean and median of these ordered lists.

a. 1 2 3 4

b. 1 2 3 4 5

c. 1 2 3 4 5 6

d. 1 2 3 4 5 . . . 97 98

e. 1 2 3 4 5 . . . 97 98 99

P18. Five 3rd graders, all about 4 feet tall, are standing together when their teacher, who is 6 feet tall, joins the group. What happens to the mean height? The median height?

P19. The stemplots in Display 2.47 show the life expectancies (in years) for the population in the countries of Africa and Europe. The means are 53.6 years for Africa and 73.6 years for Europe.

a. Find the median of each data set.

b. Is the mean or the median smaller for each distribution? Why is this so?

```
Stem-and-leaf of Life Exp Africa       Stem-and-leaf of Life Exp Europe
N = 54  Leaf Unit = 1.0                N = 39  Leaf Unit = 1.0

     1    4 1                               5    6 88999
     6    4 23333                          15    7 0000001111
    10    4 4455                          (5)    7 22233
    17    4 6666677                        19    7 455555
    23    4 888899                         13    7 6666667777777
   (5)    5 00111
    26    5 33                                   6|8 represents 68 years
    24    5 45555
    19    5 6666
    15    5 88
    13    6 011
    10    6 23
     8    6 4
     7    6 6677
     3    6 89
     1    7
     1    7 3

          6|8 represents 68 years
```

Display 2.47   Life expectancies in Africa and Europe.

Source: Population Reference Bureau, *World Population Data Sheet,* 1996.

Measuring Spread Around the Median: Quartiles and *IQR*

Pair a measure of center with a measure of spread.

If you locate the center of a distribution by dividing your data into a lower and upper half, you can use the same idea to measure spread: Find the values that divide each half in half again. These two values, the lower quartile, Q_1, and the upper quartile, Q_3, together with the median, divide your data into fourths. The distance between the upper and lower quartiles, called the **interquartile range,** or *IQR*, is a measure of spread:

$$IQR = Q_3 - Q_1$$

The next example illustrates the value of the *IQR*. San Francisco, California, and Springfield, Missouri, have about the same average temperature across the year, a little above 55 degrees Fahrenheit. In San Francisco, half the months of the year have their normal temperatures above 56.5°F, half below. For Springfield, half the months have their normal temperatures above 57°F, half below. If you judge by these medians, the difference hardly matters. But if you visit San Francisco, you had better take a jacket, no matter what month you go. If you visit Springfield, however, take your shorts and a T-shirt in the summer and a heavy coat in the winter. The difference in temperatures between the two cities is not in their centers but in their variability. In San Francisco, the middle 50% of the normal monthly temperatures lie in a narrow 9-degree interval between 52.5°F and 61.5°F, whereas in Springfield, the middle 50% of the normal monthly temperatures range widely, over a 31-degree interval, from 40.5°F to 71.5°F. In short, the *IQR* is 9 degrees for San Francisco, 31 degrees for Springfield.

Finding the Quartiles

Use quartiles as a measure of spread with the median.

If you have an even number of cases, finding the quartiles is straightforward: Order your observations, divide them into a lower and upper half, then divide each half in half. If you have an odd number of cases, the idea is still the same, but there's a question of what to do with the middle value when you form the upper and lower halves.

There is no one standard answer, and you may get a slightly different value from some computer programs, but in this book the rule is to *omit the middle value when you form the two halves.*

Example

Find the quartiles for the ages of the hourly workers before and after Round 2 of the layoffs at Westvaco.

Solution

Before: There are 10 ages: 25, 33, 35, 38, 48, 55, 55, 55, 56, 64. Because *n* is even, the median is halfway between the two middle values. The lower half of the data is made up of the first five ordered values, and the median of the lower half is the third value, so $Q_1 = 35$. The upper half of the data is the set of the five largest values, and the median of these is again the third value, so $Q_3 = 55$.

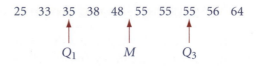

25   33   35   38   48   55   55   55   56   64

      Q_1 M Q_3

After: After the three workers are laid off, there are 7 ages: 25, 33, 35, 38, 48, 55, 56. Because *n* is odd, the median is the middle value, or 38. Omit this one number. The lower half of the data is made up of the three ordered values to the left of position 4. The median of these is the second value, so $Q_1 = 33$. The upper half of the data is the set of the three values to the right of position 4, and the median of these is again the second value, so $Q_3 = 55$.

25   33   35   38   48   55   56

      Q_1 M Q_3 ■

Discussion: Finding the Quartiles

D20. Here are the medians and quartiles for the speeds of the domestic and wild mammals:

	Q_1	Median	Q_3
Domestic	30	37	40
Wild	27.5	36	43.5

    a. Use the information in Display 2.24 to verify these numbers, and then use them to summarize and compare the two distributions.

    b. Why would the speeds of domestic mammals be less spread out than the speeds of wild mammals?

D21. The following quote comes from the mystery *The List of Adrian Messenger* by Philip MacDonald (Garden City, NY: Doubleday, 1959, page 188). Detective Firth asks Detective Seymour if eyewitness accounts have provided a description of the murderer:

> "Descriptions?" he said. "You must've collected quite a few. How did they boil down?"
>
> "To a no-good norm, sir." Seymour shrugged wearily. "They varied so much, the average was useless."

Explain what Detective Seymour means.

■ Practice

P20. Find the quartiles and *IQR*s for these ordered lists.

    a. 1 2 3 4 5 6   b. 1 2 3 4 5 6 7

    c. 1 2 3 4 5 6 7 8   d. 1 2 3 4 5 6 7 8 9

P21. Display 2.48 shows a back-to-back stemplot for the average life spans of predators and nonpredators.

```
     Predators │ │ Nonpredators
            1 │ 0 │ 34
           75 │ · │ 556788
        22222 │ 1 │ 0002222
           65 │ · │ 555555
              │ 2 │ 0000
            5 │ · │
              │ 3 │
              │ · │ 5
              │ 4 │ 1

                    1│5 stands for 15 years
```

Display 2.48 Average life spans of predators and nonpredators.

a. Use the plot to find the medians and quartiles for each group of mammals.

b. Write a pair of sentences summarizing and comparing the two distributions.

Five-Number Summaries, Outliers, and Boxplots

The visual, verbal, and numerical summaries you've seen so far tell you about the middle of a distribution but not about the extremes. If you include the minimum and maximum values, along with the median and quartiles, you get the five-number summary.

> The **five-number summary** for a set of values:
>
> **Minimum:** The smallest value in the set of data
>
> **Lower** or **first quartile, Q_1:** The median of the lower half of the values
>
> **Median:** The value that divides the data into halves
>
> **Upper** or **third quartile, Q_3:** The median of the upper half of the values
>
> **Maximum:** The largest value in the set of data

The difference of the maximum and the minimum is called the **range.**

Display 2.49 shows the five-number summary for the speeds of the mammals listed in Display 2.24.

```
1 │ 1 2              min      11
2 │ 0 5              Q1       30
3 │ 0 0 0 2 5 9      median   37
4 │ 0 0 0 2 5 8      Q3       42
5 │ 0                max      70
6 │
7 │ 0
```

Display 2.49 Five-number summary for the mammal speeds.

A boxplot is sometimes referred to as a *box and whiskers plot*.

Display 2.50 is a boxplot for the mammal speeds. A **boxplot** is a graphical display of the five-number summary. The "box" extends from Q_1 to Q_3, with a line across it at the median. The "whiskers" run from the quartiles to the most extreme values.

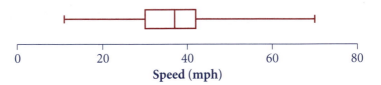

Display 2.50 Boxplot of mammal speeds.

1.5 · *IQR* rule for outliers

The maximum speed of 70 mph for the cheetah is 20 mph from the next fastest mammal (the lion) and 28 mph from the nearest quartile. It is handy to have a version of the boxplot that shows isolated cases—outliers—like the cheetah. Informally, outliers are any values that stand apart from the rest, but you can use this rule to identify them:

> A value is an **outlier** if it is more than 1.5 times the *IQR* from the nearest quartile.

Note that "more than 1.5 times the *IQR* from the nearest quartile" is another way of saying "either greater than $Q_3 + 1.5 \cdot IQR$, or less than $Q_1 - 1.5 \cdot IQR$."

Example

Use the 1.5 · *IQR* rule to identify outliers and the largest and smallest non-outliers among the mammal speeds.

Solution

From Display 2.49, $Q_1 = 30$ and $Q_3 = 42$, so the $IQR = 42 - 30 = 12$, and $1.5 \cdot IQR = 18$.

At the low end:

$$Q_1 - 1.5 \cdot IQR = 30 - 18 = 12$$

The pig, at 11 mph, is an outlier.
The squirrel, at 12 mph, is the smallest non-outlier.

At the high end:

$$Q_3 + 1.5 \cdot IQR = 42 + 18 = 60$$

The cheetah, at 70 mph, is an outlier.
The lion, at 50 mph, is the largest non-outlier. ■

A **modified boxplot** (shown in Display 2.51) is like the basic boxplot, except that the whiskers extend only as far as the largest and smallest non-outliers (sometimes called **adjacent values**) and any outliers appear as individual dots or other symbols.

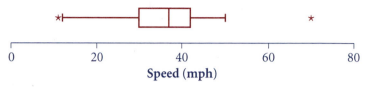

Display 2.51 Modified boxplot of mammal speeds.

Boxplots are particularly useful for comparing several distributions.

Example

Display 2.52 shows side-by-side modified boxplots of average longevity for wild and domestic mammals. Compare the two distributions.

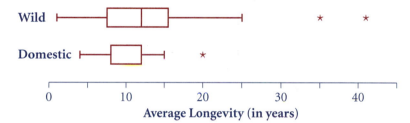

Display 2.52 Comparison of average longevity.

Solution

The boxplot for domestic animals has no median line. So many domestic animals had an average longevity of 12 years that it is both the median and the upper quartile. Keeping that in mind, these plots show that, typically, species of domestic mammals have median average life spans of about 12 years, with about half of these average life spans falling between 8 and 12 years. The average life spans for wild mammals center at about the same place, but the wild mammal averages have more variability. The unusual average life spans are on the high side; two large mammals have average life spans of more than 30 years. ■

When are boxplots most useful?

Boxplots are useful when you are plotting a single quantitative variable and

- you want to compare the shape, center, and spreads of two or more distributions
- your distribution has so many values that it would take too long, or use too much space, to show them individually in a stemplot
- you don't need to see individual values, even approximately
- you don't need to see more than the five-number summary but would like outliers clearly indicated

Discussion: Five-Number Summaries, Outliers, and Boxplots

D22. Does the five-number summary give the position of the quartiles or the value of the quartiles, or is there any difference? What is another name for the second quartile?

D23. Test your ability to interpret boxplots with these questions.

a. Approximately what percentage of the values in a data set lie within the box? Within the lower whisker, if there are no outliers? Within the upper whisker, if there are no outliers?

b. How would a boxplot look for a data set that is skewed right? Skewed left? Symmetric?

c. How can you estimate the *IQR* from a boxplot without the five-number summary? How can you estimate the range?

d. Contrast the information you can learn from a boxplot with that from a histogram. List the advantages and the disadvantages of each.

■ Practice

P22. Display 2.53 shows a boxplot of the Nielsen ratings from Display 2.20 and Display 2.21 of Section 2.1.

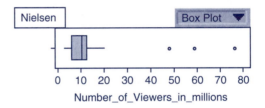

Display 2.53   Modified boxplot of Nielsen ratings.

a. Which three shows are the outliers?

b. Which show is at the top of the upper whisker (the largest non-outlier)?

c. Without looking back, sketch a histogram that could result in this boxplot.

P23. Use the medians and quartiles given in D20 and the data in Display 2.24 to construct side-by-side boxplots for the speeds of wild and domestic mammals. (Don't show outliers in these plots.)

P24. The stemplot of average mammal life spans appears in Display 2.54.

```
0 | 1 3 4                          1|5 stands for 15 years
· | 5 5 5 6 7 7 8 8
1 | 0 0 0 2 2 2 2 2 2 2 2
· | 5 5 5 5 5 5 5 6
2 | 0 0 0 0
· | 5
3 |
· | 5
4 | 1
```

Display 2.54   Average life span (in years) for 38 mammals.

a. Use it to find the five-number summary.

b. Find the *IQR*.

c. Compute $Q_1 - 1.5 \cdot IQR$. Identify any outliers (give the animal name and life span) at the low end.

d. Now identify an outlier at the high end and the largest non-outlier.

P25. Use your answers in P24 to draw a modified boxplot.

P26. Is it possible for a boxplot to be missing a whisker? If so, give an example. If not, explain why not.

Percentiles and Cumulative Frequency Plots

The first quartile, Q_1, of a distribution is the 25th percentile—the value that separates the lowest 25% of the data from the rest. The median is the 50th percentile, and Q_3 is the 75th percentile. In the same way, you can define other percentiles. The 10th percentile, for example, is the value that separates the bottom 10% of values in a distribution from the rest.

For large data sets, you may see data listed in a table or plotted in a graph like the SAT I verbal scores in Display 2.55. This plot is sometimes called a **cumulative percentage plot** or a **cumulative relative frequency plot.** The table shows that, for example, 30% of the students received a score of 450 or lower. About 14% received a score between 400 and 450.

Score	Percentile		Score	Percentile
800	99+		450	30
750	98		400	16
700	95		350	7
650	89		300	3
600	79		250	1
550	65		200	—
500	47			

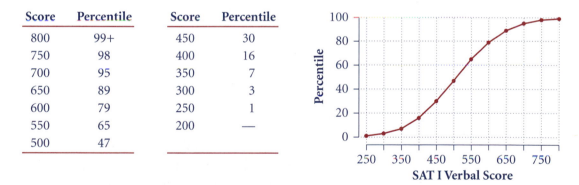

Display 2.55 Cumulative relative frequency plot of SAT I verbal scores and percentiles, 1999–2000.

Source: The College Board, *www.collegeboard.org.*

Discussion: Percentiles and Cumulative Frequency Plots

D24. Refer to Display 2.55.

a. Use the plot to estimate the percentile for an SAT I verbal score of 425.

b. What two values enclose the middle 90% of the SAT scores? The middle 95%?

c. Use the table to estimate the score that falls at the 40th percentile.

D25. What fraction of the cases lie between the 5th and 95th percentiles of a distribution? What percentiles enclose the middle 95% of the cases in a distribution?

■ Practice

P27. Estimate the quartiles and the median of the SAT I verbal scores in Display 2.55, and use those values to draw a boxplot for the distribution. What is the value of the *IQR*?

Measuring Spread Around the Mean: The Standard Deviation

There are various ways you can measure the spread of a distribution around its mean. The next activity will give you a chance to create a measure of your own.

Activity 2.3 Comparing Hand Spans: How Far Are You from the Mean?

What you'll need: a ruler

1. Spread your hand on a ruler and measure your hand span (the distance from the tip of your thumb to the tip of your little finger when you spread your fingers) to the nearest half centimeter.

2. Find the mean hand span for your group.

3. Make a dot plot of the results for your group. Write names or initials above the dots to identify the cases. Mark the mean with a wedge (▲) below the number line.

4. Give two sources of variability in the measurements. That is, give two reasons why the measurements aren't all the same.

5. How far is your hand span from that of the mean of your group? How far from the mean are the hand spans of the others in your group?

6. Make a second plot, this time a dot plot of differences from the mean. Again, label the dots with names or initials. What is the mean of these differences? Tell how to get the second plot from the first without computing any differences.

7. Using the idea of differences from the mean, invent at least two measures that give a "typical" distance from the mean.

8. Compare your measures with those of the other groups in your class. Discuss the advantages and disadvantages of each group's method.

The differences from the mean, $x - \bar{x}$, are called **deviations.** The mean is the balance point of the distribution, so the set of deviations from the mean will always add to zero.

Deviations from the mean add to zero:

$$\sum (x - \bar{x}) = 0$$

Advantages of the
standard deviation as a
measure of spread

What is a typical deviation? As you saw in the activity, there are various ways to say what you mean by "typical," but one measure, the standard deviation, abbreviated *SD,* or *s,* offers an important advantage you don't get with other measures. There is a simple relationship between the standard deviation of a list of values and the standard deviation of the averages you get when you repeatedly choose random samples from the list. This reason for using the standard deviation depends on things you won't learn about until Chapter 5. But you can get a preview of the basic idea if you turn back to Display 1.8, the simulation of the process of randomly choosing workers to lay off from Westvaco. If you'd had to do all those simulations by hand, you'd have been busy for quite a while, but there's a shortcut. Unlike other measures of spread, you can compute the value of the standard deviation for the distribution of all those sample averages *without doing any simulations.* You only need to know two things: the number of workers you were choosing in each random sample and the standard deviation for the set of 10 workers you were choosing from. This remarkable property makes the standard deviation the most useful measure of spread for working with random samples.

To get these advantages, you have to work with *squared* deviations $(x - \bar{x})^2$. To compute the standard deviation, you first square the deviations, then take the average of those squares, and then take the square root.

Dividing by *n* or *n* – 1

Two versions of the standard deviation formula are used. One divides by the sample size *n* to get the average of the squared deviations; the other divides by $n - 1$. Your calculator probably computes both of these. (On some calculators, the two versions are labeled σ_n and σ_{n-1}.) Dividing by $n - 1$ gives a slightly larger value for the standard deviation, and the larger value works better in statistical inference. If the choice makes much difference in the value of the standard deviation, however, your sample is probably too small for the standard deviation to be of much practical use anyway. For now, even though dividing by *n* may seem more natural, use $n - 1$ instead. We will come back to this in Chapter 5.

Formula for the Standard Deviation, *s*

$$s = \sqrt{\frac{\sum (x - \bar{x})^2}{n - 1}}$$

The square of the standard deviation, s^2, is called the **variance.**

Example

Compute the standard deviation for the average longevity of domesticated mammals from Display 2.24.

Solution

The table in Display 2.56 is a good way to organize the steps. First find the mean longevity $\bar{x}$, then subtract it from each observed value x to get the deviations, $x - \bar{x}$. Square each deviation to get $(x - \bar{x})^2$.

Case	Longevity x	Mean $\bar{x}$	Deviation $x - \bar{x}$	Squared Deviation $(x - \bar{x})^2$
Cat	12	11	1	1
Cow	15	11	4	16
Dog	12	11	1	1
Donkey	12	11	1	1
Goat	8	11	−3	9
Guinea pig	4	11	−7	49
Horse	20	11	9	81
Pig	10	11	−1	1
Rabbit	5	11	−6	36
Sheep	12	11	1	1
Total	**110**	**110**	**0**	**196**

Display 2.56 Computing the standard deviation.

To get the standard deviation, sum up the squared deviations, divide the sum by $n - 1$, and finally, take the square root:

$$s = \sqrt{\frac{\sum (x - \bar{x})^2}{n - 1}} = \sqrt{\frac{196}{10 - 1}} \approx 4.67 \quad \blacksquare$$

Discussion: The Standard Deviation

D26. Does 4.67 years seem like a typical distance from the mean of 11 years for the average life spans in the example?

D27. The average longevities are measured in years. What is the unit of measurement for the mean? For the standard deviation? For the variance? For the interquartile range? For the median?

D28. When you divide by $n - 1$ rather than by n, what effect does it have on the standard deviation?

D29. The standard deviation, if you look at it the right way, is a generalization of the usual formula for the distance between two points. How does the formula for the standard deviation remind you of the formula for the distance between two points?

■ Practice

P28. Verify that the sum of the deviations from the mean is 0 for the set 1, 2, 4, 6, 9. Find the standard deviation.

P29. Without computing, match each list of numbers on the left with its standard deviation in the right column. Check any answers you aren't sure of by computing.

a. 1 1 1 1   i. 0

b. 1 2 2   ii. 0.058

c. 1 2 3 4 5   iii. 0.577

d. 10 20 20   iv. 1.581

e. 0.1 0.2 0.2   v. 3.162

f. 0 2 4 6 8   vi. 3.606

g. 0 0 0 0 5 6 6 8 8   vii. 5.774

Properties of the Summary Statistics

Plot first, then look for summaries.

Which summary statistics should you use to describe a distribution? Mean and standard deviation? Median and quartiles? Something else? The right choice depends on the shape of your distribution, so you should always start with a plot. For normal-shaped distributions, the mean and standard deviation are nearly always the most suitable summaries. For skewed distributions, the median and quartiles are often the most useful summaries, in part because they have a simple interpretation based on dividing a data set into fourths.

Sometimes, however, the mean and standard deviation will be the right choices even if you have a skewed distribution. For example, if you have a representative sample of house prices for a town and you want to use your sample to estimate the total value of all the town's houses, the mean is what you want, not the median. Later, when you study statistical inference, you'll find that the standard deviation is the most useful measure of spread. This is because, as you saw in E18, the distribution of the sample means is approximately normal with a standard deviation that is easily estimated.

Choosing the right summaries is something you will get better at as you build your intuition about the properties of the summary statistics and how they behave in various situations.

Discussion: Which Summary Statistic?

D30. Explain how to determine the total amount of property taxes if you know the number of houses, the mean value, and the tax rate. In what sense is knowing the mean equivalent to knowing the total?

D31. When the "average" income of a community's residents is given, that number is usually the median. Why do you think that is the case?

D32. Which summary statistics would be most useful in the following situations?

a. You are designing airline seats and want them to be wide enough for most people.

b. You are looking for the best buy on a specific type of calculator.

c. You would like to get a job when you start college but are unsure of how many hours you will need for study time.

■ Practice

P30. A community near Los Angeles has 9751 households with a median house price of $320,000 and an average price of $392,059. Why is the mean larger than the median? The property tax rate is about 1.15%. What is the total amount of taxes that will be assessed on these houses? What is the average amount per house?

P31. A story in the *Los Angeles Times* (July 30, 1998, page W14) reported that the median age of a car in 1997 was 8.1 years, the oldest ever. The medians were 6.5 years in 1990 and 4.9 years in 1970.

a. Why were medians used in this story?

b. What reasons might there be for the increase in median age of cars?

The Effects of Recentering and Rescaling

The next example illustrates some important properties of summary statistics. It will also help you develop your intuition about how the geometry and arithmetic of working with data are related.

The lowest temperature on record for Washington, D.C., is −15°F. How does that compare with the lowest recorded temperatures for cities of other countries? Display 2.57 gives data for the few cities whose record temperatures turn out to be whole numbers in both the Fahrenheit and Celsius scales.

City	Country	Temperature (°F)
Addis Ababa	Ethiopia	32
Algiers	Algeria	32
Bangkok	Thailand	50
Madrid	Spain	14
Nairobi	Kenya	41
São Paulo	Brazil	32
Warsaw	Poland	−22

Display 2.57 Record low temperatures for seven cities.
Source: National Climatic Data Center, 2002,
www.ncdc.noaa.gov/pub/data/specal/mintemps.

The dot plot in Display 2.58 shows that the temperatures are centered at about 32 with an outlier at −22. The spread and shape are hard to determine with only seven values.

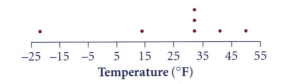

Display 2.58 Dot plot for record low temperatures in °F for seven cities.

What happens to the shape and spread of this distribution if you convert each temperature to number of degrees above or below freezing, 32°F? To find out, subtract 32 from each value, and plot the new values. Display 2.59 shows that the center of the dot plot is now at 0 rather than 32 but that the spread and shape are unchanged.

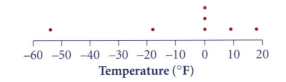

Display 2.59 Dot plot of the number of degrees Fahrenheit above or below freezing for record low temperatures for the seven cities.

Adding or subtracting a constant to each value in a set of data doesn't change the spread or the shape of a distribution but slides the entire distribution a distance equivalent to the constant. Thus, the transformation amounts to a recentering of the distribution.

What happens to the shape and spread of this distribution if you convert each temperature to °C? The Celsius scale measures temperature using the number of degrees above or below freezing, but it takes 1.8°F to make 1°C. To convert, divide each value in Display 2.59 by 1.8, and plot the new values. Display 2.60 shows that the center of the new dot plot is still at 0 and the shape is the same but the spread has decreased by a factor of 1.8.

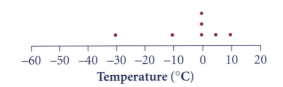

Display 2.60 Dot plot for record low temperatures in °C for the seven cities.

Multiplying or dividing each value in a set of data by a positive constant doesn't change the basic shape of the distribution. The mean and the spread both are multiplied by that number. Thus, this transformation amounts to a rescaling of the distribution.

Recentering and Rescaling a Data Set

Recentering a data set—adding the same number c to all the values—doesn't change the shape or spread but slides the entire distribution by the amount c, adding c to the median and the mean.

Rescaling a data set—multiplying all the values by the same nonzero number d—doesn't change the basic shape but stretches or shrinks the distribution, multiplying the spread (*IQR* and standard deviation) by $|d|$, and multiplying the center (median and mean) by d.

Discussion: Recentering and Rescaling Data

D33. Suppose a U.S. dollar is worth 9.4 Mexican pesos.

    a. A set of prices, in U.S. dollars, has mean $20 and standard deviation $5. Find the mean and standard deviation of the same prices expressed in pesos.

    b. Another set of prices, in Mexican pesos, has a median of 94 pesos and quartiles of 47 and 188 pesos. Find the median and quartiles for the same prices expressed in dollars.

■ **Practice**

P32. The mean height of a class of 15 children is 48 inches, the median is 45 inches, the standard deviation is 2.4 inches, and the interquartile range is 3 inches. Find the mean, standard deviation, median, and interquartile range if

    a. you convert each height to feet

    b. each child grows 2 inches

    c. each child grows 4 inches and you convert their heights to feet

P33. Compute means and standard deviations (use the formula for s) for these sets of numbers. Use recentering and rescaling wherever you can to avoid or simplify the arithmetic.

    a. 1 2 3

    b. 11 12 13

    c. 10 20 30

    d. 105 110 115

    e. −800 −900 −1000

The Influence of Outliers

A summary statistic is **resistant to outliers** if the summary statistic is not changed very much when an outlier is removed from the set of data. If the summary statistic tends to be affected by outliers, it is **sensitive to outliers.**

Display 2.61 again shows the dot plot for the Nielsen ratings from Display 2.20.

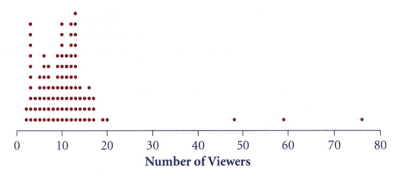

Display 2.61 Nielsen ratings of television shows from data in Display 2.20.

The three highest values—the three shows with the largest numbers of viewers—are outliers.

The printout in Display 2.62 gives summary statistics for all 101 shows.

Variable	N	Mean	Median	TrMean	StDev	SEMean
Ratings	101	11.187	10.150	9.831	9.896	0.985

Variable	Min	Max	Q1	Q3
Ratings	2.320	76.260	6.160	12.855

Display 2.62 Minitab printout of the summary statistics for all Nielsen ratings.

The second printout, in Display 2.63, gives summary statistics when the three outliers are removed from the set of ratings.

Variable	N	Mean	Median	StDev
No Outs	98	9.666	10.145	4.250

Variable	Min	Max	Q1	Q3
No Outs	2.320	20.470	6.065	12.698

Display 2.63 Summary statistics for Nielsen ratings without outliers.

Discussion: The Influence of Outliers

D34. Are these measures of center affected much by the three outliers? Explain why that is the case.

    a. Mean   b. Median

D35. Are these measures of spread affected much by the three outliers? Explain why that is the case.

    a. Range   b. Standard deviation

    c. Interquartile range

■ Practice

P34. The histogram and summary statistics in Display 2.64 and Display 2.65 show the record low temperatures for the 50 states.

    a. Hawaii has a lowest recorded temperature of 12°F. The boxplot shows Hawaii as an outlier. Verify that this is justified.

b. Suppose you exclude Hawaii from the data set. Copy the table in Display 2.65, but substitute your best estimate for the summary statistics now that Hawaii has been excluded.

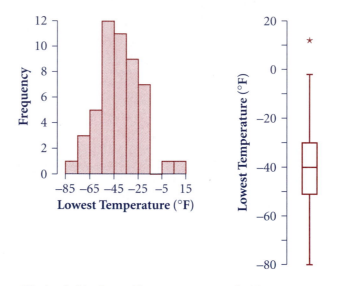

Display 2.64 Record low temperatures for the states.

Source: National Climatic Data Center, 2002, *www.ncdc.noaa.gov/pub/data/specal/mintemps.*

```
Summary of          Lowest Temperature
No Selector

Percentile        25

              Count      50
               Mean    -40.3800
             Median    -40
             StdDev     17.6946
                Min    -80
                Max     12
              Range     92
     Lower ith %tile    -51
     Upper ith %tile    -30
```

Display 2.65 Summary statistics for lowest temperatures by state.

Summaries from a Frequency Table

To find the mean of the numbers 5, 5, 5, 5, 5, 5, 8, 8, 8, you could add them and divide their sum by how many there are. However, you could get the same answer faster by taking advantage of the repetitions:

$$\bar{x} = \frac{5 \cdot 6 + 8 \cdot 3}{6 + 3} = \frac{30 + 24}{9} = \frac{54}{9} = 6$$

You can use formulas to find the mean and standard deviation of a frequency table like the one in Display 2.66.

Formulas for the Mean and Standard Deviation of a Frequency Table

If each value x occurs with frequency f, the mean of a frequency table is given by

$$\bar{x} = \frac{\sum x \cdot f}{n}$$

The standard deviation is

$$s = \sqrt{\frac{\sum (x - \bar{x})^2 \cdot f}{n - 1}}$$

where n is the sum of the frequencies or $n = \sum f$.

Example

Suppose you have 5 pennies, 3 nickels, and 2 dimes. Find the mean value per coin and the standard deviation.

Solution

The table in Display 2.66 shows a way to organize the steps for computing the mean using the formula for the mean of a frequency table.

	Value x	Frequency f	xf
Penny	1	5	5
Nickel	5	3	15
Dime	10	2	20
Sum		**10**	**40**

$$\bar{x} = \frac{\sum xf}{n} = \frac{40}{10} = 4$$

Display 2.66   Steps for computing the mean for a frequency table.

Display 2.67 gives an extended version of the table, designed to organize the steps for computing both the mean and the standard deviation.

	Value x	Frequency f	xf	$x - \bar{x}$	$(x - \bar{x})^2$	$(x - \bar{x})^2 \cdot f$
Penny	1	5	5	−3	9	45
Nickel	5	3	15	1	1	3
Dime	10	2	20	6	36	72
Sum		**10**	**40**			**120**

$$s = \sqrt{\frac{\sum (x - \bar{x})^2 \cdot f}{n - 1}} = \sqrt{\frac{120}{9}} \approx 3.65$$

Display 2.67   Steps for computing the *SD* for a frequency table.

Discussion: Summaries from a Frequency Table

D36. Display 2.68 shows the data on family size for two representative sets of 100 families, one set from 1967 and one from 1997.

a. Try to visualize the shapes of the two distributions. Are they symmetric, skewed left, or skewed right?

b. Find the median number of children per family for 1967.

c. Use the formulas to compute the mean and standard deviation for 1967.

1967		1997	
Number of Children	**Number of Families**	**Number of Children**	**Number of Families**
0	5	0	15
1	10	1	22
2	21	2	25
3	28	3	18
4	17	4	10
5	7	5	2
6	4	6	4
7	3	7	2
8	5	8	2

Display 2.68 Number of children in a sample of families, 1967 and 1997.

Source: U.S. Census Bureau, *http://www.census.gov/prod/3/98pubs/p20-509u.pdf.*

D37. Explain why the formula for the standard deviation in the boxed summary above gives the same answer as the formula on page 45.

■ Practice

P35. Refer to Display 2.68.

a. Use the formula for the mean and standard deviation of a frequency table to compute the mean number of children per family and the standard deviation for 1997.

b. Find the median number of children for 1997.

c. What are the positions of the quartiles in an ordered list of 100 numbers? Find the quartiles for 1967 and compute the *IQR*. Do the same for 1997.

d. Write a comparison of the distributions for the two years.

P36. Suppose you have 5 pennies, 6 nickels, 4 dimes, and 5 quarters.

a. Sketch a dot plot of the values of the 20 coins, and use it to estimate the mean.

b. Compute the mean using the formula for the mean of a frequency table.

c. Estimate the *SD* from your plot: Is it closest to 0, 5, 10, 15, or 20?

d. Compute the standard deviation using the formula for the standard deviation of a frequency table.

Summary 2.3: Measures of Center and Spread

Your first step in any data analysis should always be to look at a plot of your data because the shape of the distribution will help you determine what summary measures to use for center and spread.

- To describe the center of a distribution, the two most common summaries are the median and the mean. The median, or halfway point, of a set of ordered values is either the middle value (if n is odd) or halfway between the two middle values (if n is even). The mean, or balance point, is the sum of the values divided by how many there are.

- To measure spread around the median, use the interquartile range, or *IQR,* which is the width of the middle 50% of the data values and equals the distance from the lower quartile to the upper quartile. The quartiles are the medians of the lower half and upper half of the ordered list of values. To measure spread around the mean, use the standard deviation. To compute the standard deviation for a data set of size n, first find the deviations from the mean, then square them, add the squared deviations, then divide by $n - 1$, and take the square root.

A boxplot is a useful way to compare the general shape, center, and spread of two or more distributions with a large number of values. A modified boxplot shows outliers as well.

An outlier is any value more than $1.5 \cdot IQR$ from the nearest quartile. If a summary statistic doesn't depend much on whether you include or exclude outliers from your data set, then it is said to be resistant.

- The median and quartiles are resistant to outliers.

- The mean and standard deviation, on the other hand, are sensitive to outliers.

Recentering a data set—adding the same number c to all the values—slides the entire distribution. It doesn't change the shape or spread but adds c to the median and the mean. Rescaling a data set—multiplying all the values by the same nonzero number d—is like stretching or squeezing the distribution. It doesn't change the basic shape but multiplies the spread (*IQR* and standard deviation) by $|d|$ and multiplies the measure of center (median and mean) by d.

■ Exercises

E23. Discuss whether you would use the mean or the median to measure the center of the following sets of data and why you prefer the one you choose.

    a. The prices of single-family homes in your neighborhood

    b. The yield of corn (bushels per acre) for a sample of farms in Iowa

    c. The survival time, following diagnosis, of a sample of cancer patients

E24. Three histograms and three boxplots appear in Display 2.69. Which boxplot displays the same information as

a. Histogram A?

b. Histogram B?

c. Histogram C?

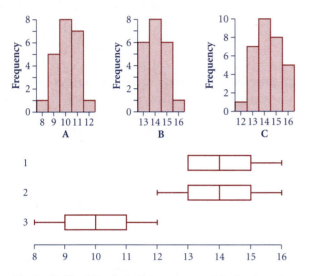

Display 2.69 Match the histograms with the boxplots.

E25. Make side-by-side boxplots for the speeds of predators and nonpredators. (The stemplot in Display 2.31 shows the values already ordered.) Are the boxplots or the back-to-back stemplot in Display 2.31 better for comparing these speeds? Explain.

E26. The test scores of 40 students in a first-period class were used to construct the first boxplot in Display 2.70, and test scores of 40 students in a second-period class were used for the second. Can the third plot be a boxplot of the scores of the 80 students in the two classes combined? Why or why not?

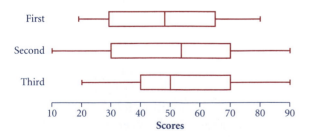

Display 2.70 Boxplots for two sets of test scores.

E27. The mean of a set of seven values is 25. Six of the values are 24, 47, 34, 10, 22, and 28. What is the seventh value?

E28. No computing should be necessary to answer these questions.

a. The mean of each of the following sets of values is 20, and the range is 40. Which set has the largest standard deviation? Which has the smallest?

I. 0 10 20 30 40

II. 0 0 20 40 40

III. 0 19 20 21 40

b. Two of the following sets of values have a standard deviation of about 5. Which two are they?

I. 5 5 5 5 5 5

II. 10 10 10 20 20 20

III. 6 8 10 12 14 16 18 20 22

IV. 5 10 15 20 25 30 35 40 45

E29. The standard deviation of the first set of values below is about 30. What is the standard deviation of the second set? Explain. No computing should be necessary.

16 23 34 56 78 92 93

20 27 38 60 82 96 97

E30. Consider the set of the heights of all female NCAA athletes and the set of heights of all female NCAA basketball players. Which distribution will have the larger mean? Which will have the larger standard deviation? Explain.

E31. *Mean versus median.*

a. You are tracing your family tree and would like to go back to the year 1700. To estimate how many generations back you will have to trace, would you need to know the median length of a generation or the mean length of a generation?

b. If a car trip takes 3 hours, do you need to know the average speed or the median speed in order to get the total distance?

c. Suppose that all trees in a forest are right circular cylinders with a radius of 3 feet. The heights vary, but the mean height is 45 feet, the median is 43 feet, the *IQR* is 3 feet, and the standard deviation is 3.5 feet. From this information, can you compute the total volume of wood?

E32. Consider the following data set: 15, 8, 25, 32, 14, 8, 25, 2. You may replace any one value with a number from 1 to 10. How would you make this replacement

a. to make the standard deviation as large as possible?

b. to make the standard deviation as small as possible?

c. to create an outlier, if possible?

E33. The histogram in Display 2.71 shows record high temperatures by state.

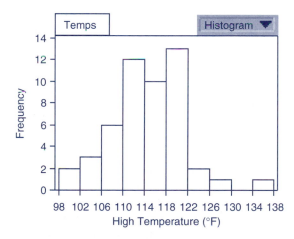

Display 2.71   Record high temperatures for the 50 U.S. states.

Source: National Climatic Data Center, 2002, *www.ncdc.noaa.gov/pub/data/special/maxtemps.pdf.*

a. Suppose each of the temperatures is converted from degrees Fahrenheit, F, to degrees Celsius, C, using the formula

$$C = \frac{5}{9}(F - 32)$$

Make a histogram of the temperatures in °C.

b. The summary statistics in Display 2.72 are for the temperatures in °F. Make a similar table for the temperatures in °C.

c. Are there any outliers in the data in °C?

Variable	N	Mean	Median	TrMean	StDev
HighTemp	50	114.10	114.00	113.95	6.69

Variable	Min	Max	Q1	Q3
HighTemp	100.00	134.00	110.00	118.00

Display 2.72   Summary statistics for record high temperatures for the 50 U.S. states.

E34. Suppose the sum of the squared deviations is 400.

a. Compare the standard deviation that would result from

i. dividing by 10 versus dividing by 9

ii. dividing by 100 versus dividing by 99

iii. dividing by 1000 versus dividing by 999

b. Does the decision to use n or $n - 1$ in the formula for the standard deviation matter very much if the sample size is large?

E35. This table shows the weights of pennies from Display 2.4 with the weights for each penny taken to be the value at the midpoint of the interval.

Weight	Frequency
2.99	1
3.01	4
3.03	4
3.05	4
3.07	7
3.09	17
3.11	24
3.13	17
3.15	13
3.17	6
3.19	2
3.21	1

a. Find the mean weight of the pennies.

b. Find the standard deviation of the weights.

c. Does the standard deviation appear to represent a typical deviation from the mean?

E36. For the countries of Europe, many of the average life expectancies are approximately the same, as you can see from the stemplot in Display 2.47. Use the formulas for a frequency table to compute the mean and standard deviation of the life expectancies for the countries of Europe.

E37. Make a back-to-back stemplot comparing the ages of those retained and those laid off among the salaried workers in the engineering department at Westvaco. Find the medians and quartiles, and use them to write a verbal comparison of the two distributions.

E38. Using only the basic boxplot in Display 2.73, show that there must be outliers in the set of average longevity.

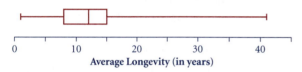

Display 2.73 Boxplot of average longevity.

E39. Display 2.74 shows the boxplot of average longevity, showing outliers. How many outliers are there?

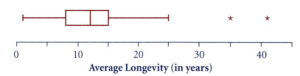

Display 2.74 Modified boxplot of average longevity, showing outliers.

How many outliers are shown in Display 2.52? How can that be, considering the boxplot shown in Display 2.74?

E40. Without computing, what can you say about the standard deviation of this set of values: 4, 4, 4, 4, 4, 4, 4, 4?

E41. Tell how you could use recentering and rescaling to simplify the computation of the mean and standard deviation for this list of numbers:

5478.1	5478.3	5478.3
5478.9	5478.4	5478.2

E42. Suppose a constant c is added to each value in a set of data, $x_1, x_2, x_3, x_4,$ and x_5. Prove that the mean increases by c by comparing the formula for the mean of the original data with the formula for the mean of the recentered data.

E43. Suppose a constant c is added to each value in a set of data, $x_1, x_2, x_3, x_4,$ and x_5. Prove that the standard deviation is unchanged by comparing the formula for the standard deviation of the original data with that for the standard deviation of the recentered data.

E44. In 1998, 32 of the 50 U.S. states either had no death penalty or executed no one. Of the states that did carry out executions, Texas led the list with 20 executions, followed by Virginia (13); South Carolina (7); Arizona, Oklahoma, and Florida (4 each); and Missouri and North Carolina (3 each). Another 10 states executed 1 person. What was the mean number of executions per state? The median number? What were the quartiles? Draw a boxplot, showing any outliers, of the number of executions.

Source: Tracy L. Snell, *Bulletin: Capital Punishment 1998* NCJ 179012 (U.S. Dept. of Justice, Bureau of Justice Statistics, 1999 [rev. Jan. 2000]).

2.4 The Normal Distribution

You have seen several reasons why the normal distribution is so important:

- It tells you how variability in measuring often behaves (tennis balls).
- It tells you how variability in populations often behaves (weights of pennies, SAT scores).
- It tells you how averages (and some other summary statistics) behave when you repeat a random process (Westvaco case, Activity 1.1).

In this section, you will learn that if you know that a distribution is normal (shape), then the mean (center) and standard deviation (spread) tell you everything else about the distribution. The reason is that, whereas skewed distributions come in many different shapes, there is only one normal shape. It's true that one normal distribution may appear tall and thin while another looks short and fat. However, the *x*-axis of the tall, thin one can be stretched out so that the two normal distributions look exactly the same.

These are both the same normal curve.

Unknown Percentage and Unknown Value Problems

The basic skills you need in order to utilize the normal distribution are illustrated by solving two related problems: the unknown percentage problem and the unknown value problem. Here's one of each type.

In a recent year, the distribution of SAT I scores for the incoming class at the University of Washington was roughly normal in shape, with mean 1055 and standard deviation 200.

Unknown percentage problem (Display 2.75): What percentage of scores were 920 or below?

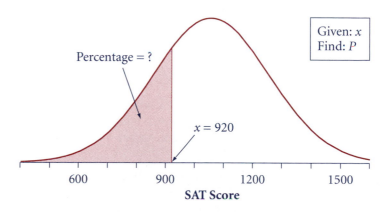

Display 2.75 The unknown percentage problem.

Unknown value problem (Display 2.76): What SAT score separates the lowest 25% of the SAT scores from the rest?

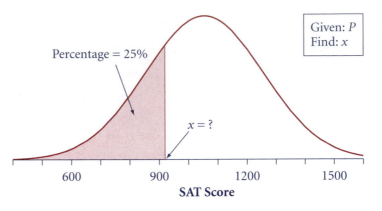

Display 2.76 The unknown value problem.

Notice how the two problems are counterparts. To find an unknown percentage, *P*, you must know the corresponding value, *x*. To find an unknown value, you must know the corresponding percentage.

Discussion: Unknown Percentage and Unknown Value Problems

D38. Which of the following situations are unknown percentage problems, and which are unknown value problems? For each, draw and label a normal curve, showing the three quantities that are given and the one quantity to find.

a. In the Westvaco simulation of Chapter 1, the averages from 1000 random samples of size 3 were roughly normal, with mean 46.9 and standard deviation 6.1. What is the chance of getting an average of 58 or more?

b. In another set of 1000 random samples, the distribution of averages was also normal, with mean 46.4 and standard deviation 6.2. For this distribution, find the age that cuts off the largest 2.5% of the values.

■ Practice

P37. Which of the following situations are unknown percentage problems, and which are unknown value problems? For each, draw and label a normal curve, showing the three quantities that are given and the one quantity to find.

a. In a recent year, students entering the University of Florida had a mean SAT I score of 1135, with standard deviation 180. The distribution was roughly normal. What percentage of SAT I scores were greater than 1300?

b. In 2000, the mean SAT I math score nationally was 514, with a standard deviation of 113. Find the upper quartile of the distribution.

The Standard Normal Distribution

Because all normal distributions have the same basic shape, you can use recentering and rescaling to change any normal distribution to the one that has mean 0 and standard deviation 1. Solving unknown percentage and unknown value problems depends on this important property.

> The normal distribution that has mean 0 and standard deviation 1 is called the **standard normal distribution.** With this distribution, we call the variable along the horizontal axis a *z*-score.

The standard normal distribution is symmetric, with total area under the curve equal to 1, or 100%. To find the percentage, *P,* that is the area to the left of the corresponding *z*-score, you can use the *z*-table or your calculator.

The next two examples show how you use the *z*-table, which is Table A in the appendix.

Example

Find the percentage, *P,* of values below $z = 1.23$.

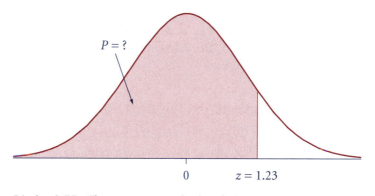

Display 2.77   The percentage of values below $z = 1.23$.

Solution

Tail probability p			
z	.02	.03	.04
1.20	.8888	**.8907**	.8925

Think of 1.23 as 1.2 + .03. In Table A in the appendix, find the row headed 1.2 and the column headed .03. Where this row and column intersect, you find the decimal .8907. So 89.07% of standard normal scores are below 1.23. ■

Example

Find the z-score that falls at the 75th percentile of the standard normal distribution; that is, the z-score that divides the bottom 75% of the values from the rest.

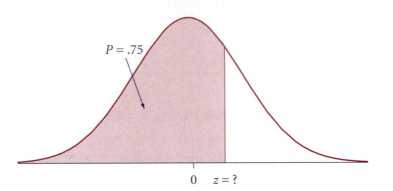

$P = .75$

$0 \quad z = ?$

Display 2.78   The z-score that corresponds to the 75th percentile.

Solution

	Tail probability p		
z	.06	.07	.08
.60	.7454	**.7486**	.7517

Look for .7500 in the body of Table A. No value is exactly equal to .7500. The closest value is .7486, which is close enough. The .7486 sits at the intersection of the row headed 0.6 and the column headed .07, so the corresponding z-score is roughly $0.6 + .07 = 0.67$.

If you have a graphing calculator, you can find the percentage or value directly. On the TI-83, for example, normalcdf (−99999,1.23) returns a value of .8907, or 89.07%. To find the 75th percentile of a standard normal, use the command invNorm(.75) to get .67449. ■

Discussion: The Standard Normal Curve

D39. What percentage of values in a standard normal distribution fall

    a. below a z-score of 1.00? 2.53?

    b. below a z-score of −1.00? −2.53?

    c. above a z-score of −1.5?

    d. between z-scores of −1 and 1?

D40. For the standard normal distribution,

    a. what is the median?

    b. what is the lower quartile?

    c. what z-score falls at the 95th percentile?

    d. what is the *IQR*?

■ Practice

P38. Find the *z*-score that has the given percentage of values below it.

    a. 32%   b. 41%   c. 87%   d. 94%

P39. Find the percentage of values below each *z*-score.

    a. −2.23   b. −1.67   c. −0.40   d. 0.80

P40. What percentage of values in a standard normal distribution fall between

    a. −1.46 and 1.46?

    b. −3 and 3?

P41. For a standard normal distribution, what interval contains

    a. the middle 90% of the *z*-scores?

    b. the middle 95% of *z*-scores?

Standard Units: How Many Standard Deviations Is It from Here to the Mean?

Converting to standard units, or **standardizing,** is the two-step process of recentering and rescaling that turns any normal distribution into the standard normal.

    First you recenter all the values of the normal distribution by subtracting the mean from each. This gives you a distribution with mean 0. Then you rescale by dividing all of the values by the standard deviation. This gives you a distribution with standard deviation 1. Now you have a standard normal distribution. You can also think of the two-step process as answering two questions: How far above or below the mean is my score? How many standard deviations is that?

> The **standard units** or **z-score** is the number of standard deviations that a given *x*-value lies above or below the mean.
>
> How far and which way to the mean?
>
> $$x - mean$$
>
> How many standard deviations is that?
>
> $$z = \frac{x - mean}{SD}$$

Example

The distribution of SAT scores for the incoming class at the University of Washington had mean 1055 and standard deviation 200. What is the *z*-score for a University of Washington student who got 912 on the SAT?

Solution

A score of 912 is 143 points below the mean of 1055. This is $\frac{-143}{200}$ or 0.715 standard deviations below the mean. Alternatively, using the formula,

$$z = \frac{x - mean}{SD} = \frac{912 - 1055}{200} = -0.715$$

so the student's z-score is −0.715. ■

> To "unstandardize," you think in reverse. Alternatively, you can solve the z-score formula for x and get
>
> $$x = mean + z \cdot SD$$

Example

What did a student at the University of Washington get on the SAT if his or her score was 1.6 standard deviations above the average?

Solution

The score that is 1.6 standard deviations above average is

$$x = mean + z \cdot SD = 1055 + 1.6(200) = 1375$$ ■

Discussion: Standard Units

D41. Standardizing is a process that is similar to others you have seen already.

   a. If you're driving at 60 mph on the interstate and are now passing the marker for mile 200, and your exit is at mile 80, how many hours from your exit are you?

   b. What two arithmetic operations did you do to get the answer in Part a? Which operation corresponds to recentering? Which one corresponds to rescaling?

D42. In the United States, heart disease kills roughly one-and-one-half times as many people as cancer. (Among 100,000 residents, there are 289 deaths per year from heart disease and 200 from cancer.) If you look at these death rates by state, the distributions are roughly normal, provided that you leave out Alaska, which is an outlier. The means and standard deviations are

	Mean	SD
Heart disease	289	54
Cancer	200	31

Alaska has 90 deaths per 100,000 residents from heart disease, 84 from cancer. Explain which death rate is more extreme compared to other states.

Source: National Center for Health Statistics, *www.cdc.gov/nchs/data*, 1996.

■ **Practice**

P42. Refer to the table in D42. California has 240 deaths from heart disease and 166 deaths from cancer per 100,000 residents. Which rate is more extreme compared to other states, and why?

P43. Refer to the table in D42.

a. Florida has 365 deaths from heart disease and 257 deaths from cancer per 100,000 residents. Which rate is more extreme?

b. Colorado has an unusually low rate of heart disease, 184 deaths per 100,000 residents. Texas has an unusually low rate of cancer, 161 per 100,000 residents. Which is more extreme?

P44. *Standardizing.* Convert each of these values to standard units, z. (Do not use a calculator. These are meant to be done in your head.)

a. $x = 12$,   *mean* 10,   *SD* 1   d. $x = 12$,   *mean* 9,   *SD* 1

b. $x = 12$,   *mean* 10,   *SD* 2   e. $x = 7$,   *mean* 10,   *SD* 3

c. $x = 12$,   *mean* 9,   *SD* 2   f. $x = 5$,   *mean* 10,   *SD* 2

P45. *Unstandardizing.* In your head, convert each of these z-scores back to the scale it came from. That is, find *x*.

a. $z = 2$,   *mean* 20,   *SD* 5

b. $z = -1$,   *mean* 25,   *SD* 3

c. $z = -1.5$,   *mean* 100,   *SD* 10

d. $z = 2.5$,   *mean* –10,   *SD* 0.2

Solving the Unknown Percentage Problem and Unknown Value Problem

Now you know all you need to solve problems involving any normal distribution.

For an unknown percentage problem:

First standardize by converting the given value to a z-score,

$$z = \frac{x - mean}{SD}$$

then look up the percentage.

For an unknown value problem, reverse the process:

First look up the z-score corresponding to the given percentage, then unstandardize,

$$x = mean + z \cdot SD$$

Example

For groups of similar individuals, heights are often approximately normal in their distribution. For example, the heights of 18- to 24-year-old males in the United States are approximately normal, with mean 70.1 inches and standard deviation 2.7 inches. What percentage of these males are more than 74 inches tall?

Source: *Statistical Abstract of the U.S. 1991.*

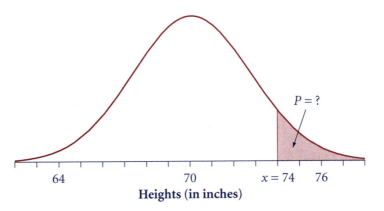

Display 2.79   The percentage of heights more than 74 inches.

Solution

Standardize:

$$z = \frac{x - mean}{SD} = \frac{74 - 70.1}{2.7} = 1.44$$

Look up the percentage: The area to the left of the *z*-score 1.44 is .9251. So the percentage taller than 74 inches is $100 - 92.51$ or 7.49%. ■

Example

The heights of females in the United States who are between the ages of 18 and 24 are approximately normally distributed, with mean 64.8 inches and standard deviation 2.5 inches. What height separates the shortest 75% from the tallest 25%?

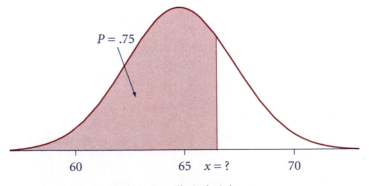

Display 2.80   The 75th percentile in height.

Solution

Look up the *z*-score: If the percentage $P = .75$, then from Table A, $z \approx 0.67$. Unstandardize:

$$x = mean + z \cdot SD \approx 64.8 + 0.67(2.5) \approx 66.475 \text{ inches} \quad \blacksquare$$

Discussion: Solving the Unknown Percentage Problem and the Unknown Value Problem

D43. The heights of 18- to 24-year-old males in the United States are approximately normal with mean 70.1 inches and standard deviation 2.7 inches.

  a. If you select a U.S. male between 18 and 24 at random, what is the approximate probability that he is less than 68 inches tall?

  b. There are roughly 13,000,000 males between 18 and 24 in the United States. About how many of them are between 67 and 68 inches tall?

  c. Find the male height that falls at the 90th percentile.

D44. If the measurements of height are transformed from inches to feet, will that change the shape of the distribution in D43? Describe the distribution of male heights in terms of feet rather than inches.

D45. For 17-year-olds in the United States, blood cholesterol levels in milligrams per deciliter have a normal distribution, approximately, with mean 176 mg/dl and standard deviation 30 mg/dl. The middle 90% of the cholesterol levels are between what two values?

■ Practice

P46. The heights of 18- to 24-year-old males in the United States are approximately normal with mean 70.1 inches and standard deviation 2.7 inches. The heights of 18- to 24-year-old females have a mean of 64.8 inches and a standard deviation of 2.5 inches.

  a. Estimate the percentage of U.S. males between 18 and 24 who are 6 feet tall or taller.

  b. How tall does a U.S. woman between 18 and 24 have to be to be at the 35th percentile?

P47. For students entering the University of Florida in a recent year, the distribution of SAT scores was roughly normal, with mean 1100 and standard deviation 180. The middle 95% of the SAT scores were between what two values?

Central Intervals for Normal Distributions

You learned in Section 2.1 that if a distribution is roughly normal, about two-thirds of the values lie within one standard deviation of the mean. (The actual percentage is closer to 68%.) It is helpful to memorize this fact as well as the others in the box that follows.

Central Intervals for Normal Distributions

68% of the values lie within 1 standard deviation of the mean.

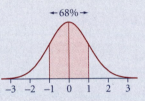

90% of the values lie within 1.645 standard deviations of the mean.

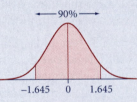

95% of the values lie within 1.96 (or about 2) standard deviations of the mean.

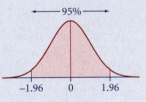

99.7% (or almost all) of the values lie within 3 standard deviations of the mean.

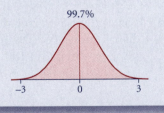

Discussion: Central Intervals

D46. Refer to the table in D42.

    a. The middle 90% of the states' death rates from heart disease fall between what two numbers?

    b. The middle 95% of the death rates from heart disease are between what two numbers?

    c. The middle 90% of the death rates from cancer are between what two numbers?

■ Practice

P48. Refer to the table in D42. Which of the following rates per 100,000 residents are outside the middle 95% of their distribution? Which of them are outside the middle 99%?

    a. California's death rate from heart disease of 240.

    b. California's death rate from cancer of 166.

    c. Alaska's death rate from heart disease of 90.

    d. Alaska's death rate from cancer of 84.

Summary 2.4: The Normal Distribution

The standard normal distribution has mean 0, standard deviation 1. The curve is symmetric, with total area of 1. All normal distributions can be converted to the standard normal by converting to standard units:

- First, recenter by subtracting the mean.

- Then rescale by dividing by the standard deviation:

$$z = \frac{x - mean}{SD}$$

Standard units z tell how far a value x is from the mean, measured in standard deviations. If you know z, you can get x using $x = mean + z \cdot SD$.

For any normal distribution,

- 68% of the values lie within 1 standard deviation of the mean

- 90% of the values lie within 1.645 standard deviations of the mean

- 95% of the values lie within 1.96 (or about 2) standard deviations of the mean

- 99.7% (or almost all) of the values lie within 3 standard deviations of the mean

■ Exercises

E45. On the same set of axes, draw two normal curves with mean 50, one having a standard deviation of 5 and the other having a standard deviation of 10.

E46. ACT scores are approximately normally distributed with mean 18 and standard deviation 6. Without using your calculator, roughly what percentage of scores are between 12 and 24? Between 6 and 30? Above 24? Below 24? Above 6? Below 6?

E47. SAT I verbal scores are scaled so that they are approximately normal; the mean is about 505, and the standard deviation is about 111.

    a. Find the probability that a randomly selected student has an SAT I verbal score

       i. between 400 and 600

       ii. over 700

       iii. below 450

    b. What SAT I verbal scores fall in the middle 95% of the distribution?

E48. SAT I math scores are scaled so that they are approximately normal and the mean is about 511 and the standard deviation is about 112. A college wants to send letters to students scoring in the top 20% on the exam. What SAT I math score should the college use as the dividing line between those who get letters and those who do not?

E49. *Height limitations for flight attendants.* To work as a flight attendant for United Airlines, you must be between 5′2″ and 6′ tall. [Source: *www.ual.com/airline.*] The mean height of 18- to 24-year-old males in the United States is about 70.1 inches, with a standard deviation of 2.7 inches. The mean height of 18- to 24-year-old females is about 64.8 inches, with a standard deviation of 2.5 inches. Both distributions are approximately normal. What percentage of men this age meet this height limitation? What percentage of women this age meet this height limitation?

E50. *Where is the next generation of male professional basketball players coming from?*

    a. Use the normal distribution to approximate the percentage of men in the United States between the ages of 18 and 24 who are as tall or taller than each basketball player below. Then, using the fact that there are about 13,000,000 men between 18 and 24 in the United States, estimate how many are as tall or taller than each player.

       i. Karl Malone, 6′9″

       ii. Michael Jordan, 6′6″

       iii. Shaquille O'Neal, 7′1″

    b. Distributions of real data that are approximately normal tend to have heavier "tails" than the ideal normal curve. Does this mean your estimates in parts i–iii are too small, too big, or just right?

E51. *Age of cars.* The cars in Clunkerville have a mean age of 12 years and a standard deviation of 8 years. What percentage of cars are more than 4 years old? (Warning: This is a trick question.)

E52. *The British monarchy.* Over the 1200 years of the British monarchy, the average reign of kings and queens lasted 18.5 years, with a standard deviation of 15.4 years.

    a. What can you say about the shape of the distribution based on the information given?

    b. Suppose you made the mistake of assuming a normal distribution. What fraction of the reigns would you estimate lasted a negative number of years?

    c. Use your work in part b to suggest a rough rule for using the mean and standard deviation of a set of positive values to check whether the distribution can possibly be normal-shaped.

E53. *NCAA scores.* The histogram in Display 2.81 was constructed from the total of the scores of both teams in all NCAA basketball play-off games between 1939 and 1995.

a. Approximate the mean of this distribution.

b. Approximate the standard deviation of this distribution.

c. Between what two values does the middle 95% of total points scored lie?

d. Suppose you choose a game at random from next year's play-offs. What is the approximate probability that the total points scored in this game will exceed 150? 190? Do you see any potential weaknesses in your approximations?

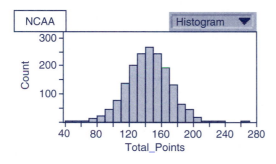

Mean	Standard Deviation	x	P
3	1	2	–?–
10	2	–?–	.18
–?–	3	6	.09
10	–?–	12	.60

E55. *More puzzle problems.* In each row below, assume the distribution is normal. Since knowing any two of the mean, standard deviation, Q_1, and Q_3 is enough to determine the other two, complete the table.

Mean	Standard Deviation	Q_1	Q_3
10	5	–?–	–?–
–?–	–?–	120	180
–?–	10	100	–?–
10	–?–	–?–	11

Display 2.81 Total points scored in NCAA play-off games.

Source: CBS SportsLine.com, *www.sportsline.com.*

E54. *Puzzle problems.* In problems that involve computations with the normal distribution, there are four quantities: mean, standard deviation, value *x*, and percentage *P* below value *x*. Any three are enough to determine the fourth. Think of each row in this table as little puzzles and find the missing value in each case. They aren't the sort of thing you are likely to run into in practice, but solving them can help you become more skilled using the normal distribution.

E56. For the comparisons below, you will be using either the SAT I verbal scores in Display 2.55 or assuming that the scores have a normal distribution with mean 505 and standard deviation 111.

a. Estimate the percentile for an SAT I verbal score of 425 using the plot. Then find the percentile for a score of 425 using a *z*-score. Are the two values close?

b. Estimate the SAT I verbal score that falls at the 40th percentile using the table in Display 2.55. Then find the 40th percentile using a *z*-score. Are they close?

c. Estimate the median from the cumulative relative frequency plot. Is this close to the median you would get by assuming a normal distribution of scores?

d. Estimate the quartiles and the interquartile range using the plot. Find the quartiles and interquartile range assuming a normal distribution of scores.

Chapter Summary: Exploring Distributions

Distributions come in various shapes, and the appropriate summaries (for center and spread) usually depend on the shape, so you should always start with a plot of your data.

Common symmetric shapes include the uniform (rectangular) distribution and the normal distribution. There are also various skewed distributions. Bimodal distributions often result from mixing cases of two kinds.

Dot plots, stemplots, and histograms show distributions graphically and let you estimate center and spread visually from the plot.

For normal-shaped distributions, you ordinarily use the mean (balance point) and standard deviation as the measure of center and spread. If you know the mean and standard deviation of a normal distribution, you can use z-scores and the z-table to find the percentage of values in any interval.

The mean and standard deviation are not resistant—their values are sensitive to outliers. So for skewed distributions, you ordinarily use the median (halfway point) and quartiles (medians of the lower and upper halves of the data) as summaries.

Later on, when making inferences about the entire population from a sample taken from that population, the sample mean and the standard deviation will be the most useful summary statistics, even if the population is skewed.

■ Review Exercises

E57. The map in Display 2.82, from the U.S. National Weather Service, gives the number of tornadoes by state, including the District of Columbia.

a. Construct a stemplot of the number of tornadoes.

b. Find the five-number summary.

c. Identify any outliers.

d. Draw a boxplot.

e. Compare the information in your stemplot with the information in your boxplot. Which plot is more informative?

f. Describe the shape, center, and spread of the distribution of the number of tornadoes.

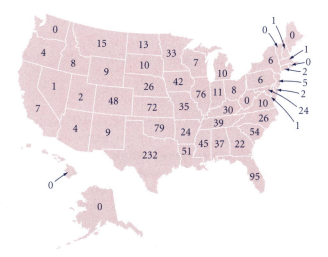

Display 2.82 The number of tornadoes per state in 1995.

Source: *www.ncdc.noaa.gov/ol/climate/severeweather/tornadoes.*

E58. The summary of SAT I scores at the University of Michigan indicates that "the middle 50% of the scores were between 1170 and 1340, with half the scores above 1210 and half below." What SAT I scores would be considered outliers for that university?

E59. The cumulative relative frequency plot in Display 2.83 shows the amount of change carried by a group of 200 students.

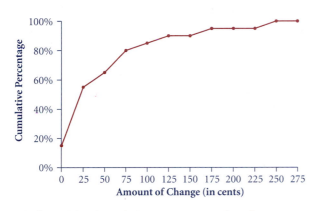

Display 2.83 Cumulative percentage plot of amount of change.

a. From this plot, estimate the median amount of change.

b. Estimate the quartiles and the interquartile range.

c. Is the original set of data skewed right, skewed left, or symmetric?

d. Does the data set look like it should be modeled by a normal distribution? Explain your reasoning.

E60. Display 2.84 shows two sets of graphs. The first set is repeated from P5, and shows smoothed histograms I–IV for four distributions. The second set shows the corresponding cumulative relative frequency plots, in scrambled order A–D. Match each plot in the first set with its counterpart in the second set.

E61. If the distributions of test scores aren't normally distributed, it's possible to have a larger z-score on Test II than on Test I yet

Distributions

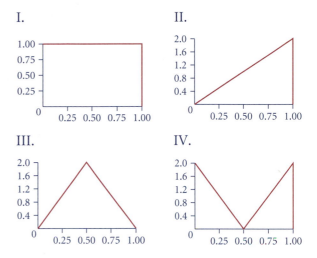

Cumulative relative frequency plots

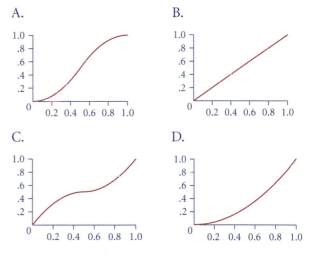

Display 2.84 Four distributions with different shapes and their cumulative relative frequency plots.

be in a lower percentile on Test II than on Test I. These computations will illustrate this point.

a. On Test I, a class got these scores: 11, 12, 13, 14, 15, 16, 17, 18, 19, 20. Compute the z-score and the percentile for the student who got a score of 19.

b. On Test II, the class got these scores: 1, 1, 1, 1, 1, 1, 1, 18, 19, 20. Compute the *z*-score and the percentile for the student who got a score of 18.

c. Do you think the student who got a score of 19 on Test I or the student who got a score of 18 on Test II did better relative to the rest of the class?

E62. The average income in dollars of people in each of the 50 states was computed for 1980 and for 1994. Summary statistics for these two distributions are given in Display 2.85.

	1980	1994
Mean	9,594	21,078
Standard deviation	1,406	3,130
Minimum	6,926	15,794
Lower quartile	8,348	18,810
Median	9,723	20,582
Upper quartile	10,612	22,542
Maximum	13,835	30,721

Display 2.85 Summary statistics of the average income in dollars by state for 1980 and 1994.

Source: *World Almanac and Book of Facts 1997.*

a. Explain the meaning of the $6,926 for the minimum in 1980.

b. Are any states outliers for either year?

c. In 1994 the average personal income in Alabama was $17,926, and in 1980 it was $7,704. Did the income in Alabama change much in relation to the other states? Explain your reasoning.

E63. The *World Almanac* records high and low temperatures by state from 1890 through 1999. Stem-and-leaf plots of the years that each state had its lowest temperature and the years that each state had its highest temperature appear in Display 2.86. What do the stems represent? What do the leaves represent? Compare the two distributions with respect to shape, center, spread, and any interesting features.

```
Stem-and-leaf of Year Low   N = 50
Leaf Unit = 1.0
    6    189   39999
   12    190   44555
   16    191   2277
   19    192   5
  (11)   193   03333446677
   21    194   0233
   17    195   14
   15    196   668
   12    197   1199
    8    198   155555
    2    199   44669

Stem-and-leaf of Year High   N = 50
Leaf Unit = 1.0
    1    188   8
    2    189   8
    3    190   0
    8    191   11135
   10    192   56
  (26)   193   00001144466666666666666667
   15    194
   15    195   24444
   10    196   1
    9    197   55
    7    198   35
    5    199   44445
```

Display 2.86 Stem-and-leaf plots of record low and high temperatures by state.

Source: *World Almanac 2001*, p. 245.

E64. Display 2.87 shows some results of the Third International Mathematics and Science study for various countries. Each case is a school.

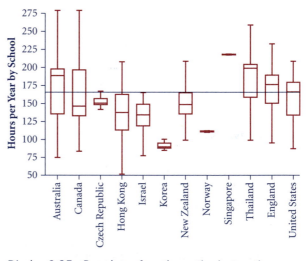

Display 2.87 Boxplots of mathematics instruction time by country for upper grade 9-year-olds.

Source: Report #8, April 1998, of the Third International Mathematics and Science Study (TIMSS), p. 6.

a. Find the boxplot for the United States. What, exactly, are the individual values that are plotted?

b. Why are there only lines and not boxes for Norway and Singapore?

c. Describe how the distribution for the United States compares to the distributions for the other countries.

E65. The side-by-side boxplots in Display 2.88 give the percentage of 4th-grade-aged children who are still in school on various continents according to the United Nations. Each case is a country. The four regions marked 1, 2, 3, and 4 are Africa, Asia, Europe, and South/Central America, not necessarily in that order.

a. Which region do you think corresponds to which number?

b. Is the distribution for any region skewed left? skewed right? symmetrical?

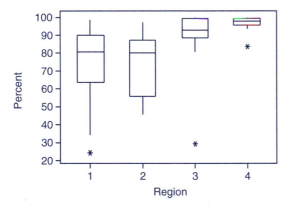

Display 2.88   Boxplots of the percentage of 4th-graders still in school in countries of the world, by continent.

Source: *The 1993 Information Please Almanac.*

Display 2.89 presents dot plots of the same data.

c. Match each of the dot plots to the corresponding boxplot.

d. In what ways do the boxplots and dot plots give different impressions? Why

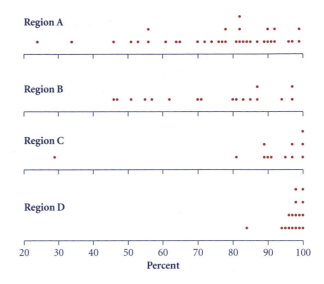

Display 2.89   Dot plots of percentage of 4th graders still in school in countries of the world, by continent.

did this happen? Which type of plot gives a better impression of the distributions?

E66. The first AP Statistics exam was given in 1997. The distribution of scores received by the 7667 students who took the exam is given in Display 2.90. Compute the mean and standard deviation of the scores.

Score	Number of Students
5	1205
4	1696
3	1873
2	1513
1	1380

Display 2.90   Scores on first AP Statistics exam.

Source: The College Board.

E67. The average number of pedestrian deaths annually for 41 metropolitan areas is given in Display 2.91.

a. What is the median number of deaths? Write a sentence explaining the meaning of this median.

Metro Area	Average Annual Fatalities	Metro Area	Average Annual Fatalities
Atlanta	84	New Orleans	47
Baltimore	66	New York	310
Boston	22	Norfolk, VA	25
Charlotte, NC	29	Orlando, FL	48
Chicago	180	Philadelphia	120
Cincinnati	23	Phoenix	79
Cleveland	36	Pittsburgh	33
Columbus, OH	20	Portland, OR	34
Dallas	76	Riverside, CA	92
Denver	28	Rochester, NY	17
Detroit	107	Sacramento, CA	37
Fort Lauderdale	58	Salt Lake City	28
Houston	101	San Antonio	37
Indianapolis	24	San Diego	96
Kansas City	27	San Francisco	43
Los Angeles	299	San Jose, CA	33
Miami	100	Seattle	37
Milwaukee	19	St. Louis	51
Minneapolis	35	Tampa	85
Nassau-Suffolk, NY	80	Washington, DC	98
Newark, NJ	51		

Display 2.91 Average annual pedestrian deaths.

Source: Environmental Working Group and the Surface Transportation Policy Project.
Compiled from National Highway Traffic Safety Administration and U.S. Census data.
USA Today, April 9, 1997.

b. Is any city an outlier in terms of the number of deaths? If so, what is the city, and what are some possible explanations?

c. Make a plot of the data you think will show the distribution in a useful way. Describe why you chose that plot and what information it gives you about the average annual pedestrian deaths.

d. In which situations might giving the death rate be more meaningful than giving the number of deaths?

E68. A game invented recently by three college students involves giving the name of an actor or actress and then trying to connect that actor or actress with actor Kevin Bacon, counting the number of steps needed. For example, Sarah Jessica Parker has a "Bacon number" of 1 because she appeared in the same movie as Kevin Bacon, *Footloose* (1984). Will Smith has a Bacon number of 2. He has never appeared in a movie with Kevin Bacon. However, he was in *Bad Boys* (1995) with Marc Macaulay, who was in *Wild Things* (1998) with Kevin Bacon. Display 2.92 gives the number of links required to connect each of the 263,484 actors and actresses in the Internet Movie Database to Kevin Bacon.

Bacon Number	Number of Actors/Actresses
0	1
1	1,267
2	78,867
3	149,018
4	32,094
5	1,903
6	299
7	34
8	1

Display 2.92 Bacon numbers.

Sources: "The Oracle of Bacon?" *www.cs.virginia.edu/oracle,* and John M. Harris and Michael J. Mossinghoff, "The Eccentricities of Actors," *Math Horizons,* Feb. 1998, pp. 23–24.

a. How many people have appeared in a movie with Kevin Bacon?

b. Who is the person with Bacon number 0?

It has been questioned whether Kevin Bacon was the best choice for the "center of the Hollywood universe." A possible challenger is Sean Connery. See Display 2.93.

Connery Number	Number of Actors/Actresses
0	1
1	1,801
2	113,571
3	135,546
4	11,591
5	824
6	130
7	20

Display 2.93 Connery numbers.

c. Do you think Kevin Bacon or Sean Connery better deserves the title of Hollywood center? Make your case using statistical evidence (as always).

d. (For movie fans.) What is Bacon's Connery number? What is Connery's Bacon number?

E69. How good are the batters in the American League of Major League Baseball? Display 2.94 shows the distribution of batting averages for all American League players who batted 100 times or more in the 1998 season. (A batting "average" is the fraction of times that a player hits safely—that is, the hit results in a player advancing to a base—usually reported to three decimal places.)

a. Does it look like batting averages are approximately normally distributed?

b. Approximate the mean and standard deviation of the batting averages from the histogram.

c. Give the interval that contains the middle 95% of the batting averages for the American League.

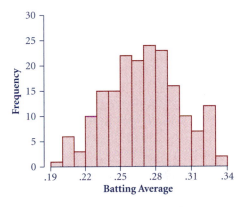

Display 2.94 American League batting averages, 1998.

Source: CBS SportsLine.com, *www.sportsline.com.*

E70. How good are batters in the National League? Display 2.95 shows the distribution of batting averages for National Leaguers who batted 100 times or more in 1998.

a. Approximate the mean and standard deviation of the batting averages from the histogram. Find the central interval that contains 95% of the batting averages. Would it be unusual to see a player hitting under .200 (or 200) in the National League?

b. Compare the distributions of batting averages for the two leagues. (See E69 for the American League.) What are the main differences between the two distributions?

c. A batter hitting .200 in the National League is traded to a team in the American League. What batting average could be expected of him in his new league if he maintains about the same position in the distribution relative to his peers?

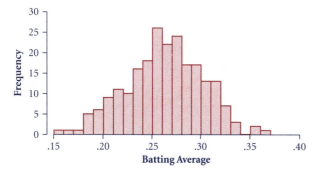

Display 2.95 National League batting averages, 1998.

Source: CBS SportsLine.com, *www.sportsline.com*.

E71. The statistics below summarize the set of Nielsen ratings from a week without any special programming. You can find a dot plot of the ratings in Display 2.22, E10, at the end of Section 2.1.

Variable	N	Mean	Median	TrMean	StDev	SEMean
Total	113	6.867	6.900	6.596	3.490	0.328

Variable	Min	Max	Q1	Q3
Total	1.400	20.700	4.550	8.250

a. Use the dot plot and summary statistics for all 113 shows to make boxplots of the ratings, showing any outliers.

b. Use the summary statistics below to make a set of side-by-side boxplots for the six networks. You will not be able to show any outliers.

Network	N	Mean	Median	TrMean	StDev	SEMean
ABC	22	7.755	7.750	7.715	2.439	0.520
CBS	25	7.856	7.500	7.765	1.728	0.346
FOX	16	6.156	6.050	6.064	1.508	0.377
NBC	28	9.157	7.100	8.892	4.308	0.814
UPN	8	2.188	2.150	2.188	0.449	0.159
WB	14	2.614	2.700	2.617	0.684	0.183

Network	Min	Max	Q1	Q3
ABC	2.900	13.400	6.025	9.125
CBS	4.900	12.900	6.950	8.550
FOX	4.000	9.600	4.850	6.900
NBC	4.500	20.700	6.800	11.100
UPN	1.700	3.100	1.825	2.375
WB	1.400	3.800	2.075	3.125

E72. A distribution is symmetric with approximately equal mean and median. Is it necessarily the case that about two-thirds of the values are within one standard deviation of the mean? If yes, explain why. If not, give an example.

E73. Construct a set of data where all values are larger than 0, but one standard deviation below the mean is less than 0.

E74. The boxplots in Display 2.96 show the life expectancies for the countries of Africa, Europe, and the Middle East. The table below the plots shows a few of the summary statistics for each of the three data sets.

a. From your knowledge of the world, match the boxplots to the correct region.

b. Match the summary statistics (for Groups A–C) to the correct boxplot (for Regions 1–3).

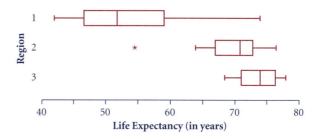

Group	Mean	Median	StDev
A	73.61	73.85	3.05
B	53.59	51.71	8.37
C	69.86	70.70	5.22

Display 2.96 Life expectancies for the countries of Africa, Europe, and the Middle East.

Source: Population Reference Bureau, World Population Data Sheet, 1996.

E75. Another measure of center that is sometimes used is the **midrange.** To find the midrange, compute the mean of the largest value and the smallest value.

The statistics in the computer output given here summarize the Nielsen data for the week of the last new *Seinfeld* episode.

```
Variable     N    Mean  Median  TrMean  StDev  SEMean
Viewers     101  11.187  10.150   9.831  9.896   0.985

Variable    Min     Max     Q1      Q3
Viewers   2.320  76.260   6.160  12.855
```

a. Using these summary statistics alone, compute the midrange both with and without the *Seinfeld* episode. (You will need to refer to Display 2.20 on page 17 to get one number.) Is the midrange sensitive to outliers, or is it resistant to outliers? Explain.

b. Compute the mean of the ratings without the *Seinfeld* episode using only the summary statistics above.

E76. In computer output like that of E75, the TrMean is the **trimmed mean.** It typically is computed by removing the largest 5% of the values and the smallest 5% of the values from the data set and then computing the mean of the remaining middle 90% of the values. (The percentage that is cut off at each end can vary depending on the software.)

a. Find the trimmed mean of the maximum longevities in Display 2.24.

b. Is the trimmed mean resistant to outliers?

CHAPTER 3
RELATIONSHIPS BETWEEN TWO QUANTITATIVE VARIABLES

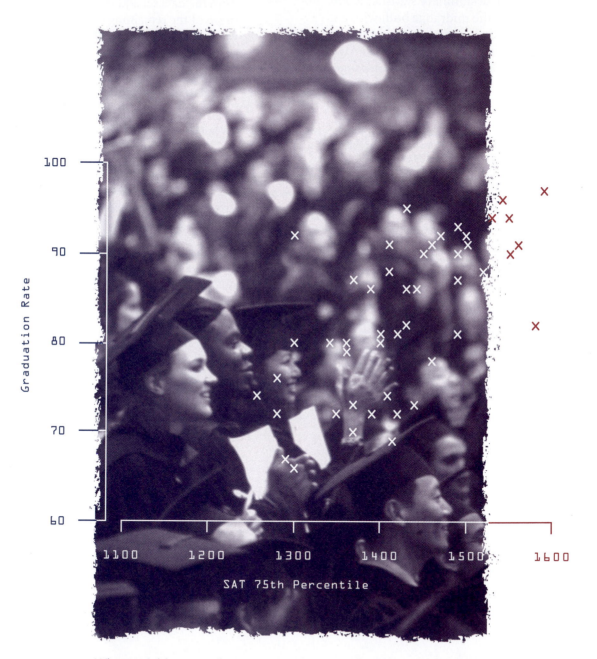

What variables contribute to a college having a high graduation rate? Scatterplots, correlation, and regression are the basic tools used to describe relationships between two quantitative variables.

In Chapter 2, you compared the speeds of predators and nonpredators. Not surprisingly, the meat-eaters were usually faster than the vegetarians. Some nonpredators, however, like the horse (48 mph) and the elk (45 mph), were faster than some predators, like the dog (39) and the grizzly (30). Because of this variability, comparing the two groups was a matter for statistics. That is, you needed suitable plots and summaries.

The comparison involved a relationship between two variables, one of them quantitative (speed) and one of them categorical (predator or not). In this chapter, you'll learn how to explore and summarize relationships between two quantitative variables. The data set in Display 2.24 on mammals raises many questions of this sort: Do mammals with longer average longevity in turn have longer maximum longevity? Is there a relationship between speed and longevity?

The approach to describing distributions in Chapter 2 boils down to "shape, center, and spread." For distributions that are approximately normal, it turns out that two numerical summaries—the mean for center, the standard deviation for spread—tell you basically all you need to know. For scatterplots with points that lie in an oval cloud, it will turn out, once again, that two summaries tell you pretty much all you need to know. The regression line tells about center. That is, what is the equation of the line that best fits the cloud of points? The correlation tells you about spread—how spread out are the points around the line?

In this chapter, you will learn to

- describe the pattern in a scatterplot, and decide what its *shape* tells you about the relationship between the two variables

- find a regression line through the *center* of a cloud of points to summarize the relationship

- use the correlation as a measure of how *spread* out the points are from this line

- use diagnostic tools to check for information the summaries don't tell you, and decide what to do with that information

- make shape-changing transformations to re-express a curved relationship so you can use a line as a summary

3.1 Scatterplots

In Chapter 1, you explored the relationship between the ages of employees at Westvaco and whether they were laid off when the company downsized. There is more to see. In the scatterplot in Display 3.1, for example, each employee is represented by a dot that shows the year when the employee was born, plotted against the year the employee was hired.

A **scatterplot** shows the relationship between two quantitative variables.

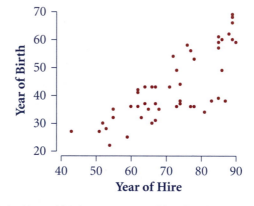

Display 3.1   *Year of birth* versus *year of hire* for the 50 employees at Westvaco Corporation's Engineering Department.

In this scatterplot, you can see a moderate positive association—employees hired in an earlier year generally were born in an earlier year and employees hired in a later year generally were born in a later year. This trend is fairly linear. You can visualize a summary line going up through the center of the data from lower left to upper right. As you go to the right, the points fan out and cluster less closely around the line.

Sometimes it's easier to think about people's ages than about the years they were born. The scatterplot in Display 3.2 shows the ages of the Westvaco employees at the time layoffs began, plotted against the year they were hired. This scatterplot shows a moderate negative association—the people who were hired in later years generally were younger at the time of the layoffs than the people who were hired in earlier years.

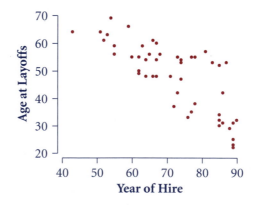

Display 3.2   *Age at layoffs* versus *year of hire* for the 50 employees at Westvaco Corporation's Engineering Department.

Discussion: Interpreting Scatterplots

D1. You will examine Displays 3.1 and 3.2 more closely in these questions.

a. Why should the two variables plotted in Display 3.1 show a positive association while the two variables plotted in Display 3.2 should show a negative association?

b. Why do all but one of the points in Display 3.1 lie below a diagonal line running from the lower left to the upper right?

c. Is this sentence a reasonable interpretation of Display 3.2? "As time passed, Westvaco tended to hire younger and younger people."

Describing the Pattern in a Scatterplot

Shape, trend, and strength

For the distribution of a single quantitative variable, "shape, center, spread" has been a useful summary. For **bivariate** (two-variable) quantitative data, the summary becomes "shape, trend, strength."

You may find it helpful to follow a set of numbered steps as you practice describing scatterplots:

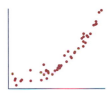

Curved and strong

1. Identify the **variables** and **cases**. On a scatterplot, each point represents a case, with the x-coordinate equal to the value of one variable and the y-coordinate equal to the value of the other variable. You should describe the scale (units of measurement) and range of each variable.

2. Describe the overall **shape** of the relationship, paying attention to

   • **linearity:** is the pattern linear (scattered about a line) or curved?

   • **clusters:** is there just one, or more than one?

   • **outliers:** are there any striking exceptions to the overall pattern?

Linear and moderate

3. Describe the **trend.** If, as x gets larger, y tends to get larger, there is a **positive trend.** (The cloud of points tends to slope up as you go from left to right.) If, as x gets larger, y tends to get smaller, there is a **negative trend.** (The cloud of points tends to slope down as you go from left to right.)

4. Describe the **strength** of the relationship. If the points cluster closely around an imaginary line, the association is **strong.** If the points are scattered farther from the line, the association is **weak.**

   If, as in Display 3.1, the points tend to fan out at one end (a tendency called heteroscedasticity), the relationship **varies in strength.** If not, it has **constant strength.**

Curved with varying strength

5. Does the pattern **generalize** to other cases, or is the relationship an instance of "what you see is all there is"?

6. Are there plausible **explanations** for the pattern? Is it reasonable to conclude that one variable causes the other? Is there a third or **lurking variable** that might be causing both?

Example: Dormitory Populations

A plot of variable *A* against (or *versus*) variable *B* shows *A* on the *y*-axis and *B* on the *x*-axis.

The plot in Display 3.3 shows, for the 50 United States, the number of people living in college dormitories versus the number of people living in cities, in thousands. Describe the pattern in the plot.

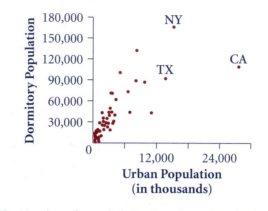

Display 3.3 Number of people living in college dormitories versus number of people living in cities for the 50 United States.

Source: U.S. Census Bureau, 1990 Census of Population, General Population Characteristics.

Solution

1. *Variables and cases.* The scatterplot shows *dormitory population* versus *urban population,* in thousands, for the 50 U.S. states. Dormitory population ranges from near zero to a high of more than 165,000 in New York. The urban population ranges from near zero to about 15 million in New York and 28 million in California.

2. *Shape.* While most states follow a linear trend, the three states with the highest urban population suggest curvature in the plot because for those states, the number of people in dormitories is proportionately lower than in the smaller states. California can be considered an outlier with respect to its urban population, which is much higher than in other states. It is also an outlier with respect to the overall pattern because it lies far below the generally linear trend.

3. *Trend.* The trend is positive—states with higher urban populations tend to have higher dormitory populations, and vice versa.

4. *Strength.* The relationship varies in strength. As *x* increases, *y* becomes more variable—the cloud of points fans out. A tight cluster in the lower left corresponds to a large number of states with small urban populations and small numbers of people living in dormitories. Toward the upper right, the points are much more spread out. There are comparatively fewer states with large urban populations, and these states show much more variation in the number of people living in college dormitories. The variation is

roughly proportional to population size. For the states with smallest urban population, the points cluster rather closely to a line. For the states with the largest urban population, the states are more scattered from the line. Overall, the strength of the relationship is moderate.

5. *Generalization.* The 50 states aren't a sample from a larger population of cases, so the relationship here does not generalize to other cases. Because both variables tend to change rather slowly, however, we can expect the relationship in Display 3.3 to be typical of other years.

6. *Explanation.* It is tempting to attribute the positive relationship to the idea that cities attract colleges. (Just pick a large city nearby and see how many colleges you can name that are located there.) The main reason for the positive relationship, however, is not nearly so interesting: both variables are related to a state's population. The more people in a state, the more people live in dormitories and the more people live in cities. (There's a moral here: Interpreting association can be tricky, in part because the two variables you see in a plot will often be related to some lurking variable that you don't see.) ■

Discussion: Describing the Pattern in a Scatterplot

D2. Display 3.4 comes from the data for Display 3.3, by converting to the *proportion* of a state's population: the proportion of the state population living in college dormitories (given as the number in dorms per 1000 state residents) and the proportion of the state population living in cities.

a. Follow steps 1–6 above to describe what you see in this new plot.

b. When you go from totals (Display 3.3) to proportion of total population (Display 3.4), the relationship changes from positive to negative and becomes weaker. Give an explanation for the differences in these two plots.

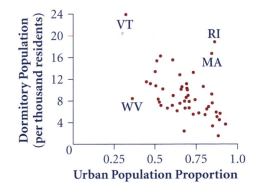

Display 3.4 Proportion of people living in college dormitories versus proportion of people living in cities for the 50 United States.

■ Practice

P1. ***Growing kids.*** The table below gives average heights of children at ages 1, 2, 3, 4, 5, and 6.

Age (yr)	Height (cm)	Age (yr)	Height (cm)
1	73.6	4	99.0
2	83.8	5	104.1
3	91.4	6	111.7

a. *Scatterplot.* Plot *height* versus *age;* that is, put height on the *y*-axis and age on the *x*-axis.

b. *Shape, trend, and strength.* Describe the shape, trend, and strength of the relationship.

c. *Generalization.* Would you expect the data to allow you to make good predictions of the average height of 7-year-olds? Of 50-year-olds?

d. *Explanation.* It doesn't quite fit to say that age "causes" height, but still, there is an underlying relationship of cause-and-effect. How would you describe it?

P2. ***Late planes and lost bags.*** The perfect way to cap off a long day of travel is to have your plane arrive late only to find that the airline has lost your luggage. As Display 3.5 shows, some airlines handle baggage better than others.

Display 3.5   On-time arrivals versus mishandled baggage.

Source: Office of Aviation Enforcement and Proceedings, U.S. Department of Transportation, June 2000, www.dot.gov/airconsumer/0008ATCR.pdf.

a. Which airline has the worst record for mishandled baggage? For being on time?

b. Where on the plot would you look to find the airline with the best on-time record and the best mishandled-baggage rate? Which airlines are best in those two categories?

c. True or false, and explain: United had a mishandled-baggage rate that was more than twice the rate of Southwest.

d. Is there a positive or negative relationship between the on-time percentage and the rate of mishandled baggage? Is it strong or weak?

e. Would you expect the relationship in this plot to generalize to some larger population of commercial airlines? Why or why not? Would you expect the relationship in this plot to be roughly the same for data from 10 years ago? For next year?

Summary 3.1: Scatterplots

A scatterplot shows the relationship between two quantitative variables. Each case is a point, with the *x*-coordinate equal to the value of the first variable and the *y*-coordinate equal to the value of the second variable.

In describing a scatterplot, be sure to cover all of the following:

- cases and variables (what exactly does each point represent?)
- shape (linear or curved, clusters, outliers)
- trend (positive, negative, or none)
- strength (strong, moderate, or weak; constant or varying)
- generalization (does the pattern generalize?)
- explanation (is there an explanation for the pattern?)

■ Exercises

E1. For each of the eight lettered scatterplots in Display 3.6, give the trend (positive or negative), strength (strong, moderate, or weak), and shape (linear or curved). Which plots show varying strength?

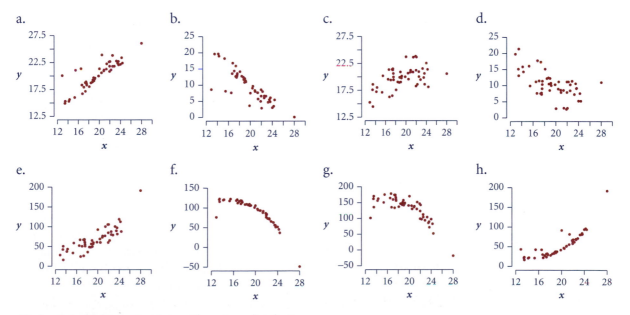

Display 3.6 Eight scatterplots with various distributions.

E2. For each set of cases, tell whether you expect the relationship to be (i) positive or negative, and (ii) strong, moderate, or weak.

Cases	Variable 1	Variable 2
a. Hens' eggs	Length	Width (diameter of cross-section)
b. High school students	SAT math score	SAT verbal score
c. Trees	Age	Number of rings
d. Squares drawn on a piece of paper	Length of side	Area
e. U.S. states	Population	Number of representatives in Congress
f. Countries of the UN	Land area	Population
g. Olympic games	Year	Winning time in the women's 100-meter race

Marion Jones races in the Women's 100-Meter Dash Event of the 2000 U.S. Olympic Trials.

E3. Match each set of cases and variables (a–d) with the short summary (i–iv) of its scatterplot.

Cases	Variable 1	Variable 2
a. Earth	Year	Human population
b. U.S. states	Area in square miles	People per square mile
c. U.S. states	Population	Number of doctors
d. Cars in the U.S.	Weight	Gas mileage

Summaries:

  i. strong negative relationship, somewhat curved

  ii. strong, curved positive relationship

  iii. moderate, roughly linear, positive relationship

  iv. moderate negative relationship

E4. ***SAT math scores.*** In 2000, the average SAT math score across the United States was 514. Minnesota students averaged 594, Iowa students averaged 600, and students from the nearby state of North Dakota did even better, with an average of 609. Why do states from the Midwest do so well? It is easy to jump to a false conclusion, but the scatterplot in Display 3.7 can help you find a reasonable explanation.

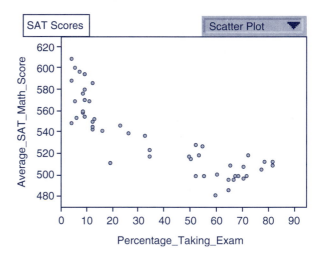

Display 3.7 Average SAT math scores by state versus percentage of high school graduates who took the test.

Source: College Board, 2000, www.collegeboard.com/press/senior00/html/table3.html.

  a. Estimate the percentage of students in North Dakota and in Iowa who took the SAT. New Jersey and Connecticut had the highest percentage of students who took the SAT. Estimate that percentage and the average SAT math scores for students in both states.

b. Describe the shape of the plot. Do you see any clusters? Any outliers? Is the relationship linear or curved? Is the overall trend positive or negative? What is the strength of the relationship?

c. Is the distribution of the percentage of students taking the SAT bimodal? Explain how the scatterplot shows this. Is the distribution of SAT scores bimodal?

d. The cases used in this plot are the 50 U.S. states. Would you expect the pattern to generalize to some other set of cases? Why or why not?

e. Suggest an explanation for the trend. (Hint: The SAT is administered from Princeton, New Jersey. An alternative, the ACT, is administered from Iowa. In the Midwest, many colleges and universities either prefer the ACT or at least accept it in place of the SAT, whereas colleges in the eastern states tend to prefer the SAT.) Is there anything in the data that you can use to help decide whether or not your explanation is correct?

E5. **Westvaco, revisited.** To determine if Westvaco discriminated by age in layoffs, you could investigate whether they may have discriminated in hiring. Examine Display 3.8 showing the age at hire plotted against the year the person was hired.

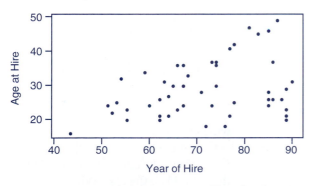

Display 3.8 Age at hire versus year of hire for the 50 employees at Westvaco Corporation's Engineering Department.

a. Describe the pattern in the plot, following the six-step format.

b. Does this plot provide evidence that Westvaco discriminated by age in hiring?

c. Display 3.9 shows the year of birth of the Westvaco employees plotted against the year they were hired. Open circles represent employees laid off, and solid circles show the employees kept. Does this scatterplot suggest a reason why older employees tended to be laid off more frequently?

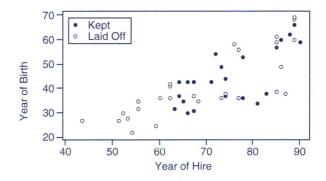

Display 3.9 Year of birth versus year hired for Westvaco employees.

E6. **Hat size.** What does hat size really measure? A group of students investigated this question by collecting a sample of hats. They recorded the size of the hat and then measured the circumference, the major axis (the length across the opening in the long direction), and the minor axis (see Display 3.10). (Hat sizes are changed to decimals; all other measurements are in inches.) Is hat size most closely related to circumference, major axis, or minor axis? Answer this question by making appropriate plots and describing the patterns in those plots.

Hat Size	Circumference	Major Axis	Minor Axis
6.625	20.00	7.00	5.75
6.750	20.75	7.25	6.00
6.875	20.50	7.50	6.00
6.875	20.75	7.25	6.00
6.875	20.75	7.50	6.00
6.875	21.50	7.25	6.25
7.000	21.25	7.50	6.00
7.000	21.00	7.50	6.00
7.000	21.00	7.50	6.25
7.000	21.75	7.50	6.25
7.125	21.50	7.75	6.25
7.125	21.75	7.75	6.50
7.125	21.50	7.75	6.25
7.125	22.25	7.75	6.25
7.250	22.00	7.75	6.25
7.250	22.50	7.75	6.50
7.375	22.25	7.75	6.50
7.375	22.25	8.00	6.50
7.375	22.50	8.00	6.50
7.375	22.75	8.00	6.50
7.375	23.00	8.00	6.50
7.500	22.75	8.00	6.50
7.500	22.50	8.00	6.50
7.625	23.00	8.25	6.50
7.625	23.00	8.25	6.50
7.625	23.25	8.25	6.75

Display 3.10 Hat size, with circumference and axes in inches.

Source: Roger Johnson, Carleton College, data from student project.

E7. **Passenger aircraft.** Airplanes vary in their size, speed, average flight length, and cost of operation. You can probably guess that larger planes use more fuel per hour and cost more to operate, but the shapes of the relationships are less obvious. You will explore the data in Display 3.11 on the 27 most commonly used passenger airplanes in the United States.
The variables are the number of seats, average cargo payload in tons, airborne speed in miles per hour, flight length in miles, fuel consumption in gallons per hour, and operating cost per hour in dollars.

a. *Cost per hour*

i. Make scatterplots with the cost per hour on the *y*-axis to explore its dependence on the other variables. Report your most interesting findings. Here are examples of some questions you could investigate: For which variable is the relationship to the cost per hour strongest? Is there any one airplane whose cost per hour, in relation to other variables, makes it an outlier?

ii. Do your results mean that larger planes are less efficient? Define your own variable, and plot it against other variables to judge the relative efficiency of the larger planes.

b. *Length of flight*

i. Make scatterplots with the length of flight on the *x*-axis to explore its relationship with the other variables. Report your most interesting findings. Here is an example of a question you could investigate: Which variable, *cargo* or *number of seats,* shows a stronger relationship with *average flight length*? Propose a reasonable explanation for why this should be so.

ii. Do planes with a longer flight length tend to use less fuel per mile than do planes with a shorter flight length?

c. *Speed, seats,* and *cargo*

i. Make scatterplots to explore the relationships between speed, seats, and cargo. Report your most interesting findings. Here are some examples of questions you could investigate: For which variable, *cargo* or *number of seats,* is the relationship with *speed* more obviously curved? Explain why that should be the case. Which plane is unusually slow for the amount of cargo it carries? Which plane is unusually slow for the number of seats it has?

ii. The plot of *cargo* against *seats* has two parts: a flat stretch on the left and a

Aircraft	Number of Seats	Cargo (tons)	Speed (mph)	Flight Length (mi)	Fuel (gal/hr)	Cost ($/hr)
B747–100	410	7.67	518	2882	3633	6567
B747–400	400	8.49	539	5063	3445	7075
B747–200/300	369	8.46	529	3321	3759	7790
L-1011–100/200	305	4.50	498	1363	2399	5081
B777	291	9.70	513	2451	2037	4194
DC10–10	286	8.54	498	1493	2233	5092
DC10–40	284	5.78	504	1963	2647	4684
DC10–30	272	6.92	516	2379	2625	5859
A300–600	266	11.63	467	1126	1671	5123
MD-11	260	10.24	524	3253	2400	6335
L-1011–500	222	5.06	523	2995	2454	4764
B767–300ER	216	7.32	495	2331	1590	3616
B757–200	187	2.39	464	1167	1048	2637
B767–200ER	181	4.52	486	2135	1432	3195
MD-90	154	0.43	441	782	817	1711
B727–200	148	0.63	440	742	1288	2396
A320–100/200	148	0.75	458	1101	816	2126
B737–400	144	0.49	414	702	792	2106
MD-80	141	0.48	432	798	924	2033
B737–300	131	0.40	416	602	836	1943
DC9–50	121	0.47	374	345	898	1925
B737–100/200	112	0.36	388	442	831	1899
B737–500	110	0.35	412	570	743	1730
DC9–40	109	0.41	387	487	837	1789
DC9–30	100	0.47	389	468	818	1749
F-100	97	0.17	384	500	705	1858
DC9–10	71	0.83	380	413	737	1614

Display 3.11 Data on passenger aircraft.

Source: Air Transport Association of America, 1996, www.air-transport.org/public/industry/67.asp.

fan on the right. Explain, in the language of airplanes, seats, and cargo, what each of the two patterns tells you.

E8. *Poverty.* What variables are most closely associated with poverty? Display 3.12 provides information on population characteristics of the 50 U.S. states, plus the District of Columbia. Each variable is measured as a percentage of the state population:

   Percentage living in metropolitan areas

   Percentage white

   Percentage of adults who have graduated from high school

   Percentage of families living under the poverty line

   Percentage of families headed by a single parent

Write a letter to your representative in Congress about poverty in America, relying only on what you find in these data. Point out the variables that appear to be most strongly associated with poverty and those that have little or no association with poverty.

State	Metropolitan Residence	White	Graduates	Poverty	Single Parent
Alabama	67.4	73.5	66.9	17.4	11.5
Alaska	41.8	75.2	86.6	9.1	14.3
Arizona	84.7	88.6	78.7	15.4	12.1
Arkansas	44.7	82.9	66.3	20.0	10.7
California	96.7	79.3	76.2	18.2	12.5
Colorado	81.8	92.5	84.4	9.9	12.1
Connecticut	95.7	89.0	79.2	8.5	10.1
Delaware	82.7	79.4	77.5	10.2	11.4
District of Columbia	100.0	31.8	73.1	26.4	22.1
Florida	93.0	83.5	74.4	17.8	10.6
Georgia	67.7	70.8	70.9	13.5	13.0
Hawaii	74.7	40.9	80.1	8.0	9.1
Idaho	30.0	96.7	79.7	13.1	9.5
Illinois	84.0	81.0	76.2	13.6	11.5
Indiana	71.6	90.6	75.6	12.2	10.8
Iowa	43.8	96.6	80.1	10.3	9.0
Kansas	54.6	90.9	81.3	13.1	9.9
Kentucky	48.5	91.8	64.6	20.4	10.6
Louisiana	75.0	66.7	68.3	26.4	14.9
Maine	35.7	98.5	78.8	10.7	10.6
Maryland	92.8	68.9	78.4	9.7	12.0
Massachusetts	96.2	91.1	80.0	10.7	10.9
Michigan	82.7	83.1	76.8	15.4	13.0
Minnesota	69.3	94.0	82.4	11.6	9.9
Mississippi	30.7	63.3	64.3	24.7	14.7
Missouri	68.3	87.6	73.9	16.1	10.9
Montana	24.0	92.6	81.0	14.9	10.8
Nebraska	50.6	94.3	81.8	10.3	9.4
Nevada	84.8	86.7	78.8	9.8	12.4
New Hampshire	59.4	98.0	82.2	9.9	9.2
New Jersey	100.0	80.8	76.7	10.9	9.6
New Mexico	56.0	87.1	75.1	17.4	13.8
New York	91.7	77.2	74.8	16.4	12.7
North Carolina	66.3	75.2	70.0	14.4	11.1
North Dakota	41.6	94.2	76.7	11.2	8.4
Ohio	81.3	87.5	75.7	13.0	11.4
Oklahoma	60.1	82.5	74.6	19.9	11.1
Oregon	70.0	93.2	81.5	11.8	11.3
Pennsylvania	84.8	88.7	74.7	13.2	9.6
Rhode Island	93.6	92.6	72.0	11.2	10.8
South Carolina	69.8	68.6	68.3	18.7	12.3
South Dakota	32.6	90.2	77.1	14.2	9.4
Tennessee	67.7	82.8	67.1	19.6	11.2
Texas	83.9	85.1	72.1	17.4	11.8
Utah	77.5	94.8	85.1	10.7	10.0
Vermont	27.0	98.4	80.8	10.0	11.0
Virginia	77.5	77.1	75.2	9.7	10.3
Washington	83.0	89.4	83.8	12.1	11.7
West Virginia	41.8	96.3	66.0	22.2	9.4
Wisconsin	68.1	92.1	78.6	12.6	10.4
Wyoming	29.7	95.9	83.0	13.3	10.8

Display 3.12   Characteristics of state populations, in percent.

Source: *Statistical Abstract of the United States, 1996.*

E9. Each of the 50 cases in the scatterplots in Display 3.13 is a top-rated university. The *y*-coordinate of a point tells the graduation rate, and the *x*-coordinate tells the value of some other variable—the percentage of alumni who gave that year, the student/faculty ratio, the 75th percentile of the SAT scores (math plus verbal) for a recent entering class, and the percentage of incoming students who ranked in the top tenth of their high school graduating class.

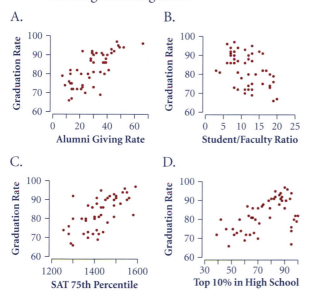

A.   B.

C.   D.

Display 3.13   Scatterplots showing the relationship between graduation rate and four other variables for 50 top universities.

Source: *U.S. News and World Report,* 2000.

a. Compare the *shapes* of the four plots.

   i. Which plots show a linear shape? Which show a curved shape?

   ii. Which show just one cluster? Which show more than one?

   iii. Which have outliers?

b. Compare the *trends* of the relationships. Which plots show a positive trend? A negative trend? No trend?

c. Compare the *strengths* of the relationships. Which variables give more accurate predictions of the graduation rate? Which variable is almost useless for predicting graduation rate?

d. *Generalization.* The cases in these plots are the 50 universities that happened to come out at the top of one particular rating scheme. Do you think the complete set of all U.S. universities would show pretty much the same relationship? Why or why not?

e. *Explanation.* Consider the two variables with the strongest relationship to graduation rates. Offer an explanation for the strength of these particular relationships. In what ways, if any, can you use the data to help decide whether your explanations are in fact correct?

3.2   Getting a Line on the Pattern

In this section, you will learn how to use a regression line to summarize the relationship between two quantitative variables. The first part of this section deals with the simple situation in which all the data points lie close to a line. In practice, however, data points are often more scattered, so the second part of this section shows how to choose a summary line when your data points form an oval cloud. But first, you will review the properties of a linear equation.

Lines as Summaries

You've seen the equation of a line before, $y = \text{slope} \cdot x + y\text{-intercept}$, so the review here will be brief. Linear relationships have the important property that for any two points on the line, the ratio

$$\frac{\text{rise}}{\text{run}} = \frac{\text{change in } y}{\text{change in } x}$$

Slope of a line is the change in *y* over the change in *x*.

is a constant. This ratio is the **slope** of the line. The rise and run are illustrated in Display 3.14. The slope is the ratio of the two sides of the right triangle. This ratio is the same for any two points on the line because the triangles formed are all similar.

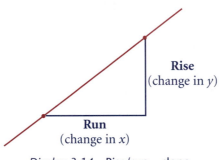

Rise
(change in *y*)

Run
(change in *x*)

Display 3.14   Rise/run = slope.

How thick is a single page of your book? One page alone is too thin to give an accurate answer if you measure directly with a ruler, but you could measure the thickness of 100 pages together, then divide. This method would give you an estimate of the thickness but no information about the precision of your measurement. The approach in the activity that follows lets you judge precision as well as thickness.

Activity 3.1   Pinching Pages

What you'll need: a ruler with a millimeter scale, a copy of your textbook

Pinch 50 *sheets* of paper, not up to page 50.

1. Pinch together the front cover and first 50 pages of your book. Then measure and record the thickness to the nearest millimeter.

2. Repeat for the cover plus 100, 150, 200, and 250 pages.

3. Plot your data on a scatterplot, with number of pages on the horizontal scale and thickness on the vertical scale.

4. Does the plot look linear? Should it? Discuss why or why not and make your measurements again if necessary. Place a straight line on the plot that fits best through the cloud of points.

(continued)

5. Find the slope and *y*-intercept of your line. What does the *y*-intercept tell you? What does the slope tell you? What is your estimate of the thickness of a page?

6. Use the information in your graph to discuss the precision of your estimate in step 5.

7. How would your line have changed if you hadn't included the front cover?

Discussion: Interpreting Slope

D3. Decide what variables *A* and *B* represent in these situations.

a. Suppose regular unleaded gasoline costs $1.20 per gallon. The number 1.20 is the slope of the line you get if you plot *A* versus *B*.

b. Suppose your car averages 30 miles per gallon. The number 30 is the slope of a line fitted to a scatterplot of *A* versus *B*.

c. "A pint's a pound the world around." This slogan summarizes the slope of a line fitted to a scatterplot of *A* versus *B* for various quantities of various kinds of liquids.

■ Practice

P3. Display 3.15 shows the cost per hour to fly various aircraft plotted against the fuel consumption per hour in gallons using data from Display 3.11. Estimate the slope of the line shown in the graph. Interpret the slope in the context of the situation.

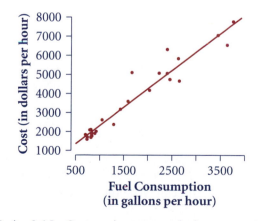

Display 3.15   Cost per hour versus fuel consumption per hour in gallons for various aircraft.

This example illustrates how to find the equation of a line when you know two points that fall on the line.

Example: Radiologists' Income

How fast did doctors' incomes increase during the decade leading up to President Clinton's health care initiative in 1993? The scatterplot in Display 3.16 shows the mean net income for radiologists (after expenses, but before taxes, in thousands of dollars) in the years from 1982 through 1992. (The data for 1983 and 1984 are not available.) The line on the plot is the least squares regression line, which you will learn more about later in this section. Estimate the slope of the line. What does the slope tell you? Estimate the equation of the line.

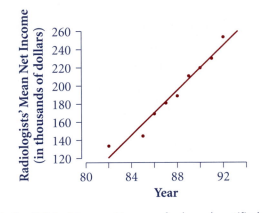

Display 3.16 Mean net income for board-certified radiologists, 1982–92.

Source: Martin L. Gonzalez, ed., *Socioeconomic Indicators of Medical Practice* (American Medical Association, 1994), p. 152.

Solution

In theory, you can find the slope from any pair of points on the line. Here, however, you have to estimate the *y*-coordinates from the graph. In such cases, using points that are farther apart gives more accurate slopes. For this plot, choosing points for the years 1982 and 1992 works well. Approximate coordinates are (82, 120) for (x_1, y_1) and (92, 245) for (x_2, y_2). So the estimated slope is

Be sure to qualify your interpretations of the slope. The average income for a radiologist didn't go up *exactly* $12,500 each year. It *tended* to go up this amount or went up *about* this amount.

$$\text{slope} = \frac{y_2 - y_1}{x_2 - x_1} = \frac{245 - 120}{92 - 82} = \frac{125}{10} = 12.5$$

This slope tells you that the mean net income for a radiologist went up *about* $12,500 per year.

You can write the equation of a line in terms of its slope and *y*-intercept:

$$y = \text{slope} \cdot x + y\text{-intercept}$$

You have the slope. You can't read the *y*-intercept from the plot. To find it, use your slope and any point (*x*, *y*) on the line, say (82, 120), and solve for the *y*-intercept:

$$y = \text{slope} \cdot x + y\text{-intercept}$$

$$120 = 12.5 \cdot 82 + y\text{-intercept}$$

$$y\text{-intercept} = -905$$

The equation of the line, then, using these values is

$$y = 12.5x - 905$$

In statistics, this equation is usually written "backwards":

$$y = -905 + 12.5x \quad \blacksquare$$

Discussion: Lines as Summaries

D4. You saw in the example that between 1982 and 1992, the net income of radiologists was going up about $12,500 per year. What about family doctors? Display 3.17 shows the mean net income (after expenses, before taxes, in thousands of dollars) for doctors who were board-certified in family practice. Fit a line by eye, and use two points to estimate its slope. What is the annual rate of increase? What is the equation of your line?

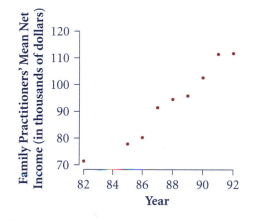

Display 3.17   Mean net income for doctors board-certified in family practice, 1982–92.

■ Practice

P4. Display 3.18 shows the data on doctors' incomes in another way. Each point corresponds to one of the years between 1982 and 1992. The *y*-coordinate is mean net income that year for a radiologist, and the *x*-coordinate is the mean for a specialist in family practice. Estimate the slope of the fitted line. What does the slope tell you? What is the equation of the line?

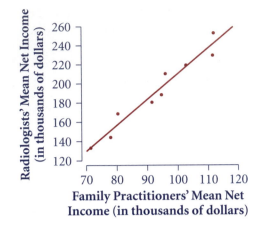

Display 3.18   Mean net income, radiology versus family practice, 1982–92.

Using Lines for Prediction

Lines are summaries and can be used to predict.

There are two main reasons that you would want to fit a line to a set of data:

- to find a summary, or model, that describes the relationship between the two variables
- to use the line to predict the value of *y* when you know the value of *x*. In cases where it makes sense to do this, the variable on the *x*-axis is called the **predictor** or **explanatory variable** and the variable on the *y*-axis is called the **response variable.**

In the last example, the equation

$$y = -905 + 12.5x$$

models the rise in radiologists' incomes for the years 1982 through 1992. Knowing this equation enables you to make general statements about radiologists' incomes throughout these years: "Roughly, the mean income went up $12,500 a year."

You might instead want to use this line to predict the income of radiologists in the missing years of 1983 and 1984 or for years before 1982 or after 1992.

Example: Predicting Radiologists' Income

Use the equation above to predict a radiologist's income in 1995.

Solution

Assuming the linear trend continues, the predicted income is

$$y = -905 + 12.5(95) = 282.5, \text{ or } \$282,500 \quad \blacksquare$$

Making the assumption that the linear trend continues can be a very risky business. This type of prediction is called **extrapolation**—making a prediction when the value of *x* falls outside the range of the actual data. **Interpolation**—making a prediction when the value of *x* falls inside the range of the data—is safer. It would be pretty safe to use the line to predict the income for 1983.

$\hat{y}$ is read "*y*-hat" and may be called the "predicted" value or the "fitted" value.

Suppose you know the value of *x* and use a line to predict the corresponding value of *y*. You know that your prediction for *y* won't be exact, but you hope that the error will be small. The **prediction error** is the difference between the observed value of *y* and the **predicted value** of *y* (called $\hat{y}$). You usually don't know what that error is, or else you wouldn't need to use the line to predict the value of *y*. You do, however, know the errors for the points used to construct the line. These are called **residuals:**

$$\text{residual} = \text{observed value of } y - \text{predicted value of } y = y - \hat{y}$$

The geometric interpretation of the residual is shown in Display 3.19.

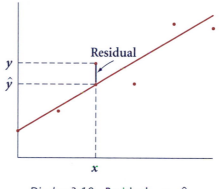

Display 3.19   Residual $= y - \hat{y}$.

Example: Finding Residuals

Display 3.20 shows the mean net income for doctors who were board-certified in family practice for the years 1982–92. The equation of the fitted line is

$$\hat{y} = -298.55 + 4.46x$$

where *x* is the number of years after 1900 and $\hat{y}$ is the income in thousands of dollars. In 1988, the mean net income was \$94,600. What is the residual for the year 1988?

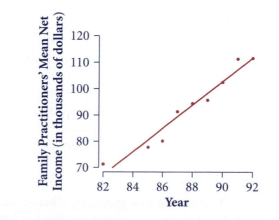

Display 3.20   Mean net income for doctors board-certified in family practice, 1982–92.

Solution

The prediction for 1988 is

$$\hat{y} = -298.55 + 4.46x = -298.55 + 4.46(88) = 93.93$$

or $93,930. The residual is

$$y - \hat{y} = 94.6 - 93.93 = 0.67$$

or $670. The residual is positive because the observed value is a bit higher than the predicted value. That is, the point lies just above the line. ■

Discussion: Using Lines for Prediction

D5. Test how well you understand residuals.

   a. If a residual is large and negative, where is the point located with respect to the line? Draw a diagram to illustrate. What does it mean if the residual is zero?

   b. If someone said that they had fit a line to a set of data points and all their residuals were positive, what would you say to them?

   c. Interpret the *y*-intercept of the regression line in the previous example. Does this make sense?

D6. What do you think of the arithmetic and the reasoning in this passage from Mark Twain's *Life on the Mississippi*?

   In the space of one hundred and seventy-six years the Lower Mississippi has shortened itself two hundred and forty-two miles. That is an average of a trifle over one mile and a third per year. Therefore, any calm person, who is not blind or idiotic, can see that in the Old Oölitic Silurian Period, just a million years ago next November, the Lower Mississippi River was upwards of one million three hundred thousand miles long, and stuck out over the Gulf of Mexico like a fishing rod.

   And by the same token any person can see that seven hundred and forty-two years from now the Lower Mississippi will be only a mile and three quarters long, and Cairo and New Orleans will have joined their streets together, and be plodding comfortably along under a single mayor and a mutual board of aldermen. There is something fascinating about science. One gets such wholesale returns of conjecture out of such a trifling investment of fact.

   Source: James R. Osgood and Company, 1883, p. 208.

   Given that the Mississippi/Missouri was about 3710 miles long in the year 2000, write an equation that Twain would say gives the length of the river in terms of the year.

■ Practice

P5. If you go to a university where class sizes tend to be small, are you more likely to give to your alumni fund after you graduate than someone who graduates from a university with large classes? Display 3.21 shows a scatterplot of a sample of 40 universities. Each university appears as a point. The vertical coordinate, *y*, tells the percentage of alumni who gave money. Each *x*-coordinate tells the student/faculty ratio (number of students per faculty member). The equation of the fitted line is approximately $\hat{y} = 55 - 2x$.

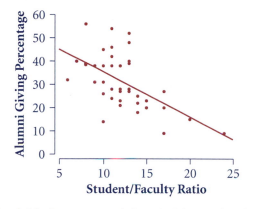

Display 3.21 Percentage of alumni giving to the alumni fund versus the student/faculty ratio for 40 highly rated U.S. universities.

a. Which is the explanatory variable and which is the response variable?

b. Explain how you can see from the graph that an increase of about four students per faculty member corresponds to an average decrease of eight percentage points in the giving rate. Explain how you can see this from the equation.

c. Interpret the *y*-intercept in this situation.

d. Use the regression line to predict the giving rate for a university with a student/faculty ratio of 16. When you use the regression line to predict the giving rate, would you expect a rather large error or a relatively small error in your prediction?

e. Use the plot to estimate the residual for the university with the highest student/faculty ratio. With the highest giving rate.

f. The university with the lowest student/faculty ratio, of 6 to 1, had a giving rate of 32%. Use the equation of the fitted line to find the residual for that university.

g. Suppose that the Alumni Association at Piranha State University boasts a giving rate of 80%. Without knowing the student/faculty ratio at PSU, can you tell whether the prediction error will be positive or negative?

P6. Examine the scatterplot in Display 3.22.

a. Which two kinds of pizza in Display 3.22 have the fewest calories? Which two have the least fat? Which region of the graph has pizzas with the most fat?

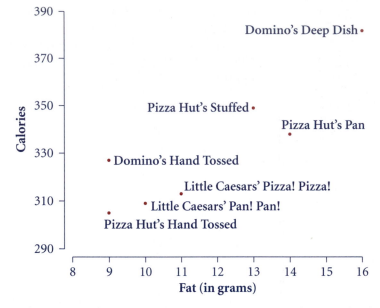

Display 3.22 Calories versus fat, per 5-ounce serving, for seven kinds
of pizza.

Source: *Consumer Reports,* Jan. 1997, p. 22.

b. Display 3.23 shows the data again, with five possible summary lines.
  Which equation (I–V) goes with which line (A–E)?

   I. *calories* = 225 + 9 *fat*   II. *calories* = 200 + 9 *fat*

 III. *calories* = 250 + 9 *fat*   IV. *calories* = 275 + 5 *fat*

  V. *calories* = 150 + 15 *fat*

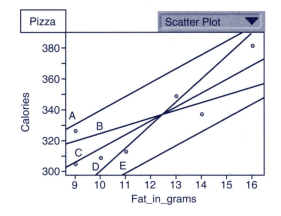

Display 3.23 Five possible fitted lines for the pizza data.

c. Each of the five lines gives a set of seven predicted values for the seven pizza brands.

   i. Which line gives predicted values for calorie content that are, on average, too high? How can you tell this from the plot?

   ii. Which line tends to give predicted calorie values that are too low?

   iii. Which line tends to overestimate calorie content for lower-fat brands and underestimate calorie content for higher-fat brands?

   iv. Which line has the opposite problem: It underestimates calorie content when fat content is lower and overestimates when fat content is higher?

   v. Which line tends to fit best overall?

Least Squares Regression Lines

The general approach to fitting lines to data is called the **method of least squares.** The method was invented about 200 years ago by Carl Friedrich Gauss (1777–1855), Adrien-Marie Legendre (1752–1833), and Robert Adrain (1775–1843), working independently of each other in Germany, France, and Ireland, respectively.

$$SSE = \sum (residuals)^2$$
$$= \sum (y - \hat{y})^2$$

The **least squares line,** or **regression line,** for a set of (x, y) data points is the line for which the **Sum of Squared Errors** (residuals), or SSE, is as small as possible.

Example: Regression Line for the DC9

This table shows *cost per hour* versus *number of seats* for three models of the DC9 jet from the data in Display 3.11. (Some of the data have been rounded.)

Aircraft	Number of Seats	Cost ($/hr)
DC9–30	100	1750
DC9–40	110	1800
DC9–50	120	1925
Average	**110**	**1825**

One of these two equations gives the least squares regression line for the DC9 data. Which one is it?

$$\hat{y} = 862.5 + 8.75x$$

$$\hat{y} = 850 + 9.00x$$

Solution

The least squares regression line minimizes the sum of the squared errors, SSE, so the equation with the smaller SSE must be the equation of the regression line.

For the equation $\hat{y} = 862.5 + 8.75x$:

x	y	$\hat{y}$	$y - \hat{y}$	$(y - \hat{y})^2$
100	1750	1737.5	12.5	156.25
110	1800	1825	−25	625
120	1925	1912.5	12.5	156.25
				SSE = 937.5

For the equation $\hat{y} = 850 + 9.00x$:

x	y	$\hat{y}$	$y - \hat{y}$	$(y - \hat{y})^2$
100	1750	1750	0	0
110	1800	1840	−40	1600
120	1925	1930	−5	25
				SSE = 1625

The first equation has the smaller SSE, so it must be the equation of the least squares regression line. Note that for this line, the sum of the residuals, $\sum(y - \hat{y})$, is equal to 0. This is always the case for the least squares regression line, but it can be true for other lines, too. ■

In addition to making the sum of the squared errors as small as possible, the least squares regression line has some other properties, given in the box.

Properties of the Least Squares Regression Line

The fact that the Sum of Squared Errors, or SSE, is as small as possible means that for this line, these properties also hold:

- The sum (and mean) of the residuals is 0.
- The variation in the residuals is as small as possible.
- The line contains the point of averages $(\bar{x}, \bar{y})$.
- The line has slope b_1 where

$$b_1 = \frac{\sum(x - \bar{x})(y - \bar{y})}{\sum(x - \bar{x})^2}$$

There are some appealing mathematical relationships among these properties, which, taken together, show that the line through $(\bar{x}, \bar{y})$ having slope b_1 does, in fact, minimize the sum of the squared errors. This gives you a way to find the equation of the least squares line $\hat{y} = b_0 + b_1 x$.

First compute the slope b_1 using the formula in the box. Then find the y-intercept, b_0, using the point $(\bar{x}, \bar{y})$ and the equation $\bar{y} = b_0 + b_1 \bar{x}$

or, solving for b_0,

$$b_0 = \bar{y} - b_1\bar{x}$$

Example: Least Squares Line for the DC9

Find the least squares line for the DC9 data given in the previous example.

Solution

Finding the line requires three main steps: find the point of averages $(\bar{x}, \bar{y})$, find the slope b_1, and use the point and slope to find the y-intercept b_0.

A convenient way to organize the computations is to work from a table:

Aircraft	Seats, x	Cost ($/hr), y
DC9–30	100	1750
DC9–40	110	1800
DC9–50	120	1925
Sum	**330**	**5475**
Mean	**110**	**1825**

Point of averages. The point of averages is $(\bar{x}, \bar{y}) = (110, 1825)$, and the least squares regression line passes through this point.

Slope. To compute the slope, first create two new columns for deviations from the mean, one for $x - \bar{x}$, one for $y - \bar{y}$:

Aircraft	Seats, x	Cost ($/hr), y	$x - \bar{x}$	$y - \bar{y}$
DC9–30	100	1750	−10	−75
DC9–40	110	1800	0	−25
DC9–50	120	1925	10	100
Sum	**330**	**5475**	**0**	**0**
Mean	**110**	**1825**	**0**	**0**

Now create another pair of columns, one for $(x - \bar{x})(y - \bar{y})$ and one for $(x - \bar{x})^2$:

Aircraft	Seats, x	Cost ($/hr), y	$x - \bar{x}$	$y - \bar{y}$	$(x - \bar{x}) \cdot (y - \bar{y})$	$(x - \bar{x})^2$
DC9–30	100	1750	−10	−75	750	100
DC9–40	110	1800	0	−25	0	0
DC9–50	120	1925	10	100	1000	100
Sum	**330**	**5475**	**0**	**0**	**1750**	**200**
Mean	**110**	**1825**	**0**	**0**		

The ratio of the sums of the last two columns gives the slope

$$b_1 = \frac{\sum(x - \bar{x})(y - \bar{y})}{\sum(x - \bar{x})^2} = \frac{1750}{200} = 8.75$$

y-intercept. Now that you have a point (110, 1825) on the line and the slope, 8.75, you can find the *y*-intercept from the equation

$$b_0 = \bar{y} - b_1\bar{x}$$

$$= 1825 - 8.75(110)$$

$$= 862.5$$

This agrees with what you found in the previous example. That is, the equation of the least squares regression line is

$$\hat{y} = 862.5 + 8.75x \quad \blacksquare$$

■ Practice: Least Squares Regression Line

P7. Here are the fat and calorie contents for 5 ounces of three kinds of pizza: (9, 305), (11, 309), (13, 316)

   a. Plot the points.

   b. Compute the equation of the least squares regression line by hand, and draw the line on your plot.

   c. Interpret the slope and *y*-intercept in the context of this situation.

   d. Verify that the least squares regression line goes through the point of averages.

   e. Verify that the sum of the residuals is 0.

P8. You should learn to use the statistical functions of your calculator to make scatterplots, find the regression equation, and compute residuals. Practice doing this for the airline data from P2. The values are given in Display 3.24.

Airline	Mishandled Baggage (per thousand passengers)	Percentage On-Time Arrivals
Northwest	5.62	75.0
Delta	4.00	73.7
Continental	6.21	73.1
Southwest	5.03	71.1
TWA	6.10	66.6
American	5.97	65.5
Alaska	6.42	65.5
USAir	5.51	63.3
America West	7.93	60.5
United	7.60	48.3

Display 3.24 Comparison, by airline, of mishandled baggage versus on-time arrival rate.

Source: Office of Aviation Enforcement and Proceedings, U.S. Department of Transportation, June 2000, www.dot.gov/airconsumer/0008ATCR.pdf.

Reading Computer Output

For real data, the best way to get the least squares line is by computer or calculator. Display 3.25 shows typical computer output for the full set of pizza data in Display 3.22.

```
Dependent variable is:    Calories
No Selector
R squared = 78.3%  R squared (adjusted) = 73.9%
s = 13.94 with 7 - 2 = 5 degrees of freedom

Source       Sum of Squares  df  Mean Square  F-ratio
Regression         3497.17   1      3497.17      18.0
Residual           971.684   5      194.337

Variable  Coefficient  s.e. of Coeff  t-ratio    prob
Constant     226.737           25.33     8.95  0.0003
Fat (g)      8.97368           2.115     4.24  0.0082
```

Display 3.25 Data Desk output giving the equation of the least squares line for the pizza data.

You can ignore most of the output for now, though you will learn how to interpret it in Chapter 11. For the time being, focus on the first two columns in the last three rows, which are reproduced in Display 3.26.

```
Variable  Coefficient
Constant     226.737
Fat (g)      8.97368
```

Display 3.26 The lower-left corner gives the slope and *y*-intercept.

The *y*-intercept is the coefficient in the row headed "Constant" and is 226.737. The slope is the coefficient for the predictor variable "Fat (g)" and is 8.97368. The SSE for the regression line is found in the "Residual" row: 971.684.

Discussion: Reading Computer Output

D7. ***Doctors' incomes.*** Display 3.27 shows Data Desk computer output for the least squares regression of the mean net income *y* for family practitioners versus year *x* for the years 1982–92. (See Display 3.17.)

   a. What is the equation of the least squares line? Estimate the SSE from Display 3.17, and then find it in the output.

```
Dependent variable is:    FamPrac
No Selector
R squared = 95.5%  R squared (adjusted) = 94.8%
s = 3.280 with 9 - 2 = 7 degrees of freedom

Source       Sum of Squares  df  Mean Square  F-ratio
Regression         1583.52   1      1583.52      147
Residual           75.3018   7      10.7574

Variable  Coefficient  s.e. of Coeff  t-ratio      prob
Constant    -298.550           32.30    -9.24   ≤0.0001
Year         4.46145          0.3677     12.1   ≤0.0001
```

Display 3.27 Data Desk output for the regression of family practitioners' income versus year.

b. The Minitab software output of this regression appears in Display 3.28. How is it different from that of Data Desk?

```
Regression Analysis

The regression equation is
Income = -299 + 4.46 Year

Predictor      Coef    Stdev   t-ratio       p
Constant    -298.55    32.30     -9.24   0.000
Year         4.4615   0.3677     12.13   0.000

s = 3.280   R-sq = 95.5%  R-sq(adj) = 94.8%

Analysis of Variance

SOURCE        DF       SS       MS       F       p
Regression     1   1583.5   1583.5  147.20   0.000
Error          7     75.3     10.8
Total          8   1658.8
```

Display 3.28 Minitab output for the regression of family practitioners' income versus year.

■ **Practice**

P9. The JMP-IN computer output below is for the pizza data in P7. Does it give the same results that you computed by hand? Where in the output is the SSE?

Linear Fit

calories = 279.75 + 2.75 fat

Summary of Fit

RSquare	0.975806
RSquare Adj	0.951613
Root Mean Square Error	1.224745
Mean of Response	310
Observations (or Sum Wgts)	3

Analysis of Variance

Source	DF	Sum of Squares	Mean Square	F Ratio
Model	1	60.500000	60.5000	40.3333
Error	1	1.500000	1.5000	Prob>F
C Total	2	62.000000		0.0994

Parameter Estimates

Term	Estimate	Std Error	t Ratio	Prob>|t|
Intercept	279.75	4.81534	58.10	0.0110
Fat	2.75	0.433013	6.35	0.0994

Summary 3.2: Fitting Lines to Scatterplots

For many quantitative relationships, it makes sense to use one variable, x, called the predictor or explanatory variable, to predict values of the other variable, y, called the response variable. When the data are roughly linear, you can use a fitted line, called the least squares regression line, as a summary, or model, that describes the relationship between the two variables. You might also use it to predict the value of an unknown y when you know the value of x.

Interpolation—using a fitted relationship to predict a response when the predictor value falls *within* the range of the data—is generally much more trustworthy than extrapolation—predicting response values based on the assumption that a fitted relationship applies *outside* the range of the observed data.

Each residual from a fitted line measures the vertical distance from a data point to the line:

$$\text{residual} = \text{observed} - \text{predicted} = y - \hat{y}$$

The least squares regression line for a set of (x, y) pairs is the line for which the Sum of Squared Errors, or SSE, is as small as possible. For this line, these properties hold:

- The sum (and mean) of the residuals is 0.
- The variation in the residuals is as small as possible.
- The line contains the point of averages $(\bar{x}, \bar{y})$.
- The line has slope b_1 where

$$b_1 = \frac{\sum(x - \bar{x})(y - \bar{y})}{\sum(x - \bar{x})^2}$$

To get the equation of the regression line,

- compute $\bar{x}$ and $\bar{y}$
- find the slope using the formula above
- compute the y-intercept: $b_0 = \bar{y} - b_1\bar{x}$

The equation then is $\hat{y} = b_0 + b_1 x$.

■ Exercises

E10. **Heights of children.** The scatterplot in Display 3.29 shows the median height in inches for boys ages 2 through 14 years.

a. Estimate the slope of the line showing the relationship between *age* and *height*. Explain the meaning of this slope with respect to boys and their height. Write the equation of this line using this slope and a point on the line. Interpret the *y*-intercept and discuss whether it makes sense in this context.

b. The next table gives the median height in inches for girls ages 2 through 14 years. Practice using your calculator by finding the equation of the least squares regression line for *height* versus *age*. Compare with the equation for boys from part a.

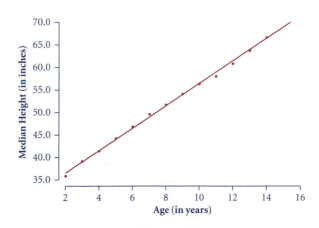

Display 3.29 Boys' median height versus age.

Source: National Health and Nutrition Examination Survey (NHANES), 2002, www.cdc.gov/nchs/about/major/nhanes/datatblelink.htm.

Age (yr)	Median Height (in.)	Age (yr)	Median Height (in.)
2	35.1	9	53.7
3	38.7	10	56.1
4	41.3	11	59.5
5	44.1	12	61.2
6	46.5	13	62.9
7	48.6	14	63.6
8	51.7		

E11. **Pizza again.** Displays 3.30 and 3.31 show the calorie and fat content of 5 ounces of various kinds of pizza.

Pizza	Calories	Fat (g)	Cost ($)
Pizza Hut's Hand Tossed	305	9	1.51
Domino's Deep Dish	382	16	1.53
Pizza Hut's Pan	338	14	1.51
Domino's Hand Tossed	327	9	1.90
Little Caesars' Pan! Pan!	309	10	1.23
Little Caesars' Pizza! Pizza!	313	11	1.28
Pizza Hut's Stuffed	349	13	1.23

Display 3.30   Food values and cost of pizza per 5-ounce serving.

Source: *Consumer Reports*, Jan. 1997, p. 22.

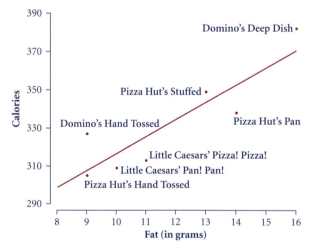

Display 3.31   Calories versus fat per 5-ounce serving, for seven kinds of pizza.

a. Use the line on the plot to predict the calorie content of a pizza with 10.5 grams of fat. Now use the line to predict calorie content for a pizza with 15 grams of fat.

b. Use the two predictions in part a to estimate the slope of the line. Write the equation of the line using this slope and a point on the line.

c. By hand, compute the equation of the least squares regression line for using fat to predict calories. How close was your estimate?

d. Now practice finding the equation of the regression line a third way—using your calculator.

e. There are 9 calories in a gram of fat. How does the slope compare?

E12. **More pizza.** Refer to E11.

a. The least squares residuals for the pizza data are, in order from smallest to largest, −14.37, −12.45, −7.47, −2.50, 5.61, 11.68, 19.50.

Match each residual with the name of its pizza.

b. What does the residual for Pizza Hut's Pan pizza say about its number of calories versus fat content?

c. For the two Domino's pizzas, are the residuals both positive or both negative? How can you tell this from the scatterplot in E11? Does this pattern suggest that Domino's pizzas may contain more carbohydrates or protein than other pizzas in the sample? Explain.

E13. **Even more pizza.** Use the pizza data in E11 to plot *cost* versus *fat* and *cost* versus *calories*. Should you model either of these relationships with a straight line?

E14. **Aircraft again.** The least squares line for the DC9 data has equation

$$cost = \$862.50 + \$8.75 \, seats$$

and the SSE is 937.5.

Aircraft	Seats, x	Cost ($/hr), y
DC9–30	100	1750
DC9–40	110	1800
DC9–50	120	1925
Sum	**330**	**5475**
Mean	**110**	**1825**

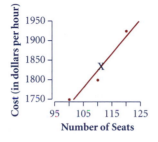

Display 3.32   Data for the three DC9 aircraft, showing the least squares line, which goes through the point of averages (shown as an X).

a. Find the SSE for the line with the same slope as the least squares line but that passes through the point for the DC9–40. Is this SSE larger or smaller than the SSE for the least squares line? According to the least squares approach, which line fits better? Do you agree?

b. Now find the slope of the line that passes through the points for the DC9–30 and the DC9–50. Then find the fitted value for the DC9–40. Finally, find all three residuals and the value of the SSE. This line passes through two of the data points. The least squares line passes through none of the data points, and yet according to SSE, that line fits better than this one. Do you agree that this line fits better than the other two? Explain why or why not.

E15. **Stopping on a dime?** In an emergency, the typical driver requires about 0.75 second to get his or her foot onto the brake. The distance that the car travels during this reaction time is called the **reaction**

distance. The table in Display 3.33 shows reaction distances for cars traveling at various speeds.

Speed (mph)	Reaction Distance (ft)
20	22
30	33
40	44
50	55
60	66
70	77

Display 3.33   Reaction distance at various speeds.

a. Plot *reaction distance* versus *speed*, with *speed* on the horizontal axis. Describe the shape of the plot.

b. What should the y-intercept be?

c. Find the slope of the line by calculating the change in y per unit change in x. What does the slope represent in this situation?

d. Write the equation of the line that fits these data.

e. Use the line of part d to predict the reaction distance for a car traveling at 55 miles per hour, and at 75 miles per hour.

f. How would the equation change if it actually took 1 second, instead of 0.75 second, for drivers to react?

E16. **Colleges again.** Refer to the scatterplot in Display 3.21 on page 121 of the percentage of alumni who give to their colleges versus the student/faculty ratio at their colleges. Part of a computer output for the regression appears in Display 3.34.

a. What equation is given for the least squares regression line?

b. Examine the table of unusual observations. What is the student/faculty ratio at the college with the largest residual (in absolute value)? Find this college in Display 3.21.

```
Predictor      Coef    Stdev   t-ratio      p
Constant     54.979    5.477     10.04  0.000
S/F Rati    -1.9455   0.4354     -4.47  0.000

s = 9.704   R-sq = 34.4%   R-sq(adj) = 32.7%

Analysis of Variance

SOURCE        DF       SS      MS       F       p
Regression     1   1880.2  1880.2   19.97   0.000
Error         38   3578.5    94.2
Total         39   5458.7

Unusual Observations
Obs.  S/F Rati  Alumni G    Fit  Stdev.Fit  Residual  St.Resid
   9      10.0     14.00  35.52       1.78    -21.52     -2.26
  14      11.0     54.00  33.58       1.60     20.42      2.13
  30      13.0     52.00  29.69       1.59     22.31      2.33
  39      20.0     15.00  16.07       3.78     -1.07     -0.12
  40      24.0      9.00   8.29       5.41      0.71      0.09
```

Display 3.34 Computer output: Regression analysis of percentage giving to alumni fund versus student/faculty ratio.

c. Verify that the fit and the value of the largest residual were computed correctly.

E17. ***More about slope.***

a. You and three friends, one right after the other, each buy the same kind of gas at the same pump. Then you make a scatterplot of your data, making one point per person, with the number of gallons on the *x*-axis and the total price paid on the *y*-axis. True or false, and explain: All four points will lie on the same line.

b. You and the same three friends each drive 80 miles but with different average speeds. Afterward, you plot your data twice, first as a set of four points with coordinates average speed, *x*, and elapsed time, *y*, and then a second time as a set of points with coordinates average speed, *x*, and with *y** defined as $\frac{1}{elapsed\ time}$. Which plot will give a straight line? Explain your reasoning. Will the other plot be a curve opening up, a curve opening down, or neither?

E18. ***A median-fit line*** is an exploratory method of fitting a line that works if you have enough data. Divide your plot into vertical slices, and find the center of the points in each slice by taking the midpoint of the horizontal interval and the median of the

y-coordinates in that slice. Then choose a line to fit the centers, as in Display 3.35.

a. Refer to the middle plot. In what sense is the median of the boxplot above 15 on the *x*-axis a reasonable predicted value for the giving rate of a university whose student/faculty ratio is 15?

b. How can you use the boxplot referred to in part a to tell how far off your prediction might be if you use the fitted line?

E19. ***Practice with the median-fit line.*** For the sample of commercial aircraft, the scatterplot in Display 3.36 gives *operating cost* (dollars per hour) versus *fuel consumption rate* (gallons per hour).

a. Divide a copy of the plot into four slices by drawing vertical lines through *x*-values 0, 1000, 2000, 3000, and 4000.

b. Create a summary point for each slice: Find the median of the *y*-values in the slice, and plot the median *y*-value against the *x*-value at the midpoint of that slice.

c. Draw a line that passes through or near your four summary points, and find its equation.

d. Do you consider your line a reasonable summary of the relationship? Why or why not?

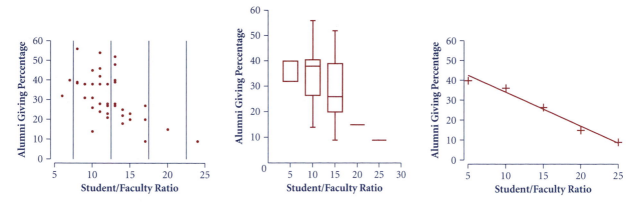

Display 3.35   Line-fitting by the method of vertical slices.

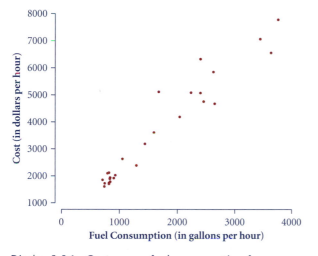

Display 3.36   Cost versus fuel consumption for commercial aircraft.

Source: Air Transport Association of America, 1996, www.air-transport.org/public/industry/67.asp.

E20. ***Sum of residuals.*** In this exercise, you will show that the mean of the residuals is equal to zero if and only if the least squares regression line passes through the point of averages.

  a. Show that for a horizontal line fitted to the DC9 data in Display 3.32, the mean of the residuals will be zero if and only if the line passes through the point of averages (110, 1825).

  b. Show that for any line fitted to the DC9 data, no matter what its slope is, the

mean of the residuals will be zero if and only if the line passes through the point of averages.

  c. Show that for any set of data points, the mean of the residuals is equal to zero if and only if the fitted line passes through the point of averages $(\bar{x}, \bar{y})$.

  d. Explain why it's not good enough to define the regression line as the line that makes the sum of the residuals zero.

E21. The data set in Display 3.37 is the pizza data set of E11 augmented by other brands of cheese pizza typically sold in supermarkets.

  a. Plot *calories* versus *fat*. Does there appear to be a linear association between *calories* and *fat*? If so, fit a least squares regression line to the data, and interpret the slope of the line.

  b. Plot *fat* versus *cost*. Does there appear to be a linear association between *cost* and *fat*? If so, fit a least squares regression line to the data, and interpret the slope of the line. Can you suggest a reason for the trend seen here?

  c. Plot *calories* versus *cost*. Does there appear to be a linear association between *cost* and *calories*?

  d. Write a synopsis of your findings.

Pizza	Calories	Fat (g)	Cost ($)
Pizza Hut's Hand Tossed	305	9	1.51
Domino's Deep Dish	382	16	1.53
Pizza Hut's Pan	338	14	1.51
Domino's Hand Tossed	327	9	1.90
Little Caesars' Pan! Pan!	309	10	1.23
Little Caesars' Pizza! Pizza!	313	11	1.28
Pizza Hut's Stuffed	349	13	1.23
DiGiorno	332	10	0.90
Tombstone's 4 Cheese	364	17	0.85
Red Baron	393	19	0.80
Boboli	347	12	1.00
Jack's	350	17	0.69
Pappalo's	353	12	0.75
Tombstone's Original	357	16	0.81
Master Choice	296	13	0.90
Celeste	358	16	0.92
Totino's	322	14	0.64
New Weight Watchers	337	10	1.54
Jeno's	323	14	0.72
Stouffer's	333	13	1.15
Ellio's	299	9	0.52
Kroger	316	7	0.72
Healthy Choice	275	4	1.20
Lean Cuisine	288	7	1.49

Display 3.37 Food values and cost per 5-ounce serving of pizza.

Source: *Consumer Reports,* Jan. 1997, p. 22.

E22. The plot in Display 3.38 shows *cost in dollars per hour* versus *number of seats* for three models of the DC9 jet from data in Display 3.32. Five lines, labeled A–E, are shown on the plot. Their equations are listed in the table, labeled I–V.

a. Match each line (A–E) with its equation (I–V).

Equation	
I.	917.5 + 8.25 *seats*
II.	975.0 + 8.25 *seats*
III.	850.0 + 8.25 *seats*
IV.	1275.0 + 5.00 *seats*
V.	505.0 + 12.00 *seats*

b. Match each line (A–E) with the appropriate verbal description:

i. This line, on average, tends to overestimate cost.

ii. This line, on average, tends to underestimate cost.

iii. This line overestimates cost for the smallest plane and underestimates cost for the largest plane.

iv. This line underestimates cost for the smallest plane and overestimates cost for the largest plane.

v. On balance, this line gives a better fit than the other lines.

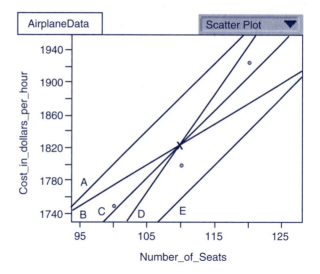

Display 3.38 Cost per hour versus number of seats for three models of the DC9 jet.

3.3 Correlation: The Strength of a Linear Trend

Some of the linear relationships you've seen in this chapter have been extremely strong, with points packed tightly about the regression line. Other linear relationships have been quite weak where although there was a general linear trend, there was a lot of variation in the values of y associated with a given value of x. Still other linear relationships were in between.

In this section, you'll learn how to measure the strength of a linear relationship numerically by using the correlation coefficient, r (which, from this point on, will be referred to simply as the correlation), where $-1 \leq r \leq 1$. Just as in the last section, you'll start by working intuitively and visually, then move to a computational approach.

The correlation coefficient measures the amount of variation from the regression line.

Examine the scatterplots and their correlations in Display 3.39. To get a rough idea of the size of a correlation, it is helpful to sketch an ellipse around the cloud of points in the scatterplot. If the ellipse has points scattered throughout and the points appear to follow a linear trend, then the **correlation** is a reasonable measure of the strength of the association. If the ellipse slants upward as you go from left to right, the correlation is **positive;** if the ellipse slants downward, the correlation is **negative.** If the ellipse is fat, the correlation is **weak** and the absolute value of r is close to 0. If the ellipse is skinny, the correlation is **strong** and the absolute value of r is close to 1.

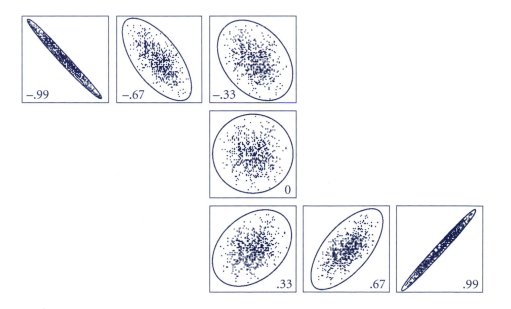

Display 3.39 Scatterplots with ellipses and their correlations.
Source: George Cobb, *Electronic Companion to Statistics* (Cogito Learning Media, Inc., 1997), p. 114.

In Activity 3.2, you will learn more about correlation and will practice finding the value of r using your calculator.

Activity 3.2 Was Leonardo Correct?

What you'll need: a measuring tape, yardstick, or meter stick

Leonardo da Vinci was a scientist and an artist who combined these skills to draft extensive instructions for other artists on how to proportion the human body in painting and sculpture. Three of Leonardo's rules were

- height is equal to the span of the outstretched arms
- kneeling height is three-fourths of the standing height
- the length of the hand is one-ninth of the height

1. Record your height. Work with a partner to measure your kneeling height, arm span, and the length of your hand. Combine your data with data from the rest of your class.

2. Check Leonardo's three rules visually by plotting the data on three scatterplots.

3. For the plots that have a linear trend, use your calculator to find the value of *r*, the correlation, and the equation of the regression line.

4. Interpret the slopes of the regression lines. Interpret the correlation.

5. Do the three relationships described by Leonardo appear to hold? Do they hold strongly?

Discussion: Estimating the Correlation

D8. Match each of the five scatterplots listed below with its correlation, choosing from

−.783   −.572   .783   .885   .999

a. *year of birth* versus *year hired* (Display 3.1 on page 102)

b. *age at layoffs* versus *year hired* (Display 3.2 on page 102)

c. *average height of boys* versus *age* (E10 on page 129)

d. *calories* in pizza versus *fat* (E11 on page 130)

e. *alumni giving percentage* versus *student/faculty ratio* (E18 on page 133)

D9. Four relationships are described below.

i. For a random sample of students from the senior class, *x* represents the day of the month of the person's birthday and *y* represents the cost of the person's most recent haircut.

ii. For a random collection of U.S. coins, *x* is the diameter and *y* is the circumference.

iii. For a random sample of bags of white socks, *x* represents the number of socks and *y* represents the price per bag.

iv. For a random sample of bags of white socks, *x* represents the number of socks and *y* represents the price per sock.

a. Which of these relationships have a positive correlation, and which a negative? Which has the strongest relationship? The weakest?

b. For each of the four relationships, pick a value of *x* and give a reasonable range of values for *y*. What is the relationship between your answers here and the strength of the correlation?

■ Practice

P10. Match each of the five scatterplots with its correlation, choosing from

$$-.95 \qquad -.5 \qquad 0 \qquad .5 \qquad .95$$

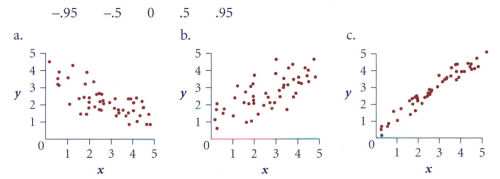

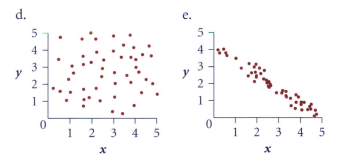

Display 3.40 Five scatterplots.

P11. This table from E11 (Display 3.30) gives the amount of fat and calories in various pizzas. Use your calculator to find the correlation, *r*.

Fat (g)	Calories
9	305
16	382
14	338
9	327
10	309
11	313
13	349

A Formula for the Correlation, *r*

A formula for the correlation, *r*, follows. It looks impressive, but the basic idea is simple—you convert *x* and *y* to standardized values (*z*-scores), then find their average product (dividing by *n* – 1).

You can think of the correlation, *r*, as the average product of the *z*-scores.

$$r = \frac{1}{n-1} \sum \left(\frac{x - \bar{x}}{s_x}\right)\left(\frac{y - \bar{y}}{s_y}\right) = \frac{1}{n-1} \sum z_x \cdot z_y$$

In this formula, s_x is the standard deviation of the *x*'s, while s_y is the standard deviation of the *y*'s. Remember that the *z*-score tells you how many standard deviations the value lies above or below its mean.

Example: Computing r for Airline Data

Compute the correlation for the relationship between the number of mishandled bags per 1000 passengers and the percentage of on-time arrivals for the airline data in Display 3.24.

Solution

When on a desert island or taking a test where you must compute *r* by hand, it is easiest to organize your work as in Display 3.41. First, compute the average value of *x* and *y*, $\bar{x}$ and $\bar{y}$. Then compute their standard deviations, s_x and s_y. These are

$$\bar{x} = \frac{60.39}{10} = 6.039 \qquad \bar{y} = \frac{662.2}{10} = 66.26$$

$$s_x = 1.146 \qquad s_y = 7.94$$

	Original Units		Standard Units (*z*-scores)		Product
Airline	**Mishandled Bags (per thousand passengers)** x	**Percentage On-Time Arrivals** y	$z_x = \dfrac{x - \bar{x}}{s_x}$	$z_y = \dfrac{y - \bar{y}}{s_y}$	$z_x \cdot z_y = \left(\dfrac{x - \bar{x}}{s_x}\right)\left(\dfrac{y - \bar{y}}{s_y}\right)$
Northwest	5.62	75.0	−0.36562	1.10076	−0.40246
Delta	4.00	73.7	−1.77923	0.93703	−1.66719
Continental	6.21	73.1	0.14921	0.86146	0.12854
Southwest	5.03	71.1	−0.88045	0.60957	−0.53670
TWA	6.10	66.6	0.05323	0.04282	0.00228
American	5.97	65.5	−0.06021	−0.09572	0.00576
Alaska	6.42	65.5	0.33246	−0.09572	−0.03182
USAir	5.51	63.3	−0.46161	−0.37280	0.17208
America West	7.93	60.5	1.65009	−0.72544	−1.19704
United	7.60	48.3	1.36213	−2.26197	−3.08109
Sum	**60.39**	**662.6**	**0**	**0**	**−6.6076**
Mean	**6.039**	**66.26**	**0**	**0**	
Standard Deviation	**1.146**	**7.94**	**1**	**1**	

Display 3.41 Calculations for the correlation.

Then convert each value of x and y to standard units or z-scores. The correlation, r, is the average product of the z-scores: the sum of the last column divided by $n - 1$, which is

$$r = \frac{1}{n-1} \sum z_x \cdot z_y$$

$$= \frac{1}{10-1}(-6.6076)$$

$$= -.734 \quad \blacksquare$$

A way to visualize the computations in this example is to look at the four quadrants formed on the scatterplot by dividing it vertically at the mean value of x and horizontally at the mean value of y. Such a plot is shown in Display 3.42. Points in Quadrant I, such as the point for Continental, have positive z-scores for both x and y, so their product contributes a positive amount to the calculation of the correlation. Points in Quadrant III, such as the point for USAir, have negative z-scores for both x and y, so their product contributes a positive amount to the calculation of the correlation. Points in Quadrants II or IV, such as Delta and America West, contribute negative amounts to the calculation for the correlation because one z-score is positive and the other is negative.

The points in Quadrants II or IV have negative products $z_x \cdot z_y$.

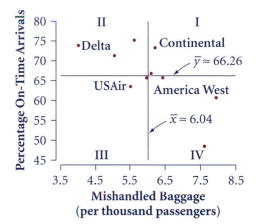

Display 3.42   Scatterplot divided into quadrants at $(\bar{x}, \bar{y})$.

The correlation, r, is a quantity without units.

Because r is the average of the products $z_x \cdot z_y$ and z-scores have no units, r has no units. In fact, r does not depend on the units of measurement in the original data. In Activity 3.2, if you measure arm spans and heights in inches and your friend measures in centimeters, you will both compute the same value for the correlation.

Discussion: A Formula for *r*

D10. Look at Display 3.41, showing the calculations of r.

    a. Confirm the calculations for the first row of data.

    b. Which point makes the largest contribution to the correlation? Where is this point on the scatterplot?

c. Which point makes the smallest contribution to the correlation? Where is it on the scatterplot?

D11. **Understanding r.**

a. Explain in your own words what the correlation measures.

b. Explain in your own words why r has no units.

c. When computing the correlation between two variables, does it matter which you select as y and which you select as x? Explain.

D12. Eight artificial "data sets" are shown here. For each one, find the value of r. Drawing a quick sketch might be helpful.

a.
x	y
−1	−1
0	0
1	1

b.
x	y
−1	−1
0	1
1	0

c.
x	y
−1	0
0	1
1	−1

d.
x	y
−1	1
0	0
1	−1

e.
x	y
99	9
100	10
101	11

f.
x	y
15	30
20	40
25	20

g.
x	y
1003	80
1006	82
1009	81

h.
x	y
9.9	1000
10.0	2000
10.1	0

D13. Refer to Display 3.42.

a. What can you say about r if there are many points in Quadrants I and III and few in II and IV?

b. What can you say about r if there are many points in Quadrants II and IV and few in I and III?

c. What can you say about r if points are scattered randomly in all four quadrants?

■ Practice

P12. The table in P11 on page 137 gives the amount of fat and calories in various pizzas. In P11, you used your calculator to find the correlation, r. This time, make a table like that in Display 3.41, and use the formula to find r. What do you notice about the products $z_x \cdot z_y$?

P13. The scatterplot in Display 3.43 is divided into quadrants by vertical and horizontal lines that pass through the point of averages $(\bar{x}, \bar{y})$.

a. Is the correlation positive or negative?

b. Give the coordinates of the point that will contribute the most to the correlation, r.

c. Consider the product

$$\left(\frac{x - \bar{x}}{s_x}\right)\left(\frac{y - \bar{y}}{s_y}\right)$$

Where are the points that have a positive product? How many of the 30 points have a positive product?

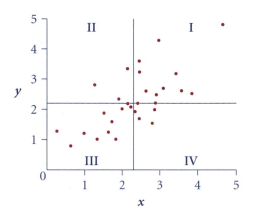

Display 3.43   Scatterplot divided into quadrants at $(\bar{x}, \bar{y})$.

d.   Where are the points that have a negative product? How many have a negative product?

The Relationship Between *r* and the Slope

By now you may have observed that the slope of the regression line, b_1, and the correlation, r, always have the same sign. But there is more:

$$b_1 = r\frac{s_y}{s_x}$$

Slope varies directly as the correlation.

where s_x is the standard deviation of the x's and s_y is the standard deviation of the y's.

Example: Verbal and Math SAT Scores

In 2000, the mean verbal score for all SAT I test takers was 505 with a standard deviation of 111. For math, the mean was 514 with a standard deviation of 113. The correlation between the two scores was not given but is known to be quite high. If you can estimate this correlation as, say, .7, you can find the equation of the regression line and use it to estimate the math score from a student's verbal score. [Source: *2000 College Bound Seniors: A Profile of SAT Program Test Takers,* The College Board, 2000, p. 7.]

Solution

Using the formula above, an estimate of the slope is

$$b_1 = r\frac{s_y}{s_x} = .7 \cdot \frac{113}{111} \approx 0.71$$

To find the y-intercept, use the fact that the point $(\bar{x}, \bar{y}) = (505, 514)$ is on the regression line:

$$y = \text{slope} \cdot x + y\text{-intercept}$$
$$514 = 0.71(505) + y\text{-intercept}$$
$$y\text{-intercept} = 155.45$$

The equation is $\hat{y} = 0.71x + 155.45$. ■

Discussion: r and b_1

D14. Find the equation of the regression line for predicting an SAT I verbal score given the student's SAT I math score.

■ Practice

P14. Imagine a scatterplot of two sets of exam scores for students in a statistics class. The score for a student on Exam 1 is graphed on the x-axis, and his or her score on Exam 2 is graphed on the y-axis. The slope of the regression line is 0.368. The mean for Exam 1 scores is 72.99, and the standard deviation is 12.37. The mean for Exam 2 scores is 75.80, and the standard deviation is 7.00.

a. Find the correlation for these scores.

b. Find the equation of the regression line for predicting an Exam 2 score from an Exam 1 score. Predict the Exam 2 score for a student who got an 80 on Exam 1.

c. Find the equation of the regression line for predicting an Exam 1 score from an Exam 2 score.

d. Sketch a scatterplot that could represent the situation described.

Correlation Does Not Imply Cause and Effect

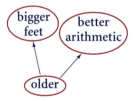

In a sample of elementary school students, there is a strong positive relationship between shoe size and scores on a standardized test of ability to do arithmetic. Does this mean that studying arithmetic makes your feet get bigger? No. Shoe size and skill at arithmetic are related to each other because they both increase as a child gets older. Age is an example of a *lurking variable*.

> Sometimes when you find a strong relationship between two variables x and y, there's a third variable z lurking in the background: x and y are related to each other because they are both consequences of the **lurking variable z**.

Two variables may be highly correlated without one causing the other.

Even if you can't identify a lurking variable, you should be careful to avoid jumping to a conclusion about cause and effect when you observe a strong relationship. *The value of r does not tell you anything about why two variables are related.* The slogan "Correlation does not imply causation" can help you remember this. To conclude that one thing causes another, you need data from a randomized experiment, as you'll learn in the next chapter.

Discussion: Correlation Does Not Imply Causation

D15. Display 3.42 on page 139 shows a negative association between the percentage of on-time arrivals and the number of mishandled bags per 1000 passengers. Discuss whether you think one of these variables might cause the other, or whether a lurking variable might account for both.

D16. For the sample of 50 top-rated universities, there's a very strong positive relationship between acceptance rate (percentage of applicants who are offered admission) and SAT scores (the 75th percentile for an entering class). Explain why these two variables have such a strong relationship: Does one "cause" the other? If not, how might you account for the strong relationship?

D17. People who argue about politics and public policy often point to relationships between quantitative variables and then assume a cause-and-effect explanation to support their points of view. For each of these relationships, first give a possible explanation by assuming a causal relationship, then give another possible explanation based on a lurking variable.

   a. Faculty positions in academic subjects with a higher percentage of male faculty tend to pay higher salaries. (For example, engineering and geology have high percentages of male faculty and high average salaries; journalism, music, and social work have much lower percentages of male faculty and much lower academic salaries.)

   b. States with larger reported numbers of hate groups tend to have more people on death row. (Here, tell also how you could adjust for the lurking variable to get a more informative relationship.)

   c. States with higher reported rates of gun ownership tend to have lower reported rates of violent crime.

■ Practice

P15. If you take a random sample of U.S. cities and measure the number of fast-food franchises in the city and the number of cases of stomach cancer per year in the city, you find a high correlation.

   a. What is the lurking variable?

   b. How would you adjust the data for the lurking variable to get a more meaningful comparison?

P16. If you take a random sample of public school students in grades K–12 and measure weekly allowance and vocabulary size, you will find a strong relationship. Explain in terms of a lurking variable why you should not conclude that raising a child's allowance will tend to increase his or her vocabulary.

P17. For the countries of the United Nations, there is a strong negative relationship between the number of TV sets per 1000 people and the birth rate. What would be a careless conclusion about cause and effect? What is the lurking variable?

Interpreting r^2

You may have noticed that computer outputs for regression analysis, like those in Display 3.25, give the value of *R*-squared, or r^2, rather than the value of *r*. The student in this discussion will show you how to think about r^2 as the fraction of the variation in the values of *y* that you can eliminate by taking *x* into account.

Student:	I've invented another way to measure the strength of the relationship between *x* and *y*. It's based on the idea that the less variation there is from the linear trend, the stronger the correlation.
Statistician:	How does it work?
Student:	Let me ask the questions for a change. What is the best way to predict the values of *y*?
Statistician:	I'd fit a least squares line and use $\hat{y} = b_0 + b_1 x$.
Student:	And how would you measure your total error?
Statistician:	Well, I'd use the sum of the squares of the residuals, just like we did in the last section:

$$\text{SSE} = \sum(y - \hat{y})^2$$

Student:	That's what I hoped you would say.
Statistician:	We consultants try to be helpful.
Student:	Don't get cocky. I'm about to change the rules. Pretend that you can't use the information about *x*. You don't have the *x*-values, and you want a single fitted value for *y*. What value would you choose?
Statistician:	I'd use $\bar{y}$.
Student:	And could you again use the sum of the squared errors to measure your total error?
Statistician:	Sure, it's almost like the standard deviation. I'd find the sum of the squares of the deviations from the mean, which I'll call the total sum of squared error, SST:

$$\text{SST} = \sum(y - \bar{y})^2$$

Student:	Okay, now for my new way to measure strength. If you have a strong relationship, the SSE will be small compared to the SST, right?
Statistician:	Right.
Student:	On the other hand, if you have a weak relationship, *x* isn't much use for predicting *y*, and SSE will be almost as big as SST. Right?
Statistician:	I see where you're going with this. You can use a ratio to measure the strength of the relationship:

Relationship	Strong	Weak
$\dfrac{\text{SSE}}{\text{SST}}$	Near 0	Near 1

Student: Exactly. Except now I've got a problem. I have two measures of strength—the correlation, r, and my new ratio. Which one should I use?

Statistician: Lucky for you—they turn out to be equivalent:

$$r^2 = \frac{\text{SST} - \text{SSE}}{\text{SST}}$$

Student: That's ugly! What happened to *my* ratio?

Statistician: It's still there, on the right. Think of it this way: If you don't know about y's relationship with x, when you want to predict y, the best you can do is to use $\bar{y}$. Your total squared error is SST. But if you do know about x, the total squared error is SSE.

Student: I can handle this from here. SST − SSE is the amount of error you get rid of. And r^2 is the proportion of error you get rid of out of the total error that you started with.

Statistician: Right! We statisticians call r^2 the **coefficient of variation.** It tells us the proportion of variation in the y's that is "explained" by x.

Student: I like it. Anything's better than those z-scores!

Example: Predicting Pizzas

Display 3.44 compares two sets of predictions of the calorie content in the seven kinds of pizza.

		Using the Mean		Using the Regression Line	
Fat (g) x	Calories y	Predicted $\bar{y}$	Error $y - \bar{y}$	Predicted $\hat{y}$	Error $y - \hat{y}$
9	305	331.86	−26.86	307.51	−2.51
16	382	331.86	50.14	370.32	11.68
14	338	331.86	6.14	352.38	−14.38
9	327	331.86	−4.86	307.51	19.49
10	309	331.86	−22.86	316.48	−7.48
11	313	331.86	−18.86	325.45	−12.45
13	349	331.86	17.14	343.40	5.60
Sum of Squared Errors		**SST = 4468.90**		**SSE = 971.68**	

Display 3.44 Predicting calorie content from fat content for seven kinds of pizza.

If you had to pick a single number as your predicted calorie content, you might choose the mean, 331.86 calories per serving. The fourth column in the table and the plot in Display 3.45 show the resulting errors.

If you use the regression equation, *calories* = 226.74 + 8.974 *fat,* to predict calories, the resulting errors are much smaller in most cases and are given in the final column of Display 3.44 and shown in the plot in Display 3.46.

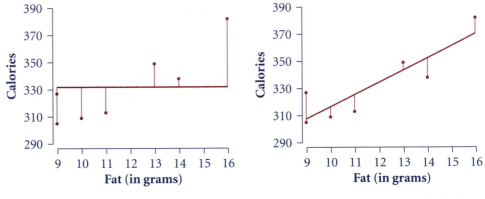

Display 3.45 Deviations around the mean of *y.*

Display 3.46 Deviations around the least squares line.

Using the student's formula,

$$r^2 = \frac{\text{SST} - \text{SSE}}{\text{SST}} = \frac{4468.9 - 971.68}{4468.9} \approx .7826$$

The coefficient of variation tells the proportion of the total variation in *y* that can be explained by the regression line.

About 78% of the variation in calories between these brands of pizza can be attributed to fat content.

Taking the square root, *r* = .885. This should be the same as the correlation you computed in P11.

Discussion: Interpreting r^2

D18. Verify the first line in Display 3.44.

D19. Use the student's formula for r^2 to explain why $-1 \leq r \leq 1$.

D20. In a study of the effect of temperature on household heating bills, an investigator said, "Our research shows that about 70% of the variability in the heating units used by a particular house over the years can be explained by outside temperature." Explain what the investigator may have meant by this statement.

■ Practice

P18. Data on the association between high school graduation rates and the percentage of families living in poverty for the 50 U.S. states were presented in E8. Display 3.47 contains the scatterplot and standard computer output of the regression analysis.

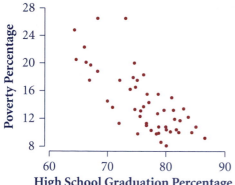

```
The regression equation is
PovR = 60.7 - 0.610 HSG

Predictor      Coef      Stdev    t-ratio        p
Constant     60.745      5.981      10.16    0.000
HSG         -0.60986    0.07826      -7.79    0.000

s = 3.095

Analysis of Variance

SOURCE        DF         SS        MS        F        p
Regression     1     581.54    581.54    60.73    0.000
Error         49     469.22      9.58
Total         50    1050.76
```

Display 3.47   Poverty rates versus high school graduation rates.

a. Under "SOURCE," the "Total" variation is the SST and "Error" variation is the SSE. From this information, find r, the correlation.

b. Write an interpretation for the slope of this regression line, in the context of these data.

c. Does the presence of a linear relationship here imply that a state that raises its graduation rate will cause its poverty rate to go down? Explain your reasoning.

d. What are the units for each of x, y, b_1, and r?

Regression Toward the Mean

Display 3.48 shows a hypothetical data set with the height of a younger sister plotted against the height of her older sister. There is a moderate positive association, $r = .337$. For both younger and older sisters, the mean height is 65 inches and the standard deviation is 2.5 inches. The line on the first plot, $y = x$, indicates where both sisters would be the same height. If you rotate your book and sight down the line, you can see that the points are scattered symmetrically about it.

Now, in the second plot, look at the vertical strip for the older sisters between 62 and 63 inches. The **X** is at the mean height of the younger sisters. It falls at about 64 inches, not between 62 and 63 inches as you would expect! Looking at the vertical strip on the right, the mean height of younger sisters with older sisters between 68 and 69 inches is only about 66 inches. If you were to use the line $y = x$ to predict the height of the younger sister, you would tend to predict a height that is too small if the older sister is shorter than average and a height that is too great if the older sister is taller than average.

The regression line as a line of means.

The flatter line through the third scatterplot of Display 3.48 is the least squares regression line. Notice that this line gets as close as it can to the center of each vertical strip. Thus, the least squares line is sometimes called the **line of means.** The predicted value for y at a given value of x, using the regression line as the model, is the estimated mean of all responses that can be produced at that particular value of x.

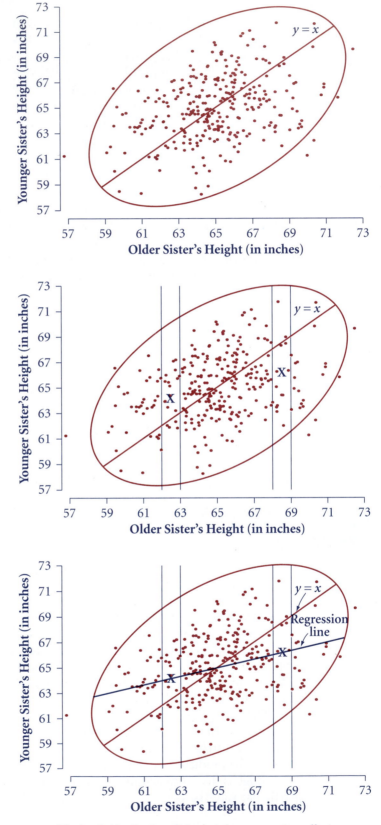

Display 3.48 Scatterplots showing regression effect.

Notice that the regression line has a smaller slope than the major axis ($y = x$) of the ellipse. This means that the predicted values are closer to the mean than you might expect. This will always be the case for positively correlated data following a linear trend. The difference between these two lines is sometimes called the **regression effect.** If the correlation is near +1 or −1, the two lines will be nearly on top of each other and the regression effect will be minimal. For a moderate correlation such as this one, the regression effect will be quite large.

"Regression toward the mean" is another term for regression effect.

The regression effect was first noticed by the British scientist Francis Galton around 1877. Galton noted that the largest sweet pea seeds tended to produce daughter seeds that were large but smaller than their parent. The smallest sweet-pea seeds tended to produce daughter seeds that were also small but larger than their parent. There was, in Galton's words, a **regression toward the mean.** This is the origin of the term "regression line." [Source: D. W. Forrest, *Francis Galton: The Life and Work of a Victorian Genius* (Taplinger, 1974).]

The regression effect is with us in everyday life whenever some element of chance is involved in a person's score. For example, in sports, there is said to be a "sophomore slump." That is, athletes who have the best rookie seasons do not tend to be the same players who have the best second year. The top students on the second exam in your class probably did not do as well, relative to the rest of the class, on the first exam. The children of extremely tall or short parents do not tend to be as extreme in height as their parents. There does, indeed, seem to be a phenomenon at work that pulls us back toward the average. As Galton noticed, this prevents the spread in human height, for example, from increasing. Look for it as you work on regression analyses of data.

Discussion: The Regression Effect

D21. Why is the regression line sometimes called the "line of means"?

D22. The equation of the regression line for the scatterplot in Display 3.48 is $\hat{y} = 43.102 + 0.337x$. Interpret the slope of this line in the context of this situation and compare it to the interpretation of the slope of the line $y = x$.

■ Practice

P19. The plot in Display 3.49 shows the heights of the older sisters plotted against the heights of their younger sisters. On a copy of this scatterplot, draw vertical lines to divide the points into six groups. Mark the mean of the y-values of each vertical strip. Sketch the regression line. Now draw an ellipse around the data and connect the two ends of the ellipse. Is the regression line "flatter" than the other line? Does this plot also show the regression effect?

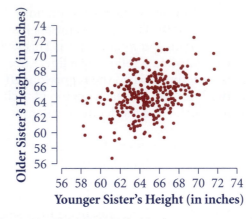

Display 3.49 The heights of older versus younger sisters.

Summary 3.3: Correlation

In your study of normal distributions in Chapter 2, you used the mean to tell the center; then you used the standard deviation as the overall measure of how much the values deviated from that center. For "well-behaved" quantitative relationships—those whose scatterplots look elliptical—you use the regression line as the center, then measure the overall amount of variation from the line using the correlation, r. A formula for the correlation, r, appears below. You can think of it as the average product of the z-scores.

$$r = \frac{1}{n-1} \sum \left(\frac{x-\bar{x}}{s_x}\right)\left(\frac{y-\bar{y}}{s_y}\right) = \frac{1}{n-1} \sum z_x \cdot z_y$$

Geometrically, the correlation measures how tightly packed the points of the scatterplot are about the regression line.

- The correlation has no units and ranges from −1 to +1. It is unchanged if you interchange x and y, or if you make a linear change of scale in x or y, such as from feet to inches or from pounds to kilograms.

- In assessing correlation, begin by making a scatterplot and then follow these steps:

  1. **Shape:** Is the plot linear, shaped roughly like an elliptical cloud, rather than curved, fan-shaped, or formed of separate clusters? If so, draw an ellipse to summarize the cloud of points. The data should spread throughout the ellipse; otherwise, the pattern may not be linear or may have unusual features that require special handling. You should not calculate the correlation for patterns that are not linear.

  2. **Trend:** If your ellipse tilts upward to the right, the correlation is positive; if it tilts downward to the right, the correlation is negative. The relationship between the correlation and the slope, b_1, of the regression line is given by

$$b_1 = r\frac{s_y}{s_x}$$

  3. **Strength:** If your ellipse is almost a circle or is horizontal, the relationship is weak and the correlation is near zero. If your ellipse is so

thin that it looks like a line, the relationship is very strong and the correlation is near 1 or −1.

- Correlation is not the same as causation. Two variables may be highly correlated without one having any causal relationship with the other. The value of *r* tells nothing about *why x* and *y* are related. In particular, a strong relationship between *x* and *y* may be due to a lurking variable.

You can interpret the value r^2 as the proportion of the total variation in *y* that can be accounted for by knowing and using *x*:

$$r^2 = \frac{\text{SST} - \text{SSE}}{\text{SST}}$$

The regression effect (or regression toward the mean) is the tendency of *y*-values to be closer to their mean than you might expect. That is, the regression line is flatter than the major axis of the ellipse surrounding the data.

■ Exercises

E23. Each scatterplot in Display 3.50 was made on the same set of axes. Match each with its correlation, choosing from −.06, .25, .40, .52, .66, .74, .85, and .90.

a.   b.   c.   d.

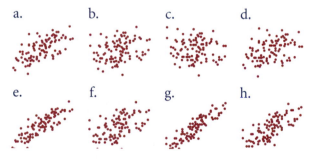

e.   f.   g.   h.

Display 3.50 Eight scatterplots with various correlations.

E24. Estimate the correlation between the variables in these scatterplots.

  a. The proportion of the state population living in dorms versus the proportion living in cities in Display 3.4 (page 105).

  b. The graduation rate versus the 75th percentile of SAT scores in E9 (page 113).

  c. The graduation rate versus the percentage of students in the top 10% of their high school in E9 (page 113).

E25. For each set of (*x, y*) pairs, compute the correlation by hand, standardizing and finding the average product.

  a. (−2, −1), (−1, 1), (0, 0), (1, 1), (2, 1)

  b. (−2, 2), (0, 2), (0, 3), (0, 4), (2, 4)

E26. For the pizza data in E11, make a scatterplot of (*cost, fat*) data and estimate the correlation. Then use a table to calculate the correlation. Can you offer a possible explanation for the strength of association between these two variables?

E27. Is it true that a high correlation, say around .9, means that the data follow a linear pattern? Give an example to illustrate your answer.

E28. Manufacturers of low-fat foods often increase the salt content in order to keep the flavor acceptable to consumers. For a sample of different kinds and brands of cheeses, *Consumer Reports* measured several variables, including the calorie content, fat content, saturated fat content, and the sodium content. Using these four variables, you can form six pairs of variables, so there are six different correlations. These correlations turned out to be either about .95 or about −.5.

a. List all six pairs, and for each pair, decide from the context whether the correlation is close to .95 or to −.5.

b. State a careless conclusion based on taking the negative correlations as evidence of cause and effect.

c. Explain the negative correlation using the idea of a lurking variable.

E29. Several biology students are working together to calculate the correlation for the relationship between air temperature and how fast a cricket chirps. They all use the same data, but some measure temperature in degrees Centigrade, and others measure in degrees Fahrenheit. Some measure chirps per second, and others measure chirps per minute. Some use x for temperature and y for chirp rate; others have it the other way around.

a. Will they all get the same value for the least squares slope? Explain why or why not.

b. Will they all get the same value for the correlation? Explain why or why not.

E30. For the sample of top-rated universities, the graduation rate has mean 82.7% and standard deviation 8.3%. The student/faculty ratio has mean 11.7 and standard deviation 4.3. The correlation is −.5.

a. Find the equation of the least squares line for predicting the graduation rate from the student/faculty ratio.

b. Find the least squares line for predicting the student/faculty ratio from the graduation rate.

E31. These questions are about the relationship between the correlation, r, and the slope, b_1, of the regression line.

a. True or false, and justify your answer: If y is more variable than x, then the least squares slope will be greater (in absolute value) than the correlation.

b. For a list of (x, y) pairs, $r = .8$, $b_1 = 1.6$, and the standard deviations for x and y are 25 and 50. (Not necessarily in that order.) Which one is the standard deviation for x? Justify your answer.

c. Students in a statistics class estimated and then measured their head circumferences in inches. The actual circumferences had SD 0.93, and the estimates had a standard deviation of 4.12. The least squares line for predicting estimated values from actual values was $\hat{y} = 11.97 + 0.36x$. What was the correlation?

d. What would be the least squares slope for predicting actual head circumferences from the estimated values?

E32. The ellipses in Display 3.51 represent scatterplots that have a basic elliptical shape.

a. Match these conditions with the corresponding pair of ellipses:

i. one of the s_y is larger than the other, the s_x are equal, and the correlations are strong

ii. one of the correlations is stronger than the other, the s_x are equal, and the s_y are equal

iii. one s_x is larger than the other, the s_y are equal, and the correlations are weak

A.   B.   C.

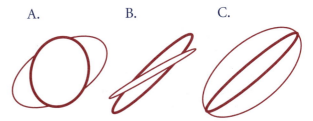

Display 3.51   Three pairs of elliptical scatterplots.

b. Draw two pairs of elliptical scatterplots to illustrate these comparisons:

i. one s_y is larger than the other, the s_x are equal, and the correlations are weak

ii. one s_x is larger than the other, the s_y are equal, and the correlations are strong

E33. **Lurking variables.** For each scenario, state a careless conclusion assuming cause and effect, then identify a possible lurking variable.

a. For a large sample of different animal species, there is a strong positive correlation between average brain weight and average life span.

b. Over the last 30 years, there has been a strong positive correlation between the average price of a cheeseburger and the average tuition at private liberal arts colleges.

c. Over the last decade, there has been a strong positive correlation between the price of an average share of stock, as measured by the S&P 500, and the number of Web sites on the Internet.

E34. Describe a scenario for which a lurking variable is responsible for a high correlation. Give the cases, the two highly correlated variables, and the lurking variable.

E35. Suppose a teacher always praises students who score exceptionally well on a test and always scolds students who score exceptionally poorly. Use the notion of regression toward the mean to explain why the results will tend to suggest the false conclusion that scolding leads to improvement, whereas praise leads to slacking off.

E36. A few years ago, a school in New Jersey tested all its 4th graders to select students for a program for the gifted. Two years later, the students were retested, and the school was shocked to find that the scores of the gifted students had dropped, whereas the scores of the other students had remained, on average, the same. What is a likely explanation for this disappointing development?

E37. A study to determine if ice cream consumption depends on the outside temperature gave the results in the table, scatterplot, and regression analysis in Displays 3.52 and 3.53.

a. Use the values for SST and SSE in the regression analysis to compute r, the correlation for the relationship between the temperature in degrees Fahrenheit and the number of pints of ice cream per person. Check your answer with the "R-sq" given.

Mean Temperature (°F)	Pints per Person
56	0.386
63	0.374
68	0.393
69	0.425
65	0.406
61	0.344
47	0.327
32	0.288
24	0.269
28	0.256
26	0.286
32	0.298
40	0.329
55	0.318
63	0.381
72	0.381
72	0.470
67	0.443
60	0.386
44	0.342
40	0.319
32	0.307
27	0.284
28	0.326
33	0.309
41	0.359
52	0.376
64	0.416
71	0.437

Display 3.52 Table of results for the effects of outside temperature on ice cream consumption.

Source: Koteswara Rao Kadiyala, "Testing for the Independence of Regression Disturbances," *Econometrica,* 38 (1970) 97–117.

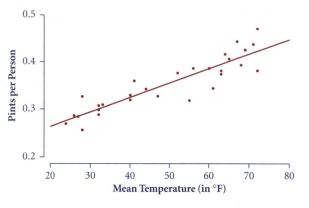

```
Regression Analysis

The regression equation is
pts/pers = 0.202 + 0.00306 Temp F

Predictor          Coef        Stdev    t-ratio         p
Constant        0.20200      0.01452      13.91     0.000
Temp F        0.0030567    0.0002791      10.95     0.000

s = 0.02457      R-sq = 81.6%      R-sq(adj) = 80.9%

Analysis of Variance

SOURCE        DF          SS          MS         F         p
Regression     1    0.072436    0.072436    119.96     0.000
Error         27    0.016304    0.000604
Total         28    0.088740
```

Display 3.53   Scatterplot and regression analysis for the effects of outside temperature on ice cream consumption.

b. Compute the value of the largest residual.

c. Is there a cause and effect relationship between the two variables?

d. What are the units for each of x, y, b_1, and r?

e. How do you think "MS" was computed? What could the letters "MS" stand for?

E38. Refer to the hat size data of E6 in Section 3.1. This exercise is best done with a computer.

a. Construct a scatterplot of *hat size* versus *head circumference*. Estimate a value for the correlation.

b. Set up a table like the one in Display 3.41. Use it to calculate a value for the correlation.

c. Identify the point that contributes the most to the correlation.

d. Construct a scatterplot of *hat size* versus *major axis*. Estimate a value for the correlation.

e. Construct a scatterplot of *hat size* versus *minor axis*. Estimate a value for the correlation.

f. Which of the three variables—circumference, major axis, or minor axis—appears to be the best predictor of hat size? Explain your reasoning.

E39. **Poverty rates.** Use what you have learned in this section to reanalyze the data on poverty from E8. As before, your goal is to identify the variable or variables most strongly related to the poverty rates of the U.S. states. This exercise is best done with a computer.

a. Use scatterplots to identify all the variables whose relationship to the poverty rate is roughly linear.

b. Then compute correlations to identify the strongest linear relationships.

c. Finally, discuss possible explanations for the relationships.

3.4   Diagnostics: Looking for Features That the Summaries Miss

As you learned in Chapter 2, summaries simplify. They are useful because they ignore a lot of detail in order to emphasize a few general features. This also makes summaries potentially misleading because sometimes the detail that is ignored has an important message to convey. Knowing just the mean and standard deviation of a distribution doesn't tell you if there are any outliers or whether the distribution is skewed. The same idea is true of the regression line and the correlation.

This section is about "diagnostics"—tools for looking beyond the summaries to see how well they describe the data and what features they leave out. There are two parts to this section. The first deals with individual cases that stand apart from the overall pattern, and how they influence the regression line and the correlation. The second part will show you how to identify systematic patterns that involve many or all of the cases—the "shape" of the scatterplot.

Which Points Have the Influence?

Just as among people, some data points have more influence than others.

Not all data points are created equal. You saw in the calculation of the correlation in Display 3.41 of Section 3.3 that some points make large contributions and some small. Some make positive contributions and some negative. Your goal is to learn to recognize the points in a data set that may have unusually large influence on where the regression line goes or on the size and sign of the correlation.

Activity 3.3 Near and Far

What you'll need: an open area in which to step off distances

This activity allows you to compare the actual distance to an object with what the distance appears to be.

1. Go to an open area, like the hall or lawn of your school, and pick a spot as your origin. Choose six objects at various distances from the origin. Five of the objects should be within ten to twenty paces of the origin, and the other should be a long way away (at least one hundred paces).

2. For each of the six objects, estimate the number of paces from the origin to the object. Record your estimates.

3. From your origin, walk to each of your objects and count the actual number of regular paces it takes you to get there. Record the actual number beside your estimate.

4. Plot your data on a scatterplot, with your estimated value on the *x*-axis and the actual value on the *y*-axis. Does the plot show a linear trend?

5. Determine the equation of the regression line, and calculate the correlation for your data.

6. Delete the data for the object that is far away from you (the largest pair of distances). Determine the equation of the regression line, and calculate the correlation for the reduced data set.

7. Did the extreme point have any influence on the regression line? On the correlation? Explain.

In Chapter 2, you learned about outliers for distributions—values that are separated by a gap from the bulk of the data. Outliers are atypical cases, and they also can exert more than their share of influence on the mean and standard deviation. For scatterplots, as you will soon see, working with two variables together means there can be outliers of various kinds. Different kinds of outliers can have different types of influence on the least squares line and correlation.

To see these ideas in action, turn again to the data on animal longevity in Display 2.24 (Section 2.2), and think about how to summarize the relationship between maximum and average life span. The average elephant lives 35 years. The oldest elephant on record lived 70 years. The average hippo lives 41 years—longer than the average elephant—but the record-holding hippo lived only 54 years. The oldest known beaver lived 50 years, almost as long as the champion granddaddy hippo, but the average beaver cashes in his wood chips after only 5 short years of making them. Other animals, however, are more predictable. If you look at the entire sample (Display 3.54), it turns out that the elephant, hippo, and beaver are the oddballs of the bunch. For the rest, there's an almost linear relationship between average age and maximum age. The least squares regression line for the entire sample has the equation

$$\hat{M} = 10.53 + 1.58A$$

where $\hat{M}$, "*M*-hat," stands for predicted maximum longevity and *A* stands for observed average longevity. For an increase of 1 year in average longevity, the model predicts a 1.58-year increase in maximum longevity. The correlation for the relationship between these two variables is .77. How much influence do the oddballs have on these summaries?

Points surrounded by white space may have strong influence.

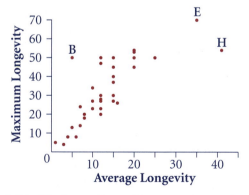

Display 3.54   Maximum longevity versus average longevity.

Points separated from the bulk of the data by white space are **outliers,** and **potentially influential.** To judge a point's influence, fit a line and compute a regression line and a correlation first with, and then without, the point in question.

The hippo has the effect of pulling the right end of the regression line downward (like putting a heavy weight on one end of a seesaw, as you can see in Display 3.55). When it is removed, that end of the regression line will spring upward and the slope will increase. One large residual has been removed, and many of the remaining residuals have been reduced in size, causing the correlation to increase. The new slope is 1.96, and the new correlation is .80. The hippo has considerable influence on the slope and some influence on the correlation.

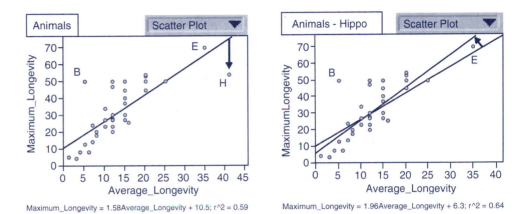

Maximum_Longevity = 1.58Average_Longevity + 10.5; r^2 = 0.59   Maximum_Longevity = 1.96Average_Longevity + 6.3; r^2 = 0.64

Display 3.55 Regression lines for the maximum longevity versus the average longevity, with and without the hippo.

Now envision the scatterplot with just the elephant, E, missing. Because E is close to the straight line fit to the data, it produces a small residual. Thus, you would think that removing E should not change the slope of the regression line much (not nearly as much as removing H did) and it should reduce the correlation just a bit. The correlation does decrease some to .72 from .77. However, the new slope is 1.53. It turns out that removing the elephant gives the hippo even more influence, and the slope decreases.

Finally, envision the scatterplot with just the beaver, B, removed. B produces a large, positive residual close to the left end of the regression line. Thus, removing B should allow the left end of the line to drop, increasing the slope, and removing a large residual should increase the correlation. The new slope is 1.69 (up from 1.58), and the new correlation is .83 (up from .77). The beaver also has considerable influence on both slope and correlation.

With a little practice, you can often anticipate the influence of certain points in a scatterplot, as in the previous example, but it is difficult to state general rules. The best rule is the one in the box: fit the line with and without the questionable point and see what happens. Then report all of the results with appropriate explanations.

Display 3.56 shows four scatterplots. These plots, known as "the Anscombe data" after their inventor, are arguably the most famous set of scatterplots in all of statistics. The Discussion questions that follow invite you to figure out why statistics books refer to them so often. In the process, you'll learn more of what this section is about.

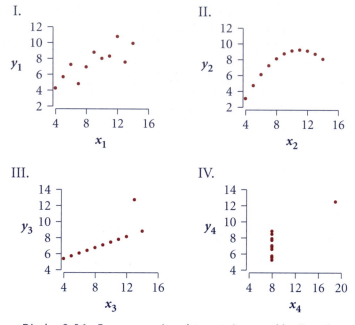

Display 3.56 Four regression data sets invented by Francis J. Anscombe.

Source: Francis J. Anscombe, "Graphs in Statistical Analysis," *American Statistician* 27 (1973): 17–21.

Discussion: Why the Anscombe Data Sets Are Important

D23. For each plot in Display 3.56, give a short verbal description of the pattern in the plot, then

   a. either fit a line by eye and estimate its slope, or else tell why you think a line is not a good summary

   b. either estimate the correlation by eye, or tell why you think a correlation is not an appropriate summary

D24. Display 3.57 shows computer output for one of the four plots. Can you tell which one? If so, tell how you know. If not, explain why you can't tell.

```
Dependent variable is:  y
No Selector
R squared = 66.6%    R squared (adjusted) = 62.9%
s = 1.297  with 11 - 2 = 9  degrees of freedom

Source        Sum of Squares   df   Mean Square   F-ratio
Regression          27.5000    1       27.5000       18.0
Residual            13.7763    9        1.53070

Variable   Coefficient   s.e. of Coeff   t-ratio     prob
Constant      3.00091           1.125       2.67    0.0258
X             0.500000          0.1180      4.24    0.0022
```

Display 3.57 Computer output for one of the Anscombe data sets.

D25. Display 3.58 gives the values for the Anscombe plots.

	Data Set I		Data Set II		Data Set III		Data Set IV	
Row	x_1	y_1	x_2	y_2	x_3	y_3	x_4	y_4
1	10	8.04	10	9.14	10	7.46	8	6.59
2	8	6.95	8	8.14	8	6.77	8	5.76
3	13	7.58	13	8.74	13	12.74	8	7.71
4	9	8.81	9	8.77	9	7.11	8	8.84
5	11	8.33	11	9.26	11	7.81	8	8.47
6	14	9.96	14	8.10	14	8.84	8	7.04
7	6	7.24	6	6.13	6	6.08	8	5.25
8	4	4.26	4	3.10	4	5.39	19	12.50
9	12	10.84	12	9.13	12	8.15	8	5.56
10	7	4.82	7	7.26	7	6.42	8	7.91
11	5	5.68	5	4.74	5	5.73	8	6.89

Display 3.58 Four data sets.

a. Which plot has a point that is highly influential both with respect to the regression slope and with respect to correlation?

b. Compared to the other points in the plot, does the influential point lie far from the least squares line or close to it?

c. How would the slope and correlation change if you were to remove this point? Discuss this without actually performing the calculations. Then carry out the calculations to verify your conjectures.

■ Practice

P20. In Activity 3.3, Near and Far, you found the influence of the farthest object on the correlation and the regression line. Now pick another point that may be influential. Calculate the slope of the regression line and the correlation with and without this point. Describe the influence of this point.

P21. The data in Display 3.59 show some interesting patterns in the relationship between domestic and international gross income from the top ten movies in domestic gross ticket sales.

a. Construct a scatterplot suitable for predicting international sales from domestic sales. Describe the pattern in the data.

b. Find the least squares regression line and the correlation for these data.

c. Now remove the most influential data point and recalculate the least squares line and correlation. Describe the influence of the removed point.

Movie	Domestic (U.S. $ millions)	International (U.S. $ millions)
Titanic	601	1235
Star Wars	461	319
Star Wars: The Phantom Menace	431	492
ET	400	305
Jurassic Park	357	385
Forrest Gump	329	350
The Lion King	313	454
Star Wars VI: Return of the Jedi	309	161
Independence Day	305	505
The Sixth Sense	294	368

Display 3.59 Ticket sales for the top ten highest grossing domestic (United States and Canada) movies of all time.

Source: Internet Movie Database, http://us.imdb.com, October 4, 2000.

Residual Plots: Putting Your Data Under a Microscope

As you can see from the Anscombe plots, there are many features of the shape of a scatterplot that you can't learn from the standard set of summary numbers. Only when the cloud of points is elliptical, as in Display 3.39, does the least squares line, together with the correlation, give a good summary of the relationship described by the plot. If the cloud of points isn't elliptical, however, these summaries aren't appropriate. How can you decide?

It turns out that a special kind of scatterplot, called a residual plot, can often help you see more clearly what's going on. For some data sets, a residual plot can even show you patterns you might otherwise have overlooked completely. Statisticians use residual plots the way a doctor uses a microscope or an X ray— to get a better look at less obvious aspects of a situation. (Plots you use in this way are called "diagnostic plots" because of the parallel with medical diagnosis.) Push the analogy just a little. You're the doctor, and data sets are your patients. Sets with elliptical point clouds are the "healthy" ones; they don't need special attention.

Residual plots may uncover more detailed patterns.

A **residual plot** is a scatterplot of residuals, $y - \hat{y}$, versus predictor values, x.

To show the details of making a residual plot, we will return to the data on percentage of on-time arrivals versus mishandled baggage for airlines, introduced in Section 3.1. Visualize each residual—the difference between the observed value of y and the predicted value, $\hat{y}$—as a vertical segment on the scatterplot in Display 3.60.

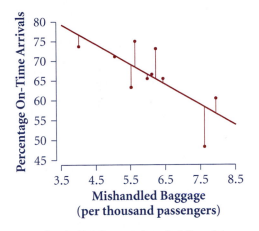

Display 3.60 Scatterplot of airline data.

The calculated residuals are shown in Display 3.61, with the list of carriers ordered from smallest to largest on the *x*-scale. This allows the size of the residuals in the right column to appear in the same order as those pictured in Display 3.60. Delta produces a negative residual of modest size, whereas Continental produces a large positive residual.

Airline	x	y	Predicted Value, $\hat{y}$	Residual, $y - \hat{y}$
Delta	4.00	73.7	76.6260	−2.9260
Southwest	5.03	71.1	71.3896	−0.2896
USAir	5.51	63.3	68.9494	−5.6494
Northwest	5.62	75.0	68.3901	6.6099
American	5.97	65.5	66.6108	−1.1108
TWA	6.10	66.6	65.9499	0.6501
Continental	6.21	73.1	65.3907	7.7093
Alaska	6.42	65.5	64.3230	1.1770
United	7.60	48.3	58.3241	−10.0241
America West	7.93	60.5	56.6464	3.8536

Display 3.61 Table showing residuals for the airline data.

The residual plot, Display 3.62, is simply a scatterplot of the residuals versus the original *x*-variable, mishandled baggage. Note that zero is somewhere around the middle of the residuals on the vertical scale.

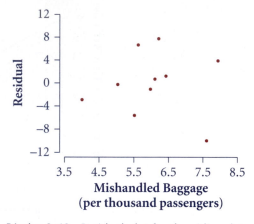

Display 3.62   Residual plot for the airline data.

Discussion: Residual Plots

D26. In Display 3.62, identify which residual belongs to Delta and which to Continental.

D27. Describe the shape of the residual plot in Display 3.62. Does it have an upward or downward trend?

D28. To see how residual plots magnify departures from the fitted line, compare the Anscombe plots of Display 3.56 with Display 3.63, which shows the four corresponding residual plots in a scrambled order.

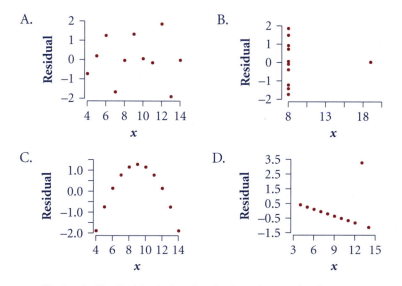

Display 3.63   Residual plots for the four Anscombe data sets.

a. Match each of the original scatterplots in Display 3.56 with its residual plot in Display 3.63.

b. Describe the overall difference between the original scatterplots and the residual plots. What do the scatterplots show that the residual plots don't? What do the residual plots show that the scatterplots don't?

■ **Practice**

P22. Display 3.64 shows four scatterplots for data from a sample of commercial aircraft. Display 3.65 shows four residual plots, numbered I, II, III, IV.

a. Match the residual plots to the scatterplots.

b. Using these plots as examples, describe how you can identify each of these shapes in a scatterplot from the residual plots:

A curve with increasing slope   Unequal variation in the responses
A curve with decreasing slope   Two linear patterns with different slopes

c. For one of the plots, two line segments joined together seem to give a better fit than either a single line or a curve. Which plot is this? Is this pattern easier to see in the original scatterplot or in the residual plot?

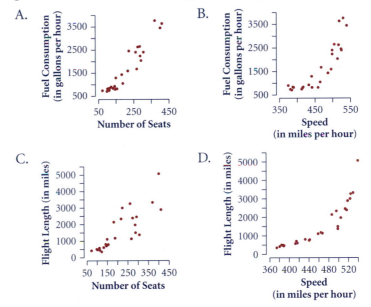

Display 3.64   Four scatterplots for the sample of commercial aircraft.

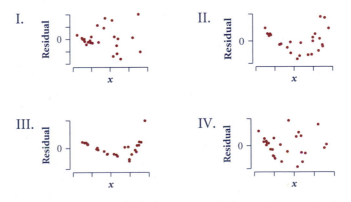

Display 3.65   Four residual plots corresponding to the scatterplots in Display 3.64.

P23. For the "data set" $(x, y) = (0, 0), (0, 1), (1, 1),$ and $(3, 2),$ the least squares line is $y = 0.5 + 0.5x.$

    a. First plot the data and graph the fitted line.

    b. Now complete a table for the predicted values and residuals, like the one in Display 3.61.

    c. Using the values in your table, plot residuals versus predictor, $x.$

    d. How does the residual plot differ from the scatterplot?

What to Look for in a Residual Plot

A careful data analyst always looks at a residual plot.

> Use residual plots to check for systematic departures from constant slope (linear trend) and constant strength (same vertical spread).

Residual plots sometimes yield surprises.

    Look in particular for plots that are curved or fan-shaped. It's true that for data sets with only one predictor value (as in this chapter), you can often get a good idea of what the residual plot will look like by carefully inspecting the original scatterplot. Once in a while, however, you get a surprise.

    E10 introduced data on the height versus age of young girls. The scatterplot of these data, with the regression line, is shown in Display 3.66.

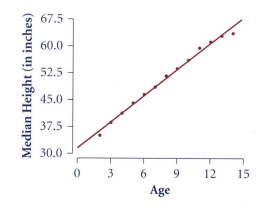

Display 3.66 Height versus age of young girls.

    The plot looks nearly linear, don't you think? But the residual plot, in Display 3.67, quite dramatically reveals that the trend is not as linear as first imagined. The overall average growth rate for the 12-year period is the slope of the regression line. The residuals show that the true growth rate is higher than this average for the first few years, about the same as the average between ages 5 and 11, and lower than the average between the ages 12 and 14.

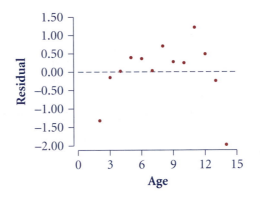

Display 3.67   Residual plot for height versus age of young girls.

Residuals may be plotted against the predicted values.

Statistical software often plots residuals against the predicted values, $\hat{y}$, rather than against the predictor values, x. For simple linear regression, both plots have exactly the same shape, as long as the slope of the regression line is positive.

Discussion: Types of Residual Plots

D29.   Display 3.68 shows a scatterplot and two residual plots for the "data set" consisting of the three number pairs $(x, y) = (0, 1), (1, 0), (2, 2)$. One plots residuals versus predictor values, x, the sort of plot you get from graphing calculators. The other plots residuals versus predicted (fitted) y-values, or $\hat{y}$, the sort of plot you get from computer packages. Explain how the residual plots were produced and how you can tell which residual plot is which. The equation of the least squares line is $\hat{y} = 0.5 + 0.5x$.

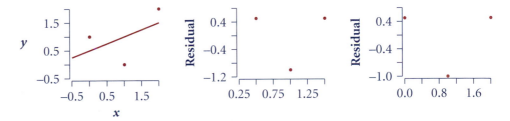

Display 3.68   A scatterplot and two residual plots.

Summary 3.4: Diagnostics for Scatterplots

For the simplest point clouds—elliptical in shape, with linear trend, and no outliers—you can summarize all the main features of a scatterplot with just a few numbers, mainly the fitted slope, y-intercept, and correlation. Not all plots are this simple, however, and a good statistician always does diagnostic checks for outliers and influential points and for departures from constant slope or constant strength.

- Points separated from the bulk of the data by white space are outliers and potentially influential. Outliers can be among the x's, y's, or the residuals.

- To judge a point's influence, fit a line and compute a correlation (or get estimates of these summaries by eye) first with, and then without, the point in question.

For some data sets, a residual plot can show you patterns you might otherwise overlook. A residual plot is a scatterplot of residuals, $y - \hat{y}$, versus predictor values, x. A residual plot can also be constructed as a scatterplot of residuals, $y - \hat{y}$, versus fitted values, $\hat{y}$. Use residual plots to check for systematic departures from linearity and constant variability in y across the values of x. If the data aren't linear, the residual plot doesn't look random. If the data have nonconstant variability, the residual plot is fan-shaped.

■ Exercises

E40. ***Extreme temperatures.*** The data in Display 3.69 provide the maximum and minimum temperatures ever recorded on the respective continents.

a. Construct a scatterplot of the data suitable for predicting minimum temperature from maximum temperature. Is a straight line a good model for the center of this cloud of points? Explain.

b. Fit a least squares regression line to the points and calculate the correlation, even if you thought in part a that a straight line was not a good model.

c. Explain, in words and numbers, what influence Antarctica has on the slope of the regression line and on the correlation. How could an account of these data be misleading if it were not accompanied by a plot?

Two climbers stand on the edge of the crater of Mount Erebus, 12,500 feet above sea level, on Ross Island, Antarctica.

E41. ***Doctors' hours.*** You've already seen that from 1982 through 1992, doctors' incomes increased rapidly. What about the hours they worked? Display 3.70 shows the average number of hours worked per week in 1992 versus hours worked in 1982 for doctors in nine specialties. (Notice that even the shortest average workweek is more than 20% longer than the conventional 40 hours.)

a. What would be the effect on the slope and correlation of removing the point for radiology? Make the calculations to see if your intuition is correct.

b. What would be the effect of removing the two points for psychiatry and

Continent	Maximum Temperature (°F)	Minimum Temperature (°F)
Africa	136	−11
Antarctica	59	−129
Asia	129	−90
Australia	128	−9
Europe	122	−67
Oceania	108	14
North America	134	−81
South America	120	−27

Display 3.69 Maximum and minimum recorded temperatures for the continents.

Source: National Climatic Data Center, 2002, www.ncdc.noaa.gov/ol/climate/globalextremes.html.

pathology? Provide an explanation with words first, and then justify the explanation with numbers.

c. Write a brief summary: What are the main features of the relationship between hours worked in 1982 and those in 1992?

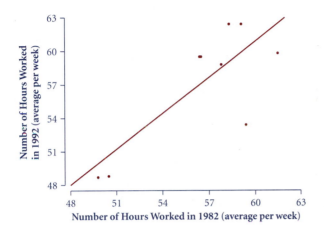

Display 3.70 Mean number of hours worked by physicians in various specialties.

Specialty	Average Hours Worked	
	1982	**1992**
Family practice	57.8	58.8
Internal medicine	59.1	62.4
Surgery	56.5	59.5
Pediatrics	56.4	59.5
Obstetrics/Gynecology	58.3	62.4
Radiology	59.4	53.4
Psychiatry	50.5	48.8
Anesthesiology	61.5	59.8
Pathology	49.8	48.7

Source: Martin L. Gonzalez, ed., *Socioeconomic Indicators of Medical Practice* (American Medical Association, 1994), p. 152.

E42. *Malpractice.* Display 3.71 shows the number of malpractice claims filed, per 100 physicians, by field, for 1985 and 1992. The correlation is .91, and the equation of the regression line is

$$\hat{y} = 2.43 + 0.58x$$

a. Describe the direction and strength of the plot.

b. Locate the point for obstetrics/gynecology. How is this point unusual? What might the explanation be?

c. Omit obstetrics/gynecology, fit a line by eye, and estimate its slope. If you omit obstetrics/gynecology, how would you describe the change in rate of malpractice claims for the other specialties between 1985 and 1992?

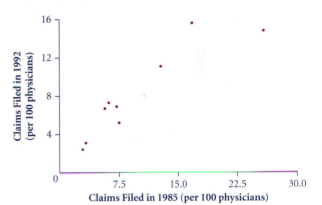

Specialty	Liability Claims per 100 Physicians	
	1985	**1992**
Family practice	5.7	6.7
Internal medicine	6.2	7.3
Surgery	16.8	15.6
Pediatrics	7.2	6.9
Obstetrics/Gynecology	25.8	14.8
Radiology	12.8	11.1
Psychiatry	2.9	2.4
Anesthesiology	7.5	5.2
Pathology	3.3	3.1

Display 3.71 Number of malpractice claims filed, per 100 physicians, by field, for 1985 and 1992.

d. "The mid-1980s saw a disproportionate number of malpractice claims against obstetricians." Do the data tend to support this statement? Explain.

e. Summarize how a least squares fitted line using all nine specialties could be misleading if your goal is to understand how the rate of malpractice claims changed between 1985 and 1992.

E43. Display 3.72 gives the little data set for the three DC9s, from the data in Display 3.11, along with a scatterplot showing the least squares line. (Values have been rounded.)

a. Use the equation of the line to find fitted values and residuals to complete the table in Display 3.72.

b. Use your numbers from part a to construct two residual plots, one with the predictor, x, on the horizontal axis and one with the predicted value, y, on the horizontal. How do the two plots differ?

Aircraft	Seats	Cost	Fitted	Residual
DC9–30	100	1750	–?–	–?–
DC9–40	110	1800	–?–	–?–
DC9–50	120	1925	–?–	–?–

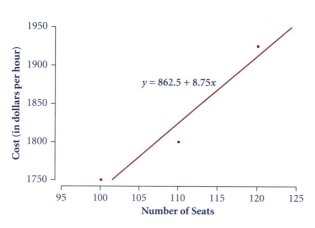

Display 3.72   Cost per hour versus number of seats for three models of the DC9 aircraft.

E44. **How effective is a disinfectant?** The data in Display 3.73 show (coded) bacteria colony counts on skin samples before and after a disinfectant is applied.

a. Plot the data, fit a regression line, and complete a table like the one in Display 3.72, filling in the fitted and residual values.

b. Plot the residuals versus x, the count before the treatment. Comment on the pattern.

c. Use the residual plot to determine for which skin sample the disinfectant was unusually effective and for which skin sample it was not very effective.

x	y
11	6
8	0
5	2
14	8
19	11
6	4
10	13
6	1
11	8
3	0

Display 3.73   Coded bacteria colony counts before and after treatment.

Source: Snedecor and Cochran, *Statistical Methods* (Iowa State University Press, 1967), p. 422.

E45. **Aircraft.** Turn back to Display 3.64, which shows a scatterplot for *flight length* versus *number of seats*.

a. Does the slope of the pattern increase, decrease, or stay roughly constant as you move across the plot?

b. Focus on the variation (spread) in flight length, y, for planes with roughly the same seating capacity, and compare the spreads for planes with few seats, a moderate number of seats, and a large number of seats. As you go from left to right across the plot, how does the spread change, if at all?

c. Suppose a friend chose a plane from the sample at random and told you the approximate number of seats. Could you guess its flight length to within

500 miles if the number of seats was between 50 and 150? If it was between 200 and 300? Explain.

d. What is the relationship between your answer in part b and the residual plot I in Display 3.65?

e. Give an explanation for why the variation in flight length shows the pattern that it does.

E46. ***Textbook prices.*** Display 3.74 compares recent prices at a college bookstore with a large online bookstore.

Type of Textbook	College Bookstore Price ($)	Online Bookstore Price ($)
Chemistry	93.40	94.18
Classic fiction	9.95	7.96
English anthology	46.70	48.75
Calculus	76.00	94.15
Biology	86.70	80.95
Statistics	7.95	6.36
Dictionary	24.00	16.80
Style manual	12.70	10.66
Art history	66.00	45.50

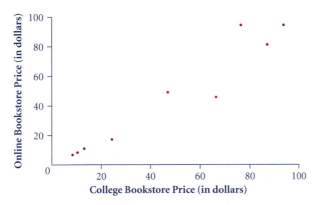

Display 3.74 Prices for a sample of textbooks at a college bookstore and an online bookstore.

a. The equation of the regression line is

online = −3.57 + 1.03 *college*

Interpret this equation in terms of textbook prices.

b. Make a residual plot. Interpret it and point out any interesting features.

c. In comparing the prices of the text, you might be more interested in a different line: $y = x$. Draw this line on a copy of the scatterplot in Display 3.74. What does it mean if a point lies above this line? Below it? On it?

d. A boxplot of the differences between the college price and the online price appears in Display 3.75. Interpret this boxplot.

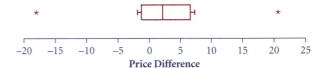
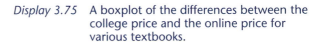

Display 3.75 A boxplot of the differences between the college price and the online price for various textbooks.

E47. Can either of the plots in Display 3.76 be a residual plot? Explain your reasoning.

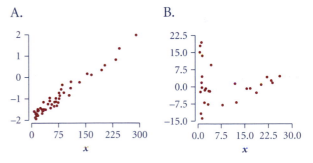

Display 3.76 A residual plot?

E48. What's wrong with this idea? "Why not use the residual plot to get a better slope for the fitted line? First, fit a least squares line to your data and compute residuals. Next, look at the plot of residuals versus the predictor. Finally, fit a least squares line to the residual plot, and use the sum of the slope from the original data and the slope from the residual plot as your new, improved slope."

E49. The plot in Display 3.77 shows the residuals from fitting a line to *female life expectancy* versus *gross national product* (GNP, in thousands of dollars per capita) for a sample of countries from around the world. The regression equation for the sample data was

$$life\ exp = 67.00 + 0.63\ GNP$$

Sketch the scatterplot of *female life expectancy* versus *GNP*.

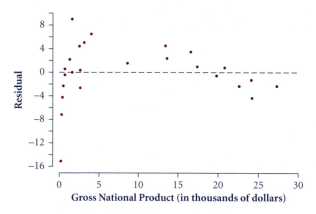

Display 3.77 Residuals of *female life expectancy* vs. *GNP*.

E50. ***Can you recapture the scatterplot from the residual plot?*** The residual plot in Display 3.78 was calculated from data showing the recommended weight (in pounds) for men at various heights over 64 inches. The fitted weights ranged from 145 to 187 pounds. Make a rough sketch of the scatterplot of these data.

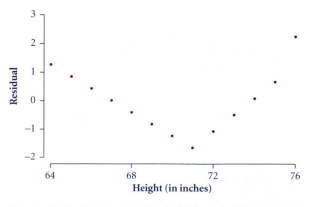

Display 3.78 Residuals from weight vs. height of men.

E51. ***Pizzas, again.*** Display 3.79 shows the pizza data used earlier in this chapter.

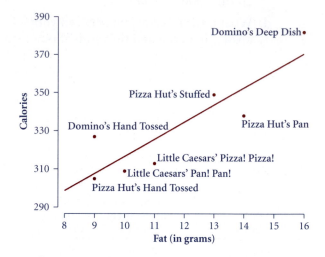

Display 3.79 Calories versus fat, per 5-ounce serving, for seven kinds of pizza.

a. Estimate the residuals from the graph, and use your estimates to sketch a rough version of a residual plot for this data set.

b. Which pizza has the largest positive residual? The largest negative residual? Are any of the residuals so extreme as to suggest those pizzas should be regarded as exceptions?

c. Is any one of the pizzas a highly influential data point? If so, specify which one(s), and describe the effect on the fitted slope and correlation of removing the influential point or points from the analysis.

E52. Explain why a residual plot of (*x*, *residual*) and one of (*fitted value*, *residual*) have exactly the same shape if the slope of the regression line is positive. What changes if the slope is negative?

3.5   Shape-Changing Transformations

For scatterplots in which the data form an elliptical cloud, the regression line and correlation tell you pretty much all you need to know. But data don't always behave so obligingly. For plots in which the points are curved, fan out, or contain outliers, the usual summaries do not tell you everything and may actually be misleading. What do you do then? This section will show you one possible remedy: Transform the data to get the shape you want.

You're already familiar with linear transformations from Section 2.3, things like changing temperatures from degrees Fahrenheit to degrees Centigrade or distances from feet to inches or times from minutes to seconds. These linear transformations—adding or subtracting a constant or multiplying or dividing by a constant—can change the center and spread of the data but don't change its basic shape.

Transforming data is sometimes called "re-expressing data."

Nonlinear transformations, such as squaring each value or taking logarithms, *do* change the basic shape of the plot. The next activity will show you how a transformation of a measurement scale can lead to simplified statistical analyses.

Activity 3.4   Finding Areas of Equilateral Triangles

What you'll need: a copy of Display 3.80

How does the area of an equilateral triangle relate to the length of a side? You may know the formula for this relationship, but even if you do, your job in this activity is to derive it from measurements. Measurements always have some "measurement error" attached to them, so you will not get an exact rule from your data; however, you should get a good approximation of that rule.

1. Find an approximation for the area of each triangle. Do this by counting the number of squares of the background grid that fall within each triangle. Make your own rule about what to do with partial squares.

2. By counting squares, find the approximate length of a horizontal or vertical side of each triangle. The triangles are nearly equilateral.

3. Construct a scatterplot in which $(x, y) = (side, area)$. Describe the pattern of this plot.

4. On the scatterplot, sketch a line or curve that appears to capture the trend.

5. What effect does each of these transformations have on the shape of your plot?

   a. dividing each area by 2

   b. taking the square root of each area

   c. taking the square of each area

   d. taking the square of each side

(continued)

Activity 3.4 (continued)

6. Pick one of the transformed data sets in 5a–d for which a line gives a good summary, and fit a line to that set.

7. From the equation of the line fit in step 6, write a statement summarizing the relationship between the length of the side of an equilateral triangle and its area.

8. There is more than one transformed data set in 5a–d for which a line fits well. Write a brief explanation of why this is so.

Power Transformations

Replacing y with $y^{1/b}$ will straighten $y = ax^b$.

In step 5, parts b–d of Activity 3.4, you replaced each value of y with a power of y, either $\sqrt{y} = y^{1/2}$ in part b or y^2 in part c, or you replaced each value of x with a power of x. These are examples of power transformations. Another common power transformation is the reciprocal, which replaces each value of y with $\frac{1}{y} = y^{-1}$, or does the same with x. Power transformations are appropriate when your data follow a power model of the form $y = ax^b$.

Power transformations of the type used to straighten out the triangle data are commonly used in data analysis because they are simple but effective nonlinear transformations; we can easily see how they expand or shrink the measurements. Thinking in terms of expanding and shrinking is the key to understanding which transformation you might want to use in a given situation. Your data from Activity 3.4 probably look something like this diagram.

Think of this as a wire that you are going to straighten by moving the upper end. To accomplish this, you have to move the upper end down and to the right. In a rectangular coordinate system, you would be moving in the negative direction on the y-axis and in the positive direction on the x-axis. Thus, you need a transformation that will shrink the y-scale (square root will work) or expand the x-scale (square will work). These are precisely the transformations you tried in Activity 3.4. The power transformation you actually use should be consistent with what you know about the physical situation underlying the measurements. It should have seemed quite natural to think of area as related to the square of a side.

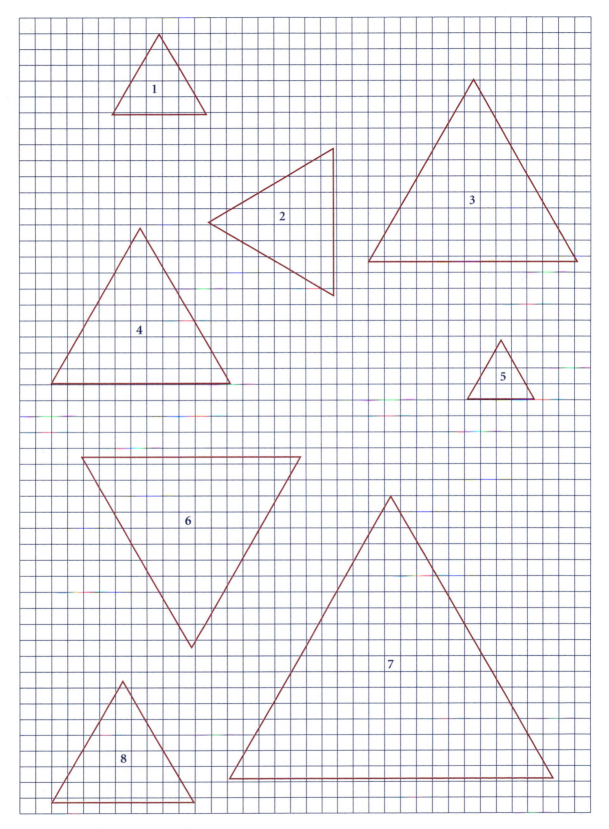

Display 3.80 Equilateral triangles.

This and similar rules for other common shapes are summarized in Display 3.81.

Shape	Desired Action	Suggested Power Transformation
	Shrink y Expand x	$y^{0.5}$; $\log y$ x^2
	Shrink y Shrink x	$y^{0.5}$; $\log y$ $x^{0.5}$; $\log x$; $\dfrac{1}{x}$
	Expand y Shrink x	y^2 $x^{0.5}$; $\log x$; $\dfrac{1}{x}$
	Expand y Expand x	y^2 x^2

Display 3.81 Summary table on power transformations.

Why is the logarithm grouped with the power transformations? The log, either natural or base ten, has the effect of shrinking a scale much like a fractional power such as a cube root or a tenth root does. (You might want to convince yourself of this by trying it on some data.) Thus the so-called ladder of powers goes, in decreasing order of strength: square, square root, log, reciprocal. Keep in mind that the reciprocal not only shrinks the large measurements but also changes the order of the measurements.

These four transformations can get you an amazingly long way in data analysis. And there is even more to this economy. Often the transformation that straightens out a curved scatterplot simultaneously makes the spread in y more constant across the values of x.

Example of a Power Transformation

The success of sustainable harvesting of timber depends on how fast trees grow, and one way to measure a tree's growth rate is to relate its diameter to its age. (If you know this relationship and the relationship between age and height of a tree, then you can estimate the growth rate for total volume of timber.) Displays 3.82 through 3.85 give data on the age (years) and diameters (inches at chest height) of a sample of oak trees, along with a scatterplot of *diameter* versus *age*, numerical summaries in the form of computer output, and a residual plot.

Age (yr)	Diameter (in.)
4	0.8
5	0.8
8	1.0
8	2.0
8	3.0
10	2.0
10	3.5
12	4.9
13	3.5
14	2.5
16	4.5
18	4.6
20	5.5
22	5.8
23	4.7
25	6.5
28	6.0
29	4.5
30	6.0
30	7.0
33	8.0
34	6.5
35	7.0
38	5.0
38	7.0
40	7.5
42	7.5

Display 3.82 Diameters and ages of oak trees.

Source: Herman H. Chapman and Dwight B. Demeritt, *Elements of Forest Mensuration,* 2nd ed. (J. B. Lyon Company, 1936 [now Williams Press]).

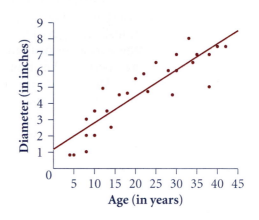

Display 3.83 Diameter versus age of oak trees (with regression line).

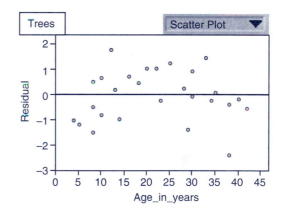

Display 3.84 Residual plot for the oak tree data.

```
The regression equation is
Diameter = 1.15 + 0.163 AGE

Predictor      Coef    Stdev   t-ratio      p
Constant     1.1507   0.4185     2.75  0.011
AGE          0.16278 0.01682     9.68  0.000

s = 1.021   R-sq = 78.9%  R-sq(adj) = 78.1%

Analysis of Variance

SOURCE          DF       SS      MS      F      p
Regression       1   97.593  97.593  93.63  0.000
Error           25   26.059   1.042
Total           26  123.652
```

Display 3.85 Numerical summary in the form of computer output of the ages and diameters (inches at chest height) of a sample of oak trees.

Inspection will reveal that the point cloud is roughly elliptical, but there is a slight curvature similar to the third shape in the summary table on power transformations, Display 3.81. Straightening this curve will require expanding the *y*-scale or shrinking the *x*-scale. If the diameter does not grow linearly with age, perhaps the cross-sectional area does. So let's try squaring the diameters.

Display 3.86 shows a scatterplot of *diameter squared* versus *age,* together with the equation of the least squares line, and Display 3.87 shows the plot of residuals versus age. Even though the value of r^2 is about the same as before, the scatterplot and residual plot are now less curved. Taking all the evidence together, it appears that a linear model fits the transformed data better than it does the original data.

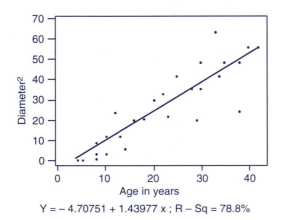

$$Y = -4.70751 + 1.43977 \, x \; ; \; R - Sq = 78.8\%$$

Display 3.86 Diameter squared versus age for the oak tree data (with regression line).

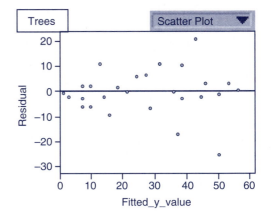

Display 3.87 Residual plot for diameter squared versus age.

Power transformations like the ones you have just seen can straighten a curved plot (or change a fan shape to a more nearly oval shape). By choosing the right powers, you can often take a data set for which a fitted straight line and correlation are not suitable and convert it to one for which those summaries work well.

Is it cheating to change the shape of your data? ("You wanted a linear cloud, but you got a curved wedge. You didn't like that, so you fiddled with the data until you got what you wanted.") In fact, as you'll see, changing scale is a matter of re-expressing the same data, not replacing the data with entirely new facts. The intelligent measurer selects a scale that is the most useful for the problem that the measurements were taken to solve.

Measurement scales are selected by the user, so select one that meets your needs.

Discussion: Power Transformations

D30. Give a plausible explanation for why a tree might grow at a rate that makes the square of its diameter proportional to age.

D31. If you need to predict the diameter of some oak trees for which you knew only the age, would you rather do the predicting for 10-year-old trees or for 40-year-old trees? Explain your reasoning.

■ **Practice**

P24. Three sets of (x, y) pairs are given below. For each set, (i) plot y versus x, (ii) find a power of x to use in place of x itself to get a linear plot, and (iii) plot y versus that power of x.

a. (0, 0), (1, 1), (2, 8), (3, 27)

b. (1, 10), (2, 5), (5, 2), (10, 1)

c. (100, 10), (25, 5), (64, 8), (9, 3), (1, 1)

P25. In P24, suppose that instead of plotting y versus a power of x, you plot a power of y versus x. For each of the three sets of (x, y) pairs, find the power of y for which the plot is a line. What is the relationship between the powers of x in P24 and the powers of y in this problem?

P26. For each of these relationships, first write the equation that relates x and y. Then use this equation to find a power of y you could plot against x in order to get a linear plot.

a. y is the area of a circle, x is the radius of the circle

b. y is the volume of a block whose sides all have equal lengths, x is the length of the side

c. y is the volume of an 8-foot-long section of log with circular cross section, x is the diameter of the log's cross section

P27. If the square of a tree's diameter is roughly proportional to its age, then you can expect the diameter itself to be roughly proportional to the square root of the tree's age. For the data in Display 3.82, use a computer or calculator to make a scatterplot of the *diameter* versus *square root of age*, fit a least squares line, and plot the residuals. Which residual plot shows more of a fan shape—the plot for diameter squared versus age or the plot for diameter versus the square root of age? If you want a plot that shows an oval point cloud, which transformation should you choose?

Power Functions and Log-Log Transformations

Display 3.88 shows one student's results from an activity similar to Activity 3.4. The student has a few questions for a statistician.

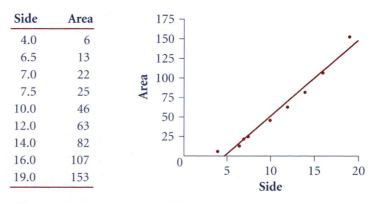

Side	Area
4.0	6
6.5	13
7.0	22
7.5	25
10.0	46
12.0	63
14.0	82
16.0	107
19.0	153

Display 3.88   Triangle data.

Student:	(*Confidently*) This activity is easy because I know the relationship between the side of an equilateral triangle and its area. It's quadratic, so I need a square-root transformation.
Statistician:	Right. See how easy this is?
Student:	(*Suspiciously*) Yes, too easy! Come to think of it, because I already know the relationship, why do I need to fit a line?
Statistician:	Good point.
Student:	And if I don't know the relationship, how do I know whether to use a cube root, a square root, or some other transformation? Does it matter?
Statistician:	It sure does. If you use a cube root transformation on your data, it will bend the data too far and you will end up with data that are curved down instead of up.
Student:	This sounds odd. If I already know the right transformation, I don't need to use it. If I don't know the right transformation, I'll probably pick the wrong one.
Statistician:	Let's explore how logs can help by doing a little algebra. If y is proportional to a power of x, then the equation has the form

$$y = ax^b$$

Taking the log of both sides,

$$\log y = \log a + \log x^b$$

$$\log y = \log a + b \log x$$

Student:	That's the equation of a straight line with slope b through a plot of $\log y$ versus $\log x$, because $\log a$ is a constant.
Statistician:	Right. If there is a power relationship between *area* and *side*, there will be a linear relationship between $\log(area)$ and $\log(side)$.
Student:	Let me try it on my data. I'll take the log of both x and y and plot the data.

Log(*side*)	Log(*area*)
0.60206	0.77815
0.81291	1.11394
0.84510	1.34242
0.87506	1.39794
1.00000	1.66276
1.07918	1.79934
1.14613	1.91381
1.20412	2.02938
1.27875	2.18469

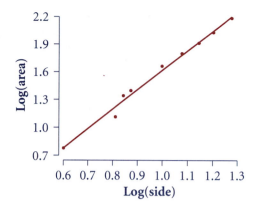

The regression equation is
logArea = -0.484 + 2.10 LogSide

Predictor	Coef	Stdev	t-ratio	p
Constant	-0.48351	0.08653	-5.59	0.000
LogSide	2.10035	0.08621	24.36	0.000

s = 0.05295   R-sq = 98.8%   R-sq(adj) = 98.7%

Analysis of Variance

SOURCE	DF	SS	MS	F	p
Regression	1	1.6641	1.6641	593.52	0.000
Error	7	0.0196	0.0028		
Total	8	1.6837			

Display 3.89 Analysis of triangle data.

Look at Display 3.89—it works! The transformation removes most of the curvature from the plot. The equation is

$$\log(area) = -0.484 + 2.10 \log(side)$$

Statistician: And the equation shows the estimated coefficient of log(*side*) to be approximately 2. This indicates that area must be proportional to the square of a side.

Student: Cool. So this is why I had to study logs!

> Replacing *x* with log *x* and *y* with log *y* will straighten a power relationship of the form $y = ax^b$.

Discussion: Log-Log Transformations

D32. Show that using a cube root transformation on the areas in the student's data results in a plot that is concave down.

D33. Use the student's data to show that it doesn't matter whether you use base 10 or base *e*.

■ Practice

P28. Display 3.90 shows the brain weights and body weights for a collection of mammals. The goal is to establish the relationship of brain weight to body weight.

   a. Assuming that there is a power relationship here, can you guess what it is from the scatterplot? If *y* is written as a function of *x* to some power, should the power be greater than 1 or less than 1?

   b. Plot log(*brain*) versus log(*body*). Describe the pattern of the plot.

   c. Fit a line to the plot of part b. Write an equation relating *y* to *x*. Does your equation support your answer to part a?

Species	Brain Weight (gm)	Body Weight (kg)	Species	Brain Weight (gm)	Body Weight (kg)
African elephant	5712	6654	Horse	655	521
African giant pouched rat	6.6	1	Human	1320	62
Arctic fox	44.5	3.385	Jaguar	157	100
Arctic ground squirrel	5.7	0.92	Kangaroo	56	35
Asian elephant	4603	2547	Lesser short-tailed shrew	0.14	0.005
Baboon	179.5	10.55	Little brown bat	0.25	0.01
Big brown bat	0.3	0.023	Mole rat	3	0.122
Cat	25.6	3.3	Mountain beaver	8.1	1.35
Chimpanzee	440	52.16	Mouse	0.4	0.023
Chinchilla	6.4	0.425	Musk shrew	0.33	0.048
Cow	423	465	Nine-banded armadillo	10.8	3.5
Desert hedgehog	2.4	0.55	North American opossum	6.3	1.7
Donkey	419	187.1	Owl monkey	15.5	0.48
Eastern American mole	1.2	0.075	Pig	180	192
European hedgehog	3.5	0.785	Rabbit	12.1	2.5
Giant armadillo	81	60	Raccoon	39.2	4.288
Giraffe	680	529	Rat	1.9	0.28
Goat	115	27.66	Red fox	50.4	4.235
Golden hamster	1	0.12	Rhesus monkey	179	6.8
Gorilla	406	207	Roe deer	98.2	14.83
Gray seal	325	85	Sheep	175	55.5
Gray wolf	119.5	36.33	Tree shrew	2.5	0.104
Ground squirrel	4	0.101	Water opossum	3.9	3.5
Guinea pig	5.5	1.04	Yellow-bellied marmot	17	4.05

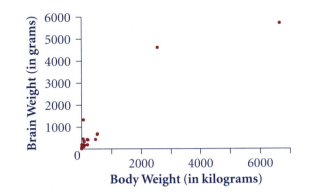

Display 3.90 Brain weights and body weights of mammals.

Source: T. Allison and D. V. Cicchetti, "Sleep in Mammals: Ecological and Constitutional Correlates," *Science* 194 (1976): 732–34.

Exponential Functions and Log Transformations

Replacing y with log y will straighten $y = ab^x$.

Taking the log of each value of y will straighten an exponential relationship of the form $y = ab^x$. How do you know if you have an exponential relationship? The points would cluster about a function similar to the ones in Display 3.91. Another clue is that you have a variable whose values range over two or more orders of magnitude (powers of 10), as in the next data set.

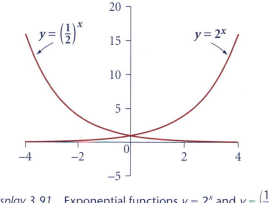

Display 3.91   Exponential functions $y = 2^x$ and $y = \left(\frac{1}{2}\right)^x$.

Example of a Logarithmic Transformation

Do aircraft with a higher typical speed also have a greater average flight length? As you might expect, the answer is yes, but the relationship is nonlinear. (See Displays 3.92 and 3.93.) Is there a simple equation that relates typical speed and flight length? The Discussion that follows leads you through one approach to these questions.

Speed (mph)	Flight Length (mi)	Speed (mph)	Flight Length (mi)
518	2882	441	782
539	5063	440	742
529	3321	458	1101
498	1363	414	702
513	2451	432	798
498	1493	416	602
504	1963	374	345
516	2379	388	442
467	1126	412	570
524	3253	387	487
523	2995	389	468
495	2331	384	500
464	1167	380	413
486	2135		

Display 3.92   Flight length and speed of aircraft.

Source: Air Transport Association of America, 1996, www.air-transport.org/public/industry/67.asp.

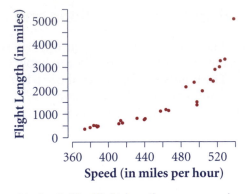

Display 3.93 Flight length versus speed.

This plot is curved much in the manner of the first row of the table in Display 3.81, and so shrinking the *y*-scale is in order. Display 3.94 shows a scatterplot of log(*flight length*) versus *speed*, with a least squares line and computer output. The transformation has improved the fit noticeably; the line fits the transformed data quite closely.

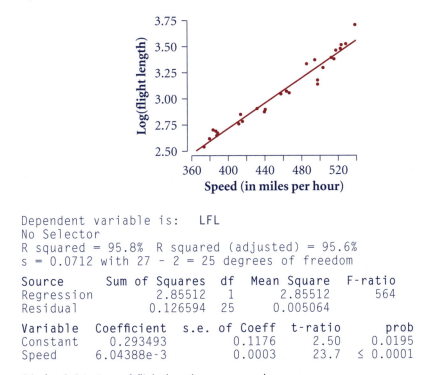

```
Dependent variable is:   LFL
No Selector
R squared = 95.8%  R squared (adjusted) = 95.6%
s = 0.0712 with 27 - 2 = 25 degrees of freedom

Source        Sum of Squares  df  Mean Square  F-ratio
Regression          2.85512   1      2.85512       564
Residual           0.126594  25     0.005064

Variable  Coefficient  s.e. of Coeff  t-ratio      prob
Constant    0.293493          0.1176     2.50    0.0195
Speed     6.04388e-3          0.0003     23.7  ≤ 0.0001
```

Display 3.94 Log of *flight length* versus *speed*.

Discussion: Log Transformations

D34. How can you tell from the table of data that the values of the flight length range over two orders of magnitude? Show how transforming from *flight length* to log(*flight length*) will shrink the larger *y*-values more than the smaller ones, and thus help straighten the plot.

D35. Although the fit is much improved and looks much closer to linear in Display 3.94, the lack of randomness in the residual plot (Display 3.95) indicates that the plot is still not really linear. Describe the pattern in the plot, and tell what it suggests as a next step in the analysis. (E57 will give you a chance to finish the detective work.)

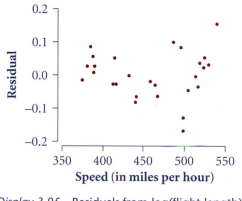

Display 3.95   Residuals from log(flight length) plotted against speed.

D36. Is there reason to expect the relationship here to generalize to some larger set of aircraft? Why or why not? In what sense, if any, is the relationship here one of cause and effect? What is your evidence?

■ Practice

P29. The logarithm of a number equals the exponent when you write the number in the form "base raised to a power." Thus, $\log_{10}1000 = 3$ means the same thing as $1000 = 10^3$. Use this fact to transform y to logs (in base 10) in this set of (x, y) pairs: (2, 1000), (1, 100), (0, 10), and (−1, 1). Then plot $\log_{10}y$ versus x and check that you get a straight line. Find the slope and y-intercept of the line.

P30. Repeat the steps of P29 for these two sets of pairs:

a. (6, 1000), (4, 100), (2, 10), (0, 1)

b. (5, 0.0001), (6, 0.01), (8, 100)

P31. Verify that if $\log_{10}y = c + dx$, then $y = ab^x$, where $a = 10^c$ and $b = 10^d$. Use this fact to rewrite each of your fitted equations in P29 and P30 in the form $y = ab^x$.

P32. Write the fitted equation in Display 3.94 in the form *flight length* $= ab^{speed}$.

Exponential Growth and Decay

Whenever a quantity changes at a rate proportional to the amount present, consider logs.

Exponential functions often arise when you study how a quantity grows or decays as time passes. Many such quantities change at a rate proportional to the amount present, resulting in a function of the form $y = ab^x$. The rate of growth of a population is proportional to population size, the rate of growth of a bank account is proportional to how much money is in the account, the rate at which a radioactive substance decays is proportional to how much is left, and the rate at which a cup of coffee cools is proportional to how much hotter it is than the air around it. For such situations, replacing *y* with log *y* often gives a plot that is much more nearly linear than the original plot.

The examples in this section illustrate exponential growth and decay. At the same time, they illustrate another important feature of measurements taken over time: There is often an "up-and-down" pattern to the residuals. This pattern cannot be removed by a simple transformation.

Data over time will have more varied patterns, sometimes difficult to see.

Whenever your measurements come in chronological order, what happens next is likely to depend on what just happened, and as a result, the patterns in your data may be more subtle than the ones you've seen up to this point. In addition, the difference between a meaningful pattern and a quirk in the data may be harder to detect because the data typically show only one observation for a single point in time.

An Example of Exponential Growth

Display 3.96 shows the population density (people per square mile) of the United States for all census years through 1990. For the years prior to 1960, only the 48 contiguous states are included. Alaska and Hawaii were added in 1960. To find a reasonable model for this situation, a scatterplot (Display 3.97) is the place to start.

Year	Density	Year	Density
1790	4.5	1900	25.6
1800	6.1	1910	31.0
1810	4.3	1920	35.6
1820	5.5	1930	41.2
1830	7.4	1940	44.2
1840	9.8	1950	50.7
1850	7.9	1960	50.6
1860	10.6	1970	57.4
1870	13.4	1980	64.0
1880	16.9	1990	70.3
1890	21.2		

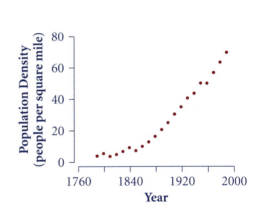

Display 3.96   Population density of the United States.

Source: *Statistical Abstract of the United States,* 1996, p. 8.

Display 3.97   U.S. population density over the census years.

Plot. Obviously, the pattern here is not linear. A curve of this type can be straightened by proportionally decreasing the large *y*-values (population densities, in this case). For variables like population growth (or growth of many other phenomena), the logarithmic transformation works well. This transformation not only solves the data analysis problem nicely, but also the resulting model has a neat interpretation.

Transform and Plot Again. This time, for practice, we will use natural logs (base *e*). You can see the transformed points in Display 3.98. For example, for the year 1800, the point (1800, ln 6.1) or (1800, 1.808) is plotted.

Fit. Although there is still some curvature, the pattern is much more nearly linear, and a straight line might be a reasonable model to fit through these data. The transformed data with the regression line and regression analysis is shown in Display 3.98.

The equation of the regression line is $\widehat{\ln y} = -25.6 + 0.0151x$. Solving for $\hat{y}$,

$$e^{\widehat{\ln y}} = e^{-25.6 + 0.0151x}$$

$$\hat{y} = e^{-25.61}(e^{0.015})^x \qquad \text{or} \qquad \hat{y} = e^{-25.61}(1.015)^x$$

This means that the population density is growing at about 1.5% per year.

```
The regression equation is
ln density = -25.6 + 0.0151 Year

Predictor          Coef       Stdev    t-ratio        p
Constant        -25.612       1.094     -23.41    0.000
Year          0.0151065   0.0005787      26.11    0.000

s = 0.1606    R-sq = 97.3%    R-sq(adj) = 97.1%

Analysis of Variance

SOURCE       DF        SS        MS        F        p
Regression    1    17.572    17.572   681.49    0.000
Error        19     0.490     0.026
Total        20    18.062
```

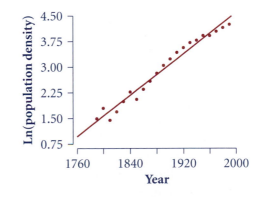

Display 3.98 Ln(*population density*) versus year with regression line and computer output.

Residuals. In the case of data over time, it is often advantageous to plot the residuals over the same time points as in Display 3.99.

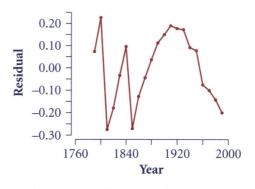

Display 3.99 Residual plot of Ln(*population density*) versus year.

Not exactly random scatter! Well, it is about the best we can do. The problem is that there are subtle patterns in the data—and in the residuals—that no simple model will adequately account for.

Discussion: Exponential Growth

D37. Relate the pattern you see in the residual plot in Display 3.99 to the pattern of the data in the original scatterplot.

D38. Why is there a huge jump from a large positive to a large negative residual as you move from 1800 to 1810? What events in U.S. history explain some of the other features of the residual plot?

D39. If you use a computer to fit a line to the (*year, density*) data, it will automatically compute a correlation.

a. Explain why a correlation is not a very useful summary for this data set.

b. In Display 3.98, the computer gave the value of "R-sq" as 97.3% for the transformed data. Statisticians are not ordinarily very interested in the size of this diagnostic measure for time-ordered data either. Can you think of any reasons?

D40. What is the estimated annual rate of growth of the population density in the United States? What is the estimated rate of growth over a decade?

■ Practice

P33. Florida is one of the fastest growing states in the United States. The population figures for each census year from 1830 through 1990 are given in Display 3.100.

a. Find a transformation that straightens the relationship.

b. Fit a line to the transformed data and use it to estimate the rate of population growth.

c. Produce a residual plot for your model and comment on the pattern in the residuals.

Year	Population
1830	34,730
1840	54,477
1850	87,445
1860	140,424
1870	187,748
1880	269,493
1890	391,422
1900	528,542
1910	752,619
1920	968,470
1930	1,468,211
1940	1,897,414
1950	2,771,605
1960	4,951,560
1970	6,791,418
1980	9,746,324
1990	12,937,926

Display 3.100   Population of Florida over the years 1830–1990.

Source: *Florida Statistical Abstract*, 1995.

Example of Exponential Decay

In the following activity, you will study a population that *decreases* exponentially over time.

Activity 3.5   Copper Flippers

What you'll need: 200 pennies, paper cup

Count your pennies. Bring 200 of them to class for an exercise to see how fast pennies "die."

A certain insect, the "copper flipper," has a life span determined by the fact that there is a 50-50 chance that a live flipper will die at the end of the day. So on average, half of any population of copper flippers will die during the first day of life. Of those that survive the first day, on average half die during the second day, and so on.

By extraordinary coincidence, these bugs behave like a bunch of tossed pennies. If you toss the 200 pennies you brought to class, about half should come up heads and half tails. The heads represent the insects that survive the first day, the tails represent those that die. You can collect the pennies that came up heads and toss them again to see how many survive the second day, and so on. This gives you a physical model for the life span distribution of the insects.

(continued)

1. Place your pennies in a cup, shake them up, and toss them on a table. Count the number of heads and record the number.

2. Set aside the pennies that came up tails. Place the pennies that came up heads back in the cup and repeat the process.

3. After each toss (day), set the tails aside, collect the heads, count them, and toss them again. Repeat this process until you have fewer than 5 heads left, but stop before you get to 0 heads.

4. Make a scatterplot of your data, with the number of the toss on the horizontal axis and the number of heads on the vertical axis. Does the pattern look linear?

5. What might you do to find an equation to summarize this pattern?

Discussion: Analyzing the Copper Flippers

These Discussion questions are based on Display 3.101, which shows the scatterplot for one student's data for Activity 3.5.

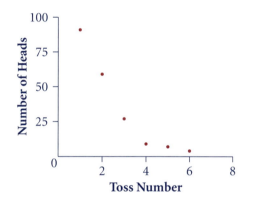

Display 3.101   Number of heads versus number of the toss.

D41. Consider whether a log transformation is appropriate.

a. Does the shape of the plot appear to be exponential?

b. Explain why the number of pennies that "die" in each time period is proportional to the amount present.

c. Do the values for the number of pennies still "alive" range over several orders of magnitude?

D42. Display 3.102 shows the natural log of the number of heads plotted against the toss number, with the regression line for the student's data from Display 3.101.

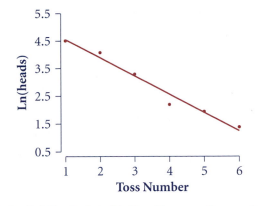

Display 3.102   A plot of ln(*heads*) versus the number
of the toss, with regression line.

Compare the scatterplot in Display 3.102 with the residual plot in
Display 3.103. (The line segments are there to help your eye follow the
time sequence.) Does the model appear to fit well if you look only at the
scatterplot? How, if at all, does the residual plot alter your judgment of
how well the line fits?

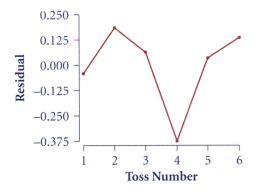

Display 3.103   Residual plot for the ln(*heads*) regression.

D43. The equation of the regression line (shown in Display 3.102) through the
transformed data is given by $\widehat{\ln y} = 5.21 - 0.66x$.

   a. Show that if you solve this equation for $\hat{y}$, you get $\hat{y} = 183.1(0.52)^x$.

   b. What is the relationship between the number under the exponent
   (or 0.52) and the chance of a coin coming up heads (or the chance
   of an insect dying on the first day)?

D44. How would the scatterplot and the least squares line change if the coins
had a probability of .6 of coming up heads? What insect death rate does
this situation model?

■ Practice: Exponential Decay

P34. ***Dying dice.*** One of your authors gathered data on dying dice, starting with 200, and used the rule that a die "lives" if it lands showing 1, 2, 3, or 4. Here are the results: 200, 122, 81, 58, 29, 19, 11, 8, 6, 4, 2, 2.

    a. Make a scatterplot of the number of "live" dice versus the roll number.

    b. Transform the number of live dice using natural logs, and make a scatterplot of ln(*dice*) versus *roll number*.

    c. Fit a line by the least squares method, and use its slope to estimate the rate of dying.

    d. Plot *residuals* versus *roll number,* and describe the pattern.

Summary 3.5: Transforming Data to Linearity

For many scatterplots that show slopes or spreads that change as x changes, you can find a shape-change transformation that brings your scatterplot much closer to the form for which the summaries of this chapter work best—a point cloud in the shape of a symmetric ellipse with linear trend. A transformation will not *always* make the pattern linear, however.

    Transformations should have some basis in reality; they should not be simply arbitrary to see what might happen. Ideally, the transformation you use will be related in a plausible way to the situation that created your data.

    The most common transformations are *powers,* in which you replace y by y^2, y by $\sqrt{y} = y^{1/2}$, or y by $\frac{1}{y} = y^{-1}$, and so forth, and *logarithms,* in which you replace y by $\log_{10} y$ or $\ln y$. You can also transform x.

    Here are some other helpful facts to consider when choosing a transformation involving logarithms:

- A log-log transformation replacing x by $\log x$ and y by $\log y$ will straighten data modeled by a power function, $y = ax^b$.

- Replacing y by $\log y$ will straighten data modeled by an exponential function of the form $y = ab^x$.

- If you have data on a quantity that changes over time, at a rate roughly proportional to the quantity at a given time, then the log of the quantity will be roughly a linear function of time.

- Consider a log transformation whenever you have a variable whose values range over two or more orders of magnitude (powers of 10).

For data collected over time (or over some other sequential ordering, such as distance along a path), there is generally one data value for each time and each data value is usually correlated quite highly with the values to either side (its close neighbors). So the data will tend to have a much more intricate pattern than a straight line can model. Careful analysis of the pattern in the residuals can often help you see what is "really" happening over time.

■ Exercises

E53. For the data of Display 3.92, try fitting a straight line to the square root of flight length as a function of speed. Does this transformation work as well as the log transformation? Explain your reasoning.

E54. *More dying dice.* Follow the same steps as in P34 for these numbers of surviving dice: 200, 72, 28, 9, 5, 2, and 1. Use your data to estimate what the probability of "dying" was in order to generate these numbers.

E55. *Growing kids.* Typical heights and weights of growing children are presented in Display 3.104. The goal is to construct a model that will allow the prediction of weight from a child's known height. What model would you choose for this prediction problem?

Age (yr)	Height (cm)	Weight (kg)
1	73.6	9.1
2	83.8	11.3
3	91.4	13.6
4	99.0	15.0
5	104.1	17.2
6	111.7	20.4
7	119.3	22.2
8	127.0	25.4
9	132.0	28.1
10	137.1	31.3
11	142.2	34.9
12	147.3	39.0
13	152.4	45.5
14	157.5	48.5

Display 3.104 Heights and weights of growing children.

E56. *Siri's equation.* Athletes and exercise scientists sometimes use the proportion of fat to overall body mass as one measure of fitness, but measuring the percentage of body fat directly poses a real challenge. Fortunately, some good statistical detective work by the scientist W. E. Siri in the 1950s provided an alternative to direct measurement that is still in use today. Siri's method lets you estimate the percentage of body fat from body density, which you can measure directly by hydrostatic (underwater) weighing. In what follows, you'll see how transformations and residual plots play a crucial role in finding Siri's model.

a. *A first model.* Use the data in Display 3.105.

Density	Percentage of Body Fat
1.053	19.94
1.053	20.04
1.055	19.32
1.056	18.56
1.048	22.08
1.040	25.81
1.030	30.78
1.064	15.33
1.001	44.33
1.052	20.74
1.014	38.32
1.076	9.91
1.063	15.81
1.023	34.02
1.046	23.15

Display 3.105 Percentage of body fat and body density of 15 women.

Source: M. L. Pollock, University of Florida, 1956.

i. Plot *percentage of body fat* versus *body density,* and use your plot to tell whether you think a line gives a poor fit, a moderately good fit, or an extremely good fit. Give the evidence for your opinion.

ii. Find the equation of the least squares line.

iii. Does the value of r^2 tend to confirm your opinion about how well the line fits?

iv. Construct a residual plot and describe the pattern. Does the plot tend to confirm or raise questions about your opinion? Explain.

b. *A new model.*

i. Explain how knowing that fat is less dense than the rest of the body might have led Siri to plot the percentage of body fat against the reciprocal of density $\left(\frac{1}{density}\right)$. Construct this plot and fit a least squares line and compare its equation with the one Siri found:

$$\% \text{ body fat} = -450 + 495 \left(\frac{1}{density}\right)$$

Plot residuals versus $\left(\frac{1}{density}\right)$. What features of this plot confirm that the transformation has improved the linear fit?

ii. Find the correlation between the percentage of body fat and body density and the correlation between the percentage of body fat and the reciprocal of body density. Comment on the use of correlation as the only criterion for assessing the usefulness of a model.

iii. Suggest another model for the percentage of body fat and body density data that might work nearly as well as Siri's.

E57. *Flight length and speed, revisited.* As you saw in Displays 3.92–3.95, the pattern relating typical flight length to average speed for commercial aircraft is not linear. Transforming flight lengths to logarithms gave a plot that is much more nearly linear. However, the residual plot strongly suggests that two line segments might provide a better model than a single line. (See Display 3.95.) Apparently, there is one relationship for slower aircraft (below 480 mph?) and another for faster aircraft. Look through the complete listing of the data (Display 3.11) to see whether any features other than speed separate the aircraft into the same two groups.

E58. The data in Display 3.106 are the population of the United States from 1830 through 1990 and the number of immigrants entering the country in the decade preceding the year indicated.

a. Find the population growth for each decade. Was the rate of growth constant from decade to decade? How can you tell?

b. Fit a model to the population data and defend your model as representative of the major trend(s) in U.S. population growth.

c. Make a plot over time of the immigration by decade. Describe the pattern you see here. Can you fit one of the models from this section to data that look like this?

Year	U.S. Population	Immigration (thousands)
1830	12,866,020	152
1840	17,069,453	599
1850	23,191,876	1713
1860	31,433,321	2598
1870	39,818,449	2315
1880	50,155,783	1812
1890	62,947,714	5247
1900	75,994,575	3688
1910	91,972,266	8795
1920	105,710,620	5736
1930	122,775,046	4107
1940	131,669,275	528
1950	150,697,361	1035
1960	179,323,175	2515
1970	203,302,031	3322
1980	226,542,199	4493
1990	248,709,873	7338

Display 3.106 Population and immigration in the United States.

Source: *Statistical Abstract of the United States, 1996.*

E59. How is the birthrate of countries related to their economic output? Do richer countries have higher birthrates, perhaps because families can afford more children? Or do poorer countries have higher birthrates, perhaps due to the need for family workers and the lack of education? The data in Display 3.107 shows the birthrates (number of births per 1000 population) and the GNP (in thousands of dollars per capita) for a selection of countries from around the world.

a. Make a scatterplot of these data and comment on the pattern you observe.

b. Fit a statistical model to these data and interpret the terms of the model in the context of these data.

Country	Birthrate	GNP
Algeria	29.0	1.6
Argentina	19.5	4.0
Australia	14.1	16.6
Brazil	21.2	2.6
Canada	13.7	20.8
China	17.8	1.3
Cuba	14.5	1.6
Denmark	12.4	24.2
Egypt	28.7	0.5
France	13.0	24.1
Germany	11.0	19.8
India	27.8	0.3
Iraq	43.6	0.7
Israel	20.4	13.6
Japan	10.7	27.3
Malaysia	28.0	2.5
Mexico	26.6	3.1
Nigeria	43.3	0.2
Pakistan	41.8	0.4
Philippines	30.4	0.7
Russia	12.6	8.6
South Africa	33.4	2.6
Spain	11.2	13.4
United Kingdom	13.2	17.4
United States	15.2	22.6

Display 3.107 Birthrates and GNP for selected countries.

Source: *Statistical Abstract of the United States,* 1995.

E60. Different body organs use different amounts of oxygen, even when you take their mass into consideration. For example, the brain uses more oxygen per kilogram of tissue than the lungs do. Scientists are interested in how oxygen consumption is related to the mass of an animal and whether that relationship differs from organ to organ. The data in Display 3.108 show typical body mass, oxygen consumption in brain tissue, and oxygen consumption in lung tissue for a selection of animals. (Oxygen consumption is often measured in milliliters per hour per gram of tissue, but the actual units were not recorded on these data.)

Animal	Body Mass (kg)	Brain Oxygen Consumption	Lung Oxygen Consumption
Mouse	0.021	32.9	12.0
Rat	0.210	26.3	8.6
Guinea pig	0.510	27.3	8.5
Rabbit	1.050	28.2	8.0
Cat	2.750	26.9	3.9
Dog	15.900	21.2	4.9
Sheep	49.000	19.7	5.4
Cattle	420.000	17.2	4.3
Horse	725.000	15.7	4.4

Display 3.108 Oxygen use by certain animal organs. The oxygen measurements are coded values (original measurements not given) but are still comparable.

Source: K. Schmidt-Nielsen, *Why Is Animal Size So Important?* (Cambridge University Press, 1984), p. 94.

a. As you can see from the table, as total body mass increases, the oxygen consumption in brain tissue tends to go down. Find a function that models this situation. Now find a way to describe the rate of decrease.

b. Repeat part a for lung tissue. How does the relationship differ from that of brain tissue?

c. It is known that the proportion of body mass concentrated in the brain decreases appreciably as the size of the animal increases, whereas the proportion concentrated in the lungs remains relatively constant. One possible theory on oxygen consumption is that the rates of consumption within organ tissue can be explained largely by the relative size of the organ within the body. Is this theory supported by the data? Explain your reasoning.

E61. According to the National Center for Health Statistics, the percentage of smokers has decreased markedly over the past 30 years, but there may still be some interesting trends to observe. Display 3.109 shows the percentage of smokers over selected years for various age groups.

Year	18–24	25–34	35–44	45–64	65+	18+
1965	54.1	60.7	58.2	51.9	28.5	51.6
1974	42.1	50.5	51.0	42.6	24.8	42.9
1979	35.0	43.9	41.8	39.3	20.9	37.2
1983	32.9	38.8	41.0	35.9	22.0	34.7
1985	28.0	38.2	37.6	33.4	19.6	32.1
1987	28.2	34.8	36.6	33.5	17.2	31.0
1988	25.5	36.2	36.5	31.3	18.0	30.1
1990	26.6	31.6	34.5	29.3	14.6	28.0
1991	23.5	32.8	33.1	29.3	15.1	27.5
1992	28.0	32.8	32.9	28.6	16.1	28.2
1993	28.8	30.2	32.0	29.2	13.5	27.5

Display 3.109 Percentages of smokers by age group and year.

Source: National Center for Health Statistics, 1996.

a. Study the trend in the percentage of smokers for the entire population aged 18 and over. What model would you use to explain the relationship between the smoking percentage and the year for this age group? Explain your reasoning.

b. Study the trend in the percentage of smokers for the group aged 18 to 24. What model would you use to explain the relationship between the smoking percentage and the year for this age group? Explain your reasoning. What feature makes this data set more difficult to model than the one in part a?

c. Study the trend in the percentage of smokers for the group aged 65 and over.

Does this group show the same kind of trend as seen in the other two groups studied? Explain.

E62. Is global warming a reality? One measure of global warming is the amount of carbon dioxide (CO_2) in the atmosphere. Display 3.110 gives the annual average carbon dioxide levels (in parts per million) in the atmosphere over Mauna Loa Observatory in Hawaii for the years 1959 through 1988.

a. Plot the data and describe the trend over the years.

b. Fit a straight line to the data and look at the residuals. Describe the pattern you see.

c. Suggest another model that might fit these data well. Fit the model and assess how well it removes the pattern from the residuals.

d. Use the model you like best to describe numerically the growth rate in carbon dioxide over Hawaii.

Year	CO_2	Year	CO_2
1959	316.1	1974	330.4
1960	317.0	1975	331.0
1961	317.7	1976	332.1
1962	318.6	1977	333.6
1963	319.1	1978	335.2
1964	*	1979	336.5
1965	320.4	1980	338.4
1966	321.1	1981	339.5
1967	322.0	1982	340.8
1968	322.8	1983	342.8
1969	324.2	1984	344.3
1970	325.5	1985	345.7
1971	326.5	1986	346.9
1972	327.6	1987	348.6
1973	329.8	1988	351.2

Display 3.110 Carbon dioxide in the atmosphere.

Source: Mauna Loa Observatory.

State	Math SAT Score	% Seniors Taking Exam	State	Math SAT Score	% Seniors Taking Exam
Alabama	558	8	Montana	547	21
Alaska	513	47	Nebraska	568	9
Arizona	521	28	Nevada	507	31
Arkansas	550	6	New Hampshire	514	70
California	511	45	New Jersey	505	69
Colorado	538	30	New Mexico	548	12
Connecticut	504	79	New York	499	73
Delaware	495	66	North Carolina	486	59
District of Columbia	473	50	North Dakota	599	5
Florida	496	48	Ohio	535	24
Georgia	477	63	Oklahoma	557	8
Hawaii	510	54	Oregon	521	50
Idaho	536	15	Pennsylvania	492	71
Illinois	575	14	Rhode Island	491	69
Indiana	494	57	South Carolina	474	57
Iowa	600	5	South Dakota	566	5
Kansas	571	9	Tennessee	552	14
Kentucky	544	12	Texas	500	48
Louisiana	550	9	Utah	575	4
Maine	498	68	Vermont	500	70
Maryland	504	64	Virginia	496	68
Massachusetts	504	80	Washington	519	47
Michigan	565	11	West Virginia	506	17
Minnesota	593	9	Wisconsin	586	8
Mississippi	557	4	Wyoming	544	11
Missouri	569	9			

Display 3.111   **Average SAT math scores by state.**   Source: *World Almanac and Book of Facts,* 1997, p. 256.

E63. How does the average SAT math score for students of a state relate to the percentage of students taking the exam? Display 3.111 shows the average SAT math score for each state in 1995, along with the percentage of high school seniors taking the exam. Find a model that looks like a good predictor of average SAT math scores from knowledge of the percentage of seniors taking the exam.

Chapter Summary

In Chapter 2, you worked with univariate data, and in Chapter 3, you learned how to uncover information for bivariate (two-variable) data, using plots and numerical summaries of center and spread. In Chapter 2, the basic plot was the histogram. For histograms that are approximately normal, the fundamental measures of center and spread are the mean and standard deviation. For bivariate data, the basic plot is the scatterplot. For scatterplots that have an elliptical shape, the fundamental summary measures are the regression line (which you can think of as the measure of center) and the correlation (which you can think of as the measure of spread).

For now, correlation and regression merely describe your set of data. But in Chapter 11, you will learn to use numerical summaries computed from a sample to make inferences about a larger population. There, using diagnostic tools such as residual plots and finding transformations that re-express a curved relationship to a linear one will come in especially handy because you won't be able to make valid inferences at all unless the points form an elliptical cloud.

■ Review Exercises

E64. ***Space shuttle Challenger.*** On January 28, 1986, because two O-rings did not seal properly, the space shuttle Challenger exploded and seven people died. The temperature predicted for the morning of the flight was from 26°F to 29°F. The engineers were concerned that the cold temperatures would cause the rubber O-rings to malfunction. There were seven previous flights when at least one of the twelve O-rings had shown some distress. The NASA officials and engineers who decided not to delay the flight had available to them data like those appearing on the scatterplot in Display 3.112 before they made that decision.

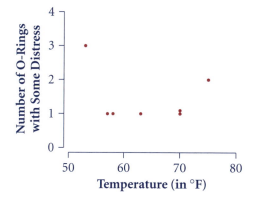
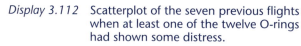

Display 3.112 Scatterplot of the seven previous flights when at least one of the twelve O-rings had shown some distress.

a. Explain why it seemed reasonable to launch the Challenger despite the low temperature.

b. Display 3.112 contains information only about the flights where there had been O-ring failures. Data for all flights were available on a table like the one in

Display 3.113. Add the missing points to a copy of the scatterplot in Display 3.112. How does this affect any trend in the scatterplot? Would you have recommended launching the shuttle if you had seen the complete plot? Why or why not?

Temperature (°F)	Number of O-Rings with Some Distress
53	3
57	1
58	1
63	1
70	1
70	1
75	2
66	0
67	0
67	0
67	0
68	0
69	0
70	0
70	0
72	0
73	0
75	0
76	0
76	0
78	0
79	0
80	0
81	0

Display 3.113 Challenger O-ring data.

Source: Siddhartha R. Dalal et al., "Risk Analysis of the Space Shuttle: Pre-Challenger Prediction of Failure," *Journal of the American Statistical Association* 84 (1989), 945–47.

E65. ***Leonardo's rules.*** A class of 15 students made these measurements for Activity 3.2. (All measurements are in centimeters.)

Student	Height	Arm Span	Kneeling Height	Hand Length
1	170.5	168.0	126.0	18.0
2	170.0	172.0	129.5	18.0
3	107.0	101.0	79.5	10.0
4	159.0	161.0	116.0	16.0
5	166.0	166.0	122.0	18.0
6	175.0	174.0	125.0	19.5
7	158.0	153.5	116.0	16.0
8	95.5	95.0	71.5	10.0
9	132.5	129.0	95.0	11.5
10	165.0	169.0	124.0	17.0
11	179.0	175.0	131.0	20.0
12	149.0	154.0	109.5	15.5
13	143.0	142.0	111.5	16.0
14	158.0	156.5	119.0	17.5
15	161.0	164.0	121.0	16.5

a. Construct scatterplots and fit least squares regression lines for each of Leonardo's rules given in Activity 3.2. Do they appear to hold?

b. Interpret the slopes of your regression lines.

c. If appropriate, find the value of r for each of the three relationships. Which is strongest? Which is weakest?

E66. ***Exam scores.*** The scores for two exams in a statistics course are given in Display 3.114 with a scatterplot with the regression line added, and a residual plot. The regression equation is *Exam 2* = 51.0 + 0.430(*Exam 1*), and the correlation is $r = .756$.

a. Is one student's point more influential than the others on the slope of the regression line? How can you tell from the scatterplot? From the residual plot?

b. How will the slope change if the scores for this one influential point are removed from the data set? How will the correlation change? Calculate the slope and correlation for the revised data to check your estimate.

c. Construct a residual plot for the revised data. Does a linear model fit the data well?

d. Refer to the scatterplot of Exam 2 versus Exam 1 scores in Display 3.114. Does this illustrate regression to the mean? Explain your reasoning.

Exam 1	Exam 2	Exam 1	Exam 2
80	88	96	99
52	83	78	90
87	87	93	88
95	92	92	92
67	75	91	93
71	78	96	92
97	97	69	73
96	85	76	87
88	93	91	91
100	93	98	97
88	86	83	89
86	85	96	83
81	81	95	97
61	73	80	86
97	92		

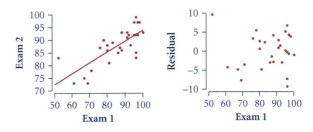

Display 3.114 Scatterplot and residual plot for exam scores in a statistics class.

A	B	C	D	E	F
−4	−2	−2	0	−6	0
−3	−1	0	−2	−5	−1
−2	0	0	0	−5	0
−1	1	0	2	−5	1
0	2	2	0	−4	0
0	2	−1	−1	4	0
1	1	−1	1	5	−1
2	0	1	−1	5	0
3	−1	1	1	5	1
4	−2	0	0	6	0

Display 3.115 A "scatterplot matrix" of all 30 possible scatterplots for the six variables listed in the table.

E67. ***More exam scores.*** Suppose you have a different set of scores on two exams.

   a. The slope of the regression line for predicting the scores on Exam 2 from the scores on Exam 1 is 0.51. The standard deviation for Exam 1 scores is 11.6, and the standard deviation for Exam 2 scores is 7.0. Use only this information to find the correlation coefficient for these scores.

   b. If you knew, in addition, that the means were 82.3 for Exam 1 and 87.8 for Exam 2, find the equation of the least squares regression line for predicting Exam 2 scores from Exam 1 scores.

E68. You are given a list of six values, −1.5, −0.5, 0, 0, 0.5, and 1.5, for *x* and the same list of six values for *y*. Note that the list has mean 0 and standard deviation 1.

   a. Match each *x*-value with a *y*-value so that the resulting six (*x, y*) pairs have correlation 1.

   b. Match again so that the points have the largest possible correlation less than 1.

   c. Match again, this time to get a correlation as close to 0 as you can.

   d. Match a fourth time to get a correlation of −1.

E69. Display 3.115 lists values for six variables, and the "scatterplot matrix" beside it shows all 30 possible scatterplots for those variables. For example, the first scatterplot in the first row has variable B on the *x*-axis and variable A on the *y*-axis. The first scatterplot in the second row has A on the *x*-axis and B on the *y*-axis.

   a. For five pairs of variables, the correlation is exactly 0, and for one other pair, it is .02, almost 0. Identify these six pairs of variables. What do they have in common?

   b. At the other extreme, there is one pair with correlation .88; the next highest correlation is .58, and the third highest is .45. Identify these three, and put them in order from strongest to weakest.

   c. Of the remaining six pairs, four have correlations around .25 (give or take a little) and two have correlations around .1 (give or take a little). Which four are around .25?

   d. Choose several scatterplots that you think best illustrate the phrase "the correlation measures direction and strength but not shape" and use them to show what you mean.

E70. Decide whether these statements are true or false, and then explain your decision.

a. The correlation is to bivariate data what the standard deviation is to univariate data.

b. The correlation measures direction and strength but not shape.

c. If the correlation is near zero, knowing the value of one variable gives you a narrow interval of likely values for the other variable.

d. No matter what data set you look at, the correlation coefficient, r, and least squares slope, b_1, will always have the same sign.

E71. Look at the scatterplot of average SAT math scores as related to the percentage of students taking the exam in Display 3.7 of Section 3.1.

a. Estimate the correlation.

b. What possibly important features of the plot are lost if you give only the correlation and the equation of the least squares line?

c. Sketch what you think the residual plot would look like if you fitted one line to all of the points.

E72. The correlation between in-state tuition and out-of-state tuition, measured in dollars, for a sample of public universities is .80.

a. Rewrite this sentence so that someone who does not know statistics can understand it.

b. Does the correlation change if you convert tuition costs to thousands of dollars and recompute the correlation? If you take logarithms of the tuition costs and recompute the correlation?

c. Does the slope of the least squares line change if you convert tuition costs to thousands of dollars and recompute the slope? If you take logarithms of the tuition costs and recompute the slope?

E73. Display 3.116 shows a scatterplot divided into quadrants by vertical and horizontal lines that pass through the point of averages, $(\bar{x}, \bar{y})$.

a. For each of the four quadrants, give the sign of z_x (the standardized value of x), z_y (the standardized value of y), and their product $z_x \cdot z_y$.

b. Which point(s) make the smallest contribution to the correlation? Explain why the contributions are small.

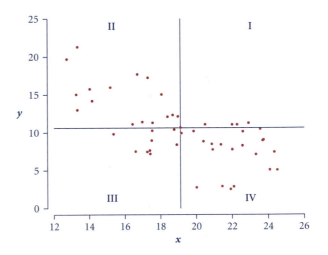

Display 3.116 A scatterplot divided into quadrants by the means.

E74. Rank these summaries for three lists of (x, y) pairs by the strength of the relationship, from weakest to strongest.

A. $\hat{y} = 90 + 100x$ $s_x = 5$ $s_y = 1000$

B. $\hat{y} = \dfrac{x}{3} - 12$ $s_x = 0.9$ $s_y = 1$

C. $\hat{y} = 1.05 + 0.01x$ $s_x = 0.05$ $s_y = 0.002$

E75. There's an extremely strong relationship between the price of books online and the price at your local bookstore.

a. Does this mean the prices are almost the same?

b. Explain why it is wrong to say that the prices online "cause" the prices at your local bookstore. Why is the relationship so strong even though neither set of prices causes the other?

E76. Describe a set of cases and two variables for which you would expect to see regression toward the mean.

E77. **Life spans.** In Display 2.24 in Section 2.2, you looked at characteristics of mammals one at a time. Now you can look at the relationship between two variables. For example, is longevity associated with the gestation period? The variables are average longevity in years, maximum longevity in years, gestation period in days, and speed in miles per hour.

   a. Make a scatterplot of the gestation period versus the maximum longevity. Describe what you see, including an estimate of the correlation.

   b. Repeat part a with the average longevity in place of the maximum longevity. Does the average longevity or the maximum longevity give a better prediction of the gestation period?

   c. Does speed appear to be associated with average longevity?

   d. Make a scatterplot of the maximum longevity versus the average longevity. What features of the plot make it hard to draw an ellipse? Does this mean that you should use only an algebraic formula to get the correlation for this pair of variables?

E78. **Spending for police.** The data in Display 3.117 give the number of police officers, the total expenditures for police officers, the population, and the violent crime rate for a sample of states.

   a. Explore and summarize the relationship between the number of police officers and expenditures for police.

   b. Explore and summarize the relationship between the population of the states and the number of police officers they employ.

   c. Is the number of police officers strongly related to the rate of violent crime in these states? Explain. Find a transformation that straightens these data. Check the linearity of your transformed data with a residual plot.

State	Number of Police Officers (thousands)	Expenditures for Police ($ millions)	Population (millions)	Violent Crime Rate (number per 100,000 of population)
California	86.2	5402	30.4	1090
Colorado	9.2	435	3.4	559
Florida	45.0	2162	13.2	1184
Illinois	39.9	1605	11.5	1039
Iowa	6.0	251	2.8	303
Louisiana	11.8	515	4.2	951
Maine	2.9	100	1.2	132
Missouri	14.6	515	5.2	763
New Jersey	30.5	1241	7.7	635
Tennessee	12.3	437	5.0	726
Texas	46.2	1840	17.3	840
Virginia	15.2	714	6.3	373
Washington	10.9	564	5.0	523

Display 3.117 Number of police officers and related variables.
Source: *Statistical Abstract of the United States,* 1994.

E79. *House prices.* Display 3.118 contains the selling prices for houses sold in Gainesville, Florida, for one month in 1995.

a. Construct a model to predict selling price from area, transforming any variables if necessary. Would you use the same model for both new and used houses?

b. Are there influential observations that have a serious effect on the model? If so, what would happen to the slope of the prediction equation and the correlation if you removed this (or these) point(s) from the analysis?

c. Predict the selling price of an old house of 1000 square feet. Do the same for a house of 2000 square feet. Which prediction do you feel more confident about? Explain.

d. Explain the effect of the number of bathrooms on the selling price of these houses. Is it appropriate to fit a regression model to price as a function of the number of bathrooms and interpret the results in the usual way? Why or why not?

Display 3.118 Selling prices of houses in Gainesville.

House	Price ($ thousands)	Area (thousands of sq ft)	Number of Bedrooms	Number of Bathrooms	New (1), Old (0)
1	48.5	1.10	3	1	0
2	55.0	1.01	3	2	0
3	68.0	1.45	3	2	0
4	137.0	2.40	3	3	0
5	309.4	3.30	4	3	1
6	17.5	0.40	1	1	0
7	19.6	1.28	3	1	0
8	24.5	0.74	3	1	0
9	34.8	0.78	2	1	0
10	32.0	0.97	3	1	0
11	28.0	0.84	3	1	0
12	49.9	1.08	2	2	0
13	59.9	0.99	2	1	0
14	61.5	1.01	3	2	0
15	60.0	1.34	3	2	0
16	65.9	1.22	3	1	0
17	67.9	1.28	3	2	0
18	68.9	1.29	3	2	0
19	69.9	1.52	3	2	0
20	70.5	1.25	3	2	0
21	72.9	1.28	3	2	0
22	72.5	1.28	3	1	0
23	72.0	1.36	3	2	0
24	71.0	1.20	3	2	0
25	76.0	1.46	3	2	0

(continued)

Display 3.118 (continued)

House	Price ($ thousands)	Area (thousands of sq ft)	Number of Bedrooms	Number of Bathrooms	New (1), Old (0)
26	72.9	1.56	4	2	0
27	73.0	1.22	3	2	0
28	70.0	1.40	2	2	0
29	76.0	1.15	2	2	0
30	69.0	1.74	3	2	0
31	75.5	1.62	3	2	0
32	76.0	1.66	3	2	0
33	81.8	1.33	3	2	0
34	84.5	1.34	3	2	0
35	83.5	1.40	3	2	0
36	86.0	1.15	2	2	1
37	86.9	1.58	3	2	1
38	86.9	1.58	3	2	1
39	86.9	1.58	3	2	1
40	87.9	1.71	3	2	0
41	88.1	2.10	3	2	0
42	85.9	1.27	3	2	0
43	89.5	1.34	3	2	0
44	87.4	1.25	3	2	0
45	87.9	1.68	3	2	0
46	88.0	1.55	3	2	0
47	90.0	1.55	3	2	0
48	96.0	1.36	3	2	1
49	99.9	1.51	3	2	1
50	95.5	1.54	3	2	1
51	98.5	1.51	3	2	0
52	100.1	1.85	3	2	0
53	99.9	1.62	4	2	1
54	101.9	1.40	3	2	1
55	101.9	1.92	4	2	0
56	102.3	1.42	3	2	1
57	110.8	1.56	3	2	1
58	105.0	1.43	3	2	1
59	97.9	2.00	3	2	0
60	106.3	1.45	3	2	1
61	106.5	1.65	3	2	0
62	116.0	1.72	4	2	1
63	108.0	1.79	4	2	1
64	107.5	1.85	3	2	0
65	109.9	2.06	4	2	1
66	110.0	1.76	4	2	0
67	120.0	1.62	3	2	1
68	115.0	1.80	4	2	1
69	113.4	1.98	3	2	0
70	114.9	1.57	3	2	0

(continued)

Display 3.118 (continued)

House	Price ($ thousands)	Area (thousands of sq ft)	Number of Bedrooms	Number of Bathrooms	New (1), Old (0)
71	115.0	2.19	3	2	0
72	115.0	2.07	4	2	0
73	117.9	1.99	4	2	0
74	110.0	1.55	3	2	0
75	115.0	1.67	3	2	0
76	124.0	2.40	4	2	0
77	129.9	1.79	4	2	1
78	124.0	1.89	3	2	0
79	128.0	1.88	3	2	1
80	132.4	2.00	4	2	1
81	139.3	2.05	4	2	1
82	139.3	2.00	4	2	1
83	139.7	2.03	3	2	1
84	142.0	2.12	3	3	0
85	141.3	2.08	4	2	1
86	147.5	2.19	4	2	0
87	142.5	2.40	4	2	0
88	148.0	2.40	5	2	0
89	149.0	3.05	4	2	0
90	150.0	2.04	3	3	0
91	172.9	2.25	4	2	1
92	190.0	2.57	4	3	1
93	280.0	3.85	4	3	0

Source: Gainesville Board of Realtors, 1995.

E80. ***Spending for schools.*** Display 3.119 provides data on spending and other variables related to public school education for 1992. The variables are defined as:

ExpPP	expenditure per pupil (dollars)
ExpPC	expenditure per capita (per person in the state, in dollars)
TeaSal	average teacher salary (thousands of dollars)
%Dropout	percentage who drop out of school
Enroll	number of students enrolled (thousands)
Teachers	number of teachers (thousands)

a. Examine the association between per-pupil expenditure and average teacher salary, with the goal of predicting per-pupil expenditure. Is this a cause-and-effect relationship?

b. Analyze the effect of average teacher salary on per-capita expenditure (spending on public schools divided by the number of people in the state). Is this association stronger than the one in part a? Explain why that should or should not be the case.

c. Are any variables good predictors of the percentage of dropouts? Explain your reasoning.

Display 3.119 Public school education by state in 1992.

State	ExpPP	ExpPC	TeaSal	%Dropout	Enroll	Teachers
Alabama	3675	684	27.0	12.6	722	41.1
Alaska	9248	1798	44.7	0.6	14	0.2
Arizona	4750	961	31.2	4.3	640	33.5
Arkansas	3770	758	26.6	10.9	433	25.9
California	4686	930	40.2	14.3	4951	221.0
Colorado	5259	961	33.1	9.6	574	33.1
Connecticut	8299	1227	47.0	9.2	469	34.8
Delaware	6080	899	34.5	11.2	106	6.1
District of Columbia	8116	1059	41.3	19.1	80	6.2
Florida	5639	934	31.1	14.2	1862	109.9
Georgia	4720	884	29.5	14.1	152	70.3
Hawaii	5453	898	34.5	7.0	172	10.0
Idaho	3528	803	26.3	9.6	221	11.6
Illinois	5248	842	36.5	10.4	1822	109.9
Indiana	5429	1003	34.8	11.4	955	55.1
Iowa	4949	882	29.2	6.5	484	31.5
Kansas	5131	941	30.7	8.4	437	29.3
Kentucky	4616	773	30.9	13.0	636	37.4
Louisiana	4378	806	27.0	11.9	785	44.0
Maine	5969	1087	30.1	8.4	215	14.8
Maryland	6273	967	39.5	11.0	715	43.3
Massachusetts	6323	863	37.3	9.5	834	56.0
Michigan	5630	987	41.1	9.9	1582	80.2
Minnesota	5510	1054	33.7	6.1	754	44.9
Mississippi	3344	670	24.4	11.7	503	27.9
Missouri	4534	780	28.9	11.2	812	52.3
Montana	5127	943	27.6	7.1	153	9.9
Nebraska	4676	866	27.2	6.6	274	18.9
Nevada	4910	1006	33.9	14.9	201	11.4
New Hampshire	5500	888	33.2	9.9	172	11.5
New Jersey	10219	1337	41.0	9.3	1090	80.5
New Mexico	4692	971	26.7	10.8	302	17.1
New York	8658	1247	43.3	10.1	2598	180.3
North Carolina	4857	835	29.2	13.2	1087	64.8
North Dakota	4119	794	24.5	4.3	118	7.8
Ohio	5451	921	33.3	8.8	1772	103.2
Oklahoma	3939	814	25.3	9.9	579	37.7
Oregon	5927	1018	34.1	11.0	485	26.7
Pennsylvania	6980	954	38.7	9.4	1668	100.5
Rhode Island	6834	903	36.0	12.9	139	9.8
South Carolina	4537	861	28.3	11.9	622	36.3
South Dakota	4255	835	23.3	7.1	129	8.6
Tennessee	3736	611	28.6	13.6	824	43.1
Texas	4651	988	29.0	12.5	3383	212.6
Utah	3092	854	26.5	7.9	448	18.3

(continued)

Display 3.119 **(continued)**

State	ExpPP	ExpPC	TeaSal	%Dropout	Enroll	Teachers
Vermont	6992	1205	33.6	8.7	96	7.0
Virginia	5487	950	31.9	10.4	998	64.7
Washington	5331	1160	34.8	10.2	840	42.9
West Virginia	5415	962	27.4	10.6	322	21.0
Wisconsin	5972	976	35.2	6.9	798	52.0
Wyoming	5333	1189	30.4	6.3	98	6.4

Source: *Statistical Abstract of the United States,* 1994.

E81. The ratings given by *Consumer Reports,* July 1997, for places to purchase eyeglasses and the corresponding prices are given in Display 3.120. A higher rating is better.

   a. Make a scatterplot of the price versus the rating. Is there a "best buy" place according to the data? Explain your thinking, using the plot you made as part of your evidence.

Source of Eyeglasses	Rating	Price ($)
Ophthalmologists, optometrists, private	77	171
Costco Wholesale	76	122
For Eyes	75	115
Shopko Optical	75	131
Lenscrafters	74	190
Frame-n-Lens	74	102
Wal-Mart Vision Center	74	136
Opti-World	73	181
Visionworks	72	183
Pearle Vision Express	71	203
Sam's Club Vision Center	70	135
Eyemasters	70	203
Empire Vision Center	69	138
Texas State Optical	69	173
Eye World	69	181
Pearle Vision Center	69	195
Sterling Optical	69	170
Cohen's Fashion Optical	67	210
NuVision	65	207
Sears Optical	64	159
J. C. Penney	63	166
America's Best Contacts & Eyeglasses	57	109

Display 3.120 Ratings and prices of eyeglasses.
Source: *Consumer Reports,* July 1997.

   b. If you fit a linear model to these data, will there be a point with strong influence? If so, compare the regression lines, fit with and without this point.

E82. On April 24, 1998, a local weekly, *The Litchfield* [Connecticut] *County Times,* published a story about the Rite Aid pharmacy chain. The chain had just built a number of stores in that part of Connecticut. The article was accompanied by data showing the number of Rite Aid stores by year from 1962 through 1997. (*Problem continues on next page.*)

Year	Number of Stores	Year	Number of Stores
1962	1	1980	800
1963	6	1981	878
1964	12	1982	963
1965	25	1983	1069
1966	36	1984	1147
1967	48	1985	1275
1968	60	1986	1392
1969	69	1987	1586
1970	132	1988	2072
1971	170	1989	2184
1972	267	1990	2352
1973	348	1991	2420
1974	426	1992	2452
1975	443	1993	2573
1976	469	1994	2690
1977	648	1995	2829
1978	663	1996	3548
1979	728	1997	3963

Display 3.121 Number of Rite Aid stores over the years 1962–1997.

a. Construct a plot of these data, and describe the pattern.

b. Find a model that will describe the nature of the growth in number of stores.

E83. *The New York Times* [Weekly Review, April 12, 1998] published the data in Display 3.122 about passengers on United Airlines flight 815, Chicago-O'Hare to Los Angeles, on October 31, 1997. There were 186 passengers, but the data concerned only the 33 passengers who had tickets for the Chicago-to-Los Angeles leg only. The variables are:

X: number of days before the flight that the ticket was purchased

Y: price of the ticket

X	Y	X	Y
11	855.97	29	87.21
7	855.97	52	154.13
0	1248.51	8	681.86
20	956.88	20	182.24
11	125.88	7	181.37
4	517.05	0	108.26
77	229.6	9	108.26
18	165.98	16	148.8
15	255.91	16	148.8
14	114.99	28	193.23
14	114.99	9	504.12
14	164.44	9	504.12
14	164.44	17	148.28
3	137.39	21	119.42
71	103.46	3	728.26
71	103.46	249	0.00
15	168.08		

Display 3.122 Number of days before the flight that the ticket was purchased and price of airline tickets.

Because it's in the airline's interest to sell tickets early, you might expect *Y* to be *negatively associated* with *X*.

It happens that the first four cases shown in Display 3.122 are for passengers who flew first class, and they pay more than other passengers no matter when they purchase their ticket. So you can justify examining the data for the 29 economy-class passengers only.

Finally, there was one passenger who paid $0 because he or she used frequent-flyer miles. You are justified in deleting this value from the data set if the goal is to find a model that relates price to time of purchase.

Can you find a model that relates cost of the fare to days in advance that the ticket was purchased? Carefully explain the problems in doing this.

E84. In E56, you explored the relationship between the percentage of body mass that is fat and body density. Display 3.123 is an extension of the data in E56, including skinfold measurements and data for men.

a. Does Siri's model for relating percentage of body fat to body density hold for men as well as it did for women? That model was

$$\% \ body \ fat = -450 + 495 \left(\frac{1}{density} \right)$$

b. The variable *skinfold* is the sum of a number of skinfold thicknesses taken at various places on the body. (The units are millimeters.) The skinfold measurements are used to predict body density. Find a good model for predicting density from skinfold measurements based on these data for the women. Do your models require re-expression?

	Women			Men	
% Fat	Density	Skinfold (mm)	% Fat	Density	Skinfold (mm)
19.94	1.053	126.5	31.60	1.029	223.0
20.04	1.053	69.0	17.99	1.058	149.5
19.32	1.055	98.0	16.10	1.062	148.0
18.56	1.056	85.5	8.68	1.079	112.5
22.08	1.048	104.5	26.87	1.038	141.5
25.81	1.040	163.0	18.06	1.058	220.5
30.78	1.030	192.0	28.71	1.034	252.5
15.33	1.064	67.0	21.71	1.049	177.5
44.33	1.001	228.0	23.38	1.046	152.0
20.74	1.052	102.0	9.09	1.078	110.0
38.32	1.014	248.5	10.77	1.074	100.5
9.91	1.076	73.0	4.58	1.089	72.0
15.81	1.063	92.5	21.93	1.049	219.5
34.02	1.023	144.5	3.82	1.091	85.5
			23.15	1.046	86.5

Display 3.123 Percentage of body fat, body density, and skinfold for 15 women and 14 men.

Source: M. L. Pollock, University of Florida, 1956.

CHAPTER 4
SAMPLE SURVEYS AND EXPERIMENTS

What prompts a hamster to prepare for hibernation? A student designed an experiment to see whether the number of hours of light in a day affects the concentration of a key brain enzyme.

Most of what you've done so far, in Chapters 2 and 3 and in the first part of Chapter 1 as well, is part of data exploration—ways to uncover, display, and describe patterns in data. Methods for exploration can help you look for patterns in just about any set of data, but they can't take you beyond the data in hand: *With exploration, what you see is all you get.* Often, that's not enough:

> *Pollster:*   I asked 100 likely voters who they planned to vote for, and 52 of them said they'd vote for you.
>
> *Politician:*   Does that mean I'll win the election?
>
> *Pollster:*   Sorry, I can't tell you. My stat course hasn't gotten to inference yet.
>
> *Politician:*   What's inference?
>
> *Pollster:*   Drawing conclusions based on your data. I can tell you about the 100 people I actually talked to, but I don't yet know how to use that information to tell you about *all* the likely voters.

Methods of *inference* can take you beyond the data you actually have, but only if your numbers come from the right kind of process. If you want to use 100 likely voters to tell you about *all* likely voters, how you choose those 100 is crucial. "The quality of your inference depends on the quality of your data," or, in other words, "Bad data, bad conclusions." This chapter tells you how to gather data in surveys and through experiments in ways that make sound conclusions possible.

Here's a simple example.[1] When you taste a spoonful of chicken soup and decide it doesn't taste salty enough, that's exploratory analysis: You've found a pattern in your one spoonful of soup. If you generalize and conclude that the whole pot of soup needs salt, that's an inference. To know if your inference is valid, you have to know how your one spoonful—the data—got taken from the pot. If there's a lot of salt sitting on the bottom, soup from the surface won't be representative, and you'll end up with an incorrect inference. But if you stir the soup thoroughly before you taste, your spoonful of data will more likely represent the whole pot. Sound methods for producing data are the statistician's way of making sure the soup gets stirred so that a single spoonful—the sample—can tell you about the whole pot. Instead of using a spoon, the statistician relies on a chance device to do the stirring and on probability theory to make the inference.

Tasting soup illustrates one kind of question you can answer using statistical methods: "Can I generalize from a small sample (the spoonful) to a larger

---

[1]The inspiration for this metaphor came from Gudmund Iversen, who teaches statistics at Swarthmore College.

population (the whole pot full of chicken soup)?" To use a sample for inference about a population, you must *randomize:* Use chance to determine who or what gets in your sample.

The other kind of question is about comparison and cause. For example, if people eat chicken soup when they get a cold, will this cause the cold to go away more quickly? When designing an experiment to determine if a pattern in the data is due to cause and effect, you also should randomize if you can. That is, you should use chance to determine which subjects get which treatments. To answer the question about chicken soup, you should use chance to decide which of your subjects eat chicken soup and which don't and then compare the duration of the colds.

The first half of this chapter is about designing surveys. A well-designed survey enables you to make inferences about a population by looking at a sample from that population. The second half of this chapter introduces experiments. An experiment enables you to determine cause by comparing the effects of two or more treatments.

In this chapter you will learn

- reasons for using samples when conducting a survey
- how to design a survey by randomly selecting participants
- how surveys can go wrong: bias
- how to design a sound experiment by randomly assigning treatments to subjects
- how experiments can determine cause
- how experiments can go wrong: confounding

4.1 Why Take Samples, and How Not To

Why samples are necessary

Taking a sample survey can help determine the percentage of people in a population who have a particular characteristic. For example, the Gallup poll periodically asks adults in the United States questions such as whether they approve of the job the President is doing. Polls or surveys such as this rely on samples to get their percentages. That is, they don't ask every adult in the United States, but only a sample of about 1500. Similarly, quality assurance methods in a manufacturing plant do not call for checking the quality of every item coming off the production line; rather, they recommend that a limited number of items (a sample) be checked carefully for quality.

Sampling saves money and time.

Cost in money or time is a primary reason to use samples. Imagine: It's Sunday night at 8:00, and Nielsen Media Research is gathering data about what proportion of TV sets are tuned to a particular program and what kinds of people are watching that program. To find out how many TV sets are tuned to the program, electronic meters have been hooked up to televisions in a sample of households. To find out who is watching it, a sample of people are filling out diaries. Why doesn't Nielsen Media Research include everyone in the United States in these surveys? To hook up a meter to every TV set would cost more than anyone would be willing to pay for this information. And to try to get a diary from every TV viewer about what they were watching at 8:00 P.M. on Sunday would take so much time that the information would no longer be very useful. So for two reasons—money and time—Nielsen ratings are based on samples.

Testing every unit can be destructive.

Sampling lets a cook know how the soup will taste without eating it all just to make sure. A lightbulb manufacturer can't test the life of every bulb produced, or they'd have none left to sell. Whenever testing destroys the things you test, your only choice is to work with a sample.

Samples can give more information than a cursory study of all individuals.

If time and money are limited, and they always are, there's a tradeoff between the number of people in your sample and the amount and quality of information you can expect to get from each person. Using a sample allows you to spend more time and money gathering information from each individual. This often produces greater accuracy in the results than you could get from a quick, but error-prone, study of every individual.

Population, units, census, sample

In statistics, the set of people or things that you want to know about is called the **population.** The individual elements of the population are sometimes called **units.** In everyday language, "population" often refers to the number of units in the set, as when you say, "In 1990, the population of Massachusetts was about six million." But in statistics, "population" refers to the set itself (for example, the people of Massachusetts). The number of units is called the **population size.** Ordinarily, you don't get to record data on all the units in the population, and so you use a sample: The **sample** is the set of units that you do get to study. In the special case that you collect data on the entire population, you have a **census.**

Discussion: Census Versus Sample

D1. In which of these situations do you think a census is used, and in which do you think sampling is used to collect data? Explain your reasoning.

    a. An automobile manufacturer inspects its new models.

    b. A cookie producer checks the number of chocolate chips per cookie.

    c. The U.S. President is determined by an election.

    d. Weekly ratings of movie attendance are released each Sunday.

    e. A Los Angeles study does in-depth interviews with teachers in order to find connections between nutrition and health.

D2. The American Statistical Association has about 17,000 members. Some members work for governments (federal, state, local), some for business and industry, and some for colleges or universities. Suppose you want to know the proportion of members in each category. Tell how to find these proportions by using a census and by using a sample.

D3. You want to estimate the average number of TV sets per household in your community.

    a. What is the population? What are the units?

    b. Explain the advantages of sampling over conducting a census.

    c. What problems do you see in carrying out this sample survey?

Bias: A Potential Problem with Survey Data

Samples offer many advantages, but not all samples are created equal: Some are much more trustworthy than others. In Activity 4.1, you'll examine the kind of problem that can lie in wait for the unwary sampler. Suppose that you have just won a contract to estimate the average length of stay in the children's ward of your local hospital. How will you gather the data?

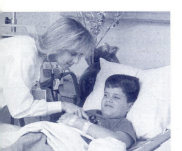

Activity 4.1   Time in the Hospital

What you'll need: a deck of cards

In this activity, you'll estimate the average length of stay in a five-bed hospital ward. You'll sample from the population of "lengths of stay" represented by the numbers on the 40 cards, not counting face cards, in an ordinary deck of cards. You'll estimate the average length from a sample of five patients.

Part A: A Random Choice of Day

1. Shuffle your 40 cards several times. Deal out a row of five cards to be your first patients. The numbers on the cards tell you how many days they will be on your ward. For example, suppose your first five cards are

A♥   2♥   8♥   8♦   3♣

That means the patient in bed 1 will be in the hospital for 1 day, the patient in bed 2 will be there for 2 days, the patient in bed 3 will be there for 8 days, and so forth. Your partner should record this information on a graph like the one in Display 4.1. Place the cards that represent patients in a separate stack throughout the Activity.

(continued)

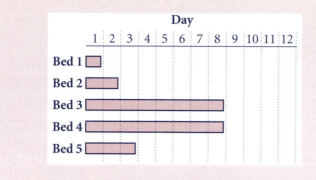

Display 4.1   Lengths of stay for the first five patients.

2. Deal out the cards one at a time, assigning each patient to a bed as it becomes available. To continue with the example, suppose the next patient is 9♣. Because the first available bed is bed 1, the patient will go into it for 9 days. The next available bed is bed 2, and the next patient, say, 9♦, will go there for 9 days. Display 4.2 shows the data at this stage. The next patient will go into bed 5.

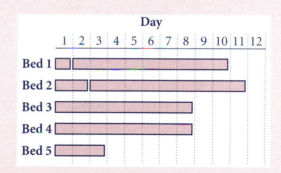

Display 4.2   Lengths of stay for the first seven patients.

3. Continue dealing out patient cards, each representing a hospital stay. Record these stays until all five beds have been filled for at least 20 days. You should end up with something like Display 4.3. (Save your chart; you'll need it later.)

4. Select a day at random from days 1–20 (or however many days it took to fill all the beds in the ward).

5. Compute the average length of stay for the patients in the beds on that day.

6. Pool your results with the rest of your class until you have about 30 estimates of the average length of a stay. Make a dot plot of the 30 averages from your class. Where is this distribution centered? Is there much variability in your estimates?

(continued)

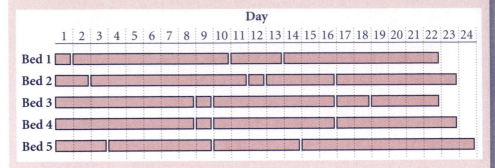

Display 4.3 Lengths of stay for the patients during the first 20 days.

7. What is the average stay for the whole population (the deck of 40 cards)?

8. Compare your results from step 6 and step 7. Are your estimates clearly too low, clearly too high, or about right?

9. What are the units in this situation? Did every unit have an equal chance of being in the sample? If so, explain. If not, which units had the greater chance?

Part B: A Random Choice of Patient

1. Shuffle the stack of cards you set aside in Part A, steps 2 and 3, and randomly choose a sample of five patients. Compute the average length of stay for your random sample.

2. Repeat step 1 and pool your results with the rest of your class until you have about 30 estimates of average length of stay. Make a dot plot of the 30 averages. Where is this distribution centered?

3. Compare your results from step 2 with the average stay for the whole population, which is 5.5 days. Are your estimates clearly too low, clearly too high, or about right?

4. Compare the dot plot from step 2 in Part B with the dot plot from step 6 in Part A. Are there any similarities? What are the main differences? How do you account for these differences?

In everyday language, we say an opinion is "biased" if it unreasonably favors one point of view over others. A biased opinion is not balanced, not objective. In statistics, bias has a similar meaning in that a biased sampling method is also unbalanced.

A sampling method is **biased** if it tends to give samples in which some characteristic of the population is underrepresented or overrepresented.

Bias is in the method, not in the sample.

There's an important distinction here between the sample itself and the method for choosing the sample.

Investigator: What makes a good sample?

Statistician: A good sample is representative; that is, it looks like a small version of the population. Proportions you compute from the sample are close to the corresponding proportions you would get if you used the whole population. The same is true for other numerical summaries, like averages and standard deviations or medians and *IQRs*.

Investigator: How can you tell if your sample is representative?

Statistician: There's the rub: in practice, you can't. You can tell only by comparing your sample with the population, and if you know that much about the population, why bother to take a sample?

Investigator: Great! First you tell me my sample should be representative, and then you tell me there's no way to know whether it is. Is that the best statisticians can do?

Statistician: Nope. Although you can't tell about any particular sample, it *is* possible to tell whether a sampling *method* is good or not. That's where bias comes in.

Investigator: I thought "biased" was just a fancy word for "nonrepresentative." Not true?

Statistician: Now we're getting to the point. Bias refers to the method, not the samples you get from it. A method is biased if it tends to give nonrepresentative samples.

Investigator: Now I get it. I may not be able to tell whether my sample is representative, but if I use an unbiased method, then I can be confident that my sample is likely to be representative. Right?

Statistician: Now you're thinking like a statistician. There's more detail to come, but you've got the big picture in focus.

Discussion: Bias

D4. Explain the difference between nonrepresentative and biased as they pertain to sampling.

D5. Which statements are possible, and which are impossible?

a. A representative sample results from a biased sample-selection method.

b. A nonrepresentative sample results from a biased sample-selection method.

c. A representative sample results from an unbiased sample-selection method.

d. A nonrepresentative sample results from an unbiased sample-selection method.

Sampling Bias

Sample selection bias, or **sampling bias,** is present in a sampling procedure if samples tend to result in numerical summaries that are systematically too high or too low. There are various forms of this selection bias that can undermine the usefulness of samples and surveys.

Size bias is one kind of selection bias.

You explored one kind of selection bias in Part A of Activity 4.1, *Time in the Hospital,* in which patients who spent more days in the hospital were more likely to be selected for the sample. In fact, the chance of selection is proportional to the length of stay. A five-day stay is five times as likely to be chosen as a one-day stay. This type of sample selection bias is called **size bias.** Suppose a wildlife biologist samples lakes in a state by dropping grains of rice at random on a map of the state and then selects for study the lakes that have rice on them. This is another example of size bias.

Voluntary response bias is another kind of selection bias.

When a television or radio program asks people to call up and take sides on some issue, those who care about the issue will be overrepresented, and those who don't care as much may not be represented at all. The resulting bias is called **voluntary response bias,** which is a second type of sample selection bias.

Convenience sampling is almost surely biased as well.

Here's a simple sampling method: Take whatever's handy. For example, what percentage of the students in your graduating class plan to go to work immediately after graduation? Rather than find a representative sample of your graduating class, it would be a lot quicker simply to ask your friends and use them as your sample—quicker, and more convenient, but almost surely biased because your friends are likely to be somewhat alike in their plans. That's a **convenience sample:** one in which the units chosen are the ones that are easy to include. The likelihood of bias of this sort makes convenience samples about as worthless as voluntary response samples.

Judgment sampling, even when taken by experts, is usually biased.

Because voluntary response sampling and convenience sampling tend to be biased, you might be inclined to rely on the judgment of an expert to choose a sample that he or she considers representative. Such samples, not surprisingly, are called **judgment samples.** Unfortunately, though, experts are sometimes biased and also may overlook important features of a population. In addition, it can be almost impossibly complicated to try to balance several features at once. In the early days of election polling, local "experts" were hired to sample voters in their locale by filling certain quotas (so many men, so many women, so many over the age of 40, so many employed workers, and so on). The polltakers used their own judgment as to whom they selected for the poll. It took a very close election (the 1948 presidential election, in which polls were wrong in their prediction) for the polling organizations to realize that quota sampling was biased.

The quality of your sample depends on having a good frame.

An unbiased sampling method requires that all units in the population have a known chance of being sampled, and so you must prepare a "list" of population units, called a **sampling frame,** or, more simply, **frame,** before you select the sample. If you think about enough real examples, you'll come to see that making this list is not something you can take for granted. For the Westvaco employees, or for the 50 United States, creating the list is not hard, but other populations can pose problems. How would you list all the people using the Internet worldwide or all the ants in Central Park or all the potato chips produced in the United States next year? For all practical purposes, you can't. This means there will often be a difference between the population—the set of units you want to know about—and the sampling frame—the list of units you use for creating your sample. A sample might represent the units in the *frame* quite well, but how well your sample represents your *population* depends on how well you've chosen your frame. Quite often, a convenient frame fails to cover the population of interest (using a telephone directory to sample residents of a neighborhood, for example), and so a bias is introduced by this incomplete coverage. If you start from a bad frame, even the best sampling methods can't save you: *bad frame, bad sample.*

Discussion: Sample Selection Bias

D6. In which of the following situations can you reasonably generalize from the sample to the population?

a. You use your statistics class to estimate the percentage of students in your school who study at least two hours a night.

b. You use the average annual income of the ambassadors to the United Nations to estimate the average per capita income for the world as a whole.

c. In 1996, a Gallup poll sampled 235 U.S. residents aged 18 to 29 to estimate the percentage of all U.S. residents in that age bracket who favored cuts in social spending.

D7. You want to know the percentage of voters who favor state funding for bilingual education. Your population of interest is the set of people likely to vote in the next election. You take as your frame the phone book listing of residential telephone numbers. How well do you think the frame represents the population? Are there important groups of individuals who belong to the population but not to the frame? To the frame but not to the population? If you think bias is likely, say what kind of bias and how it might arise.

■ Practice

P1. Describe the type of sample selection bias that would result from each of these sampling methods.

a. A student wants to determine the average size of farms in a county in Iowa. He drops some rice randomly on a map of the county and uses the farms hit by grains of rice as the sample.

b. In a study about whether valedictorians "succeed big in life," a professor "traveled across Illinois, attending high school graduations and selecting 81 students to participate. . . . He picked students from the most diverse communities possible, from little rural schools to rich suburban schools near Chicago to city schools."

Source: Michael Ryan, "Do Valedictorians Succeed Big in Life?" *Parade Magazine*, May 17, 1998, pp. 14–15.

c. To estimate the percentage of students who passed the first Advanced Placement Statistics exam, a teacher on an Internet discussion list for teachers of AP Statistics asked teachers on the list to report to him how many of their students took the test and how many passed.

d. To find the average length of the pieces of string in a bag, a student reaches in, mixes up the strings, selects one, mixes them up again, selects another, and so on.

e. In 1984, Ann Landers conducted a poll on the marital happiness of women by asking women to write to her.

P2. Suppose the Museum of Fine Arts in Boston wants to know what proportion of people who come to Boston from out of town planned their trip to Boston mainly to visit the museum. The sample will consist of all out-of-town visitors to the museum on several randomly selected days. When a person buys a ticket to the museum, he or she will be asked whether they came from out of town and, if so, what was the main reason for their trip. Do you expect the museum's estimate to be too high, too low, or just about right? Why?

Response Bias

Bias doesn't always come from the sampling method.

In all the examples so far, bias has come from the method of selecting the sample. Unfortunately, bias from other sources can contaminate data even from well-chosen samples.

Perhaps the worst case of faulty data is no data at all. It isn't uncommon for 40% of people contacted to refuse to answer a survey. These people may be different from those who agree to participate. A recent example of **nonresponse bias** came from a controversial study that found that left-handers died, on average, about nine years earlier than right-handers. The investigators sent questionnaires to the families of everyone listed on the death certificates in two counties near Los Angeles asking about the handedness of the person who died. One critic noted that only half of the questionnaires were returned. Did that change the results? Perhaps.

Nonresponse bias can occur when people do not respond to surveys.

Source: "Left-Handers Die Younger, Study Finds," *Los Angeles Times,* April 4, 1991.

Questionnaire bias

Nonresponse bias, like bias that comes from the sampling method, arises from *who* replies. **Questionnaire bias** arises from *how* you ask the questions.

The opinions people give can depend on the tone of voice of the interviewer, the appearance of the interviewer, the order in which the questions are asked, and many other factors. But the most important source of this bias is in the wording of the questions. This is so important that those who report the results of surveys should always provide the exact wording of the questions.

For example, *The Reader's Digest* commissioned a poll to determine how the wording of questions affected people's opinions. The same 1031 people were asked to respond to these two statements.

I. I would be disappointed if Congress cut its funding for public television.

II. Cuts in funding for public television are justified as part of an overall effort to reduce federal spending.

*"One final question: Do you now own or have
you ever owned a fur coat?"*

Note that agreeing with the first statement is pretty much the same as disagreeing with the second. However, 54% agreed with the first statement, 40% disagreed, and 6% didn't know; 52% agreed with the second statement, 37% disagreed, and 10% didn't know. [Source: Fred Barnes, "Can You Trust Those Polls?" *The Reader's Digest,* July 1995, pp. 49–54.]

Incorrect response

Another problem that polls and surveys have is trying to ensure that people tell the truth. Dave Barry, a newspaper columnist, reported that he was called by Arbitron, an organization that compiles television ratings. Dave reports:

So I figured the least I could do, for television, was be an Arbitron household. This involves two major responsibilities:

1. Keeping track of what you watch on TV.

2. Lying about it.

At least that's what I did. I imagine most people do. Because let's face it: Just because you watch a certain show on television doesn't mean you want to *admit* it.

Source: *Dave Barry Talks Back* by Dave Barry, copyright © 1991 by Dave Barry. Used by permission of Crown Publishers, a division of Random House, Inc.

Bias from incorrect responses may be the result of intentional lying, but it is more likely to come from inaccurate measuring devices, including inaccurate memories of people being interviewed in self-reported data. Patients in medical studies are prone to overstate how well they have followed the physician's orders, just as many people are prone to understate the amount of time they actually spend watching TV. Measuring the heights of students with a meter stick that has one end worn off leads to a measurement bias, as does weighing people on a bathroom scale that is adjusted to read on the light side.

Discussion: Response Bias

D8. Like Dave Barry, people generally want to appear knowledgeable and agreeable, and they want to present a favorable face to the world. How might that affect the results of a survey conducted by a school on the satisfaction of its graduates with their education?

D9. Another part of *The Reader's Digest* poll asked Americans if they agree with the statement that it is not the government's job to financially support television programming. The poll also asked them if they'd be disappointed if Congress cut its funding for public television. Which question do you think brought out more support for public television?

D10. How is nonresponse bias different from voluntary response bias?

■ Practice

P3. Consider this pair of questions related to gun control:

   I. Should people who want to buy guns have to pass a background check to make sure they have not been convicted of a violent crime?

   II. Should the government interfere with an individual's constitutional right to buy a gun for self-defense?

   Which question is more likely to show a higher percentage of people who favor some control on gun ownership?

P4. In one study, educators were asked to rank Princeton's undergraduate business program. Every educator rated it in the top ten in the country. Princeton does not have an undergraduate business program. What kind of bias is shown in this case?

   Source: Anne Roark, "Guidebooks to Colleges Get A's, F's," *Los Angeles Times*, November 21, 1982, Part I, p. 25.

Summary 4.1: Samples and Bias

The *population* is the set of individuals you want to know about. The *sample* is the set of individuals you choose to examine. A *census* is an examination of all individuals in the entire population. The following are important reasons to use a sample in many situations rather than taking a census:

- Testing sometimes destroys the items.
- Sampling can save money.
- Sampling can save time.
- Sampling can make it possible to collect more, or better, information on each individual.

A sampling method is biased if it tends to give results that, on average, are too low or too high. This can happen if the method of selecting the sample or the method of getting a response is flawed.

Sources of bias from the method of selecting the sample include

- using a method that gives larger individuals a bigger chance of being in the sample (size bias)
- letting people volunteer to be in the sample
- using a sampling method just because it is convenient
- selecting the sample by using "expert" judgment
- constructing an inadequate sampling frame

Types of bias derived from the method of getting the response from the sample include

- nonresponse bias
- questionnaire bias
- incorrect response or measurement bias

■ Exercises

E1. Suppose you want to know what percentage of U.S. households have children under the age of 13 living at home. Each weekday from 9 to 5, your polltakers call households in your sample. Every time they reach a person in one of the homes, they ask, "Do you have children under the age of 13 living in your household?" Eventually you give up on the households that cannot be reached.

  a. Will your estimate of the percentage of U.S. households with children under the age of 13 probably be too low, too high, or about right?

  b. How does this example help explain why polltakers are likely to call at dinnertime?

E2. Following the gun control example of P3 as a guide, make up two versions of a question about a controversial issue, with one version designed to get a higher percentage in favor, the other designed to get a higher percentage opposed. Test your questions on a sample of 20 people.

E3. Suppose you want to know the average response to the question "Do you like math?" on a scale from 1, "Not at all," to 7, "Definitely," for all students in your school. If you use your statistics class as a convenience sample, what sort of bias, if any, would be likely? Be as specific as you can. In particular, tell whether you expect the sample average to be higher or lower than the population average, and why.

E4. For a study on smoking habits, you want to know the proportion of adult males in the United States who are nonsmokers, who are cigarette smokers, and who are pipe or cigar smokers. Tell why it makes more sense to use a sample than to try to check every individual. What types of bias might appear when you attempt to collect this information?

E5. You want to estimate the average number of states that people living in the United States have visited. If you asked only those at least 40 years old, would you expect the estimate to be too high or too low? What kind of bias might you expect if you select your sample from only those living in Rhode Island?

E6. To estimate the average number of children per family in the city where you live, you use your statistics class as a convenience sample. You ask each student in the sample how many children there are in his or her family. Do you expect the sample average to be higher or lower than the population average? Explain why.

E7. In Section 1.1 on page 5, there is a list of data on the 50 employees of the Engineering Department at Westvaco where Robert Martin worked. How would you choose a sample of 20 employees that is representative of all 50 with regard to all the features listed: pay type, year of birth, year of hire, and whether laid off?

E8. "Television today is more offensive than ever, say the overwhelming majority—92%—of readers who took part in *USA Weekend*'s third survey measuring attitudes toward the small screen." More than 21,600 people responded to this write-in survey.
Source: *USA Weekend*, May 16–18, 1997, p. 20.

a. Do you trust the results of the survey? Why or why not?

b. What percentage of the entire U.S. television-watching public do you think would say that "today's shows are more offensive than ever": more than 92%, quite a bit less than 92%, or just about 92%? Why do you think that?

E9. At a meeting of local Republicans, the organizers want to estimate how well their party's candidate will do in their district in the next race for Congress. They use the people present at their meeting as their sample. What bias do you expect?

E10. Suppose you wish to estimate the average size of English classes on your campus. Compare the merits of these two sampling methods.

I. You get a list of all students enrolled in English classes, take a random sample of those students, and find out how many students are enrolled in each sampled student's English class.

II. You get a list of all English classes, take a random sample of those classes, and find out how many students are enrolled in each sampled class.

E11. Bring to class an example of a survey from the media. Identify the population and the sample, and discuss why a sample was used rather than a census. Do you see any possible bias in the survey method?

E12. A Gallup/CNN/*USA Today* poll asked this question:

As you may know, the Bosnian Serbs rejected the United Nations peace plan and Serbian forces are continuing to attack Muslim towns. Some people are suggesting that the United States conduct air strikes against Serbian military forces, while others say we should not get militarily involved. Do you favor or oppose U.S. air strikes?

On the same day, the ABC News poll worded its question this way:

> Specifically, would you support or oppose the United States, along with its allies in Europe, carrying out air strikes against Bosnian Serb artillery positions and supply lines?

Source: *Los Angeles Times,* May 20, 1993.

Explain which poll you think got a larger response favoring the air strikes.

4.2   Randomizing: Playing It Safe by Taking Chances

The best-planned surveys leave a lot to chance. The key idea is to **randomize:** Let chance choose the sample. This may seem like a paradox at first, but it makes sense once you understand a basic fact about chance-like behavior: *Choosing your sample by chance is the only method guaranteed to be unbiased.*

Relying on chance helps protect against bias.

You've seen how some common sampling methods are biased; that is, they tend to give nonrepresentative samples. Those who care enough to respond voluntarily may not be typical of others. The members of your class, although they're convenient to survey, don't represent some groups well at all; and even relying on expert judgment doesn't work as well as we'd like. Over the long run, chance beats all these methods.

Simple Random Samples

When each unit in the population has a fixed probability of ending up in the sample, we call it a **probability sample.** One type of probability sample is a simple random sample or an SRS.

In a simple random sample, each unit or group of units of the same size has the same chance of being selected.

> In **simple random sampling,** all possible samples of a given fixed size are equally likely. All units have the same chance of belonging to the sample. All possible pairs of units have the same chance of belonging to the sample; all possible triples of units have the same chance; and so on.

The simplest way to let chance choose your simple random sample amounts to essentially putting all the individuals from your population into a gigantic hat, mixing them thoroughly, and drawing individuals out, one at a time, until you have enough for your sample. Although this is exactly right in theory, in practice it's almost impossible to mix thoroughly, so it's actually better to implement the theory using random numbers.

Steps in Choosing a Simple Random Sample

1. Start with a list of all the units in the population.

2. Number the units in the list.

3. Use a random number table or generator to choose units from the numbered list, one at a time, until you have as many as you need.

Do you think you can select a representative sample as well as a random number table can? Activity 4.2 provides an opportunity for you to test yourself.

Activity 4.2   Random Rectangles

What you'll need: Display 4.4, a method of producing random digits

The goal is to choose a sample of 5 rectangles from which to estimate the average area of the 100 rectangles in the display.

1. Without studying the display of rectangles too carefully, quickly choose five that you think represent the population of rectangles on the page. This is your judgment sample.

2. Find the areas of each rectangle in your sample of five and compute the sample mean, that is, the average area of the rectangles in your sample.

3. List your sample mean with those of other students in the class. Construct a plot of the means for your class.

4. Describe the shape, center, and spread of this plot of sample means from the judgment samples.

5. Now generate five distinct random numbers between 00 and 99. (The rectangle numbered 100 can be called 00.) Find the rectangles that correspond to your random numbers. This is your random sample of five rectangles.

6. Repeat steps 2–4, this time using your random sample.

7. Discuss how the two distributions of sample means are similar and how they differ.

8. Which method of producing sample means do you think is better if the goal is to use the sample mean to estimate the population mean? (Your instructor has the value for the population mean.)

Keep your sample data on rectangle areas for future reference.

Display 4.4 Random rectangles.

Discussion: Simple Random Samples

D11. Does the judgment sampling procedure used in Activity 4.2 produce a simple random sample of rectangles? Did the method used in Activity 4.1, Part A, produce a simple random sample? How about in Part B?

D12. Use a table of random digits to select a simple random sample of six students from your class. Does the sample appear to be representative?

D13. Describe how you could obtain a simple random sample of all students enrolled in the English classes at your school.

D14. *Readability.* You've decided you simply *must* know the proportion of capital letters in this textbook because it can indicate the complexity of sentences on a page. Think of the book itself as your population, with each printed character as a unit. Describe how to set up your frame and how to use it to get a simple random sample of 10,000 characters. What are the advantages and disadvantages of this sampling method?

■ Practice

P5. Decide if the following sampling methods produce a simple random sample of students from a class of 30 students. If not, explain why.

a. Select the first 6 students on the class roll sheet.

b. Pick a digit at random and select those students whose phone number ends in that digit.

c. If the classroom has 6 rows of chairs with 5 seats in each row, choose a row at random and select all students in that row.

d. If the class consists of 15 males and 15 females, assign the males the numbers from 1 to 15 and the females the numbers from 16 to 30. Then use a random digit table to select six numbers from 1 to 30. Select the students assigned those numbers for your sample.

e. If the class consists of 15 males and 15 females, assign the males the numbers from 1 to 15 and the females the numbers from 16 to 30. Then use a random digit table to select three numbers from 1 to 15 and three numbers from 16 to 30. Select the students assigned those numbers for your sample.

f. Randomly choose a letter of the alphabet and select those students for the sample whose last name begins with that letter. If no last name begins with that letter, randomly choose another letter of the alphabet.

Stratified Random Samples

Suppose you are planning a sample survey for an international outdoor clothing manufacturer to see if their image has suffered among their retailers due to negative press coverage regarding their use of sweatshop labor. The company headquarters furnishes you with a list of retail outlets throughout the world that sell their products. Because press coverage is largely national, it will be more informative to organize this list by country and take a random sample from each country. You'll be gathering data using each country's phone or postal system in the language of that country, so it also will be more convenient to categorize the retail outlets by country.

Classifying the retail outlets by country is called **stratification.** If you can divide your population into subgroups that do not overlap and that cover the entire sampling frame, those subgroups are called **strata.** If the sample you take within each subgroup is a simple random sample, your result is a **stratified random sample.** This is a second type of probability sample.

> ### Steps in Choosing a Stratified Random Sample
> 1. Divide the units of the sampling frame into non-overlapping subgroups.
> 2. Choose a simple random sample from each subgroup.

Why stratify?

Why might you want to stratify a population? Here are the three main reasons.

- *Convenience* in selecting the sample is enhanced. It is easier to sample in smaller, more compact groups (countries) than in one large group spread out over a huge area (the world).

- *Coverage* of each stratum is assured. The company may want data from each country in which it sells products; a simple random sample from the frame does not guarantee that this would happen.

- *Precision* of the results may be improved. That is, stratification tends to give estimates that are closer to the value for the entire population than does an SRS. This is the fundamental statistical reason for stratification.

Stratification reduces variability.

Another example might help clarify the last point. A geologist has a box of rocks from a mountain stream in Mexico. Among other things, she wants to estimate the mean diameter of these discus-shaped rocks so that it can be compared to the mean diameter of similar rocks from other locations. She has a sieve with a 2-inch mesh, which allows through only rocks under 2 inches in diameter. To sieve or not to sieve before estimating, that is the question. In other words, is it better to stratify the rocks into two groups (small and large) and then sample from each stratum, or is it just as good to sample from the pile of rocks without sieving?

To simplify the problem, consider a population of 100 rocks. After sieving, 50 turn out to have diameters of less than 2 inches. Display 4.5 shows the population of diameters and that population divided into the two strata.

Twenty simple random samples, each of size 4, were taken from the population, and the sample mean was calculated for each. A plot of the 20 sample means is shown in the top half of Display 4.6. Then 20 stratified random samples were taken with each stratified sample consisting of a simple random sample of size 2 from the small rocks mixed with a simple random sample of size 2 from the large rocks. A plot of these 20 sample means is shown in the bottom half of Display 4.6. Note that in both sampling scenarios we have 20 samples, each of size 4.

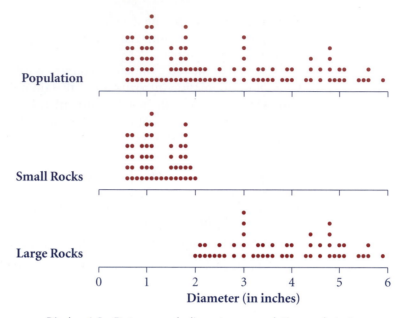

Display 4.5 Data on rock diameters; population and strata.

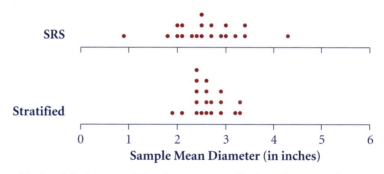

Display 4.6 Means of simple versus stratified random samples.

Make the strata as different as possible.

From this example, it certainly looks as though the stratification (sieving) pays off in producing estimates of the mean with smaller variation. This will be true generally if the stratum means are quite different. If the geologist had decided to stratify on color, and if color was not related to diameter, then the stratification would have produced little or no gain in precision of the estimates. So the guiding principle is to choose strata that have very different means, whenever that is possible.

Allocate units in the sample proportionally to the number of units in the strata.

One good way to choose the relative sample sizes in stratified random sampling is to make them proportional to the stratum sizes (the number of units in the stratum). Thus, if the strata are all of the same size, the samples should all be of the same size. If a population is known to have 65% women and 35% men, then a sample of 100 people stratified on gender should contain 65 women and 35 men. If the sample sizes are proportional to the stratum sizes, then the overall sample mean (the mean calculated from the samples of all strata mixed together) will be a good estimate of the population mean. Proportional allocation is particularly effective in reducing the variation in the sample mean if the stratum standard deviations are about equal.

Discussion: Stratification

D15. What are the main differences between simple random sampling and stratified random sampling?

D16. Suppose the geologist had a set of sieves that could divide the rocks into many small piles with essentially equal-diameter rocks in each pile. Would this process improve the precision of estimating the mean diameter over the two-strata scenario discussed earlier? Why or why not?

D17. An administrator wants to estimate the average amount of time high school students spend traveling to school. The plan is to stratify the students according to grade level and then take a simple random sample from each grade. What is potentially good and what is potentially bad about this plan for stratification?

D18. Your assignment is to estimate the mean number of hours per week spent studying by students in your school. Discuss how you would set up a stratified random sampling plan to accomplish the task.

Other Methods of Sampling

Simple random samples and stratified random samples offer many advantages, but you often run into a major practical problem: *Choosing sampling units one at a time is often too costly, too time consuming, or, if there isn't a good frame available, simply not possible.* Several compromise plans get around this problem by working with groups of individuals rather than the individuals themselves. Although choosing the individuals gives more thorough mixing (randomization) —which is the main reason to prefer the SRS—the compromise plans often work well.

Cluster Samples

A **cluster sample** is an SRS of clusters of units.

To see how well U.S. third graders do on an arithmetic test, you might choose a simple random sample of children enrolled in the third grade and give each child a standardized test. In theory, this is a reasonable plan, but it is not very practical. For one thing, how would you go about making a complete list of all the third graders in the United States? For another, imagine the work required to track down each child in your sample and get him or her to take the test. Instead of a simple random sample of third graders, it would be much better to take an SRS of elementary schools and then give the test to all the third graders in those schools. It's a lot easier to get a list of all the elementary schools in the United States than a list of all the individuals enrolled in the third grade. Moreover, once you've chosen your sample of elementary schools, it's a relatively easy organizational problem to give the test to all third graders in those schools. This example illustrates cluster sampling, in which each elementary school is a cluster of third-grade students.

> ### Steps for Choosing a Cluster Sample
>
> 1. Create a numbered list of all of the clusters in your population.
> 2. Choose a simple random sample of clusters.
> 3. Obtain data on each individual in each cluster in your SRS.

The example illustrated the two main reasons for using cluster samples: You need only a list of clusters rather than a list of individuals, and for some studies, it is much more efficient to gather data on individuals grouped by clusters than on individuals one at a time.

Two-Stage Samples

In a two-stage sample, first take a cluster sample, then take an SRS from each cluster.

A variation on the cluster sampling procedure for selecting a sample of third graders for the arithmetic test would be to start the same way, by taking an SRS of elementary schools (the clusters), but then taking an SRS of third graders at each elementary school chosen. This is an example of **two-stage sampling** because it involves two randomizations.

> ### Steps for Choosing a Two-Stage Sample
>
> 1. Create a numbered list of all the clusters in your population, and then choose a simple random sample of clusters.
> 2. Create a numbered list of all the individuals in each cluster already selected, and then choose an SRS from each cluster.

Reasons to use a two-stage sample

Two-stage samples are useful when it is much easier to list clusters than individuals but still reasonably easy and sufficient to sample individuals once the clusters are chosen.

Systematic Samples with Random Start

Systematic samples can start with counting off.

To get a quick sample of the students in your class, use the common system for choosing teams by "counting off." Count off by, say, eights, and then select a digit at random between 1 and 8. Every student who calls out that digit is in the sample. This is an example of a **systematic sample with random start.**

> ### Steps for Choosing a Systematic Sample with Random Start
>
> 1. By a method such as counting off, divide your population into groups of the size you want for your sample.
> 2. Use a chance method to choose one of the groups for your sample.

Suppose your population units are in a list, such as a list of names in a directory, and you want to select a sample of a certain size. These steps are equivalent to choosing a random starting point between units 1 and *k*, then taking every *k*th unit thereafter. The value of *k* is determined by the sample size. For example, let's say you want a systematic sample of 20 units selected from a list of 1000 units. Then *k* is 50: Selecting a random start between 1 and 50 and then taking every 50th unit thereafter will result in a systematic sample of 20 units.

When the units in the population are well mixed before the counting off, this method can be a good one to use. Systematic samples, like cluster samples, are easier to select than simple random samples of the same size.

Discussion: Other Methods of Sampling

D19. Is a systematic sample with random start a type of simple random sample?

D20. *More readability.* Reread D14, which asked about using an SRS to estimate the proportion of capital letters in this book.

    a. Now, instead of an SRS, consider taking a cluster sample. What would you use for your clusters? What are the advantages of the cluster sample? Are there any disadvantages?

    b. Suppose you want to estimate the average number of capital letters per printed line, so now your sampling unit is a line. Describe how to take a two-stage sample from your textbook.

    c. Now suppose your sampling unit is an individual printed character. Tell how to take a *three*-stage sample of characters from your book.

D21. A bank keeps its list of home mortgages in chronological order, from the earliest to the latest in terms of the date they were assumed. An auditor wants to get a quick check on the average amount of these mortgages by taking a sample. What sampling design would you suggest?

D22. Both cluster sampling and stratified random sampling involve viewing the sampling frame as a collection of subgroups. Explain the difference between these two types of sampling.

D23. Select a systematic sample of five students from your class.

■ Practice

P6. Suppose 200 people are waiting in line for tickets to a rock concert. You are working for the school newspaper and want to interview a sample of the people in line. Show how to select a systematic sample of 5% of the people in line, and then of 20% of the people in line.

P7. The American Statistical Association directory lists its roughly 17,000 members in alphabetical order. You want a sample of about 1000 names. Tell how to use the alphabetical listing to take a systematic sample with random start.

P8. A newspaper reports that about 60% of the cars in your community come from manufacturers based outside the United States. How would you design a sampling plan for getting the data to substantiate or refute this claim?

P9. The *Los Angeles Times* commissioned an analysis of St. John's wort, an over-the-counter herbal medicine. They wanted to know whether the potency of the pills matched the potency claimed on the bottle. The sampling procedure was described as follows: "For the analysis, 10 pills were sampled from each of three containers of one lot of each product."

Source: *Los Angeles Times*, August 31, 1998, p. A10.

a. What type of sampling was this?

b. Can you suggest an improvement in the sampling procedure?

P10. Suppose your population is 65% women and 35% men. In a stratified random sample of 100 women and 75 men, you find that 84 women pump their own gas and 59 men pump their own gas. What is the best estimate of the proportion of the entire population that pumps their own gas?

Summary 4.2: Randomizing

The main reason for using a chance device to choose the individuals for your sample is that randomization protects against sample selection bias. And, as you will see, random selection of the sample will make inference about the population possible.

There are several ways to choose a sample using randomization. Here are some of the most common.

- *Simple random sample*—Number the individuals and use random numbers to select those to be included in the sample.

- *Stratified random sample*—First divide your population into groups, called strata, then take a simple random sample from each group.

- *Cluster sample*—First select clusters at random, then use all the individuals in the clusters as your sample.

- *Two-stage sample*—First select clusters at random, then select individuals at random from each cluster.

- *Systematic sample with random start*—"Count off" your population by a number k determined by the size of your sample, select one of the counting numbers between 1 and k at random, and then the units with that number will be in your sample.

Display 4.7 shows schematically how five common sampling designs might look for individuals identified as points on a plane.

Simple Random Sampling

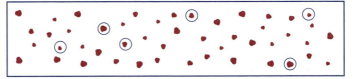

Stratified Random Sampling

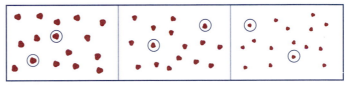

Cluster Sampling

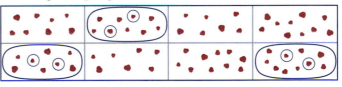

Two-Stage Sampling

Systematic Sampling

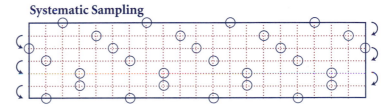

Display 4.7 Various sampling methods.

■ Exercises

E13. You want to find the percentage of people in your area with heart disease who also smoke cigarettes. The people in your area who have heart disease make up your population. You take as your frame all records of patients who were hospitalized in area hospitals within the last five years with a diagnosis of heart disease. How well do you think this frame represents the population? If you think bias is likely, say what kind of bias it would be and how it might arise.

E14. **Cookies.** Which brand of chocolate chip cookies gives you the most chips per cookie? For the purpose of this question, take as your population all the cookies now in the nearest supermarket. Each cookie is a unit in this population.

   a. Explain why it would be hard to take an SRS.

   b. Describe how to take a cluster sample of chocolate chip cookies.

   c. Describe how you would get a two-stage sample. What circumstances would make the two-stage sample better than the cluster sample?

E15. **Haircut prices.** You want to take a sample of students in your school to estimate the average amount they spent on their last haircut. Which sampling method do you think would work best: a simple random sample; a stratified random sample with two strata, male and female; or a stratified random sample with class levels as strata? Give your reasoning.

E16. Suppose that a sample consisting of 25 adults contains only women. Two explanations are possible: (1) The sampling procedure wasn't random, or (2) A non-representative sample occurred just by chance. Given no additional information, which explanation would you be more inclined to believe?

E17. An early use of sampling methods was in crop forecasting, especially in India, where an accurate forecast of the jute yield in the 1930s made some of the techniques (and their inventors) famous. Your job now is to help the crop-reporting service estimate the corn yield, right before harvest time, for a county that has five farms and a total of 1000 acres planted in corn. How would you do the sampling?

E18. The *Oxford Dictionary of Quotations,* 3rd edition, has about 600 pages of quotations. Describe how you would get a systematic sample of 30 pages to use for estimating the number of typographical errors per page.

E19. You are called upon to advise a local movie theater on designing a sampling plan for a survey of students in your school to gain information on attitudes about recent movies. The theater has time and money to interview about 50 students. What design would you use?

E20. Quality control plans in industry often involve sampling items for inspection. A manufacturer of electronic relays (switches) for the TV industry wants to set up a quality control sampling plan for these relays as they come off its production line. What sampling plan would you suggest if there is only one production line? If there are five production lines?

4.3   Experiments and Inference About Cause

Inference about cause and effect

The last two sections have shown you how to collect data that will allow you to generalize or infer from the sample you see to a larger but perhaps unseen population. A second kind of inference, perhaps even more important than the first, takes you from a pattern you observe to a conclusion about what *caused* the pattern. Does regular exercise cause your heart rate to go down? Does bilingual education increase the percentage of non-native-English-speaking students who graduate from high school? Will taking a special preparatory course raise your SAT scores? Does smoking cigarettes cause lung cancer?

Cause and Effect

Children who drink more milk have bigger feet than children who drink less milk. There are three possible explanations for this association:

1. Drinking more milk causes children's feet to be bigger.

2. Having bigger feet causes children to drink more milk.

3. A lurking variable is responsible for both.

In this case, it is number 3. The **lurking variable**—one that lies in the background that may or may not be apparent at the outset but once identified could explain the pattern between the variables—is the child's overall size. Bigger children have bigger feet, and they drink more milk because they eat and drink more of everything than do smaller children.

But suppose you think that the explanation is really number 1. How can you prove it? Can you take a bunch of children, give them milk, and then sit and wait to see if their feet grew? That won't prove anything—children's feet will grow anyway, milk or not.

Can you take a bunch of children, randomly divide them into a group that drinks milk and a group that doesn't drink milk, and then sit and wait to see if the milk-drinking group grows bigger feet? Yes! *Such an experiment is just about the only way to establish cause and effect.*

A Real Experiment: Kelly's Hamsters

The goal of every experiment is to establish cause through comparing two or more conditions, called **treatments,** using an outcome variable, called the **response.** Here's an example of a real experiment, planned and carried out by Kelly Acampora as part of her senior honors project in biology at Mount Holyoke College. What happens when an animal gets ready to hibernate? This question is too general to answer with a single experiment, but if you know a little biology, you can narrow the question enough to serve as the focus of an experiment. Kelly relied on three previously known facts:

1. Golden hamsters hibernate.

2. Hamsters rely on the amount of daylight to trigger hibernation.

3. An animal's capacity to transmit nerve impulses depends in part on an enzyme called Na^+K^+ ATP-ase.

Here are the components of Kelly's experimental design:

Kelly's Question: If you reduce the amount of light a hamster gets, from 16 hours to 8 hours per day, what happens to the concentration of Na^+K^+ ATP-ase in its brain?

Subjects: Kelly's subjects were eight golden hamsters.

Treatments: There were two treatments: being raised in long (16-hour) days or short (8-hour) days.

Random assignment of treatments: To make her study a true experiment, Kelly randomly assigned a day length of 16 hours or 8 hours to each hamster in such a way that half the hamsters were assigned to be raised under each treatment.

Replication: Each treatment was given to four hamsters.

Response variable: Because Kelly was interested in whether a difference in the amount of light causes a difference in the enzyme concentration, she chose the enzyme concentration for each hamster as her response variable.

Results: The resulting measurements of enzyme concentrations (in milligrams per 100 milliliters) for the eight hamsters were

Short days:	12.500	11.625	18.275	13.225
Long days:	6.625	10.375	9.900	8.800

You can imagine Kelly defending her design:

Kelly: I claim that the observed difference in enzyme concentrations between the two groups of hamsters is due to the difference in daylight.

Skeptic: Wait a minute. As you can see, the concentration varies from one hamster to another. Some just naturally have higher concentrations. If you happened to assign all the high-enzyme hamsters to the group that got short days, you'd get results like the ones you got.

Kelly: I agree, and I was concerned about that possibility. In fact, that's precisely why I assigned day lengths to hamsters by using random numbers. The random assignment makes it extremely unlikely that all the high-enzyme hamsters would get assigned to the same group. If you have the time, I can show you how to compute the probability.

Skeptic: (*Hastily*) That's OK for now. I'll take your word for it. But maybe you can catch me in Chapter 6.

The characteristics of the plan Kelly used are so important that statisticians try to reserve the word *experiment* for studies like hers that answer a question by comparing results of treatments assigned to subjects at random.

Discussion: A Real Experiment

D24. Plot Kelly's results in a dot plot, using different symbols for hamsters raised in short days and those raised in long days. In a later chapter, you'll see a formal method for analyzing data like these to decide whether the observed difference between the long- and short-day hamsters is too big to be due only to chance. This method will be similar to the sort of logic introduced in the *Martin v. Westvaco* case of Chapter 1. For now, just give your best judgment: Do you think the evidence supports a conclusion that the number of daylight hours causes a difference in enzyme concentration?

D25. Kelly has shown that hamsters raised in less daylight have higher hormone concentration than hamsters raised with more daylight. In order for Kelly to show that less daylight *causes* an increase in the hormone concentration, she must convince us that there is no other explanation. Has she done that?

■ Practice

P11. There are now many special courses that claim to prepare you for the SAT. Suppose you want to evaluate a particular course, using SAT scores to measure the effect of the course. You might find a reasonably large high school where students are offered the chance to take the course, then compare the SAT scores of those who completed the course with the scores of those who chose not to take it. Suppose you find that the average SAT score for students who took the course is 30 points higher than for students who didn't.

   a. Identify each of the elements in this study: population, response variable, treatments.

   b. Is this study a true experiment?

   c. Do you conclude that the course causes an increase in SAT scores? Why or why not?

P12. Each pair of variables shown here is strongly associated. Does I cause II, or does II cause I, or is there a lurking variable responsible for both?

   a. I. wearing a hearing aid

       II. dying within the next ten years

   b. I. the amount of milk a person drinks

       II. the strength of a person's bones

   c. I. the amount of money a person earns

       II. the number of years a person went to school

Confounding in Observational Studies

For sample surveys, selecting the sample at random protects against bias, which otherwise can easily mislead you into jumping to false conclusions. For comparative studies, there is a threat called *confounding* that can also lead to false conclusions. Randomizing, when it is possible, will also protect against this threat in much the same way that it protects against bias in sampling.

In everyday language, "confounded" means mixed up, confused, at a dead end, and the meaning in statistics is similar.

Confounding makes it impossible to say that one thing causes another.

> Two possible influences on an observed outcome are **confounded** if they are mixed together in a way that makes it impossible to separate their effects.

The SAT study in P11 illustrates the important concept of confounding. In the study, the effect of the review course was confounded with the fact that the course was taken only by volunteers, who would tend to be more motivated to do well on the SAT. Consequently, you can't tell if the higher scores of those who took the course were due to the course itself or to the higher motivation of the volunteers. As you read the example that follows, ask yourself which influences are being mixed together: Where's the confounding?

This thymus surgery example offers a cautionary tale about confounding.

The thymus, which is a gland in your neck, behaves in a peculiar way. Unlike other organs of the body, it doesn't get larger as you grow; it actually gets smaller. Ignorance of this fact led early 20th-century surgeons to adopt a worthless and dangerous surgical procedure. Imagine yourself in their situation: You know that many infants are dying of what seem to be respiratory obstructions, and in your search for a treatment, you begin to do autopsies on infants who die with respiratory symptoms. You've done many autopsies in the past on adults who died of various causes, so you decide to rely on these autopsy results for comparison. What stands out most when you autopsy the infants is that they all have thymus glands that look too big in comparison to body size. Aha! That must be it: The respiratory problems are caused by an enlarged thymus. It became quite common in the early 1900s for surgeons to treat respiratory problems in children by removing the thymus. In particular, in 1912, Dr. Charles Mayo, one of the two brothers for whom the Mayo Clinic is named, published an article recommending removal of the thymus. He made this recommendation even though a third of the children who were operated on died. Looking back at the study (see the table), it's easy to spot the confounded variables.

		Age	
		Child	**Adult**
Thymus Size	**Large**	Problems	No evidence
	Small	No evidence	No problems

The doctors can't know whether children with a large thymus tend to have more respiratory problems, because they have no evidence about children whose thymus was smaller. Age and size of thymus were confounded. The thymus study was an example of an observational study, not an experiment.

> In an **observational study,** no treatments get assigned to the subjects by the experimenter—the conditions of interest come already built into the units being studied.

Conditions aren't randomly assigned in an observational study.

Observational studies are less desirable than experiments, but they often are the only game in town. Suppose you want to study accident rates at a dangerous intersection to see if the rate on rainy days is higher than the rate on dry days. It turns out that most of the rainy days you have available for study are weekends and almost all of the dry days are during the week. You cannot assign rain to a weekday! The best you can do is design your observational study to cover a long enough period so that you are likely to have some weekdays and weekends in both groups. Otherwise, you have a serious confounding problem.

For drawing conclusions about cause and effect, a good randomized experiment is nearly always better than a good observational study.

A randomized comparative experiment comparing medical treatments is called a clinical trial.

These days, any new medical treatment must prove its value in a **clinical trial**—a randomized comparative experiment. Patients who agree to be part of the study know that a chance method will be used to decide whether they get the standard treatment or the new experimental treatment. If only Dr. Mayo had used such a randomized experiment to evaluate surgical removal of the thymus, he would have seen that the treatment was not effective and many lives might have been spared. However, at the time, randomized experiments were not often used in the medical profession.

Factors and Levels

The conditions that affect the response variable are called factors.

In an experiment, the conditions assigned are called treatments. In an observational study, because nothing is "treated," the word *factor* is used for the conditions. For example, in the thymus study, the factor is the size of the thymus. In this case, the factor has two **levels,** large and small. This terminology is also used for experiments, especially when several types of conditions make up the treatments. For example, suppose Kelly decided to add two different diets to her experiment (call them light and heavy). Then her experiment would have two factors, diet (light or heavy) and length of day (short or long), and therefore four possible treatments (light/short, light/long, heavy/short, and heavy/long), as diagrammed in the table.

		Factor 1 Type of Diet	
		Light	**Heavy**
Factor 2 Length of Day	**Short**	Light–Short	Heavy–Short
	Long	Light–Long	Heavy–Long

Discussion: Confounding and Observational Studies

D26. Suppose the surgeons had examined infants without respiratory problems and found that their thymus was generally small.

   a. Have they now proved that a large thymus *causes* respiratory problems in children? If so, why? If not, what is another possible explanation?

   b. Design an experiment to determine if removal of the thymus helps children with respiratory problems.

D27. What is the main difference between an experiment and an observational study?

D28. Does the type of lighting or the type of music in a dentist's waiting room have any effect on the anxiety level of the patient? Consider an experiment to study this question. There could be nine treatments in this experiment, represented by the nine cells in the table.

		Type of Music		
		Pop	**Classical**	**Jazz**
Brightness of Room	**Low**			
	Medium			
	High			

a. What are the factors? What are the levels?

b. Describe a possible response variable.

■ Practice

P13. Show the confounding of the SAT study in P11 by drawing and labeling a table, as in the thymus example.

P14–15. Decades ago, when there was less agreement than there is now about the effects of smoking on health, a large study compared death rates for three groups of men—nonsmokers, cigarette-smokers, and pipe- or cigar-smokers. The results are shown in the table.

	Deaths (per 1000 men per yr)
Nonsmokers	20.2
Cigarette-smokers	20.5
Pipe- or cigar-smokers	35.5

Source: Paul R. Rosenbaum, *Observational Studies* (New York: Springer-Verlag, 1995).

P14. The numbers seem to say that smoking a pipe or cigars or both almost doubles the death rate from about 20 to 35 per 1000, but smoking cigarettes is pretty safe.

a. Do you believe that? If not, can you suggest a possible explanation for the pattern?

b. Is this study an observational study or an experiment?

c. What is the factor? What are the levels? What is the response variable?

P15. The investigators had also recorded the ages of the men in the study so that it was possible to compare average ages for the three groups:

	Average Age (yr)
Nonsmokers	54.9
Cigarette-smokers	50.5
Pipe- or cigar-smokers	65.9

Does this information help you account for the pattern? If so, tell how. If not, tell why not. Now what are the factors in this study?

P16. ***Driving age.*** You want to know whether raising the minimum age for getting a driver's license will save lives. You compare the 1995 highway death rates for the 50 U.S. states, grouped according to the legal driving age.

  a. Is this study observational or experimental?

  b. What is the factor? What are the levels? What is the response variable?

  c. Think of a possible confounding variable.

P17. ***Popcorn.*** How much popped corn is produced by a generic popcorn as compared to a brand name? Does the type of popper make a difference? You are to compare the volume of popped corn produced by a generic and a brand-name popcorn, using both a hot air and an oil popper. What are the factors and levels? Describe the treatments. Can you think of any possible confounded variables?

The Importance of Randomizing

Randomizing can protect against confounding.

Confounding is the main threat to making reliable inferences about cause. You can think of confounding as the two groups you want to compare differing in some important way other than the relevant response variable. The nonsmokers, cigarette-smokers, and pipe- or cigar-smokers of P14 differed not only in death rate (the response variable) but also in age (a confounding variable). How can you guard against confounding? The best solution, when it is possible, is to randomize: Use a chance device to decide which people, animals, or objects to assign to each set of conditions you want to compare. *If you don't randomize, it's risky to generalize.*

Wait a minute! Are statisticians asking you to believe that you should assign people to smoke or not on the basis of a coin toss? Of course not. But it *is* true that if you can't use a chance device to assign the conditions you want to compare, like smoker and nonsmoker, it becomes very difficult to draw sound conclusions about cause and effect.

For the SAT study of P11, it *is* possible to assign treatments. You can start with a group of students who want to take the special course. Randomly assign half to take the course and the other half to a group that will not get any special preparation. Then after both groups of students take the SAT, compare their scores. This time, if the students who take the course do 30 points better on average, you can be more confident that the difference is due to the course rather than to some confounding influence.

Your goal in Activity 4.3 is to gain practice with randomization as it is used to design experiments and to see the effects of randomization.

Activity 4.3   Randomization and Its Effect

What you'll need: a box, equally sized slips of paper for a random drawing, a coin

The students in your class are to be the subjects in an experiment with two treatments, A and B. The task is to find the best of three ways to assign the treatments to the subjects.

1. Choose two leaders for the class, one for treatment A and one for treatment B. The leaders should flip a coin to decide who goes first, then alternately choose class members for their teams (much like choosing sides for a softball game).

2. For each treatment group, record in a table the number of subjects, the proportion of females, the proportion who have brothers or sisters, and the proportion who like to read novels in their spare time. Do the two groups look quite similar or quite different?

3. Next, divide the class by writing your names on the pieces of paper and putting them in a box; randomly draw out half of them (one by one), to be assigned treatment A. The names remaining in the box get treatment B.

4. Repeat step 2 for these groups.

5. Finally, divide the class by having each person flip a coin. Those getting heads go to treatment A, and those getting tails go to treatment B.

6. Repeat step 2 for these groups.

7. What are the strengths and weaknesses of each method of assigning treatments to subjects? Which method is least random?

A Control or Comparison Group Is Vital

Anecdotal evidence isn't proof.

An article on homeopathic medicine in *Time* magazine ["Is Homeopathy Good Medicine?" Sept. 25, 1995, pp. 47–48] began with an anecdote about a woman who had been having pain in her abdomen. She was told to take calcium carbonate, and after two weeks, her pain had disappeared. It was reported that her family "now turns first" to homeopathic medicine. This kind of personal evidence—"It worked for me"—is called anecdotal evidence.

Anecdotal evidence can be useful in deciding what treatments might be helpful and so should be tested in an experiment. However, anecdotal evidence such as this cannot *prove* that calcium carbonate causes abdominal pain to disappear. Why not? The pain did go away.

The problem is that pain often tends to go away anyway, especially when a person thinks he or she is getting good care. When people believe they are receiving special treatment, they tend to do better. In medicine, this is called the **placebo effect.**

Carefully conducted studies show that placebos (something that looks like a pill but contains no medicine) alleviate symptoms in 10% to 70% of people, depending on their condition. [Source: "The Placebo Effect: Gauging the Mind's Role in Healing," *USA Today,* June 15, 1993, p. 6D.] The placebo effect is another example of confounding. Do people get better because of the medical treatment or simply because they are being treated by someone they trust?

> A control group or comparison group provides the basis of comparison to evaluate the effectiveness of a treatment.

How then can scientists determine if a medication is effective or whether improvement is due to the placebo effect? They must design an experiment with a **control group** of individuals who provide a standard for comparison.

The patients in the control group are often given a placebo, a nontreatment carefully designed to be as much like the actual treatment as possible. Patients in the treatment group are given the drug to be evaluated. The control and treatment groups should be handled exactly alike, except for the treatment itself. If a new treatment is to be compared with a standard treatment, the control group, sometimes called a **comparison group,** receives the standard treatment rather than a placebo.

In order for the control group and the treatment group to be treated exactly alike, both the subjects and their doctors should be "blind." That is, the patients should not know which treatment they are receiving, and the doctors who evaluate how much the patients' symptoms are relieved should be blind as to what the patients received. If only the patients don't know, the experiment is said to be **blind.** If both the patients and the doctors don't know, the experiment is said to be **double-blind.**

> Randomized comparative experiment

To summarize, a good experiment must have both a random assignment of treatments to subjects and a control or comparison group that is compared to the group getting the treatment of interest. Such an experiment is called a **randomized comparative experiment.**

Discussion: Control Groups and Blinding

D29. Why is blinding or double-blinding important in an experiment?

D30. A report about a new study to test the effectiveness of the herb St. John's wort to treat depression says, "The subjects will receive St. John's wort, an antidepressant drug, or a placebo for at least eight weeks and as long as six months." [Source: *Los Angeles Times,* Aug. 31, 1998, p. A10.] Describe how you would design this study to compare the effects of the three treatments.

D31. *Bears in space.* Congratulations! You have just been appointed Director of Research for Confectionery Ballistics, Inc., a company that specializes in launching gummy bears from a ramp using a launcher made of tongue depressor sticks and rubber bands. The CEO has asked you to study a variety of factors that are thought to affect launch distance. Your first assignment is to study the effects of the color of the bears. Do green bears travel farther?

Display 4.8 shows actual data for bears launched by members of a statistics class at Mount Holyoke College. Each launch team did ten launches, the first set of five launches using red bears and the second set of five using green bears.

	Launch	Team 1	Team 2	Team 3	Team 4	Team 5	Team 6	Median	Mean	SD
Red Bears	1	15	44	18	13	10	125	**16.5**	**37.5**	**44.6**
	2	24	40	2	39	35	147	**37.0**	**47.8**	**50.6**
	3	48	51	10	16	24	35	**29.5**	**30.7**	**16.8**
	4	25	100	41	41	22	81	**41.0**	**51.7**	**31.7**
	5	19	37	88	41	65	125	**53.0**	**62.5**	**38.9**
Green Bears	6	17	52	46	31	45	27	**38.0**	**36.3**	**13.4**
	7	19	102	72	41	70	187	**71.0**	**81.8**	**58.9**
	8	21	86	53	55	21	84	**54.0**	**53.3**	**28.6**
	9	31	89	38	14	33	105	**35.5**	**51.7**	**36.4**
	10	74	120	92	33	37	174	**83.0**	**88.3**	**53.4**
Median		**22.5**	**69.0**	**43.5**	**36.0**	**34.0**	**115.0**			
Mean		**29.3**	**72.1**	**46.0**	**32.4**	**36.2**	**109.0**			
SD		**18.4**	**30.5**	**31.1**	**14.0**	**19.2**	**53.6**			

Display 4.8 Sample launch data for red and green bears.

a. Which color bear went farther?

b. Your CEO at Confectionery Ballistics, Inc., wants to reward you for discovering the secret of longer launches. Explain why his enthusiasm is premature.

Plot your data over time to check for confounding trends.

c. Your CEO is adamant: Color is the key to better launches. In vain you argue that color and launch order are confounded. Finally, your CEO issues an executive order: "Prove it. Show me data." Determined to meet the challenge, you remember the basic rule of statistics: *Plot your data first.* Divide your class into groups and make a time plot for the results from the team assigned to your group. Discuss what your plots show.

d. How would you change the design of the experiment to eliminate the confounding?

■ Practice

P18. In Dr. Mayo's thymus studies, what did he use as a control group? Could the placebo effect have been a factor in any successful surgeries he had? Was it possible for the study to be blind? Was it possible for the doctors who evaluated how well the patients were doing after surgery to be "blind"?

P19. In a study to see if people have a "magnetic sense"—the ability to use the Earth's magnetism to tell direction—students were blindfolded and driven around in a van over winding roads. Then they were asked to point in the direction of home. They were able to do so better than could be explained by chance alone.

Source: "Tests Point Away from 6th Sense," *Los Angeles Times*, July 19, 1982.

a. Are you satisfied with the design of this study? Are there any other possible explanations for the results other than a magnetic sense?

To improve the design, the experimenter then placed magnets on the backs of some students' heads, which were supposed to confuse their magnetic sense, and nonmagnetic metal bars on the backs of other students' heads.

b. What is the control group in the new design?

c. One very important thing wasn't mentioned in the description of the design of this experiment. What is that?

d. Are you satisfied with the new design?

Later it was determined that the magnetic bars tended to stick to the metal wall of the van.

e. Was this a blind experiment? Was this a double-blind experiment?

Experimental Units and Replication

In educational research, children come in a group—the classroom—so in order to compare two methods of teaching reading, say, a researcher must assign entire classrooms randomly to the two methods. In cases such as these, it is the classroom, not an individual student, that is the experimental unit.

Suppose that a researcher had six classrooms, each with 25 students, available for her study and assigned three classrooms to each method. The researcher might like to say that each method was replicated on 75 students because the reading ability of 75 students was measured, but alas, she can claim only that each method was replicated on three classrooms.

> **Experimental units** are the people, animals, families, classrooms, and so on to which treatments are assigned.
> **Replication** is repeating the treatment on different units.

Performing many replications in an experiment is the equivalent of a large sample size in a survey—*the more good observations you have, the more faith you have in your conclusions.* (Be careful, however, when someone uses the word *replication.* Sometimes the word is used instead to mean that your entire experiment was repeated by someone else who came to the same conclusion that you did.)

Here's a true story that shows why it is critically important to get the units right when you plan a study. Although this study happens to be observational, the lesson about units applies to experiments as well.

A Tale of Two Sponges

It could have been the best of designs. But it was the worst of designs. It looked like the biggest of samples. But in the end, it turned out to be the smallest of samples.

Once upon a time, many years ago, there lived a statistically innocent graduate student of biology who wanted to compare the lengths of cells in green sponges and white sponges.[2] After consulting his advisor—who should have known better—the graduate student went to work. For hours at a time, he sat hunched over his microscope, painstakingly measuring the lengths of tiny cells. Weeks went by, and eventually, after he had measured cell lengths for about 700 cells per sponge, the graduate student decided he had enough data. It's time, at last, to do the analysis. This dialogue between the Innocent Graduate Student and the project advisor, the Bearer of Bad News, reveals the tragic ending:

IGS: Here are my pages (and pages) of numbers. Can I conclude that cells from white and green sponges have different lengths?

BBN: That depends. We'll have to look at three things—the size of the difference in the white and green averages, the size of your sample, and the size of the natural variability from one unit to the next.

IGS: No problem. My sample size is humongous: 700 cells of each kind. And the variability from one cell to the next isn't all that big.

BBN: Seven hundred cells certainly is a lot. How many sponges did they come from?

IGS: Uh, just two, one of each color. Does that matter?

BBN: Unfortunately, it's critical. But I'll start with the good news. With 700 cells from each sponge, you have rock-solid estimates of average cell length for the two sponges you actually looked at. However, . . .

IGS: Right. And because one's green and one's white and the average lengths are different, I can say that . . .

BBN: Not so fast. You have data that lets you generalize from the cells you measured to all the cells in your particular two sponges. But that's not the conclusion you were aiming for.

IGS: Yeah. I want to say something about *all* white sponges compared to *all* green ones.

BBN: So for you, the kind of variability that matters most is the variability from one sponge to another. Your unit should be a sponge, not a cell.

IGS: You mean to tell me I've got only two units? After all that work!

BBN: I'm afraid you have samples of size one.

IGS: Woe is me! I'd rather be a character in a bad parody of a Dickens novel. But it will be a far, far better study that I do next time.

Discussion: Experimental Units and Replication

D32. Why does it seem reasonable that in the teaching-methods study, the experimental unit should be the classroom?

---

[2] The actual study was a bit more complicated than the simplified version in the story. It looked at cell length, cell width, and the ratio of width to length, for cells taken from two locations, the tip and the base, from two kinds of sponges, green and white.

D33. To study whether women released from the hospital shortly after childbirth have more problems than women who stay longer, hospitals in a large metropolitan area are placed into two groups depending on whether they release women quickly or have them stay longer. A random sample of hospitals is selected from each of the two groups. Then the health records of a random sample of women in each hospital are examined. So randomizing was done twice. What is the best unit to use in this study?

D34. Why do you have more faith in your conclusions if your experiment has many replications?

■ Practice

P20. You want to compare two different textbooks for the AP Statistics course. Ten classes with a total of 150 students will take part in the study. To judge the effectiveness of the books, you plan to use the scores of the students when they take the AP exam at the end of the year. The classes are randomly divided into two groups of five classes each. The first group of five classes, with 80 students, uses the first of the two books. The remaining five classes, with 70 students total, use the other book. Identify the treatments and the experimental units in this scenario, then tell the sample size.

P21. You have a summer job working in a greenhouse. The manager says that she has discovered a wonderful new product that will help carnations produce larger blooms, and you decide to design an experiment to check it out. What are your experimental units? How will you use randomization and replication? Do you need a control? Do you have one?

Summary 4.3: Experiments and Inference About Cause

The goal of an experiment is to compare the response to two or more treatments. The key elements in any experiment are

- randomization
- replication
- control or comparison group

Randomly assigning treatments to subjects allows you to make cause and effect statements and protects against confounding. If you can't randomize the assignment of treatments to subjects, you have an observational study and confounding will remain a threat. Confounding makes it impossible to determine whether the treatment or something else caused the response.

The amount of information you get from an experiment depends on the number of replications. Recognizing experimental units is crucial. So that there is a basis for comparison, experiments require a control group receiving no treatment or a comparison group receiving a second treatment.

■ Exercises

E21. A psychologist wants to compare children from first, third, and fifth grades to determine the relationship between grade level and how quickly a child can solve word puzzles. Would this be an observational or experimental study?

E22. ***Buttercups.*** Some buttercups grow in bright, sunny fields; others grow in woods, where it is both darker and damper. A plant ecologist wanted to know whether buttercups in one habitat had adapted in a way that made them less successful in the other habitat. For his study, he dug up 10 plants from each habitat. Five out of each 10, chosen at random, were planted in the habitat they came from; the remaining 5 were planted in the other habitat. At the end of the study, he compared the sizes of the plants. Why did he bother to dig up the plants that were just going to be replanted in the same habitat? Is this study observational or experimental?

E23. The college health service at a small residential college wants to see whether putting antibacterial soap in the dormitory bathrooms will reduce the number of visits to the infirmary. In all, 1800 students from 20 dormitories participate. Half the dormitories, chosen at random, are supplied with the special soap; the remaining ones are supplied with regular soap. At the end of one semester, the two groups of students are compared, using the average number of visits to the infirmary per person per semester.

  a. What are the units?

  b. What is the sample size?

  c. Is this an observational or an experimental study?

E24. If you've studied chemistry or biology, you may know that different kinds of sugar have different molecular structures. Simple sugars, like glucose and fructose, have six carbon atoms per molecule; sucrose, a compound sugar, has twice as many. A biologist wants to know whether complex sugars can sustain life longer than simple sugars. She prepares eight Petri dishes, each containing 10 potato leafhoppers. Two dishes are assigned to a control group (no food) and two each to glucose, fructose, and sucrose diets. The response variable is the time it takes for half the leafhoppers in a dish to die.

  a. What are the units?

  b. What is the sample size?

  c. Is this an observational or an experimental study?

E25. ***Climate and health.*** Does living in a colder climate make you healthier? To study the effects of climate on health, imagine comparing the death rates in two states that have very different climates, like Florida and Alaska. (The death rate for a given year tells how many people out of every 100,000 die during that year.) In 1990, Florida's death rate was more than twice as high as Alaska's. Why is death rate used as a response variable instead of number of deaths? Why might even death rate not be a good choice for a response variable? (Think about the kinds of people who move to the two states.)

Source: National Center for Health Statistics, www.cdc.gov/nchs/data, 1996.

E26. Suppose you have 50 subjects and you want to assign 25 to treatment A and 25 to treatment B. You flip a coin. If it is heads, the first subject goes into treatment A; if it is tails, the first subject goes into treatment B. You continue flipping and assigning subjects until you have 25 subjects in one of the treatment groups. Then the rest of the subjects go into the other treatment group. Does this method randomly divide the 50 subjects into the two treatment groups? Explain.

4.4   Designing Experiments to Reduce Variability

In the last section, you learned the basics of a good experiment: Randomly assign one of two or more treatments to each subject, and then handle them as alike as possible except for the treatment itself. However, within this framework, you have further choices about experimental design. If you can protect yourself against confounding by randomizing, then designing a good experiment becomes mainly a matter of managing variability. In the first part of this section, you will learn about two types of variability in experiments: one type that you want because it reveals differences between treatments and one type that you don't want because it obscures differences between treatments. Then you will see how a good experimental design can eliminate or minimize the "bad" kind of variability.

Differences Between Treatments Versus Variability Within Treatments

Remember Kelly's experiment with hamsters raised in short days and hamsters raised in long days? Her results are given again in Display 4.9. There is a difference in the response variable *between* the two treatments: The average enzyme concentration for the long days is 8.925, and the average concentration for the short days is 13.40625, a difference of 4.48125. There also is variability *within* each treatment—because, like all living things, hamsters will vary, even when treated exactly alike. In fact, for the short days, the difference between the largest and smallest enzyme concentrations ($18.275 - 11.625 = 6.65$) is even larger than the difference between the two treatment means. Still, because the concentration for each of the short days is larger than the concentration for each of the long days, you probably believe the treatment made a difference.

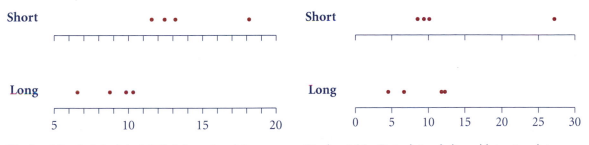

| *Display 4.9*   A dot plot of Kelly's hamster data. | *Display 4.10*   Dot plots of altered hamster data. |

But suppose Kelly had gotten results with more spread in the values but the same means. With the values more spread out, it's no longer so obvious that the treatment matters, as Display 4.10 shows.

In order to believe that the treatments make a difference, the difference **between** *the treatments has to be large enough that we are unconcerned about the variation* **within** *each treatment.*

A good experimental plan can eliminate or reduce within-treatment variability and will allow you to measure the size of any variability you can't eliminate.

A good protocol for an experiment attempts to reduce within-treatment variation.

To judge the effects of the treatments, it is important to keep all other possible influences on the response as nearly constant as possible. One important way to do this is to write out a careful **protocol** that tells exactly how the experiment is to be done. To write a good protocol, you first have to *identify the important sources of within-treatment variation.* To do this, you have to know the situation fairly well, which is one of the main reasons for relying on the bear-launch study as an example. Discovering the secret of long launches may not be quite as important as finding a cure for cancer, but, unlike examples from real science, the bears are something that everyone in your class can know firsthand. So you're going to return to the bears, but this time color won't matter. For the bear-launch activity, there are two treatments: one book and four books under the ramp. The response variable is the launch distance. Each team will do a set of 10 launches, with teams randomly assigned to use either a steeper or a flatter launch angle.

Activity 4.4   Bears in Space

What you'll need: for each team, a launch ramp and launcher, a coin, a tape measure or yardstick, a supply of gummy bears, and four copies of this textbook

1. *Form launch teams and construct your launcher.* Your class should have an even number of teams, at least four. Make your launcher from tongue depressor sticks and rubber bands. First wind a rubber band enough times around one end of a stick so that it stays firmly in place. Then place that stick and another together and wind a second rubber band tightly around the other end of the two sticks to bind them firmly together. Insert a thin pencil or small wooden dowel between the two sticks as a fulcrum.

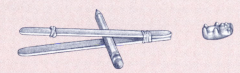

Display 4.11   How to build a bear launcher.

2. *Randomize.* Use random numbers or a random ordering of cards containing team names to assign each team to produce data for one or the other of the two treatments:

    Steeper ramp (four books)

    Flatter ramp (one book)

    The same number of teams should be assigned to each treatment.

3. *Organize your team.* Decide who will do which job or jobs: hold the launcher on the ramp, load the gummy bear, launch the bear, take measurements, and record the data. (When you launch, use a coin instead of your fingers, or you could end up with very sore fingers.)

(continued)

Display 4.12   Setting up the launch.

4. *Gather data.* Each team should do 10 launches. After each launch, record the distance. Measure the distance from the *front* of the ramp, and measure only in the direction parallel to the ramp.

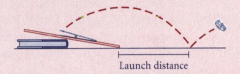

Launch distance

Display 4.13   Measuring the launch distance.

5. *Summarize.* Plot your team's 10 launch distances on a dot plot. Summarize the spread by computing the standard deviation.

Your plots from Activity 4.4 have shown variability from three sources: *the launch angle,* that is, the variability caused by the two different treatments (one book, four books); *the particular team,* that is, the variability from one team to another for teams with the same launch angle; and *the individual launches,* that is, the variability between launches, for the same team. The first is the difference you want to see—the difference between the two treatments. The second two are "nuisance" sources of variability for launches with the same angle; these are to be minimized. In the questions that follow, you examine these sources of variation.

Discussion: Within-Team Variability

D35. Why do the launch distances vary so much for your team? List as many explanations as you can, then order them, starting with the one you think has the most effect, and on down.

D36. One important source of variation between launches, as you've just seen, is the effect of practice. Can you think of an alternative strategy that would reduce or come close to eliminating the variation due to practice?

D37. Use your answers to D35–36 to write out an **experimental protocol,** a set of rules and steps to follow in order to keep launch conditions as nearly constant as possible. Your protocol should include rules for deciding when, if ever, you should not count a "bad" launch. After each team has written a protocol, your class should discuss these and combine them into a single protocol to be used for future launches.

Discussion: Between-Team Variability

D38. Which measure of center—mean or median—gives a better summary of a team's set of ten launches? Or are they about equally good? Give reasons for your judgment.

D39. Record the team summaries (mean or median) in a side-by-side dot plot, one for each launch angle. Compute the standard deviation of the means (or medians) for each launch angle. Is there more variation between teams with the same launch angle or between launches for the same team (calculated in Activity 4.4, step 5)? Why do you think the team summaries vary as much as they do?

List as many explanations as you can, then order them, starting with the one you think has the most effect, and on down. If you were going to summarize a team's results with a single-number measure of center anyway, why not do only one launch per team and use that number?

D40. What strategies would you recommend for managing the variability between the teams that had the same launch angle?

Discussion: Difference Between Treatments

D41. Use the data from the entire class to construct side-by-side dot plots showing individual launches by team. Think about ways to use the data to measure and compare the sizes of the three kinds of variability. Can you say which kind is the largest? The smallest?

D42. Without doing a formal analysis, choose a tentative conclusion from these three statements, and give your reasoning.

   I. Launch angle clearly makes a big difference.

   II. Launch angle may very well make a difference, but there's too much variation from other sources to be able to isolate and measure any effect due to launch angle.

   III. There's not much variability anywhere in the data, so it's safe to conclude that the effect due to launch angle, if any, is quite small.

D43. Combine the data from all the teams that launched with one book into a single group. Do the same for the data from the four-book teams. Make a side-by-side dot plot of one-book launches and four-book launches. Show all individual launches, ignoring teams. Compare the centers and spreads for the two sets of launches. Why might this analysis be misleading?

■ Practice

P22. Review the antibacterial soap experiment in E23.

   a. List at least two sources of within-treatment variability.

   b. Is the point of randomization to eliminate the within-treatment variability or to equalize it between the treatments?

A Design for Every Purpose

Now that you've had a chance to explore some of the general ideas for designing sound experiments, it's time to put them into practice. In Activity 4.5, you will serve as a subject for three different versions of an experiment. The goal of the experiment is to see whether there is a detectable difference in heart rate measured under two conditions or treatments:

> Treatment 1: standing, with eyes open

> Treatment 2: sitting, with eyes closed, relaxing

The goal of the activity as a whole is for you to think about the advantages and disadvantages of three different versions of the experiment.

Activity 4.5   Sit or Stand

You will take your pulse several times during this activity, as follows:

a. Get ready, sitting or standing.

b. Your teacher will time you for 30 seconds. When your teacher says "go," start counting beats until your teacher says "stop."

c. Double your count to get your heart rate in beats per minute.

Part A: Completely Randomized Design

In Part A, you and the other subjects will be randomly assigned to one treatment or the other.

1. *Random assignment.* Your teacher will pass around a box with slips of paper in it. Half say "stand" and half say "sit." When the box comes to you, mix up the slips and draw one. Depending on the instruction you get, either stand with your eyes open or sit with your eyes closed and relax.

2. *Measurement.* Take your pulse.

3. *Data record.* Record your heart rate and that of other students in your treatment group.

4. *Summaries.* Display the data using side-by-side stemplots for the two treatments. Then compute the mean and standard deviation for each group. Do you think the treatment makes a difference?

Part B: Randomized Paired Comparison Design (Matched Pairs)

In Part B, you and your classmates will first be sorted into pairs based on an initial measurement. Then within each pair, one person will be randomly chosen to stand and the other will sit.

1. *Initial measurement.* Take your pulse sitting with eyes open.

2. *Form matched pairs.* Line up in order, from fastest heart rate to slowest, and then pair off, with the two fastest in a pair, the next two after that in a pair, and so on.

(continued)

3. *Random assignment within pairs.* Either you or your partner should prepare two slips of paper, one that says "sit" and one that says "stand." One of you should then mix the two slips and let the other choose one. Thus, within each pair, one of you randomly ends up sitting and the other one standing.

4. *Measurement.* Take your pulse.

5. *Data record.* Calculate the difference, *standing* minus *sitting,* for each pair of students.

6. *Summaries.* Display the set of differences in a stemplot. Then compute the mean and standard deviation of the differences. What should the mean be if the treatment makes no difference? Do you think the observed difference is real or simply due to variation between individuals?

Part C: Randomized Paired Comparison Design (Repeated Measures)

This time each person will be his or her own matched pair: Each of you will take your pulse under both treatment I and treatment II. You'll flip a coin to decide the order: sit first, then stand; or stand first, then sit.

1. *Random assignment.* Flip a coin. If it lands heads, you will sit first, then stand. If it lands tails, you will stand first.

2. *First measurement.* Take your pulse in the position chosen by your coin flip.

3. *Second measurement.* Take your pulse in the other position.

4. *Data record.* Record your heart rates in a table with other students. Calculate the difference, *standing* minus *sitting,* for each student.

5. *Summaries.* Display the set of differences in a stemplot. Then compute the mean and standard deviation of the differences. Do you think the treatment makes a difference?

Discussion: A Design for Every Purpose

D44. Which design do you think is best for studying the effect of position on heart rate: completely randomized design, randomized paired comparison design (matched pairs), or randomized paired comparison design (repeated measures)? Explain what makes your choice better than the other two designs.

D45. Which of the three designs do you think is least suitable? Explain what makes it less effective than the other two. Make up and describe a new scenario (choose a response and two treatments to compare) for which the least suitable design would be more suitable than the other two.

D46. Describe the variation within treatments for the design in Part A. How did the other two designs eliminate variation within treatments?

The Completely Randomized Design

In a completely randomized design, treatments are randomly assigned without restriction.

In Activity 4.5, the design in Part A is an example of what statisticians call a completely randomized design. The treatments (sit or stand) get assigned to units *completely at random,* that is, without any prior sorting or restrictions. For a **completely randomized design,** use a chance device to randomly assign a treatment to each experimental unit.

Activity 4.4 was an example of a completely randomized experiment. There were two treatments, steeper and flatter launch angles. The experimental unit was a team. It is an example of a completely randomized design because you assigned treatments to individual units (teams) using a chance device.

Kelly Acampora's hamster experiment used a completely randomized design. Each hamster was randomly assigned a treatment. Kelly did this in such a way that her design was **balanced;** that is, each treatment had the same number of units assigned to it.

The Randomized Paired Comparison Design

Blocks are groups of *similar* units.

Matched pairs and repeated measures, like the designs in Parts B and C in Activity 4.5, are examples of randomized paired comparison designs. In the design in Part B (matched pairs), the units were sorted into pairs of *similar* units called **blocks** before the treatments were assigned. Each block was a pair of students with similar heart rates. Then the random assignment of treatments was done separately within each block. In the design in Part C (repeated measures), the block was a person, where the randomization determined the *order* of the treatment. For a **randomized paired comparison design,** first sort your experimental units into pairs of similar units, then randomly decide which unit in each pair gets which treatment.

In a randomized paired comparison design, pairs of similar units are randomly assigned different treatments.

In one study, investigators found seven pairs of identical twins, one of whom lived in the city and the other in the country, who were willing to participate in a study to determine how quickly their lungs cleared after inhaling a spray that contained radioactive Teflon particles. Display 4.14 gives the percentage of radioactivity remaining one hour after inhaling the spray.

In this study, a pair of twins is a block (or a matched pair). The factor is the type of environment. This was an observational study, not an experiment, because the type of environment, urban or rural, was not and could not be randomly assigned—it was a given.

Twin Pair	Environment		Difference
	Rural	Urban	
1	10.1	28.1	−18
2	51.8	36.2	15.6
3	33.5	40.7	−7.2
4	32.8	38.8	−6
5	69.0	71.0	−2
6	38.8	47.0	−8.2
7	54.6	57.0	−2.4

Display 4.14   The percentage of radioactivity remaining after one hour.

Source: Per Camner and Klas Phillipson, "Urban Factor and Tracheobronchial Clearance," *Archives of Environmental Health* 27 (1973): 82. Reprinted in Richard J. Larson and Morris L. Marx, *An Introduction to Mathematical Statistics and Its Applications,* 2nd ed. (Englewood Cliffs: N.J.: Prentice Hall, Inc., 1986).

The Randomized Block Design

In a randomized block design, treatments are randomly assigned within blocks of similar units.

A randomized paired comparison design is a special type of randomized block design. The difference is that in a block design, more than two experimental units may be in each block. For a **randomized block design,** first sort or subdivide your experimental units into groups (blocks) of similar units (this is called "blocking"), then randomly assign treatments to units separately within each block.

The term *block* comes from agriculture. One of the earliest published examples of a block design appeared in R. A. Fisher's *The Design of Experiments* (1935). The goal of the experiment was to compare five types of wheat to see which gave the highest yield. The five types of wheat were the treatments, the yield in bushels per acre was the response, and eight blocks of farm land were available for planting.

There were many possible sources of variability in the blocks of land, mainly differences in the composition of the soil: what nutrients were present, how well the soil held moisture, and so on. Because many of these possibly confounding influences were related to the soil, Fisher made the reasonable assumption that the soil within each block would be more or less uniform and that the variability to be concerned about was the variability from one block to another. His goal in designing the experiment was to keep this between-block variability from being confounded with differences between wheat types.

Fisher's plan was to divide each of the eight blocks of land "into five plots running from end to end of the block, and lying side by side, making forty plots in all." These 40 plots were his experimental units. Fisher then used a chance device to assign one type of wheat to each of the five plots in a block. Display 4.15 shows how his plan might have looked.

Suppose it turned out that one of the blocks had really poor (or really favorable) conditions for wheat. Then Fisher's blocking accomplished two things: All five types of wheat would be affected equally, and the variation within a block could be attributed to the types of wheat.

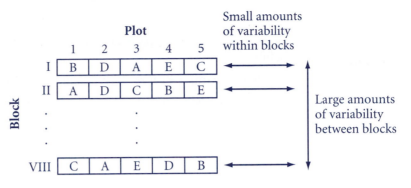

Display 4.15 A design using blocks.

Discussion: Variation Within and Between

D47. In the rural/urban twins lung-clearing study, is there more variation within the urban twins, within the rural twins, or in the differences? Do you believe the study demonstrates that environment makes a difference?

D48. Why were twins used in the lung-clearing study? What kind of variation is reduced by using identical twins in randomized paired comparison experiments?

■ Practice

P23. To test a new drug for asthma, both the new drug and the standard treatment, in random order, will be administered to each subject in the study.

  a. What kind of design is this?

  b. An observant statistician cries, "No, no! Use two different subjects in each pair, randomized to each treatment." What kind of design is this?

  c. Which design do you like, and why?

P24. In one design in the heart-rate experiment in Activity 4.5, each person provided one measurement on each treatment. Which design was that?

P25. For each of these experiments, describe the within-treatment variability that might obscure any difference between treatments. Then describe an experimental design that includes blocking, a randomization scheme, and the definition of a response variable.

  a. To determine whether studying with the radio on helps or hurts ability to memorize, there will be two treatments: listening to the radio and not listening to the radio. The subjects available are all seniors in one school.

  b. To determine whether adding MSG to soup makes customers eat more soup, a large restaurant will have two treatments: adding MSG to the soup and not adding MSG. The subjects available are all customers during one evening.

More on Blocking

In Activity 4.4 Bears-in-Space, even for the same launch angle, you found a lot of variability:

- Some teams are more skillful than others in launching the bear.

- Within a team, some launches go more smoothly than others.

These two kinds of variability within a treatment can obscure a difference caused by the launch angle.

Ordinarily, when teams carry out the bear-launch experiment, their results show that the variability between teams is so large that it is hard to tell how much difference the angle of the launch ramp makes. Following a careful protocol can reduce the variability somewhat, but there will still tend to be big differences from one team to another. One solution to the problem would be to increase the number of experimental units: Use more teams.

Although increasing the sample size is a reliable way to get more information about the difference due to treatments, it can be very costly or very time-consuming. As you'll see in Activity 4.6, sometimes there are much better ways to manage variability.

Activity 4.6   Block Those Bears!

What you'll need: the same equipment as for Activity 4.4

1. Form teams and assign jobs, as in Activity 4.4.

2. *Randomize.* Each team will conduct two sets of five launches, one set with four books under the ramp and one set with just one book. Each team should flip a coin to decide the order of the treatments:

   Heads: 4 books first, then 1 book

   Tails: 1 book first, then 4 books

3. *Gather data.* Carry out the launches, following the protocol your class developed in D37, and record the launch distances.

4. *Compute averages.* Put your team's averages for the two treatments and their difference in a table. Pool the results from all the teams. Based on the differences between the averages, do you think it is reasonable to conclude that launch angle makes a difference? Or is it impossible to tell?

Discussion: Blocked Bears

D49. What kind of experimental design is used in Activity 4.6? What are the blocks?

D50. ***Units.*** Remember that you assign a treatment to a unit. For the design of Activity 4.6, which is the unit: a single launch, a set of five launches, or a set of 10 launches by a team?

D51. ***Another block design.*** Suppose you were to redo Activity 4.6. Can you think of a way to improve on the design?

Similar units in a block make for effective blocking.

The effectiveness of blocking depends on how similar the units are in each block and how different the blocks are from each other. Here "similar units" are units that would tend to give similar values for the response if they were assigned the same treatments. *The more similar the units in a block, the more effective blocking will be.*

■ Practice

P26. What is the main difference between a completely randomized design and a randomized block design?

P27. Why is blocking sometimes a desirable feature of a design? Give an example in which you might want to block and an example in which you might not want to block.

P28. a. The dot plots in Display 4.16 show the scores on the two exams for a sample of 31 students. Does it look as if students are improving? Is it easy to answer this question from the plots given?

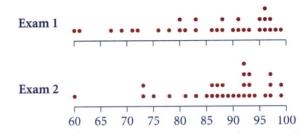

Display 4.16   Scores on two exams for a sample of students.

b. Display 4.17 shows the scatterplot for each student's pair of scores. The line on the scatterplot is the line $y = x$. Now does it look as if the students are improving from Exam 1 to Exam 2? Is it easier to answer this question from the scatterplot that shows the paired scores?

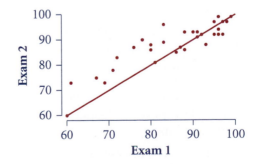

Display 4.17   A scatterplot of exam scores for a sample of students.

c. Explain why a block design (both exams recorded for a sample of students) would be better than a completely randomized design (two independent samples of students, one for Exam 1 scores and one for Exam 2 scores) for studying student improvement.

Summary 4.4: Basic Experimental Designs

Random assignment of treatments to experimental units—the fundamental principle of good design—can be accomplished through two basic plans:

- The completely randomized design is characterized by randomly assigning a treatment to each experimental unit while keeping the number of units given each treatment as equal as possible.

- The randomized block design is characterized by placing similar experimental units into different groups, called blocks, and randomly assigning the treatments to the units within each block.

The reason for the randomized block design is that in any experiment, responses vary not only because of different treatments (between-treatment variation) but also because different subjects respond differently to the same treatment (within-treatment variation). In a well-designed experiment, the experimenter should minimize the amount of variability within the treatment groups because it might obscure the variability resulting from differences between treatments. By taking the blocks into account when analyzing the data, the effect of the within-treatment variability can be minimized.

You have learned two special cases of block design:

- A randomized paired comparison (matched pairs) design involves randomly assigning two treatments within pairs of similar units, such as to twins or to left and right feet.

- A randomized paired comparison (repeated measures) design involves the assignment of all treatments, in random order, to each unit so that comparisons can be made on the same units. An example would be a study in which each patient in a clinic was assigned each of three treatments for asthma, in random order.

■ Exercises

E27. The SAT example from Section 4.3, page 241, ended with a revised plan that was truly experimental. To which of the two families does that design belong: completely randomized or randomized block? Summarize the design by giving the experimental units, factors, levels, response variable, and blocks (if any).

E28. PKU (phenylketonurea) is a disease in which the shortage of an enzyme prevents your body from producing enough dopamine, a substance that helps transmit nerve impulses. To some extent, you can relieve the symptoms by eating a restricted diet, one low in the amino acid phenylalanine. A study was designed to measure the effect of such a diet on the levels of dopamine in the body. The subjects were 10 PKU patients. Each patient was measured twice, once after a week on the low phenylalanine diet and once after a week on a regular diet. Identify blocks (if any), treatments, and units, and describe the design in this study.

E29. *Finger tapping.* Many people rely on the caffeine in coffee to get them going in the morning or to keep them going at night. But wouldn't you rather have chocolate? Chocolate contains theobromine, an alkaloid quite similar to caffeine, both in its structure and in its effects on humans. In 1944, C. C. Scott and K. K. Chen reported the results of a study designed to compare caffeine, theobromine, and a placebo. Their design used four subjects and assigned the three treatments to each subject in a random order, one treatment on each of three different days. Subjects were trained to tap their fingers in a way that allowed the rate to be measured; presumably the training got rid of any practice effect. The response was the rate of tapping two hours after taking a capsule containing either one of the drugs or the placebo. How are treatments assigned to units—completely at random or using a block design? Identify the response, units, treatments, and blocks (if any).

Source: C. C. Scott and K. K. Chen, "Comparison of the Action of 1-ethyl Theobromine and Caffeine in Animals and Man," *Journal of Pharmacologic Experimental Therapy* 82 (1944): 89–97.

E30. ***Walking babies.*** The goal of this experiment was to compare four 7-week training programs for infants, to see whether special exercises could speed up the process of learning to walk. One of the four training programs was of particular interest: It involved a daily 12-minute set of special walking and placing exercises. The second program (the exercise control group) involved daily exercise for 12 minutes but without the special exercises. The third and fourth programs involved no regular exercise (parents were given no instructions about exercise) but differed in follow-up: Infants in the third program were checked every week, whereas those in the fourth group were checked only at the end of the study. Twenty-three 1-week-old babies took part; each was randomly assigned to one of the groups. The response was the time, in months, when the baby first walked without help.

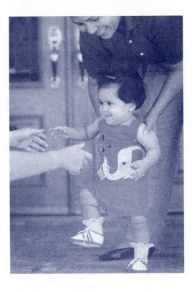

Tell how treatments are assigned to units—completely at random or using a block design. Identify the response, units, treatments, and blocks (if any).

E31. ***Sawdust for lunch?*** Twenty-four tobacco hornworms served as subjects for this experiment, which was designed to see how worms raised on low-quality food would compare with those on a normal diet. The two dozen worms were randomly divided into two groups, and the lucky half was raised on regular worm food. The unlucky half got a mixture of 20% regular food and 80% cellulose. Cellulose has no more food value for a hornworm than it has for you—neither you nor a hornworm can digest the stuff. The experimenter kept track of how much each hornworm ate and computed a response value based on the total amount eaten in relation to body weight.

Tell how treatments were assigned to units in this experiment—completely at random or using a block design. Identify the response, units, treatments, and blocks (if any).

Chapter Summary: Designing Sample Surveys and Experiments

There are two main families of chance-based methods for data collection: sampling methods and experimental designs. Sampling methods, studied in the first half of this chapter, use chance to choose the individuals to be studied. Typically, you choose the individuals in order to ask them questions, as in a Gallup poll; thus, samples and surveys often go together. Experiments, introduced in the second half of this chapter, are comparative studies that use chance to assign the treatments you want to compare. An experiment should have three characteristics: random assignment of treatments to subjects, two or more treatments to compare, and each treatment replicated on at least two subjects.

The purposes of sampling and experimental design are quite different. Sample surveys are to estimate the parameters of fixed, well-defined populations. Experiments are used to establish cause and effect by comparing treatments.

Randomization is the fundamental principle in both types of study because it allows the use of statistical inference (developed in later chapters) to generalize results. In addition, random selection of the sample reduces bias in surveys, and random assignment of treatments to subjects reduces confounding in experiments.

■ Review Exercises

E32. For each situation, tell whether it is better to take a sample or a census, and give reasons for your answer.

Characteristic of Interest	Population of Interest
a. Average life of a battery	Alkaline AAA batteries
b. Average age	Current U.S. senators
c. Average price per gallon	Purchases of regular-octane gasoline sold at U.S. stations next week

E33. Consider using your statistics class as a convenience sample in each of the following situations. For each, tell whether you think the sample will be reasonably representative, and if not, in what way(s) you expect your class to differ from the given population of interest.

Characteristic of Interest	Population of Interest
a. Percentage who can curl their tongues	U.S., age 12 or above
b. Average age	U.S., all adults
c. Average blood pressure	U.S., students your age
d. Percentage who prefer science to English	U.S., students your age
e. Average blood pressure	U.S., all adults

E34. The U.S. Bureau of Labor Statistics collects data on occupational variables from a nationwide sample of households. This paragraph is from its news bulletin "Union Members in 1996," January 31, 1997, p. 3: "The data also are subject to nonsampling error. For example, information on job-related characteristics of the worker, such as industry, occupation, union membership, and earnings, are sometimes reported by a household member other than the worker.

Consequently, such data may reflect reporting error by the respondent. Moreover, in some cases, respondents might erroneously report take-home pay rather than gross earnings, or may round up or down from actual earnings."

a. Would reporting take-home pay rather than gross earnings increase or decrease the estimate of earnings of various types of workers?

b. What other sources of nonsampling error might there be in this study?

E35. A friend who wants to be in movies is interested in how much actors earn and has decided to gather data using a simple random sample. The *World Almanac and Book of Facts* has a list of actors that your friend plans to use as a sampling frame. Would you advise against using that list as a frame? What sort of bias do you expect?

E36. What is the average number of representatives per state in the U.S. House of Representatives? If you really wanted to know, you should use a census rather than a sample, but just because you like statistics so much, you've decided to use a random sample. Which of the following two sampling methods is biased? Describe the bias: Will estimates tend to be too high or too low?

Method I. Start with a list of all the current members of the House. Choose a simple random sample of 80, and for each representative chosen, record the number of representatives from that person's state. Then take the average.

Method II. Start with a list of all 50 states. Choose a simple random sample of 5 states, and for each one chosen, record the number of representatives from that state. Then take the average.

E37. For an article on used cars, a writer wants to know the cost of repairs for a certain 1998 model of car. He plans to take a random sample of owners who bought that model from a compilation of lists supplied by all U.S. dealers of that make of car. He tells his research assistant to send letters to all the people in the sample, asking them their total repair bills for last year. His research assistant tells him, "You might want to think again about your sampling frame. I'm afraid your plan will miss an important group of owners." What group or groups?

E38. If you look up Shakespeare in just about any book of quotations, you'll find that the listing goes on for several pages. If you look up less famous writers, you find much shorter listings. What's the average number of quotations per author in, say, the *Oxford Dictionary of Quotations,* 3rd edition? Suppose you plan to base your estimate on a random sample chosen by this method: You choose a simple random sample of pages from the book. For each page chosen, you find the first quotation in the upper-left corner of that page and record the number of quotations by that author. Then you take the average of these values for all the pages in your sample. What's wrong with this sampling method?

E39. You've just gotten a job as a research assistant with the state of Maine. For your first assignment, you're asked to get a representative sample of the fish in Moosehead Lake, the state's largest and deepest lake. Your supervisor, who knows no statistics and can't tell a minnow from a muskellunge, tells you to drag a net with 1-inch mesh (hole size) behind a motor-boat, up and down the length of the lake. "No," you tell him, "that method is biased!" What is the bias? What kinds of fish will tend to be overrepresented? Underrepresented? (You *don't* have to know the difference between a minnow and muskie to answer this. You *do* have to know a little statistics.)

E40. You are asked to provide a forestry researcher with a random sample of trees in a one-acre lot. Your supervisor gives you a map of the lot, showing the location of each tree, and tells you to choose points at random on the map and for each point, to take the tree closest to the point for your sample. "Sorry, sir," you tell him. "That sampling method was once in common use, but then someone discovered it was biased. It would be better to number all the trees in the map and use random numbers to take an SRS." Draw a little map, with about 10 trees, that you could use to convince your supervisor that his method is biased. Put in several younger trees, which grow close together, and two or three older trees, whose large leaf canopies discourage other trees from growing nearby. Then explain which trees are more likely to be chosen by the method of random points, and why.

E41. Now you're working as an assistant to a psychologist. For subjects in an experiment on learning, she needs a representative sample of 20 adult residents of New York City. She tells you to run an ad in the *New York Times,* asking for volunteers, and to randomly choose 20 from the list of volunteers. What's the bias this time?

E42. Give an example where nonresponse bias is likely to distort the results of a survey.

E43. Tell how to carry out a four-stage random sample of voters in the United States, with voters grouped by precinct, precincts grouped by congressional district, and congressional districts grouped by state.

E44. Tell how to carry out a multistage sample for estimating the average number of characters per line for books on a set of shelves.

E45. *Needle threading.* With your eye firmly fixed on winning a Nobel prize, you decide to make the definitive study of the effect of background color (white, black, green, or red) on the speed of needle threading. Design two experiments, one that uses no blocks, one that creates blocks by grouping subjects, and one that creates blocks by reusing subjects. Tell which of the three plans you consider most suitable, and why.

E46. Read one of these articles from *Statistics: A Guide to the Unknown* [San Francisco: Holden-Day, 1972] and write a paragraph describing how blocks are used in the study reported in the article:

a. D. D. Reid, "Does Inheritance Matter in Disease?"

b. Elisabeth Street and Mavis G. Carroll, "Preliminary Evaluation of a New Food Product"

c. Louis J. Battan, "Cloud Seeding and Rainmaking"

E47. Design a taste-comparison experiment. Suppose you want to rate three brands of chocolate chip cookies on a scale from 1 (terrible) to 10 (outstanding). Tell how to run this experiment as a randomized block design.

E48. *Exercise bikes.* You work for a gym and have been asked to design an experiment to decide which of two types of exercise bikes are the most attractive to customers. You have space for eight bikes in your exercise bike room, which will be placed as in Display 4.18. People enter the room by the door at the top of the diagram. There is a counter on each bike that records the number of hours it has been used.

Design an experiment to compare the attractiveness to customers of the two different brands of exercise bike that takes into account the fact that some bikes are in locations that make them more likely to be used than others when the gym isn't full.

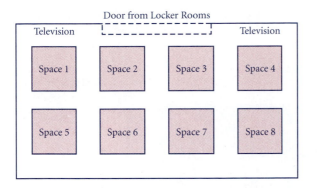
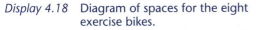

Display 4.18 Diagram of spaces for the eight exercise bikes.

E49. As the U.S. Census for the year 2000 was being planned, there was great controversy over a plan proposed by the Census Bureau to use sampling to adjust for the anticipated undercount. Here is a simplified version of the plan. The Census Bureau collects the information that is mailed in by most of the residents of a region. Some

residents, however, did not receive their forms or did not return them for some reason; these are the uncounted persons. The bureau now selects a sample of blocks (neighborhoods) from the region and sends field workers there to find all residents in the sampled blocks. The residents found by the field workers are matched to the census data, and the number of residents uncounted in the original census is noted. The census count for those blocks is then adjusted according to the proportion uncounted. (For example, if one-tenth of the residents had been uncounted, the original census figures would be adjusted upward by 11%.) In addition, the same adjustment is used for neighboring regions that have characteristics similar to the one sampled. Write a critique of this adjustment method, commenting on both strengths and weaknesses.

CHAPTER 5
SAMPLING DISTRIBUTIONS

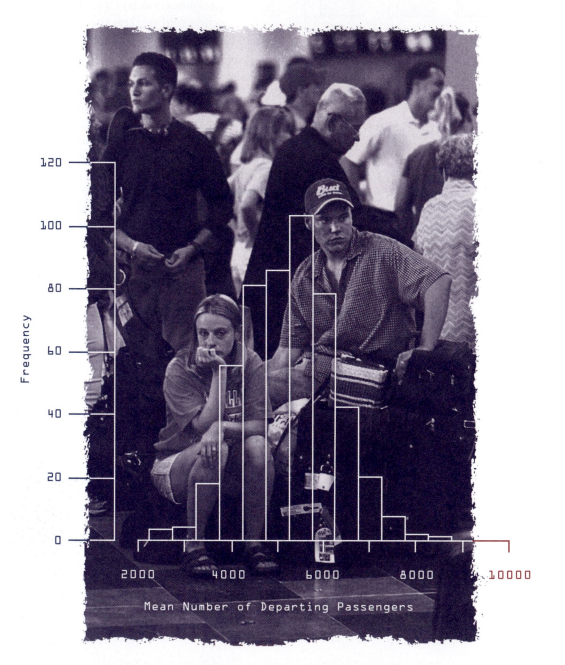

What would happen if you could take random samples over and over again from your population? Sampling distributions show how much your sample results might vary from reality, as when estimating the mean number of departures from U.S. airports for a given period of time.

The methods of Chapters 2 and 3 are good for exploration and description, but for inference—going beyond the data in hand to conclusions about the population or the process that created the data—you need to collect the data using a random sample (a survey) or by randomly assigning treatments to subjects (an experiment). The promise of Chapter 4 was that if you use those methods to produce a data set, you'd then be able to use your data to draw sound conclusions. Randomized data production not only protects against bias and confounding but also makes it possible to imagine repeating the data production process so you can estimate how the numerical summary (the statistic) you compute from the data would vary from sample to sample. To oversimplify, but only a little, a sampling distribution is what you get by repeating the process of producing the data and computing the summary.

In this chapter, you will learn

- how to use simulation to generate sampling distributions for common summary statistics such as the sample mean, the sample proportion, and the sum of two values from two populations

- to describe the shape, center, and spread of the sampling distributions for common summary statistics without actually generating them

- to use the sampling distribution to determine which results are reasonably likely and which would be considered rare

5.1 Sampling from a Population

In this section, so that you can better understand sampling distributions, you will study sampling from populations that are defined by relative frequency (proportion) tables. For example, Display 5.1 shows the proportion of households in the United States that own various numbers of motor vehicles. The "4" in Display 5.1 actually represents the category "4 or more," but the proportion of households that have more than 4 vehicles is very small.

How could you take a sample from this population? If a list of all of the millions of households in the country were available, you could write the name and address of each household on separate cards, mix them up in a box, and draw as many as you needed. Or, as you learned in Chapter 4, you could select the households by numbering them and then using a random number table.

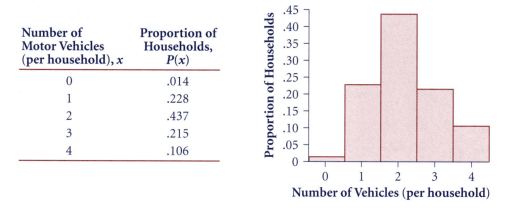

Number of Motor Vehicles (per household), *x*	Proportion of Households, *P(x)*
0	.014
1	.228
2	.437
3	.215
4	.106

Display 5.1 The number of motor vehicles per household.

Source: *Statistical Abstract of the United States,* 1996.

If you are interested only in the number of motor vehicles in a randomly selected household, there is a much easier way. All of the households that have, say, three motor vehicles are the same to you. So you could put 14 slips of paper with the number 0 on them into a box, 228 with the number 1 on them, 437 with the number 2 on them, 215 with the number 3, and 106 with the number 4. If you mix up the slips of paper and draw one at random, you have a $\frac{14}{1000}$ or .14 chance of getting a household with no motor vehicles, a $\frac{228}{1000}$ or .228 chance of getting a household with one motor vehicle, and so on.

A random digit table is even quicker because you don't have to write numbers on all those slips of paper. To use a list of random digits to select a household at random from this distribution, first assign three-digit numbers, as in Display 5.2. You need three-digit numbers because there are three decimal places in the percentages.

Number of Motor Vehicles (per household), x	Proportion of Households, $P(x)$	Random Numbers Representing This Category
0	.014	001–014
1	.228	015–242
2	.437	243–679
3	.215	680–894
4	.106	895–000

Display 5.2   Assigning random digits to represent the motor vehicle population.

For example, 14 triples out of 1000 (.014) are needed to represent households with no motor vehicles. Here the 14 triples 001 through 014 were used, although you could use 000 through 013 or any other choice as long as there are 14 of them. The next 228 triples (015–242) represent households with one vehicle, and so on.

Now that you've assigned all the triples, you're ready to take your sample. Divide a string of random digits into groups of three (because there are three decimal places in the percentages).

$$391 \mid 545 \mid 177 \mid 324 \mid 106 \mid 845 \mid 248 \mid \ldots$$

The first triple, 391, represents a household with two motor vehicles because 391 is in the interval 243–679. If you want to draw a second household, the second triple, 545, also represents a household with two motor vehicles.

Discussion: Sampling with Random Digits

D1. What numbers of motor vehicles do the remaining sets of triples represent?

D2. How would you assign the numbers in a random digit table so that they represent the distribution in Display 5.3? Use a random digit table (Table D in the appendix) to select a lung cancer patient at random and then tell whether smoking was responsible for the lung cancer.

Lung Cancer Cases	Proportion
Smoking responsible	.9
Smoking not responsible	.1

Display 5.3   The proportion of lung cancer cases for which smoking was, or was not, the cause.

Source: Lung Disease Data, American Lung Association, 1997.

■ Practice

P1. The distribution in Display 5.4 gives the number of children per family in the United States. Describe how to use the following line from a random digit table to find the number of children in a randomly selected family.

4 8 8 3 0 9 9 4 2 5 1 7 7 3 8 9 0 8 1 7

Use your process to give the number of children in three randomly selected families.

Number of Children (per family)	Proportion of Families
0	.505
1	.203
2	.191
3	.073
4 or more	.028

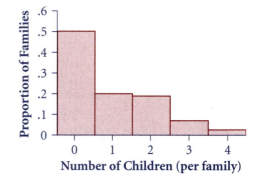

Display 5.4   The number of children per family.

Source: *Statistical Abstract of the United States,* 1999.

Sampling With and Without Replacement

Suppose the student council wants to interview a sample of thirty students from your school to find out how they think the school library could be improved. The council writes the name of every student on a slip of paper, places them in a box, and draws out thirty. Because it doesn't make sense to interview the same student twice, they don't return the slips of paper to the box after they draw them.

As you saw in Chapter 4, when you take samples in practice, it usually makes more sense to sample without replacement—don't replace the names as you draw them. This is the way a sample is selected in real life for a survey or an experiment, so this will be your usual choice when selecting a sample.

If the council allows the same student to be chosen more than once, that would be sampling *with* replacement. Although you seldom would do this in practice, the theory of sampling distributions that you'll learn in Sections 5.3 and 5.4 turns out to be simpler if you assume that you do replace. This apparent discrepancy between theory and practice doesn't make much difference as long as the sample size is relatively small compared to the population size, which is usually the case. That is, if your school is large and your sample is small, it's pretty unlikely that you will draw the same student twice even if you sample with replacement.

Sampling with replacement is about the same as sampling without replacement when the sample size is small compared to the population size.

Discussion: Sampling With and Without Replacement

D3. Suppose you are using a random digit table like the motor vehicle example on the previous page, and you come across a triple of random digits that you have used before. Should you use it again or skip it? How is this situation different from selecting students from your school?

The Mean and Standard Deviation of a Population

The mean and standard deviation for a population are denoted by μ and σ. For a sample, they are denoted by $\bar{x}$ and s.

For distributions that represent an entire population, like those in Display 5.1 and Display 5.4, statisticians use the Greek symbols μ (mu, pronounced "mew," rhyming with "few") for the mean and σ (sigma) for the standard deviation. By looking at the mound-shaped histogram in Display 5.1, you can estimate that the balance point is $\mu \approx 2$ and that an interval of $\sigma \approx 1$ on either side of the mean contains roughly two-thirds of the households.

To compute the mean and the standard deviation of populations like these, you can use formulas similar to those for frequency tables from Chapter 2:

$$\mu = \sum x \cdot P(x)$$
$$\sigma = \sqrt{\sum (x - \mu)^2 \cdot P(x)}$$

Here, $P(x)$ is the proportion of times that the value x occurs. Alternatively, you can think of $P(x)$ as the probability that the value x occurs if you pick one value at random from the population.

Example: Computing μ and σ

Use the formulas for μ and σ to compute the mean and standard deviation of the population in Display 5.1, of the number of motor vehicles per household.

Solution

The mean and standard deviation of the number of motor vehicles per household are

$$\mu = \sum x \cdot P(x)$$
$$= 0(.014) + 1(.228) + 2(.437) + 3(.215) + 4(.106)$$
$$= 2.171$$

$$\sigma = \sqrt{\sum (x - \mu)^2 \cdot P(x)}$$
$$= \sqrt{(0 - 2.171)^2(.014) + (1 - 2.171)^2(.228) + \cdots + (4 - 2.171)^2(.106)}$$
$$\approx 0.945$$

To compute the mean, it's sometimes easiest to add a third column to the table, as in Display 5.5.

To use your calculator, place the values of x in L$_1$ and the proportions $P(x)$ in L$_2$. Then 1-VarStats L$_1$, L$_2$ will give you μ and σ.

Number of Motor Vehicles (per household), x	Proportion of Households, $P(x)$	$x \cdot P(x)$
0	.014	0
1	.228	0.228
2	.437	0.874
3	.215	0.645
4	.106	0.424
		$\sum x \cdot P(x) = 2.171$

Display 5.5 Computing the mean from a relative frequency table. ■

Discussion: The Population Mean and Standard Deviation

D4. In Chapter 2, you used this formula to compute the mean, $\bar{x}$, of a frequency table, where x is the value, f is its frequency, and $n = \Sigma f$:

$$\bar{x} = \frac{\Sigma x \cdot f}{n}$$

a. Use this formula to compute the mean of the values in this frequency table.

Value, x	Frequency, f
5	12
6	23
8	15

b. Suppose that the data had been given in a relative frequency table like the one below showing the proportion of times each value occurs. Fill in the rest of the last column.

Value, x	Proportion, $P(x) = f/n$
5	.24
6	–?–
8	–?–

Show that you can find the mean μ using the formula

$$\mu = \Sigma x \cdot P(x)$$

c. Show how you can derive the formula in part b from the formula in part a.

■ Practice

P2. The distribution in Display 5.4 gives the number of children per family in the United States. Compute the mean and standard deviation of this population. Count "4 or more" as 4.

Summary 5.1: Sampling from a Population

In this section, you learned how to use a random digit table to select a value at random from a population where you're given the proportion of values in each category. The mean and standard deviation of such a population are given by the formulas

$$\mu = \Sigma x \cdot P(x)$$
$$\sigma = \sqrt{\Sigma (x - \mu)^2 \cdot P(x)}$$

where $P(x)$ is the proportion of the time that the value x occurs in the population.

■ Exercises

E1. A large class was assigned a difficult homework problem. This table shows the scores that the students received and the proportion of students who received each score.

Score	Proportion of Students
0	.45
1	.09
2	0
3	.04
4	.15
5	.27

a. Compute the mean and standard deviation of the scores.

b. Describe how to use this list of random digits to take a sample of size 4 from these scores:

30558 45957 36911 97199 08432

Note: The digits are grouped in batches of five to make them easier to read. They're to be interpreted as a single string of digits.

E2. This table gives the ages of the students enrolled at the University of Texas, San Antonio.

Source: www.utsa.edu/aviso/hilights/ student_profile.html.

Age	Percentage
17–22	44.2
23–29	31.4
30+	24.4

a. Make a reasonable estimate of the center of each age group, and compute the mean and standard deviation of the ages of these students.

b. Describe how to use this list of random digits to take a sample of six ages:

30558 45957 36911 97199 08432

E3. About 30% of parents occasionally call their child's teacher. [Source: *USA Today*, February 8, 1999.] Describe how to use the following random digits to take a sample of size 5 from this population. Take a sample of size 5. How many of the parents in your sample occasionally call their child's teacher?

24781 96045 15449 74564

E4. About 12% of teachers say it's effective for parents to email them. [Source: *USA Today*, February 8, 1999.] Describe how to use the following random digits to take a sample of size 3 from this population. Take a sample of size 3. How many of the teachers in your sample say it's effective for parents to email them?

22659 35091 36954 92166

E5. In E4, should you sample from the random digits with or without replacement?

E6. Describe how the formula for the standard deviation of a population,

$$\sigma = \sqrt{\sum (x - \mu)^2 \cdot P(x)}$$

is similar to, and how it is different from, the standard deviation formula for a frequency table,

$$s = \sqrt{\frac{\sum (x - \bar{x})^2 \cdot f}{n - 1}}$$

5.2   Generating Sampling Distributions

A sampling distribution answers the question, "How does my summary statistic behave when I repeat the process many times?"

Sometimes it is possible to imagine repeating the process of data collection. A **sampling distribution** is the distribution of the summary statistics you get from taking repeated random samples. For example, in Chapter 1 (Display 1.8), you created a simulated sampling distribution of the mean age from random samples of three employees who could have been laid off at Westvaco. This distribution is shown again in Display 5.6. There were 10 workers who could have been laid off; their ages were

$$25 \quad 33 \quad 35 \quad 38 \quad 48 \quad 55 \quad 55 \quad 55 \quad 56 \quad 64$$

You randomly selected three of these workers—without replacement because Westvaco couldn't have laid off the same worker twice.

Your summary statistic was the mean (or average) age of the three workers you chose at random. You were able to repeat this process many times in order to generate a distribution of possible values for the summary statistic.

From this sampling distribution, you could make a decision about whether it was reasonable to assume that employees were selected for layoff without respect to their age. The ages of the three workers actually laid off were 55, 55, and 64, for an average age of 58. As you can see from Display 5.6, it's rather hard to get an average age that large just by chance. So Westvaco had some explaining to do.

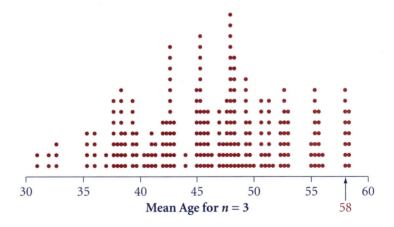

Display 5.6   A simulated sampling distribution of the mean age from random samples of three people who could have been terminated at Westvaco.

In the Westvaco analysis, you went through four steps for using simulation to generate an approximate sampling distribution of the mean age of three workers.

Random sample   I. Take a random sample of a fixed size *n* from a population.

Summary statistic   II. Compute a summary statistic.

Repetition   III. Repeat steps I and II many times.

Distribution   IV. Display the distribution of the summary statistics.

Discussion: Generating Sampling Distributions

D5. Describe how the four steps for generating a simulated sampling distribution relate to the methods used in the Westvaco simulation in Chapter 1.

Shape, Center, and Spread: Now and Forever

In Chapter 4 (Activity 4.2), you generated another sampling distribution when you selected five rectangles at random from the ones in Display 4.4 and then calculated their area. The summary statistic was the sample mean or average area of your sample's rectangles. Your simulated sampling distribution of the sample mean should resemble the one in Display 5.7, which was generated from 200 random samples of five rectangles each.

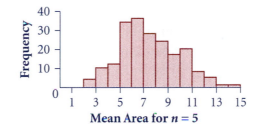

Display 5.7 A simulated sampling distribution of the sample mean ($n = 5$) for the areas of rectangles.

A good description of a sampling distribution such as this one is still the trio

SHAPE   CENTER   SPREAD

This sampling distribution is approximately normal, with a hint of a skew to the right, with mean 7.31 and standard deviation 2.39.

Compare Display 5.7 with the population of the areas of all 100 rectangles, shown in Display 5.8. It has mean 7.42 and standard deviation 5.2.

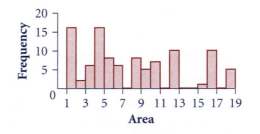

Display 5.8 The distribution of the population of rectangle areas.

Two standard deviations: population and standard error

The means of the two distributions are about the same—and in fact would be identical if all possible samples of size 5 were included in the sampling distribution. Note, however, that there are now two quite different standard deviations in the picture—the standard deviation of the population (the areas of all 100 rectangles) and the smaller standard deviation of the sampling distribution (means generated from samples of size 5). To tell them apart more easily, it helps to have different names, even though they are both standard deviations. It is natural to call the standard deviation for the population the **population standard deviation.** As you saw in Section 5.1, its symbol is σ. It is common (but not universal) to call the standard deviation of the sampling distribution the **standard error,** abbreviated *SE.*

Thus, for random samples of size 5 from the rectangle population, the standard error of the sample mean is approximately 2.39 (as seen in Display 5.7), while the population standard deviation is $\sigma \approx 5.20$ (as seen in Display 5.8).

Two outcomes: reasonably likely and rare

Values that lie in the middle 95% of an approximately normal distribution are called **reasonably likely.** Those in the outer 5% are called **rare events.** In other words, the reasonably likely values for a sampling distribution are those that lie within approximately two standard errors of the mean.

Discussion: Shape, Center, and Spread

D6. Compare Displays 5.7 and 5.8.

    a. Why is the largest value in Display 5.8 larger than the largest value in Display 5.7? Why is the smallest value smaller?

    b. Would you expect the means of the distributions in Displays 5.7 and 5.8 to be equal? Why or why not?

    c. Would you expect the standard deviations to be equal? Why or why not?

    d. How does the shape of the sampling distribution for samples of size 5 compare to the distribution of the population of the areas of all 100 rectangles?

D7. Approximately what values of the sample mean for samples of size 5 would be reasonably likely? Which would be rare events?

Once you're used to the steps for generating sampling distributions and have a calculator or computer that lets you do simulations easily, there is no end to the variety of summary statistics you can study. In the next activity, you will use the 100 rectangles from Display 4.4 of Chapter 4 to generate sampling distributions for two summary statistics other than the mean—the sample median and the sample maximum.

Activity 5.1   The Return of the Random Rectangles

What you'll need: a copy of Display 4.4, a method of producing random digits

1. Generate five distinct random numbers between 00 and 99.

2. Find the rectangles in Display 4.4 that correspond to your random numbers. (The rectangle numbered 100 can be called 00.) This is your random sample of five rectangles.

3. Determine the areas of the rectangles in your random sample and find the sample median. (Save these five areas for step 7.)

(continued)

4. Combine your sample median with those of other students in the class and repeat steps 1 through 3 until you have 200 sample medians. Construct a plot showing the distribution of the 200 sample medians.

5. Describe the shape, mean, and standard error of this plot of sample medians. How do these compare to the shape, mean, and standard deviation of the population of areas in Display 5.8?

6. How does this sampling distribution compare to the simulated sampling distribution for the sample mean in Display 5.7?

7. Repeat steps 4 and 5, but this time use the maximum area of the five rectangles in each sample as the summary statistic. For example, if the areas in your sample are 1, 8, 4, 8, and 3, the maximum area is 8. (You can use the same samples as before.) Keep your data. You will need it for E14.

Discussion: Return of the Random Rectangles

D8. Compare the population in Display 5.8 to the sampling distributions you plotted in Activity 5.1.

a. What is the median of the population of 100 rectangle areas? Is the sample median a good estimate of the median of the areas of all 100 rectangles?

b. What is the maximum of the population of 100 rectangle areas? Is the sample maximum a good estimate of the largest area of the 100 rectangles?

D9. ***Reasonably likely and rare events.*** Use the plots you constructed in Activity 5.1 to answer these questions.

a. What values of the sample median are reasonably likely? What values would be rare events?

b. What values of the sample maximum are reasonably likely? What values would be rare events?

D10. You could use either the sample mean or the sample median to estimate the "center" of the population of the areas of the rectangles. Discuss the difference between the two sampling distributions. In what circumstances would each be preferable as an estimate of center?

■ Practice

P3. The histogram in Display 5.9 shows the distribution of exam scores for 192 students in an introductory statistics course at the University of Florida.

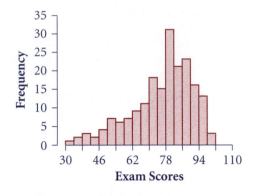

Display 5.9 A distribution of exam scores.

a. Match the histograms A, B, and C in Display 5.10 to their descriptions, I, II, or III.

I. The individual scores for one random sample of 30 students

II. A simulated sampling distribution of the mean of the scores of 100 random samples of 30 students

III. A simulated sampling distribution of the mean of the scores of 100 random samples of 4 students

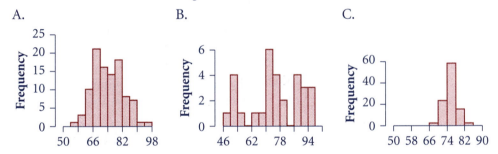

Display 5.10 Histograms for exam scores.

b. The second-hour class of 30 students, whose exam scores are part of the data set, had a class average of 86 on this exam. The instructor says that this class is just a random assortment of typical students taking this course. Do you agree?

The midrange is another measure of center:
$\frac{\text{min} + \text{max}}{2}$.

P4. The mean and median are only two of many possible measures of center. Another is the **midrange**, defined as the midpoint between the minimum and the maximum in a data set. The rectangles in Display 4.4 have areas ranging from 1 to 18, so the population midrange is $\frac{1+18}{2}$ or 9.5. The histogram in Display 5.11 shows a simulated sampling distribution of the midrange based on the same 200 samples used in Display 5.7. The mean is 7.84 and the standard error is 2.22.

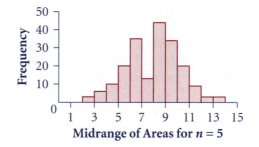

Display 5.11 A sampling distribution for the sample midrange (*n* = 5).

    a. Take a random sample of 5 rectangles; compute the midrange of their areas.

    b. Describe the sampling distribution of the midrange. How does it compare with the sampling distributions of the sample mean and the sample median?

    c. Would you use the midrange as a measure of center for the rectangle area population? Why or why not?

P5. From the exam scores of P3, the sampling distributions of different summary statistics in Display 5.12 were generated for the minimum, maximum, midrange, and lower quartile of random samples of size 10.

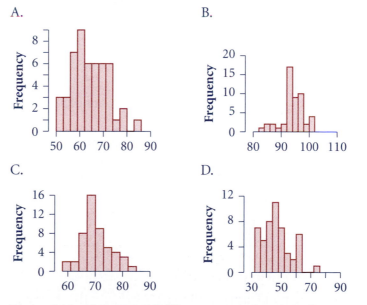

Display 5.12 Histograms of different summary statistics from random samples of size 10 of statistics exam scores.

    a. Match each histogram to its summary statistic:

        I. The minimum   II. The maximum

       III. The midrange   IV. The lower quartile

    b. Which sampling distribution has the largest standard error? The next largest?

Summary 5.2: Sampling Distributions

A sampling distribution answers the question, "How would my summary statistic behave if I could repeat the process of collecting data using random samples?" By showing you the summary statistics from random samples, a sampling distribution can help you decide whether the same process created the result from your actual sample.

You can generate a simulated sampling distribution for any sample statistic by following these steps:

Random sample
    I. Take a random sample of size *n* from a population.

Summary statistic
    II. Compute a summary statistic.

Repetition
    III. Repeat steps I and II many times.

Distribution
    IV. Display the distribution of the summary statistics.

You can sometimes create an exact sampling distribution by listing all the possible samples. In many more situations, you will be able to describe the shape, mean, and standard error of the sampling distribution without simulation and without listing samples. You will learn how to do this for sample means and sample proportions in Sections 5.3 and 5.4.

There are several key facts about sampling distributions.

- Sampling distributions are useful only for random samples and are generated by repeating the random sampling process many times.

- Sampling distributions, like data distributions, are best described by shape, center, and spread.

- The standard deviation of a sampling distribution is called the standard error.

- Many, but not all, sampling distributions are approximately normal. For normal distributions, reasonably likely outcomes are those that fall within approximately two standard errors of the mean.

■ Exercises

E7. Match each histogram in Display 5.13 to its description.

    I. The population of random digits

    II. The distribution of one random sample of 20 digits taken from this population

    III. A simulated sampling distribution of the sample mean for random samples of size 2 taken from this population

    IV. A simulated sampling distribution of the sample mean for random samples of size 20 taken from this population

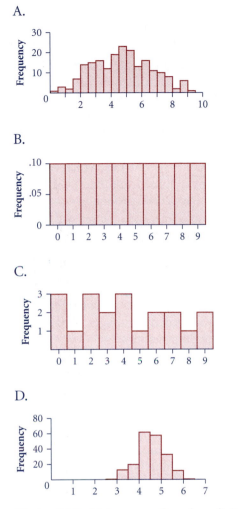

A.

B.

C.

D.

Display 5.13   Histograms of random digits.

State	Acres (millions)	State	Acres (millions)
AL	33.1	MT	94.1
AK	366.0	NB	49.5
AZ	73.0	NV	70.8
AR	34.0	NH	5.9
CA	101.6	NJ	5.0
CO	66.6	NM	77.8
CT	3.2	NY	31.4
DE	1.3	NC	33.7
FL	37.5	ND	45.2
GA	37.7	OH	26.4
HI	4.1	OK	44.8
ID	53.5	OR	62.1
IL	36.1	PA	29.0
IN	23.1	RI	0.8
IA	36.0	SC	19.9
KS	52.6	SD	49.4
KY	25.9	TN	27.0
LA	30.6	TX	170.8
ME	21.3	UT	54.3
MD	6.7	VT	6.1
MA	5.3	VA	26.1
MI	37.4	WA	43.6
MN	54.0	WV	15.5
MS	30.5	WI	35.9
MO	44.6	WY	62.6

E8. The areas of all fifty U.S. states, in millions of acres, are given in Display 5.14.

a. Describe the shape of the distribution.

b. Use a random digits table to select a random sample of five areas. Do these values appear to be representative of the population? What is the area of your five values?

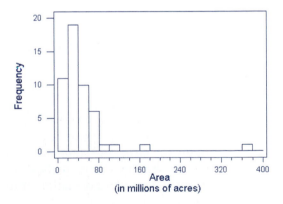

Display 5.14   Areas of U.S. states in millions of acres (not including inland bodies of water).

Source: *World Almanac and Book of Facts 2001,* p. 621.

c. The plot in Display 5.15 shows the mean area for 25 random samples, each of size 5, just as you generated in part b. Describe this simulated sampling distribution.

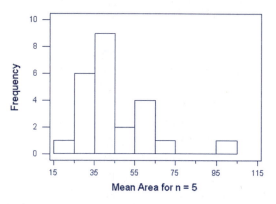

Display 5.15 A simulated sampling distribution of the sample mean of state areas.

d. The largest value in the plot above could have come from what particular sample?

E9. The five tennis balls in a can have diameters 62, 63, 64, 64, and 65 mm. Suppose you select two of the tennis balls at random without replacing the first before selecting the second.

a. Construct a dot plot of the population values.

b. List all possible sets of size 2 that can be chosen from the five balls. There are 5C2 ("5 choose 2") or 10 possible sets of two balls.

c. List all 10 possible sample means, and construct a dot plot of the sampling distribution of the sample mean. Compute the mean and standard error of this distribution. Compare these to the mean and standard deviation of the population.

d. Construct a dot plot displaying the sampling distribution of the maximum diameter of a sample of size 2.

e. Construct a dot plot of the sampling distribution of the range (maximum minus minimum) of a sample of size 2.

f. Which of your sampling distributions in parts c–e look closer to normal in shape than the dot plot of the population?

E10. As part of a statistics project at Iowa State University, a student tested how well a bike with treaded tires stopped on concrete. In six trials, these lengths of skid marks (in cm) were produced: 365, 374, 376, 391, 401, 402. Suppose instead the student had done only three trials, which could have been any three from this population of six, each with equal probability.

Source: Stephen B. Vardeman, *Statistics for Engineering Problem Solving.* (Boston: Prindle, Weber & Schmidt, 1994), p. 349.

a. Construct a dot plot of the population values.

b. List all possible sets of three skid lengths the student could get. There are 6C3 or 20 possible sets.

c. Construct a dot plot of the sampling distribution of the sample mean. Find its mean and standard error. Is the mean equal to the mean of the population?

d. Construct a dot plot of the sampling distribution of the sample median. Find its median. Is this equal to the median of the population?

e. Construct a dot plot of the sampling distribution of the sample minimum. Describe the shape of this distribution. Is the sample minimum a good estimator of the population minimum of all six trials? Explain.

f. Which of your sampling distributions in parts c–e look closer to normal in shape than the dot plot of the population?

E11. An inspector is called to see how much damage has been caused by a recent power failure to a warehouse full of frozen fish. The fish are stored in cartons containing 24 one-pound packages. There are hundreds of cartons in the warehouse.

The inspector decides he needs a sample of 48 one-pound packages of fish in order to assess the damage. He selects two cartons at random and treats this as a sample of 48 one-pound units.

The inspector doesn't realize it, but if 1 package of fish in a carton spoils, very rapidly the whole carton is spoiled. So it is safe to assume that each carton in the warehouse is either completely spoiled or not spoiled at all.

a. If the inspector is using pounds of spoiled fish in the sample as a statistic, what will the sampling distribution of such a statistic look like? Describe this distribution as completely as you can.

b. Did the inspector choose a good sampling plan? If not, explain how he could have done better.

E12–13. Back in Chapter 2, you may have wondered why the denominator of the sample standard deviation is $n - 1$ rather than n. Now that you know about sampling distributions, you can use them to answer this question in true statistical fashion. If a sample standard deviation is to be a good estimator of the population standard deviation σ, its sampling distribution should be centered at σ, or at least near that value. In E12 and E13, you will see if that turns out to be the case.

E12. Suppose you have a population that consists of only the three numbers 2, 4, and 6. You take a sample of size 2, replacing the first number before you select the second. Use parts a–d to complete the table.

a. In the first column, list all nine possible samples. (Count 2, 4 as different from 4, 2; that is, order matters.)

Sample	Mean	Variance, Dividing by $n = 2$	Variance, Dividing by $n - 1 = 1$
2, 2	2	0	0
2, 4	3	1	2
⋮	⋮	⋮	⋮
Average			

b. Compute the mean of each sample and write it in the second column.

c. Compute the variance of each sample, dividing by $n = 2$, and enter it in the third column. (You should be able to do this in your head.)

d. Compute the variance of each sample, dividing by $n - 1 = 1$, for the fourth column.

e. Compute the average of each column.

f. Compute the variance of the population {2, 4, 6} by using the formula given in Section 5.1 and assuming each of the three values is equally likely to be sampled. Compare it to your answers in part e.

g. On the average, do you get a better estimate of the variance of the population if you divide by n or by $n - 1$? Explain how you know.

h. If you divide by n rather than by $n - 1$, does the variance tend to be too big or too small on average?

E13. Look again at the rectangle areas of Display 5.8. Take a random sample of five rectangles. Compute the standard deviation of the areas of these rectangles using the regular formula, dividing by $n - 1$ or 4. Then compute the standard deviation again, but divide by n or 5. Which is larger?

The mean area of the population is 7.42, and the population standard deviation is σ or 5.20. The first histogram in Display 5.16 shows the sample standard deviations from the same samples of size 5 whose means are plotted in Display 5.7. The mean of these standard deviations is 5.03.

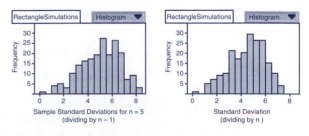

Display 5.16   Simulated sampling distributions of sample standard deviations ($n = 5$).

If you divide by 5 rather than by 4 (that is, n instead of $n - 1$), the results produce the distribution shown in the second histogram. The mean of this distribution is 4.50. Explain why dividing by $n - 1$ gives a better estimate of the population standard deviation.

E14. As you found in Activity 5.1, step 7, if you use the sample maximum as an estimate of the population maximum, it tends to be too small. Here is one possible rule for adjusting it:

$$population\ maximum \approx \frac{n + 1}{n} \cdot sample\ maximum$$

As always, n is the sample size.

a. Use the dotplot to explain why this rule is reasonable. This dotplot shows a sample of size 4 from a uniform population with minimum 1.

b. Test how well this rule works using your results from Activity 5.1, step 7.

E15. You have seen a number of examples in which the variation in the sampling distribution of the sample mean is smaller than the variation in the sampling distribution for the sample median. However, these examples arose from populations that were symmetric, or nearly so. Suppose the distribution for your population is highly skewed. (For example, the areas of the U.S. states.) Will the sampling distribution for the sample median have less variation than the sampling distribution for the sample mean? Use simulation to provide evidence for your answer.

5.3   Sampling Distribution of the Sample Mean

For a finite population, the sampling distribution of the sample mean is the set of means from all possible samples of a specific size.

You have seen that the sample mean (average) is a very important statistic, especially for symmetric distributions. In fact, you knew that the mean is important before you ever took a statistics course because you see averages used all around you—average exam score, average income, average age, batting average, and on and on. In this section, you will learn to predict the mean, standard deviation, and shape of the sampling distribution of the sample mean. Because its properties are so well understood, the sample mean will be used almost exclusively as the measure of center in the remainder of this book.

Some of the distributions of data that you have studied so far have had a roughly normal shape, but many others were not normal at all. Part of what makes the normal distribution important is its tendency to emerge when you create sampling distributions. The next activity shows how the normal distribution comes up unexpectedly when you work with the sampling distribution of a mean.

Activity 5.2   Cents and Center

What you'll need: 25 pennies collected from recent day-to-day change

1. Make a list of the dates on your random sample of 25 pennies. Next to each date, write the age of the penny by subtracting its date from the current year. If you were to make a histogram of the ages of all the

(continued)

pennies from all the students in your
class, what do you think the shape of
the distribution would look like?

2. Make that histogram of the ages of all
the pennies in the class.

3. Estimate the mean and the standard
deviation of the distribution. Now confirm these estimates by actual
computation.

4. Take a random sample of size 5 from the ages of your class' pennies, and
compute the mean age of your sample.

5. If you were to make a histogram of all the mean ages computed by
the students in your class in step 4, do you think the mean of the values
in this histogram will be larger than, smaller than, or the same as the one
for the population of the ages of all the pennies? Regardless of your
choice, try to make an argument to support each choice. Estimate what
the standard deviation of the distribution of the mean ages will be.

6. Make the histogram of the mean ages, and determine its mean and standard
deviation. Which of the three choices in step 5 appears to be correct?

7. Repeat steps 4–6 for samples of size 10 and size 25.

8. Look at the four histograms that your class has constructed. As *n*
increases, what can you say about the shape of the histogram? About the
center? About the spread?

Shape, Center, and Spread

In the rest of this section, you'll see if the results from Activity 5.2, step 8, are true
for other distributions. Display 5.17 shows the data from Section 5.1 on the
number of motor vehicles per household in the United States. This isn't a sampling
distribution; it's the population distribution you will use to build a sampling
distribution. The population has mean 2.171 and standard deviation 0.945.

Number of Motor Vehicles (per household), x	Proportion of Households, $P(x)$
0	.014
1	.228
2	.437
3	.215
4	.106

Display 5.17 The number of motor vehicles per household.

Source: *Statistical Abstract of the United States, 1996.*

These four steps describe how to construct a simulated sampling distribution of the mean for samples of size 4.

Random sample

I. Take a random sample from a population.

    In your model, there should be a .014 chance that a randomly selected household will have no motor vehicles, and so on. Use a list of random digits, as you did in Section 5.1, to select a random sample of four households from this distribution.

Summary statistic

II. Compute a summary statistic for your random sample.

    Suppose that your random sample represents households with 2, 2, 1, and 2 motor vehicles. The sample mean is then

$$\bar{x} = \frac{(2 + 2 + 1 + 2)}{4} = 1.75$$

Repetition

III. Repeat steps I and II many times.

Distribution

IV. Display the distribution of the summary statistics.

    Display 5.18 gives the results from hundreds of repetitions. This simulated sampling distribution has mean 2.17 and standard error 0.47.

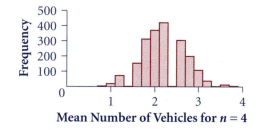

Display 5.18   A histogram of the means of samples of size 4 for number of vehicles per family.

The simulated sampling distribution in Display 5.19 is constructed like that in Display 5.18 except that larger samples, of size 10, are used. It has mean 2.17 and standard error 0.30.

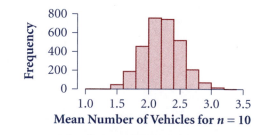

Display 5.19   A histogram of the means of samples of size 10 for the number of vehicles per family.

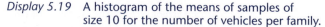

Discussion: Shape, Center, and Spread

D11. Compare the shape, center, and spread of the three distributions in Displays 5.17, 5.18, and 5.19 (population, means of samples of size 4, and means of samples of size 10).

D12. Explain why there are gaps in Display 5.18.

■ Practice

P6. This distribution is from Display 5.4 in Section 5.1; it gives the number of children per family in the United States.

Number of Children (per family)	Proportion of Families
0	.505
1	.203
2	.191
3	.073
4 or more	.028

a. Describe how to construct a simulated sampling distribution of the mean from random samples of size 10 from this distribution.

b. Describe how to construct a simulated sampling distribution of the mean from random samples of size 1 from this distribution. What would this sampling distribution look like?

Shape, Center, and Spread

The plot of the population of the number of motor vehicles in Display 5.17 is fairly symmetric. The distribution of the number of children per family in Display 5.4 is highly skewed. The plots in Display 5.20 show simulated sampling distributions of the sample mean for samples of size 1, 10, 20, 40, and 80 from the population of the number of children per family. In each case, a thousand sample means were generated.

Discussion: Sampling from a Skewed Population

D13. *The Mean*

a. How does the mean of each sampling distribution in Display 5.20 compare to the mean of the population?

b. Does the mean of the sampling distribution appear to depend on n, the sample size?

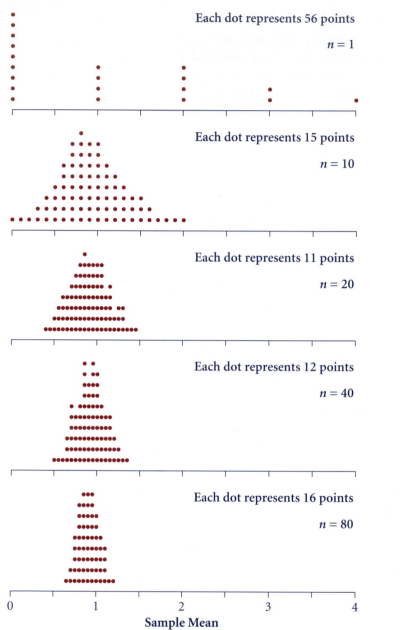

Sample Size, n	Mean	Standard Error, SE
1	0.92	1.11
10	0.92	0.35
20	0.92	0.25
40	0.92	0.18
80	0.92	0.12

Display 5.20 Sampling distributions of the sample mean for samples of size 1, 10, 20, 40, and 80 from the population of the number of children per family.

D14. *The Standard Error*

    a. How do the standard errors of the sampling distributions in Display 5.20 compare to the standard deviation of the population?

    b. Make a scatterplot of the standard error plotted against n. Does the standard error depend on the sample size?

    c. Try to find a rule that gives the standard error in terms of σ and n.

    d. Check your rule with the sampling distributions for $n = 4$ and $n = 10$ for the number of motor vehicles in Displays 5.18 and 5.19.

D15. *The Shape*

    a. Describe what happens to the shapes of the sampling distributions in Display 5.20 as the sample size increases.

    b. Is this pattern consistent with those for the sampling distributions constructed for motor vehicles per household?

It is time to collect the information you have found in D13, D14, and D15 into a set of rules for describing the behavior of sampling distributions of the sample mean. First the symbols commonly used are organized in this table.

	Population	Sample	Sampling Distribution
Mean	μ	$\bar{x}$	$\mu_{\bar{x}}$
Standard Deviation	σ	s	$\sigma_{\bar{x}}$
Size	N	n	

Properties of the Sampling Distribution of the Sample Mean

If a random sample of size n is selected from a population with mean μ and standard deviation σ, then

Center

- the mean $\mu_{\bar{x}}$ of the sampling distribution of $\bar{x}$ equals the mean of the population μ:

$$\mu_{\bar{x}} = \mu$$

Spread

- the standard deviation $\sigma_{\bar{x}}$ of the sampling distribution of $\bar{x}$, sometimes called the **standard error of the mean,** equals the standard deviation of the population σ divided by the square root of the sample size n:

$$\sigma_{\bar{x}} = \frac{\sigma}{\sqrt{n}}$$

Central Limit Theorem

- the shape of the sampling distribution will be approximately normal if the population is approximately normal; for other populations, the sampling distribution becomes more normal as n increases. This property is called the **Central Limit Theorem.**

All three properties are of great importance in statistics, and all three depend on *random* samples. The first property, $\mu_{\bar{x}} = \mu$, says that the means of random samples are centered at the population mean. This may seem obvious, and it is obvious for symmetric populations. But as you saw in Activity 5.2, it's also true for skewed populations.

The second property, $\sigma_{\bar{x}} = \sigma/\sqrt{n}$, is the main reason for using the standard deviation to measure spread—you can find the standard error for the sample mean *without simulation*. This result validates our intuitive feeling of why large samples are better: The larger the sample, the closer its mean tends to be to the population mean.

The third property, the Central Limit Theorem, which deals with the shape of the sampling distribution, will be central to your work in the rest of this course. It helps you decide which outcomes are reasonably likely and which are not. Notice that the normality may depend on the sample size being large, but the first two results on center and spread are true for any sample size.

Discussion: Properties of a Sampling Distribution

D16. Justify the comment, "Large samples are better; the sample mean tends to be closer to the population mean."

■ Practice

P7. From 1910 through 1919, the single-season batting averages of individual major league baseball players had a distribution that was approximately normal, with a mean of 266 and standard deviation of 37. [Source: Stephen Jay Gould, *The Spread of Excellence from Plato to Darwin* (New York: Harmony Books, 1996).] Suppose you construct the sampling distribution of the mean batting average of a random sample of 15 players. What is the shape, mean, and standard error of this distribution?

Finding Probabilities for Sample Means

As you will see in the next example, you can solve problems involving the sample mean using the three properties of the sampling distribution in combination with what you learned about normal distributions in Chapter 2.

Example: Average Number of Children

What is the probability that a random sample of 20 families in the United States will have an average of 1.5 children or fewer?

Solution

By looking at the simulated distribution in Display 5.20 for $n = 20$, you can see that the sampling distribution of the sample mean is approximately normal.

The mean of the population is 0.916 (from **Section 5.1**), and the standard deviation of the population is 1.11. So for **the sampling distribution**, we have

$$\mu_{\bar{x}} = \mu = 0.916$$

$$\sigma_{\bar{x}} = \frac{\sigma}{\sqrt{n}} = \frac{1.11}{\sqrt{20}} \approx 0.2482$$

Display 5.21 shows a sketch of this situation.

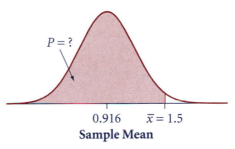

Sample Mean

Display 5.21 Sampling distribution of $\bar{x}$ when $\mu_{\bar{x}} = 0.916$, $\sigma_{\bar{x}} = 0.2482$, and $n = 20$. Shaded area shows the probability that the sample mean is less than 1.5.

The *z*-score for the value 1.5 is

Because this is a *z*-score for a sampling distribution, the mean is $\mu_{\bar{x}}$ and the standard deviation is the standard error or $\sigma/\sqrt{n}$.

$$z = \frac{\bar{x} - mean}{standard\ deviation} = \frac{\bar{x} - \mu_{\bar{x}}}{\sigma_{\bar{x}}} = \frac{\bar{x} - \mu}{\sigma/\sqrt{n}} \approx \frac{1.5 - 0.916}{0.2482} \approx 2.35$$

The area under a standard normal curve to the left of $z = 2.35$ is .9906, which represents the probability that the sample mean will fall below 1.5. In a random sample of 20 families, it is almost certain that the mean number of children per family will be less than 1.5. ■

Example: Reasonably Likely Averages

What average numbers of children are reasonably likely in a random sample of 20 families?

Solution

From the preceding example, the distribution of the mean number of children is approximately normally distributed with mean 0.916 and standard error 0.2482. Therefore, reasonably likely outcomes are those within approximately two standard errors of the mean. This interval is

$$0.916 \pm 2(0.2482)$$

or an average number of children between 0.420 and 1.412. ■

Discussion: Finding Probabilities for Sample Means

D17. What is the probability that a random sample of nine U.S. families will have an average of 1.5 children or fewer? Can this problem be done like the *Average Number of Children* example? Why or why not?

■ Practice

P8. Suppose a television network selects a random sample of 1000 families in the United States for a survey on TV viewing habits.

   a. Describe the distribution of the possible values of the mean number of children per family.

   b. What is the probability that the mean number of children per family will be .9 or less?

Finding Probabilities for Sample Totals

Sometimes situations are stated in terms of the total number in the sample rather than the average number: "What is the probability that there are 30 or fewer children in a random sample of 20 families in the United States?" You have the choice of two equivalent ways to do this problem.

- **Method I:** Find the equivalent average number of children, $\bar{x}$, by dividing the total number of children, 30, by the sample size, 20:

$$\bar{x} = \frac{30}{20} = 1.5$$

Then you can use the same formulas and procedure as in the previous examples.

- **Method II:** Convert the formulas from the previous examples to equivalent formulas for the sum, then proceed as in the next example.

The sum (or total) of a sample is equal to the sample size times the sample mean,

$$\sum x = n\bar{x}$$

You can get the properties of the sampling distribution of the sum from those of the sampling distribution of the mean simply by multiplying by n.

Properties of the Sampling Distribution of the Sum of a Sample

If a random sample of size n is selected from a distribution with mean μ and standard deviation σ, then

- the mean of the sampling distribution of the sum is

$$\mu_{sum} = n\mu$$

(continued)

- the standard error of the sampling distribution of the sum is

$$\sigma_{sum} = n\sigma_{\bar{x}}$$

$$= n\frac{\sigma}{\sqrt{n}}$$

$$= \sqrt{n} \cdot \sigma$$

- the shape of the sampling distribution will be approximately normal if the population is approximately normal; for other populations, the sampling distribution becomes more normal as n increases.

Example: The Probability of 30 or Fewer Children

What is the probability that a random sample of 20 families in the United States will have a total of 30 children or fewer?

Solution

The sampling distribution of the sum is approximately normal because it has the same shape as the sampling distribution of the mean. It has mean $\mu_{sum} = n\mu$ or $20(0.916) = 18.32$ and standard error $\sigma_{sum} = \sqrt{n} \cdot \sigma = \sqrt{20}(1.11) \approx 4.96$. The z-score for a total of 30 children is

$$z = \frac{sample\ sum - \mu_{sum}}{\sigma_{sum}} = \frac{sample\ sum - n\mu}{\sqrt{n} \cdot \sigma}$$

$$= \frac{30 - 20(0.916)}{\sqrt{20}(1.11)} \approx 2.35$$

This is the same z-score as in the previous example, so again, the probability that z is less than 2.35 is about .9906. ■

Example: Reasonably Likely Totals

In a random sample of 20 families, what total numbers of children are reasonably likely?

Solution:

From the preceding example, the sampling distribution of the total number of children is approximately normal with mean 18.32 and standard error 4.96. The reasonably likely outcomes are those within approximately two standard errors of the mean. This is the interval

$$18.32 \pm 2(4.96)$$

or a total number of children between 8.40 and 28.24. ■

Discussion: The Sampling Distribution of the Sample Total

D18. How do the shapes of the sampling distributions of the mean and of the total compare? Show how to convert the plots for the sample mean in Display 5.20 to plots for the sample total.

■ Practice

P9. As in P8, suppose a television network selects a random sample of 1000 families for a survey on TV viewing habits.

a. Do you think it is reasonably likely that a sample of 1000 households will produce at least 1000 children? Explain your reasoning.

b. Suppose the network changes to a random sample of 1200 households. Does this dramatically improve the chance of seeing at least 1000 children in the sampled households? Explain.

P10. Suppose the network in P9 repeatedly uses one of the sample sizes given in parts a–d and computes the mean number of children per family. What interval should contain the middle 95% of the sample means? That is, what sample means are reasonably likely?

a. 25

b. 100

c. 1000

d. 4000 (This is the approximate sample size for the Nielsen television ratings.)

Sample Size and Population Size

Suppose you want a random sample of four households from the population given in Display 5.17. If a list of the millions of households in the country were available, you could number the households on the list and select a random sample by drawing numbers from a box or by using a random number table. Because you don't want to sample the same household twice, you sample without replacement.

No list of households in the United States is available, so random digits are used to represent households. This model of sampling was presented earlier in the chapter on page 274. Listen to Pop and Sam discuss this method of sampling:

Pop: That sampling model in step I was all wrong. Say there are 1000 households in the United States. Then 228 of them have one motor vehicle. Now I take a random sample of size 200. Suppose the first 100 households have one motor vehicle each. Out of the 900 households left, only 128 have one motor vehicle—that's now only $\frac{128}{900}$, or roughly 14.2%!

Sam: Hmmm. I see your point. The probabilities actually change depending on what you get first in your sample. In the sampling model in step I, they stayed the same on each draw.

Pop:	Right, you're quick.
Sam:	Wait a minute. There aren't 1000 households in the United States! There are closer to 100,000,000 households. About 22,800,000 of them have one motor vehicle.
Pop:	And if we take a random sample of size 200 and the first 100 have one motor vehicle . . .
Sam:	Then there are 22,799,900 out of the 999,999,900 households remaining that have one motor vehicle. That's still 22.79992%!
Pop:	So with a large population, it doesn't seem to matter if we sample with or without replacement.

Pop and Sam are right. As long as the sample size n is relatively small compared to the population size N, it doesn't matter if you sample with or without replacement. They have discovered one of the realities of doing statistics: A model may be useful even if it's not 100% correct.

A Rule for Relating Sample and Population Sizes

If the sample size is a small percentage (around 10% or less) of the population size, it doesn't matter much if you sample with or without replacement, and, in fact, the population size will have little effect on the statistical analysis.

Population size has little effect as long as it is large compared to the size of the sample.

The second part of this rule says something surprisingly strong. For example, a sample of size 20 provides about as much information from a population of size 200 as it does from a population of size 2000 or even 20,000,000. If, on the other hand, you had a sample of size 20 from a population of only 30, then the population size must be taken into account in the analysis by using a different formula for the standard error, one that is more complex and gives a smaller value. A sample that is more than 10% of the population rarely occurs in practice, so you won't study this case in this book.

	Pop and Sam are interrupting again.
Sam:	Are they trying to tell us that a random sample of 100 households out of the United States is just as good as a random sample of 100 households out of a country with only 5000 households?
Pop:	That's exactly what they are saying.
Sam:	But 100 out of 100,000,000 is a *very* small percentage of the households. How can that be as representative as 100 out of 5000?
Pop:	Well, technically you're right. One hundred out of 5000 gives you a little more information, but not enough to bother about.
Sam:	That's better. The population size *does* matter. Apparently not as much as I thought it might, though.

Discussion: Sample Size and Population Size

D19. If you select 100 households at random, would the standard error of the sampling distribution of the mean number of children be larger if the population size N is 1,000 or if it is 10,000?

D20. Consider the sampling distribution for the mean of a random sample of size n taken from a population of size N with mean μ and standard deviation σ.

 a. For a fixed N, how does the mean of the sampling distribution change as n increases?

 b. For a fixed N, how does the standard error change as n increases?

 c. If $N = n$, find the mean and the standard error of the sampling distribution.

■ Practice

P11. Describe how to select five households at random from 1000 households, using a list of random digits, if you want to sample:

 a. with replacement

 b. without replacement

P12. A typical political opinion poll in the United States questions about 1500 people. How would the analysis change if this was a random sample of 1500 people from Arizona rather than from the entire United States?

Summary 5.3: Sampling Distribution of the Sample Mean

Because averages are so important, it seems fortunate that the mean and standard error of the sampling distribution of $\bar{x}$ have such simple formulas. It seems doubly fortunate that for large sample sizes, the shape is approximately normal. This allows you to use z-scores and the standard normal distribution to determine whether a given sample mean is reasonably likely or a rare event. But fortunate has nothing to do with it! It's precisely because these formulas are so simple that the mean and standard deviation are so important.

Here is a summary of the characteristics of the sampling distribution of $\bar{x}$.

- The samples must be selected at random.

- The mean of the sampling distribution equals the mean of the population, $\mu_{\bar{x}} = \mu$, and this fact does not depend on the sample size.

- The standard error of the sampling distribution equals the population standard deviation divided by the square root of the sample size, or $\sigma_{\bar{x}} = \sigma/\sqrt{n}$, and this fact does not depend on the sample size.

- If the population is normally distributed, the sampling distribution is normally distributed for all sample sizes n. If the population is not normal, then as the sample size increases, the sampling distribution will become more normal. This result is called the Central Limit Theorem.

- The population size does not have much effect on the analysis unless the sample size is greater than 10% of the population size.

If you have a situation involving the sum or total in the sample, you can find a sample mean by dividing the total by n. Alternatively, the sampling distribution of the sample sum has mean

$$\mu_{sum} = n\mu$$

and standard error

$$\sigma_{sum} = \sqrt{n} \cdot \sigma$$

What rule can you use to decide whether the shape of your sampling distribution is approximately normal? Unfortunately, there isn't one—the required size of n depends on how close the population itself is to being normal and how accurate you want your approximations to be. There will be more on this topic later.

■ Exercises

E16. Display 5.22 shows the three simulated sampling distributions of the sample mean for samples of sizes 2, 4, and 25 selected from the skewed population shown.

a. Match the summary information with the correct sampling distribution (A, B, or C) and the correct sample size (2, 4, or 25).

   I. mean 2.50; standard error 0.50

  II. mean 2.50; standard error 1.25

  III. mean 2.50; standard error 1.80

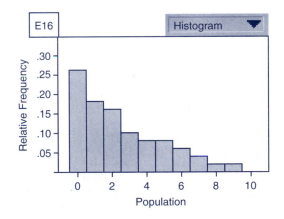

A.   B.   C.

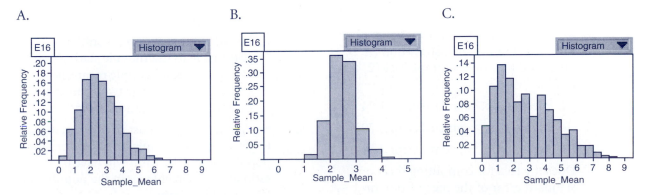

Display 5.22 Three sampling distributions of the sample mean ($n = 2, 4,$ and 25, not necessarily in that order).

b. Does the rule for computing the standard error of the mean from the standard deviation of the population appear to hold in all three situations?

c. Does the sampling distribution appear to be approximately normal in all cases? If not, explain how the given shape came about.

d. For which sample sizes would it be reasonable to use the rule stating that about 95% of all sample means lie within approximately two standard errors of the population mean?

E17. Explain how to use a list of random digits to get a random sample of size 5 from the population in E16. (You will have to estimate the percentages in the population from the histogram.)

E18. Display 5.23 shows three simulated sampling distributions of the sample mean for samples of sizes 2, 4, and 25 selected from this M-shaped population.

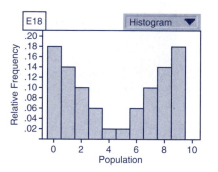

a. Match the summary information with the correct sampling distribution (A, B, or C) and the correct sample size (2, 4, or 25).

   I. mean 4.39;   standard error 1.84

   II. mean 4.53;   standard error 0.70

   III. mean 4.47;   standard error 2.39

b. Does the rule for computing the standard error of the mean from the standard deviation of the population appear to hold in all three situations?

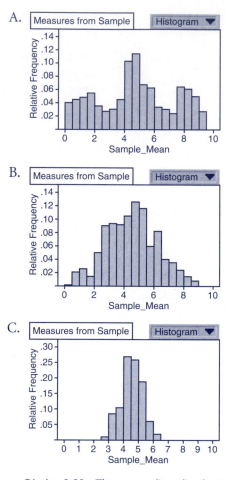

Display 5.23   Three sampling distributions ($n = 2$, 4, and 25, in some order).

c. Does the sampling distribution appear to be approximately normal in all cases? If not, explain how the given shape came about.

d. For which sample sizes would it be reasonable to use the rule stating that about 95% of all sample means lie within approximately two standard errors of the population mean?

E19. Explain how to use a list of random digits to get a random sample of size 5 from the population in E18. (You will have to estimate the percentages in the population from the histogram.)

E20. The distribution of grades on the 1998 AP Statistics exam are given in the table.

Grade	Percent Receiving Grade
5	13.5
4	21.3
3	24.6
2	18.7
1	21.9

Source: College Board, 1998, www.apcentral. collegeboard.com.

a. Make a plot of the distribution of the population of scores.

b. The plots in Display 5.24 show simulated sampling distributions of the mean for random samples of sizes 1, 5, and 25. Match each histogram to its sample size.

c. A teacher reports that her class averaged 3.6 on this exam. Do you think she has a class of 5 students or a class of 25 students? Explain your reasoning.

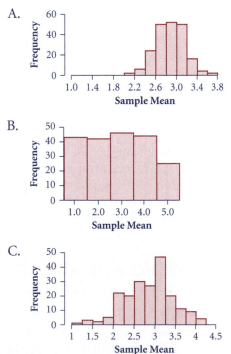

A.

B.

C.

Display 5.24   Simulated sampling distributions: $n = 1$, 5, and 25 (not necessarily in that order).

E21. Suppose that police records show that the number of automobile accidents per day in a small city has the frequency distribution shown in Display 5.25.

a. Would it be unusual to see two accidents per day in this city?

The two histograms in Display 5.26 show simulated sampling distributions of the mean number of accidents per day for random samples of four days and eight days.

b. Which histogram is for samples of four days. Which is for samples of eight days?

c. Would it be unusual to see an average of two accidents per day over a four-day period in this city?

d. Would it be unusual to see an average of two accidents per day over an eight-day period in this city?

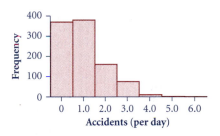
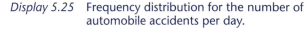

Display 5.25   Frequency distribution for the number of automobile accidents per day.

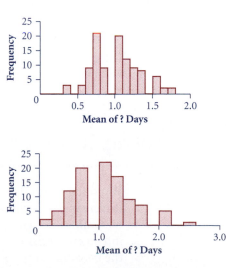

Display 5.26   Simulated sampling distributions ($n = 4$ and $n = 8$, not necessarily in that order).

e. When you use the distributions in Display 5.26 to make your decisions in parts c and d, what assumptions are you making? Do they seem reasonable for this situation?

E22. College entrance examination scores have an approximately normal distribution with mean 500 and standard deviation 100.

a. If you select 1 score at random, what is the probability that it is 510 or greater?

b. If you select 4 scores at random, what is the probability that their average is 510 or greater?

c. If you select 25 scores at random, what is the probability that their average is 510 or greater?

d. Explain how to use simulation to find the probability that each of four randomly selected scores is 510 or greater.

E23. College entrance examination scores have an approximately normal distribution with mean 500 and standard deviation 100.

a. How large a random sample of scores would you need in order to be 95% sure that the sample mean would be within 10 points of the population mean of 500?

b. Within 5 points?

c. Within 1 point?

E24. The process for manufacturing a ball bearing results in weights that have an approximately normal distribution with mean 0.15 g and standard deviation 0.003 g.

a. If you select 1 ball bearing at random, what is the probability that it weighs less than 0.148 g?

b. If you select 4 ball bearings at random, what is the probability that their mean weight is less than 0.148 g?

c. If you select 10 ball bearings at random, what is the probability that their mean weight is less than 0.148 g?

E25. As the sample size n increases, describe what happens to the mean and standard error of the sampling distribution of

a. the sample mean

b. the sample total

E26. Suppose a survey organization wants a sample of families that contains a total of at least 30 children. Use Display 5.20 to estimate the probability of at least 30 children in a random sample of size

a. 10   b. 20   c. 40   d. 80

Check your answer to part c using the properties of the normal distribution.

E27. Use Display 5.4 to find the interval that should contain the middle 95% of possible values for the *total* number of children in random samples of U.S. households of size

a. 25   b. 100   c. 1000   d. 4000

E28. Refer to Display 5.4. Suppose you will take a random sample of U.S. households. How large should the sample size be in order to give a chance of .025 that the total number of children in the sample is less than 1000?

5.4 Sampling Distribution of the Sample Proportion

You often hear reports of percentages or proportions: About 60% of automobile drivers in Kentucky use seat belts. (The national average is about 71%.) [Source: www.nhtsa.dot.gov/people/ncsa/ResearchNotes/20000SBURNR.pdf.] To make intelligent decisions based on data that is reported this way, you must understand the behavior of proportions that arise from random samples. As you will see, the properties of sample proportions parallel the properties of sample means that you learned in Section 5.3. (What luck!)

Suppose you take a random sample of 40 Kentucky drivers and count the number who wear seat belts. You would "expect" to get 60% of 40, or 24, who wear seat belts, but you might get a lot fewer or a lot more. In this section, you will learn to determine which outcomes are reasonably likely and which are rare events.

Some notation will make your work more efficient. Suppose you count the number of successes in a random sample of size *n* from a population in which the true proportion of successes is given by *p*. The symbol for the sample proportion is $\hat{p}$ (read as "p-hat"). That is,

$$\hat{p} = \frac{\text{number of "successes"}}{\text{sample size}}$$

For example, suppose your sample of automobile drivers contains 26 who use seat belts. Then $n = 40$, $p = .60$, and $\hat{p} = \frac{26}{40} = .65$.

In Activity 5.3, you will construct a simulated sampling distribution for the proportion of drivers wearing seat belts in samples of size 10, 20, and 40.

Activity 5.3   Buckle Up!

What you'll need: a table of random digits

1. Describe how to use a table of random digits to select a random sample of size 10 from a population with 60% "successes."

2. Use your method to select a random sample of size 10. Count the number of "successes," and compute the sample proportion, $\hat{p}$.

3. Repeat step 2 until your class has 100 samples of size 10.

4. Plot your 100 sample proportions on a dot plot.

5. Calculate the mean and standard error of your simulated sampling distribution of the sample proportion.

6. Repeat steps 2 through 5 for samples of size 20 and 40.

The histograms in Display 5.27 show the exact sampling distributions for samples of size 10, 20, and 40 drawn from a population with 60% "successes." They should be similar in shape, center, and spread to your dot plots from Activity 5.3.

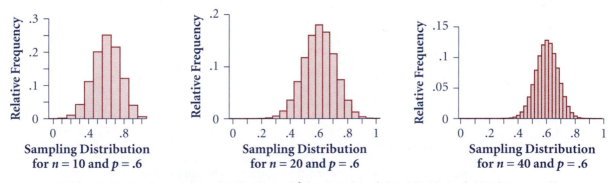

Display 5.27   Exact sampling distributions of $\hat{p}$ for samples of size 10, 20, and 40 when $p = .60$.

Discussion: Sampling Distribution for *p* = .6

D21. Use Display 5.27 to answer these questions.

a. How many drivers, on the average, would you expect to be wearing seat belts in a sample of size 40?

b. Is it reasonably likely that in a sample of 40 drivers, more than 75% would be wearing seat belts?

c. Would it be unusual to find fewer than 10 drivers wearing seat belts in a sample of 40 drivers?

d. Which would be more likely, to find that 75% or more of the drivers use seat belts in a sample of size 10, in one of size 20, or in one of size 40?

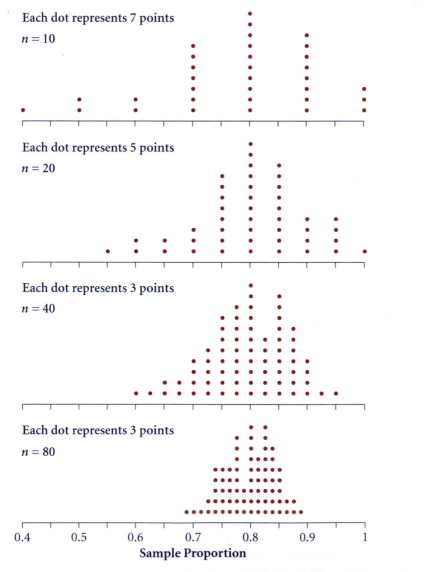

Sample Size	Mean	SE
10	.788	.125
20	.793	.085
40	.796	.063
80	.802	.042

Display 5.28 Simulated sampling distributions of the sample proportion when *p* = .8 (*n* = 10, 20, 40, and 80).

D22. Using Display 5.27, compare the shape, center, and spread of the distribution of possible values of $\hat{p}$ for samples of sizes 10, 20, and 40.

D23. Let's see what happens to the sampling distribution of the sample proportion if the percentage of successes changes to, say, 80%. The simulated sampling distributions in Display 5.28 show the results for simulations from sample sizes of 10, 20, 40, and 80 when $p = .8$.

   a. Compare the shape, center, and spread of these four distributions.

   b. How do the distributions for $p = .8$ compare with those for $p = .6$?

D24. Do the rules for sample means found on page 289 hold for sample proportions?

Center and Spread for Sample Proportions

To find formulas for the mean and the standard error of the sampling distribution of the sample proportion, let's go back to the seat belt example.

First note that the formula for $\hat{p}$ is equivalent to assigning the number 1 to every driver who uses seat belts and a 0 to every driver who doesn't. So for a sample that contains 26 drivers who wear seat belts and 14 who don't, the sample total is $26(1) + 14(0)$ or 26. Dividing this total by the sample size of 40, you get $\frac{26}{40}$. This looks familiar: $\hat{p}$ is the same thing as the mean of the sample. That is,

$\hat{p}$ is the mean of a sample taken from a population of 0's and 1's.

$$\hat{p} = \frac{\text{sum of values}}{\text{sample size}} = \bar{x}$$

You can describe the population in the same way: 40% of it is 0's and 60% of it is 1's. The summary table and histogram are shown in Display 5.29.

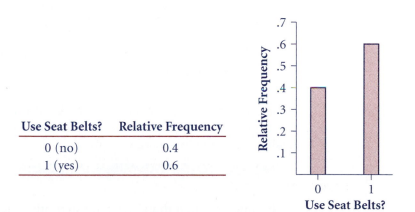

Use Seat Belts?	Relative Frequency
0 (no)	0.4
1 (yes)	0.6

Display 5.29   Relative frequency histogram for a population with $p = .6$.

You can find the mean of this population in the usual way:

$$\mu = \sum x \cdot P(x) = 0(.4) + 1(.6) = .6$$

which is equal to p.

The standard deviation is

$$\sigma = \sqrt{\sum (x - \mu)^2 \cdot P(x)}$$
$$= \sqrt{(0 - .6)^2(.4) + (1 - .6)^2(.6)}$$
$$= \sqrt{(.6)^2(.4) + (.4)^2(.6)}$$
$$= \sqrt{(.6)(.4)(.6 + .4)}$$
$$= \sqrt{(.6)(.4)}$$

which is equal to $\sqrt{p(1 - p)}$. This is true for any value of p: The mean and the standard deviation of the population of 0's and 1's are

$$\mu = p \qquad \text{and} \qquad \sigma = \sqrt{p(1 - p)}$$

Now you can use the same three rules as for the sampling distribution of the sample mean.

Properties of the Sampling Distribution of the Sample Proportion

If a random sample of size n is selected from a population with proportion of successes p, then the sampling distribution of $\hat{p}$ has these properties:

- The mean of the sampling distribution is equal to the mean of the population, or $\mu_{\hat{p}} = p$.

- The standard error of the sampling distribution is equal to the standard deviation of the population divided by the square root of the sample size:

$$\sigma_{\hat{p}} = \frac{\sigma}{\sqrt{n}} = \frac{\sqrt{p(1 - p)}}{\sqrt{n}} = \sqrt{\frac{p(1 - p)}{n}}$$

- As the sample size gets larger, the shape of the sampling distribution gets more normal and will be approximately normal if n is large enough.

As a guideline, if both np and $n(1 - p)$ are at least 10, then using the normal distribution as an approximation for the shape of the sampling distribution will give reasonably accurate results.

It's not often that things wrap up so neatly. The reason the properties of the sampling distributions of sample proportions are the same as those for sample means is that a proportion is a special kind of mean. You can use the same strategy to solve a problem about a proportion as you did when the problem was about a mean.

Now go back to the distributions in Display 5.28. The formula for $\mu_{\hat{p}}$ says that the mean of the sampling distribution of $\hat{p}$, for $p = .8$, should be $\mu_{\hat{p}} = p = .8$, no matter what the sample size is. This is fairly close to the means from the simulation. The standard errors predicted from the theory, for samples of size 10, 20, and 40, are

For $n = 10$: $\sigma_{\hat{p}} = \sqrt{\dfrac{p(1-p)}{n}} = \sqrt{\dfrac{.8(1-.8)}{10}} \approx .126$

$n = 20$: $\sigma_{\hat{p}} = \sqrt{\dfrac{.8(1-.8)}{20}} \approx .089$

$n = 40$: $\sigma_{\hat{p}} = \sqrt{\dfrac{.8(1-.8)}{40}} \approx .063$

Again, these are close to the standard errors from the simulation.

Discussion: Properties for Proportions

D25. Find the theoretical means and standard errors of the three sampling distributions in Display 5.27.

D26. Which sampling distributions in Display 5.28 look approximately normal? Does your visual impression agree with the guideline that you can consider the distribution approximately normal if both $np \geq 10$ and $n(1-p) \geq 10$?

■ Practice

P13. In the 2000 United States census, 53% of the population over the age of 30 were women.

Source: U.S. Bureau of the Census, www.census.gov/population/estimates/nation/intfile2-1.txt.

a. Describe the shape, mean, and standard error of the sampling distribution of $\hat{p}$ for random samples of size 100 taken from this population. Make an accurate sketch of this distribution with a scale on the x-axis.

b. To be a member of the U.S. Senate, you must be at least 30 years old. In 2000, there were 100 members of the U.S. Senate, and 9 of them were women. Is this a reasonably likely event to occur if gender plays no role in whether a person becomes a U.S. Senator?

Finding Probabilities for Proportions

You can now find answers to specific questions by using your knowledge of the properties of the sampling distribution of a proportion, and, in fact, you no longer have to refer to simulated sampling distributions.

Example: Using the Properties to Find Probabilities

About 60% of Kentucky drivers use seat belts. Suppose your class conducts a survey of 40 randomly selected drivers from Kentucky.

a. What is the chance that 75% or more of the drivers are wearing seat belts?

b. Would it be quite unusual to find fewer than 25% of the drivers wearing seat belts?

Solution

From the developments in this section, you know that $\hat{p}$ has a sampling distribution with mean and standard error

$$\mu_{\hat{p}} = p = .6$$

$$\sigma_{\hat{p}} = \sqrt{\frac{p(1-p)}{n}} = \sqrt{\frac{.6(1-.6)}{40}} \approx .0775$$

Furthermore, you can consider the sampling distribution to be approximately normal because np and $n(1-p)$ are both 10 or greater:

$$np = 40(.6) = 24$$

$$n(1-p) = 40(1-.6) = 16$$

This distribution is shown in Display 5.30.

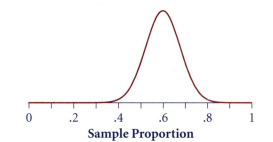

Sample Proportion

Display 5.30   Approximate sampling distribution of $\hat{p}$ for $p = .6$ and $n = 40$.

Using the normal distribution as an approximation

a. The *z*-score for the value $\hat{p} = .75$ in this distribution is

$$z = \frac{\hat{p} - \mu_{\hat{p}}}{\sigma_{\hat{p}}} = \frac{\hat{p} - p}{\sqrt{\frac{p(1-p)}{n}}} \approx \frac{.75 - .6}{.0775} \approx 1.935$$

The proportion below this value of *z* is .9735. It follows that the probability of getting 75% or more of drivers who wear seat belts in a sample of size 40 is only $1 - .9735$, or .0265.

b. The sample proportion of .25 is about 4.5 standard deviations below the mean of the sampling distribution, so such a result would be unusual indeed! ■

■ Practice: Finding Probabilities for Proportions

P14. In 1999, about 28% of the citizens of the United States were Republicans.
      Source: Gallup Poll of April 9, 1999, www.gallup.com/poll/releases/pr990409c.asp.

   a. If 435 citizens are selected at random, what is the probability of getting 51% or more who are Republicans?

b. In 2001, the U.S. House of Representatives had 435 members; 221 were Republicans. Is this percentage higher than could reasonably occur by chance? If so, what are some possible reasons for this unusually large sample percentage?

Source: www.clerkweb.house.gov/mbrcmtee/stats.htm.

P15. About 60% of Kentucky drivers wear seat belts. Determine what proportions of seat belt users would be reasonably likely to occur in a random sample

a. of 40 drivers   b. of 100 drivers   c. of 400 drivers

Finding Probabilities for the Number of Successes

Sometimes situations are stated in terms of the number of successes rather than the proportion of successes: "In a random sample of 40 Kentucky drivers, what is the probability that 30 or more use seat belts?" As with sample means in Section 5.3, you have the choice of two equivalent ways to do this problem.

- **Method I:** Convert the number of successes, 30, to the proportion of successes by dividing by the sample size: $\hat{p} = \frac{30}{40} = .75$. Then proceed as in the preceding example.

- **Method II:** Use the equivalent formulas for the sum and proceed as in the next example.

As in Section 5.3, you can get the formulas for the total number of successes if you multiply those for the proportion of successes by the sample size n.

Properties of the Sampling Distribution of the Number of Successes

If a random sample of size n is selected from a population with proportion of successes p, then the sampling distribution of the number of successes

- has mean $\mu_{sum} = np$
- has standard error $\sigma_{sum} = \sqrt{np(1-p)}$
- will be approximately normal as long as n is large enough

As a guideline, if both np and $n(1-p)$ are at least 10, then using the normal distribution as an approximation to the shape of the sampling distribution will give reasonably accurate results.

Example: Probability of 30 or More Wearing Seat Belts

In a random sample of 40 Kentucky drivers, what is the probability that 30 or more use seat belts?

Solution

Because 60% of the population use seat belts, both np and $n(1-p)$ are 10 or greater:

$$np = 40(.6) = 24$$

$$n(1-p) = 40(1-.6) = 16$$

Thus, the sampling distribution of the number of successes is approximately normal. It has mean and standard error

$$\mu_{sum} = np = 40(.6) = 24$$

$$\sigma_{sum} = \sqrt{np(1-p)} = \sqrt{40(.6)(1-.6)} \approx 3.098$$

This distribution is shown in Display 5.31.

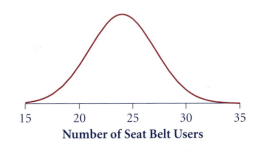

Number of Seat Belt Users

Display 5.31   Approximate sampling distribution of the number of successes when *p* = .6 and *n* = 40.

The *z*-score for 30 successes is

$$z = \frac{sample\ sum - \mu_{sum}}{\sigma_{sum}} = \frac{sample\ sum - np}{\sqrt{np(1-p)}} \approx \frac{30-24}{3.098} \approx 1.937$$

So the probability of getting 30 or more successes is 1 − .9737, or .0263, the same result (except for round-off error) as in the previous example. ■

Discussion: Finding Probabilities for the Number of Successes

D27. Show how the formulas for the mean and the standard error of the sampling distribution of the number of successes are derived from the formulas for the proportion of successes.

■ Practice

P16. About 64% of two-year college students in the United States are part-time. What is the probability a random sample of 500 two-year college students will have 316 or more who go to college part time? Do this problem using both methods.

Source: American Association of Community Colleges, 1995, www.aacc.nche.edu/allaboutcc/snapshot.htm.

Summary 5.4: Sampling Distribution of the Sample Proportion

Sample proportions, like sample means, are among the most common summary statistics in practical use, and knowledge of their behavior is fundamental to the study of statistics. A sample proportion is just another form of a sample mean, so the formulas and procedures in this section derive from those in Section 5.3.

These facts about the sampling distribution of the sample proportion parallel those in Section 5.3. So, they enable you to use *z*-scores and the normal distribution to approximate the probability of getting specified results from a random sample.

- The samples must be selected randomly.
- The mean $\mu_{\hat{p}}$ of the sampling distribution of $\hat{p}$ is equal to the proportion p of successes in the population, or $\mu_{\hat{p}} = p$.
- The standard deviation $\sigma_{\hat{p}}$ of the sampling distribution of $\hat{p}$, also called the standard error of a sample proportion, is given by

$$\sigma_{\hat{p}} = \sqrt{\frac{p(1-p)}{n}}$$

- Alternatively, the sampling distribution of the number of successes has mean μ_{sum} and standard error σ_{sum}, where

$$\mu_{sum} = np \quad \text{and} \quad \sigma_{sum} = \sqrt{np(1-p)}$$

- The shape of the sampling distribution becomes more normal as *n* increases. You may assume it is normal for computing probabilities as long as *np* and $n(1-p)$ are both 10 or larger.

■ Exercises

E29. The ethnicity of about 92% of the population of China is Han Chinese. Suppose you will take a random sample of 1000 Chinese.

Source: *World Almanac and Book of Facts 2001*, p. 775.

a. Make an accurate sketch, with a scale on the *x*-axis, for the sampling distribution of the proportion of Han Chinese.

b. Make an accurate sketch, with a scale on the *x*-axis, for the sampling distribution of the number of Han Chinese.

c. What is the probability of getting 90% or fewer Han Chinese in the sample?

d. What is the probability of getting 925 or more Han Chinese?

e. What numbers of Han Chinese would be rare events? What proportions?

E30. In the fall of 2000, 46% of the almost 34,000 first-year students attending one of the California State Universities needed remedial work in mathematics.

Source: California State University, www.asd.calstate.edu/performance/remediation.htm.

a. Suppose you will select 136 students at random from this population of students. Make an accurate sketch, with a scale on the *x*-axis, of the sampling distribution of the number of students who need remedial work.

b. What is the probability that 72 or fewer in a random sample of 136 students need remedial work?

c. Suppose you will select 2075 of the students at random. Make an accurate sketch, with a scale on the *x*-axis, of the sampling distribution of the proportion who need remediation.

d. What is the probability of getting 63% or more who need remediation in a random sample of 2075 students?

e. Of the 2658 students entering Cal State University Northridge, 57% needed remediation. Is this result about what you would expect from a random sample, or should you conclude that this group is special in some way?

E31. In 1991, the median age of residents of the United States was 33.1 years.

Source: U.S. Bureau of the Census, www.census.gov/population/estimates/nation/intfile2−1.txt.

a. What is the probability that one person, selected at random, was under the median age?

b. In a random sample of 50 people from the United States, what is the probability of getting 10 or fewer under the median age?

c. Of the 50 Westvaco employees listed in Display 1.1 of Chapter 1, 10 were under the median age. Is this about what you would expect from a random sample of 50 residents, or should you conclude that this group is special in some way? If you think the group is special, what is special about it ?

E32. Suppose that 80% of a certain brand of computer disk contain no bad sectors. If 100 such disks are inspected, what is the approximate chance that 15 or fewer contain bad sectors? What assumptions underlie this approximation?

E33. The guideline states that it is appropriate to use the normal approximation for the sampling distribution of sample proportions if np and $n(1 - p)$ are both greater than or equal to 10. To check this out, generate simulated sampling distributions for values of p equal to .1 and .98. Use sample sizes of 50, 100, and 500 with each value of p. Does the guideline appear to be reasonable?

E34. The histograms in Display 5.32 are sampling distributions for $\hat{p}$ for samples of size 5, 25, and 100, first for a population with $p = .2$ and then for a population with $p = .4$.

a. Do the means of the sampling distributions depend on p? On n?

b. How do the spreads of the sampling distributions depend on p and n?

c. How do the shapes of the sampling distributions depend on p and n?

d. For which of the combinations of p and n would you be willing to use the rule that roughly 95% of the values lie within two standard errors of the mean?

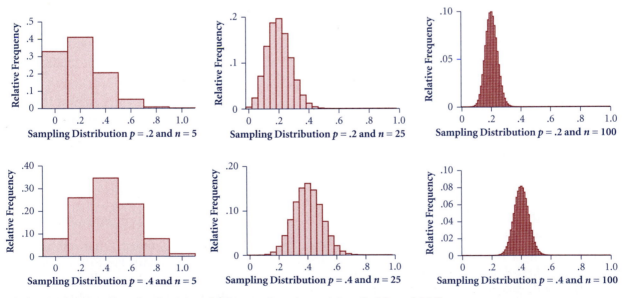

Display 5.32 Sampling distributions of $\hat{p}$ for $p = .2$ and $p = .4$ ($n = 5$, 25, and 100).

5.5 Sampling Distribution of the Sum and Difference

Because the sample mean is equal to the sample sum divided by the sample size, you were able to answer questions in Sections 5.3 and 5.4 by using either the sampling distribution of the mean or the sampling distribution of the sum. You will continue that idea in this section.

You will also study the sampling distribution of a difference. Why are differences important? In Chapter 4, Kelly compared the enzyme level of hamsters raised in short days with the enzyme level of hamsters raised in long days. It would be natural for Kelly to use, as her summary statistic, the size of the difference in the mean enzyme levels of the two groups of hamsters. To evaluate this summary statistic, she would have to know what differences might occur simply by chance if the length of day does not affect the enzyme levels.

Imagine the familiar situation of rolling a balanced die and observing the number of dots on the upper face. Each of the six possible outcomes should occur with equal frequency in the long run; the distribution of possible outcomes is shown in Display 5.33.

Outcome	Probability
1	1/6
2	1/6
3	1/6
4	1/6
5	1/6
6	1/6

Display 5.33 Distribution of outcomes of a roll of a single die.

This distribution has mean

$$\mu = \sum x \cdot P(x) = 1\left(\frac{1}{6}\right) + 2\left(\frac{1}{6}\right) + \cdots + 6\left(\frac{1}{6}\right) = 3.5$$

and variance

$$\sigma^2 = \sum (x - \mu)^2 \cdot P(x)$$

$$= (1 - 3.5)^2 \frac{1}{6} + (2 - 3.5)^2 \frac{1}{6} + \cdots + (6 - 3.5)^2 \frac{1}{6}$$

$$\approx 2.917$$

In Activity 5.4, you will look at the distributions of the sum and the difference of two values taken at random from this population.

Activity 5.4 The Sum and Difference of Two Rolls of a Die

What you'll need: a die

You will roll your die twice. One partner will record the sum of the numbers on the two rolls. The other partner will record the difference. For example, if the first roll is 3 and the second roll is 5, the first partner records 8 and the second records –2.

1. What do you think the distribution of sums (first roll plus second roll) and the distribution of differences (first roll minus second roll) will look like after many rolls? Where will they be centered? Which will have the larger spread?

2. Roll until your class has recorded 100 sums and 100 differences.

3. Plot the 100 sums and the 100 differences on separate dot plots. Do these distributions have the same shape as that of Display 5.33 for the outcome of a single die toss? Should they?

4. What is the mean of your distribution of sums? Of differences? How do these means compare to the mean of the distribution of a single roll of a die?

5. What is the variance of your distribution of sums? Of differences? How do the variances of your distributions of sums and differences compare with the variance of the outcomes from a single die toss? How do they compare with each other?

While doing Activity 5.4, you may have realized that the formulas in Section 5.3 for the sampling distribution of a sum should apply to the sum of two dice:

$$\mu_{sum} = n\mu = 2\mu = 2(3.5) = 7$$

$$\sigma^2_{sum} = n\sigma^2 = 2\sigma^2 = 2(2.917) = 5.834$$

Discussion: The Sum and Difference of Two Dice

D28. How close are your results from step 4 of the activity to the results given by the formulas for μ_{sum} and σ^2_{sum}? Do these formulas or similar ones appear to work for differences as well as sums?

D29. Provide an intuitive argument as to why the variation in a sum should be larger than the variation in either variable making up the sum. Do the same for the variation in a difference.

The rolls of a die create a situation simple enough that you can check the formulas for a sum by constructing the exact distributions of the sum and of the difference. The set of all possible outcomes when rolling a die twice is listed in Display 5.34. Each of these rolls is just as likely as any other.

$$
\begin{array}{cccccc}
1,1 & 1,2 & 1,3 & 1,4 & 1,5 & 1,6 \\
2,1 & 2,2 & 2,3 & 2,4 & 2,5 & 2,6 \\
3,1 & 3,2 & 3,3 & 3,4 & 3,5 & 3,6 \\
4,1 & 4,2 & 4,3 & 4,4 & 4,5 & 4,6 \\
5,1 & 5,2 & 5,3 & 5,4 & 5,5 & 5,6 \\
6,1 & 6,2 & 6,3 & 6,4 & 6,5 & 6,6
\end{array}
$$

Display 5.34   The possible outcomes from rolling a die twice.

The only way to get a sum of 2 is to roll (1, 1). So the probability that the sum is 2 is $\frac{1}{36}$. There are two rolls that give a sum of 3, namely, (1, 2) and (2, 1), so the probability of getting a sum of 3 is $\frac{2}{36}$. Continuing, you get the distribution of the sum in Display 5.35: The shape is triangular.

Sum	Probability	Sum	Probability
2	1/36	8	5/36
3	2/36	9	4/36
4	3/36	10	3/36
5	4/36	11	2/36
6	5/36	12	1/36
7	6/36		

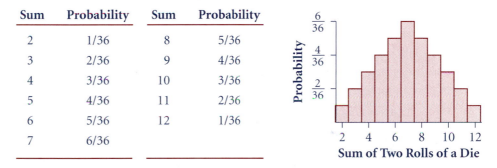

Display 5.35   The sampling distribution of the sum of two rolls of a die.

The mean is given by

$$
\mu = \sum x \cdot P(x) = 2\left(\frac{1}{36}\right) + 3\left(\frac{2}{36}\right) + \cdots + 12\left(\frac{1}{36}\right) = 7
$$

and the variance is given by

$$
\sigma^2 = \sum (x - \mu)^2 \cdot P(x)
$$

$$
= (2 - 7)^2 \frac{1}{36} + (3 - 7)^2 \frac{2}{36} + \cdots + (12 - 7)^2 \frac{1}{36}
$$

$$
\approx 5.834
$$

These are the same results as the results from the formulas for the mean and variance of a sum following Activity 5.4.

■ **Practice: Sampling Distribution for Two Rolls of a Die**

P17. Complete a sampling distribution of the difference when a die is rolled twice (first roll minus second roll).

a. What is the shape of this distribution?

b. Compute the mean and the variance of this distribution using the standard formulas

$$\mu = \sum x \cdot P(x) \quad \text{and} \quad \sigma^2 = \sum (x - \mu)^2 \cdot P(x)$$

c. Verify that the mean of this sampling distribution is equal to $\mu - \mu$ or 0 and the variance is $\sigma^2 + \sigma^2$ or $2\sigma^2$, where μ and σ are the mean and the standard deviation of the distribution of a single roll of a die.

A Regular Die and a Tetrahedral Die

In the dice activity, the two variables were taken from the same population (the outcomes of the roll of a six-sided die). The formulas, however, still apply even if the two populations aren't the same.

Example: The Sum of a Regular and a Tetrahedral Die

A regular die is rolled and then a tetrahedral (four-sided) die is rolled. The distribution of the outcomes of the six-sided die was given in Display 5.33 and has mean $\mu_1 = 3.5$ and variance $\sigma_1^2 = 2.917$.

There are four outcomes in the roll of a four-sided die: 1, 2, 3, or 4, and each has probability $\frac{1}{4}$. This distribution has mean $\mu_2 = 2.5$ and variance $\sigma_2^2 = 1.25$.

There are 24 possible outcomes if both dice are rolled, listed in Display 5.36.

Six-Sided Die

		1	2	3	4	5	6
	1	1, 1	1, 2	1, 3	1, 4	1, 5	1, 6
Four-Sided Die	**2**	2, 1	2, 2	2, 3	2, 4	2, 5	2, 6
	3	3, 1	3, 2	3, 3	3, 4	3, 5	3, 6
	4	4, 1	4, 2	4, 3	4, 4	4, 5	4, 6

Display 5.36 Possible outcomes from the roll of a six-sided die and a four-sided die.

The sampling distribution of the sum of the two dice is given in Display 5.37. It has mean 6 and variance 4.167. Once again, the means and the variances add:

$$\mu_{sum} = \mu_1 + \mu_2 = 3.5 + 2.5 = 6$$
$$\sigma_{sum}^2 = \sigma_1^2 + \sigma_2^2 = 2.917 + 1.25 = 4.167$$

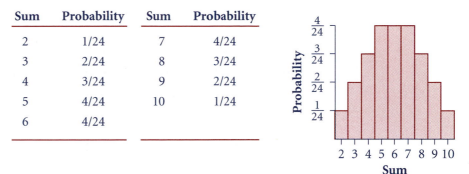

Sum	Probability	Sum	Probability
2	1/24	7	4/24
3	2/24	8	3/24
4	3/24	9	2/24
5	4/24	10	1/24
6	4/24		

Display 5.37   Sampling distribution of the sum of the rolls of a four-sided die and a six-sided die. ■

■ **Practice: A Regular Die and a Tetrahedral Die**

P18. a. Construct the sampling distribution of the difference when a regular die is rolled and then a tetrahedral die.

  b. Verify that the mean of the sampling distribution is $\mu_1 - \mu_2$ and the variance is $\sigma_1^2 + \sigma_2^2$ where μ_1 and σ_1^2 are, respectively, the mean and variance of the distribution of the roll of a single six-sided die and μ_2 and σ_2^2 are, respectively, the mean and variance of the distribution of the roll of a single four-sided die.

The results with dice apply to any sampling distribution of a sum or a difference.

Properties of the Sampling Distribution of the Sum and Difference

Suppose two values are taken randomly from two populations with means μ_1 and μ_2 and variances σ_1^2 and σ_2^2, respectively. Then the sampling distribution of the sum of the two values has mean

$$\mu_{sum} = \mu_1 + \mu_2$$

If the two values were selected independently, the variance is

$$\sigma_{sum}^2 = \sigma_1^2 + \sigma_2^2$$

The sampling distribution of the difference of the two values has mean

$$\mu_{difference} = \mu_1 - \mu_2$$

If the two values were selected independently, the variance is

$$\sigma_{difference}^2 = \sigma_1^2 + \sigma_2^2$$

The shapes of the sampling distributions of the sum and the difference depend on the shapes of the two original populations. If both populations are normal, so are the sampling distributions of the sum and the difference.

Sam: There they go again. Why did they try to sneak that word "independently" into the part about the variance? We haven't seen that before.

Pop: It hasn't come up before because in our dice examples the rolls *were* independent—the result of the first roll didn't change the probabilities for the second roll. This doesn't have to be the case.

Sam: Huh? How can two rolls of a die be random and not independent?

Pop: Well, suppose you got lazy and used the results of the first roll twice. The only pairs of rolls you could get are $(1, 1)$, $(2, 2)$, $(3, 3)$, $(4, 4)$, $(5, 5)$, and $(6, 6)$, each with equal probability. The rolls are random but not independent.

Sam: I see. Let's try the computations for the sum. The possible sums are 2, 4, 6, 8, 10, and 12. Each has probability $\frac{1}{6}$. This distribution has mean 7 and variance 11.7.

Pop: Right. See, it's still the case that the mean of all possible sums is $3.5 + 3.5 = 7$, but the variance is not $2.917 + 2.917 = 5.934$.

Example: Using the Sampling Distribution of a Difference

Bottle caps are manufactured so that their inside diameters have a distribution that is approximately normal with mean 36 mm and standard deviation 1 mm. The distribution of the outside diameters of the bottles are approximately normal with mean 35 mm and standard deviation 1.2 mm. If a bottle cap and a bottle are selected at random (and independently!), what is the probability that the cap will fit on the bottle?

Solution

The cap fits on the bottle if $d_c > d_b$, where d_c is the inside diameter of the cap and d_b is the outside diameter of the bottle. So the problem asks for the probability that $d_c - d_b > 0$. Because both sampling distributions are approximately normal, the sampling distribution of $d_c - d_b$ is approximately normal. This sampling distribution has mean and variance

$$\mu_{difference} = \mu_1 - \mu_2 = 36 - 35 = 1 \text{ mm}$$

$$\sigma^2_{difference} = \sigma^2_1 + \sigma^2_2 = 1^2 + 1.2^2 = 2.44 \text{ mm}^2$$

Taking the square root, the standard deviation is about 1.562 mm. The area to the right of zero is shaded in the sampling distribution for $d_c - d_b$, shown in Display 5.38 to indicate where $d_c - d_b > 0$.

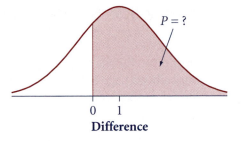

Display 5.38 Sampling distribution of $d_c - d_b$.

The z-score for a difference of zero is

$$z = \frac{sample\ difference - \mu_{difference}}{\sigma_{difference}} = \frac{0 - 1}{1.562} \approx -0.640$$

The area to the right of this z-score is .739. The chance the cap fits on the bottle is only .739—not very good for a manufacturing process. ■

Discussion: Using the Properties

D30. How are these two sets of formulas related?

$$\mu_{sum} = \mu_1 + \mu_2 \qquad \mu_{sum} = n\mu$$
$$\sigma^2_{sum} = \sigma^2_1 + \sigma^2_2 \qquad \sigma^2 = n\sigma^2$$

D31. Suppose a die is rolled and the same number is used twice, as in the dialogue between Pop and Sam.

a. Find the sampling distribution of the difference of the numbers on the dice.

b. Find the mean and variance of this sampling distribution using the standard formulas,

$$\mu = \sum x \cdot P(x)$$

and

$$\sigma^2 = \sum (x - \mu)^2 \cdot P(x)$$

c. Is the mean equal to $\mu_{difference}$ or $\mu_1 - \mu_2$?

d. Is the variance equal to $\sigma^2_{difference}$ or $\sigma^2_1 + \sigma^2_2$?

■ Practice

P19. Refer to the preceding example. Suppose a cap is too loose if it is at least 1.1 mm larger than the bottle. If a cap and a bottle are selected at random, what is the probability the cap is too loose?

Summary 5.5: Sampling Distribution of the Sum and Difference

The properties for the sampling distribution of the sum and difference are stated symbolically on page 315. It may help you to remember them if you say them occasionally using words:

- If two values are selected randomly from two populations (which may be the same), then the mean of the sampling distribution of the sum of the values is equal to the sum of the population means.

- The mean of the sampling distribution of the difference of the values is equal to the difference of the population means.

- The variance of the sampling distribution of the sum, or of the difference, is equal to the sum of the variances—but only if the values are picked independently.

- Finally, if the two populations are normally distributed, so are the distributions of the sum and the difference.

As you will see in E39, the rule for sums extends to the case of picking more than two values.

■ Exercises

E35. In 1998, SAT verbal scores had a mean of 505 and a standard deviation of 111. The SAT math scores had a mean of 512 and a standard deviation of 112. Each distribution was approximately normal.

Source: *1998 College Bound Seniors: A Profile of SAT Program Test Takers* (The College Entrance Examination Board and Educational Testing Service, 1998), p. 9.

a. Suppose one SAT verbal score is selected at random and one SAT math score is independently selected at random and the scores are added. Find the mean and the standard error of the sampling distribution of this sum.

b. Find the probability that the sum of the scores from part a is less than 800.

c. What total SAT scores are reasonably likely?

d. What is the probability that the verbal score is at least 100 points higher than the math score?

e. Suppose now that one student is selected at random and his or her SAT verbal score and and math score are added. Describe the mean and standard error of the sampling distribution of this sum.

E36. The table gives the National League West baseball standings on July 4, 1999.

Team	Won	Lost
Giants	46	35
Diamondbacks	44	37
Padres	39	40
Rockies	36	42
Dodgers	34	44

Source: *Los Angeles Times,* July 4, 1999, p. D6.

a. What is the mean and variance of the numbers of games won? Of the numbers of games lost?

b. If you select a team at random and find the total number of games played (games won plus games lost), what is the mean and variance of the sampling distribution of this total number of games?

c. Are your answers in part b equal to the sum of those in part a? Explain why they should or shouldn't be equal.

E37. A student drives to school in the morning and drives home in the afternoon. She finds that her commute time depends on the day of the week and whether it is morning or afternoon.

Day	Morning Commute Time (in minutes)	Afternoon Commute Time (in minutes)
Monday	16	9
Tuesday	14	8
Wednesday	13	5
Thursday	11	7
Friday	10	11

a. What is the mean and variance of the morning commute times? Of the afternoon commute times?

b. If the student selects a day at random and finds the total commute time for that day, what is the mean and variance of the sampling distribution of this total commute time?

c. Are your answers in part b equal to the sum of those in part a? Explain why they should or shouldn't be equal.

E38. Is it true that standard deviations add like variances? Try it with the numbers from the situation in Display 5.34.

E39. The rule for sums extends to the sum of more than two values, taken either from the same population or from different populations. In the case where n values are taken at random and independently from the same population with mean μ and standard deviation σ, the sampling distribution of their sum has mean $\mu + \mu + \cdots + \mu = n\mu$ and variance $\sigma^2 + \sigma^2 + \cdots + \sigma^2 = n\sigma^2$.

a. Where have you seen these formulas before?

b. If you roll a die three times, what is the mean of the sampling distribution of the sum? The variance?

c. If you roll a die seven times, what is the mean of the sampling distribution of the sum? The variance?

d. In 1998, SAT verbal scores had a mean of 505 and a standard deviation of 111. If 20 scores are selected at random, what is the mean and variance of their sum? What is the probability the sum is less than 10,000?

E40. W. J. Youden (Australian, 1900–1971) weighed many new pennies and found that the distribution is approximately normal with mean 3.11g and standard deviation 0.43g. What are the reasonably likely weights of a roll of 50 randomly selected new pennies? (Use the rules in E39.)

Source: W. J. Youden, *Experimentation and Measurement* (National Bureau of Standards, 1984), pp. 107–109.

E41. In 1977, F. Paca weighed rams in Lima, Peru, and found that the distribution was approximately normal with mean 75.4 kg and standard deviation 7.38 kg.

Source: F. Paca (New Mexico State University, 1977).

a. Suppose two rams are selected at random. Describe the sampling distribution of the sum of their weights.

b. What is the probability that the sum of their weights is less than 145 kg?

c. What is the probability that the sum of the weights of ten randomly selected rams is more than 750 kg? (Use the rules in E39.)

d. What is the probability that the first ram selected is more than 5 kg heavier than the second ram?

E42. Suppose n draws are made from a population with proportion of successes p.

a. Using the rules in E39, show why the mean and standard error for the sampling distribution of the number of successes are equal to np and $\sqrt{np(1-p)}$, respectively.

b. Using the rules of E39, show why the standard error for the sampling distribution of the mean $\bar{x}$ is $\sigma/\sqrt{n}$.

E43. Suppose you and your partner each roll two dice. Each of you computes the average of your two rolls. The summary statistic is the difference between your average and your partner's average. Describe the sampling distribution of these differences.

Chapter Summary

In this chapter, you have learned to use simulation to generate an approximate sampling distribution of a summary statistic. First you define a process for taking a random sample from the given population. You take such a random sample and compute the summary statistic that you are interested in. Finally, you generate a distribution of values for the summary statistic through repetitions of the process. With some simple situations, such as rolling dice, you can list all possible outcomes and get the exact distribution of the summary statistic.

Some summary statistics have easily predictable sampling distributions.

- If a random sample of size n is taken from a population with mean μ and variance σ^2, the sampling distribution of the *sample mean $\bar{x}$* has mean and standard error

$$\mu_{\bar{x}} = \mu \qquad \text{and} \qquad \sigma_{\bar{x}} = \frac{\sigma}{\sqrt{n}}$$

The sampling distribution of the *sum* of the values in the sample has mean and standard error

$$\mu_{sum} = n\mu \qquad \text{and} \qquad \sigma_{sum} = \sqrt{n} \cdot \sigma$$

These formulas are summarized in Display 5.39.

- If a random sample of size n is taken from a population with percentage of successes p, the sampling distribution of the *sample proportion $\hat{p}$* has mean and standard error

$$\mu_{\hat{p}} = p \qquad \text{and} \qquad \sigma_{\hat{p}} = \sqrt{\frac{p(1-p)}{n}}$$

The sampling distribution of the *number* of successes has mean np and standard error $\sqrt{np(1-p)}$. These formulas are summarized in Display 5.39.

- Finally, suppose two values are taken randomly from two populations with means μ_1 and μ_2 and variances σ_1^2 and σ_2^2, respectively. Then the sampling distribution of the sum of the two values has mean

$$\mu_{sum} = \mu_1 + \mu_2$$

If the two values were selected independently, the variance is

$$\sigma_{sum}^2 = \sigma_1^2 + \sigma_2^2$$

The sampling distribution of the difference of the two values has mean

$$\mu_{difference} = \mu_1 - \mu_2$$

If the two values were selected independently, the variance is

$$\sigma^2_{difference} = \sigma^2_1 + \sigma^2_2$$

In the case of means and sums, the sampling distribution gets more and more normal in shape as the sample size increases. This is called the Central Limit Theorem. If the population is normally distributed to begin with, the sampling distributions of means, sums, and differences are normal for all sample sizes. For proportions, you can use the normal distribution as an approximation of the sampling distribution as long as np and $n(1 - p)$ are both 10 or larger.

It may still be unclear to you why sampling distributions are so important. If that's the case, it may help you to review the Westvaco case at the beginning of Section 5.2. That sampling distribution helped you decide if the ages of the workers laid off could reasonably be attributed to the variation that occurs from random sample to random sample or whether you should investigate whether some other mechanism is at work.

There's another reason why sampling distributions are so important. Suppose a pollster takes a random sample of 1500 voters to find out what percentage p of the voters will vote for the incumbent in the next election. The proportion $\hat{p}$ in the sample who will vote for the incumbent almost certainly won't be exactly equal to the proportion p in the population who will vote for the incumbent—that's called sampling error. How large is this error likely to be? That's what sampling distributions can tell us. That may seem a bit backward. In this chapter, you always knew p and then found the reasonably likely values of $\hat{p}$. But the pollster has $\hat{p}$ and wants to find plausible values of p. That won't be hard to do now that you know what values of $\hat{p}$ tend to come from which values of p. But the details will have to wait until Chapter 8.

Sampling Distribution

		Of the Mean	$\rightarrow$	**Of the Sum**
Center	**All Populations**	$\mu_x = \mu$	times n	$\mu_{sum} = n\mu$
	Special Case of Success/Failure Population	$\mu_{\hat{p}} = p$	times n	$\mu_{sum} = np$
Spread	**All Populations**	$\sigma_{\bar{x}} = \dfrac{\sigma}{\sqrt{n}}$	times n	$\sigma_{sum} = \sqrt{n} \cdot \sigma$
	Special Case of Success/Failure Population	$\sigma_{\hat{p}} = \sqrt{\dfrac{p(1 - p)}{n}}$	times n	$\sigma_{sum} = \sqrt{np(1 - p)}$

Display 5.39 Formulas for the mean and standard error of sampling distributions.

■ Review Exercises

E44. On March 13, 1999, *China Daily*, an English-language newspaper published in the People's Republic of China, reported the following Air Pollution Index (API) for 32 major Chinese cities.

City	API	City	API
Beijing	123	Nanchang	56
Changchun	74	Nanjing	61
Changsha	77	Nanning	75
Chengdu	89	Shanghai	75
Chongqing	77	Shengyang	114
Fuzhou	86	Shenzhen	42
Guangzhou	98	Shijiazhuang	218
Guiyang	85	Taiyuan	346
Haikou	41	Tianjin	156
Hangzhou	62	Urumqi	94
Harbin	138	Wuhan	83
Hefei	42	Xi'an	224
Hohhot	164	Xining	500
Jinan	128	Yantai	76
Kunming	119	Yinchuan	328
Lanzhou	500	Zhengzhou	157

The mean of this distribution, shown in the dot plot in Display 5.40, is 140.9, and its standard deviation is 119.4.

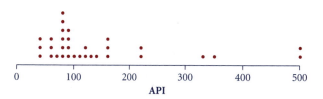

Display 5.40 Dot plot of the mean and standard deviation for the Air Pollution Index (API) for 32 major Chinese cities.

Suppose China had the resources to monitor air quality in only three of these cities.

a. Describe how to use simulation to construct the sampling distribution of the sample mean $\bar{x}$ for randomly selected samples of three cities, without replacement. Take one such random sample and compute $\bar{x}$.

b. Find the mean and the approximate standard error of the sampling distribution of the sample mean $\bar{x}$ for samples of size 3.

c. Can the sampling distribution be considered approximately normal?

E45. Refer to the situation in E44 of selecting three Chinese cities at random from the list. Air quality is considered "good" if the API is less than 100.

a. Describe how to use simulation to construct the sampling distribution of the number of cities with good air quality for randomly selected samples of three cities.

b. Find the mean and approximate standard error of the sampling distribution of the number of cities with good air quality for samples of size 3.

c. Can the sampling distribution be considered approximately normal?

E46. About 68% of the people in China live in rural areas. Suppose a random sample of 200 Chinese people is selected.

a. Describe the shape, mean, and standard error of the sampling distribution of $\hat{p}$, the proportion of people in the sample who live in rural areas.

b. What is the probability that 75% or more live in rural areas?

c. What values of $\hat{p}$ would be rare events?

d. What is the probability that 130 or more in the sample live in rural areas?

E47. Suppose that the spine widths of the books in a particular library have a distribution that is skewed slightly to the right with mean 4.7 cm and *SD* 2.1 cm.

a. If you select 50 books at random from this library, what is the probability that they will fit on a shelf that is 240 cm long?

b. Does your answer to part a imply that the 50 books in the philosophy section probably will fit on one 240 cm shelf?

E48. *Forbes Magazine* (November 8, 1993) found that the mean age of the CEOs (chief executive officers) of America's best small companies was about 51.5 years with a standard deviation of 8.9 years. The distribution of ages was approximately normal. If you select 10 of these CEOs at random, what is the probability their average age is between 50 and 53 years?

E49. Suppose you roll a dodecahedral (12-sided) die twice. Your summary statistic will be the sum of the two rolls.

a. What is the probability the sum is 3 or less?

b. Compute the mean and standard deviation of the distribution of outcomes of a single roll.

c. Compute the mean and standard error of the sampling distribution of the sum of two rolls.

E50. Young men have an average height of about 68 in. with a standard deviation of 2.7 in. Young women have an average height of about 64 in. with a standard deviation of

2.5 in. Both distributions are approximately normal. A young man and a young woman are selected at random.

a. Describe the sampling distribution of the difference of their heights.

b. Find the probability that he is 2 in. or more taller than she is.

c. What is the probability that she is taller than he is?

E51. According to the Caltrans Web site [www.dot.ca.gov (1999)], on an average day about 113,500 vehicles travel east on the Santa Monica freeway in Los Angeles to the interchange with the San Diego freeway. Assume that a randomly selected vehicle is equally likely to go straight through the interchange, go south on the San Diego freeway, or go north on the San Diego freeway.

a. What is the single best estimate of the number that will go straight through on an average day?

b. What numbers of vehicles are reasonably likely to go straight through?

c. On an average day, Caltrans found that 63,800 vehicles went straight through the interchange, 21,600 went north on the San Diego Freeway, and 28,100 went south. What can you conclude?

E52. Two buses always arrive at a bus stop between 11:13 and 11:23 each day. Over a sample of five days, the average time between the times the two buses arrived was 7 minutes. Could the buses be arriving at random in the 10-minute interval? Follow these steps to arrive at an answer you think is reasonable.

a. The population of interest is the set of possible times between the two arrivals (interarrival times). Describe a way to generate five of these interarrival times, assuming buses arrive at random and independent times during the 10-minute interval. Then get such a sample.

b. What was the mean interarrival time for your sample of size 5?

c. The dot plot in Display 5.41 shows the results of 100 repetitions of the process you did in parts a and b. What is the largest mean interarrival time recorded? Give an example of five pairs of arrival times that would give that mean.

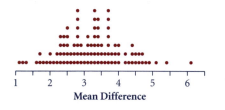

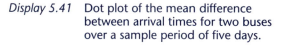

Display 5.41   Dot plot of the mean difference between arrival times for two buses over a sample period of five days.

d. Based upon the sampling distribution, do you think the buses in question were arriving at random?

E53. You select two random integers from 0 through 9 (with replacement) and take their average. Your opponent also selects two random integers and takes their average. You win an amount equal to the difference between your average and your opponent's average. (If the difference is negative, you lose that amount.)

a. Describe how to simulate a sampling distribution for the amount you win per play of the game. Do this simulation three times and record the differences.

b. Describe the sampling distribution of all possible outcomes.

E54. You estimate that the people using an elevator in an office building have an average weight of roughly 150 lbs. with a standard deviation of approximately 20 lbs. The elevator is designed for a 2000-lb. weight maximum. This maximum can be exceeded on occasion but should not be exceeded on a regular basis. Your job is to post a sign in the elevator stating the maximum number of people for safe use.

Keep in mind that it is inefficient to make this number too small but dangerous to make it too large.

a. What number would you use for maximum occupancy? Explain your reasoning and assumptions.

b. Otis Elevator Company sells a Gearless Low Rise Hydraulic Elevator. Its capacity is 2000 lbs. or 13 people in the United States and 12 people in Canada. [Source: www.otis.com (2002).] Are these numbers close to the one you decided on in part a? Explain why you would or would not expect that to be the case.

E55. Refer to the four sampling distributions in Display 5.28 (all with $p = .8$, remember).

a. Find the approximate fraction of simulated values of $\hat{p}$ that are within two standard deviations of p. Do this separately for each of the distributions in the display.

b. For a distribution that is approximately normal, what fraction of values should be within two standard deviations of the mean? How does this answer compare to those you got in part a?

E56. In Chapter 4, you read about Kelly Acampora's hamster experiment. The four hamsters raised in short days had enzyme levels of

| 12.500 | 11.625 | 18.275 | 13.225 |

The hamsters raised in long days had enzyme levels of

| 6.625 | 10.375 | 9.900 | 8.800 |

In this exercise, you'll use a sampling distribution to decide if this is persuasive evidence that hamsters raised in short days have higher enzyme levels than hamsters raised in long days. Your summary statistic for Kelly's experiment will be d, the difference in the mean enzyme level of hamsters raised in short days and the mean enzyme level of hamsters raised in long days.

a. What is the value of this summary statistic *d* for Kelly's hamsters?

b. Now suppose that the length of a day makes no difference in enzyme levels. That is, suppose the eight numbers would have been the same if the hamsters had all received the opposite treatment. Use simulation to construct an approximate sampling distribution of all possible values of *d*. In other words, assign the hamsters at random so that four get each treatment but the enzyme level for the hamster is the same no matter what treatment it gets.

c. Is Kelly's value of *d* a reasonably likely outcome if the length of a day makes no difference? What can you conclude?

E57. Suppose you take a sample of size *n* from some large population. As *n* increases, describe what happens to the shape, mean, and standard error of each of these sampling distributions.

a. The sample mean

b. The sample total

E58. The histogram in Display 5.42 shows the number of airline passengers (in thousands) departing from major U.S. airports during a recent year. The four outliers are Chicago, Dallas, Atlanta, and Los Angeles with 25,636; 22,899; 22,666; and 18,438, respectively. The mean of this population is 5190, and the standard deviation is 5273.

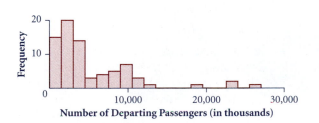

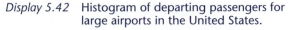

Display 5.42 Histogram of departing passengers for large airports in the United States.

Source: *Journal of Statistics Education*, Data Archives (1990), www.amstat.org/publications/jse/datasets/airport.dat.

The histograms and summary statistics in Display 5.43 show simulated distributions of sample means for samples of size 5, 10, and 30, selected without replacement from the numbers of departing passengers.

Check how well the shapes, means, and standard errors of the simulated sampling distributions agree with what the theory says they should be. Do you see any reason why the theory you have learned should not work well in any of these cases?

Sample Size	Mean	Standard Error
5	5592	2300
10	5063	1585
30	5228	705

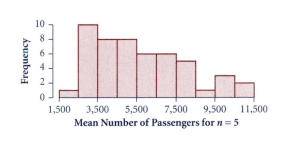

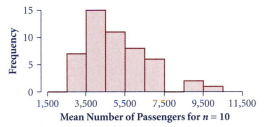

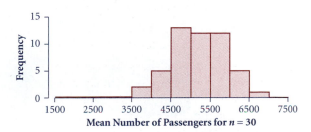

Display 5.43 Summary statistics and histograms of the distribution of sample means (in thousands) for samples of size 5, 10, and 30.

CHAPTER 6
PROBABILITY MODELS

Were some types of people on the Titanic more likely to go down with the ship than others? Basic concepts of probability can help you decide.

In your study of *Martin v. Westvaco* back in Chapter 1, a key question was this: If you draw three numbers at random from the set of ages 25, 33, 35, 38, 48, 55, 55, 55, 56, 64, what is the probability that the average of the three numbers will be 58 or more? This is a typical question for one kind of inference: You have a model, which lets you predict how things are supposed to behave. You also have data, which tells you how things actually did behave. Are the data consistent with the model? Or should you scrap the model and look for some other way to account for the data? The answer to the question on average ages is given by the sampling distribution of the mean—its set of values, together with how often they occur.

As you saw in Chapter 5, you can approximate sampling distributions using simulation, and some sampling distributions can be approximated by a normal distribution. Sometimes you can construct exact sampling distributions using probability theory, and that's what you'll learn to do in this chapter and the next.

In this chapter, you will learn

- to list all possible outcomes for a chance process in a systematic way
- to use the Addition Rule to compute the probability that event *A* or event *B* (or both) occurs
- to use the Multiplication Rule to compute the probability that event *A* and event *B* both occur
- to compute conditional probabilities: What is the probability that event *B* occurs given that event *A* occurs?

As you work through the chapter, you will see a number of problems involving coins, cards, and dice. These familiar devices effectively demonstrate fundamental principles that apply to many probabilistic situations. As you gain understanding, you'll be able to apply what you've learned to new statistical problems.

6.1 Sample Spaces with Equally Likely Outcomes

Taste tests are done routinely by food companies to make sure that new products taste at least as good as the competing brand. Jack and Jill, co-owners of Downhill Research, just won a contract to determine if people can tell tap water from bottled water. They will give each person in their sample both kinds of water, in random order, and ask which is the tap water.

Jack and Jill know they have a long climb ahead of them, but they're prepared to take it a step at a time. Before they start, they want to know what might happen if people can't identify tap water. For example, they want to know: If we ask *n* people which is the tap water and none of the people really know, what is the probability they all will guess correctly?

Jack and Jill know these basics about probability:

Fundamental Facts About Probability

Probability is a number between 0 and 1 (or between 0% and 100%) that tells how likely it is for something to happen. At one extreme, events that can't happen have probability 0. At the other extreme, events that are certain to happen have probability 1.

If you have a list of all possible outcomes and all outcomes are equally likely, then the probability of a specific outcome is

$$\frac{1}{\text{the number of equally likely outcomes}}$$

and the probability of an event is

$$\frac{\text{the number of outcomes in the event}}{\text{the number of equally likely outcomes}}$$

They begin by formulating a model that specifies that if a person can't really identify tap water, then they will choose the tap water, *T*, with a probability of .5 and the bottled water, *B*, with a probability of .5. If Jack and Jill have only one taster, that is, $n = 1$, then the probability that the person will guess correctly is written $P(T) = .5$.

Onward and upward: What about two tasters, or $n = 2$? Assuming that they can't identify tap water, what is the probability that both people will guess correctly and choose *T*? Here, the research stumbles:

Jack: There are three possible outcomes: Neither person chooses *T*, one chooses *T*, or both choose *T*. These three outcomes are equally likely, so each outcome has probability $\frac{1}{3}$. In particular, the probability that both choose *T* is $\frac{1}{3}$.

Jill: Jack, did you break your crown already? I say there are *four* equally likely outcomes: The first taster chooses *T* and the second also chooses *T* (*TT*); the first chooses *T* and the second chooses *B* (*TB*);

the first chooses B and the second chooses T (BT); or both choose B (BB). Because these four outcomes are equally likely, each has probability $\frac{1}{4}$. In particular, the probability that both choose T is $\frac{1}{4}$, not $\frac{1}{3}$.

A **probability distribution** gives all possible values resulting from a random process and gives the probability of each. Display 6.1 summarizes Jill's probability distribution.

Number Who Choose T	Probability
0	1/4
1	2/4
2	1/4

Display 6.1   Jill's probability distribution for $n = 2$.

Discussion: Probability Distributions

D1. Copy Jill's table, leaving out the probabilities. Fill in new probabilities so they represent Jack's probability distribution.

D2. What is the sum of the probabilities in a probability distribution? Why must this be so?

D3. Who do you think is right, Jack or Jill? Give your reason(s).

D4. How could Jack and Jill decide which of them is right?

Where Do Probabilities Come From?

To determine which one of them has assigned the correct probabilities, Jack and Jill decide to be philosophical for a moment and think about how probabilities can be assigned.

Probabilities come from three different sources.

- *Observed data* (long-run relative frequencies). For example, observation of thousands of births has shown that about 51% of newborns are boys. You can use these data to say that the probability of the next newborn being a boy is about .51.

- *Symmetry* (equally likely outcomes). If you flip a fair coin, there is nothing about the two sides of the coin to suggest that one side is more likely than the other to land facing up. Relying on symmetry, it is reasonable to think that heads and tails are equally likely. So the probability of heads is .5.

- *Subjective estimates*. What's the probability that you'll get an A in this statistics class? That's a reasonable, everyday kind of question, and the use of probability is meaningful, but you can't gather data or list equally likely outcomes. However, you *can* make a subjective judgment.

Based on symmetry, Jack and Jill agree that if $n = 1$, then $P(T) = .5$. They both used symmetry to justify their models for $n = 2$, saying their outcomes were equally likely, but they know that one of them must be wrong. To decide which, they decide to collect some data. They use two flips of a coin to simulate the taste test experiment with two tasters who can't identify tap water. Two tails counted as neither person choosing the tap water. One head and one tail counted as one person choosing the tap water and the other choosing the bottled water. Two heads counted as both people choosing the tap water. Hours later, they had their results. These are given in Display 6.2.

Number Who Choose T	Frequency	Relative Frequency
0	782	.26
1	1493	.50
2	725	.24
Total	**3000**	**1.00**

Display 6.2 Results of 3000 simulations for two tasters when $P(T) = .5$.

Jack: Whew! Now I see you must be right. The relative frequencies from the simulation match your probabilities fairly well. I admit I fell down a bit here. I forgot that there are two ways to get one person choosing correctly—the first chooses correctly and the second doesn't or the second chooses correctly and the first doesn't. There is only one way for two people to choose correctly: They both choose the tap water.

Jill: Now that this has been settled, let's try to construct the probability distribution for three people, so $n = 3$.

Jack: Let me redeem myself. Following your reasoning that different orders should be listed separately, there are eight possible outcomes:

First Person	Second Person	Third Person
T	T	T
T	T	B
T	B	T
B	T	T
T	B	B
B	T	B
B	B	T
B	B	B

Jill: I'll count the number of people in each of the eight equally likely outcomes who picked the tap water. That gives us the probability distribution in Display 6.3. The probability that all three people choose correctly is $\frac{1}{8}$.

Number Who Choose T	Probability
0	1/8
1	3/8
2	3/8
3	1/8

Display 6.3 Jack and Jill's probability distribution for $n = 3$.

Jack: This makes me think we need a larger sample size. Even if nobody can identify tap water, there is a $\frac{1}{8}$ chance everyone will choose correctly! The reputation of Downhill Research is at stake!

■ Practice: Jack, Jill, and Probability

P1. Suppose Jack and Jill use a sample of four people who can't identify tap water. Construct the probability distribution for the number of people in the sample who would choose the tap water just by chance. What is the probability that all four people will choose correctly? Is four people a large enough sample to ease Jack's concern about the reputation of Downhill Research?

Sample Spaces

Both Jack and Jill used the same principle: Start by making a list of possible outcomes. Over the years, mathematicians realized that such a list of possible outcomes, called a sample space, must satisfy two requirements.

A **sample space** for a chance process is a *complete* list of *disjoint* outcomes.

Complete means that no outcomes are left off the list, that is, every possible outcome is on the list. **Disjoint** means that no two outcomes can occur at once. Sometimes the term **mutually exclusive** is used instead of "disjoint." This book will alternate between the terms so you can get used to both of them.

Deciding whether two outcomes are disjoint sounds easy enough, but be careful. For example, suppose your sample space consists of all possible outcomes from two flips of a coin. Then getting two heads, *HH*, and getting two tails, *TT*, are disjoint outcomes: If you get two heads, you do not get two tails. The outcomes *HT* (heads on the first flip, tails on the second) and *HH* are also disjoint because if you get one, you can't get the other on the same two flips, even though the first flip is heads in both cases.

Jack's sample space {neither person chooses *T*, one chooses *T*, two choose *T*} has outcomes that are complete and disjoint, but they aren't equally likely. He has a legitimate sample space; he just assigned the wrong probabilities.

Once an appropriate listing of outcomes is specified, an **event** is any set of these outcomes—in other words, any subset of the sample space.

Discussion: Sample Spaces

D5. In a slightly different form, Jack and Jill's argument holds an important place in the early history of probability. One famous mathematician, Jean d'Alembert (French, 1717–1783), wrote in a 1750s textbook that the probability of getting two heads in two flips of a fair coin is $\frac{1}{3}$. Here's his list of outcomes, which he said were equally likely:

- The first flip is tails.
- The first flip is heads, the second is tails.
- The first flip is heads, the second is heads.

Is his list of outcomes complete? Are the outcomes disjoint? Are the outcomes equally likely?

D6. Suppose you were to select a student at random from your school and look at the books that student is carrying. Construct a sample space that is complete but not disjoint. Construct one that is disjoint but not complete.

■ Practice

P2. D'Alembert was co-author (with another Frenchman, Denis Diderot [1713–1784]) of a 35-volume *Enclyclopédie*. In it, he wrote that the probability of getting at least one head in two flips of a fair coin is $\frac{2}{3}$. This time, he said these three outcomes were equally likely:

- heads on the first flip
- heads on the second flip
- heads on neither flip

Is this sample space complete? Are the outcomes disjoint? Are the three outcomes equally likely? Is d'Alembert's probability of getting at least one head correct?

P3. You randomly choose two people to be laid off from a group whose ages are 28, 35, 41, 47, and 55. List a sample space that has outcomes that are disjoint and complete. What is the probability that the two youngest people are the ones laid off?

P4. You flip a penny four times and note how it lands each time: heads or tails. List a sample space. Check that your list is complete and that the outcomes are disjoint. Make a table of the probability distribution of the number of heads. What is the probability of getting exactly two heads?

P5. Roll two dice and note the total number of dots on the top. List a sample space, and check that it is complete and that the outcomes are disjoint. Make a table of the probability distribution for the sum of the two dice. What is the probability that the sum is 7?

Data and Symmetry Working Together

Data and symmetry work together in constructing useful probability models.

How can you tell if the outcomes in your sample space are equally likely? There's no magic answer, only two useful principles—data and symmetry. A die has six sides, and the only thing that makes one side different from another is the number of dots. It's hard to imagine that the number of dots would have much effect on which side lands up, so it's reasonable to assume that the six sides are equally likely. This seems like a good model for a die, but to verify this, you also need data: Roll the die many times and compare your model's predictions with actual data to see if you have a good fit.

In Jack and Jill's model, they are temporarily *assuming* that symmetry applies—that people are as likely to choose the tap water as the bottled water. In fact, it's probably not the right model and when Jack and Jill collect data with enough real people, they may be able to reject it. Or perhaps the model is correct and they won't be able to reject it. In either case, they will have done what they have been hired to do—decide if people can distinguish the tap water.

Here is an activity that will help you see some of the difficulties in coming up with a realistic probability model.

Activity 6.1   Spinning Pennies

What you'll need: one penny per student

For flipping a penny, heads and tails have the same probability. You might think that the same probability model is true for spinning a penny.

1. Use one finger to hold your penny on edge on a flat surface, with Lincoln's head right side up, facing you. Flick the penny with the index finger of your other hand so that it spins around many times on its edge. When it falls over, record whether it lands heads up or tails up.

2. Repeat and combine results with the rest of your class until you have a total of at least 500 spins.

3. Are the data consistent with the model that heads and tails are equally likely outcomes, or do you think that the model can safely be rejected?

Discussion: Data and Symmetry

D7. Suppose you spin a penny three times and record whether it lands heads or tails.

   a. How many possible outcomes are there?

   b. Are these outcomes equally likely? If not, which one is the most likely? the least likely?

D8. Suppose that you pick four students at random from your school and check if they are left-handed or right-handed. Can you list a sample space? Can you determine the probability that all four are right-handed?

The Law of Large Numbers

Why was Jack convinced that his probability model was wrong after the simulation of 3000 flips of two coins (Display 6.2)? In Activity 6.1, why might you be convinced that the equal probability model does not work for coin spinning? Results from observed data seem to preclude the proposed model, and that is as it should be, according to a mathematical principle—the law of large numbers—established in the 1600s by Jacob Bernoulli (Swiss, 1654–1705).

The law of large numbers guarantees that sample proportions converge to true probabilities when sampling is random.

The **law of large numbers** says that in random sampling, the larger the sample, the closer the proportion of successes in the sample tends to be to the proportion in the population. In other words, the difference between a sample proportion and a population proportion must get smaller (except in rare instances) as the sample size gets larger. You can see a demonstration of this law in Display 6.4. Think of flipping a coin in such a way that it comes up heads with probability .4. The graph in Display 6.4 shows a total of 150 flips with the proportion of heads accumulated after each flip plotted against the number of the flip. Notice that the coin came up heads for the first few flips so that the proportion of heads starts out at 1. This proportion quickly decreases, however, as the number of flips increases, and it ends up at about .377 after 150 flips.

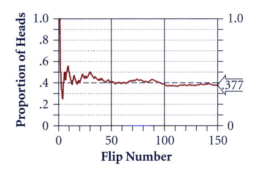

Display 6.4 Proportion of heads for the given number of flips of a coin with *P*(*heads*) = .4.

Most people intuitively understand the law of large numbers. If they want to estimate a proportion, they know it is better to take a larger sample than a smaller one. After Jack saw the results from 3000 pairs of coin flips, he immediately rejected his model that $p = \frac{1}{3}$. If there had been only 10 pairs of coin flips, he couldn't have been so positive his model was wrong.

Discussion: Law of Large Numbers

D9. An opinion pollster says, "All I need to ensure accuracy of the results of my polls is to make sure I have a large sample."

A casino operator says, "All I need to ensure that the house will win most of the time is to keep a large number of people flocking into my casino."

A manufacturer says, "All I need to keep my proportion of defective products low is to manufacture a lot of them."

Comment on the correctness of each of these statements. In particular, do the people speaking understand the law of large numbers?

D10. Flip a coin 20 times, keeping track of the cumulative number of heads and the cumulative proportion of heads after each flip. Plot the cumulative number of heads versus the flip number on one graph. Plot the cumulative proportion of heads versus the flip number on another graph. On each graph, connect the points. Repeat this process five times so that you have five sets of connected points on each plot. Comment on the differences in patterns formed by the five lines on each graph.

The Fundamental Principle of Counting

Jack and Jill are now ready to tackle samples of size 5. They are beginning to realize that using Jill's method will result in a long list of possible outcomes, and they have become a bit discouraged. The key is for them to notice that

1. they can always think of asking their question in stages (one person at a time).

2. they can list the possible outcomes at each stage.

These two observations will let them count outcomes by multiplying. To see how, note that Jack and Jill could also have used a tree diagram like the one in Display 6.5 to count the number of outcomes when $n = 2$.

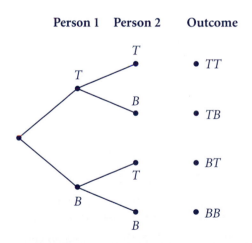

Display 6.5 A tree diagram of all possible outcomes when $n = 2$.

There are two main branches, one for each of the two ways the first person can answer. Each main branch has two secondary branches, one for each way the second person can answer once the first person has answered. In all, there are $2 \cdot 2$, or 4, outcomes.

For $n = 3$, you would add two branches to the end of each of the four branches in Display 6.5. This would give a total of $2 \cdot 2 \cdot 2$, or 8, outcomes, as Jack and Jill already discovered. This idea is called the **Fundamental Principle of Counting.**

Fundamental Principle of Counting

For a two-stage process, with n_1 possible outcomes for stage 1 and n_2 possible outcomes for stage 2, the number of possible outcomes for the two stages together is $n_1 n_2$.

More generally, if there are k stages, with n_i possible outcomes for stage i, then the number of possible outcomes for all k stages taken together is $n_1 n_2 n_3 \cdots n_k$.

So with five people, or $n = 5$, and two possible outcomes for each person, Jack and Jill have $2 \cdot 2 \cdot 2 \cdot 2 \cdot 2$, or 32, possible outcomes. If their model is correct—people are just guessing which is the tap water—these 32 outcomes are all equally likely. So the probability that each of the five people will correctly choose the tap water is $\frac{1}{32}$.

When there are only two stages, it is often more convenient to list them using a two-way table.

Example: Rolling Two Dice

Make a two-way table that shows all possible outcomes when you roll two fair dice. What is the probability you get doubles (both dice show the same number)?

Solution

Because there are six faces on each die and the dice are fair, there are $6 \cdot 6$, or 36, equally likely outcomes, as shown in Display 6.6. Six of these are doubles, so the probability of rolling doubles is $\frac{6}{36}$.

Second Roll

		1	2	3	4	5	6
	1	1, 1	1, 2	1, 3	1, 4	1, 5	1, 6
	2	2, 1	2, 2	2, 3	2, 4	2, 5	2, 6
First	3	3, 1	3, 2	3, 3	3, 4	3, 5	3, 6
Roll	4	4, 1	4, 2	4, 3	4, 4	4, 5	4, 6
	5	5, 1	5, 2	5, 3	5, 4	5, 5	5, 6
	6	6, 1	6, 2	6, 3	6, 4	6, 5	6, 6

Display 6.6 The 36 outcomes when rolling two dice. ■

Discussion: The Fundamental Principle of Counting

D11. Suppose Jack and Jill have four people taste three well-known brands of cola. The people are asked if they prefer the first cola, the second cola, the third cola, or if they can't tell any difference. How many possible outcomes will be in Jack and Jill's list for this experiment? Are these outcomes equally likely?

D12. Suppose you flip a fair coin five times.

    a. How many possible outcomes are there?

    b. What is the probability you get five heads?

    c. What is the probability you get four heads and one tail?

■ Practice

P6. Make a tree diagram to show all possible outcomes if you ask a person to taste a particular brand of strawberry ice cream and then evaluate it as good, okay, or poor on flavor and as acceptable or unacceptable on price. Are the outcomes equally likely?

P7. A dental clinic has three dentists and seven dental hygienists. If you are assigned a dentist and a dental hygienist at random when you go in, how many different pairs could you end up with? What is the probability that you get your favorite dentist and your favorite dental hygienist? Make a two-way table to illustrate your answer.

Summary 6.1: Sample Spaces

A **probability model** is a sample space together with an assignment of probabilities. The sample space is a complete list of disjoint outcomes for which these properties hold:

- Each outcome is assigned a probability between 0 and 1.
- The sum of all of the probabilities must equal 1.

Often you can rely on symmetry to recognize that outcomes are equally likely. If they are, then you can compute probabilities just by counting outcomes: The probability of an event is the number of outcomes that make up the event divided by the total number of possible outcomes. In statistics, the main practical applications of equally likely outcomes are in the study of random samples and randomized experiments. As you saw in Chapter 4, in a survey, all possible simple random samples are equally likely. Similarly, in a completely randomized experiment, all possible assignments of treatments to units are equally likely.

The only way to know whether a probability model fits a real situation is to compare probabilities from the model with observed data.

The Fundamental Principle of Counting says that if you have a process of k stages, with n_i outcomes for stage i, the number of outcomes for all k stages taken together is $n_1 n_2 n_3 \cdots n_k$.

■ Exercises

E1. Suppose you flip a fair coin five times and count the number of heads. Make a table that gives the probability distribution for the number of heads.

E2. Refer to the sample space in Display 6.6 for rolling two dice. Determine each of these probabilities.

    a. not getting doubles

b. getting a sum of 5

c. getting a sum of 7 or 11

d. a 5 occurring on the first die

e. getting at least one 5

f. a 5 occurring on both dice

g. the larger number is a 5 (if you roll doubles, the number is both the smaller and the larger number)

h. the smaller number is a 5

i. the difference of the larger number and the smaller number is 5

E3. *Data or symmetry?*

a. Is there any way to tell from symmetry (as opposed to data) that in a taste test of two foods, food A will be preferred over food B half of the time when, in fact, both foods are the same? Give an argument in support of your answer.

b. Is there any way to tell from symmetry (as opposed to data) that a randomly selected mathematics major will be female with a probability of about .5? Give an argument in support of your answer.

E4. *Equally likely outcomes?*

a. Use the Fundamental Principle of Counting to find the number of outcomes for the situation of flipping a fair coin six times. Can you find the probability that you will get heads all six times?

b. Use the Fundamental Principle of Counting to find the number of outcomes for the situation of rolling a fair die six times. Can you find the probability that you will get a 3 all six times?

c. Use the Fundamental Principle of Counting to find the number of outcomes for the situation of picking six U.S. residents at random and asking if they go to school. Can you find the probability that all six will say "yes"?

E5. Michael is designing an experiment to test which combination of colors makes the type on computer screens the easiest to read. His two factors are the color of the background (blue, green, red, yellow, white, or beige) and the color of the type (brown, black, navy, or gray).

a. How many different possible treatments are there?

b. Make a tree diagram and a two-way table showing them.

c. Suppose that Michael adds a third factor of brightness to his experiment. He will have two levels of brightness on the screen: high and low. How many possible distinct treatments are there?

d. Make a tree diagram of this new situation.

E6. Draw a tree diagram showing all possible outcomes for flipping four fair coins. What is the probability of getting exactly three heads?

E7. Pat has six shirts (blue, green, red, yellow, and two white—one long-sleeved, one short) and four pairs of pants (brown, black, blue, and gray).

a. Use the Fundamental Principle of Counting to find the number of possible outfits Pat can wear.

b. Make a two-way table to show all possible outfits.

c. If Pat selects a shirt and pants at random, determine the probability of Pat wearing

- a white shirt

- the gray pants

- a white shirt and the gray pants

- a white shirt or the gray pants

E8. A tetrahedral die has four sides with the numbers 1, 2, 3, and 4.

a. How many equally likely outcomes are there when two fair tetrahedral dice are rolled?

b. Show all possible equally likely outcomes using a two-way table.

c. What is the probability of getting doubles?

d. What is the probability that the sum is 2?

e. What is the probability that the larger number is a 2? (If you roll doubles, the number is both the smaller and the larger number.)

f. What is the probability that the smaller number is a 2?

g. What is the probability that the larger number minus the smaller number is equal to 2?

h. What is the probability that the smaller number is less than 2?

E9. Suppose you roll a fair tetrahedral die twice (see E8). Determine which of the proposed sample spaces are disjoint and complete. If the sample space is disjoint and complete, assign probabilities to the given outcomes. If it is not, explain why.

a. {no 4 on the two rolls, one 4, two 4's}

b. {the first roll is a 4, the second roll is a 4, neither roll is a 4}

c. {the first 4 comes on the first roll, the first 4 comes on the second roll}

d. {the sum of the two rolls is less than 2, the sum is more than 2}

e. {the first die is a 4, neither die is a 4}

E10. How many three-digit numbers can you make from the digits 1, 2, and 7? You can use the same digit more than once. If you choose the digits at random, what is the probability that the number is less than 250?

6.2 The Addition Rule and Disjoint Events

In mathematics, "A or B" means A or B or both.

One of the shortest words in the English language—*or*—is often misunderstood. That's because *or* can have two different meanings. First, suppose you know your course grade will be either an A or a B. You know it can't be both—this meaning of *or* doesn't allow A *and* B as a possibility. On the other hand, in statistics, the meaning of *or* allows *both* as a possibility: Either the first flip landed heads or the second flip landed heads (or both landed heads). Either the first taster chose the tap water or the second taster chose the tap water (or both chose the tap water). There are times in statistics where *or* is used even though *both* cannot occur (either the first flip lands heads or it lands tails), but *or* always forces you to check on the possibility of both.

Statistical data are often presented in tables like the one in Display 6.7, which gives basic figures for employment in the United States for all nonmilitary adults who were employed or seeking employment.

Nonmilitary Employable Adults	Number of People (in thousands)
Employees on farm payrolls	4,639
Employees on nonfarm payrolls	128,911
Seeking employment	5,836
Total civilian labor force	**139,386**

Display 6.7 The U.S. labor force, as of September 1999.
Source: Bureau of Labor Statistics, www.bls.gov.

If two events are disjoint, the probability of either event occurring is the sum of their individual probabilities.

The categories are disjoint, and this makes computing probabilities easy. For example, if you pick one of the people in the civilian labor force at random, you can find the probability that he or she is employed by adding the employees on farm payrolls to those on nonfarm payrolls and then dividing by the total number in the civilian labor force:

$$P(employed) = P(on\ farm\ payroll\ or\ on\ nonfarm\ payroll)$$

$$= \frac{4{,}639 + 128{,}911}{139{,}386} = \frac{133{,}550}{139{,}386} \approx .958$$

Display 6.8 is a little more complex. This table tells where the 35,578,000 people in the United States who fish do their fishing. If you randomly select a person who fishes and you want to find the probability they fish in freshwater or in saltwater, you can't add the numbers in these two categories because you'd get more than the total number of people who fish! People who fish in *both* saltwater and freshwater must have been included in both categories.

Type of Fishing	Number (thousands)
All freshwater fishing	31,041
Freshwater other than Great Lakes	30,186
Great Lakes	2,552
Saltwater fishing	8,885
Total, all fishing	**35,578**

Display 6.8 Number of people who fish in the United States by location of fishing.

Source: *Statistical Abstract of the United States, 1996.*

Discussion: *A or B*

D13. Are the categories in the table of Display 6.8 complete? Are they disjoint?

D14. What is the probability that a randomly selected person who fishes does their fishing in freshwater or in saltwater? How many people fish in both freshwater and saltwater?

D15. Consider the data in Display 6.8 on where people fish.

a. Suppose you randomly select one of the 35,578,000 people who fish. Find the probability that the person fishes only in freshwater or only in saltwater (but not both).

b. Suppose you randomly select a freshwater fisherman. Find the probability that the person fishes only in the Great Lakes or only in freshwater other than the Great Lakes (but not both).

■ Practice

P8. Display 6.9 categorizes the child support received by divorced women with children under 21 in the United States.

Child Support Received by Divorced Women	Number (in thousands)
Supposed to receive payments in 1991	4,884
Received full amount	2,552
Received partial amount	1,176
Did not receive payments	1,156
Not supposed to receive payments in 1991	5,035
Total divorced women with children under 21	**9,919**

Display 6.9 Mothers receiving court-ordered child support payments from fathers, 1991.

Source: *Statistical Abstract of the United States*, 1996, Table 604.

a. Are the five categories in the table complete for divorced women with children under 21? Are they disjoint?

b. Revise the table so that the categories are complete and disjoint.

c. If one of the 9,919,000 women is selected at random, what is the probability that she was supposed to receive payments and received the full amount or received a partial amount?

Addition Rule for Disjoint Events

While doing this activity, notice when you can find $P(A \text{ or } B)$ just by adding.

Activity 6.2   Exploring "or"

1. For each of these questions, record the number of students in your class who say "yes."

   • Is blue your favorite color?

   • Is red your favorite color?

2. From the information in step 1 alone, can you determine the number of students in your class who say blue is their favorite color or who say red is their favorite color? If so, explain how to do it. If not, what additional information do you need?

3. Now for each of these questions, record the number of students in your class who say "yes."

   • Are you female?

   • Is blue your favorite color?

4. From the information in step 3 alone, can you determine the number of students in the class who are female or who have blue as their favorite color? If so, explain how to do it. If not, what additional information do you need?

In Activity 6.2, you can get the answer to one question by adding because the categories are disjoint. For the other question, adding doesn't work because the categories are not disjoint.

Two useful rules have emerged from all of this. First, if two events are disjoint, the probability of their occurring together is 0.

Two events are disjoint if they can't happen on the same outcome.

A Property of Disjoint (Mutually Exclusive) Events

If event A and event B are disjoint, then

$$P(A \text{ and } B) = 0$$

Second, if two events are disjoint, then you can add probabilities to find $P(A \text{ or } B)$.

Using set notation, this rule is written $P(A \cup B) = P(A) + P(B)$ where $P(A \cup B)$ is read "the probability of A union B."

The Addition Rule for Disjoint (Mutually Exclusive) Events

If event A and event B are disjoint, then

$$P(A \text{ or } B) = P(A) + P(B)$$

The Addition Rule for Disjoint Events can be generalized. For example, if each pair of events A, B, and C are disjoint, then

$$P(A \text{ or } B \text{ or } C) = P(A) + P(B) + P(C)$$

Example: The Addition Rule with Dice

Use the Addition Rule to find the probability that if you roll two dice, you get a sum of 7 or a sum of 8.

Solution

Let A be the event of getting a sum of 7. Let B be the event of getting a sum of 8. These two events are mutually exclusive on the same roll, so $P(A \text{ and } B) = 0$. Then $P(A \text{ or } B) = P(A) + P(B)$. Using the sample space of equally likely outcomes from Display 6.6 in Section 6.1,

$$P(sum\ of\ 7\ or\ sum\ of\ 8) = P(sum\ of\ 7) + P(sum\ of\ 8)$$

$$= \frac{6}{36} + \frac{5}{36} = \frac{11}{36} \quad \blacksquare$$

Example: The Addition Rule with Tap Water

Use the Addition Rule to compute the probability that exactly one person out of two correctly chooses the tap water in Jack and Jill's model.

Solution

$$P(exactly\ one\ T\ among\ two\ tasters)$$

$$= P[(1st\ chooses\ T\ and\ 2nd\ chooses\ B)\ or$$
$$(1st\ chooses\ B\ and\ 2nd\ chooses\ T)]$$

$$= P(TB \text{ or } BT) = P(TB) + P(BT) = \frac{1}{4} + \frac{1}{4} = \frac{1}{2}$$

You can add the probabilities because the two outcomes *TB* and *BT* are disjoint. ■

Discussion: The Addition Rule for Disjoint Events

D16. Suppose you select a person at random from your school. Which of these pairs of events must be disjoint?

   a. the person has ridden a roller coaster; the person has ridden a Ferris wheel

   b. owns a classical music CD; owns a jazz CD

   c. is a senior; is a junior

   d. has brown hair; has brown eyes

   e. is left-handed; is right-handed

   f. has shoulder-length hair; is a male

D17. If you select a person at random from your classroom, what is the probability that the person is a junior or a senior?

D18. Suppose there is a 20% chance of getting a certain disease each time you are exposed. Can you use the Addition Rule for Disjoint Events to compute the probability that you will get the disease if you are exposed three times?

■ Practice

P9. If you roll two dice, which of these pairs of events are mutually exclusive?

   a. doubles; sum is 8

   b. doubles; sum is odd

   c. there is a 3 on one die; sum is 10

   d. there is a 3 on one die; doubles

The Addition Rule

What about "or" when you don't have a list of disjoint categories? In that situation, you have an extra step.

Activity 6.3   More on "or"

1. For each of these questions, record the number of students in your class who say "yes."

   • Are you female?

   • Is blue your favorite color?

   • Is it true that you are female and blue is your favorite color?

(continued)

2. From the information in step 1 alone, can you determine the number of students in your class who are female or have blue as their favorite color? (Remember, when we use the word *or* in probability, we mean *one or the other or both.*) If so, explain how to do it. If not, what additional information do you need?

3. Make a two-way table showing the number of members of your class whose responses fall into each of the cells of the table.

4. If you select a person at random from your class, what is the probability that the person is female or has blue as a favorite color? What is the probability that a randomly selected student in your class is male?

Discussion: More on "or"

D19. Use the table you filled out as part of Activity 6.3. In parts a–b, use the given method to find the number of students who are female or have blue as their favorite color:

    a. as a row total plus a column total minus a cell count

    b. as the sum of three cell counts

D20. Let *B* be the event that the favorite color of the randomly chosen student is blue. Let *F* be the event that the randomly chosen student is female. Write two formulas, one corresponding to part a and one to part b in D19, for computing $P(B \text{ or } F)$.

The ideas you explored in the activity and discussion questions can be stated formally as the Addition Rule.

Using set notation, this rule is written $P(A \cup B) = P(A) + P(B) - P(A \cap B)$ where $P(A \cap B)$ is read "the probability of *A* intersect *B*."

Addition Rule

For any two events *A* and *B*,

$$P(A \text{ or } B) = P(A) + P(B) - P(A \text{ and } B)$$

Example: The Addition Rule for Events that Aren't Disjoint

Use the Addition Rule to find the probability that if you roll two dice, you get doubles or a sum of 8.

Solution

Let *A* be the event of getting doubles, and let *B* be the event of getting a sum of 8. These two events are not mutually exclusive, because they have a common outcome $(4, 4)$, which occurs with probability $\frac{1}{36}$. Using the Addition Rule, $P(A \text{ or } B) = P(A) + P(B) - P(A \text{ and } B)$,

$$P(doubles \text{ or } sum \text{ of } 8)$$

$$= P(doubles) + P(sum \text{ of } 8) - P(doubles \text{ and } sum \text{ of } 8)$$

$$= \frac{6}{36} + \frac{5}{36} - \frac{1}{36} = \frac{10}{36}$$

Another way to solve problems like this one is to organize the given information in a table.

Doubles?

		Yes	No	Total
Sum of 8?	**Yes**	1	4	5
	No	5	26	31
	Total	6	30	36

Now any question can be answered. The probability of getting doubles or a sum of 8 is

$$\frac{1 + 5 + 4}{36} = \frac{10}{36} \quad \blacksquare$$

You can also apply the Addition Rule "backward"; that is, you can compute $P(A \text{ and } B)$ when you know the other probabilities.

Example: Computing P(A *and* B)

In a local school, 80% of the students carry a backpack, B, or a wallet, W. Forty percent carry a backpack and 50% carry a wallet. If a student is selected at random, find the probability that the student carries both a backpack and a wallet.

Solution

Categories B and W aren't disjoint, so this Addition Rule applies:

$$P(B \text{ or } W) = P(B) + P(W) - P(B \text{ and } W)$$

$$.80 = .40 + .50 - P(B \text{ and } W)$$

Solving, $P(B \text{ and } W) = .10$. The probability that the student carries both is .10. $\blacksquare$

Discussion: Addition Rule

D21. Use the Addition Rule to find the probability that if you roll a pair of dice, you do not get doubles or you do get a sum of 8.

D22. What happens if you use the general form of the Addition Rule,

$$P(A \text{ or } B) = P(A) + P(B) - P(A \text{ and } B)$$

in a situation when A and B are mutually exclusive?

■ Practice

P10. Use the information about fishing in Display 6.8 to complete this table of the people who fish in freshwater. Then, from the table, find the probability that a person fishes in the Great Lakes and in other freshwater.

		Fish in the Great Lakes?		
		Yes	**No**	**Total**
Fish in Other Freshwater?	**Yes**	–?–	–?–	–?–
	No	–?–	–?–	–?–
	Total	–?–	–?–	31,041

P11. Use the Addition Rule to compute the probability that if you roll two dice,

  a. you get doubles or a sum of four

  b. you get doubles or a sum of seven

  c. you get a 5 on the first die or a 5 on the second die

P12. Use the Addition Rule to compute the probability that if you flip two fair coins, you get heads on the first coin or heads on the second coin.

Summary 6.2: The Addition Rule

In statistics, $P(A \text{ or } B)$ indicates the probability that event A occurs, event B occurs, or they both occur in the same random outcome. The word *or* always allows for the possibility that the two events occur together. In this section, you learned the Addition Rule, namely, for any events A and B,

$$P(A \text{ or } B) = P(A) + P(B) - P(A \text{ and } B)$$

Two events are disjoint (mutually exclusive) if they cannot occur together. In the case of disjoint events, $P(A \text{ and } B) = 0$, so the Addition Rule simplifies to

$$P(A \text{ or } B) = P(A) + P(B)$$

Two-way tables are often helpful in seeing the structure of a probability problem and in answering questions about the joint behavior of two events.

■ Exercises

E11. Suppose you flip a fair coin five times. Which of these pairs of events are disjoint?

  a. you get five heads; you get four heads and a tail

  b. the first flip is a head; the second flip is a head

  c. you get five heads; the second flip is a tail

d. you get three heads and two tails; the second flip is a tail

e. the first four flips are heads; the first three flips are heads

f. the first head occurs on the third flip; the fourth flip is a head

g. the first head occurs on the third flip; three of the flips are heads

E12. Display 6.10 shows the U.S. population broken down by the categories of age and race as given by the U.S. Bureau of the Census.

You are working for a polling organization that is about to select a random sample of U.S. residents. What is the probability that the first person selected will be

a. white? white and 40 or over?

b. nonwhite and under the age of 10? nonwhite or under age 10?

c. under the age of 10? under the age of 10 and black?

d. black? black or under 10?

E13. Fill in the rest of this table describing students in a large class.

Owns a Beeper?

Has a Dog?		Yes	No	Total
	Yes	–?–	46	72
	No	112	–?–	–?–
	Total	–?–	–?–	216

What is the probability that a randomly selected student doesn't have a dog or doesn't own a beeper?

E14. Suppose 80% of people can swim. Suppose 70% of people can whistle. Suppose 55% of people can do both. What percentage of people can swim or whistle?

a. Construct a table to help you answer this question.

b. Use the Addition Rule to answer this question.

			Race		
Age (yr)	White	Black	American Indian, Eskimo, Aleut	Asian, Pacific Islander	Total
Under 5	15,184	2,892	202	872	**19,150**
5–9	15,560	3,147	226	805	**19,738**
10–14	15,093	2,937	239	770	**19,040**
15–19	15,151	2,963	219	735	**19,068**
20–24	13,970	2,598	186	758	**17,512**
25–29	15,163	2,615	191	900	**18,869**
30–34	16,903	2,762	184	892	**20,741**
35–39	18,710	2,858	183	874	**22,625**
40 or older	95,600	11,175	692	3,427	**110,893**
Total	**221,334**	**33,947**	**2,322**	**10,033**	**267,636**

Display 6.10 Resident population by race and age, 1997 (in thousands). (Prior to the 2000 census, Hispanics were required to classify themselves into these categories.)

Source: U.S. Census Bureau, www.census.gov/population/www/estimates/popest.html.

E15. A 1995 survey found that two-year college mathematics programs had 7,578 full-time faculty members and 14,266 part-time faculty members. Of the full-time faculty, 1% had a bachelor's degree as their highest degree, 82% had a master's degree, and 17% had a doctorate. The corresponding percentages for part-time faculty members were 18%, 75%, and 7%. [Source: Don O. Loftsgaarden, Donald C. Rung, and Ann E. Watkins, *Statistical Abstract of Undergraduate Programs in the Mathematical Sciences in the United States: Fall 1995 CBMS Survey* (Washington, D.C.: Mathematical Association of America, 1997), pp. 96 and 106.] If a two-year college mathematics faculty member is selected at random, what is the probability that his or her highest degree is a master's or a doctorate? Make a two-way table to answer this question.

E16. Polls of registered voters often report the percentage of Democrats and the percentage of Republicans who approve of the job the president is doing. Suppose in a poll of 1500 randomly selected voters, 860 are Democrats and 640 are Republicans. Overall, 937 approve of the job the president is doing, and 449 of those are Republicans. Assuming the people in this poll are representative, what percentage of registered voters are Republicans or approve of the job the president is doing? First answer this question by using the Addition Rule. Then make a two-way table showing the situation.

E17. Jill computes the probability that she gets at least one head in two flips of a fair coin:

P(*at least one head in two flips*)

= *P*(*heads on first* or *heads on second*)

= *P*(*heads on first*) + *P*(*heads on second*)

$= \frac{1}{2} + \frac{1}{2} = 1$

She defends her use of the Addition Rule because getting a head on the first flip and a head on the second flip are mutually exclusive—they can't both happen at the same time. What would you say to her?

E18. Next, Jill computes the probability that she gets exactly one head in two flips of a fair coin:

P(*exactly one head in two flips*)

= *P*((*tails on first flip* and *heads on second*) or (*heads on first* and *tails on second*))

= *P*(*tails on first flip* and *heads on second*) + *P*(*heads on first* and *tails on second*)

$= \frac{1}{4} + \frac{1}{4} = \frac{1}{2}$

She defends her use of the Addition Rule because *HT* and *TH* are mutually exclusive. What would you say to her?

E19. When is this statement true?

P(*A* or *B* or *C*) = *P*(*A*) + *P*(*B*) + *P*(*C*)

E20. In a local school, 80% of the students carry a backpack, *B*, or a wallet, *W*. Forty percent carry just a backpack and 30% carry just a wallet. If a student is selected at random, find the probability that the student carries both a backpack and a wallet. (Note how the word *just* makes this exercise different from the example on page 345.)

6.3 Conditional Probability

The Titanic sank in 1912 without enough lifeboats for the passengers and crew. Almost 1500 people died, most of them men. Was that because a man was less likely than a woman to survive? Or did more men die simply because men outnumbered women by more than 3 to 1 on the Titanic? For an answer, you might turn to an old campfire song, which suggests what may have happened. One line is about loading the lifeboats: "And the captain shouted, 'Women and children first.'" Statisticians aren't opposed to songs. Some of us have even been known to sing! But statisticians also know that stories and theories, even stories set to music, are no substitute for data. Do the data in Display 6.11 support the lyrics of the song?

		Gender		
		Male	**Female**	**Total**
Survived?	**Yes**	367	344	711
	No	1364	126	1490
	Total	**1731**	**470**	**2201**

Display 6.11   Titanic survival data.
Source: *Journal of Statistics Education* 3, no. 3 (1995).

Although the numbers alone can't tell you who got to go first on the lifeboats, they do show that $\frac{344}{470}$, or roughly 73%, of the females survived, while only $\frac{367}{1731} \approx 21\%$ of the males survived. Thus, the data are fully consistent with the hypothesis that the song was right about what happened.

Overall, $\frac{711}{2201}$, or approximately 32%, of the people survived, but the survival rate for females was much higher and for males much lower. The chance of surviving depended on the *condition* of whether the person was male or female. This commonsense notion that probability can change if you are given additional information is called **conditional probability.**

Conditional Probability from the Sample Space

Often you can calculate conditional probabilities directly from the sample space.

Example: The Titanic and Conditional Probability

Suppose you pick a person at random from the list of people aboard the Titanic. Let S be the event that this person survived, and let F be the event that the person was female. Find $P(S\,|\,F)$, read "the probability of S given that F is known to have happened."

Solution

The symbol for the conditional probability that A happens given that B happens is $P(A|B)$.

The conditional probability $P(S\,|\,F)$ is the probability that the person survived *given the condition* that the person was female. To find $P(S\,|\,F)$, restrict your sample space to only the 470 females (the outcomes for whom the condition is true). Then compute the probability of survival as the number of favorable outcomes, 344—the women surviving—divided by the total number of outcomes, 470, in the restricted sample space.

$$P(S\,|\,F) = P(survived\,|\,female) = \frac{344}{470} \approx .732$$

Overall, only 32.3% of the people survived, but 73.2% of the females survived. ■

Example: Sampling Without Replacement

When you sample without replacement from a small population, the result of the first draw affects the probabilities for the second draw. For example, imagine a population of four recent graduates (that is $N = 4$), two of them from the west, W, and two from the east, E. Suppose you randomly choose one person from the population and he or she is from the west. What is the probability that the second person selected is also from the west? In symbols, that is

$$P(W \text{ selected 2nd} \mid W \text{ selected 1st})$$

Solution

If you start with a population $\{W, W, E, E\}$ and the first selection is W, that leaves the restricted sample space $\{W, E, E\}$ to choose from on draw two. The conditional probability that the second selection is from the west is

$$P(W \text{ selected 2nd} \mid W \text{ selected 1st}) = \frac{1}{3} \quad \blacksquare$$

Discussion: Conditional Probability

D23. For the Titanic data in Display 6.11, let S be the event a person survived and F be the event a person was female. Find and interpret these probabilities.

  a. $P(F)$ and $P(F \mid S)$

  b. $P(\text{not } F)$, $P(\text{not } F \mid S)$, and $P(S \mid \text{not } F)$

D24. When you compare sampling with and without replacement, how does the size of the population affect the comparison? Conditional probability lets you answer the question quantitatively. Imagine two populations of students, one large ($N = 100$) and one small ($N = 4$), with half of the students in each population being male, M. Draw random samples of size $n = 2$ from each population.

  a. First, consider the small population. Find the probability $P(2\text{nd is } M \mid 1\text{st is } M)$, assuming you sample without replacement. Then calculate it again, but this time assume you sample with replacement.

  b. Now do the same for the larger population.

  c. Based on these calculations, how would you describe the effect of population size on the difference between the two sampling methods?

■ Practice

P13. Suppose Jack draws marbles at random without replacement from a population of three red and two blue marbles. Find these conditional probabilities:

  a. $P(2\text{nd draw is red} \mid 1\text{st draw is red})$

  b. $P(2\text{nd draw is red} \mid 1\text{st draw is blue})$

  c. $P(3\text{rd is blue} \mid 1\text{st is red and } 2\text{nd is blue})$

  d. $P(3\text{rd is red} \mid 1\text{st is red and } 2\text{nd is red})$

P14. Suppose Jill draws a card out of a standard 52-card deck. Find the probability that

    a. it is a club, given that it is black

    b. it is a jack, given that it is a heart

    c. it is a heart, given that it is a jack

The Multiplication Rule for *P(A and B)*

You will now investigate a general rule for finding $P(A \text{ and } B)$ for any two events A and B. Suppose, for example, you choose one person at random from the passenger list of the Titanic. What is the probability that the person was female, F, *and* survived, S? You already know how to answer this question by directly using the data in Display 6.11. Out of 2201 people, 344 were women and survived, so

$$P(F \text{ and } S) = \frac{344}{2201}$$

You can also find the answer by using conditional probability. It sometimes helps to show the order of events in a tree diagram such as Display 6.12, where the condition comes first.

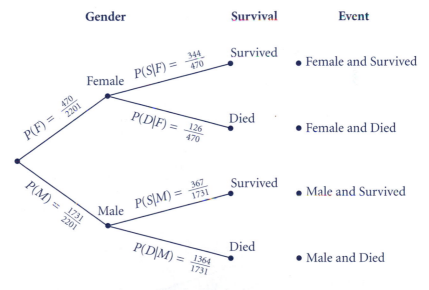

Display 6.12 Tree diagram for the Titanic data.

You can interpret the top branch of the tree diagram, for example, as

$$P(F) \cdot P(S \mid F) = \frac{470}{2201} \cdot \frac{344}{470} = \frac{344}{2201} = P(F \text{ and } S)$$

Discussion: Computing *P(A and B)*

D25. Look again at the Titanic data, this time thinking of whether or not the person survived as the condition.

   a. Make a tree diagram to illustrate this situation.

   b. Write these probabilities as unreduced fractions.

      i. *P(survived)*

      ii. *P(female | survived)*

      iii. *P(survived and female)*

   c. Now write a formula that tells how these three probabilities are related. Compare it with the computation on the bottom of page 351.

   d. Find two formulas to compute *P(male and survived)*.

The formulas you just used in D25, parts c and d, are equivalent to this rule for finding *P(A and B)*.

P(A and B) is sometimes written as P(A ∩ B) and sometimes as P(AB).

The Multiplication Rule

The probability that event *A* and event *B* both happen is given by

$$P(A \text{ and } B) = P(A) \cdot P(B \mid A)$$

Alternatively,

$$P(A \text{ and } B) = P(B) \cdot P(A \mid B)$$

Display 6.13 shows the Multiplication Rule on the branches of a tree diagram.

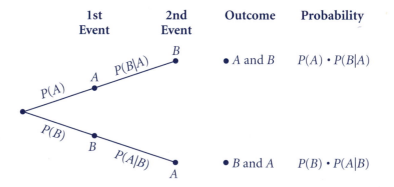

Display 6.13   The general Multiplication Rule, shown on a tree diagram.

From past work with fractions, you know that *of* translates to multiplication. Three-fourths of twenty means $\frac{3}{4} \cdot 20$, or 15. The Multiplication Rule is simply an extension of this same logic. For example, 21% of all Titanic passengers were female, and 73% of female passengers survived. So 73% of 21% of all passengers were females and survived:

$$P(female \text{ and } survived) = (.21)(.73) = P(female) \cdot P(survived \,|\, female)$$

Example: Hunting Moose

This article appeared in the *Los Angeles Times* on July 18, 1978.

> **Man, Wife Beat Odds in Moose-Hunt Draw**
>
> The Washington State Game Department conducted a public drawing last week in Olympia for three moose permits. There were 2,898 application cards in the wire mesh barrel. It was cranked around several times before the first name was drawn: Judy Schneider of East Wenatchee, Wash. The barrel was spun again. Second name drawn: Bill Schneider, Judy's husband. Result: groans of disbelief from hopeful moose hunters in the auditorium.

a. What is the probability that Judy's name would be the first drawn and her husband Bill would be second?

b. What assumptions are you making when you answer this question?

c. According to the article, a Game Department spokesman said that "Mathematicians at the University of Washington told us that the odds of that happening are astronomical." (Assume that the spokesman meant that the odds *against* this happening are "astronomical," and use the term in that context.) Comment.

Solution

a. *P(Judy first* and *Bill second)*

$$= P(Judy \; first) \cdot P(Bill \; second \,|\, Judy \; first)$$

$$= \frac{1}{2898} \cdot \frac{1}{2897}$$

$$\approx .000000119$$

b. The assumptions are that Bill and Judy had only one card each in the barrel and that the cards were well mixed.

c. The chance that Judy's name and then her husband's name will be called is indeed astronomical. The probability is only .000000119. However, there is also the chance that Bill's name will be called first and then Judy's. This doubles their chances of being the first two names to .000000238—still astronomical. However, suppose that the only names in the barrel were those of couples. Then the probability that the first two names drawn will be those of a couple is $\frac{1}{2897} \approx .000345$ because the first name can be anyone and there is one partner out of the 2897 left for the second draw. This is still small but no longer astronomical. Finally, this lottery happens every year, so the chances that something like this would happen eventually and get in the paper are much larger still. ■

Discussion: The Multiplication Rule

D26. Use the tree diagram for the Titanic data in Display 6.12 to find $P(male$ and $survived)$ and $P(male$ and $died)$.

D27. Use the Titanic data in Display 6.11 to show that

$$P(M \text{ and } S) = P(S) \cdot P(M \mid S) = P(M) \cdot P(S \mid M)$$

D28. Use the frequencies in the table below to show that

$$P(A \text{ and } B) = P(A) \cdot P(B \mid A) = P(B) \cdot P(A \mid B)$$

is true in general.

		Event A Present?	
		Yes	**No**
Event B	**Yes**	c	d
Present?	**No**	e	f

■ Practice

P15. Use the Multiplication Rule to find the probability that if you draw two cards from a deck without replacing the first before drawing the second, both the cards will be hearts. What is the probability if you replace the first card before drawing the second?

P16. Suppose you take a random sample of size $n = 2$, without replacement, from the population $\{W, W, M, M\}$. Find these probabilities: $P(W \text{ chosen 1st})$ and $P(W \text{ chosen 2nd} \mid W \text{ chosen 1st})$. Now find $P(W \text{ chosen 1st and } W \text{ chosen 2nd})$.

P17. Use the Multiplication Rule to find the probability of getting a sum of 8 and doubles when you roll two dice.

The Definition of Conditional Probability

It's time to define conditional probability formally.

> ### The Definition of Conditional Probability
>
> For any two events A and B such that $P(B) > 0$,
>
> $$P(A \mid B) = \frac{P(A \text{ and } B)}{P(B)}$$

The Multiplication Rule is a consequence of the formal definition of conditional probability. Just solve the equation for $P(A \text{ and } B)$.

Example: Conditional Probability

Suppose you roll two dice. Use the definition of conditional probability to find the probability that you get a sum of 8 given that you rolled doubles.

Solution

Thinking in terms of the restricted sample space, you already know that the answer is $\frac{1}{6}$ because there are six equally likely ways to roll doubles and only one of them is a sum of 8. However, using the definition, the arithmetic is a little different:

$$P(sum\ 8 \mid doubles) = \frac{P(sum\ 8\ and\ doubles)}{P(doubles)}$$

$$= \frac{\frac{1}{36}}{\frac{6}{36}} = \frac{1}{6} \quad \blacksquare$$

Example: Elizabeth Dole

In 1999, Elizabeth Dole was a candidate to become the first woman president in U.S. history, and many observers assumed that she would have particular strength among women. According to a Gallup poll, "She did slightly better among Republican women than among Republican men, but this strength was not nearly enough to enable her to challenge Bush. In the October poll, Dole received the vote of 16% of Republican women, compared to 7% of Republican men." [Source: www.gallup.com (2000).] With the additional information that the Republican party is about 60% male, find the probability that a Republican randomly selected from the October survey would have voted for Dole.

Solution

You can interpret the two percentages given by Gallup as the approximate conditional probabilities that a randomly selected Republican woman, W, or Republican man, M, respectively, would have voted for Dole, D. That is,

$$P(D \mid W) = .16 \quad \text{and} \quad P(D \mid M) = .07$$

You can write the probability of a Republican voting for Dole in terms of two disjoint events as follows:

$$P(D) = P[(D\ and\ W)\ or\ (D\ and\ M)]$$
$$= P(D\ and\ W) + P(D\ and\ M)$$
$$= P(W) \cdot P(D \mid W) + P(M) \cdot P(D \mid M)$$
$$= (.40)(.16) + (.60)(.07)$$
$$= .11$$

Only about 11% of the Republicans appeared to be willing to vote for Elizabeth Dole. ■

Discussion: Definition of Conditional Probability

D29. Fill in a copy of this table with percentages derived from the example about Elizabeth Dole. Then use the definition of conditional probability to find the probability that a randomly selected Republican is a woman, given that the person would vote for Dole.

		Would Vote for Dole?		
		Yes	**No**	**Total**
Gender	**Male**	–?–	–?–	–?–
	Female	–?–	–?–	–?–
	Total	–?–	–?–	100%

■ Practice

P18. Suppose you know that in a class of 30 students, 10 have blue eyes and 20 have brown eyes. Twenty-four of the students are right-handed and 6 are left-handed. Of the left-handers, 2 have blue eyes. Make and fill in a table showing this situation. Then use the definition of conditional probability to find the probability that a randomly selected person from this classroom is right-handed, given that they have brown eyes.

P19. As of July 1, the Los Angeles Dodgers had won 53% of their games in 2000. Eighteen percent of their games had been played against left-handed starting pitchers. The Dodgers won 36% of the games against left-handed starting pitchers. What percentage of their games did they win against right-handed starting pitchers?

Source: *Los Angeles Times,* July 1, 2000, p. D8.

Conditional Probability and Medical Tests

In medicine, *screening tests* give a quick indication of whether or not a person is likely to have a particular disease. (For example, the ELISA test is a screening test for HIV, and a chest X ray is a screening test for lung cancer.) Because screening tests are intended to be relatively quick and noninvasive, they are often not as accurate as other tests that take longer or are more invasive. (For example, a biopsy is more accurate than a chest X ray if you want to know if a lung tumor is cancerous.)

A two-way table like the one in Display 6.14 is often used to show the four possible outcomes of a screening test.

		Test Result		
		Positive	**Negative**	**Total**
Disease	**Present**	*a*	*b*	*a + b*
	Absent	*c*	*d*	*c + d*
	Total	*a + c*	*b + d*	*a + b + c + d*

Display 6.14   Possible results of a screening test.

The effectiveness of tests like these are judged using conditional probabilities. Here are definitions of four terms in common use:

$$\text{False positive rate} = P(no\ disease \,|\, test\ positive) = \frac{c}{a + c}$$

$$\text{False negative rate} = P(disease\ present \,|\, test\ negative) = \frac{b}{b + d}$$

$$\text{Sensitivity} = P(test\ positive \,|\, disease\ present) = \frac{a}{a + b}$$

$$\text{Specificity} = P(test\ negative \,|\, no\ disease) = \frac{d}{c + d}$$

Relatively rare diseases have tables somewhat like that in Display 6.15.

Test Result

		Positive	Negative	Total
Disease	**Present**	9	1	10
	Absent	50	9,940	9,990
	Total	59	9,941	10,000

Display 6.15 Hypothetical results of a screening test for a rare disease.

In some ways, this is a pretty good test. It finds 9 out of the 10 people who have the disease for a sensitivity of .9 and correctly categorizes 9940 out of the 9990 people who don't have the disease for a specificity of .99. The false negative rate is only 1 out of 9941, or .0001. However, notice that only 9 out of 59, or 15%, of the people who test positive for the disease actually have it! The false positive rate is extremely high, 85%.

Even with a test of high specificity, this is what can happen when the population being screened is mostly disease free. Most of the people who test positive do not, in fact, have the disease. On the other hand, if the population being screened has a high incidence of the disease, then the false negative rate tends to be high.

Because the rates of false positives and false negatives depend on the population as well as the test, statisticians prefer to judge a test using the other pair of conditional probabilities: sensitivity and specificity.

Discussion: Interpreting Screening Tests

D30. A laboratory technician is being tested on her ability to detect contaminated blood samples. Among 100 samples given to her, 20 are contaminated, each with about the same degree of contamination. Suppose the technician makes the correct decision 90% of the time. Make a table showing what you would expect to happen. What is her false positive rate? What is her false negative rate? How would these rates change if she were given 100 samples with 50 contaminated?

D31. Most patients who take a screening test find the false positive rate and false negative rate easier to interpret than sensitivity and specificity. Explain why.

■ Practice

P20. This table presents hypothetical data on 130 people for an inkblot test for bureaucratic pomposity disorder (BPD). Use the data to compute the four conditional test probabilities defined in the screening tests example. Is the test a good one, in your judgment? Why or why not?

Test Result

		Positive	Negative	Total
BPD	**Present**	60	40	100
	Absent	20	10	30
	Total	80	50	130

Conditional Probability and Statistical Inference

In statistics, probabilities are computed under the conditions imposed by a model. For example, you can calculate the probability of observing an even number of dots on a roll of a die under the condition that the die is fair. That is, you know that

$$P(even\ number \mid fair\ die) = \frac{3}{6}$$

If, however, you roll a die and get an even number, you cannot calculate $P(fair\ die \mid even\ number)$. So how can you discredit a model?

If you keep rolling until you have rolled 20 times and you get an even number every time, you can compute (by the method of the next section),

$$P(even\ number\ all\ 20\ rolls \mid fair\ die) = \left(\frac{3}{6}\right)^{20} \approx .000001$$

This is so unlikely that you can feel justified in abandoning the model that the die is fair. But still, you didn't and can't compute

$$P(fair\ die \mid even\ number\ all\ 20\ rolls)$$

The following dialog is invented and did not actually occur in the Westvaco case, but it is based on real conversations one of your authors had on several occasions with a number of different lawyers as they grappled with conditional probabilities.

Statistician: Suppose you draw 3 workers at random from the set of 10 hourly workers. This establishes random sampling as the model for the study.

Lawyer: Okay.

Statistician:	It turns out that there are $\binom{10}{3}$, or 120, possible samples of size 3, and only 6 of them give an average age of 58 or more.
Lawyer:	So the probability is $\frac{6}{120}$, or .05.
Statistician:	Right.
Lawyer:	There's only a 5% chance the company didn't discriminate and a 95% chance that they did.
Statistician:	No, that's not true.
Lawyer:	But you said . . .
Statistician:	I said that *if* the age-neutral model of random draws is correct, then there's only a 5% chance of getting an average age of 58 or more.
Lawyer:	So the chance the company is guilty must be 95%.
Statistician:	Slow down. If you start by assuming the model is true, you can compute the chances of various results. But you're trying to start from the results and compute the chance that the model is right or wrong. You can't do that.

Here is the analysis: The lawyer is having trouble with conditional probabilities. The statistician has computed

$$P(average\ age \geq 58 \mid random\ draws) = .05$$

The lawyer wants to know

$$P(no\ discrimination \mid average\ age = 58)$$

There are two things that make the lawyer's conditional probability impossible to compute. One is that "random draws" and "no discrimination" are not the same thing. Random selection is an age-neutral process, but it is not the only way to select people without regard to age. The other is that it makes a tremendous difference which is the condition. The model, drawing at random without replacement, tells how to compute the statistician's conditional probability—if you know the model, you can find $P(data \mid model)$. But the lawyer wants $P(model \mid data)$, and there's not enough information in the data to compute this.

Discussion: Conditional Probability and Statistical Inference

D32. Jill has two pennies. One is an ordinary fair penny, but the other came from a magic shop and has two heads. Jill chooses one of these coins, but she does not choose it randomly. She also doesn't tell you how she chose it, so you don't know the probabilities $P(coin\ is\ two\text{-}headed)$ or $P(coin\ is\ fair)$. She flips the coin once. For each of these probabilities, give a numerical value if it is possible to find one. If it is not possible to compute a probability, explain why.

a. $P(coin\ lands\ heads \mid coin\ is\ fair)$

b. $P(coin\ lands\ heads \mid coin\ is\ two\text{-}headed)$

   c. $P(coin\ is\ fair \mid coin\ lands\ heads)$

   d. $P(coin\ is\ two\text{-}headed \mid coin\ lands\ heads)$

   e. $P(coin\ is\ fair \mid coin\ lands\ tails)$

   f. $P(coin\ is\ two\text{-}headed \mid coin\ lands\ tails)$

D33. Compare the situations of the confused lawyer and the two-headed coin in D32. In the coin example, tell what corresponds to "model" and what corresponds to "data."

■ Practice

P21. There's a parallel between statistical testing and medical testing. Write expressions for conditional probabilities—in terms of finding a company guilty or not guilty of discrimination—that correspond to a false positive rate and false negative rate. (You might think of the true medical status as being comparable to the true state of affairs within the company.)

Summary 6.3: Conditional Probability

In this section, you saw these important uses of conditional probability in statistics:

- To compare sampling with and without replacement
- To study the effectiveness of medical tests
- To describe probabilities of the sort used in statistical tests of hypotheses

You have learned the definition of conditional probability and the Multiplication Rule and how to use these two key concepts in practical problems. When solving conditional probability problems, by use of the Multiplication Rule or otherwise, tree diagrams and two-way tables are a big help in organizing the data and seeing the structure of the problem.

- The definition of conditional probability is that for any two events A and B, where $P(B) > 0$, the probability of A given the condition B is

$$P(A \mid B) = \frac{P(A \text{ and } B)}{P(B)}$$

- The Multiplication Rule is

$$P(A \text{ and } B) = P(A) \cdot P(B \mid A)$$

■ Exercises

E21. Display 6.16 gives a breakdown of the U.S. population by race and age as given by the U.S. Bureau of the Census.

You are working for a polling organization that is about to select a random sample of U.S. residents. What is the probability that the first person selected will be

a. age 40 or older, given that the person is Asian/Pacific islander?

	Race				
Age (yr)	White	Black	American Indian, Eskimo, Aleut	Asian, Pacific Islander	Total
Under 5	15,184	2,892	202	872	**19,150**
5–9	15,560	3,147	226	805	**19,738**
10–14	15,093	2,937	239	770	**19,040**
15–19	15,151	2,963	219	735	**19,068**
20–24	13,970	2,598	186	758	**17,512**
25–29	15,163	2,615	191	900	**18,869**
30–34	16,903	2,762	184	892	**20,741**
35–39	18,710	2,858	183	874	**22,625**
40 or older	95,600	11,175	692	3,427	**110,893**
Total	**221,334**	**33,947**	**2,322**	**10,033**	**267,636**

Display 6.16 Resident population by race and age, 1997 (in thousands). (Prior to the 2000 census, Hispanics were required to classify themselves into these categories.)

Source: U.S. Census Bureau, www.census.gov/population/www/estimates/popest.html.

b. black, given that the person is under the age of 10?

c. under the age of 10, given that the person is white?

d. black or under the age of 10, given that the person is under the age of 30?

E22. Joseph Lister (British, 1827–1912), surgeon at the Glasgow Royal Infirmary, was one of the first to believe in Pasteur's germ theory of infection. He experimented with using carbolic acid to disinfect operating rooms during amputations. When carbolic acid was used, 34 of 40 patients lived. When carbolic acid was not used, 19 of 35 patients lived. If a patient is selected at random, find

a. $P(patient\ died \mid carbolic\ acid\ used)$

b. $P(carbolic\ acid\ used \mid patient\ died)$

c. $P(carbolic\ acid\ used\ and\ patient\ died)$

d. $P(carbolic\ acid\ used\ or\ patient\ died)$

E23. As you have seen, a useful way to work with the Multiplication Rule is to record the probabilities along the branches of a tree diagram. To practice this, suppose you draw two marbles at random, without replacement, from a bucket containing three red and two blue marbles. Compute the probabilities in parts a–f and then use these probabilities to fill in and label the branches in a copy of the tree diagram in Display 6.17.

a. $P(R\ on\ 1st\ draw)$

b. $P(B\ on\ 1st\ draw)$

c. $P(R\ on\ 2nd\ draw \mid R\ on\ 1st\ draw)$

d. $P(B\ on\ 2nd\ draw \mid R\ on\ 1st\ draw)$

e. $P(R\ on\ 2nd\ draw \mid B\ on\ 1st\ draw)$

f. $P(B\ on\ 2nd\ draw \mid B\ on\ 1st\ draw)$

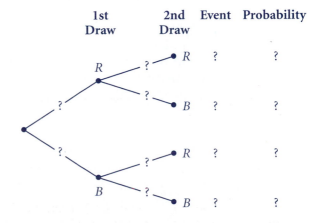

Display 6.17 Tree diagram for drawing marbles.

E24. Suppose that if rain is predicted, there is a 60% chance that it will actually rain, and if rain is not predicted, there is a 20% chance that it will rain. Rain is predicted on 10% of the days.

　a. On what percentage of the days does it rain?

　b. Describe a simulation that you could use to check your answer.

E25. A Harris poll released on July 28, 1997, found that 36% of all adults said they would be interested in going to Mars. Of those who want to go to Mars, 62% were under 25 years old. Write a problem for which the solution would be to compute (.36)(.62).

E26. Suppose the only information you have about the Titanic disaster appears in the two-way table below. Is it possible to fill in the rest of the table? Is it possible to find the probability that a randomly selected person aboard the Titanic was female and survived? Either calculate the answers or explain why it is not possible without additional information, and tell what information you would need.

Gender

Survived?	Male	Female	Total
Yes	–?–	–?–	711
No	–?–	–?–	1490
Total	1731	470	2201

E27. A class collects information on their gender and handedness. The next two-way table gives the proportions of the class members who fall into each category.

　a. If possible, fill in the rest of this table of proportions.

Gender

Handedness	Male	Female	Total
Right	.44	–?–	–?–
Left	–?–	–?–	.08
Total	–?–	.48	1.00

　b. If a student is chosen at random from this class, what is the probability of the student being left-handed, given that the student is male?

　c. What is the probability that the randomly selected student is female and left-handed?

E28. A 1995 survey of mathematics programs in two-year colleges found 7578 full-time permanent faculty members and 14,266 part-time faculty. Overall, part-time faculty taught 38% of the classes but tended to teach lower-level courses, teaching 47% of the classes of remedial mathematics. Remedial mathematics made up 53% of the total number of classes. Fill in a copy of this two-way table with the percentage of courses that fall into each category.

Source: Don O. Loftsgaarden, Donald C. Rung, and Ann E. Watkins, *Statistical Abstract of Undergraduate Programs in the Mathematical Sciences in the United States: Fall 1995 CBMS Survey* (Washington, D.C.: Mathematical Association of America, 1997), pp. 96 and 106.

Course Taught By

Type of Course	Full-time Faculty	Part-time Faculty	Total
Remedial	–?–	–?–	–?–
Non-remedial	–?–	–?–	–?–
Total	–?–	–?–	100%

E29. A laboratory test for the detection of a certain disease gives a positive result 6% of the time for people who do not have the disease. The test gives a negative result 0.5% of the time for people who do have the disease. Large-scale studies have shown that the disease occurs in about 3% of the population.

a. Fill in this two-way table showing the results expected for every 100 people.

Test Result

		Positive	Negative	Total
Disease	**Yes**	–?–	–?–	–?–
	No	–?–	–?–	–?–
	Total	–?–	–?–	100

b. What is the probability that a person selected at random would test positive for this disease?

c. What is the probability that a person selected at random who tests positive for the disease does not have the disease?

E30. Find the probability of each of these events involving drawing two cards from a deck.

a. Both cards are aces; the first card is replaced before drawing the second.

b. Both cards are aces; the first card is not replaced before drawing the second.

c. The first is an ace and the second is a king; the first card is replaced before drawing the second.

d. The first is an ace and the second is a king; the first card is not replaced before drawing the second.

e. Both cards are of the same suit; the first card is not replaced before drawing the second.

E31. Make a table and do a computation to illustrate that if the population being screened has a high percentage of people who have a certain disease, the false negative rate will tend to be high.

E32. A screening test for bureaucratic pomposity disorder (BPD) has reasonably good sensitivity and specificity, say 90% for both. (This means that, on average, 9 out of every 10 people who have BPD give a positive result on the test, and also 9 out of every 10 who don't have BPD give a

negative result on the test.) Now consider testing two different populations.

a. Officiousville has a population of 1000. Half the people in Officiousville have BPD. Suppose everyone in Officiousville gets tested. Fill out a table like the one in Display 6.14. Then use it to compute the false positive and false negative rates for the test on the Officiousville population.

b. Mellowville also has a population of 1000, but only 10 people in the whole town have BPD. Fill out another table to show how the test would perform if you used it to screen Mellowville. Then compute the false positive and false negative rates.

E33. Return to the situation in D32 on page 359 where Jill has two pennies. One is an ordinary fair penny, but the other one came from a magic shop and has two heads. Jill chooses one of these coins, this time *at random*. She flips the coin once. For each of these probabilities, give a numerical value if it is possible to find one. If it is not possible to compute a probability, explain why not.

a. $P(coin\ lands\ heads \mid coin\ is\ fair)$

b. $P(coin\ lands\ heads \mid coin\ is\ two\text{-}headed)$

c. $P(coin\ is\ fair \mid coin\ lands\ heads)$

d. $P(coin\ is\ two\text{-}headed \mid coin\ lands\ heads)$

e. $P(coin\ is\ fair \mid coin\ lands\ tails)$

f. $P(coin\ is\ two\text{-}headed \mid coin\ lands\ tails)$

E34. One of these two statements is true and one is false. Which is which? Make up an example using hypothetical data to illustrate your decision.

$P(A) = P(A \mid B) + P(A \mid not\ B)$

$P(A) = P(A\ and\ B) + P(A\ and\ not\ B)$

(The event *not B* is the event that *B* did not occur, sometimes called the **complement** of *B*. It consist of all outcomes in the sample space that are not in *B*.)

6.4 Independent Events

Jack and Jill have finished conducting taste tests with 100 adults from their neighborhood. They found that 60 of them correctly identified the tap water. But how well the person did depended on whether the person regularly uses bottled water for drinking. Of the people who regularly drink bottled water, $\frac{24}{30}$, or 80%, correctly identified the tap water. (See Display 6.18.) Of people who drink tap water, only $\frac{36}{70}$, or 51.4%, correctly identified the tap water.

Identified Tap Water?

		Yes	No	Total
Drinks Bottled Water?	**Yes**	24	6	30
	No	36	34	70
	Total	60	40	100

Display 6.18 Results of taste test depend on whether a person drinks bottled water.

On the other hand, as Display 6.19 shows, men and women did equally well in identifying the tap water. In each case, 60% correctly identified the tap water.

Identified Tap Water?

		Yes	No	Total
Gender	**Male**	21	14	35
	Female	39	26	65
	Total	60	40	100

Display 6.19 Results of taste test don't depend on gender.

Based on these data, the events *drinks bottled water* and *correctly identifies tap water* are **dependent events**. If a person drinks bottled water, he or she is more likely to correctly identify tap water. The events *is a male* and *correctly identifies tap water* are **independent events**. Knowing a person is male doesn't change the probability that the person correctly identifies tap water.

More formally, events A and B are independent if the probability of event A happening doesn't depend on whether event B happens. That is, knowing that B happens doesn't change the probability that A happens.

Definition of Independent Events

Events A and B are independent if and only if $P(A \mid B) = P(A)$.
Equivalently, events A and B are independent if and only if $P(B \mid A) = P(B)$.

In E56, you will show that $P(A \mid B) = P(A)$ if and only if $P(B \mid A) = P(B)$. In E47, you will show that if events A and B are independent, then so are events A and *not B* (event *not B* is the complement of event B, consisting of all possible outcomes that are not in B).

You can write event *not B* as $\bar{B}$, so $P(\bar{B}) = 1 - P(B)$.

Example: Water, Gender, and Independence

Show that the events *is a male* and *correctly identifies tap water* are independent and that the events *drinks bottled water* and *correctly identifies tap water* are not independent.

Solution

First, consider the events *is a male* and *correctly identifies tap water*.

$$P(correctly\ identifies\ tap\ water) = \frac{60}{100} = .60$$

$$P(correctly\ identifies\ tap\ water \mid is\ a\ male) = \frac{21}{35} = .60$$

Because these two probabilities are equal, the events are independent.

Now consider the events *drinks bottled water* and *correctly identifies tap water*.

$$P(correctly\ identifies\ tap\ water) = \frac{60}{100} = .60$$

$$P(correctly\ identifies\ tap\ water \mid drinks\ bottled\ water) = \frac{24}{30} = .80$$

Because these two probabilities are not equal, the events are not independent. ■

Discussion: Definition of Independent Events

D34. Show in another way that the events *is a male* and *correctly identifies tap water* are independent and that the events *drinks bottled water* and *correctly identifies tap water* are not independent. That is, use *correctly identifies tap water* as the condition in both cases.

D35. Suppose you choose a student at random from your school. In each case, does knowing that event A happened increase the probability of event B, decrease the probability of event B, or leave the probability of event B unchanged?

a. *A:* The student is a football player.

B: The student weighs less than 120 pounds.

b. *A:* The student has long fingernails.

B: The student is female.

c. *A:* The student is a freshman.

B: The student is male.

d. *A:* The student is a freshman.

B: The student is a senior.

■ Practice

P22. Refer to the Titanic data in Display 6.11 on page 349. Use the definition of independent events to determine if the events *didn't survive* and *male* are independent. Are any pair of events in this table independent?

P23. Suppose you draw a card at random from a standard deck. Use the definition of independent events to determine which of these pairs of events are independent.

    a. getting a heart; getting a jack

    b. getting a heart; getting a red card

    c. getting a seven; getting a heart

Multiplication Rule for Independent Events

In Section 6.3, you learned how to use the Multiplication Rule, $P(A \text{ and } B) = P(A) \cdot P(B \mid A)$. If A and B are independent, then $P(B \mid A) = P(B)$, so the Multiplication Rule reduces to a simple product.

Multiplication Rule for Independent Events

Two events A and B are independent if and only if

$$P(A \text{ and } B) = P(A) \cdot P(B)$$

More generally, events $A_1, A_2, \ldots, A_n$ are independent, if and only if

$$P(A_1 \text{ and } A_2 \text{ and } \ldots \text{ and } A_n) = P(A_1) \cdot P(A_2) \cdots P(A_n)$$

You can also use this rule to decide if two events are independent.

    If $P(A \text{ and } B) = P(A) \cdot P(B)$, then A and B are independent.

For example, if you flip a fair coin four times, the outcomes of the first flips don't change the probabilities on the remaining flips. The flips are independent, and so for a sequence of four flips of a fair coin,

$P(heads\ on\ 1st\ flip$ and *heads on 2nd* and *heads on 3rd* and *heads on 4th*)

    = $P(heads\ on\ 1st) \cdot P(heads\ on\ 2nd) \cdot P(heads\ on\ 3rd) \cdot P(heads\ on\ 4th)$

$$= \left(\frac{1}{2}\right)\left(\frac{1}{2}\right)\left(\frac{1}{2}\right)\left(\frac{1}{2}\right) = \frac{1}{16}$$

Example: Multiplication Rule and Gas Grills

According to the U.S. Department of Energy, about 30% of households in the United States have an outdoor gas grill. [Source: *World Almanac and Book of Facts 1997*, p. 237.] What is the chance that if you choose two families at random, the first has a gas grill and the second does not?

Solution

Because the sample is selected from the large number of families in the United States, you can model the outcomes of the two trials as independent events so that

Probabilities still multiply even if the independent events are not equally likely.

P(*1st family has grill* and *2nd family has no grill*) = (.3)(.7) = .21

You can conveniently list the probabilities of all four possible outcomes for two families in a "multiplication" table, as in Display 6.20.

Second Family

		Grill	No Grill	Total
First Family	**Grill**	(.3)(.3) = .09	(.3)(.7) = .21	**.30**
	No Grill	(.7)(.3) = .21	(.7)(.7) = .49	**.70**
	Total	**.30**	**.70**	**1.00**

Display 6.20 Two-way table for the gas grill problem. ■

In practice, computing P(*A* and *B*) for independent events is straightforward—use the definition, which tells you to multiply. What's not always straightforward in practice is to recognize whether or not independence gives a good model for a real situation. The basic idea is to ask whether one event has a bearing on the likelihood of another, and this is something you try to decide by thinking about the events in question. As always, models are simplifications of the real thing, and you'll often use independence as a simplifying assumption to be able to calculate probabilities, at least approximately.

Discussion: Multiplication Rule

D36. About 42% of people have type O blood. Suppose you select two people at random and check if they have type O blood or not. Make a table like the one in Display 6.20 to show all possible results. What is the probability that exactly one of the people has type O blood? Make a tree diagram that illustrates the same situation.

■ Practice

P24. In the state of Idaho, about 80% of 9th grade students graduate from high school three years later. [Source: National Center for Education Statistics, *World Almanac and Book of Facts 1997*, p. 253.] Suppose you select two Idaho 9th graders at random and check three years later to see if they have graduated from high school.

a. Find the probability that both students have graduated from high school.

b. Show how to use a table like the one in Display 6.20 to find the probability that exactly one student has graduated from high school.

c. Show how to use a tree diagram to illustrate this situation.

Independence with Real Data

You have seen that data often serves as a basis for establishing a probability model or for checking whether an assumed model is reasonable. In using data to check for independence, however, you have to be careful. For example, it would be fairly unusual for Jack and Jill to come up with exactly the same percentage of men as women who can identify tap water even if men and women are equally likely to be able to identify tap water. That example was "cooked up"; it's almost impossible to find such perfection in practice.

Example: Independence and Baseball

As of July 1, 2000, the Los Angeles Dodgers had won a total of 41 games and lost 37 games. The breakdown by whether the game was played during the day or at night is shown in Display 6.21. Are the events *win* and *day game* independent?

		Won the Game?		
		Yes	**No**	**Total**
Time of Game	**Day**	11	10	**21**
	Night	30	27	**57**
	Total	**41**	**37**	**78**

Display 6.21   Dodger record by time of game.
Source: *Los Angeles Times*, July 1, 2000, p. D8.

Solution

First check the probabilities:

$$P(win) = \frac{41}{78} \approx .526$$

$$P(win \mid day\ game) = \frac{11}{21} \approx .524$$

Because these two probabilities aren't exactly equal, you must say that the events *win* and *day game* aren't independent. And yet, given that the Dodgers played 21 day games, the percentage of day games won couldn't be any closer to .526. A statistician would conclude that if these results can be considered a random sample of their games, the probability that the Dodgers will win doesn't depend on whether they play a day or night game.   ■

Activity 6.4 will further illustrate the difficulty of establishing independence with real data.

Activity 6.4   Independence with Real Data

What you'll need: one penny per student

1. Collect and study data on eyedness and handedness:

   a. Are you right-handed or left-handed?

   b. Determine if you are right-eyed or left-eyed:

      Hold your hands together in front of you at arm's length. Make a space between your hands that you can see through. Through the space, look at an object at least 15 feet away. Now close your right eye. Can you still see the object? If so, you are left-eyed. Now close your left eye. Can you still see the object? If so, you are right-eyed.

   c. Would you expect being right-handed and right-eyed to be independent?

   d. Complete a two-way table for members of your class, showing frequencies of eyedness and handedness.

   e. For the students in your class, are being right-handed and right-eyed independent?

2. Collect and study data on the results of two coin flips:

   a. With other members of your class, flip a coin twice until you have 100 pairs of flips. Place your results in a two-way table.

   b. Would you expect the results of the first flip and the second flip to be independent?

   c. For your data, are the results of the first flip and the second flip independent?

3. Did you get the results you expected in steps 1 and 2? Explain.

In the general population, being right-handed and being right-eyed aren't independent. That's probably what you found for your class in Activity 6.4. On the other hand, two flips of a coin are independent, even though for real data—actual flips—it almost never happens that $P(A)$ *exactly* equals $P(A|B)$, as the definition of independence requires.

Sampling with replacement yields independent selections.

There is another connection between independence and data, and that has to do with sampling. As you saw in Chapters 4 and 5, there are two kinds of simple random samples. For one kind, sampling with replacement, you draw the first unit, then *put it back* and re-mix the population before drawing the next unit. For this method of sampling, the population you draw from always stays the same and the probabilities for the various draws never change. For the second kind of simple random sampling, called sampling *without* replacement, you draw the first unit, then draw the next one *without putting the first unit back*. Each draw changes the population, and, therefore, the outcome on the first draw changes the probabilities for the second and later draws. Sampling with replacement from {*H*, *T*} is just like flipping a coin repeatedly. Sampling with replacement from {1, 2, 3, 4, 5, 6} is just like rolling a die over and over. But if you don't replace, the probabilities change from trial to trial and independence is lost.

In Chapter 10, you will learn how to examine tables like the ones of your coin flips in Activity 6.4 and the Dodger record and then do a statistical test of independence. The key question, just as in Chapter 1, will be whether the data are too extreme (too far from the strict definition of independence) to be consistent with that hypothesis.

Probability and Mendel's Laws of Inheritance

The modern science of genetics began around 1860 when an Austrian monk, Gregor Mendel (1823–1884), started experimenting with pea plants. Mendel noticed that when he bred short pea plants to other short pea plants, all plants in the next generation were short. However, when he bred tall pea plants to other tall plants, some of the plants in the next generation were short and some were tall.

Today we know that for some physical traits, each individual has two genes that determine that trait. The individual receives one gene from each parent. For Mendel's peas, the gene can exist in only two forms (called alleles), tall and short. A tall gene is dominant, and a short gene is recessive. This means that if the offspring receives a tall gene from one parent plant and a short gene from the other, the offspring is tall. If the offspring receives a tall gene from each parent, the offspring is tall. If the offspring receives a short gene from each parent, the offspring is short.

Suppose each parent plant is heterozygous—that is, each parent has a dominant tall gene, *T*, and a recessive short gene, *t*. To visualize this situation, around the turn of the 20th century, the geneticist Reginald C. Punnett (English, 1875–1967) invented the Punnett square, as in Display 6.22, essentially just a two-way table with rows for the alleles of one parent and columns for the alleles of the other parent.

Male Parent

		T	*t*
Female Parent	*T*	*TT*	*Tt*
	t	*tT*	*tt*

Display 6.22 Punnett square for the offspring of two pea plants, both heterozygous, with one dominant gene for tall plants.

Three-fourths of the next generation will be tall plants (*TT*, *Tt*, *tT*) and one-fourth will be short (*tt*).

Discussion: Probability and Genetics

D37. What assumptions were made about inheritance of genes in order to conclude that three-fourths of the next generation will be tall and one-quarter will be short?

D38. When Mendel crossed two heterozygous plants, he found that in the next generation, 787 were tall and 277 were short. How well do these data fit his theory?

■ Practice

P25. Make a Punnett square for the case of one parent plant that has two tall genes and one parent plant that has two short genes. What proportion of the next generation of peas will be tall?

Summary 6.4: Independence

The word *independent* is used in several related ways in statistics. First, two events are independent if knowledge that one event occurs does not affect the probability of the other event occurring. Second, in sampling units from a population with replacement, the outcome of the second selection is independent of what happened on the first selection (and the same is true for any other pair of selections). Sampling without replacement leads to dependent outcomes on successive selections. (However, if the population is large compared to the sample, you can compute probabilities as if the selections were independent without much error.) In both situations, the same definition of independence applies:

> Two events A and B are independent if and only if $P(A) = P(A\,|\,B)$; the probability of A does not change with knowledge that B has happened. If $P(A) = P(A\,|\,B)$, then it is also true that $P(B) = P(B\,|\,A)$.

You also learned in this section that the Multiplication Rule simplifies to

$$P(A \text{ and } B) = P(A) \cdot P(B)$$

if and only if A and B are independent events.

■ Exercises

E35. Which of these pairs of events, A and B, do you expect to be independent? Give a reason for your answer.

a. For a test for tuberculosis antibodies:

  A: The test is positive.

  B: The person has a relative with tuberculosis.

b. For a test for tuberculosis antibodies:

  A: A person's test is positive.

  B: The last digit of the person's social security number is 3.

c. For a randomly chosen state in the United States:

  A: The state lies east of the Mississippi River.

  B: The state's highest elevation is more than 8000 feet.

E36. Use the definition of independent events to determine which of these pairs of events are independent when you roll two dice.

  a. rolling doubles; rolling a sum of 8

  b. rolling a sum of 8; getting a 2 on the first die rolled

  c. rolling a sum of 7; getting a 1 on the first die rolled

  d. rolling doubles; rolling a sum of 7

  e. rolling a 1 on the first die; rolling a 1 on the second die

E37. When a baby is expected, there are two possible outcomes: a boy and a girl. However, they aren't equally likely! About 51% of all babies born are boys.

  a. List the four possible outcomes for a family that has two babies (no twins).

  b. Which of the outcomes is the most likely? Which is the least likely?

  c. What is the probability that both children will be boys in a two-child family? What assumption are you making?

  d. Sometimes people say that "girls run in our family" or "boys run in our family." What do they mean? If they are right, how would this affect your answers?

E38. Marianne has six shirts (blue, green, red, yellow, and two identical white shirts) and four pairs of pants (brown slacks, black slacks, and two identical pair of jeans). Suppose Marianne selects a pair of pants at random and a shirt at random. Make a tree diagram showing Marianne's choices. Use the tree diagram to compute the probability that she wears a white shirt with black pants.

E39. Jill computes the probability that if two dice are rolled, then exactly one die will show a 4 as follows:

$P(4 \text{ on one die and not 4 on the other})$

$$= \left(\frac{1}{6}\right)\left(\frac{5}{6}\right) = \frac{5}{36}$$

Jill is wrong. What did she forget?

E40. A family has two girls and two boys. To select two children to do the dishes, the mother puts the names of the children on separate pieces of paper, mixes them up, and draws two different pieces of paper at random. What is the probability that they are both girls?

E41. Mendel also looked at the shape of the pea seeds and at their color. The shape was either round (dominant) or wrinkled (recessive). The color was either yellow (dominant) or green (recessive). Suppose that each parent is heterozygous with respect to both shape and color. What proportion of the offspring will be round and yellow? Round and green? What assumption(s) are you making?

E42. Suppose you cross two heterozygous pea plants—each parent plant has one dominant (tall) gene and one recessive (short) gene. You then cross all of the plants of the new generation.

  a. Make a Punnett square to represent the offspring of the "grandchild" generation.

  b. What proportion of the "grandchild" plants will be tall?

E43. Construct a two-way table that represents this situation:

$$P(A) = \frac{1}{4} \quad P(B \mid A) = \frac{1}{2} \quad P(B \mid not\ A) = \frac{1}{4}$$

Are events *A* and *B* independent?

E44. Forty-two percent of people in a town have type O blood and 5% are Rh-positive. Assume blood type and Rh type are independent.

a. Find the probability that a randomly selected person has type O blood and is Rh-positive.

b. Find the probability that a randomly selected person has type O blood or is Rh-positive.

c. Make a table that summarizes this situation.

E45. In the "Ask Marilyn" column of *Parade* magazine, the following question appeared:

> Suppose a person was having two surgeries performed at the same time. If the chances of success for surgery A are 85% and the chances of success for surgery B are 90%, what are the chances that both would fail?

Source: *Parade*, November 27, 1994, p. 13.

Write an answer to this question for Marilyn.

E46. In *Dave Barry Talks Back*, Dave relates that he and his wife bought an oriental rug "in a failed attempt to become tasteful." Before going out to dinner, they admonished their dogs Earnest and Zippy to stay away from the rug.

"NO!!" we told them approximately 75 times while looking very stern and pointing at the rug. This proven training technique caused them to slink around the way dogs do when they feel tremendously guilty but have no idea why. Satisfied, we went out to dinner. I later figured out, using an electronic calculator, that this rug covers approximately 2 percent of the total square footage of our house, which means that if you (not you personally) were to have a random . . . [Dave's word omitted because this is a textbook] attack in our home, the odds are approximately 49 to 1 against your having it on our Oriental rug. The odds against your having four random attacks on this rug are more than five million to one. So we had to conclude that it was done on purpose.

Source: *Dave Barry Talks Back* by Dave Barry, copyright © 1991 by Dave Barry. Illustrations copyright © 1991 by Jeff MacNelly. Used by permission of Crown Publishers, a division of Random House, Inc.

Do you agree with Dave's arithmetic? Do you agree with his logic?

E47. Show that if events *A* and *B* are independent, then so are events *A* and *not B*.

Chapter Summary

The probabilities used in statistical investigations often come from observed data (the past being used to approximate the future) or from a model based on symmetry (the possible events being broken down into equally likely outcomes).

Two important concepts were introduced in this chapter: disjoint (mutually exclusive) events and independent events.

- Two events are disjoint if they can't happen on the same outcome. If *A* and *B* are disjoint events, then $P(A \text{ and } B) = 0$.

- Two events are independent if the occurrence of one doesn't change the probability that the other will happen. That is, events *A* and *B* are independent if and only if $P(A) = P(A \mid B)$.

There were two important rules given in this chapter.

- The Addition Rule is

$$P(A \text{ or } B) = P(A) + P(B) - P(A \text{ and } B)$$

If the events A and B are disjoint (mutually exclusive), this simplifies to

$$P(A \text{ or } B) = P(A) + P(B)$$

- The Multiplication Rule is given by

$$P(A \text{ and } B) = P(A) \cdot P(B \mid A)$$

If A and B are independent events, this simplifies to

$$P(A \text{ and } B) = P(A) \cdot P(B)$$

■ Review Exercises

E48. Suppose you roll a fair four-sided (tetrahedral) die and a fair six-sided die.

   a. How many equally likely outcomes are there?

   b. Show all of them in a table or in a tree diagram.

   c. What is the probability of getting doubles?

   d. What is the probability of getting a sum of 3?

   e. Are the events *getting doubles* and *getting a sum of 4* disjoint? Are they independent?

   f. Are the events *getting a 2 on the tetrahedral die* and *getting a 5 on the six-sided die* disjoint? Are they independent?

E49. Jorge has a CD player attached to his alarm clock. He has set the CD player so that when it's time for him to wake up, it randomly selects one song to play on the CD. Suppose there are nine songs on the CD in the player and Jorge's favorite song is the third one. Describe a simulation to estimate the solution to this probability problem: What is the probability that Jorge will hear his favorite song at least once in the next week?

E50. The next table gives probabilities for a randomly selected adult in the United States.

		Age	
		18 to 25	**25 or Older**
Has a Cell Phone?	**Yes**	p	q
	No	r	s

Write an expression for the probability that a randomly selected adult

   a. has a cell phone

   b. is 25 or older, given that the person has a cell phone

   c. has a cell phone, given the person is age 18 to 25

   d. is 25 or older and has a cell phone

   e. is 25 or older or has a cell phone

E51. If you select a student at random from the school being described, what is the probability that the student has ridden a merry-go-round or a roller coaster, given the information about that school? Make a table to illustrate each situation.

   a. In a particular school, 80% of the students have ridden a merry-go-round and 45% have ridden a roller coaster. Only 15% have done neither.

   b. In another school, 30% of the students have ridden a merry-go-round but not a roller coaster. Forty-five percent have ridden a roller coaster but not a merry-go-round. Only 20% have done neither.

E52. Backgammon is one of the world's oldest games. Players move counters around the board in a race to get "home" first. The number of spaces moved is determined by a roll of two dice. If your counter "hits" your opponent's counter, your opponent has to move his or her counter back to the beginning. For example, suppose your opponent's counter is five spaces ahead of yours. You can "hit" that counter by rolling a sum of 5 with both dice or by getting a 5 on either die.

  a. Use the sample space for rolling a pair of dice to find the probability of being able to hit your opponent's counter on your next roll if you are five spaces behind.

  b. Can you use the Addition Rule for Disjoint Events to compute the probability that you roll a sum of 5 with both dice or get a 5 on either die? Either do the computation or explain why you can't.

E53. "So they stuck them down below, where they'd be the first to go." This line from the song about the Titanic refers to the third-class passengers. The first two-way table below gives the survival rate for Titanic passengers in various classes of travel. The members of the ship's crew have been omitted from this table. The second and third tables break down survival by class

and by gender. Use the ideas about conditional probability and independence that you have learned to analyze these tables.

Source: *Journal of Statistics Education* 3, no. 3 (1995).

E54. This problem appeared in the "Ask Marilyn" column of *Parade* magazine on July 27, 1997, and stirred up a lot of controversy. Marilyn devoted at least four columns to it.

  A woman and a man (unrelated) each have two children. At least one of the woman's children is a boy, and the man's older child is a boy. Do the chances that the woman has two boys equal the chances that the man has two boys?

  a. Design a simulation to answer this question. (The process of designing a simulation will help you clarify your assumptions.)

  b. What is the answer?

E55. Which of these situations are possible? Explain your answers.

  a. Events *A* and *B* are disjoint and independent.

  b. Events *A* and *B* are not disjoint and are independent.

  c. Events *A* and *B* are disjoint and dependent.

  d. Events *A* and *B* are not disjoint and are dependent.

Class

Survived?	First	Second	Third	Total
Yes	203	118	178	499
No	122	167	528	817
Total	325	285	706	1316

Females: Class

Survived?	First	Second	Third	Total
Yes	141	93	90	324
No	4	13	106	123
Total	145	106	196	447

Males: Class

Survived?	First	Second	Third	Total
Yes	62	25	88	175
No	118	154	422	694
Total	180	179	510	869

E56. Two events *A* and *B* are independent if any of these statements hold true:

$$P(A) = P(A \mid B)$$
$$P(B) = P(B \mid A)$$
$$P(A \text{ and } B) = P(A) \cdot P(B)$$

Use the definition of conditional probability to show that these three conditions are equivalent as long as $P(A)$ and $P(B)$ aren't zero. That is, show that if any one is true, so are the others.

E57. Among recent graduates of mathematics departments, half intend to teach high school. A random sample of size 2 is to be selected from the population of recent graduates.

 a. If mathematics departments had only four recent graduates altogether, what is the chance that the sample will consist of two graduates who intend to teach high school?

 b. If mathematics departments had 10 million recent graduates, what is the chance that the sample will consist of two graduates who intend to teach high school?

 c. Are the selections technically independent in part a? Are they in part b? In which part can you assume independence anyway? Why?

E58. a. Prove that

 $$P(A) = P(A \mid B) \cdot P(B) + P(A \mid \bar{B}) \cdot P(\bar{B})$$

 This result is sometimes called the law of total probability. Use it to find $P(Dodgers\ win)$ from the information in Display 6.21.

 b. Prove that

 $$P(B \mid A) = \frac{P(A \mid B) \cdot P(B)}{P(A \mid B) \cdot P(B) + P(A \mid \bar{B}) \cdot P(\bar{B})}$$

 This result is called Bayes' Theorem. Use it to find $P(Dodgers\ win \mid day\ game)$.

E59. Two screening tests are used on patients from a certain population. Test I, the less expensive of the two, is used about 60% of the time and produces false positives in

about 10% of the cases. Test II produces false positives only about 5% of the time.

 a. Fill in this table of percentages.

Results		Test I	Test II	Total %
	False Positives	–?–	–?–	–?–
	Other Results	–?–	–?–	–?–
	Total %	60	40	100

 b. A false positive is known to have occurred in a patient tested by one of these two tests. Find the approximate probability that Test I was used by direct observation from the table and then by use of the formula in E58.

E60. Suppose a randomly selected Republican was known to have favored Elizabeth Dole in the October poll described in Section 6.3. Use the formula in E58 to find the conditional probability that the Republican was a woman.

E61. DNA contains coded information that tells living organisms how to build proteins by combining 20 different amino acids. The code is based on the sequence of building blocks called nucleotides. There are four kinds of nucleotides grouped into triplets called *codons*. Each codon corresponds to an amino acid. In a codon, order doesn't matter and the same nucleotide can be used more than once. So, for example, AAB is the same as ABA. Show that there is exactly one codon for each of the 20 amino acids.

E62. A telephone area code is a 3-digit number. With only a few exceptions, current area codes

 • start with any digit from 2 through 9

 • have 0 or 1 as a second digit

 • have any third digit, except that area codes ending in 11 are not allowed

 How many area codes are possible, according to these rules?

E63. A few years ago, the U.S. Post Office added the "zip plus 4" to the original 5-digit zip codes. For example, the zip code for the national headquarters of the American Red Cross is now 20006-5310.

For this problem, assume that a zip code can be any 5-digit number except 00000 and that the 4-digit "plus 4" can be any number except 0000.

  a. Assume that the U.S. population is roughly 270 million. If all possible 5-digit zip codes are used, what is the average number of people per zip code?

  b. If all possible "zip plus 4" codes are used, what is the average number of people per "zip plus 4"?

E64. Jack and Jill found that 60 out of 100 people correctly identified the tap water. They are uncertain what to conclude in their report. They are concerned that perhaps everyone was just guessing and that 60 of them got the right answer just by chance. So Jack and Jill flipped a fair coin 100 times and counted the number of heads. They did this over and over again. Out of thousands of trials, they got 60 or more heads about 2.5% of the time. Write a conclusion for their report.

CHAPTER 7
PROBABILITY DISTRIBUTIONS

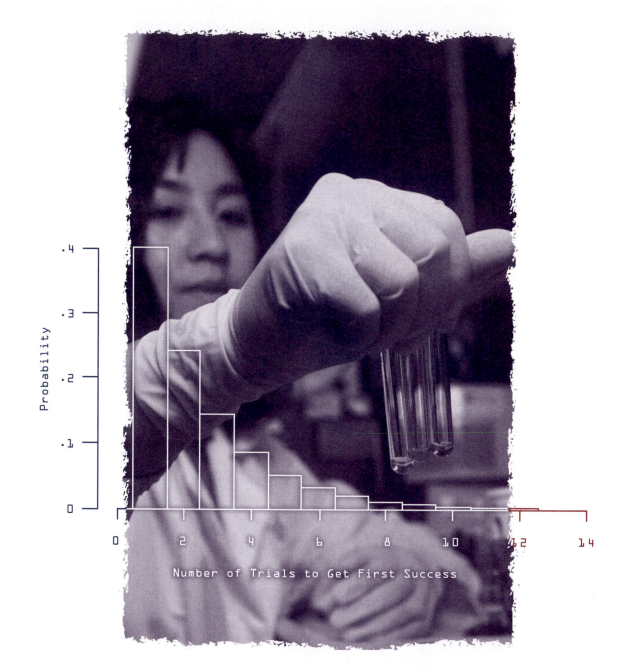

Forty percent of people have type A blood. A blood bank has 10 donors scheduled for today. How likely is the bank to get no type A's, all type A's, or any number in between? Probability distributions describe the likelihood of all possible outcomes in situations like this.

A **probability distribution** describes the possible numerical outcomes of a chance process and allows you to find the probability of any set of possible outcomes. Sometimes a probability distribution is defined by a table, like the ones Jack and Jill made in Chapter 6. Sometimes a probability distribution is defined by a curve, like that for the normal curve in Chapter 2. If you select a male at random, all possible heights he could have are given by the values on the x-axis, while the probability of getting someone whose height is between two specified x-values is given by the area under the curve between those two x-values. And, as you will learn in this chapter, sometimes a probability distribution is defined by a formula.

Some types of probability distributions occur so frequently in practice that it is important to know their names and formulas. Among them are the binomial and geometric distributions, which closely (but not perfectly) reflect many real-world situations and so are used as models for many applications that have similar characteristics.

In this chapter, you will learn

- the terminology of probability distributions
- the uses of expected value
- to recognize and apply the binomial distribution
- to recognize and apply the geometric distribution

7.1 Random Variables and Expected Value

In Chapter 6, you constructed a table like Display 7.1, which lists all possible outcomes when two dice are rolled. From this table, you can make different probability distributions, depending on the numerical summary that interests you.

Second Die

		1	2	3	4	5	6
	1	1, 1	1, 2	1, 3	1, 4	1, 5	1, 6
	2	2, 1	2, 2	2, 3	2, 4	2, 5	2, 6
First **3**		3, 1	3, 2	3, 3	3, 4	3, 5	3, 6
Die **4**		4, 1	4, 2	4, 3	4, 4	4, 5	4, 6
	5	5, 1	5, 2	5, 3	5, 4	5, 5	5, 6
	6	6, 1	6, 2	6, 3	6, 4	6, 5	6, 6

Display 7.1 The 36 outcomes when rolling two dice.

For example, if you are playing monopoly, you need to find the sum of the two dice. If you are playing backgammon, you might be more interested in the larger of the two numbers. Display 7.2 shows these two probability distributions, one for the sum of two dice and one for the larger number on the two dice.

Sum of Two Dice, x	Probability, p		Larger Number, x	Probability, p
2	1/36		1	1/36
3	2/36		2	3/36
4	3/36		3	5/36
5	4/36		4	7/36
6	5/36		5	9/36
7	6/36		6	11/36
8	5/36		**Total**	**1**
9	4/36			
10	3/36			
11	2/36			
12	1/36			
Total	**1**			

Display 7.2 Two different probability distributions from rolling two dice.

When you roll two dice, the probability that the sum of the numbers is 3 is $\frac{2}{36}$, while the probability that the larger number is a 3 is $\frac{5}{36}$. (In the case of doubles, the larger number and smaller number are both the same.)

The language of probability distributions reflects the chance involved in random outcomes. The variable of interest, X, is called a **random variable** because its value varies depending on the result of a particular trial.

The mean of a probability distribution has a special name, although continuing to call it the "mean" isn't wrong.

Expected Value

The mean of a probability distribution for the random variable X is called its expected value, and is usually denoted by μ_X, or $E(X)$.

The formulas for the expected value (mean) and standard deviation of a probability distribution table are exactly parallel to those for a relative frequency table.

Formulas for the Mean and Variance of a Probability Distribution Table

The expected value (mean) μ_X and variance σ_X^2 of a probability distribution listed in a table are

$$E(X) = \mu_X = \sum x_i \cdot p_i \qquad \text{and} \qquad Var(X) = \sigma_X^2 = \sum (x_i - \mu_X)^2 \cdot p_i$$

where p_i is the probability that the random variable takes on the specific value x_i. To get the standard deviation, take the square root of the variance.

Example: CDs Purchased by USC Students

The probability distribution shown in Display 7.3 was constructed from a recent survey at the University of Southern California (USC). The random variable is the number of CDs purchased by a student in the previous year. So, for example, if you select a USC student at random, the probability is .06 that he or she purchased no CDs in the previous 12 months. Estimate the expected value and standard deviation for this distribution.

Number of CDs Purchased Within 12 Months, x_i	Probability, p_i
0	.06
1–5	.28
6–10	.25
11–20	.21
21–50	.17
Over 50	.03
Total	**1.00**

Display 7.3 CDs purchased by USC students.

Source: Survey of usage for MP3, the online music service. Mark Latonero, June 2000, www.entertainment.usc.edu/publications/mp3.pdf.

Student: I remember how to do this. First I have to make a reasonable estimate of the center of each interval, like for the interval of 1–5 CDs, I'll use 3. Then I simply use the formulas:

$$\mu_X = \Sigma x_i \cdot p_i$$

$$= 0(.06) + 3(.28) + 8(.25) + 15.5(.21) + 35.5(.17) + 55(.03)$$

$$= 13.78$$

and

$$\sigma_X^2 = \Sigma (x_i - \mu_X)^2 \cdot p_i$$

$$= (0 - 13.78)^2(.06) + (3 - 13.78)^2(.28) + (8 - 13.78)^2(.25)$$
$$+ (15.5 - 13.78)^2(.21) + (35.5 - 13.78)^2(.17)$$
$$+ (55 - 13.78)^2(.03)$$

$$= 184.0766$$

so, $\sigma_X \approx 13.57$

Statistician: Your computations are perfect. Now, what is the expected number of CDs purchased by a USC student?

Student: Easy. My computations give 13.78, but because we can't have part of a CD, we say the expected number of CDs purchased by a USC student is 13 or 14.

Don't round expected values to whole numbers.

Statistician: And you were doing so well. It's true you can't have part of a CD, but expected value is the same thing as the *average,* and you know that the average doesn't have to be one of the values in the distribution. So you should say that the expected number of CDs is 13.78. ∎

Discussion: Random Variables and Expected Value

D1. Define a random variable for Display 7.1 that is different from the two random variables given in Display 7.2. Give its probability distribution and compute its expected value and standard deviation.

D2. This sentence once appeared in the British humor magazine *Punch:*

The figure of 2.2 children per adult female was felt to be in some respects absurd, and a Royal Commission suggested that the middle classes be paid money to increase the average to a rounder and more convenient number. Who would find the figure absurd—the student or the statistician?

Source: M. J. Moroney, *Facts From Figures* (Baltimore: Penguin Books, 1951).

■ Practice

P1. Describe the shape of the distributions of the two random variables in Display 7.2. Find their expected values and standard deviations.

Expected Value in Everyday Situations

In many real-life situations, you can find the best course of action by considering expected value.

Example: The Wisconsin Lottery

The Wisconsin lottery recently had a scratch-off game called "Big Cat Cash." It cost $1 to play. The probabilities of winning various amounts are listed in Display 7.4. Find and interpret the expected value when playing one game.

Winnings, x_i	Probability, p_i
$1	1/10
$2	1/14
$3	1/24
$18	1/200
$50	1/389
$150	1/20,000
$900	1/120,000

Display 7.4 Probabilities for Wisconsin scratch-off game.
Source: Sept. 2001, www.wilottery.com.

Solution

First, note that the probabilities don't add up to 1—it's not even close. That's because the most likely outcome is winning nothing. So imagine another row with $0 for "Winnings" and .7793 for "Probability." Using the expected-value formula, you can verify that the expected value for this probability distribution is 0.601. This means that if you spend $1 to play this game, you "expect" to get 60.1 cents back in winnings. Of course, you can't get this amount on any one play, but in the long run, that would be the average return. Another way to understand this is to imagine playing the game 1000 times. You expect to get back $601, but you will have spent $1000. The standard deviation of the winnings per game is $4.040, which is quite large because of the possibility of winning the larger prizes. ∎

Example: Burglary Insurance

A person who lives in a large city comes to your insurance agency and asks you to insure her household so that if it is burgled, you pay her $5000. What should you charge per year for such insurance?

Solution

How much you charge per year depends on how likely it is that her household will be burgled. Insurance companies and the mathematicians they employ, called actuaries, keep careful records of various crimes and disasters so they can know the probabilities that these will occur. Unfortunately, you neglected to hire an actuary, and all the information you can find is the nationwide rate of 40.9 burglaries per 1000 households.

Display 7.5 gives the two possible outcomes, their payouts, and their probabilities.

Outcome	Payout	Probability
No burglary	$0	.9591
Burglary	$5000	.0409

Display 7.5 Table of payouts and their probabilities.

Source: U.S. Bureau of Justice Statistics, *Criminal Victimization, 2000,* www.ojp.usdoj.gov/bjs/pub/pdf/cv00.pdf.

The expected payout per policy is

$$E(X) = \mu_X = 0(.9591) + 5000(.0409) = 204.50$$

You expect to break even if you charge $204.50 for the insurance. But as a good businessperson, you will actually charge more to cover your costs of doing business and to give yourself some profit. ■

Discussion: Expected Value in Everyday Situations

D3. A few years ago, Taco Bell had a scratch-off card promotion that contained four separate games. The player could play only one of them and could choose which one. The probability of winning game A was $\frac{1}{2}$, and the prize was a drink worth 55 cents. The probability of winning game B was $\frac{1}{4}$, and the prize was a food item worth 69 cents. The probability of winning game C was $\frac{1}{8}$, and the food prize was worth $1.44. The probability of winning game D was $\frac{1}{16}$, and the food prize was worth $1.99. Assuming that you like all of the prizes, which is the best game to play? This question is open to different interpretations, so let's be specific: On which game are your expected winnings the greatest?

You can organize the information for each game into a table. For example, for game A, the table looks like this.

Value of Prize	Probability of Winning
$0.55	.5
$0.00	.5

The expected winnings if you play game A are

$$\mu_X = (0.55)(.5) + (0)(0.5) = 0.275$$

or 27.5 cents.

a. Compute the expected winnings from games B, C, and D. On which game are your expected winnings the greatest?

b. For what reasons would people choose to play the other games, even if they knew the expected values were lower?

■ Practice

P2. For a raffle, 500 tickets will be sold.

   a. What is the expected value of a ticket if the only prize is worth $600?

   b. What is the expected value of a ticket if there is one prize worth $1000 and two prizes worth $400 each?

P3. Refer to the example on burglary insurance. The burglary rate for a suburban location is 27.2 burglaries per 1000 households, so if the home is in a suburban area, what should you charge for the insurance in order to break even?

Rules for Means and Variances

Adding a constant recenters, or shifts, a distribution, while multiplying by a constant stretches or shrinks it.

In Section 2.3, you learned that if you recenter a data set—add the same number *c* to all the values—it doesn't change the shape or standard deviation but adds *c* to the mean. If you rescale a data set—multiply all the values by the same nonzero number *d*—it doesn't change the basic shape but multiplies the mean by *d* and the standard deviation by $|d|$. These rules for distributions of data also apply to probability distributions.

> ### Linear Transformation Rule:
> ### The Effect of a Linear Transformation on μ and σ
>
> Suppose you have a probability distribution with random variable X, mean μ_X, and standard deviation σ_X. If you transform each value by multiplying it by d and then adding c, where c and d are constants, then the mean and variance of the transformed values are given by
>
> $$\mu_{c+dX} = c + d\mu_X$$
> $$\sigma^2_{c+dX} = d^2\sigma^2_X$$

In Section 5.5, you learned rules for sampling distributions that are equivalent to these rules for probability distributions.

> ### Addition and Subtraction Rules for Random Variables
>
> If X and Y are random variables, then
>
> $$\mu_{X+Y} = \mu_X + \mu_Y$$
> $$\mu_{X-Y} = \mu_X - \mu_Y$$
>
> and if X and Y are independent, then
>
> $$\sigma^2_{X+Y} = \sigma^2_X + \sigma^2_Y$$
> $$\sigma^2_{X-Y} = \sigma^2_X + \sigma^2_Y$$

In the following examples, you will see how to apply these rules.

Example: Buying Three Lottery Tickets

Refer to the example about the Wisconsin lottery game with an expected value of $0.601 and standard deviation $4.040. Suppose a person buys three tickets. What is the expected total winnings and what is the standard deviation of this total?

Solution

Let $X + Y + Z$ represent the person's total winnings on three tickets. Then μ_{X+Y+Z} is the expected value of the total winnings on three tickets. This problem requires a generalization of the Addition Rule:

$$\mu_{X+Y+Z} = \mu_X + \mu_Y + \mu_Z$$

$$= 0.601 + 0.601 + 0.601$$

$$= 1.803$$

Because the winnings on the three tickets can be considered independent, the variance of $X + Y + Z$ is

$$\sigma^2_{X+Y+Z} = \sigma^2_X + \sigma^2_Y + \sigma^2_Z$$

$$= 4.040^2 + 4.040^2 + 4.040^2$$

$$\approx 48.965$$

and the standard deviation is about $6.997. The person expects winnings of $1.803 with a standard deviation of $6.997. ∎

Example: Expected Savings

Suppose you earn $12 an hour for tutoring but spend $8 an hour for dance lessons. You save the difference between what you earn and the cost of your lessons. The number of hours you spend on each activity in a week varies independently according to the probability distributions in Display 7.6. Find your expected weekly savings and the standard deviation of your weekly savings.

Hours of Dance Lessons, x_i	Probability, p_i	Hours of Tutoring, y_i	Probability, p_i
0	.4	1	.3
1	.3	2	.3
2	.3	3	.2
		4	.2

Display 7.6   Probability distributions for the number of hours per week taking dance lessons and tutoring.

Solution

Take X to be the number of hours per week of dance lessons and Y the number of hours per week tutoring. The expected number of hours you take dance lessons, or μ_X, is $0(.4) + 1(.3) + 2(.3) = 0.9$ with a standard deviation, σ_X, of 0.831 hours.

Similarly, the expected number of hours you tutor, or μ_Y, is 2.3 hours with a standard deviation, σ_Y, of 1.1 hours. You spend \$8 for each hour of dance lessons, so by the Linear Transformation Rule, your mean expenditure for the week of dance lessons is

$$\mu_{8X} = 8 \cdot \mu_X = (8) \cdot (0.9) = 7.20$$

or \$7.20. Similarly, if you earn \$12 for each hour of tutoring, then by the Linear Transformation Rule, your mean earnings for the week from tutoring is

$$\mu_{12Y} = 12 \cdot \mu_Y = (12) \cdot (2.3) = 27.60$$

or \$27.60. Now, using the Subtraction Rule, your expected weekly savings is

$$\mu_{12Y} - \mu_{8X} = 27.60 - 7.20 = 20.40$$

or \$20.40. Using the Linear Transformation Rule, the standard deviation of the cost of your weekly lessons is $\sigma_{8X} = 8(0.831)$, or \$6.648, and the standard deviation of your weekly earnings from tutoring is $\sigma_{12Y} = 12(1.1)$, or \$13.20. Because the amounts are independent, you can now use the Subtraction Rule to find that the variance of your weekly savings is

$$\sigma^2_{12Y-8X} = \sigma^2_{12Y} + \sigma^2_{8X} = 13.2^2 + 6.648^2 = 218.435904$$

So the standard deviation is about \$14.78. ■

Discussion: Rules for Means and Variances

D4. If you expect to work 10 hours next week with a standard deviation of 2 hours and you expect to study 15 hours next week with a standard deviation of 3 hours, what is the total number of hours you expect to spend working or studying? Use the rule for adding variances to estimate the standard deviation of the total number of hours spent studying or working. Is it reasonable to use this rule in this case? Explain.

D5. This question demonstrates one reason why the mean and variance are considered so important. Suppose you select one book at random to read from List A and one from List B. The numbers given are the number of pages in each book.

List A	List B
200	100
225	250
600	700

a. Find the total number of pages in each of the nine different possible pairs of books. Compute the standard deviation σ and variance σ^2 of these nine totals.

   b. Compute the variance σ^2 of the number of pages in the three books in List A. Compute the variance σ^2 of the number of pages in the three books in List B. What relationship do you see between these two variances and the one in part a?

   c. Does that same relationship hold for the three standard deviations?

   d. Can you add the mean of the numbers in List A to the mean of the numbers in List B to get the mean of the nine totals?

   e. Can you add the median of the numbers in List A to the median of the numbers in List B to get the median of the nine totals?

■ Practice

P4. Refer to the example on expected savings. Find the amount you expect to save in a year and the standard deviation of the yearly savings.

Summary 7.1: Random Variables and Expected Value

A probability distribution describes the possible numerical outcomes, x_i, of a chance process and allows you to find the probability, $P(x)$, of any set of possible outcomes. A probability distribution table for a given random variable lists each of these outcomes in one column and their associated probability in another column. You can compute the expected value (mean) and variance of a probability distribution table using the formulas

$$E(X) = \mu_X = \Sigma x_i \cdot p_i$$

$$Var(X) = \sigma_X^2 = \Sigma(x_i - \mu_X)^2 \cdot p_i$$

Estimating the expected value has many real-world applications. For example, you can figure out your expected weekly savings or the break-even price for insurance.

   If you have random variables X and Y and constants c and d, then

- the mean and variance of a linear transformation of X is

$$\mu_{c+dX} = c + d\mu_X$$

$$\sigma_{c+dX}^2 = d^2\sigma_X^2$$

- if you add or subtract random variables X and Y, the mean is

$$\mu_{X+Y} = \mu_X + \mu_Y$$

$$\mu_{X-Y} = \mu_X - \mu_Y$$

- if X and Y are independent, then the variance is

$$\sigma_{X+Y}^2 = \sigma_X^2 + \sigma_Y^2$$

$$\sigma_{X-Y}^2 = \sigma_X^2 + \sigma_Y^2$$

■ Exercises

E1. Refer to the burglary insurance **example.**

    a. If you charge an urban householder $204.50 for insurance, what is the largest profit you could earn on that one policy for the year? What is the largest possible loss for the year?

    b. If you want to earn an expected yearly profit of $5000 per 1000 urban customers, how much should you charge per customer?

    c. There are other factors besides the location of the household that affect the probability of a burglary. What other factors might an insurance company take into account?

E2. Suppose you roll two tetrahedral dice, each with faces numbered 1, 2, 3, and 4.

    a. Make a probability distribution table for the sum of the numbers on the two dice. What is the probability that the sum is 3?

    b. Make a probability distribution table for the absolute value of the difference of the numbers on the two dice. What is the probability that the absolute value of the difference is 3?

    c. What is the expected value for the probability distribution in part b?

E3. For each million tickets sold, the original New York Lottery awarded one $50,000 prize, nine $5000 prizes, ninety $500 prizes, and nine hundred $50 prizes.

    a. What was the expected value of a ticket?

    b. The tickets sold for 50 cents each. How much could the state of New York expect to earn for every million tickets sold?

    c. What percentage of the income from the lottery was returned in prizes?

E4. The passenger vehicle with the highest theft loss is the two-door Acura Integra. There are 21.6 claims for theft per 1000 insured vehicles per year and an average payment of $10,676 per claim. How much would you charge an owner of a two-door Acura Integra for theft insurance per year if you simply want to expect to break even?

Source: Highway Loss Data Institute news release, May 23, 2001, www.hwysafety.org/news_releases/2001/pr052301.htm.

E5. A scratch-off card at Burger King gave the information contained in the table. All cards were potential winners. In order to actually win the prize, the player had to scratch off a winning path without a false step. The probability of doing this was given as $\frac{1}{5}$ for the food item and $\frac{1}{10}$ for any nonfood items. So, for example, if you got the card with the prize worth $1,000,000, you would still have to scratch off a path without making a mistake. The probability of doing that and actually winning the $1,000,000 was $\frac{1}{10}$.

Value of Prize (in dollars)	Number of Game Cards
1,000,000	1
500,000	2
(house) 200,000	1
100,000	10
(car) 16,619	75
10,000	50
(cruise) 2,900	150
1,000	200
100	1,000
10	10,000
(food item) 0.73	196,000,000

    a. What is the total value of all the potential prizes?

    b. What is the total amount that Burger King expected to pay out?

    c. What is the expected value of a card?

E6. You can't decide whether to buy Brand A or Brand B as your new dishwasher. Each brand is expected to last about 10 years. Brand A costs $950 with an unlimited number of repairs at $150 each. Brand B costs more ($1200) but comes with an unlimited number of free repairs. Which dishwasher you buy depends on the number of repairs you expect Brand A to require. You investigate and find out this information about Brand A.

Number of Repairs	Probability
0	.40
1	.30
2	.15
3	.10
4	.05

a. Find the expected cost of each brand, including original price and repairs.

b. What would be the advantage of buying Brand A? Brand B?

E7. A recent publication from the New York City Department of Consumer Affairs, "The Buyer's Guide to Better Cellular Service," gave this information about the four most commonly used local service plans for the New York/New Jersey calling area.

Service	Monthly Charge	Monthly Minutes	Each Additional Minute
AT&T	$49.99	600	$0.45
Verizon	$49.00	500	$0.45
Voicestream	$39.99	500	$0.25
Sprint PCS	$49.99	500	$0.30

Source: April 2002, www.nyc.gov/html/dca/pdf/cellguide.pdf.

If you talk for 500 minutes or less each month, Voicestream will be the cheapest. But some months there is a lot to talk about, and you estimate that the number of minutes you talk each month follows this probability distribution.

Number of Minutes	Probability
500 or less	.4
550	.3
600	.2
700	.1

Rank these plans according to how much you would expect them to cost per month, knowing that your only calls will be within the New York/New Jersey calling area.

7.2   The Binomial Distribution

Notation: p is the probability of a success on any one trial, and n is the number of trials.

In Section 5.4, you used the normal approximation to answer questions like this: If 60% of drivers in Kentucky wear seat belts and you take a random sample of 1200 drivers, what is the probability that 800 or more of them wear seat belts?

It doesn't work well, however, to use a normal approximation if the sample size is small—if you took a sample of only 12 drivers, for example, and wanted to find the probability that 8 or more wear seat belts. That's because the distribution of the number of drivers wearing seat belts isn't approximately normal unless both np and $n(1-p)$ are at least 10.

In this section, Jack and Jill will come to your rescue and show you how to find such probabilities exactly, using a formula that works even when the number of trials is small.

Binomial Probabilities

Many random variables seen in practice amount to counting the number of successes in n independent trials, such as

- the number of doubles in 4 rolls of a pair of dice (on each roll either you get doubles or you don't),
- the number of patients with type A blood in a random sample of 10 patients (either each person has type A blood or they don't),
- the number of defective items in a sample of 20 items (either each item is defective or it isn't).

These situations are called "binomial" because each trial has two possible outcomes: "success" or "failure."

Back in Chapter 6, Jack and Jill were asking questions about a binomial situation when they wanted to know the answer to a question like this: "If we ask four people which is the tap water and none of them can tell bottled water from tap water, what is the probability that two of the four will guess correctly?"

Jack: That's exactly what we were doing—looking at n people and counting the number who correctly identified the tap water.

Jill: Do they mean there is an easier way than making a list of all those outcomes?

Jack: Sounds like it. But I bet we can figure it out for ourselves now that we know there is a formula. When we asked four people who can't tell the difference to identify the tap water, we got a list of 16 possible outcomes.

Jill: Yeah, look, I've organized them in this table. Let's try to figure out the probability of getting exactly two people who select the tap water.

Number Who Select Tap Water	Outcomes, B (Bottled Water), T (Tap Water)						Number of Outcomes
0	BBBB						1
1	BBBT	BBTB	BTBB	TBBB			4
2	BBTT	BTBT	BTTB	TBBT	TBTB	TTBB	6
3	BTTT	TBTT	TTBT	TTTB			4
4	TTTT						1

Jack: Now that we have the Multiplication Rule for Independent Events, we know that the probability of getting a particular outcome when two people guess correctly, say, *BTTB*, is $\frac{1}{2} \cdot \frac{1}{2} \cdot \frac{1}{2} \cdot \frac{1}{2} = \frac{1}{16}$. Then, from the last column, there are 6 outcomes like this when two people guess correctly. So the probability of getting two people who correctly select tap water is $6 \cdot \frac{1}{2} \cdot \frac{1}{2} \cdot \frac{1}{2} \cdot \frac{1}{2} = \frac{6}{16}$. But I wish we could find the numbers in the last column without trying to list all the possible outcomes!

Jill: I've seen those numbers before—in the fifth row of Pascal's triangle.

Jack: That's it! The number of ways we can choose exactly two people out of four people to identify the tap water is:

$$\binom{4}{2} = \frac{4!}{2!2!} = \frac{4 \cdot 3 \cdot 2 \cdot 1}{2 \cdot 1 \cdot 2 \cdot 1} = 6$$

```
          1
        1   1
      1   2   1
    1   3   3   1
  1   4   6   4   1
```
Pascal's triangle

Discussion: Binomial Probabilities

D6. Explain why each of the 16 outcomes in Jill's table has probability $\frac{1}{2} \cdot \frac{1}{2} \cdot \frac{1}{2} \cdot \frac{1}{2}$.

D7. Now suppose Jack and Jill ask six people who can't tell the difference between tap and bottled water to identify the tap water. Use their method to make a table of the probability distribution for six people, as Jill did, but omit the "outcomes" column. Then make a graph of the distribution.

Reminder:

$$\binom{n}{k} = \frac{n!}{k!(n-k)!}$$
$$\binom{n}{n} = 1$$
$$\binom{n}{0} = 1$$

■ Practice

P5. Suppose you flip a coin eight times. What is the probability that you'll get exactly three heads? Exactly 25% heads? At least 7 heads?

Example: College Graduates

The proportion of adults 25 and older in the United States with at least a bachelor's degree is .25. Suppose you pick seven adults at random. What is the probability that exactly three will have a bachelor's degree or higher?

Source: *Statistical Abstract of the United States,* U.S. Census Bureau, 2000, p. 159, www.census.gov/prod/www/statistical-abstract-us.html.

Jack: Now what do we do? They've changed the problem so that the outcomes aren't equally likely. That is *so* like them.

Jill: There is nothing about our method that requires equally likely outcomes—all that matters is independence and that the probability stays the same. For example, one outcome with three grads is

grad, not, grad, grad, not, not, not

So by the Multiplication Rule, the probability of this particular outcome is

$$(.25)(.75)(.25)(.25)(.75)(.75)(.75) = (.25)^3(.75)^4$$

Another outcome with three grads is

not, grad, grad, not, not, not, grad

The probability of this particular outcome is

$$(.75)(.25)(.25)(.75)(.75)(.75)(.25) = (.25)^3(.75)^4$$

which is exactly the same.

Jack: That's because the probabilities for the outcomes with exactly three grads all have the same factors but in a different order.

Jill: And there are 35 of them because

$$\binom{7}{3} = \frac{7!}{3!4!} = 35$$

Good thing we didn't have to list all of them!

Jack: So the probability of getting exactly three college grads is

(the number of ways to get 3 grads) · (probability of each way)

or

$$\binom{7}{3}(.25)^3(.75)^4 \approx .1730$$

Jill: Yeah! Now we can do any problem they throw at us.   ■

On the next page, you will find a summary of Jack and Jill's method.

The Binomial Probability Distribution

Suppose you have a series of trials that satisfy these conditions:

B: They are binomial—each trial must have one of two different outcomes, one called a "success" and the other a "failure."

I: Each trial is independent of the others. That is, the probability of a success doesn't change or depend on what has happened before.

N: There is a fixed number, *n*, of trials.

S: The probability, *p*, of a success is the same on each trial, $0 < p < 1$.

Then the distribution of the random variable *X* that counts the number of successes is called a **binomial distribution.** Further, the probability that you get exactly $X = k$ successes is

$$P(X = k) = \binom{n}{k} p^k (1 - p)^{n-k}$$

and $\binom{n}{k}$ is the binomial coefficient, calculated as

$$\binom{n}{k} = \frac{n!}{k!(n - k)!}$$

where $n! = n(n - 1)(n - 2) \cdots 3 \cdot 2 \cdot 1$.

You may have noticed that the trials in Jack and Jill's example aren't really independent. The first adult they selected who was 25 or older has a probability of .25 of being a college graduate. If that person is a college graduate, the probability that the next person is a college graduate is a bit less. However, the change in probability is so small that Jack and Jill can safely ignore it. If there are 150,000,000 adults who are 25 and older in the United States, then there would be 37,500,000 college graduates. The probability that the first adult selected is a college graduate is $\frac{37,500,000}{150,000,000} = .25$. If that person turns out to be a college graduate, the probability that the second adult selected is a college graduate is $\frac{37,499,999}{149,999,999}$ or .249999995, which is very close to .25.

You can treat your random sample as a binomial situation as long as the sample size *n* is small compared to the population size *N*. The rule is simple: *n* should be less than 10% of the size of the population, or $n < .10N$.

It it generally safe to assume independent trials if $n < .10N$.

Discussion: The Binomial Probability Formula

D8. Make a probability distribution table for the number of college graduates in a random sample of seven adults if 25% of adults are college graduates.

D9. Make a probability distribution table for the number of people who *aren't* college graduates. Make a histogram of this distribution and one of the distribution in D8 and compare them.

D10. Show why the population must be "large" to use the binomial probability formula by pretending that the total adult population of the United States

In a famous episode of *I Love Lucy*, Lucille Ball takes her passion for chocolate to comic extremes.

is 12, of which 4 are "chocoholics." Compute the exact probability that in a random sample of 7 adults, none will be chocoholics; compare your results to those from using the binomial formula.

D11. From a class of 20 students, half of whom are seniors, the teacher will select 10 students at random to check for completion of today's paper. If X denotes the number of papers belonging to seniors among those checked, will X have a binomial distribution? Explain why or why not.

■ Practice

P6. About 11.2% of people aged 16 to 19 are "dropouts," a person who is not in regular school and who has not completed the 12th grade or received a general equivalency degree. Suppose you pick five people at random from this age group.

Source: *Statistical Abstract of the United States*, U.S. Census Bureau, 2000, p. 159, www.census.gov/prod/www/statistical-abstract-us.html.

a. What is the probability that none are dropouts?

b. What is the probability that at least one is a dropout?

c. Fill out a table for this situation like you did in D8.

P7. Describe how to use simulation to construct an approximate binomial distribution for the situation described in P6. Conduct two trials of your simulation using these random digits.

| 89254 | 99538 | 18315 | 45716 | 36270 | 79665 | 49830 | 06226 |
| 88863 | 02322 | 36630 | 07176 | 04011 | 70959 | 23449 | 62572 |

Activity 7.1   Can People Identify the Tap Water?

What you'll need: small paper or plastic cups, a container of tap water, a container of bottled water, about 20 volunteer subjects

1. Design a study to see if people can identify which of two cups of water contains the tap water. Refer to the principles of good experimental design in Chapter 4.

2. Suppose that none of your 20 subjects actually can tell the difference and so their choice is equivalent to selecting one of the cups at random. Make a probability distribution of the number who choose the tap water.

3. If only 10 people correctly select the tap water, you don't have any reason to conclude that people actually can tell the difference. Why not?

4. Decide how many people will have to make the correct selection before you are reasonably convinced that people can tell the difference.

5. Conduct your study and write a conclusion.

The Shape, Center, and Spread of a Binomial Distribution

You can always compute the mean and the standard deviation of a binomial distribution using the basic formulas

$$E(X) = \mu_X = \sum x_i \cdot p_i \qquad \text{and} \qquad \sigma_X = \sqrt{\sum (x_i - \mu_X)^2 \cdot p_i}$$

But the good news is you won't have to. You can use the simpler formulas from Section 5.4:

$$\mu_{sum} = np \qquad \text{and} \qquad \sigma_{sum} = \sqrt{np(1-p)}$$

These formulas apply here, too, because although the terminology is different, the sampling distribution of the number of successes in Section 5.4 is just like the binomial distributions in this section. In both cases, you are finding the probabilities for given numbers of successes.

> ### The Characteristics of a Binomial Distribution
>
> *These formulas hold for all sample sizes n.*
>
> A binomial distribution of n trials with probability of success p has expected value (mean) and standard deviation given by
>
> $$E(X) = \mu_X = np \qquad \text{and} \qquad \sigma_X = \sqrt{np(1-p)}$$
>
> The shape of the distribution becomes more normal as n increases.

For example, if you select 100 adults at random from the United States, you expect to get $E(X) = 100(.25) = 25$ college graduates with a standard deviation of $\sigma_X = \sqrt{100(.25)(1-.25)} = 4.33$.

Plots of binomial distributions for various values of n and p are shown in Display 7.7.

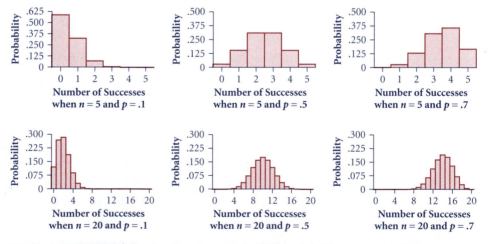

Display 7.7 Binomial distributions for sample sizes n and different probabilities p.

Discussion: Shape, Center, and Spread

D12. How do the shapes, centers, and spreads of the binomial distributions change as *p* increases for fixed values of *n*? How do they change as *n* increases for fixed values of *p*?

D13. Find the expected value and standard deviation for the number of college graduates, *X*, observed in random samples of four adults (*p* = .25). Repeat the calculations for samples of size 10, 20, and 40.

    a. As the sample size increases, what do you notice about the change in the expected value for the counts, *X*, that differs from that for the change in the expected value for the *proportion* studied in Section 5.4?

    b. As the sample size increases, what do you notice about the change in the standard deviation for the counts, *X*, that differs from that for the change in the standard deviation for the *proportion* studied in Section 5.4?

D14. Ten CD players of a discontinued model are selling for $100 each with a "double your money back" guarantee if they fail in the first month of use. Suppose the probability of such a failure is .08. What is the expected net gain for the retailer after all ten CD players are sold? Ignore the original cost of the CD players to the retailer.

■ Practice

P8. The median annual household income for U.S. households is about $39,000.

Source: *Statistical Abstract of the United States*, U.S. Census Bureau, 2000, p. 467, www.census.gov/prod/www/statistical-abstract-us.html.

    a. Among five randomly selected U.S. households, find the probability that four or more have incomes exceeding $39,000 per year.

    b. Consider a random sample of 16 U.S. households.

      i. What is the expected number of households with income below $39,000?

      ii. What is the standard deviation of the number of households with incomes under $39,000?

      iii. What is the probability of seeing at least 10 of the 16 households with incomes under $39,000 annually?

    c. Suppose in a sample of 16 U.S. households, none had incomes below $39,000. What might you suspect about this sample?

P9. You neglected to study for a true–false quiz on the government of Botswana, so you will have to guess on all of the questions. You need to have at least 60% of the answers correct in order to pass. Would you rather have a 5-question quiz or a 20-question quiz? Explain your reasoning, referring to Display 7.7.

Summary 7.2: The Binomial Distribution

The random variable X is said to have a binomial distribution if X represents the number of "successes" in n independent trials, where the probability of a success is p on each trial. For example, X may represent the number of successes in a random sample of size n from a large population, with probability of success p on each selection.

A binomial distribution has these important characteristics.

- The probability of getting exactly $X = k$ successes is given by the formula

$$P(X = k) = \binom{n}{k} p^k (1 - p)^{n-k}$$

and, unlike the normal approximation in Section 5.4, this formula may be used with any sample size n.

- The expected value or mean is $E(X) = \mu_X = np$.
- The standard deviation is $\sigma_X = \sqrt{np(1-p)}$.

■ Exercises

E8. If you roll a pair of dice five times, find the probability of each outcome.

   a. You get doubles exactly once.

   b. You get exactly three sums of 7.

   c. You get at least one sum of 7.

   d. You get at most one sum of 7.

E9. Now suppose you roll a pair of dice 60 times. What is the probability you get doubles fewer than five times?

   a. Find this probability exactly by using the method of this section.

   b. Find an estimate of this probability by using the normal approximation, as you did in Section 5.4. Compare your answer with part a.

E10. According to a recent Census Bureau report, 34.5 million Americans, which is 12.7% of the population, live below the poverty level. Suppose these figures hold true for the region in which you live. You plan to randomly sample 25 Americans from your region.

Source: *Statistical Abstract of the United States,* U.S. Census Bureau, 2000, p. 475, www.census.gov/prod/www/statistical-abstract-us.html.

   a. What is the probability that your sample will include at least two people with incomes below the poverty level?

   b. What is the expected value and standard deviation of the number of people in your sample with incomes below the poverty level?

E11. As stated in E10, 12.7% of Americans live below the poverty level. Suppose you plan to sample at random 100 Americans and count the number who live below the poverty level. What is the probability that you count 10 or fewer?

   a. Find this probability exactly by using the method of this section.

   b. Find an estimate of this probability by using the normal approximation, as you did in Section 5.4. Compare your answer with part a.

E12. You buy 15 lottery tickets for $1 each. With each ticket, you have a .05 chance of winning $10. Taking into account the cost of the tickets,

   a. what is your expected earnings (or loss) on this purchase?

   b. what is the probability you will gain $10 or more?

E13. An oil exploration firm is to drill 10 wells, each in a different location. Each well has a probability of .1 of producing oil. It will cost the firm $50,000 to drill each well. A successful well will bring in oil worth $1,000,000. Taking into account the cost of drilling,

   a. Find the firm's expected gain from the 10 wells.

   b. Find the standard deviation of the firm's gain for the 10 wells.

   c. Find the probability that the firm will lose money on the 10 wells.

   d. Find the probability that the firm will gain $1.5 million or more from the 10 wells.

E14. A home alarm system has detectors covering n zones of the house. Suppose the probability is .7 that a detector sounds an alarm when an intruder passes through its zone, and this probability is the same for each detector. The alarms operate independently. An intruder enters the house and passes through all the zones.

   a. What is the probability that an alarm sounds if $n = 3$?

   b. What is the probability that an alarm sounds if $n = 6$? Does the probability of part a get doubled?

   c. How large must n be in order to have the probability of an alarm sounding be about .99?

E15. A potential buyer will sample videotapes from a large lot of new videotapes. If she finds at least one defective one, she'll reject the entire lot. In each case, find the sample size n so that the probability of detecting at least one defective tape is .50.

   a. Ten percent of the lot is defective.

   b. Five percent of the lot is defective.

E16. If X represents the number of successes in a binomial distribution of n trials, then $\frac{X}{n}$ represents the proportion of successes in the n trials. Show that if p represents the probability of a success, then

$$E\left(\frac{X}{n}\right) = p \quad \text{and} \quad \sigma\left(\frac{X}{n}\right) = \sqrt{\frac{p(1-p)}{n}}$$

Is this consistent with what you learned in Section 5.4?

E17. For a random sample of three adults, the expected number of college graduates is $(3)(.25) = 0.75$, less than one whole person. Provide a meaningful interpretation of this number.

E18. The sum of the probabilities in a binomial distribution must equal 1. To show this is true, first let $q = 1 - p$ so that $p + q = 1$. Then the table shows that for a binomial distribution with $n = 1$ trial, the sum of the probabilities is 1.

Number of Successes	Probability
0	q
1	p
Total	**1**

   a. Show that the sum of the probabilities when $n = 2$ is $q^2 + 2qp + p^2$. Factor this sum to show that it is equal to 1.

   b. Find the sum of the probabilities when $n = 3$. Factor this sum to show that it is equal to 1.

   c. Show that the terms of the expansion of $(q + p)^n$ are the probabilities for $x = 0, 1, 2, \ldots, n$ in the binomial distribution with n trials and probability of success p.

7.3 The Geometric Distribution

In waiting-time problems, you count the number of trials it takes to get the first success.

Your favorite sports team has made the finals. You have a good estimate of the probability that it will win any one game against its opponent. In the last section, you answered questions like "What is the probability your team will win exactly three out of seven games?" The number of trials was fixed. In this section, you will answer questions like "What is the probability that your team will play at least three games before it gets its first win?" Problems like this are called **waiting-time problems** because the variable in question is the number of trials you have to wait before the event of interest happens. The number of trials isn't fixed—you simply count the number of trials until you get the first success.

The goal of the next activity is to design a simulation for a waiting-time situation and use it to construct an approximate probability distribution.

Activity 7.2   Waiting for Type A Blood

What you'll need: a device for generating random outcomes

In the general population, about 40% of people have type A blood. Suppose that a worker in a blood bank needs type A blood today and wants to know about how many blood donations he might have to process before he finds the first donation that is type A.

Source: April 2002, www.redcross.org/services/biomed/blood/supply/usagefacts.html.

1. Describe how to use random digits to simulate the blood type of the first donation he checks if he selects one at random.

2. Simulate the outcome for the first donation he checks. Was it type A blood? In other words, was his waiting time to success just one trial?

3. If he was successful on the first donation, then stop this run of the simulation. If he did not have success on the first trial, then continue generating outcomes until he gets his first success. Record the value of *x*, the number of the trial on which the first success occurred.

4. Repeat steps 2 and 3 at least ten times, recording a value for *x* each time.

5. Combine your values of *x* with others from the class and construct a plot to represent the simulated distribution of *X*. Describe the shape of this distribution and find its mean.

6. What is your estimated probability that
   a. the first donation with type A blood is the second one he checks?
   b. he will have to check five or fewer donations to find the first that was type A?
   c. he has to check at least two donations before getting one that is type A?
   d. he has to check at most four donations to get one that is type A?

7. Which donation is the most likely to be the first that is type A?

The probability distribution you constructed in the activity is called a **geometric distribution.** Display 7.8 shows geometric distributions for p equal to .1, .3, .5, and .8, where p is the probability of "success" on each trial. For example, when $p = .3$, the probability of getting a success on the first trial is .3, as the height of the first bar indicates. The probability of getting the first success on the second trial is .21, the height of the second bar. This is because you must fail on the first try and then succeed on the second, and the probability of this is

$$(\text{probability of failure}) \cdot (\text{probability of success}) = (.7)(.3) = .21$$

Note that in each graph, each bar is the same fixed proportion of the height of the bar on its left.

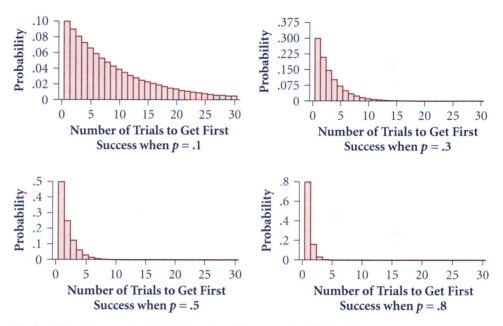

Display 7.8 Geometric distributions for different probabilities of a success.

Discussion: Waiting-Time Situations

D15. Refer to the graphs in Display 7.8.

    a. What is the shortest possible waiting time? What is the longest? What is the most likely waiting time? Why is this the case?

    b. What are the exact heights of the first and second bars in the graph for $p = .8$? Explain.

    c. How are the shapes of the four distributions in Display 7.8 similar? How are they different?

D16. What real-life situation could the graph for $p = .5$ represent?

■ Practice

P10. Suppose you are rolling a pair of dice and waiting for a sum of 7 to occur.

  a. What is the probability you get a sum of 7 for the first time on your first roll? For the first time on your second roll?

  b. Using the graphs in Display 7.8 as a guide, sketch an approximate graph for this situation.

  c. Is the sum of the probabilities in your graph less than 1, equal to 1, or more than 1? Explain why this is the case.

The Geometric Distribution

As you may already have discovered, you can derive a formula for the probability distribution for a geometric (waiting-time) distribution quite easily.

Jack: Yep, we can do it. This is easier than our problem from Section 7.2 about the binomial distribution.

Jill: It sure is. Let's do it for the blood bank example, where about 40% of people have type A blood. There the probability that the first donation checked is type A is .4.

Jack: Right, the probability that the waiting time, *X*, is 1 is $P(X = 1) = .4$.

Jill: Now for $P(X = 2)$. For the second donation to be the first that is type A . . .

Jack: What kind of nonsense is that? "For the second donation to be the first . . ."?

Jill: Sorry, let me say it with more words. Suppose the first donation checked isn't type A and the second donation is, then we have our first success with the second donation.

Jack: That's better. The probability of this sequence of outcomes is *P(first donation isn't type A* and *second is type A)* = (.6)(.4) or .24.

Always check conditions!   *Jill:* But you get to multiply like that only if the events are independent. What if a whole family had donated blood? Then their blood types might not be independent.

Jack: Yeah, we will have to be careful about things like that. The probability can't change depending on who else has come in.

Jill: If we can assume independence, the probability that the third donation checked will be the first that is type A is $P(X = 3) = (.6)(.6)(.4)$.

Jack: Because we have to have two "failures" and then our first success.

Jill: This could go on forever. So we better get started and make one of our tables.

Number of Trials to Get First Success	Probability
1	.4
2	(.6)(.4) = .24
3	(.6)(.6)(.4) = .144
4	(.6)(.6)(.6)(.4) = .0864
5	(.6)(.6)(.6)(.6)(.4) = .05184

Jack: This is making me thirsty. Five rows is enough. Let's go up the hill and get some water.

Display 7.9 summarizes Jack and Jill's reasoning, this time for a general geometric distribution with probability of success p and probability of failure $q = 1 - p$.

Number of Trials to Get First Success	Probability
1	p
2	qp
3	$q^2 p$
4	$q^3 p$
⋮	⋮
k	$q^{k-1} p$
⋮	⋮

Display 7.9   The table for the general geometric distribution.

The Geometric Probability Distribution

Suppose you have a series of trials that satisfy these conditions:

- They are binomial—each trial must have one of two different outcomes, one called a "success" and the other a "failure."
- Each trial is independent of the others. That is, the probability of a success doesn't change or depend on what has happened before.
- The trials continue until the first success.
- The probability, p, of a success is the same on each trial, $0 < p < 1$.

Then the distribution of the random variable X that counts the number of trials needed until the first "success" is called a geometric distribution. The probability that the first success occurs on the $X = k$th trial is

$$P(X = k) = (1 - p)^{k-1} p$$

for $k = 1, 2, 3, \ldots$

Discussion: The Formula for a Geometric Distribution

D17. About 10% of the U.S. population has type B blood. Suppose our medical technician is checking donations that may be considered independent with respect to blood type.

Source: April 2002, www.redcross.org/services/biomed/blood/supply/usagefacts.html.

a. What is the probability that the first donation that is type B is the third one checked?

b. What is the probability that he will have to check at most three donations to get the first that is type B? At least three donations?

c. Suppose the first donation that is type B is the third one checked. What is the probability that at least three *more* donations will have to be checked to obtain a second type B?

d. Part c refers to the *memoryless* property of the geometric distribution. Explain why that term is appropriate.

D18. In Activity 3.5, Copper Flippers, you started by flipping 200 pennies. You then removed the ones that "died" (landed tails) and continued flipping with a decreasing number of "live" copper flippers. Make a plot with the number of the flip on the horizontal axis and the theoretical number of copper flippers left after that flip on the vertical axis. Compare with your results from Activity 3.5 if you still have them.

D19. Give at least two ways of finding the probability that it will take three or more rolls of a pair of dice to get doubles.

■ Practice

P11. Suppose 9% of the engines manufactured on a certain assembly line have at least one defect. Engines are randomly sampled from this line one at a time and tested. What is the probability that the first nondefective engine is found

a. on the third trial?   b. before the fourth trial?

P12. About 60% of the time, the telephone lines coming into a concert ticket agency are all busy. If you are calling this agency, what is the probability that it takes you

a. only one try to get through?   b. two tries?   c. four tries?

d. What assumption are you making in computing these probabilities?

Expected Value and Standard Deviation

In this activity, you will discover the formula for the mean of a geometric distribution.

Activity 7.3   The Expected Value of a Geometric Distribution

Divide your class into nine groups. Each group will be assigned a different value of p, for p equal to .1, .2, .3, .4, .5, .6, .7, .8, and .9.

(continued)

1. Make a table that shows the probabilities for the values of X from 1 through 30 for the geometric distribution with your group's assigned value of p.

2. Use this table to compute an approximation of the expected value of this distribution.

3. Will your approximation in part a be a bit larger or a bit smaller than the theoretical expected value? Explain.

4. Using the results from all nine groups, make a scatterplot of the expected values plotted against the value of p. What is the shape of this plot?

5. Now make a scatterplot of the expected values plotted against the reciprocals of the values of p. Give a hypothesis about the formula for the expected value of a geometric distribution.

Like the formulas for the binomial distribution, those for the expected value and standard deviation of the geometric distribution turn out to be surprisingly simple. Although using the mean as the measure of center and the standard deviation as the measure of spread seemed complicated in Chapter 2, it pays off now.

The Characteristics of a Geometric Distribution

A geometric distribution with probability of success p has an expected value (mean) and standard deviation of

$$\mu_X = \frac{1}{p} \quad \text{and} \quad \sigma_X = \frac{\sqrt{1-p}}{p}$$

The shape is skewed right, with bars decreasing proportionally.

You can find a proof of the formula for the expected value in E25. The proof of the formula for the standard deviation is a bit more involved.

Example: Mean and Standard Deviation for Blood Donations

The probability that a random blood donation is type A is .4. Thus, the expected number of blood donations that the technician would have to check in order to get the first one that is type A is

$$\mu_X = \frac{1}{p} = \frac{1}{.4} = 2.5$$

with a standard deviation of

$$\sigma_X = \frac{\sqrt{1-p}}{p} = \frac{\sqrt{1-.4}}{.4} \approx 1.94$$

The expected value is the expected number of trials needed to get the first success.

Note that the expected value is the expected number of donations that must be checked to get one that is type A. It is *not* the expected number checked *before* getting one that is type A. So in this case, the expected number includes 1.5 donations that aren't type A and one donation that is. ■

Example: The Expected Number for the Second Success

The expected number of blood donations that are checked to obtain two of type A is 2(2.5), or 5, because the technician must check, on average, 2.5 donations to get the first success. Then the process starts again to get the second success, checking an average of 2.5 donations more, or five in all. ■

Discussion: Expected Values and Standard Deviation

D20. The phone line for a 24-hour ticket office is busy about 60% of the time.

   a. If you dial at random times throughout the week, what is the expected number of tries it will take you to get through? What is the standard deviation of the number of tries?

   b. What is the expected total number of times you will have to dial if you forget to ask a question after getting through the first time and have to call again?

   c. Suppose you have been trying for three days and so your friend starts dialing as well, at random times, while you continue. Who do you expect to get through first, you or your friend?

D21. If p doubles, how does the expected value change?

■ Practice

P13. The probability that a random blood donation is type B is .1.

   a. What is the expected number of donations that have to be checked to obtain the first one that is type B?

   b. What is the standard deviation of the number of donations checked to obtain the first one of type B?

   c. What is the expected number of donations checked to obtain two that are type B? Three?

Summary 7.3: The Geometric Distribution

For a sequence of independent trials in which p, the probability of "success," stays the same for each trial, the number X of the trial on which the first success occurs has a geometric distribution.

• The probabilities for this distribution are given by the formula

$$P(X = k) = (1-p)^{k-1}p$$

for $k = 1, 2, 3, \ldots$ and $0 < p < 1$.

- The mean, or expected value, of the distribution is

$$E(X) = \mu_X = \frac{1}{p}$$

- The standard deviation is

$$\sigma_X = \frac{\sqrt{1-p}}{p}$$

- The expected number of trials until the *n*th success is $n \cdot \frac{1}{p}$.

■ Exercises

E19. You are participating in a "question bee" in history class. You remain in the game until you give your first incorrect answer. The questions are all multiple choice with four possible answers, exactly one of which is correct. Unfortunately, you have not studied for this bee, and you simply guess randomly on each question.

    a. What is the probability that you are still in the bee after the first round?

    b. What is the probability you are still in the bee after the third round?

    c. What is the expected number of rounds you will be in the bee?

    d. If an entire class of 32 is simply guessing on each question, how many do you expect to be left after the third round?

E20. Suppose 12% of the engines manufactured on a certain assembly line have at least one defect. Engines are randomly sampled from this line one at a time and tested.

    a. What is the expected number of engines that need to be tested before finding the first engine *without* a defect? Before finding the third?

    b. What is the standard deviation of the number of engines that need to be tested before finding the first engine without a defect?

    c. If it costs $100 to test one engine, what are the expected value and standard deviation of the cost of inspection up to and including the first engine without a defect?

    d. Will the cost of inspection often exceed $200? Explain.

E21. An oil exploration firm is to drill wells at a particular site until it finds one that will produce oil. Each well has a probability of .1 of producing oil. It costs the firm $50,000 to drill each well.

    a. What is the expected number of wells to be drilled?

    b. What are the expected value and standard deviation of the cost of drilling to get the first successful well?

    c. What is the probability that it will take at least 5 tries to get the first successful well? At least 15?

E22. An engineer is designing a machine to find tiny cracks in airplane wings that are difficult to detect. The specifications state that a section of wing known to have these cracks may be tested by the machine up to five times, but the probability of locating the cracks in five trials or fewer must be at least .99. Therefore, what probability for the machine's locating cracks should the engineer be trying to attain in any one trial?

E23. A *geometric sequence* is a sequence of the form

$$a, ar, ar^2, ar^3, ar^4, \ldots$$

where *a* is the first term and *r* is called the *common ratio*. Explain why the probability distribution developed in this section is called the *geometric* distribution.

E24. The sum of the probabilities in a geometric distribution form an infinite series:

$$p + pq + pq^2 + pq^3 + pq^4 + \cdots = \sum_{k=1}^{\infty} pq^{k-1}$$

(See Display 7.9.) Use the next formula for the sum of an infinite geometric series to prove that the sum of the probabilities of a geometric distribution is equal to 1.

$$\sum_{k=1}^{\infty} ar^{k-1} = a + ar + ar^2 + ar^3 + \cdots = \frac{a}{(1-r)}$$

as long as $|r| < 1$.

E25. In this exercise, you will use the formula in E24 twice to prove that the expected value of the geometric distribution is $\frac{1}{p}$.

a. Use the formula for the expected value of a probability distribution to write the formula for the mean of a geometric distribution with probability of success *p*. This will be an infinite series.

b. Show that you can write the series from part a in this form:

$$E(X) = p(1 + q + q^2 + q^3 + \cdots$$
$$+ q + q^2 + q^3 + \cdots$$
$$+ q^2 + q^3 + \cdots$$
$$+ q^3 + \cdots$$
$$+ \cdots)$$

c. Ignore *p* for now and sum each row using the formula in E24.

d. Find the sum of all the rows, again using the formula in E24.

Chapter Summary: Probability Distributions

In this chapter, you have learned that random variables are variables with probabilities attached to their possible outcomes. Probability distributions are characterized by their shape, center, and spread, just as data distributions and sampling distributions are. And you can use the same formulas for the mean and standard deviation.

The binomial distribution is a model for the situation where you count the number of successes in a random sample of size *n* from a large population. A typical question is, "If you perform 20 trials, what is the probability of getting exactly 6 successes?"

The geometric distribution is a model for the situation where you count the number of trials needed to get your first success. A typical question is, "What is the probability that it will take you exactly five trials to get the first success?"

■ Review Exercises

E26. Construct the probability distribution table for the situation of flipping a coin five times and counting the number of heads. Compute the mean and standard deviation of the distribution, and make a probability distribution graph.

E27. Display 7.10 gives the chances that an adult (18 or over) living in the United States will become a victim of the given events.

Event	Rate per 1000 Adults per Year
Accidental injury	242.0
Personal theft	72.0
Violent victimization	31.0
Injury in motor vehicle accident	17.0
Death	11.0
Injury from fire	0.1

Display 7.10   Rates of insurable causes of accidents or death to 1000 adults per year.

Source: *Report to the Nation on Crime and Justice,* 2nd ed., U.S. Bureau of Justice Statistics, 1988, p. 24.

a. What should a randomly selected U.S. adult be charged yearly for a $100,000 life insurance policy if the insurance company expects to break even?

b. What should a randomly selected U.S. adult be charged yearly for a policy that pays him or her $10,000 if he or she is the victim of a violent crime?

c. An insurance company determines that people are willing to pay $50 a year for a policy that pays them a set amount if they are injured by fire. What set amount should this be if the insurance company expects to break even?

E28. A blood bank knows that only about 10% of its regular donors have type B blood.

a. The technician will check 10 donations today. What is the chance that at least one will be type B? What assumptions are you making?

b. The technician will check 100 donations this month. What is the approximate probability that at least 10% of them will be type B?

c. This bank needs 16 type B donations. If the technician checks 100 donations, does the blood bank have a good chance of getting the amount of type B blood it needs? What recommendation would you have for the blood bank managers?

d. What is the probability that the technician will have to check at least 4 donations before getting the first that is type B?

E29. "One in 10 high school graduates in the state of Florida sends an application to the University of Florida," says the director of admissions there. If Florida has approximately 120,000 high school graduates next year, what is the mean and standard deviation of the number of applications the University will receive from Florida high school students? Do you see any weaknesses in using the binomial model here?

E30. "Girls Get Higher Grades" is a headline from *USA Today,* August 12, 1998. Eighty percent of the girls surveyed said it was important to them to do their best in all classes, whereas only 65% of the boys responded this way. Suppose you take a random sample of 20 female students and another sample of 20 male students in your region of the country.

a. What is the probability that more than half of the girls in the sample want to do their best in every class? What assumptions are you making?

b. What is the probability that more than half of the boys in the sample want to do their best in every class? What assumptions are you making?

c. Suppose you are to take one sample of 40 students at random, instead of 20 girls and 20 boys. What is the approximate probability that more than half of those sampled will want to do their best in every class? What assumptions are you making?

E31. It is estimated that 14% of Americans have no health insurance. A polling organization randomly samples 500 Americans to ask questions about their health.

Source: Current Population Survey, U.S. Census Bureau, March 2000 and 2001.

a. What is the probability that more than 430 of those sampled will have health insurance?

b. If it costs $40 to interview a respondent with health insurance and $20 to interview a respondent without health insurance, what is the expected cost of conducting the 500 interviews?

c. If the interviews are conducted sequentially, what is the expected number of interviews to be conducted before the second person without health insurance is found?

E32. Airlines routinely overbook popular flights because they know that not all ticket holders show up. If more passengers show up than there are seats available, an airline offers passengers $100 and a seat on the next flight. A particular 120-seat commuter flight has a 10% no-show rate.

a. If the airline sells 130 tickets, how much money does it expect to have to pay out per flight?

b. How many tickets should the airline sell per flight if they want their chance of giving any $100 amounts to be about .05?

E33. Suppose X has a geometric distribution with probability of success denoted by p.

a. Find the conditional probability that $X = 6$ given that $X > 5$.

b. Find the probability that $X > 6$ given that $X > 5$.

c. Find the probability that X exceeds $k + m$ given that it exceeds m. That is, find an expression for

$$P(X > k + m \mid X > m)$$

This is referred to as the "memoryless" property of the geometric distribution. Explain why this is an appropriate name.

E34. Random sampling from a large lot of a manufactured product yields a number of defectives, X, that has an approximate binomial distribution with p being the true proportion of defectives in the lot. A sampling plan consists of specifying the number, n, of items to be sampled and an acceptance number, a. After n items are inspected, the lot is accepted if $X \le a$ and it is rejected if $X > a$.

a. For $n = 5$ and $a = 0$, calculate the probability of accepting the lot for values of p equal to 0, .1, .3, .5, and 1.

b. Graph the probability of lot acceptance as a function of p for this plan. This is called the *operating characteristic curve for the sampling plan.*

E35. Refer to E34. A quality control engineer is considering two different lot acceptance sampling plans: $(n = 5, a = 1)$ and $(n = 25, a = 5)$.

a. Construct operating characteristic curves for both plans.

b. If you were a seller producing lots with proportions of defectives between 0% and 10%, which plan would you prefer?

c. If you were a buyer wishing high protection against accepting lots with proportions of defectives over 30%, which plan would you prefer?

E36. Each box of a certain brand of cereal contains a coupon that can be redeemed for a poster of a famous sports figure. There are five different coupons, representing five different sports figures.

a. Suppose you are interested in a particular sports figure. Set up a simulation that would produce an approximate probability distribution of the number of boxes of cereal you would have to buy in order to get one coupon for that particular poster. (You stop buying cereal when you get the poster you want.)

b. What is the probability that you would get the coupon for the particular sports figure in four or fewer boxes of cereal? (You stop buying cereal when you get the poster you want.)

c. What is the expected number of boxes of cereal you would have to buy to get the one coupon you want?

d. Suppose you want to get two particular posters out of the five available. What is the expected number of boxes of cereal you would have to buy to get the two specific ones? (*Hint:* At the outset, the probability of getting a coupon you want in any one box is $\frac{2}{5}$. After you get one of them, what is the probability of getting the other coupon you want? Think of the expected number of boxes of cereal being purchased in terms of these two stages.)

e. What is the expected number of boxes of cereal you would have to buy to get a full set of all five posters? Set up a simulation for this event, and compare your theoretical result to the one obtained from the simulation.

E37. Refer to the copper flipper description in D18. This time suppose you start with 10 pennies. Design and carry out a simulation to answer these questions.

a. On average, how many flips does it take for a specified copper flipper to "die"?

b. On average, how many flips will it take for all 10 to die?

c. On average, how many flips of the entire set of 10 pennies does it take to get at least one copper flipper that dies?

CHAPTER 8
INFERENCE FOR PROPORTIONS

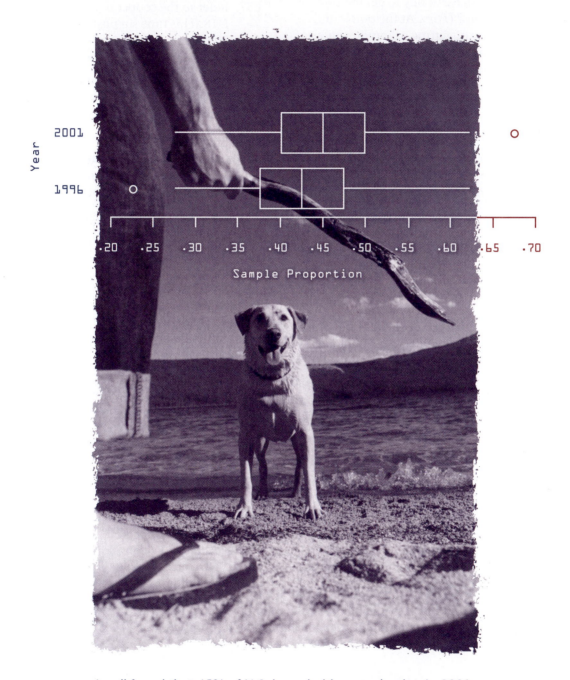

A poll found that 45% of U.S. households owned a dog in 2001, up from 43% in 1996. Are you confident that there was an increase in the percentage of *all* U.S. households that own a dog? Your answer will depend on how many households were included in the poll.

Open the daily newspaper, newsmagazine, or your own favorite magazine and you are likely to see the results of a public opinion poll. For example, a recent Phi Delta Kappa/Gallup poll reported that a record 51% of the American public assigns a grade of A or B to the public schools in their community and that this survey had a margin of error of 3%. [Source: 2001, www.gallup.com/poll/releases/pr010823.asp.]

In this chapter, you will learn the two basic techniques of statistical inference—confidence intervals and significance testing. If you are familiar with polls, you have seen the idea of confidence interval: Gallup's result of 51% ± 3% is equivalent to a confidence interval of 48% to 54%. In your work in Chapter 1 on the Westvaco case, you used the idea behind significance testing when you saw that if you randomly selected 3 of the 10 hourly workers to be laid off, you would be unlikely to get a group with an average age as large as those actually laid off by Westvaco.

The ideas developed here that are behind confidence intervals and significance testing are fundamental to work in statistical inference. They will form the basis of the inferential procedures in the chapters to follow.

In this chapter, you will learn

- to construct and interpret a confidence interval to estimate the proportion of successes in a binomial population

- to use a significance test (hypothesis test) to decide if it is reasonable to conclude that your sample might have been drawn from a binomial population with a specified proportion of successes

- to construct and interpret a confidence interval for the difference between the proportion of successes in one population and the proportion of successes in another population

- to use a test of significance to decide if it is reasonable to conclude that two samples might have been drawn from two binomial populations that have the same proportion of successes

8.1 Estimating a Proportion with Confidence

The Phi Delta Kappa/Gallup poll of public approval of the public schools reported that

> These results are based on telephone interviews with a randomly selected national sample of 1108 adults, 18 years and older, conducted May 23–June 6, 2001. For results based on this sample, one can say with 95 percent confidence that the maximum error attributable to sampling and other random effects is plus or minus 3 percentage points. In addition to sampling error, question wording and practical difficulties in conducting surveys can introduce error or bias into the findings of public opinion polls.

The Gallup organization is disclosing that they didn't ask all adults in the United States, only 1108. Even so, unless there are some special difficulties such as problems with the wording of the question, they are 95% confident that the error is less than 3% either way in the percentages they report. That is, they are 95% confident that if they were to ask all adults in the United States to give a grade to the public schools, 51% ± 3%, or between 48% and 54%, would give a grade of A or B. How can the Gallup organization possibly make such a statement?

You can quantify how confident you are in a result from a sample.

Random sampling is the key.

In Section 5.4, you learned how much variability to expect in the proportions of successes $\hat{p}$ from repeated random samples taken from a given binomial population. About 95% of all sample proportions $\hat{p}$ will fall within about two standard errors of the population proportion p, that is, within the interval

$$p \pm 1.96\sqrt{\frac{p(1-p)}{n}}$$

where n is the sample size. The proportions in this interval are called **reasonably likely.**

Reasonably Likely Outcomes and Rare Events

Reasonably likely outcomes are those in the middle 95% of the distribution of all possible outcomes. The outcomes in the upper 2.5% and the lower 2.5% of the distribution are rare events—they happen, but rarely.

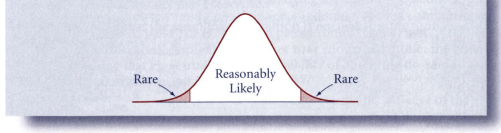

Example: Reasonably Likely Results from Coin Flips

Suppose that you will flip a fair coin 100 times. What are the reasonably likely values of the sample proportion $\hat{p}$? What numbers of heads are reasonably likely?

In Section 5.4, you
learned $\mu_{\hat{p}} = p$ and
$\sigma_{\hat{p}} = \sqrt{\dfrac{p(1-p)}{n}}$.

Solution

Here the actual value of p for each trial is .5 and you have 100 trials—a sample
size of 100. From the previous formula, 95% of all sample proportions $\hat{p}$ should
fall in the interval

$$p \pm 1.96\sqrt{\frac{p(1-p)}{n}} = .5 \pm 1.96\sqrt{\frac{.5(1-.5)}{100}}$$

or about $.5 \pm .1$. So the reasonably likely values of $\hat{p}$ are in the interval from
.4 to .6. Values of $\hat{p}$ outside this interval can happen, but they happen rarely.

Equivalently, in about 95% of the samples, the number of successes x in the
sample will be in the interval

$$np \pm 1.96\sqrt{np(1-p)} = 100(.5) \pm 1.96\sqrt{100(.5)(1-.5)}$$

or about 50 ± 10 successes. ■

Even though Gallup doesn't know p, the actual proportion of people who
would give their schools an A or a B, it can use the idea in the preceding example.
For each possible value of p, Gallup can compute how close to p most sample
proportions will be. By knowing the variability expected from random samples
taken from populations with different values of p, Gallup can estimate how close
$\hat{p}$ should be to the "truth."

Discussion: Reasonably Likely Outcomes

D1. Suppose 35% of a population think they pay too much for car insurance.
A polling organization takes a random sample of 500 people in this
population and computes the sample proportion $\hat{p}$ of people who think
that they pay too much for car insurance.

a. There is a 95% chance that $\hat{p}$ will be between what two values?

b. Is it reasonably likely to get 145 people in the sample who think they pay
too much for car insurance?

D2. In parts a through d, you will review the method of using simulation to
construct a sampling distribution.

a. Using a calculator, computer, or random digit table, take a random
sample of size 40 from a population with 60% successes. Count the
number of successes in your sample.

b. Continue taking samples of size 40 until your class has the results from
100 samples.

c. Plot your 100 values on a dot plot or a histogram.

d. From your plot in part c, what numbers of successes are reasonably likely?

■ Practice

P1. Suppose 40% of your graduating class plans to go on to higher education.
You survey a random sample of 50 of your classmates and compute the
sample proportion $\hat{p}$ of students who plan to go on to higher education.

    a. There is a 95% chance that $\hat{p}$ will be between what two numbers?

    b. Is it reasonably likely to find that 25 students in your sample plan go on to higher education?

P2. Describe how to use simulation to find the reasonably likely outcomes for a random sample of size 40 taken from a population with $p = .3$.

In Activity 8.1, you will collect data about the proportion of students in your class who are right-eye dominant. Later in this section, you will learn to find a confidence interval for the percentage of *all* students who are right-eye dominant.

Activity 8.1   Determining Eye Dominance

What you'll need: an $8\frac{1}{2}'' \times 11''$ piece of paper

1. Determine which eye is your dominant eye by following these instructions:

   Tear a $1'' \times 1''$ square in the middle of an $8\frac{1}{2}'' \times 11''$ piece of paper. Hold the paper at arm's length. Look through the square at a relatively small object across the room. Without changing your position, close your right eye. If you can still see the object, your left eye is your dominant eye. If not, your right eye is your dominant eye.

2. Gather the results about eye dominance from exactly 40 students. What proportion of students in your sample have a dominant right eye?

3. Suppose you can reasonably consider your sample of 40 students a random sample of all students. From the information you gathered in the activity, is it plausible that 10% of all students are right-eye dominant? Is it plausible that the true percentage is 60%? What percentages do you find plausible?

The Meaning of a Confidence Interval

In the following activity, you will make a chart that allows you to see the reasonably likely outcomes for all population proportions p when the sample size is 40. From there, it is a short step to confidence intervals.

Activity 8.2   Constructing a Chart of Reasonably Likely Outcomes

What you'll need: a copy of Display 8.1

1. Suppose you take repeated random samples of size 40 from a population with 60% successes. What proportions of successes would be reasonably likely in your sample?

2. Compare the proportion of students you found to be right-eye dominant from Activity 8.1 to your answer in step 1. Is it reasonable to assume that 60% of all students are right-eye dominant? Explain your reasoning.

(continued)

3. On a copy of the chart in Display 8.1, draw a horizontal line segment aligned with .6 on the vertical axis, "Proportion of Successes in the Population." The line segment should stretch in length over your interval from step 1.

Your instructor will give you one of the population proportions whose line segments are missing in Display 8.1.

4. Compute the reasonably likely outcomes for your population proportion p.

5. On your copy of the chart in Display 8.1, draw a horizontal line segment showing the reasonably likely outcomes for your group's proportion p.

6. Get the reasonably likely outcomes from the other groups in your class and complete the chart with the line segments from those values of p.

7. Refer to your results from Activity 8.1 about eye dominance. For which population proportions is your sample proportion reasonably likely?

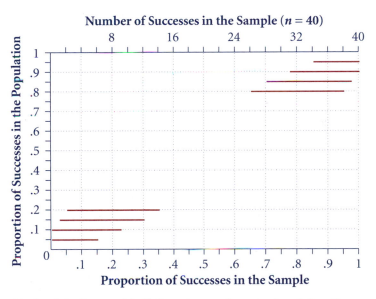

Display 8.1   Reasonably likely outcomes for samples of size 40.

The line segments you drew on your copy of Display 8.1 show reasonably likely outcomes to get when taking random samples of size 40 from a population with a given proportion of successes p.

Example: Plausible Percentages of Right-Eye Dominance

In a group of 40 adults doing Activity 8.1, exactly 30 were right-eye dominant. Assuming this can be considered a random sample of all adults, is it plausible that if you tested *all* adults, you would find that 50% are right-eye dominant? Is it plausible that 80% are right-eye dominant? What percentages are plausible?

Solution

No, it's not plausible that 50% of all adults are right-eye dominant. That's because 30 out of 40, or $\hat{p} = .75$, isn't a reasonably likely outcome for a random sample taken from a population with 50% successes. You can see this from your chart or from Display 8.2 by going to the horizontal line segment next to the population proportion of .5. This line segment goes from about .35 to .65, and that does not include a sample proportion of $\hat{p} = .75$. Of course, it is *possible* that the true percentage is 50% and that the adults were an unusual sample, but it's not very *plausible*.

It is plausible that 80% is the actual percentage of all adults who are right-eye dominant. The horizontal line segment for a population percentage of 80% goes from .65 to .95, and that does include the sample proportion $\hat{p} = .75$. If the true percentage is 80%, it is reasonably likely to get 30 adults who are right-eye dominant from a random sample of 40 adults.

To find the plausible percentages from your chart, draw a vertical line upward from $\hat{p} = .75$, as in Display 8.2. The line segments it crosses, in addition to 80%, represent the other plausible population percentages: 60%, 65%, 70%, 75%, and 85%. In all these populations, a sample proportion of $\hat{p} = .75$ would be a reasonably likely result.

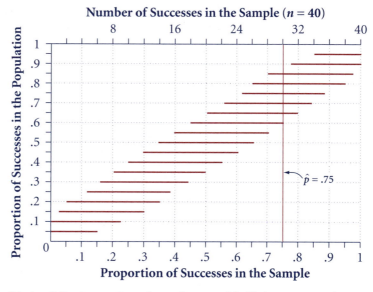

Display 8.2 A complete chart of reasonably likely outcomes for samples of size 40.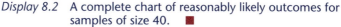

The percentages listed in the last part of the preceding example are in the 95% confidence interval for the percentage of all adults who are right-eye dominant. That is, you are 95% confident that if you tested all adults, the percentage who are right-eye dominant is somewhere in the interval from about 60% to 85%.

95% Confidence Interval for a Proportion *p*

p is the unknown parameter.

A **95% confidence interval** consists of those population percentages *p* for which the sample proportion $\hat{p}$ is reasonably likely.

Discussion: Confidence Intervals

Use your completed chart or the one in Display 8.2 to answer these questions.

D3. According to the 2000 U.S. Census, about 60% of Hispanics in the United States are of Mexican origin. Would it be reasonably likely in a survey of 40 randomly chosen Hispanics to find that 27 are of Mexican origin?

Source: www.census.gov/prod/2001pubs/c2kbr01-3.pdf.

D4. According to the 2000 U.S. Census, about 30% of people over age 85 are men. In a random sample of 40 people over age 85, would it be reasonably likely to get 60% who are men?

Source: www.census.gov/prod/2001pubs/c2kbr01-10.pdf.

D5. Suppose that in a random sample of 40 toddlers, 34 know what color Barney is. What is the 95% confidence interval for the percentage of toddlers who know what color Barney is?

D6. Polls usually report a margin of error. Suppose a poll of 40 randomly selected statistics majors finds that 20 are female. The poll reports that 50% of statistics majors are female, with a margin of error of 15%. Use your completed chart to explain where the 15% came from.

D7. Why do we say we want a confidence interval for p rather than saying we want a confidence interval for $\hat{p}$?

■ Practice

Use your completed chart or the one in Display 8.2 to answer these questions.

P3. Suppose you flip a fair coin 40 times. How many heads is it reasonably likely for you to get?

P4. According to the *2000 Statistical Abstract of the United States*, about 60% of 18- and 19-year-olds are enrolled in school. If you take a random sample of 40 randomly chosen 18- and 19-year-olds, would it be reasonably likely to find that 32 were in school?

Source: www.census.gov/prod/2001pubs/statab/sec04.pdf (Table 245).

P5. According to the *2000 Statistical Abstract of the United States*, about 80% of people in the United States aged 25 or over have graduated from high school. In a random sample of 40 people aged 25 or over, how many high school graduates is it reasonably likely to get? What proportions of high school graduates is it reasonably likely to get?

Source: www.census.gov/prod/2001pubs/statab/sec04.pdf (Table 249).

P6. In a random sample of 40 adults, 25% know what color Barney is. What is the 95% confidence interval for the percentage of all adults who know what color Barney is?

P7. Suppose that in a random sample of 40 retired women, 45% of the women travel more than they did while they were working. Find the 95% confidence interval for the proportion of all retired women who travel more.

From the Chart to a Formula

If you have noticed that Display 8.2 can be used to find a confidence interval only when the sample size is 40, you may be thinking, "Constructing a new chart each time they give me a different sample size is going to be a lot of work. Perhaps it's not too late to transfer to that snowboarding class." But don't leave yet. There is a quick and easy formula you can use.

To develop this formula, think again about the geometry behind a confidence interval. Suppose in a random sample of size 40, the sample proportion turns out to be $\hat{p} = .6$. As in Display 8.3, you can approximate the 95% confidence interval by drawing a vertical line from $\hat{p} = .6$ and seeing which horizontal line segments it intersects. These horizontal line segments give the population proportions, p, for which the sample proportion $\hat{p}$ is a reasonably likely result. These populations are $p = .45$ up through $p = .70$ almost to $p = .75$. This is the confidence interval.

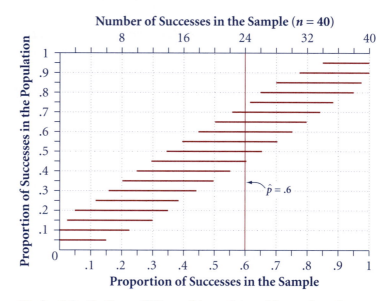

Display 8.3   Finding a 95% confidence interval for p when the sample proportion is .6 and the sample size is 40.

How can you get this confidence interval without the chart? The key is to notice that the two bold line segments in Display 8.4 cross at the point (.6, .6) and are approximately the same length. (They are approximately the same length because the two sets of endpoints of the horizontal line segments almost lie on two parallel lines with slope 1.)

The bold horizontal line segment represents the reasonably likely outcomes or the middle 95% of all sample proportions from a population with $p = .6$. You know how to find its endpoints:

$$p \pm 1.96\sqrt{\frac{p(1-p)}{n}} = .6 \pm 1.96\sqrt{\frac{.6(1-.6)}{40}} \approx .6 \pm .15$$

The bold vertical line segment is the confidence interval, and it has the same endpoints, namely, .6 ± .15, or .45 to .75. So to get the (vertical) confidence interval, all you have to do is find the endpoints of the horizontal line segment by substituting the known value of $\hat{p}$ for the unknown value of p in the formula.

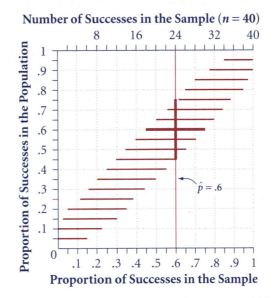

Number of Successes in the Sample (n = 40)

Display 8.4 The length of the 95% confidence interval (the vertical line segment) is the same as the length of the horizontal line segment of reasonably likely outcomes for $p = .6$.

Discussion: Toward a Formula

D8. Suppose you observe a sample proportion of $\hat{p} = .5$ in a random sample of size 40.

   a. Use Display 8.2 to find an approximate 95% confidence interval for the true population proportion.

   b. On a copy of Display 8.2, sketch in the two bold line segments for the proportion .5 as in Display 8.4.

   c. Explain the meaning of the horizontal bold line segment. Use a formula to find its endpoints.

   d. Explain the meaning of the vertical bold line segment. Find the endpoints of the vertical bold line segment using your result from part c.

   e. What is the 95% confidence interval?

■ Practice

P8. Suppose that in a random sample of 40 students, 25 are wearing sneakers. Find the 95% confidence interval for the percentage of all students who wear sneakers,

   a. using the chart in Display 8.2

   b. without using the chart

Using the Formula

You can use the reasoning described above to find a confidence interval for the percentage of successes p in a population, no matter what the sample size or level of confidence you desire. The general formula and conditions for using it are given in this box.

A Confidence Interval for a Proportion

Confidence intervals have the form estimate ± z · SE.

A confidence interval for the proportion of successes p in the population is given by the formula

$$\hat{p} \pm z \cdot \sqrt{\frac{\hat{p}(1 - \hat{p})}{n}}$$

Here, n is the sample size and $\hat{p}$ is the proportion of successes in the sample. The value of z depends on how confident you want to be that p will be in the confidence interval. If you want a 95% confidence interval, use $z = 1.96$; for a 90% confidence interval, use $z = 1.645$; for a 99% confidence interval, use $z = 2.576$; and so on.

"Conditions" are sometimes called "assumptions."

This confidence interval is reasonably accurate when three conditions are met:

- The sample was a simple random sample from a binomial population.
- Both $n\hat{p}$ and $n(1 - \hat{p})$ are at least 10.
- The size of the population is at least 10 times the size of the sample.

The first two conditions are necessary to be able to use the normal distribution (and z-scores) as an approximation to the binomial distribution. If the last condition isn't met, your confidence interval will be longer than it needs to be.

The quantity

$$E = z \cdot \sqrt{\frac{\hat{p}(1 - \hat{p})}{n}}$$

is called the **margin of error**. It is half the length of the confidence interval.

Example: Safety Violations

Suppose you have a random sample of 40 buses from a large city and find that 24 have a safety violation. Find the 90% confidence interval for the proportion of all buses that have a safety violation.

Solution

Check conditions.

You will check the three conditions in D9.

Do computations.

For a 90% confidence interval, use $z = 1.645$. The confidence interval for the proportion p of all buses that have a safety violation based on a sample proportion of $\hat{p} = \frac{24}{40} = .6$ can be written

$$\hat{p} \pm z \cdot \sqrt{\frac{\hat{p}(1 - \hat{p})}{n}} = .6 \pm 1.645 \sqrt{\frac{.6(1 - .6)}{40}}$$

$$\approx .6 \pm 1.645(.077)$$

$$\approx .6 \pm .13$$

or about $(.47, .73)$.

Conclusion in context.   You are 90% confident that the percentage of all of this city's buses that have a safety violation is between 47% and 73%. The margin of error for this survey is .13.   ■

Discussion: Using the Formula

D9. In the bus example, are the three conditions met for constructing a confidence interval? Explain. (You should always verify that the three conditions are met before reporting a confidence interval.)

D10. What, in words, does $n\hat{p}$ represent in the bus example? What, in words, does $n(1 - \hat{p})$ represent in the bus example? What do these two quantities represent in general?

■ Practice

P9. A survey of 744 teenagers in grades 7–12 found that 82% believe that computer skills are necessary to make a good living.

Source: CNN/USA Today/Gallup poll sponsored by the National Science Foundation, April 22, 1997.

a. Check to see if the three conditions for computing a confidence interval are met in this case.

b. Find a 95% confidence interval for the percentage of all teenagers who believe this. What is the margin of error?

c. Find a 90% confidence interval for the percentage of all teenagers who believe this. What is the margin of error?

d. Which of your confidence intervals is longer? Why should that be the case?

The Capture Rate

If you construct one hundred 95% confidence intervals, you expect that the population proportion p will be in 95 of them. This statement is not as obvious as it may seem at first. You will test this statement in Activity 8.3.

Activity 8.3   The Capture Rate

What you'll need: a calculator or one table of random digits per student; a copy of Display 8.5

1. Your instructor will assign you a group of 40 random digits, or you will generate 40 random digits on your calculator.

(continued)

2. Count the number of even digits in your sample of 40.

3. Use the formula to construct a 95% confidence interval for the proportion of random digits that are even.

4. Each member of your class should draw his or her confidence interval as a vertical line segment on a copy of Display 8.5.

5. What is the true proportion of all random digits that are even?

6. What percentage of the confidence intervals captured the true proportion? Is this what you expected? Explain.

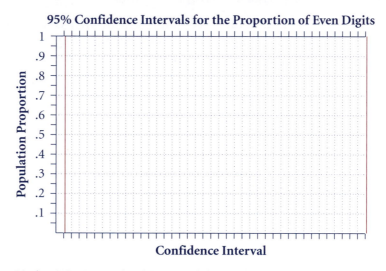

95% Confidence Intervals for the Proportion of Even Digits

Display 8.5   A sample of 95% confidence intervals.

Student: In the activity, about 95% of the 95% confidence intervals captured the true population proportion of .5. That was no surprise. But I don't see *why* that happened. Just calling something a 95% confidence interval doesn't make it one.

Teacher: You're right; this isn't obvious. For me to explain the logic to you, you'll have to answer some questions as we go along.

Student: Okay.

Teacher: Go back and ask those of your classmates who had confidence intervals that captured the true proportion of even digits ($p = .5$) what values of $\hat{p}$ they got. We'll talk again tomorrow.

Student: (*The next day*) They had values of $\hat{p}$ between .35 and .65.

Teacher: Right again. Now look at Display 8.6. What do you notice about these values of $\hat{p}$?

Student: They all lie on the horizontal line segment for $p = .5$.

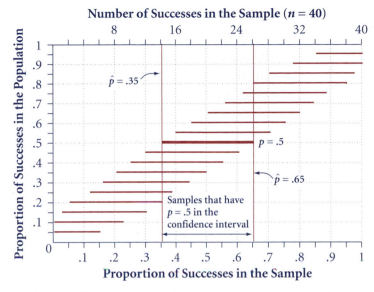

Display 8.6 The proportion *p* is in the confidence interval if and only if $\hat{p}$ is a reasonably likely outcome for *p*.

Teacher: Yes, the values of $\hat{p}$ that give a confidence interval that captures *p* = .5 are the reasonably likely outcomes for *p* = .5. What is the chance of getting one of these "good" values of $\hat{p}$?

Student: 95%!

Discussion: The Capture Rate

D11. Was every student's confidence interval in Activity 8.3 the same? Is this how it should be? Explain.

D12. Suppose you know that a population proportion is *p* = .60. Now, 80 different students are going to select independent random samples of size 40 from this population. Each student then constructs his or her own 90% confidence interval. How many of the resulting confidence intervals would you expect to include the true population proportion of *p* = .60?

A note about correct language

It is correct to say that we expect the true value of *p* to be in 95 out of every 100 of the 95% confidence intervals we construct. However, it is not correct to say, after we have found a confidence interval, that there is a 95% chance that *p* is in that confidence interval. Here is an example that shows why. If you pick a date in the next millennium at random, it is reasonable to say that there is a $\frac{1}{7}$ chance you will pick a Tuesday. However, say that the date you pick turns out to be March 3, 3875. It sounds a bit silly to say that there is a $\frac{1}{7}$ chance that March 3, 3875, is a Tuesday. Once the date is selected, there is no randomness left. March 3, 3875, is either a Tuesday or it isn't. All you can say is that the process you have used to select a date gives you a Tuesday $\frac{1}{7}$ of the time. (This example is due to Wes White, an AP Statistics teacher in Los Angeles.)

Margin of Error and Sample Size

You may have noticed that the 95% confidence intervals for large sample sizes are shorter than those for small sample sizes. This makes sense—the larger the sample size, the closer $\hat{p}$ should be to p. Let's see how that works for a specific example. Suppose you take a survey and get $\hat{p} = .7$. If your sample size is 100, your 95% confidence interval would be

$$\hat{p} \pm 1.96\sqrt{\frac{\hat{p}(1-\hat{p})}{n}} = .7 \pm 1.96\sqrt{\frac{.7(1-.7)}{100}}$$

$$\approx .7 \pm 1.96(.0458)$$

$$\approx .7 \pm .0898$$

The margin of error for this survey is approximately .0898.

If you quadruple your sample size to 400, your 95% confidence interval would be

$$\hat{p} \pm 1.96\sqrt{\frac{\hat{p}(1-\hat{p})}{n}} = .7 \pm 1.96\sqrt{\frac{.7(1-.7)}{400}}$$

$$\approx .7 \pm 1.96(.0229)$$

$$\approx .7 \pm .0449$$

Quadrupling the sample size cuts the margin of error in half.

The margin of error is approximately .0449, or half what it was before.

Discussion: Margin of Error and Sample Size

D13. All else being equal, will the confidence intervals for samples of size 80 be longer or shorter than those for samples of size 40? Explain your reasoning.

D14. You can use Display 8.2 only for samples of size 40 and for 95% confidence intervals.

  a. How would a chart for 99% confidence intervals be different from Display 8.2?

  b. How would the lengths of the confidence intervals be different?

■ Practice

P10. All else being equal, how would a chart for samples of size 100 be different from Display 8.2? How would the lengths of the confidence intervals be different?

P11. What value of z should you use for an 80% confidence interval? Will 80% confidence intervals be longer or shorter than 95% confidence intervals? Explain your answer.

P12. Suppose you take a survey with sample size n and get a margin of error that is three times larger than you would like. What sample size should you have used? Prove your answer using algebra.

Back to the Opinion Polls

At the beginning of this section, you read about a recent Phi Delta Kappa/Gallup poll that reported that a record 51% of the American public assigns a grade of A or B to the public schools in their community. The sample size was 1108. This survey had a margin of error of 3%, and so their 95% confidence interval is 48% to 54%. (The procedure Gallup uses to select a sample is more complicated than simple random sampling, but you can use your formula for a confidence interval to approximate Gallup's margin of error.)

You now should be able to answer these questions:

- What is it that you are 95% sure is in the confidence interval?

  The proportion of *all* Americans who would assign a grade of A or B to their local public schools.

- What is the interpretation of the confidence interval of 48% to 54%?

  We are 95% confident that if we could ask all Americans to give a grade to their local public schools, between 48% and 54% of them would give an A or B.

- What is the meaning of "95% confidence"?

  If we were to take 100 random samples of Americans and compute the 95% confidence interval from each sample, then we expect that 95 of them will contain the true proportion of all Americans that would assign a grade of A or B (whatever that proportion is).

Discussion: Back to the Opinion Polls

D15. Verify the margin of error for the Phi Delta Kappa/Gallup poll.

D16. Are the three conditions for constructing a confidence interval met in the Phi Delta Kappa/Gallup poll?

D17. In the report at the beginning of this section on the Phi Delta Kappa/Gallup poll, the phrase "error attributable to sampling" is used. Explain what this means.

■ Practice

P13. A report of a recent survey about Internet use on homework gave this information.

Seventy-one percent of middle school and high school students with Internet access said they relied the most on the Internet in completing a project, according to a survey conducted by the Pew Internet and American Life Project. Of students aged 12 to 17, the Pew study found that 73 percent, or 17 million children, have Internet access. The Pew study surveyed 754 of those students.
Source: www.siliconvalley.com/docs/news/svfront/063319.html, Sept. 1, 2001.

a. Check if the three conditions are met for computing a confidence interval for the percentage of students with Internet access who would say they rely the most on the Internet in completing a project.

b. Regardless of your answer to part a, compute the 95% confidence interval for the percentage of middle and high school students with Internet access who rely on the Internet the most in completing a project.

c. What is it that you are 95% sure is in the confidence interval?

d. Interpret this confidence interval.

e. Explain the meaning of "95% confidence."

P14. Most legitimate polling organizations publish details on how their polls are conducted. Many include a statement similar to this: "In addition to sampling error, the practical difficulties of conducting a survey of public opinion may introduce other sources of error." List at least three other sources of error and explain why they are not included in the sampling error measured by statistical procedures.

What Sample Size Should You Use?

How large the sample should be depends on how small a margin of error you want.

People who conduct surveys and experiments often ask statisticians, "What sample size should I use?" The simple answer is that the larger the sample size n, the more accurate the results will probably be. When everything else is held constant, the margin of error is smaller with a larger sample size than with a smaller one. However, researchers always have limits of time and money and so for practical reasons have to limit the size of their samples. For a confidence interval, the margin of error, E, is given by

$$E = z \cdot \sqrt{\frac{p(1-p)}{n}}$$

You can solve this formula for n:

$$E = z \cdot \sqrt{\frac{p(1-p)}{n}}$$

$$E^2 = z^2 \cdot \left(\frac{p(1-p)}{n}\right)$$

$$n = z^2 \cdot \left(\frac{p(1-p)}{E^2}\right)$$

To use this formula, you have to know the margin of error, E, that is acceptable. You have to decide on a level of confidence so you know what value of z to use. And you have to have an estimate of p. If you have a good estimate of p, use it in this formula. If you do not have a good estimate of p, use $p = .5$. Using this value for p may give you a sample size that is a bit too large, but never one that is too small. (This is because the largest possible value of $p(1-p)$ occurs when $p = .5$. See E13.)

Example: Estimating Needed Sample Size

What sample size should you use for a survey if you want the margin of error to be at most 3% with 95% confidence but you have no estimate of p?

Solution

Because you do not have an estimate of p, use $p = .5$, then

$$n = z^2 \cdot \left(\frac{p(1-p)}{E^2}\right) = 1.96^2\left(\frac{.5(1-.5)}{.03^2}\right)$$

$$\approx 1067.111$$

Because you must have a sample size of at least 1067.111, round up to 1068. To get a margin of error around 3% is one reason that national polling organizations use sample sizes around 1000. ∎

Discussion: Sample Size Needed

D18. Examine the formula for a confidence interval and explain why is it true that when everything else is held constant, the margin of error is smaller with a larger sample size than with a smaller one.

D19. Suppose it costs $1 to survey each person in your sample. You judge that p is about .5. What will your survey cost if you want a margin of error of about 10%? 1%? .1%?

■ Practice

P15. The accuracy of opinion polls is examined closely in presidential election years. Suppose you are running a large polling organization and want to have a margin of error of 2% for your final poll before the presidential election. What sample size should you use with a 95% confidence level? What sample size should you use if you want a margin of error of 1% with a 99% confidence level? What sample size should you use if you want a margin of error of 0.5% with 90% confidence?

P16. A Gallup poll found that 30% of adult Americans report that drinking has been a source of trouble in their families. Gallup asks this question every year. What sample size should Gallup use next year to get a margin of error of 3% and be as economical as possible?

Source: www.gallup.com/poll/news/970927b.html, Oct. 10, 1997.

Summary 8.1: Confidence Intervals for a Proportion

A 95% confidence interval consists of those population percentages p for which the observed sample proportion $\hat{p}$ is reasonably likely.

The formula for a confidence interval has the form

estimate $\pm z \cdot$ (*standard error of the estimate*)

In the case of estimating a proportion, the confidence interval is

$$\hat{p} \pm z \cdot \sqrt{\frac{\hat{p}(1-\hat{p})}{n}}$$

where $\hat{p}$ is the observed sample proportion and n is the sample size.

With $z = 1.645$, the method will capture the true proportion 90% of the time; with $z = 1.96$, it will do so 95% of the time; and with $z = 2.576$, it will do so 99% of the time.

This confidence interval is reasonably accurate when three conditions (or "assumptions") are met.

- The sample was a simple random sample from a binomial population.
- Both $n\hat{p}$ and $n(1 - \hat{p})$ are at least 10.
- The size of the population is at least 10 times the size of the sample.

There are two parts to giving an interpretation of a confidence interval: Describe what is in the confidence interval and describe what is meant by "confidence." For a 95% confidence interval, for example, this interpretation would include:

- You are 95% confident that if you could examine each unit in the population, the proportion of successes, p, in this population would be in this confidence interval.
- If you were able to repeat this process 100 times and construct the 100 resulting confidence intervals, you'd expect that 95 of them will contain the population proportion of successes, p. In other words, the process results in a confidence interval that captures the true value of p 95% of the time.

Remember that there can be sources of error other than sampling error. For example, if the samples aren't randomly selected or if there is a problem such as a poorly worded questionnaire, the capture rates don't apply.

To estimate the sample size n needed for a given margin of error E, use this formula:

$$n = z^2 \cdot \left(\frac{p(1-p)}{E^2} \right)$$

with a rough estimate of p if you have it, and if not, use $p = .5$.

■ Exercises

E1. Review the use of the chart in Display 8.2 by answering these questions.

a. The proportion of successes in a random sample of size 40 is .90. Is this sample proportion reasonably likely if the population has 75% successes?

b. In the multicandidate 2000 presidential election, Al Gore received 48% of the votes cast, "winning" the popular vote but losing in the electoral college. In a random sample of 40 voters in this election, what is the largest proportion of people who voted for Vice President Gore that is reasonably likely? The smallest?

c. Suppose that a poll of a random sample of 40 shoppers finds that 16 prefer open-air malls to enclosed malls. What is the 95% confidence interval for the percentage of all shoppers who prefer open-air malls?

E2. A recent nationwide survey of 19,441 teens about their attitudes and behaviors toward epilepsy found that only 51% knew that epilepsy is not contagious.

The Executive summary from the Epilepsy Foundation for this survey gave this technical information:

The two-page survey was distributed to teens nationally by 20 affiliates of the Epilepsy Foundation from March 2001 through July 2001 in schools selected by each affiliate. A total of 19,441 valid surveys were collected. Mathew Greenwald & Associates, Inc., edited the surveys and performed the data entry. Greenwald & Associates was also responsible for the tabulation, analysis, and reporting of the data. The data were weighted by age and region to reflect national percentages.

The margin of error for this study (at the 95% confidence level) is plus or minus approximately 1%.

The survey was funded by the Centers for Disease Control and Prevention.

Source: www.efa.org/blurt/etr/survey.pdf, April 16, 2002.

a. Describe the survey design.

b. Are the conditions for constructing a confidence interval met?

c. Does a computation of the margin of error using the formula in this section agree with that given in the press release?

d. Assuming the conditions have been met to construct a confidence interval, what is it that you are 95% sure is in the confidence interval?

e. Interpret this confidence interval.

f. Explain the meaning of 95% confidence.

E3. A *U.S. News & World Report* survey of 1000 adults from the general public (published on April 15, 1996) reported that 81% thought TV contributed to a decline in family values. If the sample was randomly selected, what can you say about the proportion of all adults who think TV contributes to a decline in family values? (As always, discuss whether the three conditions are met, give the confidence interval itself, and give an interpretation of this interval, stating clearly what it is that is supposed to be in the confidence interval.)

E4. Another part of the *U.S. News & World Report* survey went to Hollywood leaders. Of 6059 mailed surveys, only 570 were returned. Among the returned surveys, 46% thought TV contributed to a decline in family values. The magazine does not report a margin of error for this part of the survey. Should a margin of error be reported? Explain.

E5. Observe 40 students on your campus. Find the 98% confidence interval for the percentage of students who carry backpacks. (As always, discuss whether the three conditions are met, give the confidence interval, and give an interpretation of this interval, stating clearly what it is that is supposed to be in the confidence interval.)

E6. Suppose a random sample of size 40 produces a sample proportion of $\hat{p} = .80$.

a. Find an approximate 95% confidence interval from Display 8.2.

b. Find a 95% confidence interval using the formula.

c. Compare the answers from parts a and b. Give reasons for any discrepancy between the two.

E7. Suppose a random sample of size 100 produces a sample proportion of .60. Find a 95% confidence interval estimate of the population proportion. Can Display 8.2 be used in this case? Why or why not?

E8. If everything else in your sample is left unchanged,

a. what happens to the length of a confidence interval as the sample size increases?

b. what happens to the length of a confidence interval as the degree of confidence you have in the answer increases?

E9. Explain the difference between p and $\hat{p}$. Which is the parameter? Which one is always in the confidence interval?

E10. If you want a margin of error that is one-quarter what you estimate it will be with your current sample size, by what factor should you increase the sample size?

E11. Supply a reason for each step in parts a–d in this alternative explanation for why the 95% confidence interval for a population proportion "works."

a. Ninety-five percent of all sample proportions $\hat{p}$ are within two standard errors of p.

b. There is a 95% chance that the proportion $\hat{p}$ from a specific random sample will be within two standard errors of p.

c. There is a 95% chance that p will be within two standard errors of $\hat{p}$.

d. Thus, the formula for a 95% confidence interval is approximately

$$\hat{p} \pm 2\sqrt{\frac{\hat{p}(1-\hat{p})}{n}}$$

E12. A survey was conducted to determine what adults prefer in cell phone services. The results of the survey showed that 73% of the people wanted email service, with a margin of error of plus or minus 4%. Which of these sentences explains most accurately what is meant by "plus or minus 4%"?

A. They estimate that 4% of the population that was surveyed may change their minds between the time the poll is conducted and the time the survey is published.

B. There is a 4% chance that the true percentage of adults who want email service will not be in the confidence interval of 69% to 77%.

C. Only 4% of the population was surveyed.

D. It would be unlikely to get the observed sample proportion of 73% unless the actual percentage of all adults who want email service is between 69% and 77%.

E. The probability that the sample proportion is in the confidence interval is .04.

E13. If you don't have an estimate for p when computing the needed sample size, you use $p = .5$. In this exercise, you will show that this is the safest value to use because $p = .5$ requires the largest sample size, all other things being equal. Use these prompts to prove that the largest possible value of $p(1 - p)$ is achieved when $p = .5$.

a. Let $y = x(1 - x)$, where $0 \le x \le 1$. Why are we letting $y = x(1 - x)$? Why are we restricting x to $0 \le x \le 1$?

b. Sketch the graph of $y = x(1 - x)$. What is the name of this type of function?

c. Where does the largest possible value of y occur for this function? At what value of x does this occur? What is the largest possible value of y?

8.2 Testing a Proportion

People often make decisions with data by comparing the results from a sample to some predetermined standard. These kinds of decisions are called **tests of significance** because the goal is to test the significance of the difference between the sample and the standard. If the difference is small, there is no reason to conclude that the standard doesn't hold. But when the difference is large enough that it can't reasonably be attributed to chance, you can conclude that the standard no longer holds.

Here is an example of this type of reasoning. About 2% of barn swallows have white feathers in places where the plumage is normally blue or red. The white feathers are caused by genetic mutations. In 1986, the Russian nuclear reactor at Chernobyl leaked radioactivity. Researchers continue to be concerned that the radiation may have caused mutations in the genes of humans and animals that were passed on to offspring. In a sample of barn swallows captured around Chernobyl in 1991 and 1996, about 14% had white feathers in places where the plumage is normally blue or red. Researchers compared the proportion $\hat{p} = .14$ in the sample of captured barn swallows to the standard of .02. If the overall percentage was still only 2%, it is not reasonably likely to get 14% in their sample. So they came to the conclusion that there was an increased probability of genetic mutations in the Chernobyl area. [Source: *Los Angeles Times*, October 9, 1997, page B2.]

Tests of significance look for departures from a standard.

In Activity 8.4, you will record the results of 40 flips of a penny and 40 spins of a penny. You will use the data later to perform tests of significance.

Activity 8.4   Spinning and Flipping Pennies

What you'll need: pennies (at least one per student)

People tend to believe that pennies are balanced. They generally have no qualms about flipping a penny to make a fair decision. Is it really the case that penny flipping is fair? What about spinning pennies?

1. Begin spinning pennies. To spin, hold the penny upright on a table or the floor with the forefinger of one hand and flick the side edge with a finger of the other hand. The pennies should spin freely, without bumping into things before they fall. Spin pennies until your class has a total of 40 spins. Count the number of heads and compute $\hat{p}$.

2. Do you believe heads and tails are equally likely when spinning pennies, or can you reject this standard? Explain.

3. Begin flipping pennies. Let the penny fall and record whether it is heads or tails. Flip the pennies until your class has a total of 40 flips. Count the number of heads and compute $\hat{p}$.

4. Do you believe heads and tails are equally likely when flipping pennies, or can you reject this standard? Explain.

Informal Significance Testing

The logic involved in deciding whether or not to reject the standard that spinning a penny results in heads 50% of the time makes use of the same logic as that involved in estimating a proportion in Section 8.1.

Example: Jenny and Maya's Spins

Jenny and Maya wonder if heads and tails are equally likely when a penny is spun. They spin pennies 40 times and get 17 heads. Should they reject the standard that pennies fall heads 50% of the time even if heads and tails are equally likely?

Solution

Of course, Jenny and Maya shouldn't expect to get exactly 20 heads (50% of 40). There is **variation in sampling**—you don't get exactly the same result each time you take a sample from a population. Jenny and Maya consult the chart in Display 8.2. They draw a vertical line upward from their sample proportion of $\hat{p} = \frac{17}{40} = .425$, as shown in Display 8.7.

Variation in sampling means that different samples usually give different results.

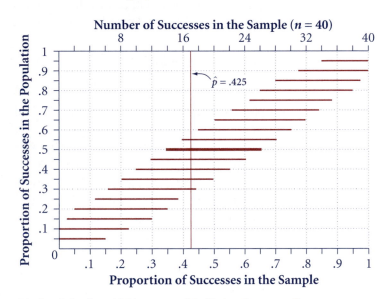

Display 8.7 $\hat{p} = .425$ is reasonably likely when $p = .5$.

The population with 50% successes is in the 95% confidence interval. This means that if the percentage of heads is actually 50% when a penny is spun, it is reasonably likely to get 17 heads out of 40 spins. The population with 50% successes is a plausible one, so Jenny and Maya don't reject the standard that a penny lands heads 50% of the time when spun. ■

Discussion: Informal Significance Testing

D20. Consider the question of whether spinning a penny is fair. You will need your penny-spinning results from Activity 8.4.

   a. When investigating whether spinning a penny is fair, what is the standard hypothesis? An alternate hypothesis?

b. What is your sample proportion $\hat{p}$ of heads from Activity 8.4?

c. Locate your sample proportion $\hat{p}$ on the chart in Display 8.7. If spinning a penny is fair, is your sample proportion $\hat{p}$ reasonably likely?

d. What is your conclusion? Do your data support the standard?

■ Practice

P17. Do the four steps in D20. Use your penny-flipping results from Activity 8.4.

Statistical Significance

A sample proportion is said to be **statistically significant** if it isn't a reasonably likely outcome when the proposed standard is true. Jenny and Maya's result of 17 heads out of 40 spins wasn't statistically significant. Their value of $\hat{p} = .425$ is a reasonably likely outcome if spinning a penny is fair.

Example: Miguel and Kevin's Spins

Miguel and Kevin also spun pennies and got 10 heads out of 40 spins for a sample proportion of $\hat{p} = .25$. Is this a statistically significant result?

Solution

As shown in Display 8.8, if .5 is the true proportion of heads when a penny is spun, it would be a rare event to get only 10 heads in a sample of 40 spins. Miguel and Kevin have a statistically significant result. This leads Miguel and Kevin to conclude that .5 is probably not the true proportion of heads when a penny is spun.

Another way of seeing this is to note that in this case, $p = .5$ isn't included in the 95% confidence interval for the true proportion of heads when a penny is spun. That is, the sample proportion $\hat{p} = .25$ is far enough from the standard of .5 to reject .5 as a plausible value for the proportion of times heads appear when a penny is spun.

> Reject a value of the standard that is not likely to have produced the sample proportion.

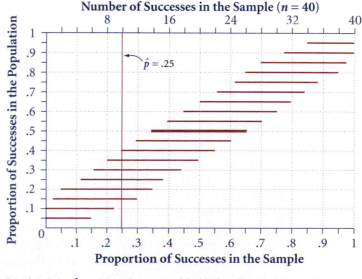

Display 8.8 $\hat{p} = .25$ isn't reasonably likely when $p = .5$. ■

When you spin a penny, there are three different proportions to keep straight:

p: the population proportion

- The true proportion of heads, p, when a penny is spun. No one knows this value for sure. You may suspect from your own results that $p \neq .5$.

$\hat{p}$: the sample proportion

- The proportion of heads, $\hat{p}$, in your sample when you spin 40 times. Jenny and Maya's sample proportion was $\hat{p} = \frac{17}{40}$, or .425. Miguel and Kevin's sample proportion was $\hat{p} = \frac{10}{40}$, or .25.

p_0: the hypothesized value of the population proportion

- The proportion of heads that you hypothesize will come up before you do the spinning. The symbol used for this standard value is p_0. You have been testing the standard that spinning a penny is fair, so you have been using $p_0 = .5$.

Discussion: Statistical Significance

D21. A 1997 article reported that two-thirds of teens in grades 7–12 want to study more about medical research. You wonder if this proportion still holds today and decide to test it. You take a random sample of 40 teens and find that only 23 want to study more about medical research.

Source: CNN Interactive Story Page, www.cnn.com/tech/9704/22/teentech.poll/, April 22, 1997.

a. What is the standard (the hypothesized value, p_0, of the population proportion)?

b. What is an alternate hypothesis?

c. What is the sample proportion, $\hat{p}$?

d. Is the result statistically significant? That is, is there evidence leading you to believe that the proportion today is different from the proportion in 1997?

■ Practice

P18. A student took a 40-question true–false test and got 30 correct. The student says, "That proves I was not guessing at the answers."

a. What is the standard, p_0?

b. What is the student's alternate hypothesis?

c. What is the sample proportion, $\hat{p}$?

d. Is the result statistically significant? Does this prove that the student was not guessing?

The Test Statistic

Miguel and Kevin could see from Display 8.8 that .50 is not one of the population proportions from which it is reasonably likely to get $\hat{p} = .25$. So they rejected the hypothesis that a spun penny comes up heads half the time. How could they have determined this without the chart?

All Miguel and Kevin need to do is to determine whether their value of $\hat{p} = .25$ is a reasonably likely value from a population with $p = .5$. That is, they need to find out if their value of $\hat{p}$ lies in the middle 95% of all possible values of $\hat{p}$.

Display 8.9 shows the distribution of $\hat{p}$ when $n = 40$ and $p = .5$.

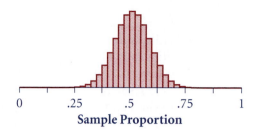

Sample Proportion

Display 8.9 The sampling distribution of $\hat{p}$ for $n = 40$ and $p = .5$.

The distribution is approximately normal, so Miguel and Kevin can compute a *z*-score for their sample proportion of $\hat{p} = .25$:

$$z = \frac{estimate\ from\ sample - mean}{standard\ error}$$

$$= \frac{\hat{p} - p_0}{\sqrt{\dfrac{p_0(1 - p_0)}{n}}}$$

$$= \frac{.25 - .5}{\sqrt{\dfrac{.5(1 - .5)}{40}}}$$

$$\approx -3.16$$

Having a sample proportion that is 3.16 standard errors below the hypothesized mean of $p_0 = .5$ would definitely be a rare event if it is true that half of spun pennies land heads. The value of $\hat{p}$ is so far from .5 that they reject the hypothesis that spinning a penny is fair.

The *z*-score that Miguel and Kevin computed is called a *test statistic*.

A *z*-score in disguise!

> To determine if $\hat{p}$ is a reasonably likely outcome or a rare event for a given standard p_0, you need to check the value of the **test statistic:**
>
> $$z = \frac{\hat{p} - p_0}{\sqrt{\dfrac{p_0(1 - p_0)}{n}}}$$
>
> This tells you how many standard errors the sample proportion $\hat{p}$ lies from the standard p_0.

Critical Values and Level of Significance

A large value of $|z|$ indicates that $\hat{p}$ is far from p_0.

You have been rejecting the standard when $\hat{p}$ would be a rare event if that standard were true. This is equivalent to rejecting the standard when the value

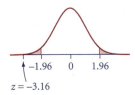

−1.96   0   1.96

z = −3.16

A result of z = −3.16 is statistically significant.

of the test statistic z is less than −2 or greater than 2, the dividing points between reasonably likely outcomes and rare events. Such dividing points are called **critical values** (denoted z^*). Other z^*-values commonly used as critical points are $z^* = \pm1.645, \pm1.96$, and ±2.576. If the value of the test statistic is more extreme than the critical values you have chosen, you reject the standard and say that the result is statistically significant.

A larger critical value makes it harder to reject the standard. For example, if you use $z^* = \pm1.96$, then to reject the standard, the test statistic z must fall in the outer 5% of the standard normal distribution. If you use $z^* = \pm1.645$, the value of z must fall in only the outer 10% of the distribution. In this way, each critical value is associated with a corresponding percentage, α (alpha), called the **level of significance.** If a level of significance isn't specified, it is usually safe to assume that $\alpha = .05$ and $z^* = \pm1.96$.

Discussion: The Test Statistic

D22. Find the test statistic from Jenny and Maya's data on spinning pennies. What do you conclude if $\alpha = .10$?

D23. Use your z-table to answer these questions.

    a. What level of significance is associated with critical values of $z^* = \pm2.576$?

    b. What critical values are associated with a level of significance of 2%?

D24. If everything else remains constant, what will happen to the test statistic z when

    a. the sample size n increases?

    b. $\hat{p}$ gets farther from p_0?

D25. Could Miguel and Kevin be wrong when they decided that spinning a penny isn't fair? Explain.

■ Practice

P19. If your standard (which is also called your null hypothesis) is that spinning a coin is fair, find the test statistic z for each of these situations and give your conclusion if $\alpha = .05$.

    a. Your results from *spinning* 40 pennies in Activity 8.4

    b. A class spun quarters 500 times and got 194 heads.

P20. If your standard (null hypothesis) is that flipping a coin is fair, find the test statistic for each of these situations and give your conclusion if $\alpha = .05$.

    a. Your results from *flipping* 40 pennies in Activity 8.4

    b. A class flipped quarters 500 times and got 265 heads.

P21. Use your z-table to answer these questions.

    a. What critical value is associated with a level of significance of .12?

    b. What level of significance is associated with critical values of $z^* = \pm1.73$?

The Formal Language of Tests of Significance

You have been using informal language to describe tests of significance. More formal terminology appears in the box. You will examine each component of a significance test thoroughly in the rest of this section.

Components of a Significance Test for a Proportion

1. **Give the name of the test and check the conditions for its use.** For a significance test for a proportion, three conditions must be met.

   - The sample is a simple random sample from a binomial population.
   - Both np_0 and $n(1 - p_0)$ are at least 10.
   - The population size is at least 10 times the sample size.

2. **State the hypotheses, defining any symbols.** When testing a proportion, the null hypothesis H_0 is

   H_0: The percentage of successes p in the population from which the sample came is equal to p_0.

   The alternate hypothesis, H_a, can be of three forms:

   H_a: The percentage of successes p in the population from which the sample came is not equal to p_0.

   H_a: The percentage of successes p in the population from which the sample came is greater than p_0.

   H_a: The percentage of successes p in the population from which the sample came is less than p_0.

   For now, we will use just the first H_a.

3. **Compute the test statistic z and compare it to the critical values z^*** (or find the *P*-value—as explained later in this section).

   The test statistic is

$$z = \frac{\hat{p} - p_0}{\sqrt{\dfrac{p_0(1 - p_0)}{n}}}$$

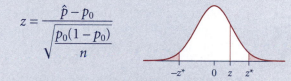

   Compare the value of z to the predetermined critical values. Include a sketch that illustrates the situation.

4. **Write a conclusion.** There are two parts to stating a conclusion.

   - Say whether you reject the null hypothesis or don't reject the null hypothesis, linking your reason to the results of your computations.
   - Tell what your conclusion means in the context of the situation.

To illustrate the components of a significance test, look again at Jenny and Maya's results from spinning a penny 40 times and getting 17 heads.

Always check conditions first.

1. Apparently Jenny and Maya spun the pennies carefully and independently, so we can consider the number of heads they got a binomial random variable. Both np_0 and $n(1 - p_0)$ are at least 10 because $np_0 = 40(.5) = 20$ and $n(1 - p_0) = 40(.5) = 20$. The population of all possible spins is infinitely large, so it definitely is at least 10 times their sample size of 40. So it is appropriate to do a significance test for a proportion.

State your hypotheses next.

2. H_0: The proportion of heads when a penny is spun is equal to .5.

H_a: The proportion of heads when a penny is spun is not equal to .5.

Or you could write,

H_0: $p = .5$, where p is the proportion of heads when a penny is spun.

H_a: $p \neq .5$.

Compute the test statistic and draw a sketch.

3. As you computed in D22, the test statistic is

$$z = \frac{\hat{p} - p_0}{\sqrt{\dfrac{p_0(1 - p_0)}{n}}}$$

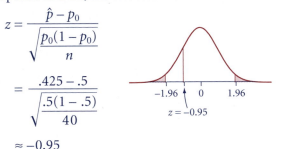

$$= \frac{.425 - .5}{\sqrt{\dfrac{.5(1 - .5)}{40}}}$$

$$\approx -0.95$$

Write a conclusion that is linked to the value of z or the P-value and is stated in the context of the situation.

4. Because $z = -0.95$ is less than one standard error from the mean and is not in the shaded critical region, this isn't a statistically significant result.

If spinning a penny results in heads 50% of the time, it is reasonably likely to get only 17 heads out of 40 flips. Thus, you cannot reject the null hypothesis that spinning a penny is a fair process. (You should never say that you "accept" the null hypothesis, because if you constructed a confidence interval for p, all of the other values in it, not just .5, are also plausible as the true value of p.)

Discussion: Formal Language of Tests of Significance

D26. Using the data in P19 that there were 194 heads out of 500 spins, carry out the four steps in a test of the null hypothesis that spinning a quarter is fair.

D27. What is the difference in the meaning of the symbols p, p_0, and $\hat{p}$? Of the three, which varies depending on the sample you select? Which of the three is unknown?

D28. A large study of births in the United States found that about 51.2% of all babies born in the United States are boys and about 48.8% are girls. This is a statistically significant difference, meaning that the difference can't reasonably be attributed to chance.

Sources: U.N. World Fertility Patterns, 1997. Joyce A. Martin, Brady E. Hamilton, Stephanie J. Ventura, Fay Menacker, and Melissa M. Park, Division of Vital Statistics, "Births: Final Data for 2000," *National Vital Statistics Reports* 50, no. 5 (Feb. 12, 2002), p. 7.

Just because a result is statistically significant doesn't mean the difference is of any practical importance.

a. Is this result of any practical significance to a pair of prospective parents who were hoping for a girl?

b. Give an example where this difference might have some practical significance.

D29. Suppose the null hypothesis is true and you have a random sample of size n.

a. What is the approximate probability that the test statistic will be greater than 1.96? What is the probability it will either be larger than 1.96 or be less than −1.96?

b. What is the probability that the test statistic will exceed 1.645? What is the probability it will either be larger than 1.645 or be less than −1.645?

c. Why do you need to know H_0 is true to answer parts a and b?

■ **Practice**

P22. Suppose in a random sample of 500 bookstores across the United States, 265 also sell CDs. Carry out the four steps in a test of the null hypothesis that half of the bookstores in the United States sell CDs.

P23. What is an appropriate null hypothesis in each of these situations?

a. You wonder if a student is guessing on a multiple-choice test of 60 questions where there are 5 possible answers for each question.

b. You wonder if there is an equal proportion of male and female newscasters on local television stations.

c. You wonder if people who wash their cars once a week are more likely to wash them on Saturday.

P24. The United States 2000 Census found that 4% of all households were multigenerational (consisting of three or more generations of parents and their children). You want to test the null hypothesis that this percentage is the same this year as it was in 2000. You write the null hypothesis as $p = .04$. Which of these statements is the best description of what p stands for?
Source: www.census.gov/press-release/www/2001/cb01cn182.html.

A. the proportion of all households that were multigenerational in 2000

B. the proportion of all households that are multigenerational this year

C. the proportion of multigenerational households in the sample in 2000

D. the proportion of multigenerational households in the sample this year

E. the proportion of all multigenerational households in both 2000 and this year

Types of Errors

The reasoning of significance tests is often compared to that of a jury trial. The possibilities in such a trial are given in this diagram.

		Defendant Is Actually	
		Innocent	**Guilty**
Jury's Decision	**Not Guilty**	Correct	Error
	Guilty	Worse error	Correct

In the same way, there are two types of errors in significance testing.

		Null Hypothesis Is Actually	
		True	**False**
Your Decision	**Don't Reject H_0**	Correct	Type II error
	Reject H_0	Type I error	Correct

Miguel and Kevin got a test statistic with a large absolute value and so concluded that spinning a penny is not fair. However, Jenny and Maya got a test statistic with a value that was close to zero; the result from their sample was quite consistent with the idea that spinning a penny is fair. Who is right? Thousands of statistics students have contributed to the solution of this problem by spinning pennies. Their instructors have reported that the true proportion of heads seems to be around .4. So Miguel and Kevin made the correct decision. Jenny and Maya have made the error of not rejecting a null hypothesis that is false. They made a **Type II error.**

If, like Miguel and Kevin, your test statistic is large in absolute value, then there are several possibilities to consider.

- The null hypothesis is true and a rare event occurred. That is, it was just bad luck that resulted in $\hat{p}$ being so far from p_0.

- The null hypothesis isn't true, and that's why the sample proportion $\hat{p}$ was so far from p_0.

- The sampling process was biased in some way, and so the value of $\hat{p}$ is itself suspicious.

If you can rule out the last possibility, then the usual decision is to reject the null hypothesis. However, you may be making a **Type I error**—rejecting H_0 even though H_0 is actually true.

If, like Jenny and Maya, your test statistic is small in absolute value, then there are several possibilities to consider.

- The null hypothesis is true, and you got just about what you would expect in the sample.

- The null hypothesis isn't true, and it was just by chance that $\hat{p}$ turned out to be close to p_0.

- The sampling process was biased in some way, and so the value of $\hat{p}$ is itself suspicious.

If you can rule out the last possibility, then the usual decision is to not reject the null hypothesis. However, you may be making a Type II error, like Jenny and Maya.

Minimizing the Probability of an Error

Suppose the null hypothesis is false. If you don't reject it, you will make a Type II error. The probability you will make a Type II error is larger when the sample size is smaller. If Jenny and Maya had spun more pennies, they soon would have realized that the proportion of heads is less than .5. Their sample size was too small for their test to have much *power*—they were unable to see that the null hypothesis was false. *To avoid a Type II error, your best strategy is to have a large sample size.*

Now suppose the null hypothesis is true. What is the chance you will reject it, making a Type I error? The only time you can make a Type I error is if you get a rare event from your sample, that is, the value of the test statistic turns out to be larger than 1.96 or smaller than −1.96. This happens only 5% of the time. Thus, if the null hypothesis is true, the probability of a Type I error is .05. If you had used a critical value of ±2.576, which has a significance level of .01, then the probability of making a Type I error would be only 1%. *If the null hypothesis is true, the probability of a Type I error is equal to the level of significance.*

To avoid a Type I error, then, your best strategy is to have a large critical value or, equivalently, a low level of significance.

The only way you can decrease the probability of making a Type I error (rejecting a true null hypothesis) is to lower the level of significance. However, if the null hypothesis is actually false, this strategy results in a higher probability of a Type II error! For example, suppose Miguel and Kevin had decided to have only a .001 chance of making a Type I error. With $\alpha = .001$, the critical values from your calculator are ±3.29. (If you are using Table A from the Appendix, you will find that −3.27 to −3.32 all cut off areas of approximately .0005.) In this case, the value of their test statistic, $z = -3.16$, is inside the critical values, so they wouldn't have rejected the false null hypothesis that spinning a penny is fair. They would have made a Type II error.

Catch-22: Attempting to decrease the chance of a Type I error can increase the chance of a Type II error.

Type I Error

Suppose the null hypothesis is true. If you reject it, you have made a Type I error. The probability of making a Type I error is equal to the significance level of the test. To decrease the probability of a Type I error, decrease the significance level. Changing the sample size has no effect on the probability of a Type I error.

Type II Error

Suppose the null hypothesis is false. If you fail to reject it, you have made a Type II error. To decrease the probability of making a Type II error, you can take a larger sample or you can increase the significance level.

(continued)

(continued)

Power of a Test

The **power** of a test is the probability of rejecting the null hypothesis. (Usually, you can't know this probability because you don't know if the null hypothesis is true or false.) If the probability of a Type II error is small, then the power of your test is large. To increase the power of your test for a fixed α, you have to take a larger sample.

Discussion: Types of Errors

D30. Jeffrey and Taline each want to test if the proportion of adults in their state who have graduated from college is .6, as claimed in the newspaper. Jeffrey takes a random sample of 200 adults and uses $\alpha = .01$. Taline takes a random sample of 200 adults and uses $\alpha = .05$. Suppose the newspaper's percentage is actually right.

　　a. Is it possible for Jeffrey or for Taline to make a Type I error? If so, who is more likely to do so?

　　b. Is it possible for Jeffrey or for Taline to make a Type II error? If so, who is more likely to do so?

D31. Jeffrey and Taline each want to test if the proportion of adults in their state who have gone to nursery school is .55, as claimed in the newspaper. Each takes a random sample of 200 adults. Jeffrey uses $\alpha = .01$. Taline uses $\alpha = .05$. Suppose the newspaper's percentage is actually wrong.

　　a. Is it possible for Jeffrey or for Taline to make a Type I error? If so, who is more likely to do so?

　　b. Is it possible for Jeffrey or for Taline to make a Type II error? If so, who is more likely to do so?

D32. Suppose you use a critical value of ± 2.576 (so $\alpha = .01$) and the null hypothesis is actually true. What is the probability that you will get a sample that results in rejecting the null hypothesis? Explain. What type of error is this?

D33. In order to avoid Type I errors, why not always use a very large critical value?

■ Practice

P25. Hila is rolling a pair of dice to test if they land doubles $\frac{1}{6}$ of the time. She doesn't know it, but the dice are actually fair. She will use a significance level of $\alpha = .05$. Hila rolls the dice 100 times and gets doubles 22 times.

　　a. What conclusion should Hila come to?

　　b. Did Hila make an error? If so, which type?

P26. Jeffrey and Taline each want to test if the proportion of adults in their neighborhood who have graduated from high school is .95, as claimed in the newspaper. Jeffrey takes a random sample of 200 adults and uses $\alpha = .05$. Taline takes a random sample of 500 adults and uses $\alpha = .05$. Suppose the newspaper's percentage is actually right.

a. Is it possible for Jeffrey or Taline to make a Type I error? If so, who is more likely to do so?

b. Is it possible for Jeffrey or Taline to make a Type II error? If so, who is more likely to do so?

P27. Jeffrey and Taline each want to test if the proportion of adults in their neighborhood who took chemistry in high school is .25, as claimed in the newspaper. Jeffrey takes a random sample of 200 adults and uses $\alpha = .05$. Taline takes a random sample of 500 adults and uses $\alpha = .05$. Suppose the newspaper's percentage is actually wrong.

a. Is it possible for Jeffrey or Taline to make a Type I error? If so, who is more likely to do so?

b. Is it possible for Jeffrey or Taline to make a Type II error? If so, who is more likely to do so?

P-Values

Instead of just reporting that you either have or have not rejected the null hypothesis, it has become common practice also to report a *P*-value.

> The **P-value** for a test is the probability of seeing a result from a random sample that is as extreme as or more extreme than the one you got from your random sample *if the null hypothesis is true.*

Example: P-Value for Jenny and Maya

Find the *P*-value for Jenny and Maya's test that spinning a penny is fair.

Solution

Jenny and Maya got 17 heads out of 40 spins, so the value of the test statistic was

$$z = \frac{\hat{p} - p_0}{\sqrt{\dfrac{p_0(1 - p_0)}{n}}} = \frac{.425 - .5}{\sqrt{\dfrac{.5(1 - .5)}{40}}} \approx -0.95$$

Assume for now that the null hypothesis is true: Spinning a penny gives you heads 50% of the time. The probability of getting a value of z less than -0.95 can be found in Table A in the Appendix or from a calculator. This value is approximately .171, as shown in Display 8.10.

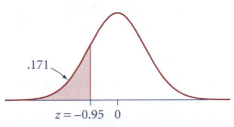

.171

$z = -0.95$ 0

Display 8.10   The probability of 17 or fewer heads with 40 spins of a coin if spinning a coin is fair.

However, Jenny and Maya notice that getting 23 or more heads is just as extreme as getting 17 heads or fewer. Now their sketch looks like Display 8.11. The probability of getting a result as far out in the tails of the distribution as they did is $2(.171) = .342$. This is their *P*-value. *If the probability of getting heads is .5, there is a .342 chance of getting 17 heads or fewer or 23 heads or more in 40 spins.*

The phrase in italics is crucial in describing the meaning of a *P*-value.

Always draw a picture of the *P*-value.

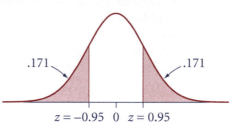

.171   .171

$z = -0.95 \quad 0 \quad z = 0.95$

Display 8.11 The probability of at most 17 or at least 23 heads with 40 spins of a coin, if spinning a coin is fair. ■

Discussion: *P*-Values

D34. Suppose that 22 students out of a random sample of 40 students carry a backpack to school. Follow steps a–d to test the claim that exactly half of the students in the school carry backpacks to class.

    a. Name the test and check the conditions needed for it.

    b. State the hypotheses in words and symbols.

    c. Calculate the value of the test statistic. Calculate the *P*-value for the test. Use this *P*-value in a sentence that explains what it represents.

    d. What is your conclusion? Explain in the context of this problem.

D35. Which of statements A–E is the best explanation of what is meant by the *P*-value of a test of significance?

    A. Assuming that you had a random sample and the other conditions for a significance test are met, it is the probability that H_0 is true.

    B. Assuming that H_0 is true, it is the probability of observing a value of a test statistic at least as far out in the tails of the sampling distribution as the *z* from your sample.

    C. It is the probability that H_0 is false.

    D. Assuming that the sampling distribution is normal, it is the probability that H_0 is true.

    E. Assuming that H_0 is true, it is the probability of observing the same value of *z* that you got in your sample.

■ Practice

P28. Follow the four steps of D34 for a random sample of 40 students, of which 28 students carry backpacks to class.

P29. Find the *P*-value for Miguel and Kevin's test statistic of −3.16. Write a sentence explaining what this *P*-value means in the context of their situation.

P30. At the beginning of this section, you read that 2% of barn swallows have white feathers in places where the plumage is normally blue or red, but about 14% of the barn swallows captured around Chernobyl in 1991 and 1996 had such genetic mutations. Whether the researchers believe the difference can reasonably be attributed to chance or whether the radioactivity is associated with the mutations depends on how many swallows were examined. That sample size was not reported in the article. Suppose that the sample size was 500. What is the *P*-value? What should the researchers say to the press?

One-Tailed Tests of Significance

When testing the effectiveness of a new drug, the investigator must establish that the new drug has a *better* cure rate than the older treatment (or that there are *fewer* side effects). He or she isn't interested in simply rejecting the null hypothesis that the new drug has the same cure rate as the older treatment. He or she needs to know if it is *better*. In such situations, the alternate hypothesis should state that the new drug cures a larger proportion of people than does the older treatment. This is called a **one-tailed test of significance.** Tests of significance can be one-tailed if the investigator has an indication of which way any change from the standard should go. This must be decided before looking at the data.

The tests of significance discussed previously were two-tailed (or two-sided); the investigator was interested in detecting a change from the standard in either direction. For example, Jenny and Maya were testing to see if the percentage of heads when a penny is spun is higher or lower than .5.

When testing a proportion, the alternate hypothesis can take one of three forms.

H_a: The percentage of successes p in the population from which the sample came is not p_0. Or, in symbols, $p \neq p_0$.

H_a: The percentage of successes p in the population from which the sample came is greater than p_0. Or $p > p_0$.

H_a: The percentage of successes p in the population from which the sample came is less than p_0. Or $p < p_0$.

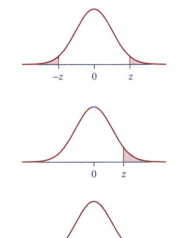

The first alternate hypothesis is for a two-tailed test; the latter two are for one-tailed tests. For a one-tailed test, the *P*-value is found on one side only.

Example: One-Sided Test of Significance

The editors of a magazine have noticed that people seem to believe that a successful life depends on having good friends. They would like to have a story about this and use a headline such as "Most adults believe friends are important for success." So they commissioned a survey to ask a random sample of adults whether a successful life depends on having good friends. In a random sample of 1027 adults, 53% said yes. Should the editors go ahead and use their headline?

Solution

As always, the answer must include statistical evidence, in this case a test of significance. You will use $p_0 = .5$ because "most" means greater than .5.

The conditions are the same for a one-tailed test as for a two-tailed test.

1. The sample was a simple random sample of adults in the United States. Both $np_0 = 1027(.5) = 513.5$ and $n(1 - p_0) = 1027(.5) = 513.5$ are at least 10. The population of all adults in the United States is much larger than $10(1027) = 10{,}270$.

2. The claim (the alternate hypothesis) is that "most" adults believe this. This is the same as "more than 50%." Let p represent the proportion of all adults in the United States who believe that a successful life depends on having good friends. Then the null hypothesis is that the percentage who believe this is 50% (or less):

The null hypothesis also is the same for a one-tailed test as for a two-tailed test.

$$H_0:\ p = .5$$

The alternate hypothesis is different for a one-tailed test and a two-tailed test.

The alternate hypothesis is that this percentage is more than 50%:

$$H_a:\ p > .5$$

The test statistic is the same for a one-tailed test as for a two-tailed test.

3. The test statistic is

$$z = \frac{\hat{p} - p_0}{\sqrt{\dfrac{p_0(1 - p_0)}{n}}} = \frac{.53 - .5}{\sqrt{\dfrac{.5(1 - .5)}{1027}}} \approx 1.92$$

The P-value will be half as large for a one-tailed test as it is for a two-tailed test.

As shown in Display 8.12, the *P*-value is the probability of getting a test statistic greater than 1.92 if the null hypothesis is true. This probability is .0274.

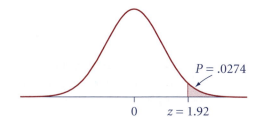

$$P = .0274$$

$$0 \qquad z = 1.92$$

Display 8.12 The *P*-value for the successful life/friends problem.

Conclusion in context with link to computations

4. The *P*-value is fairly small. If the percentage of all adults who believe a successful life depends on having good friends is 50% or less, then the probability of getting a sample proportion of 53% or more is only .0274.

This is unlikely. Thus, you should reject the null hypothesis. This is quite strong evidence that the true percentage must be greater than 50%.

The editors should feel free to use their headline. ∎

Discussion: One- and Two-Sided Tests

D36. What would have been the conclusion if 40% of the sample in the preceding example believed that friends are important for success?

D37. A psychologist was struck by the fact that at square tables many pairs of students sat on adjacent sides rather than across from one another and wondered if people prefer to sit that way. The psychologist collected some data, observing 50 pairs of students seated in the student cafeteria of a California university. The tables were square and had one seat available on each of the four sides. The psychologist observed that 35 pairs sat on adjacent sides of the table, while 15 sat across from one another.

Source: Joel E. Cohen, "Turning the Tables," in *Statistics by Example: Exploring Data*, edited by Frederick Mosteller et al. (Reading, Mass.: Addison-Wesley, 1973), pp. 87–90.

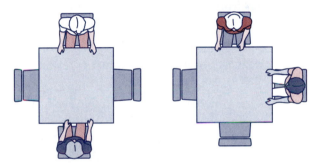

a. What is the probability that if two students sit down randomly at a square table, they sit on adjacent sides?

b. What is an appropriate null hypothesis to test whether students have a preference?

c. What is the psychologist's alternate hypothesis?

d. Does this situation meet the conditions for a significance test?

e. What is the test statistic?

f. Find the *P*-value for this test.

g. Write a conclusion for the psychologist.

D38. If everything else remains the same, is it easier to reject a false null hypothesis with a one-sided or with a two-sided test?

∎ Practice

P31. You claim that the percentage of teens in your community who know that epilepsy is not contagious is larger than the national percentage of 51%. (See E2.) You take a random sample of 169 teens in your community and find that 55% know epilepsy is not contagious. Carry out the steps of a test of your claim.

P32. Determine whether each statement is true or false.

   a. Always look at the data before writing the null and alternate hypotheses.

   b. All else being equal, using a one-sided test will result in a larger *P*-value than using a two-sided test.

   c. The *P*-value is the probability that the null hypothesis is true.

   d. A statistically significant result means the *P*-value is "small."

Summary 8.2: Tests of Significance

Suppose you want to decide if it is reasonable to assume that your random sample comes from a population with proportion of successes, *p*. A test of significance for a proportion *p* tells you whether the results from your sample are so different from what you would expect that you should reject that value of *p* as plausible. If the sample proportion $\hat{p}$ is relatively close to *p*, you can reasonably attribute the difference to chance variation and you should not reject *p* as the possible proportion of successes in the population. If the sample proportion $\hat{p}$ is relatively far from *p*, you cannot reasonably attribute the difference to chance variation and you should reject *p* as the possible proportion of successes in the population.

The four components of a test of significance are listed here, with some suggestions about how to interpret each one.

- *Checking conditions.* In real surveys, it is almost always the case that the sample was not a simple random sample from the population. If the sample comes from a more complicated design that uses randomization, inference is possible. In other cases, convenience samples are used—people are chosen for a sample because they were easy to find. In such cases, confidence intervals and significance tests should not be used.

- *Writing hypotheses.* Remember that when you write the hypotheses, you are not supposed to have seen the sample yet. Thus, the value from the sample, $\hat{p}$, does not appear in the null hypothesis.

  If you use the symbol *p* in the null hypothesis, always say in words what it represents in the context of the problem.

  Using a one-sided alternate hypothesis makes it easier to reject a false null hypothesis—if you have the correct side. But you can use a one-sided alternate hypothesis only if, before looking at the sample, you have an indication of which direction from the standard the true proportion falls. When in doubt about which type of test to use, use a two-tailed test.

- *Doing computations.* Test statistics typically come in this form:

$$\frac{\textit{estimate from the sample} - \textit{standard value from } H_0}{\textit{standard error of the sampling distribution if } H_0 \textit{ is true}}$$

In the case of testing a proportion, this becomes

$$z = \frac{\hat{p} - p_0}{\sqrt{\dfrac{p_0(1 - p_0)}{n}}}$$

- *Writing a conclusion.* If the *P*-value is small (smaller than some significance level α that you have decided on in advance), this is evidence against the null hypothesis and so you should reject it. If it is larger than the predetermined value α, do not reject the null hypothesis.

If you reject the null hypothesis, you are saying that the difference between $\hat{p}$ and p_0 is so large that you can't reasonably attribute it to chance variation. If you don't reject the null hypothesis, you are saying that the difference between $\hat{p}$ and p_0 is small enough so that it looks like the kind of variation you would expect from a binomial situation where the proportion of successes is p_0.

Always write your conclusion in the context of the situation. Do not simply write "Reject the null hypothesis" or "Do not reject the null hypothesis." Explain what this means in terms of the context of the problem.

You should never write "Accept the null hypothesis." The reason is that the sample can tell you only if the null hypothesis is reasonable or not. If you get a very small *P*-value, then it isn't reasonable to assume that the null hypothesis is true and you should reject it. However, if you get a large *P*-value, then the null hypothesis is plausible, but it still may not give the exact value of *p*.

Finally, just because a result is statistically significant doesn't mean the difference is of any practical importance in the real world. In other words, the difference may exist but may be so small that it doesn't have an impact.

■ Exercises

E14. A random sample of dogs was checked to see how many wore a collar. The 95% confidence interval for the percentage of all dogs that wear a collar turned out to be .82 to .96. Which of the following is *not* a true statement?

A. You can reject the hypothesis that 75% of all dogs wear a collar.

B. You cannot reject the hypothesis that 90% of all dogs wear a collar.

C. If 90% of all dogs wear a collar, then you are reasonably likely to get a result like the one from this sample.

D. If 75% of all dogs wear a collar, then you are reasonably likely to get a result like the one from this sample.

E. The true proportion of dogs that wear a collar may or may not be between .82 and .96.

E15. A Gallup poll asked 506 adult Americans this question: "Increased efforts by business and industry to reduce air pollution might lead to higher prices for the things consumers buy. Would you be willing to pay $500 more each year in higher prices so that industry could reduce air pollution, or not?" The report said, "Almost two-thirds (63%) claim they would be willing to pay $500 more per year for this purpose." Use a test of significance to determine if it is plausible that if Gallup had asked all Americans, exactly two-thirds of them would have said they are willing to pay $500 more per year for this purpose.
Source: www.gallup.com/poll/releases/pr010316.asp, March 16, 2001.

E16. In March 2001, a Gallup poll asked, "How would you rate the overall quality of the environment in this country today—as excellent, good, only fair, or poor?" Of

1060 adults nationwide, 46% gave a rating of excellent or good. Use a test of significance to determine if this is convincing evidence that fewer than half of the nation's adults would give a rating of excellent or good. Use $\alpha = .01$.

Source: www.gallup.com/poll/indicators/indenvironment.asp.

E17. The Gallup Organization asked 1003 adults to answer the question "If you could pick only one specific problem for science to solve in the next 25 years, what would it be?" Thirty percent responded with a cure for cancer. Use a test of significance to determine if it is plausible that if you could ask all adults in the United States this question, at most 25% would pick a cure for cancer.

Source: www.publicagenda.org/issues/pcc_detail.cfm?issue_type=medical_research&list=1, June 2001.

E18. Suppose that when flipping a penny, the probability of getting heads is .5. Suppose 100 people do a test of significance with this penny. Each person plans to reject the null hypothesis that $p = .5$ if their *P*-value is less than .05. How many of these people do you expect to make a Type I error?

E19. It has been said that tests of significance are meaningless because if you get a large enough sample size, you can reject any null hypothesis. Explain why this is the case.

E20. A friend wants to test whether spinning pennies results in heads 50% of the time by doing 20 spins. Determine whether this test has much power. (You know that the probability really is about .4.) What should you suggest to your friend?

E21. Suppose that 42 out of 80 randomly selected students prefer hamburgers to hot dogs and 38 prefer hot dogs to hamburgers.

a. Test the null hypothesis that the percentage of all students who prefer hamburgers is .55.

b. Test the null hypothesis that the percentage of all students who prefer hot dogs is .55.

c. How can you reconcile your conclusions to parts a and b?

Now suppose that you go out and get a larger random sample of 800 students. Suppose that 420 of them prefer hamburgers to hot dogs and 380 prefer hot dogs to hamburgers.

d. Test the null hypothesis that the percentage of all students who prefer hamburgers is .55.

e. Test the null hypothesis that the percentage of all students who prefer hot dogs is .55.

f. How can you reconcile your conclusions in parts a, b, d, and e?

E22. **Investigative task.** In an election between two men, does the taller man tend to win more often? Display 8.13 gives the results of the U.S. presidential elections in the 20th century.

a. Display these data graphically.

b. Do these data meet the conditions of a significance test?

c. If you write a slightly different type of null hypothesis, you can proceed with a test of significance and write a reasonable conclusion. Begin by writing a null hypothesis to test whether a sequence of coin flips is likely to result in as many heads as there were times when the taller man won. That is, you will test whether the fact that the taller man has won more often can be reasonably attributed to chance. If so, there is nothing to explain. If not, an explanation is called for. See Display 8.13 on next page.

Year	Winning Candidate Candidate	Height	Runner-Up Candidate	Height	Year	Winning Candidate Candidate	Height	Runner-Up Candidate	Height
1900	McKinley	5'7"	Bryan	6'0"	1952	Eisenhower	5'10.5"	Stevenson	5'10"
1904	T. Roosevelt	5'10"	Parker	6'0"	1956	Eisenhower	5'10.5"	Stevenson	5'10"
1908	Taft	6'0"	Bryan	6'0"	1960	Kennedy	6'0"	Nixon	5'11.5"
1912	Wilson	5'11"	T. Roosevelt	5'10"	1964	Johnson	6'3"	Goldwater	6'0"
1916	Wilson	5'11"	Hughes	5'11"	1968	Nixon	5'11.5"	Humphrey	5'11"
1920	Harding	6'0"	Cox	NA	1972	Nixon	5'11.5"	McGovern	6'1"
1924	Coolidge	5'10"	Davis	6'0"	1976	Carter	5'9.5"	Ford	6'0"
1928	Hoover	5'11"	Smith	NA	1980	Reagan	6'1"	Carter	5'9.5"
1932	F. Roosevelt	6'2"	Hoover	5'11"	1984	Reagan	6'1"	Mondale	5'10"
1936	F. Roosevelt	6'2"	Landon	5'8"	1988	Bush	6'2"	Dukakis	5'8"
1940	F. Roosevelt	6'2"	Wilkie	6'1"	1992	Clinton	6'2"	Bush	6'2"
1944	F. Roosevelt	6'2"	Dewey	5'8"	1996	Clinton	6'2"	Dole	6'2"
1948	Truman	5'9"	Dewey	5'8"	2000	Bush	5'11"	Gore	6'1"

Display 8.13   Heights of presidential candidates.

Source: Paul M. Sommers, "Presidential Candidates Who Measure Up," *Chance* 9, no. 3 (1996): 29–31.

8.3   A Confidence Interval for the Difference of Two Proportions

In Sections 8.1 and 8.2, you estimated the proportion of successes in a single population. The more common and important situation involves taking two samples independently from two different populations with the goal of estimating the size of the difference between the proportion of successes in one population and the proportion of successes in the other.

Many decisions are based on comparisons.

A February 2001 poll of 1016 U.S. households found that 45% owned a dog. The percentage five years earlier was 43%. [Source: www.gallup.com/poll/releases/pr010307.asp.] The two populations are the households in the United States in 1996 and those in 2001. The question you will investigate is "What was the change in the percentage of U.S. households that own a dog?" The obvious answer is that the percentage increased by 2 percentage points. But this is only an estimate because 2% is the difference of two *sample* percentages and they probably are not exactly equal to the *population* percentages. In this section, you will learn how to find a margin of error for the 2%.

A confidence interval for the difference of two proportions $p_1 - p_2$, where p_1 is the proportion of successes in the first population and p_2 is the proportion of successes in the second population, has this familiar form:

$$(\hat{p}_1 - \hat{p}_2) \pm z^* \cdot (standard\ error\ of\ \hat{p}_1 - \hat{p}_2)$$

Here, $\hat{p}_1$ and $\hat{p}_2$ are the proportions of successes in the two samples. Substituting in what you have so far, a 95% confidence interval for the difference between the proportion of U.S. households that owned dogs in 2001 and the proportion in 1996 is

$$.02 \pm 1.96 \cdot (standard\ error\ of\ \hat{p}_1 - \hat{p}_2)$$

The hard part, as always, is estimating the size of the standard error. You will first see how to estimate how much $\hat{p}_1 - \hat{p}_2$ tends to vary from $p_1 - p_2$ using simulation, and then you'll develop a formula.

A Sampling Distribution for the Difference of Two Proportions

Like all statistics computed from samples, the difference of two sample proportions, $\hat{p}_1 - \hat{p}_2$, has a sampling distribution. In Activity 8.5, you will construct a sampling distribution for the difference of two proportions.

Pouring out the first 20 M&M's in a bag gives you a random sample. So does pouring out the next 20.

You will be sampling from the population of plain M&M's® (now called "Milk Chocolate"), which are 30% brown, and from the population of peanut M&M's, which are 20% brown. Because bags of M&M's are filled from a huge vat with the colors already mixed in designated proportions, you may consider any sample that you pour from a bag of M&M's a random sample from the population of all M&M's of that type.

Activity 8.5 Brown Plain and Peanut M&M's

What you'll need: a large bag of plain M&M's, a large bag of peanut M&M's

1. As a class, pour a sample of 20 M&M's out of the bag of plain M&M's and find the proportion that are brown.

2. Pour a sample of 20 M&M's out of the bag of peanut M&M's and find the proportion that are brown.

3. Subtract the proportion in step 2 from the proportion in step 1.

4. When the bags are passed to you and your partner, repeat steps 1 through 3 until your class has at least 100 differences $\hat{p}_1 - \hat{p}_2$. When you run out of M&M's, you may use your calculator or computer to simulate the remaining differences.

5. Construct a histogram or dot plot of your approximate sampling distribution of $\hat{p}_1 - \hat{p}_2$. Estimate its mean and standard error. What is the shape?

6. What differences are you reasonably likely to get?

Discussion: Simulating the Sampling Distribution for the Difference of Two Proportions

D39. Now suppose your two populations are two bags of plain M&M's, each of which has 30% browns. You take a sample of size 40 from each bag and compute $\hat{p}_1 - \hat{p}_2$. Display 8.14 is a typical simulated sampling distribution for 200 trials.

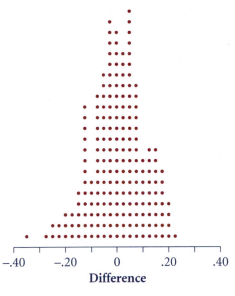

Display 8.14   A simulated sampling distribution for $\hat{p}_1 - \hat{p}_2$,
where $p_1 = p_2 = .3$ and $n = 40$.

a. Estimate the mean and standard error of this distribution. What is its shape?

b. What differences are reasonably likely to occur?

c. How does this distribution compare to the one you got in Activity 8.5?

■ Practice

P33. Use your calculator to construct an approximate sampling distribution of $\hat{p}_1 - \hat{p}_2$, where $\hat{p}_1$ is computed from a sample of size 30 taken from a population with 50% successes and $\hat{p}_2$ is computed from a sample of size 25 taken from a population with 60% successes.

Conduct 200 trials. What are the shape, mean, and standard error of this distribution? What differences are you reasonably likely to get?

The Formula for the Confidence Interval

From your simulations, you should have observed that the mean of the sampling distribution of $\hat{p}_1 - \hat{p}_2$ falls at $p_1 - p_2$ and that the shape is approximately normal if the sample sizes are large enough. You still need a formula for the standard error.

The standard error of $\hat{p}_1 - \hat{p}_2$

To find the standard error, you can use the result from Section 5.5, which says that the variance of the difference of two *independent* random variables is equal to the sum of their two variances:

$$\sigma^2_{\hat{p}_1-\hat{p}_2} = \sigma^2_{\hat{p}_1} + \sigma^2_{\hat{p}_2}$$

Equivalently, the standard deviation is

$$\sigma_{\hat{p}_1 - \hat{p}_2} = \sqrt{\sigma_{\hat{p}_1}^2 + \sigma_{\hat{p}_2}^2}$$

From Section 5.4, the standard error of the distribution of the sample proportion $\hat{p}_1$ is

$$\sigma_{\hat{p}_1} = \sqrt{\frac{p_1(1 - p_1)}{n_1}} \quad \text{which can be estimated by} \quad \sqrt{\frac{\hat{p}_1(1 - \hat{p}_1)}{n_1}}$$

where n_1 is the sample size. Similarly, the standard error of the distribution of the sample proportion $\hat{p}_2$ is

$$\sigma_{\hat{p}_2} = \sqrt{\frac{p_2(1 - p_2)}{n_2}} \quad \text{which can be estimated by} \quad \sqrt{\frac{\hat{p}_2(1 - \hat{p}_2)}{n_2}}$$

where n_2 is the sample size.

Substituting these into the formula for the standard deviation of a difference, the standard error of the difference of two proportions is

$$\sigma_{\hat{p}_1 - \hat{p}_2} \quad \text{which can be estimated by} \quad \sqrt{\frac{\hat{p}_1(1 - \hat{p}_1)}{n_1} + \frac{\hat{p}_2(1 - \hat{p}_2)}{n_2}}$$

Now you can write the complete confidence interval for the difference between two proportions.

Confidence Interval for the Difference of Two Proportions

The confidence interval for the difference $p_1 - p_2$ of the proportion of successes in one population and that of a second population is

$$(\hat{p}_1 - \hat{p}_2) \pm z^* \cdot \sqrt{\frac{\hat{p}_1(1 - \hat{p}_1)}{n_1} + \frac{\hat{p}_2(1 - \hat{p}_2)}{n_2}}$$

Here, $\hat{p}_1$ is the proportion of successes in a random sample of size n_1 taken from the first population and $\hat{p}_2$ is the proportion of successes in a random sample of size n_2 taken from the second population. (The sample sizes don't have to be equal.)

The conditions that must be met in order to use this formula are

- the two samples are taken randomly and independently from two populations
- each population is at least 10 times as large as its sample size
- $n_1 \hat{p}_1$, $n_1(1 - \hat{p}_1)$, $n_2 \hat{p}_2$, and $n_2(1 - \hat{p}_2)$ are all 5 or more

Example: A Difference in Dog Ownership?

Find a 95% confidence interval for the difference between the proportion of U.S. households that owned a dog in 2001 and the proportion of U.S. households that owned a dog in 1996.

Solution

The sample proportions for 2001 and 1996 are $\hat{p}_1 = .45$ and $\hat{p}_2 = .43$, and the sample size in 2001 was given as 1016. We'll assume that was the sample size in 1996 also.

Check conditions.

The two samples may be considered random samples. They were taken independently from the population of U.S. households in 2001 and in 1996. The number of U.S. households in each year is larger than 10 times 1016. Finally, each of

$$n_1\hat{p}_1 = 1016(.45) = 457.2, \qquad n_1(1 - \hat{p}_1) = 1016(.55) = 558.8,$$

$$n_2\hat{p}_2 = 1016(.43) = 436.88, \qquad n_2(1 - \hat{p}_2) = 1016(.57) = 579.12$$

is more than 5.

Do computations.

The 95% confidence interval for the difference of the two population proportions, p_1 and p_2, is

$$(\hat{p}_1 - \hat{p}_2) \pm z^* \cdot \sqrt{\frac{\hat{p}_1(1 - \hat{p}_1)}{n_1} + \frac{\hat{p}_2(1 - \hat{p}_2)}{n_2}}$$

$$= (.45 - .43) \pm 1.96 \sqrt{\frac{(.45)(1 - .45)}{1016} + \frac{(.43)(1 - .43)}{1016}}$$

$$= .02 \pm .043$$

Alternatively, you can write this confidence interval as $(-.023, .063)$.

Write your interpretation in context.

You are 95% confident that the difference of the 2001 and 1996 rates of dog ownership is between $-.023$ and $.063$. This means that it is plausible that 2.3% fewer households owned dogs in 2001 than in 1996, and it is also plausible that 6.3% more households owned dogs in 2001 than in 1996. Note that a difference of 0 lies within the confidence interval. This means that if the difference in the proportion of dog owners in 1996 and in 2001 is actually 0, getting a difference of .02 in the samples is reasonably likely. Thus, you aren't convinced that there was a change in the percentage of households that own a dog. ■

Discussion: The Formula for the Confidence Interval

D40. Thirty percent of plain M&M's are brown, whereas 20% of peanut M&M's are brown.

a. Use the formula to find the standard error of the distribution of $\hat{p}_1 - \hat{p}_2$, where $\hat{p}_1$ is the proportion of brown M&M's in a sample of size 20 of plain M&M's and $\hat{p}_2$ is the proportion of brown M&M's in a sample of size 20 of peanut M&M's.

b. Compare your result in part a with your result from Activity 8.5, step 5.

D41. What does it suggest if the confidence interval for the difference of two proportions

a. includes zero?

b. does not include zero?

D42. Statements I and II are interpretations of a 95% confidence interval for a single proportion. Write similar statements for a 95% confidence interval for a difference of two proportions.

I. A 95% confidence interval consists of those population proportions p for which the proportion from the sample, $\hat{p}$, is reasonably likely to occur.

II. If you construct one hundred 95% confidence intervals, you expect that the population proportion p will be in 95 of them.

■ Practice

P34. The UCLA Center for Communication Policy found that 66.9% of Americans used the Internet in 2000 and 72.3% used the Internet in 2001. Assume the samples were independently and randomly selected and that the sample size was 2006 in both years. Find a 99% confidence interval for this situation. Include an interpretation of this confidence interval.

Source: *Los Angeles Times*, November 29, 2001, pp. T1 and T8.

P35. What is the effect of an increase in the sample sizes on the length of a confidence interval for the difference of two proportions?

Experiments to Compare Two Treatments

The example comparing dog ownership in 1996 and 2001 is typical of surveys: Independent random samples are taken from two different populations. Medical experiments are an equally important setting for inference about the difference of two proportions. For example: What is the difference when AIDS-related complex (ARC) patients are given the drug AZT combined with acyclovir rather than AZT alone? As you will see, with experiments the conditions are different, but the form of the confidence interval is the same.

From 1986 to 1988, an important study on the treatment of patients with ARC was run by a collaborative group of researchers from eight European countries and Australia. (AIDS-related complex is a condition that generally leads to a diagnosis of AIDS.) A total of 134 patients with ARC agreed to participate in the study, which was a double-blind, randomized clinical trial. Two treatments were to be compared. Each patient was given either zidovudine (commonly known as AZT) by itself or a combination of AZT and acyclovir (ACV). One of the outcome measures was the number of ARC patients who developed AIDS during the one year of the study (Display 8.15).

Treated With

		AZT	AZT + ACV	Total
Developed AIDS?	**Yes**	12	10	22
	No	55	57	112
	Total	67	67	134

Display 8.15 ARC patients progressing to AIDS during the study.

Source: David A. Cooper, et al., "The efficacy and safety of zidovudine alone or as cotherapy with acyclovir for the treatment of patients with AIDS and AIDS-related complex: a double blind, randomized trial," *AIDS* 7 (1993): 197–207.

The sample of ARC patients weren't randomly selected from the population of all ARC patients. They were volunteers who had to give "informed consent." Thus, they might be quite different from the population of all ARC patients and should not be regarded as a random sample from any specific population.

Don't we have to have a random sample?

However, the ARC patients who were available were assigned randomly to the treatments. This type of randomization is typical in medical research, agricultural research, and many other kinds of experiments where getting a simple random sample from the entire population of interest is impossible. Although this type of randomization does not satisfy the conditions for constructing a confidence interval for the difference of two proportions, statisticians have shown that the formula

$$(\hat{p}_1 - \hat{p}_2) \pm z^* \cdot \sqrt{\frac{\hat{p}_1(1 - \hat{p}_1)}{n_1} + \frac{\hat{p}_2(1 - \hat{p}_2)}{n_2}}$$

gives close-to-optimal results. This means, for example, that you can expect ninety-five out of every hundred 95% confidence intervals to capture the true difference due to the treatments. But because you will consider the group of 134 volunteers as the entire population, you have to make a few changes in your wording of the question and interpretation of the confidence interval, as shown in the next example.

Example: Confidence Interval for an Experiment

During the life of the study, 12 of 67 (or 17.9%) patients using AZT progressed to AIDS and 10 of 67 (or 14.9%) patients using AZT + ACV progressed to AIDS. Find a 95% confidence interval for the difference between p_1, the proportion of patients *in this study* who would have developed AIDS if they all had taken the AZT treatment, and p_2, the proportion who would have developed AIDS if they all had taken the AZT + ACV treatment.

Solution

Check conditions.

The first condition that needs to be met by an experiment is that the subjects were randomly assigned to treatments. Second, $n_1 \hat{p}_1 = 12$, $n_1(1 - \hat{p}_1) = 55$, $n_2\hat{p}_2 = 10$, and $n_2(1 - \hat{p}_2) = 57$ are all 5 or more. So both conditions are met here.

Do computations.

The 95% confidence interval for the difference between the two proportions p_1 and p_2 is

$$(\hat{p}_1 - \hat{p}_2) \pm z^* \cdot \sqrt{\frac{\hat{p}_1(1 - \hat{p}_1)}{n_1} + \frac{\hat{p}_2(1 - \hat{p}_2)}{n_2}}$$

$$= (.179 - .149) \pm 1.96 \sqrt{\frac{(.179)(.821)}{67} + \frac{(.149)(.851)}{67}}$$

$$= .03 \pm .125$$

You also can write this confidence interval as $(-.095, .155)$.

Write your interpretation in context.

If all of these patients could have been given each treatment, we estimate that the difference between the rate of progression to AIDS if all had been given the AZT alone and the rate of progression if all had been given AZT + ACV would be someplace between a 9.5% difference in favor of AZT alone to a 15.5% difference in favor of AZT + ACV. Note that a difference of 0 lies well within the confidence interval. This means that if the difference in the proportions of patients who would progress to AIDS is actually 0, getting a difference of .03 in our samples is reasonably likely. Thus, we aren't convinced that the two therapies differ much with respect to the rate of progression to AIDS in this group of patients. ■

Discussion: A Medical Experiment to Compare Two Treatments

D43. The ARC data came from a "double-blind, randomized clinical trial." What is the meaning of each part of this phrase?

D44. Suppose the difference in the proportion of patients who improve with Treatment A and the proportion of patients who improve with Treatment B is .05. Does this mean that the majority of patients improve with Treatment A? Does this mean that Treatment B isn't very good?

■ Practice

P36. The landmark study of the effect of low-dose aspirin on the incidence of heart attacks checked to see if the aspirin use was associated with increased risk of ulcers. This was a randomized, double-blind, placebo-controlled clinical trial. Male physician volunteers with no previous important health problems were randomly assigned to an experimental group ($n_1 = 11,037$) or a control group ($n_2 = 11,034$). Those in the experimental group were asked to take a pill containing 325 mg of aspirin every second day, and those in the control group were asked to take a placebo pill every second day. Of those who took the aspirin, 169 got an ulcer, compared to 138 in the placebo group. Find and interpret a 95% confidence interval for the difference of two proportions.

Source: The Steering Committee of the Physicians' Health Study Research Group, "The Final Report on the Aspirin Component of the Ongoing Physicians' Health Study," *New England Journal of Medicine* 321, no. 3 (1989): 129–135; www.content.nejm.org/cgi/content/abstract/321/3/129.

Summary 8.3: Confidence Intervals for a Difference in Proportions

To get a simulated sampling distribution of the difference $\hat{p}_1 - \hat{p}_2$, you

- take random samples from two different populations with proportions of success p_1 and p_2
- compute the proportion of successes in each sample, $\hat{p}_1$ and $\hat{p}_2$
- subtract the proportions of successes in the samples, $\hat{p}_1 - \hat{p}_2$
- repeat this process many times

When the sample sizes, n_1 and n_2, are large enough, the sampling distribution is approximately normal for all sample sizes, it is centered at $p_1 - p_2$, and has standard error

$$\sigma_{\hat{p}_1 - \hat{p}_2} = \sqrt{\frac{p_1(1 - p_1)}{n_1} + \frac{p_2(1 - p_2)}{n_2}}$$

In practice, you estimate $\sigma_{\hat{p}_1 - \hat{p}_2}$ using the two sample proportions. This leads to a confidence interval for the difference of the two population proportions of

$$(\hat{p}_1 - \hat{p}_2) \pm z^{\star} \cdot \sqrt{\frac{\hat{p}_1(1 - \hat{p}_1)}{n_1} + \frac{\hat{p}_2(1 - \hat{p}_2)}{n_2}}$$

where $\pm z^{\star}$ are the standardized values that enclose an area under a normal curve equal to the confidence coefficient, which typically is .95. Don't forget to check the conditions that must be met to use this confidence interval, especially that the two samples were taken independently from two different populations.

When interpreting, say, a 95% confidence interval, you should say that you are 95% confident that the difference between the proportion of successes in the first population and the proportion of successes in the second population is in this confidence interval. (However, say this in the context of the situation.) If the confidence interval includes 0, you can't conclude that there is any difference in the two proportions.

In an experiment with random assignment, the available subjects are divided randomly into two treatment groups. Although this doesn't satisfy the condition that two samples are taken independently from two different populations, you may still use the methods of this section. Just think of the population for each treatment as being all of the subjects in the experiment.

■ Exercises

E23. In a recent national survey, 16,262 students in 151 schools completed questionnaires about physical activity. Male students (55.5%) were significantly more likely than female students (42.3%) to have played on sports teams run by their school during the 12 months preceding the survey. Check the accuracy of the phrase "significantly more likely," assuming that there were equal numbers of male and female students in this survey and that the samples are equivalent to simple random samples.

Source: Youth Risk Behavior Surveillance System, National Centers for Disease Control and Prevention, 1997; www.cdc.gov/nccdphp/dash/yrbs/natsum97/supa97.html.

E24. The survey in E23 found that overall, 49.5% of the students had played on sports teams run by their school during the 12 months preceding the survey. Use this information along with the information in E23 to find the number of male students and the number of female students in the survey.

E25. A Harris poll of November 14, 2001, asked about 425 men and 425 women, "When you get a sales or customer service phone call from someone you don't know, would you prefer to be addressed by your first name or by your last name, or don't you care one way or the other?" Twenty-three percent of the men and 34% of the women said they would prefer their last name. Find and interpret a 99% confidence interval for the difference of two proportions. Although this was not quite the case, you may assume that the samples were random.

Source: www.harrisinteractive.com/harris_poll/index.asp?PID=267, 2001.

E26. A Harris on-line poll asked 99 on-line respondents, "Would you support the proposed bill that would eliminate the U.S. penny?" There were 22 (or 22.2%) "yes" votes and 77 (or 77.8%) "no" votes.

Source: www.harriszone.com, Feb. 25, 2002.

A student computed a 95% confidence interval for the difference in the proportion of the population answering "yes" and the proportion answering "no" as follows:

$$(\hat{p}_1 - \hat{p}_2) \pm z^* \cdot \sqrt{\frac{\hat{p}_1(1 - \hat{p}_1)}{n_1} + \frac{\hat{p}_2(1 - \hat{p}_2)}{n_2}}$$

$$= (.222 - .778)$$

$$\pm 1.96 \sqrt{\frac{(.222)(.778)}{99} + \frac{(.778)(.222)}{99}}$$

$$= -.556 \pm .116$$

Assuming that the respondents can be considered a random sample from some population, is the student's method correct? If so, write an interpretation of this interval. If not, use a more appropriate method.

E27. A famous medical experiment was conducted by the Nobel Laureate Linus Pauling (United States, 1901–1994), who believed that vitamin C prevents colds. His subjects were 279 French skiers who were randomly assigned to receive vitamin C or a placebo. Of the 139 given vitamin C, 17 got a cold. Of the 140 given the placebo, 31 got a cold.

Source: L. Pauling, "The Significance of the Evidence about Ascorbic Acid and the Common Cold," *Proceedings of the National Academy of Sciences* 68 (1971): 2678–81.

a. Find and interpret a 95% confidence interval for the difference of two proportions.

b. Find a 99% confidence interval for the difference of two proportions. Does your conclusion change?

E28. In 1954, the largest medical experiment of all time was carried out to test whether the newly developed Salk vaccine was effective in preventing polio. This study incorporated all three characteristics of an experiment: use of a control group of children who received a placebo injection (an injection that felt like a regular immunization but contained only salt water), random assignment of children to either the placebo injection group or the Salk vaccine injection group, and assignment of several hundred thousand children to each treatment. Of the 200,745 children who received the Salk vaccine, 82 were diagnosed with polio. Of the 201,229 children who received the placebo, 162 were diagnosed with polio. Find and interpret a 95% confidence interval for the difference of two proportions.

Source: Paul Meier, "The Biggest Public Health Experiment Ever," in *Statistics: A Guide to the Unknown*, 3rd ed., edited by Judith M. Tanur et al. (Pacific Grove, CA: Brooks/Cole, 1989).

E29. The formula for the confidence interval for the difference of two proportions involves z. Why is it okay to use z in this case?

E30. The survey reported in this section found that 45% of households own a dog, 34% own a cat, and 20% own both.

a. What percentage own neither?

b. Are owning a dog and owning a cat independent events?

c. Should you use the techniques of this section to estimate the difference in the percentage of U.S. households that own a dog and the percentage that own a cat?

8.4   A Significance Test for the Difference of Two Proportions

In the previous section, you learned how to estimate the size of the difference of two proportions. But sometimes you must decide between two alternatives. For example:

- Is snowboarding or skiing more likely to result in a serious injury?

- Does a new treatment for AIDS result in fewer deaths than an old treatment?

- Is Reggie Jackson's World Series record so much better than his play during the regular season that the difference can't reasonably be attributed to chance?

The size of any difference involved is not the issue, as it was in Section 8.3. All you care about is whether you have enough evidence to conclude that there is a difference. In this section, you will learn to perform a test of significance in order to decide if the observed difference can reasonably be attributed to chance alone or if it is so large that something other than chance variation must be causing it.

Could the observed difference be due to chance alone?

Differences Occur Even When There Is No Difference

Before learning how to use a test of significance to answer questions like those above, you will review the idea of a sampling distribution for the difference of two proportions.

In the M&M's activity of the previous section, you simulated a sampling distribution of $\hat{p}_1 - \hat{p}_2$, where $\hat{p}_1$ was the proportion of browns in a sample of plain M&M's and $\hat{p}_2$ was the proportion of browns in a sample of peanut M&M's. The mean of this distribution of differences turned out to be .10 because 30% of plain M&M's are brown and 20% of peanut M&M's are brown. There *is* a difference between the proportion of plain M&M's that are brown and the proportion of peanut M&M's that are brown.

However, as M&M's fans know, there is no difference between the proportion of plain M&M's and the proportion of peanut M&M's that are red. Each type is 20% red. In the next activity, you will examine what a sampling distribution of $\hat{p}_1 - \hat{p}_2$ looks like when $p_1 = p_2$.

This activity demonstrates chance variation.

Activity 8.6   Differences When There Is No Difference

What you'll need: graph paper for constructing dot plots

1. Use your calculator or computer to simulate taking a sample of 10 M&M's out of a bag of plain M&M's. Find the proportion of M&M's that are red.

2. Use your calculator or computer to simulate taking a sample of 10 M&M's out of a bag of peanut M&M's. Find the proportion of M&M's that are red.

3. Subtract the proportion in step 2 from the proportion in step 1.

(continued)

4. Repeat steps 1 through 3 so that you have 1000 differences of two proportions.

5. Construct a dot plot or histogram of the (simulated) sampling distribution of the difference. Describe its shape, mean, and standard deviation. How might you have predicted what these would be?

6. Repeat steps 1–5, this time using samples of size 40.

7. Compare the shape, center, and spread of your distribution of $\hat{p}_1 - \hat{p}_2$ for samples of size 10 with that for samples of size 40.

The dot plots in Display 8.16 show typical results from Activity 8.6. The first dot plot is for samples of size 10, while the second is for samples of size 40. These dot plots, like yours, demonstrate that chance variation usually results in a difference between $\hat{p}_1$ and $\hat{p}_2$ even if the two populations from which the samples are drawn have the same proportion of successes. Further, the set of differences $\hat{p}_1 - \hat{p}_2$ is centered at 0, and the variation in the differences tends to be smaller when the sample size is larger.

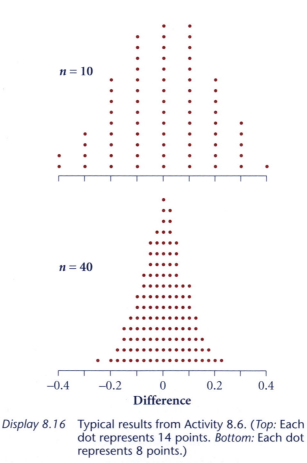

Display 8.16 Typical results from Activity 8.6. (*Top:* Each dot represents 14 points. *Bottom:* Each dot represents 8 points.)

Discussion: A Difference When There Really Is No Difference

D45. As you saw in Section 8.3, the standard error for the difference of two proportions $\hat{p}_1 - \hat{p}_2$ can be estimated by

$$\sigma_{\hat{p}_1-\hat{p}_2} \approx \sqrt{\frac{\hat{p}_1(1-\hat{p}_1)}{n_1} + \frac{\hat{p}_2(1-\hat{p}_2)}{n_2}}$$

a. Compute the standard error of the sampling distribution of $\hat{p}_1 - \hat{p}_2$ from Activity 8.6 for samples of size 10. Compare your result to your estimate from the activity or from Display 8.16.

b. Compute the standard error of the sampling distribution of $\hat{p}_1 - \hat{p}_2$ from Activity 8.6 for samples of size 40. Compare your result to your estimate from the activity or from Display 8.16.

D46. Suppose you take two independent samples from two different populations that have the same proportion of successes. Answer true or false to each statement.

a. It is always the case that $\hat{p}_1 = \hat{p}_2$.

b. It is true that $p_1 = p_2$.

c. Sometimes the difference between the two sample proportions can be relatively large.

d. Theoretically, the dot plots from Activity 8.6 should be centered exactly at 0.

e. To have a bigger chance of having the difference $\hat{p}_1 - \hat{p}_2$ nearer 0, have larger sample sizes.

■ Practice

P37. Use your calculator or computer to take 200 pairs of random samples, both of size 25 and both from populations with 35% successes.

a. Make a dot plot of the differences of the two sample proportions.

b. From your dot plot, estimate the mean and standard error of the sampling distribution of $\hat{p}_1 - \hat{p}_2$.

c. Find the theoretical value of the mean and standard error of the sampling distribution of $\hat{p}_1 - \hat{p}_2$. Compare your answers to your estimates in part b.

The Theory of a Significance Test for the Difference of Two Proportions

To see how to apply these ideas to an experiment, let's return to the clinical trial experiment on comparing the two treatments for AIDS-related complex (ARC) introduced in Section 8.3. The investigators first wanted to find the difference between the proportion of patients who progress to AIDS when given AZT and the proportion of patients who progress to AIDS when given AZT + ACV. The confidence interval of Section 8.3 answers that question. They are 95% confident that the difference between these two proportions is between −.095 and .155.

The investigators' next question dealt with the survival of patients who had already developed AIDS. These data are shown in Display 8.17.

Treated With

		AZT	AZT + ACV	Total
Survived?	**No**	28	13	41
	Yes	41	49	90
	Total	69	62	131

Display 8.17 Treatment results among AIDS patients.

Here, $\hat{p}_1 = \frac{41}{69} \approx .594$, where $\hat{p}_1$ is the proportion who survived in the group treated with AZT, and $\hat{p}_2 = \frac{49}{62} \approx .790$, where $\hat{p}_2$ is the proportion who survived in the group treated with AZT + ACV.

You could use a confidence interval or a significance test.

You could analyze these data in the same way you did in Section 8.3. That is, you could construct a confidence interval for the difference in the proportion of patients, $p_1 - p_2$, who will survive with each treatment. You will be confident that AZT alone is worse than AZT + ACV if the entire confidence interval lies below zero. Generally, constructing such a confidence interval is the best procedure to use. In this section, however, you will learn how to do a test of significance, which you often see in reports of all types of research.

In order for a new therapy (in this case, AZT + ACV) to become accepted practice among physicians, the research must show that the new therapy is an improvement over the old. This leads to a one-sided test of significance. The hypotheses are

H_0: The new therapy is not better than the old, or $p_1 = p_2$, where p_1 is the proportion of the patients who would have survived if all 131 patients could have been given AZT alone and p_2 is the proportion of the patients who would have survived if they all could have been given AZT + ACV.

H_a: The new therapy is better than the old, or $p_1 < p_2$.

The general form of a test statistic for testing hypotheses is

$$\frac{estimate - parameter}{standard\ deviation\ of\ the\ estimate}$$

The estimate is $\hat{p}_1 - \hat{p}_2$, or $.594 - .790 = -.196$. The parameter is 0, the value under the null hypothesis: $p_1 - p_2 = 0$. The standard deviation of the estimate (or the standard error) is given exactly by

$$\sigma_{\hat{p}_1 - \hat{p}_2} = \sqrt{\frac{p_1(1 - p_1)}{n_1} + \frac{p_2(1 - p_2)}{n_2}}$$

You could do as you did in Section 8.3 and estimate p_1 with $\hat{p}_1$ and estimate p_2 with $\hat{p}_2$. However, you can do even better. The null hypothesis states that $p_1 = p_2$. You can estimate this common value of p_1 and p_2 by combining the data from both samples into a **pooled estimate $\hat{p}$**. In Display 8.17, there are a total of 90 surviving patients out of 131 cases. The null hypothesis says that the two

$\hat{p}$ is called the pooled estimate of the common proportion of successes.

treatments are equivalent, so the best estimate of the probability that patients will survive given either treatment is $\hat{p} = \frac{total\ survived}{grand\ total} = \frac{90}{131} \approx .687$. The standard error of the difference then is approximately

$$\sqrt{\frac{p_1(1-p_1)}{n_1} + \frac{p_2(1-p_2)}{n_2}} \approx \sqrt{\frac{\hat{p}(1-\hat{p})}{n_1} + \frac{\hat{p}(1-\hat{p})}{n_2}}$$

$$= \sqrt{\hat{p}(1-\hat{p})\left(\frac{1}{n_1} + \frac{1}{n_2}\right)}$$

$$= \sqrt{.687(1-.687)\left(\frac{1}{69} + \frac{1}{62}\right)}$$

$$\approx .081$$

The test statistic then takes on the value

$$z = \frac{estimate - parameter}{standard\ deviation\ of\ the\ estimate}$$

$$= \frac{(\hat{p}_1 - \hat{p}_2) - (p_1 - p_2)}{\sqrt{\hat{p}(1-\hat{p})\left(\frac{1}{n_1} + \frac{1}{n_2}\right)}}$$

$$= \frac{(.594 - .790) - 0}{\sqrt{.687(1-.687)\left(\frac{1}{69} + \frac{1}{62}\right)}}$$

$$\approx -2.415$$

The test statistic is normally distributed.

This test statistic is based on the difference of two approximately normally distributed random variables, $\hat{p}_1$ and $\hat{p}_2$. You learned in Chapter 5 that such a difference itself has a normal distribution. Thus, you can use Table A to find the *P*-value. The test statistic $z = -2.42$ has a (one-sided) *P*-value of .0078, as illustrated in Display 8.18. This *P*-value is very small, so reject the null hypothesis; if all subjects in the experiment could have been given the AZT + ACV treatment, you are confident that there would have been a larger survival rate than if they all had received only AZT. The difference between the survival rates for these two treatments is too large to attribute to chance variation alone.

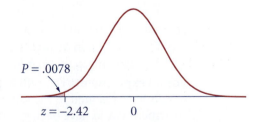

$P = .0078$

$z = -2.42$ 0

Display 8.18   The *P*-value for a one-sided test with test statistic $z = -2.42$.

As you have seen, the significance test for the difference of two proportions proceeds along the same lines as the test of a single proportion. The steps in the significance test for the difference of two proportions are given in this box.

Components of a Significance Test for the Difference of Two Proportions

1. *Check conditions.*

   For a survey: A random sample of size n_1 is taken from a large binomial population with proportion of successes p_1. The proportion of successes in this sample is $\hat{p}_1$. A second and independent random sample of size n_2 is taken from a large binomial population with proportion of successes p_2. The proportion of successes in this sample is $\hat{p}_2$. All of these quantities must be at least 5:

   $$n_1\hat{p}_1 \quad n_1(1-\hat{p}_1) \quad n_2\hat{p}_2 \quad n_2(1-\hat{p}_2)$$

   Each population size must be at least 10 times the sample size.

   For an experiment: Subjects must be randomly assigned to treatments, and all of these must be at least 5:

   $$n_1\hat{p}_1 \quad n_1(1-\hat{p}_1) \quad n_2\hat{p}_2 \quad n_2(1-\hat{p}_2)$$

2. *Write a null and an alternate hypothesis.*

   For a survey: You can write the null hypothesis in several different ways as long as you define p_1 and p_2:

   H_0: The proportion of successes p_1 in the first population is equal to the proportion of successes p_2 in the second population.

   H_0: $p_1 = p_2$, where p_1 is the proportion of successes in the first population and p_2 is the proportion of successes in the second population.

   H_0: $p_1 - p_2 = 0$, where p_1 is the proportion of successes in the first population and p_2 is the proportion of successes in the second population.

   The form of the alternate hypothesis depends on whether you have a two-sided or a one-sided test:

   H_a: The proportion of successes p_1 in the first population is not equal to the proportion of successes p_2 in the second population. Equivalently, $p_1 \neq p_2$ or $p_1 - p_2 \neq 0$.

   H_a: The proportion of successes p_1 in the first population is greater than the proportion of successes p_2 in the second population. Equivalently, $p_1 > p_2$ or $p_1 - p_2 > 0$.

   H_a: The proportion of successes p_1 in the first population is less than the proportion of successes p_2 in the second population. Equivalently, $p_1 < p_2$ or $p_1 - p_2 < 0$.

   For an experiment: The null hypothesis is that the two treatments would have resulted in the same proportion of successes if each treatment could have been given to all subjects. The alternate hypothesis for a two-sided test is that the two treatments would *not* have resulted in the same proportion of successes if each treatment could have been given to all subjects. The alternate hypothesis for a

Null hypothesis: $p_1 = p_2$

Alternate hypothesis: $p_1 \neq p_2$, $p_1 > p_2$, or $p_1 < p_2$

(continued)

one-sided test is that a specific treatment would have resulted in a higher proportion of successes than the other if each treatment could have been given to all subjects.

Computations with a sketch

3. **Compute the test statistic and a P-value.**

$$z = \frac{(\hat{p}_1 - \hat{p}_2) - (p_1 - p_2)}{\sqrt{\hat{p}(1 - \hat{p})\left(\frac{1}{n_1} + \frac{1}{n_2}\right)}}$$

where $\hat{p} = \dfrac{\textit{total number of successes in both samples}}{n_1 + n_2}$

The P-value is the probability of getting a value of z as extreme as or even more extreme than that from your samples if H_0 is true. See Display 8.19.

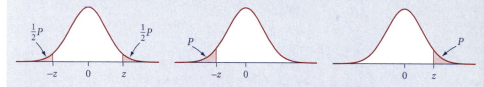

Display 8.19 Critical regions for a two-sided test (left) and a one-tailed test (center and right).

Conclusion

4. **Write a conclusion.**

State whether you reject or do not reject the null hypothesis. Link this conclusion to your computations by comparing z to the critical value z^* or by appealing to the P-value. You reject the null hypothesis if z is more extreme than z^* or, equivalently, if the P-value is smaller than the level of significance, α. Then write a sentence giving your conclusion in the context of the situation.

Example: Significance Test for a Difference

A survey found that half of men age 20 to 24 live with at least one parent, whereas a third of women age 20 to 24 live with at least one parent. The sample size wasn't stated, but it's a reasonable estimate that about 150 men this age and 150 women this age were polled. Perform a test of significance to determine if there is a difference in the proportion of men and women age 20 to 24 who live with at least one parent. Use $\alpha = .01$.

Source: "USA Snapshots," *USA Today*, Aug. 8, 1997, p. A1.

Solution

Conditions

1. Although the situation probably is actually more complicated, you can assume that you have two independent random samples. You know that the proportion of men age 20 to 24 in the sample who are living with at

least one parent is $\hat{p}_1 = .50$ and the proportion of women age 20 to 24 in the sample who are living with at least one parent is $\hat{p}_2 = \frac{1}{3}$. Using the estimates of $n_1 = 150$ and $n_2 = 150$, all of

You should *show* these
computations worked
out, not just say they
are true, even if that
is obvious.

$$n_1\hat{p}_1 = 150(.5) = 75 \quad n_1(1 - \hat{p}_1) = 150(1 - .5) = 75,$$

$$n_2\hat{p}_2 = 150\left(\frac{1}{3}\right) = 50 \quad n_2(1 - \hat{p}_2) = 150\left(1 - \frac{1}{3}\right) = 100$$

are larger than 5. Both the number of men and the number of women in this age group in the United States are much larger than 10 times the sample size for both samples.

Hypotheses

2. H_0: The proportion p_1 of men age 20 to 24 who live with one parent is equal to the proportion p_2 of women age 20 to 24 who live with one parent.

H_a: $p_1 \neq p_2$. (Note that you can use the symbols p_1 and p_2 because you defined exactly what they stand for in the null hypothesis.)

*Computations with
a diagram*

3. The pooled estimate, $\hat{p}$, is

$$\hat{p} = \frac{\textit{total number of successes in both samples}}{n_1 + n_2} = \frac{75 + 50}{150 + 150} \approx .417$$

So the test statistic is

$$z = \frac{(\hat{p}_1 - \hat{p}_2) - (p_1 - p_2)}{\sqrt{\hat{p}(1 - \hat{p})\left(\frac{1}{n_1} + \frac{1}{n_2}\right)}} \approx \frac{(.5 - .33) - (0)}{\sqrt{.417(1 - .417)\left(\frac{1}{150} + \frac{1}{150}\right)}} \approx 2.99$$

With a test statistic of 2.99, the P-value for a two-tailed test is .0014(2) = .0028. See Display 8.20.

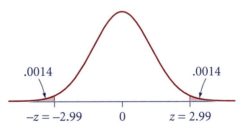

Display 8.20 A two-sided test with $z = 2.99$.

(Using your calculator's two-proportion z-test, which carries more decimal places, you get the more accurate $z = 2.928$ and a P-value of .0034.)

Conclusion in context

4. This difference is statistically significant, and you reject the null hypothesis. If the proportions of men and women age 20 to 24 who live with at least one parent were equal, then there would be only 28 chances in 10,000 repeated samples of size 150 of getting a difference in proportions of 17% or larger. Because the P-value of .0028 is less than $\alpha = .01$, you cannot reasonably attribute the difference to chance variation. You conclude that the proportions of men and women who live with at least one parent are different.

Display 8.21 shows a printout for this example.

Test of Men vs. Women	Compare Proportions ▼

Attribute (categorical): Man_Lives_With_Parent

Attribute (categorical): Woman_Lives_With_Parent

In **Man_Lives_With_Parent 75** out of **150**, or **0.5**, are **True**
In **Woman_Lives_With_Parent 50** out of **150**, or **0.333333**, are **True**
Alternative hypothesis: The population proportion for **True** in
Man_Lives_With_Parent is **not equal to** that for **True** in
Woman_Lives_With_Parent

The test statistic, **z**, is **2.928**.

If it were true that the two population proportions were equal (the null
hypothesis), and the sampling process were performed repeatedly, the
probability of getting a value of **z with an absolute value this great or
greater** would be **0.0034**.

Display 8.21   A printout for the significance test.   ■

Discussion

D47. In the AZT/AZT + ACV example, the null hypothesis was rejected.

   a. Which type of error might have been made in the significance test?

   b. How can you check quickly that the condition is met that all of
      $n_1\hat{p}_1$, $n_1(1-\hat{p}_1)$, $n_2\hat{p}_2$, and $n_2(1-\hat{p}_2)$ are at least 5?

D48. What does it suggest about the two proportions if the significance test for
   a difference of two proportions

   a. fails to reject the null hypothesis?

   b. rejects the null hypothesis?

D49. Look again at the distributions you generated in Activity 8.6 (or at
   Display 8.16) for samples of size 10 and 40. Explain why they illustrate
   the need for the assumption that all of the following quantities must be
   at least 5:

$$n_1\hat{p}_1 \quad n_1(1-\hat{p}_1) \quad n_2\hat{p}_2 \quad n_2(1-\hat{p}_2)$$

■ Practice

P38. A random sample of 600 probable voters was taken three weeks before the
   start of a campaign for mayor; 321 of the 600 said they favored the new
   candidate over the incumbent. However, it was revealed the week before the
   election that the new candidate had dozens of outstanding parking tickets.
   Subsequently, a new random sample of 750 probable voters showed that
   382 voters favored the new candidate.

   Do these data support the conclusion that there was a decrease in voter
   support for the new candidate after the parking tickets were revealed? Give
   appropriate statistical evidence to support your answer.

P39. What type of error might have been made in the significance test in the example about young men and women living with a parent?

Summary 8.4: Testing for a Significant Difference Between Proportions

In this section, you learned how to test whether two samples were drawn from populations that have the same proportion of successes. Or, in the case of an experiment, you saw how to test whether two treatments have the same results. You follow the same steps as with any significance test.

- *Check the conditions.* For a survey, you need two sufficiently large random samples selected independently from two different populations. For an experiment, the treatments must have been assigned randomly to subjects.

- *Write a null and an alternate hypothesis.* The null hypothesis for a survey is that the two populations contain the same proportion of "successes." For an experiment, the null hypothesis is that the two treatments would have resulted in the same proportion of successes if each treatment could have been given to all subjects. The alternate hypothesis can be one-sided or two-sided.

- *Compute the test statistic and a P-value.* The test statistic is

$$z = \frac{(\hat{p}_1 - \hat{p}_2) - (p_1 - p_2)}{\sqrt{\hat{p}(1 - \hat{p})\left(\frac{1}{n_1} + \frac{1}{n_2}\right)}}$$

  where $\hat{p}$ is computed by combining both samples.

  The P-value is the probability of getting a value of z as extreme as or even more extreme than that from your samples *if H_0 is true.*

- *Write a conclusion.* State whether you reject or do not reject the null hypothesis, linking this decision to your value of z or to your P-value. Then restate your conclusion in the context of the situation.

■ Exercises

E31. A poll of 256 boys and 257 girls age 12 to 17 asked, "Do you feel like you are personally making a positive difference in your community?" More girls (76%) than boys (63%) answered "yes."

Source: www.ncpc.org/rwesafe/2001/rwesafe11.html, April 17, 2002.

a. Using a one-sided test, is this a statistically significant difference? That is, if all teens were asked, are you

confident that a larger proportion of girls than boys would say "yes"? Assume that the samples were selected randomly.

b. The report says, "Participants were selected through random digit dialing." Do you have any concerns that such a procedure would give a random sample?

c. Find a 95% confidence interval for the proportion of all teens who would say "yes." What additional assumption do you need to make to do this?

E32. An annual poll of the sleeping habits of Americans found that of a random sample of 158 Americans age 18 to 29, 35% slept eight hours or more on a weekday. Of a random sample of 435 people age 30 to 64, 36% slept eight hours or more on a weekday. Is this a statistically significant difference? (When answering this question, include all four steps in your test of significance.)

Source: www.sleepfoundation.org/publications/2001poll.html.

E33. The Physicians' Health Study conducted a famous experiment on the effect of low-dose aspirin on heart attacks. This was a randomized, double-blind, placebo-controlled clinical trial. Male physician volunteers with no previous important health problems were randomly assigned to an experimental group ($n_1 = 11,037$) or a control group ($n_2 = 11,034$). Those in the experimental group were asked to take a pill containing 325 mg of aspirin every second day, and those in the control group were asked to take a placebo pill every second day. After about five years, there were 139 heart attacks in the aspirin group and 239 in the placebo group. Is the difference statistically significant at the 1% level?

Source: The Steering Committee of the Physicians' Health Study Research Group, "The final report on the aspirin component of the ongoing Physicians' Health Study," *New England Journal of Medicine* 321, no. 3 (1989): 129–135, www.content.nejm.org/cgi/content/abstract/321/3/129.

E34. A related part of the landmark study discussed in E33 dealt with the question of whether aspirin reduces the seriousness of a heart attack, should one occur. Of those 139 in the aspirin group who had a heart attack, 10 died. Of those 239 in the placebo group who had a heart attack, 26 died. Of those who had heart attacks, is the death rate less for those who take aspirin?

E35. The great baseball player Reggie Jackson batted .262 during regular season play and .357 during World Series play. He was at bat 9864 times in regular season play and 98 times in the World Series. Can this difference reasonably be attributed to chance, or did Reggie earn his nickname "Mr. October"? Use $\alpha = .05$. Note that you do not have two random samples from different populations. You have Reggie's entire record, so all you can do is decide if the difference is about the size you would expect from chance variation or whether you should look for some other explanation.

Source: Major League Baseball: www.mlb.com/NASApp/mlb/mlb/history/mlb_history_event.jsp?event=worldseries&story=worldseries11 and www.mlb.com/NASApp/mlb/mlb/stats_historical/mlb_individual_stats_player.jsp?playerID=116439&HS=True, Feb. 23, 2002.

E36. At Bunker Hill, near Boston, in one of the first battles of the American Revolution, 2400 British troops were engaged, of which 1054 were wounded. Of the 1054 wounded, 226 died. Out of 1500 American participants in the battle, losses were estimated at 140 killed and an additional 271 wounded who didn't die of their wounds. (The number of casualties varies somewhat depending on the source.) You have all of the necessary information about the two populations. Can the difference in the pairs of proportions in I and II reasonably be attributed to chance variation, or should you look for another explanation? Use $\alpha = .05$.

Source: Christopher Ward, *The War of the Revolution* (New York: Macmillan, 1952), p. 96.

I. The proportion of the American troops who were wounded (including those killed) and the proportion of the British troops who were wounded (including those killed)

II. The proportion of the American troops who were killed and the proportion of the British troops who were killed

E37. Tests of significance are based upon randomness somewhere in the design of the study and collection of the data. Explain how randomness would be used if you want to compare two proportions and the study is

a. a sample survey   b. an experiment

E38. There are 35% successes in each of two different large populations. A random sample of size 40 is drawn from the first population, and a random sample of size 50 is drawn from the second. What is the probability that the difference in the sample proportions will be 10% or greater?

E39. There are 45% successes in one large population and 65% successes in a second large population. A random sample of size 30 is taken from the first population, and a random sample of size 25 is taken from the second population. What is the probability that the difference in the sample proportions will be 20% or greater in favor of the second population?

E40. Explain how knowledge of sampling distributions for differences in sample proportions is used in the development of the significance test for the difference of two proportions.

E41. You can take either Bus A or Bus B to school, and you're concerned about being late to class. You check out Bus A on 100 randomly selected mornings and find that you would be late to class 5 mornings. You check out Bus B on 64 randomly selected mornings and find that you would be late 5 times. Because you live a bit closer to the stop for Bus B, you will use Bus A only if you can reject the null hypothesis that the difference in the percentages is more than 2%. Which bus will you use?

Chapter Summary

To use the confidence intervals and significance tests of this chapter, you need either

- a random sample from a population that is made up of "successes" and "failures," or

- two random samples taken independently from two distinct such populations. If the study is an experiment, you should have two treatments randomly assigned to the available subjects.

(When there is no randomness involved, you proceed with the test only if you state loudly and clearly the limitations of what you have done. If you reject the null hypothesis in such a case, all you can conclude is that something happened that can't reasonably be attributed to chance.)

You use a confidence interval if you want to find a range of plausible values for

- p, the proportion of successes in a population

- $p_1 - p_2$, the difference between the proportion of successes in one population and the proportion of successes in another population

Both of the confidence intervals you studied have the same form:

$$(\textit{estimate from the sample}) \pm z^* \cdot (\textit{estimated standard error})$$

And both of the significance tests you have studied include the same steps:

1. Justify your reasons for choosing this particular test. Discuss whether the conditions are met and decide whether it is okay to proceed if they are not strictly met.

2. State the null hypothesis and the alternate hypothesis.

3. Compute the test statistic and find the *P*-value. Draw a diagram of the situation.

4. Use the computations to decide whether to reject or not to reject the null hypothesis by comparing the test statistic to $z^{\star}$ or by comparing the *P*-value to α, the level of significance. Then state your conclusion in terms of the context of the situation. (Just saying "Reject H_0" is not sufficient.) Mention any doubts you have about the validity of that conclusion.

■ Review Exercises

E42. A 6th grade student, Emily Rosa, performed an experiment to test the validity of "therapeutic touch." According to an article written by her RN mother, a statistician, and a physician, "Therapeutic Touch (TT) is a widely used nursing practice rooted in mysticism but alleged to have a scientific basis. Practitioners of TT claim to treat many medical conditions by using their hands to manipulate a 'human energy field' perceptible above the patient's skin." To investigate whether TT practitioners can actually perceive a "human energy field," 21 experienced TT practitioners "were tested under blinded conditions to determine whether they could correctly identify which of their hands was closest to the investigator's hand. Placement of the investigator's hand was determined by flipping a coin. Practitioners of TT identified the correct hand in 123 of 280 trials.

Source: *Journal of the American Medical Association* 279 (1998): 1005–10.

a. What does "blinded" mean in the context of this experiment? How might it have been done? Why wasn't double-blinding necessary?

b. Perform a statistical analysis of this experiment and write a conclusion. Use $\alpha = .05$.

c. The article says, "The statistical power of this experiment was sufficient to conclude that if TT practitioners could reliably detect a human energy field, the study would have demonstrated this." What does this sentence mean?

E43. In P13, a report of a recent survey about Internet use on homework was described. The survey also reports that 94% of the 754 students age 12 to 17 with Internet access said the Internet helped them with their homework.

Source: www.siliconvalley.com/docs/news/svfront/ 063319.html, Sept. 1, 2001.

a. Compute the 92% confidence interval.

b. Interpret this confidence interval, making it clear exactly what it is that you are 92% sure is in the confidence interval.

c. Explain the meaning of 92% confidence.

E44. Refer to the 2001 Gallup poll at the beginning of Section 8.1 in which 51% of the American public assigned a grade of A or B to the public schools in their community.

In 2000, the comparable figure was 47%. Assuming a sample size of 1108 in both 2000 and 2001, find and interpret a 90% confidence interval for the difference of two proportions.

E45. A study of all injuries from the two winter seasons 1999–2000 and 2000–01 at the three largest ski areas in Scotland found that of the 531 snowboarders who were injured, 148 had fractures. Of the 952 skiers who were injured, 146 had fractures. (For both groups, most of the other injuries were sprains, lacerations, or bruising.)

Source: www.ski-injury.com/accstat9901.html, March 29, 2002.

a. Is this difference statistically significant at the .05 level?

b. There are about twice as many skiers as snowboarders. Can you use this fact and the data from this study to determine if snowboarders are more likely to be injured than are skiers?

E46. As part of a study of the brain's response to placebos, researchers at UCLA gave 51 patients with depression either an antidepressant or a placebo. The article reports that the "51 subjects then were randomly assigned to receive 8 weeks of double-blind treatment with either placebo or the active medication." "Overall, 52% of the subjects (13 of 25) receiving antidepressant medication responded to treatment, and 38% of those receiving placebo (10 of 26) responded." Perform a test to determine if the difference in the proportion who responded is statistically significant. Use $\alpha = .02$.

Source: A. F. Leuchter, I. A. Cook, E. A. Witte, M. Morgan, and M. Abrams, "Changes in Brain Function of Depressed Subjects During Treatment with Placebo," *American Journal of Psychiatry* 159, no. 1 (January 2002): 122–29.

E47. Suppose that a null hypothesis is tested at the .05 level of significance in 100 different studies. Also suppose that in each study, the null hypothesis is true.

a. How many Type I errors do you expect?

b. Find the probability that no Type I error is made in any of the 100 studies.

E48. To become a lawyer, a person must pass the bar exam for his or her state. Suppose that you are searching for possible inequities that result in unequal proportions of males and females passing the bar exam. Each state has its own bar exam, so you check a random sample of the bar exam records in each of the 50 states. For each state, you plan to perform a test of the significance of the difference of the proportion of males passing the exam and the proportion of females passing the exam, using $\alpha = .05$. If the difference is statistically significant for any state, you will conclude that there has been some inequity. Do you see anything wrong with this plan?

E49. Suppose you use the formula for a 90% confidence interval for many different random samples from the same population. What fraction of the times should the confidence intervals produced actually capture the population proportion? Explain your reasoning.

E50. What is the best explanation of the use of the term "95% confidence"?

A. We can never be 100% confident in statistics; we can only be 95% confident.

B. The sample proportion will fall in 95% of the confidence intervals we construct.

C. The population proportion will fall in 95% of the confidence intervals we construct.

D. We are 95% confident that the range of reasonably likely outcomes contains the population proportion.

E. We are 95% confident that we have made a correct decision when we reject the null hypothesis.

E51. The following explanation about sample size was posted on the "Frequently Asked Questions" page of the Gallup Organization's Web site on March 17, 2002.

Read it carefully and write a brief report on how it could be improved.

Source: Frank Newport, Lydia Saad, and David Moore, "How Polls Are Conducted," in *Where America Stands* (New York: John Wiley & Sons, Inc., © 1997); www.gallup.com/help/FAQs/poll1.asp. This material is used by permission of John Wiley & Sons, Inc.

The Number of Interviews or Sample Size Required

One key question faced by Gallup statisticians: How many interviews does it take to provide an adequate cross-section of Americans? The answer is, not many—that is, if the respondents to be interviewed are selected entirely at random, giving every adult American an equal probability of falling into the sample. The current U.S. adult population in the continental United States is 187 million. The typical sample size for a Gallup poll which is designed to represent this general population is 1,000 national adults.

The actual number of people which need to be interviewed for a given sample is to some degree less important than the soundness of the fundamental equal probability of selection principle. In other words—although this is something many people find hard to believe—if respondents are not selected randomly, we could have a poll with a million people and still be significantly less likely to represent the views of all Americans than a much smaller sample of just 1,000 people—if that sample is selected randomly.

To be sure, there is some gain in sampling accuracy which comes from increasing sample sizes. Common sense—and sampling theory—tell us that a sample of 1,000 people probably is going to be more accurate than a sample of 20. Surprisingly, however, once the survey sample gets to a size of 500, 600, 700 or more, there are fewer and fewer accuracy gains which come from increasing the sample size. Gallup and other major organizations use sample sizes of between 1,000 and 1,500 because they provide a solid balance of accuracy against the increased economic cost of larger and larger samples. If Gallup were to—quite expensively—use a sample of 4,000 randomly selected adults each time it did its poll, the increase in accuracy over and beyond a well-done sample of 1,000 would be minimal, and generally speaking, would not justify the increase in cost.

Statisticians over the years have developed quite specific ways of measuring the accuracy of samples—so long as the fundamental principle of equal probability of selection is adhered to when the sample is drawn.

For example, with a sample size of 1,000 national adults (derived using careful random selection procedures), the results are highly likely to be accurate within a margin of error of plus or minus three percentage points. Thus, if we find in a given poll that President Clinton's approval rating is 50%, the margin of error indicates that the true rating is very likely to be between 53% and 47%. It is very unlikely to be higher or lower than that.

To be more specific, the laws of probability say that if we were to conduct the same survey 100 times, asking people in each survey to rate the job Bill Clinton is doing as president, in 95 out of those 100 polls, we would find his rating to be between 47% and 53%. In only five of those surveys would we expect his rating to be higher or lower than that due to chance error.

As discussed above, if we increase the sample size to 2,000 rather than 1,000 for a Gallup poll, we would find that the results would be accurate within plus or minus 2% of the underlying population value, a gain of 1% in terms of accuracy, but with a 100% increase in the cost of conducting the survey. These are the cost value decisions which Gallup and other survey organizations make when they decide on sample sizes for their surveys.

E52. Some people complain that election polls cannot be right, because they personally were not asked how they were going to vote. Write an explanation to such a person about how polls can get a good idea of how the entire population will vote by asking a relatively small number of voters.

CHAPTER 9
INFERENCE FOR MEANS

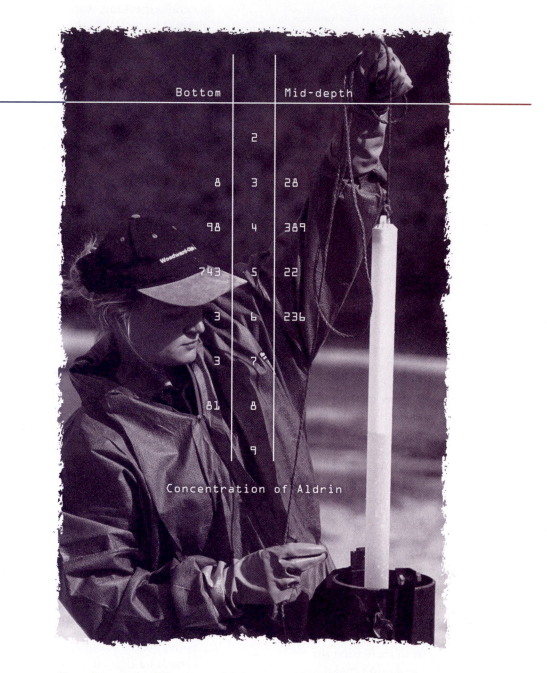

Does it make a difference whether you take water samples at mid-depth or near the bottom when studying pesticide levels in a river? Inference for differences is fundamental to statistical applications because many surveys and all experiments are comparative in nature.

You now know how to make inferences for proportions based on categorical data from random samples. But much of the data that arise in everyday contexts is measurement data rather than categorical data. Measurement data, as you know, are most often summarized by using the mean to represent a typical value. Mean income, mean scores on exams, mean waiting times at checkout lines, and mean heights of people your age are all commonly used to make decisions that affect your life.

In Chapter 8, you learned how to construct confidence intervals and do significance tests for proportions. Although the formulas you'll use for the inferential procedures in this chapter will change a little from those you just learned, the basic concepts remain the same. These concepts are all built around the question "What are the reasonably likely outcomes from a random sample?" This chapter will follow the same outline as Chapter 8 except that the methods are for means rather than proportions.

In this chapter, you will learn how to

- construct and interpret a confidence interval for estimating an unknown mean

- perform a significance test for a single mean

- construct and interpret a confidence interval for estimating the difference between two means

- perform a significance test for the difference between two means

9.1 Toward a Confidence Interval for a Mean

In this section, you will learn the logic behind a confidence interval for a population mean. Activity 9.1 will help you get into the spirit of making an inference about a mean.

Activity 9.1   What Makes Strong Evidence?

Consider the four data sets of Display 9.1, each coming from an experiment that measured the differences between sitting and standing pulse rates.

1. Which experiment gives the strongest evidence in favor of a larger mean pulse rate when people are standing than when they are sitting?

2. Rank the four experiments, from weakest to strongest, in terms of the evidence provided. Explain your reasoning.

3. What else would you like to know about these data, in addition to what you see in the boxplots, in order to improve your inference-making ability?

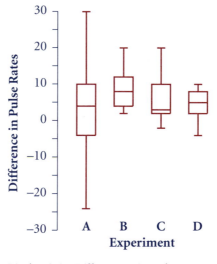

Display 9.1   Differences in pulse rates, standing minus sitting.

Reasonably Likely Outcomes

In Chapter 5, you saw that the sampling distribution of a sample proportion, $\hat{p}$, depends on just two numbers, the population proportion, p, and the sample size, n. In Chapter 8, you knew n and $\hat{p}$ and you wanted to estimate p. But now there are three numbers, not two, that determine the sampling distribution of the sample mean, $\bar{x}$: the population mean, μ, the population standard deviation, σ, and the sample size, n. You know n and $\bar{x}$ and you want to estimate μ—but usually you won't know σ. The examples in this section will be artificially simple. They will assume that you have a normal population and that you *do* know its standard deviation, σ. Sections 9.3 and 9.4 will show you how to adjust things to cover unknown standard deviations and non-normal shapes.

So let's begin with a situation in which you will actually know the shape and population standard deviation. Colleges and universities that rely on SAT scores as part of their admissions process typically provide summary information on the scores of first-year students. For many institutions, the distribution of scores is roughly normal, although the mean and standard deviation vary from one school to another. Here are the mean combined SAT scores (math plus verbal) for several universities.

University	Mean Combined SAT Score
Notre Dame	1345
Wake Forest	1300
North Carolina State	1175
Hofstra	1085
Albright	1015

Source: *U.S. News and World Report*, April 2002, www.usnews.com/usnews/edu/college/rankings/rankindex.htm.

The standard deviation σ of the distribution of SAT scores was about 150 for each of these universities. You will use these known populations to review the idea of "reasonably likely outcomes" and then use this idea to reach conclusions based on samples.

Example: SAT Scores

A first-year student is picked at random from one of the universities on the list.

   a. Suppose her combined SAT score is 1200. Is this score a reasonably likely outcome for a student from Wake Forest? From Albright?

   b. Suppose her combined score had been 980 instead of 1200. Is 980 a reasonably likely outcome for a randomly chosen student from Wake Forest? From Albright?

Solution

Using the "two standard deviation" rule, reasonably likely outcomes for Wake Forest students would lie between $1300 - 2(150)$ and $1300 + 2(150)$, or between 1000 and 1600. Similarly, the reasonably likely outcomes of scores for Albright students lie between $1015 - 300$ and $1015 + 300$, or in the interval 715 to 1315.

   a. The observed score of 1200 is in the interval 1000 to 1600, so it is a reasonably likely outcome for Wake Forest. It is a reasonably likely outcome for Albright as well.

   b. A score of 980 is in the interval (715, 1315) but not in (1000, 1600), so it is a reasonably likely outcome for Albright but not for Wake Forest.

You can see this reasoning more clearly by using a graph, much as in Chapter 8. In Display 9.2, the true mean for each of the five universities is plotted along the vertical axis and the SAT scores along the horizontal axis. The horizontal line segments represent the middle 95% of the SAT scores for that university. For example, Albright is represented by a horizontal line segment plotted at 1015 on the vertical axis (the bottom line on the graph), with x-values ranging from 715 to 1315.

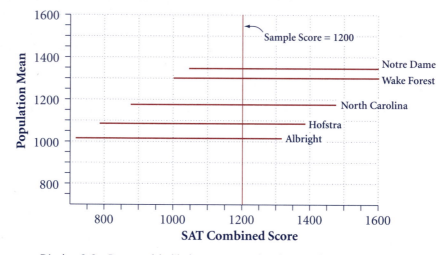

Display 9.2 Reasonably likely outcomes for the combined SAT scores.

The vertical line segment placed at a score of 1200 intersects all five horizontal segments, making it easy to see that a randomly chosen first-year student with a combined SAT score of 1200 could reasonably have come from any of the five universities. The use of this plot continues in the following discussion. ■

Discussion: Reasonably Likely Outcomes

D1. Suppose you randomly select a first-year student from a university with $\sigma = 150$.

  a. Her combined SAT score is 1390. Use Display 9.2 to indicate which universities could have produced a student with this score as a reasonably likely outcome.

  b. Her combined SAT score is x. To be a reasonably likely outcome from a university with mean μ, what is the farthest x can be from μ? Write an interval that gives the plausible values of μ. In other words, what values of μ have x as a reasonably likely outcome?

D2. Let the student scores along the horizontal axis in Display 9.2 be denoted by x and the university means along the vertical axis be denoted by μ. Let σ denote the population standard deviation.

  a. Write the equations of the parallel lines that intersect the endpoints of the horizontal segments.

b. Find the intersection of these lines with a vertical line at an arbitrary point, x.

c. Explain how these intersection points relate to the statement "For an observed score, x, the plausible values of μ are those between $x - 2\sigma$ and $x + 2\sigma$."

D3. All universities so far have had $\sigma = 150$. Now you will generalize to other standard deviations.

a. Suppose a randomly selected first-year student from a university with $\sigma = 100$ has a score of x. What is the set of means for which x is a reasonably likely outcome (in the middle 95% of the data)?

b. Repeat part a for a randomly selected student from a university with $\sigma = 210$.

You are already well aware of the fact that inferences should not be drawn from a single observation. Most often inference builds upon a random sample of n observations. In the next example, you will see how to handle cases when the sample size is larger than 1.

Example: Reasonably Likely Sample Means

Suppose you have a random sample of SAT scores for 25 first-year students from a university whose scores are normally distributed with $\sigma = 150$. Your sample has a mean of $\bar{x} = 1000$.

a. Is a sample mean of 1000 a reasonably likely outcome for Albright? For Hofstra?

b. Let μ represent the mean SAT score of all students at this university. For what values of μ would the sample mean of 1000 be reasonably likely?

Solution

The standard error of the sample mean is $\sigma/\sqrt{n} = 150/\sqrt{25}$, or 30.

The standard error of the sampling distribution of $\bar{x}$ is $SE = \sigma_{\bar{x}} = \sigma/\sqrt{n}$.

a. For Albright, then, reasonably likely sample means lie between $1015 - 2(30)$ and $1015 + 2(30)$, or from 955 to 1075. For Hofstra, the corresponding interval is 1025 to 1145. Therefore, the observed sample mean of 1000 is a reasonably likely outcome for Albright but not for Hofstra.

If the population is normal, so is the sampling distribution of $\bar{x}$.

b. If $\bar{x} = 1000$ is a reasonably likely sample mean for this unknown university, it must lie in the interval from $\mu - 2(30)$ to $\mu + 2(30)$:

$$\mu - 2(30) < 1000 < \mu + 2(30)$$

Solving this inequality for μ,

$$1000 - 2(30) < \mu < 1000 + 2(30)$$

or $940 < \mu < 1060$. Any population mean in this interval plausibly could have produced a sample mean of 1000. ∎

Display 9.3 shows the reasonably likely outcomes for means of samples of 25 SAT scores from the five universities. Notice that the horizontal lines here are much shorter than those in Display 9.2, reflecting the decrease in the standard error from 150 to $150/\sqrt{25}$. You can find the answer to part a of the last example by drawing a vertical line upward from a sample mean of 1000 and seeing for which universities it is a reasonably likely outcome.

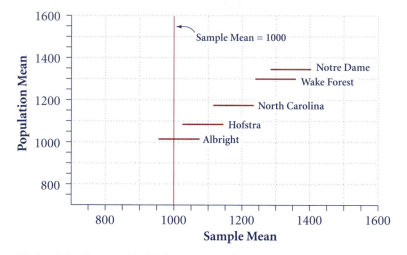

Display 9.3   Reasonably likely outcomes for sample means ($n = 25$), with a sample mean of 1000 indicated.

Confidence Intervals

The goal of statistical inference for means is to decide which populations plausibly could have produced the observed sample mean $\bar{x}$. We call the means of these populations a "confidence interval" for μ.

> A **confidence interval** for a population mean μ is the set of values for μ that make the observed mean of the random sample, $\bar{x}$, a reasonably likely outcome. The definition of reasonably likely will depend on the **confidence level** desired (typically 95%, as you have been using).

A formula follows from the idea in the last example. The mean of a random sample is a reasonably likely outcome from a normally distributed population if it falls within two standard errors of the mean of the population:

$$\mu - 2 \cdot SE < \bar{x} < \mu + 2 \cdot SE$$

where $SE = \sigma_{\bar{x}} = \sigma/\sqrt{n}$.

As in part b of the last example, you can write this in another algebraically equivalent way:

$$\bar{x} - 2 \cdot SE < \mu < \bar{x} + 2 \cdot SE$$

The first interval, $\mu \pm 2 \cdot SE$, is the one you're used to. Here you start with a population mean, μ, and find the interval within two standard errors of μ. This interval contains the values of $\bar{x}$ that are reasonably likely.

For the second interval, $\bar{x} \pm 2 \cdot SE$, you start with a sample mean, $\bar{x}$, and find the interval that contains all values of μ that fall within two standard errors of $\bar{x}$. This interval contains all the values of μ that are plausible as the population mean. In other words, each population mean in the interval $\bar{x} \pm 2 \cdot SE$ plausibly could have produced the observed $\bar{x}$.

If you want a level of confidence different from 95%, you would use a value of z different from $z \approx 2$.

A Confidence Interval for a Mean: Normal Population, σ Known

A confidence interval for the mean μ of a normally distributed population is given by

$$\bar{x} \pm z^* \cdot \frac{\sigma}{\sqrt{n}}$$

Here n is the sample size, σ is the population standard deviation, and z^* depends on the confidence level desired. Use

$$z^* = \begin{cases} 1.645 & \text{for 90\% confidence} \\ 1.960 & \text{for 95\% confidence} \\ 2.576 & \text{for 99\% confidence} \end{cases}$$

If the sample size is large, the Central Limit Theorem guarantees that this procedure also gives good results for populations that aren't normal.

Notice that, as in Chapter 8, the formula has the form

$$estimate \pm z^* \cdot (standard\ error\ of\ the\ estimate)$$

The quantity

$$E = z^* \cdot \frac{\sigma}{\sqrt{n}}$$

is called the **margin of error.** It is half the width of the confidence interval.

Example: Confidence Interval for a Mean, σ Known

A random sample of 16 students taken from an unspecified university whose SAT scores have a standard deviation of 100 has a mean of $\bar{x} = 942$. Find a 95% confidence interval for the university mean.

Solution

The 95% confidence interval is

$$\bar{x} \pm z^* \cdot \frac{\sigma}{\sqrt{n}} = 942 \pm 1.96 \cdot \frac{100}{\sqrt{16}}$$

$$= 942 \pm 49$$

or 893 to 991. We are 95% confident that the mean SAT score of all students in this university is in the interval 893 to 991. ∎

Discussion: Confidence Interval for a Mean

D4. A first-year student is chosen at random from a university where the combined SAT scores have a normal distribution with $\sigma = 150$, but the mean is unknown.

    a. If that student's score is 1000, find three confidence intervals for the unknown mean, with confidence levels of 90%, 95%, and 99%, respectively.

    b. Does the mean for Notre Dame belong to the 90% interval? The 95% interval? The 99% interval?

D5. A random sample of 16 first-year students is chosen at random from a university where the SAT scores have a normal distribution with $\sigma = 150$.

    a. If the mean score for that sample is 1000, find and interpret three confidence intervals for the unknown mean, with confidence levels of 90%, 95%, and 99%, respectively.

    b. Does the mean for Notre Dame belong to the 90% interval? The 95% interval? The 99% interval?

D6. "Let's be sure to capture μ by producing an interval with a 100% confidence level," a student says. Is this possible? Would it be wise to shoot for a 99.5% confidence level?

D7. Explain why the logic of confidence intervals as described in the explanation of part b of the *Reasonably Likely Sample Means* example on page 483 breaks down if you try to apply it to a set of universities with different standard deviations, such as Cal Tech ($\sigma = 100$), Wake Forest ($\sigma = 150$), University of the Pacific ($\sigma = 210$), and so on.

D8. Explain why the population must be normal to construct and interpret the confidence intervals in D4 and D5.

D9. What happens to the margin of error if all else remains the same except that

    a. the standard deviation of the population is increased?

    b. the sample size is increased?   c. the confidence level is increased?

■ Practice

P1. You take an SRS of 9 SAT scores from first-year students from a university with standard deviation 150 and get a sample mean of 1110.

 a. Find and interpret 90%, 95%, and 99% confidence intervals for the unknown mean.

 b. Does the mean for Albright belong to the 90% interval? The 95% interval? The 99% interval?

P2. You have produced a confidence interval for your supervisor and she says, "This interval is too long to be of any practical value." What are your options for producing a shorter interval in the next study of this same type? Which option would you choose if the study may have consequences that are vital to the future of the firm?

P3. Suppose there is a population of test scores on a large, standardized exam for which the mean is unknown but the standard deviation is known. Two different random samples of 50 scores are taken from the population. A 95% confidence interval is constructed for each sample. Which of the following statements most likely would be true of these two confidence intervals?

 A. They would have identical values for the lower and upper limits of the confidence interval.

 B. The confidence intervals would have the same width but different lower and upper limits.

 C. The confidence intervals would have different widths.

 D. There is no way to compare the two intervals.

Capture Rate

The capture rate of this method for producing confidence intervals is the chance that the result will contain the population mean.

Just as in Chapter 8, the **capture rate** for a confidence interval is the proportion of random samples for which the resulting confidence interval captures the true (population) value. You can think of this as the chance that the method used to produce a confidence interval will actually work correctly.

Consider this scenario, which is unrealistic but helpful in understanding capture rate. An admissions official is to choose a random sample of 25 first-year students from Wake Forest and use their SAT scores to construct a 95% confidence interval estimate of the population mean score. The distribution of the SAT scores for Wake Forest, with its standard deviation of 150, actually has a mean of 1300. For $n = 25$, $\sigma/\sqrt{n} = 30$, and the interval of reasonably likely values for $\bar{x}$ is 1240 to 1360. If the random sample of scores turns out to have a mean of 1200, the confidence interval of $1200 \pm 2(30)$ will not capture the true mean. If the sample mean is 1320, then the interval of $1320 \pm 2(30)$ will capture the true mean. In fact, the confidence interval will capture the true mean as long as the observed sample mean is between 1240 and 1360. Therefore, the chance of the confidence interval capturing the true mean is .95, the same as the confidence level.

Remember that the capture rate is a property of the method, not of the sample. In the scenario above, the admissions official had a .95 chance of capturing the population mean in her confidence interval before the sample was selected. After the sample was selected, either the resulting interval captured the population mean or it didn't. So all she can say is that she is 95% confident that μ is in her interval, meaning that the method produces a confidence interval that captures μ, the true mean, 95% of the time.

> ### The Capture Rate for Confidence Intervals for the Mean
>
> The proportion of intervals of the form $\bar{x} \pm z^* \cdot \sigma/\sqrt{n}$ that capture μ is equal to the confidence level. That is, a 95% interval ($z^* = 1.96$) will have a 95% capture rate, a 90% interval ($z^* = 1.645$) will have a 90% capture rate, and so on. This will be true *provided that*
>
> - your sample is random (or experimental treatments are randomly assigned)
> - the population is normal
> - the standard deviation is known
>
> The capture rate will be approximately correct for non-normal populations as long as the sample size is large enough.

When assumptions are met, the capture rate is the same as the confidence level.

In practice, samples may not be random, distributions are often not normal, and you almost never know the standard deviation. All the same, the result about the capture rate is important because it allows you to estimate, at least, how often the method will produce a "correct" result.

Discussion: Capture Rate

D10. Is the capture rate affected by

    a. changes in the sample size?

    b. the size of the population standard deviation?

    c. the confidence level?

D11. "The capture rate equals the confidence level." Explain why this statement depends on

    a. the randomness of the sampling

    b. the normality of the population

What Sample Size Should You Use?

Provided your samples are random, larger samples provide more information than smaller ones, and this fact is reflected in the relationship between the margin of error and the sample size. As the sample size increases, the margin of error decreases at a rate proportional to the square root of the sample size. To cut the margin of error in half, you have to quadruple the sample size. Here is the rule that tells you the sample size needed in order to have a specified margin of error.

To have a margin of error, E, you need a sample size of

$$n = \left(\frac{z^* \cdot \sigma}{E}\right)^2$$

Example: Enzyme Concentrations

A biology student plans an honors project that will require her to measure enzyme concentrations. Each independent measurement takes her about a half hour. Assume the measurements behave like a random sample from a normal population and that the standard deviation of the measurements is 0.1 moles/liter.

a. Find the margin of error for a 90% confidence interval based on four independent measurements.

b. If the student wants a 95% confidence interval to have width 0.01, how many independent measurements will she need?

Solution

The margin of error is $E = z^* \cdot \sigma/\sqrt{n}$.

a. For $n = 4$ and 90% confidence,

$$E = 1.645 \cdot \frac{0.1}{\sqrt{4}} = 0.08225$$

b. The margin of error, E, is half the width of the confidence interval, so the student wants $E = \frac{0.01}{2} = 0.005$. For a 95% interval, $z^* \approx 2$. Thus,

$$n = \left(\frac{z^* \cdot \sigma}{E}\right)^2 \approx \left(\frac{2 \cdot 0.1}{0.005}\right)^2 = 1600$$

At a half hour per measurement, 1600 measurements would take 20 weeks of full-time work. (The student can use this information to rethink her plans for the project.) ■

■ Practice: Choosing a Sample Size

P4. You want to estimate the mean resting pulse rate for athletes, and you know that for many populations the distribution of resting pulse rates is roughly normal with $\sigma \approx 10$. How many athletes would you need in an SRS if you want a 95% confidence interval to have width 1?

P5. You want to estimate the mean number of hours of sleep of students who hold part-time jobs. Assume that the distribution of hours of sleep is roughly normal with $\sigma \approx 2$ hours. What size SRS would give an interval with a margin of error of 10 minutes?

Summary 9.1: The Logic Behind a Confidence Interval for a Mean

In this section, you have learned how to construct a confidence interval for a population mean μ, but you can use it only if you know σ and only if either the population is normal or the sample size is large.

Suppose $\bar{x}$ is the mean of a simple random sample of measurements from a normal population with known standard deviation σ and unknown mean μ. Then the 95% confidence interval for μ is

$$\bar{x} \pm 1.96 \cdot \frac{\sigma}{\sqrt{n}}$$

Here is what you need to know to interpret this confidence interval:

- The set of values in the confidence interval are those values of μ that make $\bar{x}$ a reasonably likely outcome.

- The method of constructing a 95% confidence interval ensures that the chance of getting a value of $\bar{x}$ whose interval captures μ is equal to 95%. This is called the *capture rate* or *confidence level*.

If you want 90% confidence, then the interval takes the form $\bar{x} \pm 1.645 \cdot \sigma/\sqrt{n}$; if you want 99% confidence, the interval takes the form $\bar{x} \pm 2.576 \cdot \sigma/\sqrt{n}$.

To obtain a specified margin of error, E, the required sample size is

$$n = \left(\frac{z^{*} \cdot \sigma}{E} \right)^{2}$$

■ Exercises

E1. A random sample of four colleges from the *U.S. News and World Report* list of the top 50 U.S. liberal arts colleges has a mean acceptance rate of 52%. (The colleges were Scripps, Davidson, Bryn Mawr, and Hamilton.) The acceptance rates for all 50 colleges range from 19% to 90% and are mound-shaped.

a. Based on your knowledge of distributions, approximate a standard deviation for these percentages. Then use it to construct a 95% confidence interval estimate of the true mean acceptance rate of the 50 colleges.

b. What is the variable and what are the cases in this problem?

c. Explain carefully what the interval found in part a represents. (What is the population being studied here?)

E2. Going back to the *SAT Scores* example, Dartmouth has a mean SAT score for first-year students of about 1490, with a standard deviation of about 150. We intentionally left Dartmouth out of the discussion of confidence intervals for mean SAT score. Why?

E3. Suppose a large sample, with $n = 100$, is going to be taken from a population of weights of children. A 90% confidence interval will be constructed to estimate the mean weight of children. A smaller sample, with $n = 50$, will also be taken from the same population of weights, and a 99% confidence interval will be constructed to estimate the mean weight. Which of these methods will have a better chance of producing a confidence interval that captures the population mean weight? Explain the reasoning you used in making your choice.

A. The 90% confidence interval based on a sample of 100 weights.

B. The 99% confidence interval based on a sample of 50 weights.

C. Both methods have an equal chance.

D. I can't determine which will have a better chance.

E4. A 95% confidence interval for a population mean is calculated for a sample of weights, and the resulting confidence interval is 42 to 48 lb. For each of these statements, indicate whether it is a true or false interpretation of the confidence interval. Then explain why the false statements are false.

a. Ninety-five percent of the weights in the population are between 42 and 48 lb.

b. Ninety-five percent of the weights in the sample are between 42 and 48 lb.

c. The probability that the interval includes the population mean, μ, is 95%.

d. The sample mean, $\bar{x}$, may not be in the confidence interval.

e. If 200 confidence intervals were generated using the same process, about 10 of the confidence intervals would not include the population mean.

E5. Heights of women in the United States are normally distributed and have a standard deviation of about 2 inches. Suppose you want to estimate the mean height of all women in your city from a random sample

of size n at the 95% confidence level. What sample size would you need in order to estimate the mean height to within

a. two inches?

b. one inch?

c. one-half inch?

E6. This is the formula for the required sample size for a desired margin of error E:

$$n = \left(\frac{z^* \cdot \sigma}{E} \right)^2$$

Show how it follows from the formula for a confidence interval.

E7. Jack and Jill have opened a water-bottling factory. The machine that fills the bottles is fairly precise. The distribution of the number of ounces of water in the bottles is normal with $\sigma = 0.06$ oz. The mean μ is supposed to be 16 oz, but the machine slips away from that occasionally and has to be readjusted. Jack and Jill take a random sample of 10 bottles from today's production and weigh the water in each. The weights are

15.91	16.08	16.08	15.94	16.02
15.94	15.96	16.03	15.82	15.96

Should Jack and Jill readjust the machine? Use a statistical argument to support your advice.

E8. A 95% confidence interval is to be constructed from a random sample selected from a population with $\sigma = 5.11$. A 90% confidence interval is to be constructed from a random sample of the same size selected from a population with $\sigma = 6.09$. If these two constructions are made over and over, explain how the performances of these intervals differ. Sketch a picture to illustrate your explanation.

9.2   Toward a Significance Test for a Mean

Going back to the SAT score scenario of Section 9.1, suppose a student file shows up with a combined SAT score of 1010 and a sticky note attached to the file says that the student has entered Notre Dame. Looking back at Display 9.2, you can see that a score of 1010 is not a reasonably likely outcome for Notre Dame. Thus, you would be within your statistical rights to check to see if the sticky note might really belong on another file.

This reasoning is a simple example of a significance test. Significance tests and confidence intervals rely on the same basic idea: Be suspicious of any model that assigns low probability to what actually happened. In the language of sampling, be suspicious of any claimed population value (for example, μ or p) if it puts the summary statistic ($\bar{x}$ or $\hat{p}$) outside the range of reasonably likely outcomes.

Now, suppose you have a random sample of 25 SAT scores for first-year students from a single university that is among the set of universities with $\sigma = 150$ but the label that identified which university these scores came from was lost. The sample mean is 1010. Is it plausible that the sample came from Albright, which has a population mean score of 1015?

You can answer this question by computing a z-score or by looking at Display 9.4. The z-score in this case involves a sample mean:

$$z = \frac{\bar{x} - \mu}{\sigma/\sqrt{n}} = \frac{1010 - 1015}{150/\sqrt{25}} = \frac{-5}{30} = -0.1667$$

A sample mean that is only one-sixth of a standard error from Albright's mean is well within the two-standard-error boundary for reasonably likely outcomes, so it is quite plausible that this sample of scores came from Albright. As shown in Display 9.4, a vertical line at a sample mean of 1010 is well within the range of reasonably likely outcomes for Albright.

Is it plausible that the observed sample came from Hofstra? Hofstra has a mean score of 1085, so the z-score is

$$z = \frac{\bar{x} - \mu}{\sigma/\sqrt{n}} = \frac{1010 - 1085}{150/\sqrt{25}} = -2.5$$

Now we have a sample mean that is 2.5 standard errors away from Hofstra's mean, and this is beyond the two-standard-error boundary, so it does not seem plausible that this sample came from Hofstra. Again, Display 9.4 provides a graphical view of the situation. The vertical line at 1010 just misses the range of reasonably likely outcomes for Hofstra.

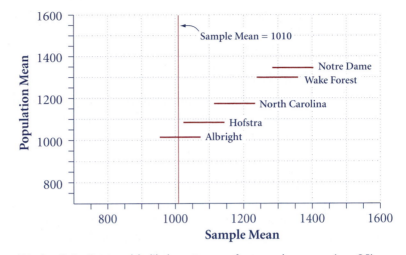

Display 9.4 Reasonably likely outcomes for sample means ($n = 25$), with a sample mean of 1010 indicated.

Discussion: Plausible Values for Means

D12. A random sample of 25 SAT scores for first-year students from a single university yielded a mean score of 1300. The university is from the set of universities with $\sigma = 150$.

    a. Using Display 9.4, is it plausible that the sample came from Hofstra? From Wake Forest?

    b. Verify your decision by computing the z-statistic for Hofstra and then for Wake Forest.

Fixed-Level Testing

With a confidence interval, you were looking for the set of population means for which the observed sample mean is a reasonably likely outcome. In a significance test, you have a single hypothesized population mean in mind and you are asking the question "Is it plausible that a population with this hypothesized mean produced the observed sample mean?" If your hypothesized population mean is in the confidence interval, the answer to the question is "yes." Otherwise, the answer is "no."

    As you saw previously and in Chapter 8, you do not have to calculate the confidence interval to answer the question about the hypothesized mean—you can answer it by calculating the z-score. Once you have this z-score, you compare it to a predetermined critical value z^*.

z^* is the same value that would be used as a multiplier in the corresponding confidence interval.

    When you are asked to compute a confidence interval, you are given the desired confidence level or capture rate, typically 95%. You then find the critical value z^* to use in the formula for a confidence interval. When you are asked to do a test of significance, you will be given the desired level of significance, α, typically 5%. This is because a significance test's performance is usually measured by its chance of *rejecting* a hypothesized value rather than the chance of capturing it. Thus, the **level of significance** (or simply the **level**) of a significance test is the complement of the corresponding confidence level, and you use the same value of z^* for both the confidence interval and a two-tailed test. This is illustrated in the next example.

Example: Finding Critical Values

If you are doing a two-tailed fixed-level test of significance with $\alpha = .02$, what is your critical value? Do you reject the null hypothesis if your test statistic is $z = 2.01$?

Solution

z	.02	.03
−2.30	.0102	.0099

The values of z that cut off upper and lower areas of $\frac{.02}{2} = .01$ of a normal distribution are closest to −2.33 and 2.33. So the critical value is $z^* \approx 2.33$. You would not reject the null hypothesis, because $z = 2.01$ does not fall in the critical region. ■

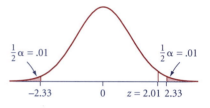

Discussion: Fixed-Level Testing

D13. What are the critical values in Table A for rejecting hypotheses at the 10% level, the 5% level, and the 1% level for a two-tailed test? For a one-tailed test?

D14. The University of the Pacific (UOP) has a mean combined SAT score of 1100 for first-year students, with $\sigma = 210$.

    a. Suppose a random sample of 25 SAT scores for first-year students from a single university has a mean of 1170. Would you reject the claim that this sample came from UOP students at the 10% level? At the 5% level? If $\alpha = .01$?

    b. In answering part a, did you have to imagine that UOP belonged to a whole set of different universities, all with $\sigma = 210$, and visualize a plot similar to that in Display 9.4?

The Formal Structure of Tests of Significance

Tests of significance for means have the same general structure as tests for proportions, although some of the details are a bit different.

Components of a Fixed-Level Test for a Mean

1. ***Name the test and check conditions.*** For a test of significance of a single mean, the methods of this section require that

- the sample was selected at random (or, in the case of an experiment, treatments were randomly assigned to units)

- the population is normal (or your sample size is large) with unknown mean μ

- the standard deviation σ of the population is known

(continued)

2. ***State your hypotheses.*** The null hypothesis is that the population mean μ has a particular value μ_0. This is typically abbreviated $H_0: \mu = \mu_0$.

   The alternate hypothesis can have any one of three forms:

   $$H_a: \mu \neq \mu_0 \quad H_a: \mu < \mu_0 \quad H_a: \mu > \mu_0.$$

3. ***Compute the test statistic, compare it to the critical values, and draw a sketch.*** The test statistic is the distance from the sample mean, $\bar{x}$, to the hypothesized value μ_0 measured in standard errors:

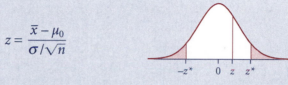

$$z = \frac{\bar{x} - \mu_0}{\sigma / \sqrt{n}}$$

   Compare the value of z to the critical value z^*. The critical value z^* for a two-tailed test is

   $$z^* = \begin{cases} 1.645 & \text{for } \alpha = .10 \\ 1.960 & \text{for } \alpha = .05 \\ 2.576 & \text{for } \alpha = .01 \end{cases}$$

4. ***Write your conclusion linked to your computations and in the context of the problem.*** If the alternate hypothesis is two-sided ($H_a: \mu \neq \mu_0$), you reject H_0 if $z > z^*$ or $z < -z^*$. Then give a conclusion that relates to the situation.

Example: Mean Weight of Pennies

Newly minted U.S. pennies are targeted to have a mean weight of 3.11 g, but there is a manufacturing "tolerance" of 0.13 g. That is, about 95% of all weights are within 0.13 g of the mean. As you saw in Chapter 2, the weights of manufactured pennies are approximately normally distributed, so $0.13 \approx 2\sigma$. A sample of 9 pennies taken from a production line had a mean weight of 3.16 g. Using a 5% level of significance, does it look like this production line has a mean that has moved off target?

Solution

Check conditions.

The problem asks a question about the mean of a normal distribution and provides data from a random sample with which to study the claim. The standard deviation of the population is known to be half the tolerance, or 0.065 g.

State H₀ and Hₐ.

The hypotheses are

H_0: $\mu = 3.11$, where μ denotes the mean weight of all pennies on this production line

H_a: $\mu \neq 3.11$ because the mean could have moved off target in either direction

Compute the value of the test statistic and draw a sketch.

The test statistic is

$$z = \frac{\bar{x} - \mu_0}{\sigma / \sqrt{n}}$$

$$= \frac{3.16 - 3.11}{.065 / \sqrt{9}} = 2.31$$

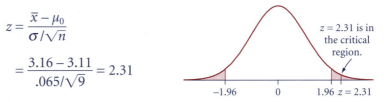

$z = 2.31$ is in the critical region.

Write a conclusion linked to your computations and in the context of the situation.

For a 5% significance level, the critical value is $z^* = 1.96$. Because $z = 2.31$ falls in the critical region, there is sufficient evidence to reject the hypothesis that the manufacturing process is still on target. ■

The smaller the significance level (the larger the critical value) you choose, the stronger you require the evidence to be in order to reject H_0. The stronger the evidence you require, the less likely you are to make a Type I error, that is, to reject H_0 when it is true.

In the practical problems you will see in later sections, the alternate (research) hypothesis is usually described first and the null hypothesis is the standard against which the alternate is compared. For example, in Activity 4.5, you collected data on change in pulse rates as people moved from sitting to standing positions. The research hypothesis is that "the mean pulse rate changes as people change position." The standard against which this would be compared—the null hypothesis—is that "there is no change in the mean pulse rate." The term *null* is used here to reflect the idea that for these baseline standards, "nothing is changing."

Discussion: Formal Structure of Fixed-Level Tests

D15. Suppose you randomly choose four first-year students from an unknown university, with $\sigma = 150$. The four students have a mean combined SAT score of 1173. Using $\alpha = .05$, carry out the four steps of a test of the hypothesis that the students are from Notre Dame, which has a mean of 1345. Do you come to the same conclusion at the 10% level? At the 1% level?

D16. A first-year student is chosen at random from one of the universities with $\sigma = 150$. Based on her SAT score, a 5% test rejects the hypothesis that she is from North Carolina State, which has a mean of 1175, but a 1% test fails to reject the same hypothesis. Tell as much as you can about her score.

■ Practice

P6. A random sample of four colleges from the *U.S. News and World Report* list of the top 50 U.S. liberal arts colleges showed a mean acceptance rate of 52%. (The colleges were Scripps, Davidson, Bryn Mawr, and Hamilton.)

The acceptance rates for all 50 colleges range from 19% to 90% and are mound-shaped. Test the hypothesis that the mean acceptance rate for all of the colleges differs from 50%. Write a statement summarizing your conclusion that could be understood by a newspaper reader who has never taken a statistics course.

P7. A random sample of 9 first-year students is chosen at random from one of the universities with $\sigma = 150$. The mean SAT score for the sample is 1350. The null hypothesis is that the sample is from Wake Forest. Find the value of the test statistic z, and state your conclusion. Use a fixed-level test with $\alpha = 5\%$.

P8. A random sample of 36 first-year students is chosen from an unknown university with $\sigma = 150$, and the sample mean is 1150. The null hypothesis is that the sample comes from Wake Forest. Find the value of the test statistic, and state your conclusion using a 10% fixed-level test.

P9. You take an SRS of 9 first-year students from a university with $\sigma = 150$ and get a sample mean of 1000. Do you reject the hypothesis that the students are from a university with mean 1085 at the 10% level? At the 5% level? At the 1% level?

The Meaning of *Reject*

Just as in Chapter 8, a large value of the test statistic z tells you there is a large distance between your sample statistic $\bar{x}$ and the standard μ_0. There are three ways you can get a large value of z: The model is wrong, a rare event has occurred, or the null hypothesis is false.

- **Wrong model:** The model you used (that you have a random sample from a normal population or a random sample with a large enough sample size) doesn't fit the data.

- **Rare event:** The model fits and the null hypothesis is true, but just by chance an unlikely event occurred and gave you an unusually large value of the test statistic.

- **Reject is right:** The model fits but the true population mean is not μ_0. You are right to reject H$_0$.

In general, if you know you have the right model so that either a rare event occurred or the null hypothesis is false, standard practice is to reject the null hypothesis.

There's only one way you can be confident of having the right model: Use sound methods of data production, either random samples or randomized experiments. At the other extreme, there are many ways you can be confident that the model doesn't apply. For example, voluntary-response samples are worthless when it comes to inference. Many situations fall between the two extremes. What then? For example, what if you use your class as a sample instead of taking a random sample? There are no formal rules; the value of the inference methods is rarely all-or-nothing. The more reasonable it is to regard your data as coming from a random sample or randomized experiment, the more reasonable it is to trust your conclusions based on the inference methods. For many data sets, making a careful judgment about this issue is the hardest aspect of a statistician's job.

Discussion: The Meaning of *Reject*

D17. Compare the lists of components of significance tests for proportions (page 439) and for means (page 494). List all the similarities you find. What are the main differences?

D18. Suppose you are doing fixed-level testing at the 1% level. Which of these questions can be answered?

   I. What percentage of tests will reject the null hypothesis?

   II. If the null hypothesis is true, what is the probability it will be rejected? That it will not be rejected?

   III. If the null hypothesis is false, what is the probability it will be rejected? That it will not be rejected?

P-Values: The Observed Significance Level

Fixed-level testing has a major drawback: It boils all the data down to the tiny residue of "reject" or "don't reject." To see what gets lost, think about a court trial, where "not guilty" can mean anything from "This guy is so innocent, he should never have been brought to trial" to "Everyone on the jury thinks the defendant did it, but the evidence presented isn't quite strong enough." In the same way, "do not reject" reports only a small fraction of the information in the data. A standard way to improve on fixed-level testing is to report a *P*-value, sometimes called an **observed significance level,** just as in Chapter 8.

Finding the *P*-Value

A **P-value** is the probability of seeing a result as extreme as or even more extreme than the one observed, when the null hypothesis is true.

1. Compute the value of the test statistic *z*.

2. Find the area under a standard normal curve that corresponds to outcomes as extreme as or even more extreme than the test statistic, as shown in Display 9.5 for a two-tailed test.

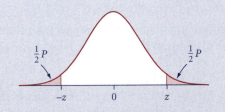

Display 9.5   A standard normal curve with *P*-value shaded.

Example: P-Value for the Pennies

Consider once again the pennies in the previous example. Find the *P*-value for testing to see whether the population mean has moved off target.

Solution

The test statistic z was 2.31.

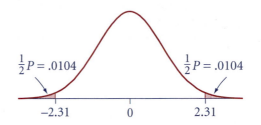

$\frac{1}{2}P = .0104$ $\frac{1}{2}P = .0104$

−2.31   0   2.31

The smaller the P-value, the stronger the evidence against the null hypothesis.

The P-value is the area under a normal curve above $z = 2.31$ and below $z = -2.31$, which turns out to be $2(.0104)$, or $.0208$. Such a small P-value indicates that the sample is quite inconsistent with the null hypothesis. If the null hypothesis is true, there is only a 2.08% chance of getting a sample mean this far from 3.11. It looks like the population mean has moved off target. ◼

Discussion: *P*-Values

D19. Study the formula for the test statistic when testing a hypothesis about a population mean, and think about the relationship of the test statistic to the P-value.

    a. What happens to the P-value if the population standard deviation increases but everything else remains the same?

    b. What happens to the P-value if the sample size increases but everything else remains the same?

◼ Practice

P10. Refer to P6, P7, and P8. Find the P-values for each of the three significance tests. Relate the P-values to the conclusion you came to when you worked these problems in the fixed-level significance test scenario.

P11. Suppose a random sample of 100 SAT scores is to be taken from a university with $\sigma = 150$. The null hypothesis is that the university has a mean of 1175. Find the values of the sample mean that would make the P-value for this significance test

    a. .10   b. .05   c. .01

One-Sided Alternatives, One-Tailed Tests

Sometimes it makes sense to consider alternatives that depart from the standard in only one direction.

So far in this section, every alternate hypothesis has been two-sided: The true mean, μ, is not equal to the standard, μ_0. But there are two other possible alternatives—μ might be less than the standard, or μ might be greater than the standard. In real applications, you can sometimes use the context to rule out one of those two possibilities as meaningless, impossible, uninteresting, or irrelevant.

In *Martin v. Westvaco* (in Chapter 1), a statistical analysis compared the ages of the workers who were laid off with the ages of those in the population of employees working for Westvaco at the time of the layoff. An average age for the laid-off workers that was greater than the population mean would tend to support a claim of age discrimination. On the other hand, the opposite inequality is not relevant: If the average age of those laid-off had been less than the overall average, that would not be evidence of discrimination, because younger workers aren't protected under the law.

When the context tells you to use a one-sided alternate hypothesis, you need to adjust your *P*-value accordingly. These adjusted *P*-values are called one-tailed or one-sided *P*-values, and the corresponding test is called a one-tailed or one-sided test, just as in Chapter 8.

Example: One-Sided P-Values

A magazine reports that the mean number of minutes students at a particular university study each week is 1015 minutes. A dean says she believes the mean is greater than 1015 minutes. To test her claim, she checks a random sample of 64 students and finds the sample mean is 1050 minutes. Is this strong evidence in favor of her claim? Assume the study times at this particular university have *SD* = 150 minutes.

Solution

Check conditions. A random sample of size 64 would produce a nearly normal sampling distribution for the sample mean even if the population was not normal. This requires a one-tailed test because the official is interested only in the alternate (research) hypothesis that the population mean may be larger than reported.

State your hypotheses.

    H_0: $\mu = 1015$ where μ denotes the mean number of minutes studied for all of the university's students

  H_a: $\mu > 1015$

Compute the test statistic and draw a sketch.

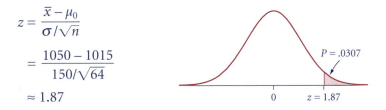

$$z = \frac{\bar{x} - \mu_0}{\sigma/\sqrt{n}}$$

$$= \frac{1050 - 1015}{150/\sqrt{64}}$$

$$\approx 1.87$$

$P = .0307$

$z = 1.87$

The *P*-value is the area to the right of 1.87 under the standard normal curve, which turns out to be about .0307.

Conclusion in context. This is fairly strong evidence in favor of her claim that the population mean is larger than 1015. You would tend to reject the null hypothesis with a *P*-value this small. Note that in this one-sided scenario, the *P*-value is well below .05 even though the *z*-score is smaller than 2. ■

Discussion: One-Tailed *P*-Values and One-Tailed Tests

D20. Suppose the value of your test statistic is 1.645.

    a. Draw a picture of a standard normal curve and shade the area that corresponds to the *P*-value for a two-tailed test.

    b. Now identify the part of the shaded area that corresponds to evidence that $\mu > \mu_0$. What is the relationship between the two-tailed and one-tailed *P*-values for a given value of the test statistic?

    c. How big must the test statistic be to give a one-tailed *P*-value equal to the two-tailed *P*-value for $z = 1.645$?

    d. If all else is equal, will the *P*-value be larger for a one-tailed or for a two-tailed test?

For D21 and D22, do these steps.

    a. Decide whether the alternate hypothesis should be one-sided or two-sided.

    b. Tell what $\bar{x}$ and μ are for the problem.

    c. State the null and alternate hypotheses.

D21. The board of the Gainesville Home Builders Association knows the selling price of all new houses sold in the city last month and believes the mean selling price has gone up this month. To check this claim quickly, they obtain a random sample of the selling prices for houses sold this month.

D22. Theory says that if you dissolve salt in water, the freezing point will be lower than it is for pure water. To test the theory, you plan to dissolve a teaspoon of salt in a bowl of water, put a thermometer in the mixture, and then put the bowl in your freezer, checking periodically so you can observe the temperature at which the solution begins to freeze. (If you actually try this, you need to check often enough to be sure of removing the thermometer once ice begins to form, or you may end up with a broken thermometer.)

D23. Invent two situations, one for which a two-tailed test is more appropriate and one for which a one-tailed test is more appropriate.

■ Practice

P12. Theory says that at high altitude, water boils at a lower temperature than it does at sea level. You plan to test this theory by flying to the mile-high Denver airport and observing the boiling point of water there. Decide whether the alternate hypothesis should be one-sided or two-sided. Define $\bar{x}$ and μ, and then state the null and alternate hypotheses in words and in symbols.

P13. Gas mileage for a certain brand of compact car is reported by the manufacturer to be 35 mpg. Describe a scenario in which the appropriate alternate hypothesis would be $\mu > 35$. Now describe a scenario in which the appropriate alternate hypothesis would be $\mu < 35$.

Summary 9.2: Toward a Significance Test for a Mean

Suppose $\bar{x}$ is the mean of a simple random sample of measurements from a normal population (or, if not normal, you have a large enough sample size) with known standard deviation σ and unknown mean μ. A claim is made about a value for μ, and you want to conduct a test of the significance of this claim.

- The null hypothesis gives the hypothesized, or standard, value for μ.

- The alternate (research) hypothesis conjectures how the true mean differs from the standard given in the null hypothesis.

- The test statistic gives the distance from the summary statistic, $\bar{x}$, to the standard, μ_0, measured in standard errors:

$$z = \frac{\bar{x} - \mu_0}{\sigma_{\bar{x}}} = \frac{\bar{x} - \mu_0}{\sigma/\sqrt{n}}$$

- If the null hypothesis is true, the sampling distribution of the test statistic, z, will be standard normal.

- In a fixed-level test, you compare the observed test statistic z to a critical value of z^*, which depends on the preassigned significance level α of the test. For a two-tailed test, if $|z| > z^*$, you reject the null hypothesis. Otherwise, you do not reject it. The level of significance gives the chance that the test will reject a true null hypothesis.

- The observed significance level (P-value) gives the probability of seeing a result as extreme as or even more extreme than the one observed, when the null hypothesis is true. It serves as a measure of the evidence against the null hypothesis. The smaller the P-value, the greater the evidence in the sample against the null hypothesis.

■ Exercises

E9. Select the best answer. A P-value measures

    A. the probability that the null hypothesis is true

    B. the probability that the null hypothesis is false

    C. the probability that an alternate hypothesis is true

    D. the probability of seeing a result at least as far from μ_0 as the one observed, given that the null hypothesis is true

E10. Select the best answer. The confidence level measures

    A. the fraction of times the confidence interval will capture the parameter it is estimating in repeated usage on different samples

    B. the fraction of times the confidence interval will fail to capture the parameter it is estimating in repeated usage on different samples

    C. the fraction of times that the confidence interval captures the sample statistic on which it is based

    D. none of the above

E11. A sample of 25 values for monthly rent of two-bedroom apartments was selected from a recent edition of the *Gainesville Sun*. The distribution of the 25 values is symmetric, with no outliers. The mean of this sample was $575, and you may estimate that the population standard deviation was $165.

a. A group of students thinks the true mean rent for this type of apartment is over $650. Is there statistical evidence in the data to reject their claim?

b. A newspaper reports that the average rent for two-bedroom apartments last year was $500. Is there statistical evidence of a change in mean rent from last year to this?

c. Could you have made the decision in part b based only on a confidence interval? Explain.

E12. The distribution of annual incomes of employees of a large firm is highly skewed toward the larger values because of the high salaries of the upper-level managers. A random sample of size n is to be selected for the purpose of testing the claim that the mean salary is over $50,000, using the methods of this section. A skeptic says this is not appropriate. Do you agree? Explain.

E13. Refer to the Jack and Jill water-bottling factory scenario of E7. Use a test of significance to help them make a decision.

E14. Heights of women in the United States have a mean of about 65 inches and a standard deviation of about 2 inches. Suppose 36 randomly selected female runners have a mean height of 66 inches. Does it appear that female runners have the same mean height as all women?

a. Answer this through a test of hypothesis.

b. Suppose part a leads to a rejection of the null hypothesis that the true mean is 65, yet the problem states that the true mean *is* 65. What is it that we are rejecting?

9.3 When You Estimate σ: The *t*-Distribution

True standard error (*SE*):
$$\sigma_{\bar{x}} = \sigma/\sqrt{n}$$
Estimated standard error:
$$\sigma_{\bar{x}} \approx s/\sqrt{n}$$

In real applications, you almost never know the true population standard deviation σ; you have to use the sample standard deviation s as an estimate. How will making that change—substituting s for σ—affect your inferences? Some samples give an estimate that's too small: $s < \sigma$. Others give an estimate that's too big: $s > \sigma$. On average, though, the small and large values even out so that the sampling distribution of s has its center very near σ. So if you replace σ by s and change nothing else in constructing a confidence interval for a mean, the *average* width of the confidence intervals produced by your new method should be about equal to the width of the intervals produced by the old method.

That's nice, but the key feature of interest for confidence intervals is their capture rate, not their average width. Does replacing σ by s change your chance of capturing the unknown population mean? You will investigate this matter in the following activity.

Activity 9.2 The Effect of Estimating the Standard Deviation

SAT math scores are approximately normally distributed, with a mean of 511 and a standard deviation of 112. Suppose you don't know that and you're going to estimate the mean score from a random sample of 5 scores.

1. Use this calculator command to generate a random sample of 5 SAT math scores and place them in list 1.

randNorm(511, 112, 5)   STO   L1

(continued)

2. Use the 1VarStat command to find the mean $\bar{x}$ and standard deviation s.

3. In Section 9.1, you learned to construct a 95% confidence interval using the formula $\bar{x} \pm 1.96 \cdot \sigma/\sqrt{n}$. What percentage of the time should this confidence interval capture the true value of μ? Does the capture rate depend on the sample size?

4. This time you aren't supposed to know σ, so substitute s for σ (because that's all you have and all you'll ever have in a real situation) and calculate a "95%" confidence interval. Did your confidence interval capture the true value of μ?

5. Working with a partner, follow the procedure in steps 1–4 to get 10 samples. For each sample, calculate and record the sample mean $\bar{x}$, the sample standard deviation s, the confidence interval using $z = 1.96$, and whether or not the confidence interval captures the true population mean μ.

6. Collect the values of s and the confidence intervals from the other groups. Calculate the capture rate for the class. What is your conclusion about the effect of replacing σ by s?

7. Make a dot plot for the values of s from your class and describe its shape, mean, and standard error. How close is the mean value of s to σ? How many times is s smaller than σ? Larger?

You learned in the activity that although s is about equal to σ on average, it tends to be smaller than σ more often than it is larger. This is because the sampling distribution of s is skewed right. The sampling distribution of s becomes less skewed and more approximately normal as the sample size increases.

Discussion: The Effect of Estimating σ

D24. Use the information in the preceding paragraph to answer these questions.

   a. Explain why the fact that the sampling distribution of s is skewed right means that the capture rate is less than advertised if you simply substitute s for σ.

   b. Consider two extreme situations, $n = 10$ and $n = 1000$. If you use s for σ in your confidence intervals $\bar{x} \pm z^* \cdot \sigma/\sqrt{n}$, which sample size, 10 or 1000, do you expect to give capture rates closest to the confidence level of 95%? Why?

D25. When you use s to estimate σ, the capture rate is too small unless you make a further adjustment. If an interval's true capture rate is lower than what you want it to be, do you need to use a wider or narrower interval to get the capture rate you want? If the true capture rate is higher than you want it to be, do you need to use a wider or narrower interval to get the capture rate you want?

■ Practice

P14. *Aldrin in the Wolf River.* Aldrin is a highly toxic organic compound that can cause various cancers and birth defects. Ten samples taken in Tennessee's Wolf River, downstream from a toxic waste site once used by the pesticide industry, gave aldrin concentrations, in nanograms per liter, of 5.17, 6.17, 6.26, 4.26, 3.17, 3.76, 4.76, 4.90, 6.57, and 5.17. The sample mean is 5.0, so if we knew the standard deviation, a 95% confidence interval would be $5.0 \pm (1.96)\sigma/\sqrt{10}$. Although we don't know σ, we can estimate it from the sample and then use the estimated value $s = 1.10$ in place of the unknown true value.

a. Calculate the confidence interval using s for σ.

b. In general, how do you expect this change—using s as an estimate for σ—to affect the centers of confidence intervals? The spreads? The capture rate? Explain your reasons for each answer.

How to Adjust for Estimating σ

Replacing σ with s tends to make the interval too short. Compensate by replacing z^* with t^*.

Estimating the standard deviation doesn't affect the center of a confidence interval, because the center is at the sample mean $\bar{x}$. But substituting s for σ does lower the overall capture rate unless you compensate by increasing the interval widths by replacing z^* with a larger value, t^*.

Confidence Intervals for an Unknown Mean When Populations Are Normal

$$\sigma \text{ known: } \quad \bar{x} \pm z^* \cdot \frac{\sigma}{\sqrt{n}}$$

$$\sigma \text{ unknown: } \quad \bar{x} \pm t^* \cdot \frac{s}{\sqrt{n}}$$

Student: Where does the value of t^* come from?

Statistician: In principle, you could find it using simulation. Set up an approximately normal population, take a random sample, compute the mean and standard deviation. Do this thousands of times. Then use the results to figure out the value of t^* that gives a 95% capture rate for intervals of the form $\bar{x} \pm t^* \cdot s/\sqrt{n}$.

Student: Wouldn't that take a lot of work?

Statistician: Yes, especially if you went about it by trial and error. Fortunately, this work has already been done, long ago. A statistician, W. S. Gosset (English, 1876–1937), who worked for the Guinness Brewery, actually did this back in 1915. Four years later, the geneticist and statistician R. A. Fisher (English, 1890–1962) figured out how to find values of t^* using probability theory. It turns out that the value of t^* depends on just two things—how many observations you have and the capture rate you want.

Student: So t^* doesn't depend on the unknown mean or unknown standard deviation?

Statistician: No it doesn't, which is very handy because in practice you don't know these numbers. Suppose, for example, you have a sample of size $n = 5$ and you want a 95% interval. Then you can use $t^* = 2.776$ no matter what the values of μ and σ are.

Student: Where did you get that value for t^*?

Statistician: From a t-table, although I could have gotten it from a computer. A brief version of the table is shown in Display 9.6. Table B in the Appendix is more complete. The confidence level tells you which column to look in. For example, for a 95% interval, you want a tail area of .025 (half of .05) on either side, so you look in the column headed .025. For the row, you need to know the degrees of freedom, or df for short.

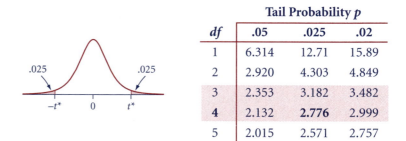

		Tail Probability p	
df	.05	.025	.02
1	6.314	12.71	15.89
2	2.920	4.303	4.849
3	2.353	3.182	3.482
4	2.132	**2.776**	2.999
5	2.015	2.571	2.757

Display 9.6 An abbreviated version of a t-table with a sketch showing the shaded tail areas.

Student: Degrees of freedom? What's that?

Statistician: There's a short answer, a longer answer, and a very long answer. The longer answer will come in E19. The very long answer is for another course. For the moment, here's the short answer: The **degrees of freedom** is the number you use for the denominator when you calculate the sample standard deviation. So for these confidence intervals, $df = n - 1$, where n is your sample size. If $n = 5$, for example, then $df = 4$ and you look in that row. If you turn to Table B in the Appendix and look in the row with $df = 4$ and the column with tail probability .025, you'll find the value 2.776 for t^*.

Example: Average Body Temperature

What is the average body temperature under normal conditions? Is it the same for both men and women? Medical researchers interested in this question collected data from a large number of men and women. Two random samples from that data, each of size 10, are recorded in Display 9.7.

a. Use a 95% confidence interval to estimate the mean body temperature of men.

b. Use a 95% confidence interval to estimate the mean body temperature of women.

Body Temperatures (°F)

Male	Female
96.9	97.8
97.4	98.0
97.5	98.2
97.8	98.2
97.8	98.2
97.9	98.6
98.0	98.8
98.1	98.8
98.6	99.2
98.8	99.4

Display 9.7   Two random samples of body temperatures.

Source: *Journal of Statistics Education Data Archive,* www.amstat.org/publications/jse/datasets/normtext.txt, April 15, 2002.

Solution

First, you need to check that these are random samples from a distribution that is approximately normal. The problem states that these are both random samples from a group of men and women examined by the researchers, so your final result generalizes only to this select population. As to the shape of the distributions, body temperatures cannot stray too far from the mean (or else the person would not have been available to have his or her temperature taken!). So the populations cannot be very skewed. You should still plot the data, however, and the stem-and-leaf plots in Display 9.8 give no reason to suspect that the populations are not normal.

With a sample size of 10 (9 degrees of freedom) and a confidence level of 95%, Table B gives $t^* = 2.262$, a little larger than the corresponding z-value.

Male Temperatures		Female Temperatures	
9 6		9 7	
·	9	·	8
9 7	4	9 8	0 2 2 2
·	5 8 8 9	·	6 8 8
9 8	0 1	9 9	2 4
·	6 8		

97 | 8 represents 97.8°

Display 9.8 Stemplots of body temperature data.

a. For the males, the sample mean is 97.88 and the sample standard deviation is 0.555. That results in a 95% confidence interval of

$$\bar{x}_m \pm t^* \cdot \frac{s_m}{\sqrt{n}} = 97.88 \pm 2.262 \cdot \frac{0.555}{\sqrt{10}}$$

or (97.48, 98.28). Any population of body temperatures with a mean in this interval could have produced the sample mean seen here as a reasonably likely outcome.

b. For the females, the sample mean is 98.52 and the sample standard deviation is 0.527. You are 95% confident that the mean body temperature of the women in this population is in the interval

$$\bar{x}_f \pm t^* \cdot \frac{s_f}{\sqrt{n}} = 98.52 \pm 2.262 \cdot \frac{0.527}{\sqrt{10}}$$

or (98.14, 98.90). From the sample means, it may appear that the average body temperature of females is higher than that of males, but beware! There is some overlap of the confidence intervals, so that conclusion may not be valid. You'll come back to the question of comparing means in a later section. ∎

Checking Conditions

As always, whenever you construct a confidence interval based on t, there are two conditions you must check. Officially, you need a random sample from a normal population. Although you must always have a random sample (or, in the case of an experiment, a random assignment of treatments to experimental units), you can get away with less than a normal population if the sample size is large enough, which will be explained in Section 9.4.

Just as musicians have to practice scales to improve their technical skills even though they never play scales at a concert, statisticians need certain mechanical skills in order to be able to analyze real data sets. The real data sets will come later, but you've reached a point where it makes sense to practice some of the mechanical skills.

Discussion: Confidence Intervals Based on *t*

D26. Which interval is the widest? (You should be able to rule out some of the possibilities without using a *t*-table; for others, you'll need to use the table.)

a. A 95% interval with $n = 4$ and $s = 10$, a 90% interval with $n = 5$ and $s = 9$, or a 99% interval with $n = 4$ and $s = 10$?

b. A 95% interval with $n = 3$ and $s = 10$, a 95% interval with $n = 4$ and $s = 12$, or a 95% interval with $n = 5$ and $s = 10$?

c. A 90% interval with $n = 10$ and $s = 5$, a 95% interval with $n = 10$ and $s = 5$, or a 95% interval with $n = 10$ and $s = 10$?

D27. ***Walking Babies.*** Some babies start walking well before they are a year old, whereas others still haven't taken their first unassisted steps at 18 months or even later. The data in Display 9.9 come from an experiment designed to see whether a program of special walking and placing exercises for 12 minutes each day could speed up the process of learning to walk. In all, 23 male infants (and their parents) took part in the study and were randomly assigned either to the special exercise group or to one of three control groups.

In order to isolate the effects of interest, the scientists used three different control groups. In the "exercise control" group, parents were told to make sure their infant sons exercised at least 12 minutes per day, but they were given no special exercises and no other instructions about exercise. In the "weekly report" group, parents got no instructions about exercise, but each week they were called to find out about their progress. Parents in the "final report" group also got no instructions about exercise, nor did they get weekly check-in calls. Instead, they gave a report at the end of the study. The response variable is the age, in months, when the baby first walked without help.

a. Find a 95% confidence interval estimate of the true mean time to first unaided steps for each group.

b. Attempt to order the four groups with respect to mean time to first steps based on the analysis in part a. Do you see any problems with your method of ordering?

Group	Age (months) of First Unaided Steps					
Special exercises	9	9.5	9.75	10	13	9.5
Exercise control	11	10	10	11.75	10.5	15
Weekly report	11	12	9	11.5	13.25	13
Final report	13.25	11.5	12	13.5	11.5	

Display 9.9 Age (in months) when 23 male infants first walked without support.

Source: Phillip R. Zelazo, Nancy Ann Zelazo, and Sarah Kolb, "Walking in the Newborn," *Science* 176 (1972): 314–315 (via Larsen and Marx).

■ Practice

P15. *Using a* t-*table.* Find the correct value of t^* to use for each of these situations.

    a. a 95% confidence interval based on a sample of size 10

    b. a 95% confidence interval based on a sample of size 7

    c. a 99% confidence interval based on a sample of size 12

    d. a 90% confidence interval with $n = 3$

    e. a 99% confidence interval with $n = 2$

    f. a 99% confidence interval with $n = 4$

P16. For each of these situations, construct and interpret a 95% confidence interval for the unknown mean.

    a. $n = 4$ $\bar{x} = 27$ $s = 12$

    b. $n = 9$ $\bar{x} = 6$ $s = 3$

    c. $n = 16$ $\bar{x} = 9$ $s = 48$

P17. Using the data provided in P14, construct and interpret a 95% confidence interval for the mean level of aldrin.

Hypothesis Tests for a Mean When σ Is Unknown

In Section 9.2, with σ known, you saw how to test hypotheses about the mean by computing a value of the test statistic

$$z = \frac{\bar{x} - \mu_0}{\sigma/\sqrt{n}}$$

and comparing it with a critical value z^*. What will change if the standard deviation is unknown? Perhaps you can guess what's coming: If you don't know σ, you can test H_0: $\mu = \mu_0$ by substituting s for σ and t^* for z^*.

Example: Distance from the Sun

"Earth is 93 million miles from the Sun." You have probably heard or read that often. But is it true? Display 9.10 shows 15 measurements made of the average distance between Earth and the Sun (called the Astronomical Unit, abbreviated A.U.), including Newcomb's original measurement from 1895. Use these data to test the hypothesis that they are consistent with the true mean A.U. being 93 million miles. Use a .05 significance level.

Astronomical Unit (millions of miles)	Astronomical Unit (n = 15), Leaf Unit = 0.010
93.28	9 2 8 | 1 3 4
92.83	· | 7 7 8
92.91	9 2 9 | 1 1 1 2
92.87	· | 6 6 8
93.00	9 3 0 | 0
92.91	·
92.84	9 3 1
92.98	·
92.91	9 3 2
92.87	· | 8
92.88	
92.92	928 | 1 represents 92.81 million miles
92.96	
92.96	
92.81	

Display 9.10 Measurements of the Astronomical Unit.

Source: W. J. Youden, *Experimentation and Measurement* (National Science Teachers Association, 1985), p. 94.

Solution

Check conditions. This type of problem is commonly referred to as a *measurement error* problem. There is no actual population of measurements from which this sample was selected. There are, however, measurements independently determined by different scientists, which can be thought of as a "random" sample from a conceptual population of all such measurements that could be made.

The stemplot of the measurements in Display 9.10 shows that one measurement is extremely large compared to the others, which makes the normality assumption suspect. We will keep this in mind in the analysis.

State your hypotheses. Letting μ denote the true value of the Astronomical Unit, the hypotheses are

$$H_0: \ \mu = 93 \text{ versus } H_a: \ \mu \neq 93$$

because the question asks only if the commonly reported value of 93 million miles is true or not.

Compute the test statistic and draw a sketch. From the data, $\bar{x} = 92.93$ and $s = 0.112$, the test statistic is then given by

$$t = \frac{\bar{x} - \mu_0}{s/\sqrt{n}} = \frac{92.93 - 93.00}{0.112/\sqrt{15}} \approx -2.42$$

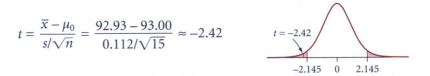

For $\alpha = .05$ and 14 degrees of freedom, the critical *t*-value is $t^* = 2.145$.

Write a conclusion in context, linked to the computations. The observed *t*-value lies farther away from zero than the critical *t*-value, so the null hypothesis is rejected. There is sufficient evidence to conclude that the true mean value of the Astronomical Unit differs from 93 million miles. What about the influence of the outlier? The large measurement of 93.28 is Newcomb's original measurement from 1895, so there may be good scientific reason to remove it from the data set. If we do that, the new observed *t*-value is

$$t = \frac{\bar{x} - \mu_0}{s/\sqrt{n}} = \frac{92.90 - 93.00}{0.057/\sqrt{14}} = -6.56$$

which gives even stronger evidence to reject the hypothesized value of 93 million miles. The A.U. is not 93 million miles, but the true value does not deviate much from this commonly accepted value. ■

Significance Test for a Mean When σ Is Unknown

Follow the same procedure as the test for μ when σ is known (Section 9.2), except use the test statistic

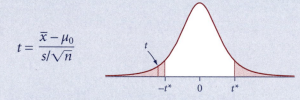

Compare this value of *t* to the appropriate value of t^* found in Table B, using $n - 1$ degrees of freedom.

■ Practice: Fixed-Level Testing

P18. For the aldrin data of P14, carry out a two-sided test of the null hypothesis that the true mean is 4 nanograms per liter. Use $\alpha = .05$.

P-Values Instead of Fixed-Level Tests

A P-value is the probability of seeing a result more unlikely than the one observed, under the conditions of the null hypothesis.

Just as before, instead of doing a fixed-level test, you can report a *P*-value in order to provide more information about the strength of evidence than you give by simply saying "reject" or "don't reject." And, as always, the *P*-value is a conditional probability, computed assuming the null hypothesis is true. It tells the chance of getting a random sample whose value of the test statistic is as extreme as or more extreme than the value computed from the data. The value of t^* depends on the degrees of freedom, so it is a little more difficult to find the *P*-value here than it was for the normal case. You can get almost exact *P*-values from a graphing calculator or computer. You can get only approximate *P*-values from a table.

Example: Body Temperatures

The question about mean body temperature addressed with the data in Display 9.7 on page 507 could have been phrased differently: "Does mean body temperature differ from 98.6 degrees?" The standard that is often reported is 98.6 degrees, but do the data provide evidence to suggest that this might not be true? Test the hypothesis that the mean of the population from which the male sample was drawn differs from 98.6 degrees. Then test the same hypothesis for the female population.

Solution

Check conditions. This was done in Display 9.8.

State your hypotheses. The research claim here is that the mean may differ from the advertised standard of 98.6. This claim forms the alternate (or research) hypothesis. The standard against which the sample mean is compared forms the null hypothesis.

H_0: $\mu = 98.6$, where μ is the mean body temperature of all persons in the population under study

H_a: $\mu \neq 98.6$

Compute the test statistic and draw a sketch.

$$t = \frac{\bar{x} - \mu_0}{s/\sqrt{n}}$$

$$= \frac{97.88 - 98.6}{0.555/\sqrt{10}}$$

$$\approx -4.10$$

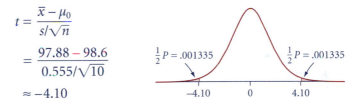

The actual *P*-value as produced by a graphing calculator or a computer is about $P = .00267$. If you have to use the table to find the *P*-value, all you can say is that it is between $2(.001)$ and $2(.0025)$, or $.002 < P\text{-value} < .005$. This partial *t*-table shows these values.

	Tail Probability p	
df	.0025	.001
9	3.690	4.297

Write a conclusion in context, linked to the computations. The small *P*-value implies that if the null hypothesis were true, the chance of seeing a sample mean this unusual is extremely small. Thus, you should reject 98.6 as a plausible value for the mean body temperature of all males in this population.

To test the same hypothesis for the females, everything is the same except that the observed *t*-statistic is

$$t = \frac{98.52 - 98.6}{0.527/\sqrt{10}}$$

$$\approx -0.48$$

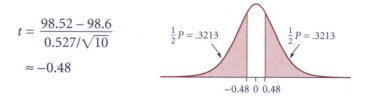

and the resulting *P*-value is .6426. This sample mean is quite likely to occur under the null hypothesis, so you cannot reject the null hypothesis; the claim that the mean body temperature for females is 98.6 is plausible. ■

■ Practice: *P*-Values

P19. For the aldrin data of P14 and P18, find a *P*-value appropriate for these alternative hypotheses.

  a. H_a: $\mu \neq 4$

  b. H_a: $\mu < 4$

  c. H_a: $\mu > 4$

P20. Suppose H_0 is true. Which gives a larger chance of a Type I error, a 5% one-tailed test or a 5% two-tailed test?

P21. Suppose $\mu > 0$. Which has the greatest chance of rejecting H_0: $\mu = 0$?

  A. a 5% one-tailed test of H_0 versus H_a: $\mu > 0$

  B. a 5% two-tailed test

  C. a 5% one-tailed test of H_0 versus H_a: $\mu < 0$

Summary 9.3: Inference Based on *t*

When you use *s* to estimate σ, you must use values of *t* rather than values of *z* when constructing a confidence interval for a mean or when doing a significance test for a mean. You must check that you have a random sample (or a random assignment of treatments to units) and that the sample came from a distribution that is approximately normal or, if not, that the sample size is large enough.

- To construct a confidence interval for a mean, use the formula $\bar{x} \pm t^* \cdot s/\sqrt{n}$. Look up t^* in Table B using $df = n - 1$.

- You can test the null hypothesis H_0: $\mu = \mu_0$ using the test statistic

$$t = \frac{\bar{x} - \mu_0}{s/\sqrt{n}}$$

For example, suppose you are using $\alpha = .05$, with $df = n - 1$.

If your alternate hypothesis is H_a: $\mu \neq \mu_0$, reject H_0 if $|t| > t^*$, using a tail probability of .025 in the *t*-table.

If your alternate hypothesis is H_a: $\mu > \mu_0$, reject H_0 if $t > t^*$, using a tail probability of .05 in the *t*-table.

If your alternate hypothesis is H_a: $\mu < \mu_0$, reject H_0 if $t < -t^*$, using a tail probability of .05 in the *t*-table.

A *P*-value, or observed significance level, tells the proportion of random samples (from a population with H_0 true) that give a value of the test statistic as extreme as (or more extreme than) the value computed from the actual data. To find a right-tailed *P*-value, find the chance of a value greater than *t*, where *t* is the value computed from your data. For a left-tailed *P*-value, find the chance of a value less than *t*. For a two-tailed *P*-value, double the one-tailed *P*-value.

■ Exercises

E15. The data in Display 9.11 come from a survey taken in an introductory statistics class during the first week of a term. Assume this class is a random sample from all students taking this course. These data are the number of hours of study per week, classified by females (F) and males (M).

Study Hours	Gender	n	Mean	Median	StDev
	F	46	10.93	10.00	6.22
	M	15	8.20	7.00	5.94

```
Stem-and-leaf of Study        Stem-and-leaf of Study
Gender = Female N = 46        Gender = Male N = 15
   Leaf Unit = 1.0               Leaf Unit = 1.0

    5  0  22233                     2  0  23
   10  0  55555                     6  0  4445
   16  0  666777                  (3)  0  677
   19  0  888                       6  0  8
  (8)  1  00000011                  5  1  00
   19  1  22233                      3  1  3
   14  1  555555                     2  1  5
    8  1                             1  1
    8  1  88                         1  1
    6  2  0000                       1  2
    2  2  3                          1  2
    1  2                            1  2  5
    1  2
    1  2
    1  3  0
```

Display 9.11 Two stemplots of study hours per week, classified by gender.

a. Estimate the mean study hours per week for all the females taking this course, with confidence level .90. Do you think the estimate would be valid for all females in the university? Explain your reasoning. Be sure here and in parts b and c to state any doubts you have about the validity of the conditions.

b. Suppose the true mean for the population of females taking this course is 11 hours. Test the claim that males study less than females, on average.

c. Estimate the mean study hours for all students taking this course.

E16. Display 9.12 shows the manufacturer's suggested retail price (MSRP), the highway miles per gallon (mpg), and the weight for each case in two different samples of car models. The top half of the table shows a random sample of five models of family sedans (which rules out luxury cars, sports cars, and convertibles); the bottom half shows a random sample of five models of sports-utility vehicles.

Sedans	MSRP ($)	Highway mpg	Weight (lb)
Buick Century Custom	20,020	29	3,368
Chevrolet Malibu	17,150	29	3,051
Chrysler Concorde LX	22,510	28	3,488
Ford Taurus LX	18,550	27	3,354
Toyota Camry LE	20,415	32	3,120

SUVs	MSRP ($)	Highway mpg	Weight (lb)
Blazer 4WD LX	26,905	20	4,049
Explorer AWD XLT	30,185	19	4,278
Jimmy 4WD SLT	30,225	20	4,170
Trooper S 4×4	27,920	19	4,465
Grand Cherokee 4WD	35,095	20	4,024

Display 9.12   Facts on samples of vehicles from the 2001 model year.

Source: www.autoweb.com.

a. Estimate the mean model price of family sedans in a 95% confidence interval.

b. Estimate the mean model weight of the SUVs in a 90% confidence interval.

c. Estimate the mean model highway mpg of each group using 98% confidence.

d. Which of the confidence intervals in parts a–c is the least informative? What would you do to improve the situation if you were designing a new survey of this type?

E17. The National Football League (NFL) reports that the average hot dog price across the 31 league teams for the 2001 season is $3.04. Suppose you call three stadiums chosen at random and find the hot dog prices in Display 9.13.

Team	Hot Dog Price ($)
Baltimore Ravens	3.75
New England Patriots	3.00
Chicago Bears	3.00

Display 9.13   A sample of hot dog prices.

Source: *USA Today,* Sept. 6, 2001.

a. Is a population mean of $3.04 consistent with these data? (Construct an appropriate hypothesis test.)

b. Before looking at these data, a friend says, "I believe the league average hot dog price must be greater than $3.04." Construct an appropriate hypothesis test to see whether your friend's claim is plausible.

c. There are only 31 teams in the NFL. If we had sampled 10 teams instead of 3, would the inferential methods of this section still have worked properly? If not, where do the potential problems lie?

E18. When you do not know σ and have to estimate it with s, the capture rate will be too small if you use the z-interval. Complete this exercise to see part of the reason why.

a. In Activity 9.2, you generated sample standard deviations, s, from samples of size 5 taken from a population with mean 511 and SD 112. The histogram and summary statistics shown here show 100 sample standard deviations from that activity. What is the shape of this distribution? On average, how well does s approximate σ?

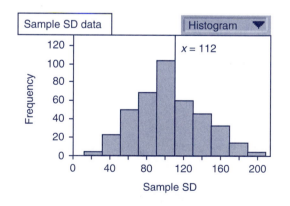

Descriptive Statistics

Variable	N	Mean	Median	TrMean	StDev	SEMean
SampleSD	405	105.80	103.38	105.27	36.33	1.81

Variable	Min	Max	Q1	Q3
SampleSD	18.17	206.87	81.66	130.42

b. Which happens more often, s is smaller than σ or s is larger than σ?

c. How does your answer to part b explain why the capture rate of a z-interval tends to be too small?

E19. Degrees of freedom tell how much information in your sample is available for estimating the standard deviation of the population. In a t-test, you use s to estimate σ. The more deviations from the mean, $(x - \bar{x})$, you have available to use in the formula for s, the closer s should be to σ. But, as you will see in this exercise, not all of these deviations give you independent information.

a. For a sample of size n, how many deviations from the mean are there? What is their sum? (See Section 2.3.)

b. Suppose you have a random sample of size $n = 1$ from a completely unknown population. Call the sample value x_1. Does the value of x_1 by itself give you information about the spread of the population?

c. Next suppose you have a sample of size $n = 2$, with values x_1 and x_2. If one deviation is $x_1 - \bar{x} = 3$, what is the other deviation?

d. Now suppose you have a sample of size $n = 3$. Suppose you know two of the deviations, $x_1 - \bar{x} = 3$ and $x_2 - \bar{x} = -1$. What is the third deviation?

e. Finally, suppose you have a sample of size n. Show how to find the final deviation once you know all of the others. With a sample size of n, how many deviations give you independent information about the size of σ, and how many are redundant?

f. Explain what your answers to parts a–e have to do with degrees of freedom.

E20. Histograms A and B in Display 9.14 were generated from 200 random samples of size 4, each taken from a normal distribution with mean 100 and SD 20. One histogram shows the 200 z-values (one for each sample, using σ) for testing the hypothesis that the population mean is, in fact, 100; the other shows the 200 t-values (using s).

a. Compare the shapes, centers, and spreads of these two distributions.

b. Choose which one is the distribution of t-values (**the t-distribution**, for short), and give the reason for your choice.

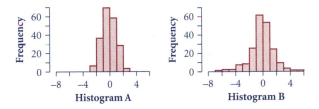

Display 9.14 Two histograms of the t-values and z-values computed from 200 random samples of size 4, each taken from a normal distribution with mean 100 and standard deviation 20.

9.4 The Effect of Long Tails and Outliers

If you think about what you've done so far in terms of "the big three" for distributions—center, spread, and shape—you'll see that there's more work to do.

1. **Center:** The mean is the target of our inferences. You don't know the mean, but that's the whole point: If you did know the mean, you wouldn't be trying to estimate it.

2. **Spread:** For the confidence intervals of Section 9.1, you knew the population standard deviation. In real applications, that's almost never true

and you have to use *s* as an estimate of σ. In Section 9.3, you saw that for small samples, using *s* in place of σ has a pretty big effect on confidence intervals. Unless you make a suitable adjustment to the width of the interval by using t^* instead of z^*, its capture rate will be way off.

3. **Shape:** So far, the shape of the population has been approximately normal, but that's often not the case.

For example, Display 9.15 gives the brain weights of 68 species of animals under study by a certain zoo; Display 9.16 shows a histogram of these weights. Obviously, this distribution is highly skewed toward the larger values. What effect will that have on capture rates?

Species	Brain Weight (g)	Species	Brain Weight (g)
African elephant	5712	Nine-banded armadillo	10.8
African giant pouched rat	6.6	North American opossum	6.3
Arctic fox	44.5	Owl monkey	15.5
Arctic ground squirrel	5.7	Pig	180
Asian elephant	4603	Rabbit	12.1
Baboon	179.5	Raccoon	39.2
Big brown bat	0.3	Rat	1.9
Cat	25.6	Red fox	50.4
Chimpanzee	440	Rhesus monkey	179
Chinchilla	6.4	Roe deer	98.2
Cow	423	Sheep	175
Desert hedgehog	2.4	Tree shrew	2.5
Donkey	419	Water opossum	3.9
Eastern American mole	1.2	Yellow-bellied marmot	17
European hedgehog	3.5	Canary	0.85
Giant armadillo	81	Crow	9.3
Giraffe	680	Flamingo	8.05
Goat	115	Loon	6.12
Golden hamster	1	Ostrich	42.11
Gorilla	406	Pheasant	3.29
Gray seal	325	Pigeon	2.69
Gray wolf	119.5	Stork	16.24
Ground squirrel	4	Vulture	19.6
Guinea pig	5.5	Barracuda	3.83
Horse	655	Brown trout	0.57
Jaguar	157	Catfish	1.84
Kangaroo	56	Mackerel	0.64
Lesser short-tailed shrew	0.14	Northern trout	1.23
Little brown bat	0.25	Salmon	1.26
Man	1320	Tuna	3.09
Mole rat	3	Blue whale	6800
Mountain beaver	8.1	Porpoise	1735
Mouse	0.4	Seal	442
Musk shrew	0.33	Walrus	1126

Display 9.15 Brain weights for a selection of species.

Source: T. Allison and D. V. Cicchetti, "Sleep in Mammals: Ecological and Constitutional Correlates." *Science* 194 (1976): 732–734.

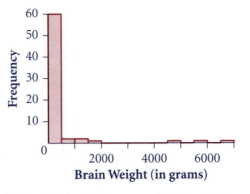

Display 9.16 A histogram of brain weights.

Discussion: Departures from Normality

D28. Think of the 68 species in Display 9.15 as the population of interest. Imagine taking a random sample of size 5 and using $\bar{x} \pm t^* \cdot s/\sqrt{n}$ to create a 95% confidence interval for the true mean weight of all 68 animal species. What do you think the real capture rate will be? (You will check your speculation in Activity 9.3.)

D29. Which is more cautious, to use an interval that is longer than that for the advertised rate or an interval that is shorter?

Activity 9.3 The Effect of Skewness on Confidence Intervals

What you'll need: the data set of Display 9.15

Your goal is to see what happens to the capture rate when you construct many confidence intervals based on random samples from the distribution of brain weights of species given in Display 9.15.

1. Decide on a way to get a random sample of five of the species. Select the sample and record the brain weights of the five species in your sample.

2. Construct a 95% confidence interval estimate of the mean brain weight of all the species using the sample mean, the sample standard deviation, and the appropriate value for t^*.

3. Does your confidence interval include the true mean weight of 394.49 grams? If not, does the interval lie below the true mean or above it?

4. Repeat steps 1 through 3 with a second sample of size 5.

5. Combine your answers with those from other groups in the class to compute an estimate of the overall capture rate. Is it close to 95%?

6. Looking at the capture rate from your simulation and the locations of the intervals that do not capture the population mean, write a brief statement on what happens to confidence intervals when techniques based on the normal distribution are used with distributions highly skewed toward large values.

Plotting to Keep Your Intervals Honest

The lesson from Activity 9.3 is that if you have data from a distribution that is far from normal, confidence intervals and hypothesis tests may not behave the way they are supposed to.

> ### The Effect of Long Tails and Outliers
>
> If the sample size is small and the underlying distribution is highly skewed rather than normal or if it has gross outliers, then the capture rate for an interval of the form $\bar{x} \pm t^* \cdot s/\sqrt{n}$ may be substantially lower than the advertised capture rate, for example, 95%, and a test of significance based on the normal distribution will falsely reject H_0 substantially more often than the advertised rate, for example, 5%.

Skewness implies that the capture rate will be too small.

What can you do about skewness, heavy tails, and outliers? Most important, before you do any tests or construct any intervals, *always plot your data* to see its shape. If the plot looks like the data came from a normal distribution, you don't need to worry about shape. On the other hand, if your plot shows any major deviations from a normal shape, you should be cautious.

Always plot your data to check for skewness and outliers.

What If My Population Is Not Normal?

I. *Try a Transformation*

Fortunately, skewed distributions can almost always be made much more nearly symmetric by transforming to a new scale—often, for example, by taking the logarithm of all the numbers in the sample. If you can find a change of scale that makes your data look roughly normal, once again you don't have to worry about shape. Sometimes a change of scale will also take care of what at first looked liked outliers, making them look much more like "just part of the herd." For a distribution that is skewed right (a common shape), try a log transformation.

Example: The Log Transformation

Take the logarithm of each of the brain weights in Display 9.15. Is it now safe to proceed with a confidence interval based on the *t*-procedure?

Solution

Display 9.17 shows a histogram of the brain weights after taking the natural log of each value. Notice how the outliers have been drawn in toward the center of the data in a way that makes the distribution much more nearly symmetric. This distribution is still not normal, but it is symmetric enough to make the distribution of the sample means for 100 samples of size 5 in Display 9.18 look nearly symmetric. Thus, the confidence interval using *t* will work well on the transformed data.

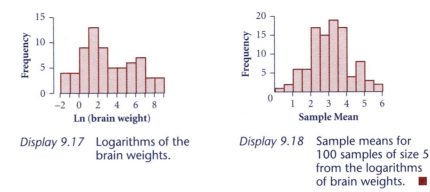

Display 9.17 Logarithms of the brain weights.

Display 9.18 Sample means for 100 samples of size 5 from the logarithms of brain weights.

Example: The Reciprocal Transformation

Gasoline mileage for cars is typically computed as miles per gallon. But why not use gallons per mile? Display 9.19 shows the typical miles per gallon (*mpg*) figures for a sample of small-car models along with the gallons per mile (*gpm*) figures. The stem-and-leaf plots show that the *gpm* distribution is far less skewed than the *mpg* distribution and therefore may be better suited to inference.

Display 9.19 Stemplots of gas mileage for a sample of small-car models.

Source: *Journal of Statistics Education Data Archive,* www.amstat.org/ publications/jse/datasets/93cars.dat.

Manufacturer	Model	mpg	gpm
Acura	Integra	31	0.032
Dodge	Colt	33	0.030
Dodge	Shadow	29	0.034
Ford	Festiva	33	0.030
Ford	Escort	30	0.033
Geo	Metro	50	0.020
Honda	Civic	46	0.022
Hyundai	Excel	33	0.030
Hyundai	Elantra	29	0.034
Mazda	323	37	0.027
Mazda	Protege	36	0.028
Mitsubishi	Mirage	33	0.030
Nissan	Sentra	33	0.030
Pontiac	LeMans	41	0.024
Saturn	SL	38	0.026
Subaru	Justy	37	0.027
Subaru	Loyale	30	0.033
Suzuki	Swift	43	0.023
Toyota	Tercel	37	0.027
Volkswagen	Fox	33	0.030

(continued)

*Display 9.19 (**continued**)*

Miles per Gallon Leaf Unit = 1.0		Gallons per Mile Leaf Unit = 0.0010	
2	9 9	2	0
3	0 0 1	.	2 3
.	3 3 3 3 3 3	.	4
.		.	6 7 7 7
.	6 7 7 7	.	8
.	8	3	0 0 0 0 0 0
4	1	.	2 3 3
.	3	.	4 4
.			
.	6		
.			
5	0		

    a. Use a 95% confidence interval to estimate the mean *mpg*.

    b. Use a 95% confidence interval to estimate the mean *gpm*.

Solution

The 95% confidence interval for the sample mean when $n = 20$ ($df = 19$) is

$$\bar{x} \pm t^{\star} \cdot \frac{s}{\sqrt{n}} = \bar{x} \pm 2.093 \cdot \frac{s}{\sqrt{20}}$$

    a. The sample mean *mpg* is 35.6 and $s = 5.725$, so we are 95% confident that the mean *mpg* is in the interval 35.6 ± 2.679 or $32.92 < \mu(mpg) < 38.28$.

    b. The sample mean *gpm* is 0.0285 and $s = 0.004$, so we are 95% confident that the mean *gpm* is in the interval 0.0285 ± 0.001872 or $0.0266 < \mu(gpm) < 0.0304$.

The methodology of this chapter applied to the *gpm* data will, over the long run, produce intervals with a capture rate closer to the nominal 95%. One moral of the story is that whenever data come in the form of a ratio, think about what might happen if you inverted the ratio. ■

II. *Do the Analysis With and Without the Outliers*

Don't remove outliers without a good reason.

If changing scales doesn't take care of the outliers, do two parallel analyses, one with all the data and the other with the outliers removed. If both analyses lead you to the same conclusion, you're all set. But if the two conclusions differ, you need more data: You don't want your conclusions to depend on what you assume about one or two observations!

III. *Get a Large Sample Size*

A robust procedure works okay even if certain conditions are not met.

The worst cases are small samples with extreme skewness or with gross outliers. For moderate sample sizes, you can rely on the **robustness** of the *t*-procedures: The *t*-procedures are comparatively insensitive to departures from normality, especially for larger samples. What this means in practice is that the true capture rates and significance levels will be close to the advertised values except in extreme situations. For large samples, the Central Limit Theorem says that the sampling distribution of $\bar{x}$ will be approximately normal, so it is safe to use the *t*-procedures. As a rough guide, you can rely on the 15/40 rule.

The 15/40 Rule for Inferences Using the *t*-Procedures

- If your sample size is less than 15: Be very careful. Your data or transformed data must look like they came from a normal distribution—little skewness, no outliers.

- If your sample size is between 15 and 40: Proceed with caution. Strongly skewed distributions should be transformed to a scale that makes them more nearly symmetric before using the *t*-procedures. If you have gross outliers, a transformation may be in order. If you don't transform or if the outliers remain even after a change of scale, do two versions of your test or interval, one with and one without the outliers. Don't rely on any conclusions that depend on whether you include the outliers.

- If your sample size is over 40: You're in good shape. Your sample size is large enough that skewness will not reduce capture rates or alter significance levels enough to matter. Still, if your sample shows strong skewness, it is worth asking whether a change of scale would make the usual summary statistics (especially the standard deviation) more meaningful. Even though outliers may not have much effect on capture rates or significance levels, you should still check by doing two versions of your *t*-procedure.

Small samples:	Moderate-sized samples:	Large samples:
Inferences are quite sensitive to shape.	Transform if data are strongly skewed.	Transforming for symmetry is less important.

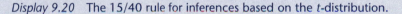

Sample Size

Display 9.20   The 15/40 rule for inferences based on the *t*-distribution.

Discussion: The 15/40 Rule

D30. Discuss what approach you might use to estimate the population mean in each scenario.

 a. You are taking a sample of size 10 from the population of prices of single-family houses in your city.

 b. You are taking a sample of size 100 from the population of prices of single-family houses in your city.

 c. You are taking a sample of size 5 from the SAT scores of students entering a college as freshmen.

 d. You are taking a sample of size 20 from the waiting times of customers at a bank drive-up window.

■ Practice

P22. Pretend that each data set described in parts a–d is an SRS and that you want a confidence interval for the unknown mean. Use the sample size and the shape of the distribution (you may need to make a plot) to decide which of these descriptions (I. through IV.) best fits each data set.

I. There are no outliers, and there is no evidence of skewness. Methods based on the normal distribution are suitable.

II. The distribution is not symmetric, but the sample is large enough that it is reasonable to rely on the robustness of the *t*-procedures and go ahead with a confidence interval, without transforming to a new scale.

III. The shape suggests transforming. With a larger sample, this might not be necessary, but for a skewed sample of this size, transforming is worth trying.

IV. It would be a good idea to analyze this data set twice, once with the outliers and once without.

a. Let *x* be the number of times you have to flip a penny until you get heads. Create a dozen or so *x*-values, recording the data in a plot.

b. Use the data in Display 9.7 on page 507 on body temperatures.

c. Use the data in the study-hours-per-week problem in E15 on page 515.

d. Use the data on the manufacturer's suggested retail prices for SUVs in E16 on page 516.

Summary 9.4: The Effect of Long Tails and Outliers

In this section, you learned that small samples from a skewed population will tend to produce confidence intervals that are too short and off-center. The capture rate of such intervals will be less than the confidence level. If the sample size is small, be careful! If the sample size is large, however, the skewness of the population is of little consequence. If the sample size is moderate, then the sampling distribution may be slightly skewed, but *t*-procedures are robust to slight departures from normality; that is, they still will work well.

Always plot your data first and then proceed with a confidence interval or test of significance using the 15/40 rule:

• If the sample size is less than 15, make sure your data or transformed data have little skewness and no outliers.

• If the sample size is between 15 and 40, your data or transformed data should be nearly symmetric with no gross outliers. (If you don't transform, or if the outliers remain even after a change of scale, do two versions of your test or confidence interval, one with and one without the outliers. Don't rely on any conclusions that depend on whether you include the outliers.)

- If the sample size is over 40, it's safe to proceed. Still, if your sample shows strong skewness, it is worth asking whether a change of scale would make the usual summary statistics (especially the standard deviation) more meaningful.

With small samples, in addition to making plots, think carefully about the nature of the population before concluding that approximate normality is a reasonable assumption.

■ Exercises

E21. Fish absorb mercury as water passes through their gills, and too much mercury makes the fish unfit for human consumption. In 1994, the state of Maine issued a health advisory warning that people should be careful about eating fish from Maine lakes because of the high levels of mercury. Before the warning, data on the status of Maine lakes were collected by the U.S. Environmental Protection Agency (EPA) working with the state. Fish were taken from a random sample of lakes, and their mercury content was measured in parts per million (ppm). Display 9.21 shows a subset of that data from a random sample of 35 lakes.

Mercury (Hg) (ppm)		
1.05	1.22	0.77
0.23	0.24	0.67
0.1	0.9	0.6
0.77	2.5	0.68
0.91	0.34	0.22
0.25	0.4	0.47
0.13	0.45	0.37
0.29	1.12	0.29
0.41	0.32	0.43
0.21	0.37	0.16
0.94	0.54	0.49
0.36	0.86	

Display 9.21 Mercury content of fish in Maine lakes.

Source: R. Peck, L. Haugh, and A. Goodman, *Statistical Case Studies,* 1998, ASA-SIAM, 1–14.

a. Explore the data on mercury levels to see if there are any problems with computing a standard confidence-interval estimate of the mean mercury levels in fish for the lakes of Maine. If so, how do you recommend handling the analysis?

b. One newspaper headline proclaimed, "Mercury: Maine Fish Are Contaminated by This Deadly Poison." Most states consider mercury levels of 0.5 ppm as the borderline for issuing a health advisory. Does it appear that, on the average, Maine lakes deserved the headline? Justify your answer statistically.

E22. Health insurance companies are constantly looking for ways to lower costs, and one way is to shorten the length of stay in hospitals. A study to compare two insurance companies on length of stay (LOS) for pediatric asthma patients randomly sampled 393 cases from Insurer A. Summary statistics and a histogram for the data are shown in Display 9.22.

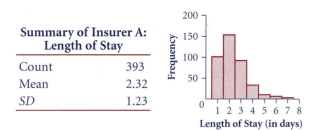

Summary of Insurer A: Length of Stay	
Count	393
Mean	2.32
SD	1.23

Display 9.22 Data plot and summary for lengths of stay in hospitals for Insurer A.

Source: R. Peck, L. Haugh, and A. Goodman, *Statistical Case Studies,* 1998, ASA-SIAM, 45–64.

a. Is it appropriate to use the standard methods to construct a confidence interval from these data? Explain.

b. Use the standard techniques, regardless of your answer to part a, to estimate the mean LOS for Insurer A in a 90% confidence interval.

c. Would you be more concerned about using standard techniques for answering part b if the sample size was 40 instead of nearly 400? How about a sample size of 4 instead of nearly 400?

E23. According to the popular press, some kinds of thinking (visual tasks) are "right-brained," whereas others (verbal tasks) are "left-brained." There is a fair amount of scientific support for this theory. For example, here is an experiment invented by L. R. Brooks, who devised two kinds of tasks.

Source: L. R. Brooks, "Spatial and verbal components of the act of recall," *Canadian Journal of Psychology* 22 (1968): 349–368.

The verbal task was to scan a sentence such as "The pencil is on the desk" and to decide whether or not each word was a noun. (The correct response is "No Yes No No No Yes.") The visual task was to scan a block letter like the F shown below, starting at the arrow and moving clockwise, deciding whether or not each corner is an outside corner. (The correct response is "Yes Yes Yes Yes No No Yes Yes No Yes.")

Brooks also devised two ways to report, one verbal and one visual. To report verbally, you would simply say "Yes" or "No" out loud; to report visually, you

would point in sequence to a "Yes" or "No" on a piece of paper.

If visual and verbal tasks are handled independently, then a visual task with verbal report or a verbal task with visual report would be easier and so would take less time than verbal task/verbal report or visual task/visual report. The theory predicts that when the task and the report were of the same kind, they would interfere with each other in memory and slow the response time.

Display 9.23 shows the data from a version of the experiment run by a psychology lab at Mount Holyoke College.

	Visual Task		Verbal Task	
Subject	**Visual Report Time (sec)**	**Verbal Report Time (sec)**	**Visual Report Time (sec)**	**Verbal Report Time (sec)**
1	11.60	13.06	16.05	11.51
2	22.71	6.27	13.16	23.86
3	20.96	7.77	15.87	9.51
4	13.96	6.48	12.49	13.20
5	14.60	6.01	14.69	12.31
6	10.98	7.60	8.64	12.26
7	21.08	18.77	17.24	12.68
8	15.85	10.29	11.69	11.37
9	15.68	9.18	17.23	18.28
10	16.10	5.88	8.77	8.33
11	11.87	6.91	8.44	10.60
12	17.49	5.66	9.05	8.24
13	24.40	6.68	18.45	8.53
14	23.35	11.97	24.38	15.85
15	11.24	7.50	14.49	10.91
16	20.24	11.61	12.19	11.13
17	15.52	10.90	10.50	10.90
18	13.70	5.74	11.11	9.33
19	28.15	9.32	13.85	10.01
20	33.98	12.64	15.48	28.18
Totals	**363.46**	**180.24**	**273.77**	**256.99**

Display 9.23 Data on visual/verbal report time (in seconds) from the Mount Holyoke experiment.

a. Compute averages for the four combinations of task and report. Then discuss the meaning of the pattern in light of the two predictions.

b. Carry out an exploratory analysis of the data.

c. Estimate each of the four treatment means in 90% confidence intervals. Based on your exploratory analysis and the description of the experiment, tell whether you think your estimates are valid. If not, tell why not.

E24. Go back to the brain weight data in Display 9.15. Consider it a random sample from all such species.

a. Compute the 95% confidence interval for the mean brain weight of all such species, using the 68 species in the sample.

b. Now omit the three outliers and construct the 95% confidence interval for the mean.

c. How much difference is there between the two confidence intervals? What does this tell you about how reliable your confidence intervals are?

d. Take the log of each data value and construct a 95% confidence interval for the population mean on this transformed scale. Transform the endpoints of this interval back to the original scale. Compare the result to the intervals you found in parts a and b.

E25. How should you measure the relationship between housing units (H) and the number of people living in them (P): as persons per housing unit or as housing units per person? The data in Display 9.24

show the population and number of housing units for a random sample of 10 counties from Florida.

a. Make a stemplot for the variable *persons per housing unit* (P/H) and another for *housing units per person* (H/P).

b. Construct and interpret a 95% confidence interval for the mean number of persons per housing unit.

c. Construct and interpret a 95% confidence interval for the mean number of housing units per person.

d. Turn the confidence interval in part c into one for the mean number of persons per housing unit by taking the reciprocals of the endpoints. How does this interval compare with the one found in part b?

e. Looking at parts a, b, and d, which variable do you think is better for measuring the relationship between housing units and their occupants?

County	Population (P) (thousands)	Housing units (H) (thousands)
Alachua	217.9	95.1
Collier	251.4	144.5
Duval	778.9	329.8
Hernando	430.8	62.7
Polk	483.9	226.4
Seminole	365.2	147.1
St. Lucie	192.7	91.3
Suwanee	34.8	15.7
Volusia	443.3	212.0
Walton	40.6	29.1

Display 9.24 Population and housing units in a sample of Florida counties.

Source: U.S. Census Bureau, *Census 2000*.

9.5   Inference for the Difference Between Two Means

Most scientific studies involve comparisons. For example, for purposes of studying pesticide levels in a river, does it make a difference whether you take water samples at mid-depth, near the bottom, or at the surface of the river? Do special exercises help babies learn to walk sooner? Inference about comparisons is more often used in scientific investigations than is inference about a single parameter. In fact, almost all experiments are comparative in nature. The study of inference for differences, then, is fundamental to statistical applications.

Warm-up: How Strong Is the Evidence?

Before you begin to develop formal methods of inference for differences between means, Activity 9.4 will help you develop your intuition and ability to make informal judgments based on comparing means. It revisits the Wolf River data of P14 in Section 9.3 on page 505.

Activity 9.4   The Strength of Evidence

The Wolf River in Tennessee flows past an abandoned site once used by the pesticide industry for dumping wastes, including hexachlorobenzene (chlordane), aldrin, and dieldren. These highly toxic organic compounds can cause various cancers and birth defects. The standard method to test whether these poisons are present in a river is to take samples at six-tenths depth, that is, six-tenths of the way from the surface to the bottom. Unfortunately, there are good reasons to worry that six-tenths is the wrong depth. The organic compounds in question don't have the same density as water, and their molecules tend to stick to particles of sediment. Both these facts suggest you'd be likely to find higher concentrations near the bottom than near mid-depth.

1. Display 9.25 shows eight back-to-back stemplots. The first plot is for actual data. It compares concentrations of aldrin (in nanograms per liter) for 20 water samples taken from the Wolf River downstream from the dumpsite. Ten of the samples were taken at mid-depth, and 10 were taken at the bottom. The other seven plots, numbered 1 through 7, show hypothetical data.

   Compare the plot of each hypothetical data set, 1 through 7, with the plot for the actual data, and evaluate the strength of the evidence of a difference in concentration at the two depths. Is the evidence stronger in the actual data, stronger in the hypothetical data, or about the same for both? State your choice, and also give the reason for your answer.

2. With everything else the same, is the evidence of a difference in concentration stronger when the difference in the means is smaller or larger? When the spreads are smaller or larger? When the sample sizes are smaller or larger?

Actual Data

Bottom		Mid-Depth
	2	
8	3	2 8
9 8	4	3 8 9
7 4 3	5	2 2
3	6	2 3 6
3	7	
8 1	8	
	9	

Hypothetical Data Set 1

Bottom		Mid-Depth
	2	
	3	2 8
8	4	3 8 9
9 8	5	2 2
7 4 3	6	2 3 6
3	7	
3	8	
8 1	9	

Hypothetical Data Set 2

Bottom		Mid-Depth
	2	
	3	8
9 8 8	4	2 3 8 9
7 4 3	5	2 2 6
3	6	2 3
8 3	7	
1	8	

Hypothetical Data Set 3

Bottom		Mid-Depth
8	2	8
	3	2
9 8	4	3
7 4 3	5	8 9
3	6	2 2
3	7	2 6
8	8	3
1	9	

Hypothetical Data Set 4

Bottom		Mid-Depth
	2	
88	3	2288
9988	4	338899
774433	5	2222
33	6	223366
33	7	
8811	8	
	9	

Hypothetical Data Set 5

Bottom		Mid-Depth
	2	
8	3	6
9	4	6 8
4 5	5	2
3	6	3 4
8	7	
8	8	

Hypothetical Data Set 6

Bottom		Mid-Depth
	2	
	3	
8	4	2 8
9 8	5	3 8 9
7 4 3	6	2 2
3	7	2 3 6
3	8	
8 1	9	

Hypothetical Data Set 7

Bottom		Mid-Depth
	2	
8	3	
9 8	4	2 8
7 4 3	5	3 8 9
3	6	2 2
3	7	2 3 6
8 1	8	
	9	

Key

$2|8$ = 2.8 nanograms

Display 9.25 Concentrations of aldrin (nanograms per liter) for samples from the Wolf River: actual data and seven hypothetical data sets.

Source: P. R. Jaffe, F. L. Parker, and D. J. Wilson, "Distribution of Toxic Substances in Rivers," *Journal of Environmental Engineering Division* 108 (1982): 639–649 (via Robert V. Hogg and Johannes Ledolter, *Engineering Statistics* (1987), New York: Macmillan).

Confidence Intervals When Shapes Are Normal and σ_1 and σ_2 Are Known

Most people can just look at hypothetical data set 1 in Display 9.25 and correctly conclude that the mean concentration at the bottom probably isn't equal to the mean concentration at mid-depth. However, with the actual data and the other hypothetical data sets, it's not clear if there really is a difference. And to determine the *size* of the difference, you will need to compute a confidence interval. A confidence interval for the difference of two means has the standard form

$$estimate \pm z^\star \cdot (SE\ of\ the\ estimate)$$

In the unusual situation where you know the standard deviation of both populations, the formula for the standard error comes directly from page 315 of Chapter 5.

Confidence Interval for the Difference Between Two Means, σ_1 and σ_2 Known

A confidence interval for the difference $\mu_1 - \mu_2$ between the means of two populations has the form

$$(\bar{x}_1 - \bar{x}_2) \pm z^\star \cdot \sqrt{\frac{\sigma_1^2}{n_1} + \frac{\sigma_2^2}{n_2}}$$

where $\bar{x}_1$ is the mean of a random sample of size n_1 taken from the first population with known standard deviation σ_1 and $\bar{x}_2$ is the mean of a random sample of size n_2 taken independently from a second population with known standard deviation σ_2. Use

$$z^\star = \begin{cases} 1.645 & \text{for 90\% confidence} \\ 1.960 & \text{for 95\% confidence} \\ 2.576 & \text{for 99\% confidence} \end{cases}$$

These are three conditions that must be met:

- Samples have been randomly and independently selected from two different populations (or two treatments were randomly assigned).

- The two populations are approximately normally distributed or the sample sizes are large enough. (The 15/40 rule in Section 9.4 can be applied to each sample, although it is a bit conservative.)

- The standard deviations σ_1 and σ_2 are known.

Just as for any confidence interval, unless the conditions are satisfied, there is no automatic guarantee that the capture rate will be equal to the advertised confidence level.

Generally, you do not know the population standard deviations in situations where you want to compare two means. An exception occurs when there are data available from past records that can help you in the design of new studies. The next example illustrates such a situation.

Example: Farm Acreage

The U.S. Department of Agriculture records data on farm acreage and number of farms by county for every county in the country. This takes considerable time and energy, and the results for each update (usually occurring about every five years) are a long time in process. Suppose you are assigned the task of using the 1992 data (presented in Display 9.26) to design a survey of counties to get a quick estimate of the difference between mean acreage for the North Central and Northeast regions for the current year.

    a. You're told to use independent random samples of 100 counties from each of the two regions. What is the margin of error for a 95% confidence interval?

b. If you want a margin of error of 40 acres and the sample sizes are to be equal for each region, what sample size should you use?

Region	Number of Counties	Mean (thousands of acres)	Standard Deviation (thousands of acres)
North Central	1052	326	271
Northeast	210	95	79
South	1376	200	244
West	418	730	837

Display 9.26 Summary of acreage devoted to farming in 1992 by county.

Source: U.S. Department of Agriculture, www.nass.usda.gov/census/census92/agrimenu.htm.

Solution

a. The 1992 standard deviations are for the population of counties and are the best approximations you have for the current standard deviations. Thus, they can be used in the formula for approximating the margin of error:

$$E = z^* \cdot \sigma_{\bar{x}_1 - \bar{x}_2} = z^* \cdot \sqrt{\frac{\sigma_1^2}{n_1} + \frac{\sigma_2^2}{n_2}} = 2\sqrt{\frac{271^2}{100} + \frac{79^2}{100}} \approx 56$$

You can expect the margin of error to be around 56 acres if you estimate the difference in means with samples of 100 counties from each region. (This is an approximation already, so you can use 2 instead of 1.96).

b. Setting $n_1 = n_2$ because the two sample sizes are supposed to be equal and $E = 40$ because the margin of error you want is 40 acres, you get

$$E = z^* \cdot \sqrt{\frac{\sigma_1^2}{n_1} + \frac{\sigma_2^2}{n_2}}$$

$$40 \approx 2\sqrt{\frac{271^2 + 79^2}{n_1}}$$

$$\sqrt{n_1} \approx \frac{2}{40}\sqrt{271^2 + 79^2}$$

$$n_1 \approx \left(\frac{2}{40}\right)^2 (271^2 + 79^2)$$

$$n_1 \approx 199.2$$

You will need about 200 observations from each region, for a total sample size of 400, in order to meet your margin of error requirement. ∎

Discussion: When σ_1 and σ_2 Are Known

D31. Why is it more appropriate to use z rather than t in the *Farm Acreage* example?

D32. What happens to the margin of error in part a of the *Farm Acreage* example if you distribute the 200 observations so that 150 go to the North Central states and 50 go to the Northeastern states?

D33. When population standard deviations are known to be equal, find a general rule for allocating a total sample of size n to two populations so as to minimize the margin of error when estimating a difference between two means.

■ Practice

P23. SAT math scores are standardized so that the standard deviation of the scores for any one exam is about 100.

   a. Suppose independent random samples of 25 students each from two different school districts have mean SAT math scores of 510 and 470, respectively. With a confidence level of 95%, estimate the true difference between mean scores for the two school districts. Be sure to check conditions, and specify which, if any, you are worried about.

   b. Could a difference as large as the one given in part a reasonably be attributed to chance, or should the districts involved look for another explanation?

P24. Suppose you need to design a study similar to the one described in P23. Assuming the two samples are of equal size, how large a sample should you choose to get a 40-point margin of error in a 95% confidence interval for the difference in means?

P25. Suppose you are assigned the task of using the 1992 data in Display 9.26 to design a survey of counties to get an estimate of the difference between mean acreage for the South and the West for this year. You are expected to sample a total of 200 counties. What will be the margin of error for a 95% confidence interval? What will be the margin of error for a 95% confidence interval if sample sizes are allocated to the regions in direct proportion to the standard deviations of the regions?

Confidence Intervals for Differences When σ_1 and σ_2 Are Unknown

You've seen how to find a confidence interval for $\mu_1 - \mu_2$ provided you have random samples from normal populations with known σ_1 and σ_2. In practice, you almost never know the standard deviations. What then? If you think back to what you did in Section 9.3 for inferences about μ with σ unknown, you may be able to guess what to do: Substitute s_1 for σ_1 and s_2 for σ_2, and use t^* instead of z^* to adjust the interval width by just the right amount to make the capture rate equal to the confidence level.

Confidence Interval for the Difference Between Two Means, σ_1 and σ_2 Unknown

A confidence interval for the difference $\mu_1 - \mu_2$ between the means of two populations with unknown standard deviations has the form

$$(\bar{x}_1 - \bar{x}_2) \pm t^* \cdot \sqrt{\frac{s_1^2}{n_1} + \frac{s_2^2}{n_2}}$$

(continued)

where $\bar{x}_1$ is the mean of a random sample of size n_1 taken from the first population with standard deviation estimated by s_1 and $\bar{x}_2$ is the mean of a random sample of size n_2 taken independently from a second population with standard deviation estimated by s_2. It is best to use your graphing calculator to find this confidence interval because the value of t^* depends on a complicated calculation. (See D35.)

These are the conditions that must be met:

- Samples have been randomly and independently selected from two different populations (or two treatments were randomly assigned).

- The two populations are approximately normally distributed or the sample sizes are large enough. (The 15/40 rule in Section 9.4 can be applied to each sample, although it is a bit conservative.)

- The standard deviations σ_1 and σ_2 are unknown.

Example: Walking Babies 1

We'll use the same scenario and experimental data as in D27 of Section 9.3 on page 509. The effects of special exercises to speed up infant walking were isolated with three different control groups. Display 9.27 shows the data again.

Group	Age (months) at First Unaided Steps					
Special exercises	9	9.5	9.75	10	13	9.5
Exercise control	11	10	10	11.75	10.5	15
Weekly report	11	12	9	11.5	13.25	13
Final report	13.25	11.5	12	13.5	11.5	

Display 9.27 Age (in months) when 23 male infants first walked without support.

Source: Phillip R. Zelazo, Nancy Ann Zelazo, and Sarah Kolb, "Walking in the Newborn," *Science* 176 (1972): 314–315 (via Larsen and Marx).

Use these data to find a 95% confidence interval estimate of the difference between mean walking times for the special exercises group and the exercise control group if all babies could have had each treatment.

Solution

Check conditions. Two treatments were randomly assigned to the babies. As you will see in the next section, you should plot the data from the two samples to see if there is any reason to doubt that the two populations are approximately normally distributed. Here there is some doubt because each sample has one relatively large value. (See Display 9.28.) Further, we were told that some babies are 18 months old before they walk, so the distribution of ages must be somewhat skewed right. However, the sampling distribution of the difference of means tends to be more symmetric than that of either sample mean by itself, so the standard procedures should work for this amount of skewness. The standard deviations σ_1 and σ_2 are unknown but may be estimated by the sample standard deviations, s_1 and s_2.

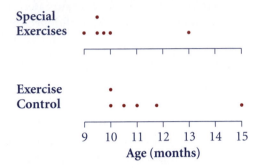

Display 9.28   Dot plots of two samples.

Computations.

Special exercises:	$\bar{x}_1 = 10.125$	$n_1 = 6$	$s_1 = 1.447$
Exercise control:	$\bar{x}_2 = 11.375$	$n_2 = 6$	$s_2 = 1.896$

If you are using the two-sample *t*-interval function of a calculator, you will get $df \approx 9.35$ and a confidence interval of -3.44 to 0.94.

To justify this interval by hand, round the *df* from your calculator to 9 and go to Table B to find $t^\star = 2.262$ for 95% confidence. The confidence interval is about the same:

$$(\bar{x}_1 - \bar{x}_2) \pm t^\star \cdot \sqrt{\frac{s_1^2}{n_1} + \frac{s_2^2}{n_2}} = (10.125 - 11.375) \pm 2.262 \sqrt{\frac{1.447^2}{6} + \frac{1.896^2}{6}}$$

$$= -1.25 \pm 2.20$$

resulting in the interval $(-3.45, 0.95)$.

Conclusion in context. You are 95% confident that if all of the babies could have been in the special exercises group and all of the babies could have been in the exercise control group, the difference in the mean age at which they would learn to walk is in the interval -3.44 months to 0.94 months. Because 0 is in the confidence interval, you have no evidence that there would be any difference if you were able to give each treatment to all of the babies. However, it is important to note that any conclusion is subject to doubt about the appropriateness of this procedure due to skewness of the population, so it would be good to confirm this finding with more data.

Display 9.29 shows the printouts for the two-sample *t*-interval for the walking babies from three commonly used statistical software packages.

Data Desk

```
2-Sample t-Interval for μ1-μ2
No Selector
Individual Confidence 95.00%
Bounds: Lower Bound < μ1-μ2 < Upper Bound

With 95.00% Confidence, -3.4399682 < μ(Special)-μ(Exercise) < 0.93996820
```

Fathom

Estimate of Babies	Difference of Means ▼

First attribute (continuous): Specialexercises

Second attribute (continuous or categorical): Exercisecontrol

Interval estimate for the population mean of **Specialexercises** minus that of **Exercisecontrol**

	Specialexercises	**Exercisecontrol**
Count:	**6**	**6**
Mean:	**10.125**	**11.375**
Std dev:	**1.44698**	**1.89572**
Std error:	**0.590727**	**0.773924**

Confidence level: 95

Using unpooled variances

Estimate: **-1.25** +/- **2.18997**

Range: **-3.43997 to 0.939967**

Minitab

```
TWOSAMPLE T FOR Special VS Exercise
             N    Mean   StDev   SEMean
Special      6   10.12    1.45     0.59
Exercise     6   11.38    1.90     0.77

95 PCT CI FOR MU Special - MU Exercise: (-3.45, 0.95)

TTEST MU Special = MU Exercise (VS NE): T= -1.28 P=0.23 DF= 9
```

Display 9.29 Printouts of estimation from three statistical software packages. ■

Discussion: Confidence Intervals Using $t*$

D34. Here are means for the actual and hypothetical data sets of Activity 9.4. Assume samples are random, observations are independent, populations are normal, and σ_1 and σ_2 are unknown.

   i. Construct a 95% confidence interval for the difference between the two means, *bottom – mid-depth*, for each data set.

   ii. Are the results consistent with the intuitive judgments you made about the strength of the evidence in Activity 9.4?

a. **Actual Data**

Bottom:	$\bar{x}_1 = 6.04$	$n_1 = 10$	$s_1 = 1.6$
Mid-depth:	$\bar{x}_2 = 5.05$	$n_2 = 10$	$s_2 = 1.1$

b. **Hypothetical Data Set 1**

Bottom:	$\bar{x}_1 = 7.04$	$n_1 = 9$	$s_1 = 1.6$
Mid-depth:	$\bar{x}_2 = 5.05$	$n_2 = 10$	$s_2 = 1.1$

c. **Hypothetical Data Set 2**

Bottom:	$\bar{x}_1 = 6.04$	$n_1 = 10$	$s_1 = 1.3$
Mid-depth:	$\bar{x}_2 = 5.05$	$n_2 = 10$	$s_2 = 0.8$

d. **Hypothetical Data Set 3**

Bottom:	$\bar{x}_1 = 6.04$	$n_1 = 10$	$s_1 = 1.9$
Mid-depth:	$\bar{x}_2 = 5.75$	$n_2 = 10$	$s_2 = 1.8$

e. **Hypothetical Data Set 4**

Bottom:	$\bar{x}_1 = 6.04$	$n_1 = 20$	$s_1 = 1.5$
Mid-depth:	$\bar{x}_2 = 5.05$	$n_2 = 20$	$s_2 = 1.1$

f. **Hypothetical Data Set 5**

Bottom:	$\bar{x}_1 = 6.07$	$n_1 = 7$	$s_1 = 1.7$
Mid-depth:	$\bar{x}_2 = 5.15$	$n_2 = 6$	$s_2 = 1.1$

g. **Hypothetical Data Set 6**

Bottom:	$\bar{x}_1 = 7.04$	$n_1 = 10$	$s_1 = 1.6$
Mid-depth:	$\bar{x}_2 = 6.05$	$n_2 = 10$	$s_2 = 1.1$

h. **Hypothetical Data Set 7**

Bottom:	$\bar{x}_1 = 6.04$	$n_1 = 10$	$s_1 = 1.6$
Mid-depth:	$\bar{x}_2 = 6.05$	$n_2 = 10$	$s_2 = 1.1$

D35. Your calculator or software uses this formula to find *df* when doing a two-sample *t*-procedure. You may be curious why we need this complicated rule for degrees of freedom. The simple answer is that using *t* in place of *z* does not provide quite the right adjustment in the two-sample case, as it does in the one-sample case, unless we make this additional adjustment on the degrees of freedom. (The whole theoretical story is a little complicated.)

$$\frac{[variance(\bar{x}_1 - \bar{x}_2)]^2}{df} = \frac{[variance(\bar{x}_1)]^2}{df_1} + \frac{[variance(\bar{x}_2)]^2}{df_2}$$

$$\frac{\left(\frac{s_1^2}{n_1} + \frac{s_2^2}{n_2}\right)^2}{df} = \frac{\left(\frac{s_1^2}{n_1}\right)^2}{n_1 - 1} + \frac{\left(\frac{s_2^2}{n_2}\right)^2}{n_2 - 1}$$

a. Verify the *df* given in the *Walking Babies* example.

b. If $n_1 = n_2$, derive a simplified version of the formula for *df*.

c. If $n_1 = n_2$ and, in addition, $s_1 = s_2$, derive an even simpler rule for *df*.

■ Practice

P26. The data shown in Display 9.30 come from a survey done in an introductory statistics class during the first week of a term. (See E15 in Section 9.3.) Assume this class is a random sample from all students taking this course. These data are the number of hours of study per week, classified by females (F) and males (M).

```
              Gender   N    Mean   Median   StDev
Study Hours      F    46   10.93   10.00    6.22
                 M    15    8.20    7.00    5.94

Stem-and-leaf of Study              Stem-and-leaf of Study
Gender = Female N = 46              Gender = Male N = 15
Leaf Unit = 1.0                     Leaf Unit = 1.0
      5  0  22233                        2  0  23
     10  0  55555                        6  0  4445
     16  0  666777                      (3) 0  677
     19  0  888                          6  0  8
    (8)  1  00000011                     5  1  00
     19  1  22233                        3  1  3
     14  1  555555                       2  1  5
      8  1                               1  1
      8  1  88                           1  1
      6  2  0000                         1  2
      2  2  3                            1  2
      1  2                               1  2  5
      1  2
      1  2
      1  3  0
```

Display 9.30 Two stemplots of study hours per week, classified by gender.

Estimate the difference in mean study hours per week for all the males and females taking this course, with confidence level .90.

Significance Tests for Differences When σ_1 and σ_2 Are Unknown

As you know, the other way to approach an inference problem is through significance testing. Here's a summary of significance testing for the difference between two means when you don't know σ_1 and σ_2.

Components of a Test for the Difference Between Two Means

1. *Name the test and check conditions.* For a test of significance of a difference between two means, the methods of this section require the same conditions as for a confidence interval:

   • Samples have been randomly and independently selected from two different populations (or two treatments were randomly assigned).

(continued)

Components of a Test for the Difference Between Two Means (continued)

- • The two populations are approximately normally distributed or the sample sizes are large enough. (The 15/40 rule can be applied to each sample, although it is a bit conservative.)
- • The standard deviations σ_1 and σ_2 are unknown.

2. ***State your hypotheses.*** The null hypothesis is, ordinarily, that the two population means are equal. In symbols, $H_0: \mu_1 = \mu_2$, or, in terms of the difference between the means, $H_0: \mu_1 - \mu_2 = 0$. Here μ_1 is the mean of the first population and μ_2 is the mean of the second. There are three forms of the alternate or research hypothesis:

$$H_a: \mu_1 \neq \mu_2 \quad \text{or} \quad H_a: \mu_1 - \mu_2 \neq 0$$
$$H_a: \mu_1 < \mu_2 \quad \text{or} \quad H_a: \mu_1 - \mu_2 < 0$$
$$H_a: \mu_1 > \mu_2 \quad \text{or} \quad H_a: \mu_1 - \mu_2 > 0$$

3. ***Compute the test statistic and draw a sketch.*** Compute the difference between the sample means (because the hypothesized mean difference is zero), measured in estimated standard errors:

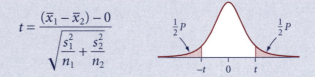

$$t = \frac{(\bar{x}_1 - \bar{x}_2) - 0}{\sqrt{\dfrac{s_1^2}{n_1} + \dfrac{s_2^2}{n_2}}}$$

Use the two-sample t-test function of your calculator or computer to get the P-value.

4. ***Write a conclusion that is linked to your computations and is stated in context.*** Report the P-value. Reject H_0 if the P-value is less than the level of significance, α. Alternatively, if the alternate hypothesis is two-sided ($H_a: \mu \neq \mu_0$), you may reject H_0 if $t > t^*$ or $t < -t^*$. If it is one-sided, reject H_0 if t falls in the appropriate critical region.

Example: Wolf River

You've been given the responsibility to analyze the Wolf River data in Display 9.25 to test whether the true mean aldrin concentrations at the bottom and mid-depth might differ. That is, you want to set up a test of significance for the difference between two population means based on data from independent random samples. Use $\alpha = .10$.

Solution

Check conditions. From the discussion at the beginning of this section, it seems reasonable to assume that these were independent random samples. The stemplot of the actual data in Display 9.25 gives no indication that the populations aren't normal. The standard deviations σ_1 and σ_2 are unknown but can be estimated by the sample standard deviations, s_1 and s_2.

State your hypotheses. In terms of the difference between two means, your null hypothesis is

$$H_0: \mu_{bottom} = \mu_{mid\text{-}depth} \quad \text{or} \quad H_0: \mu_{bottom} - \mu_{mid\text{-}depth} = 0$$

where μ_{bottom} is the mean aldrin concentration at the bottom of the Wolf River and $\mu_{mid\text{-}depth}$ is the mean concentration at six-tenths depth.

You are looking for a difference in either direction, so the alternate hypothesis is two-sided:

$$H_a: \mu_{bottom} \neq \mu_{mid\text{-}depth} \quad \text{or} \quad H_a: \mu_{bottom} - \mu_{mid\text{-}depth} \neq 0$$

Compute the test statistic and draw a sketch. Here are the summary statistics for the aldrin concentrations:

Bottom: $\bar{x}_1 = 6.04 \qquad n_1 = 10 \qquad s_1 = 1.579$
Mid-depth: $\bar{x}_2 = 5.05 \qquad n_2 = 10 \qquad s_2 = 1.104$

The value of the test statistic is

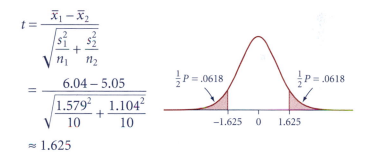

$$t = \frac{\bar{x}_1 - \bar{x}_2}{\sqrt{\dfrac{s_1^2}{n_1} + \dfrac{s_2^2}{n_2}}}$$

$$= \frac{6.04 - 5.05}{\sqrt{\dfrac{1.579^2}{10} + \dfrac{1.104^2}{10}}}$$

$$\approx 1.625$$

If you use a calculator, you get $df \approx 16.1$ and a P-value of .1236. The calculator does not tell you the value of t^*, but some software programs will provide it.

If you justify the calculator's interval with the table using $df \approx 16$, you find that the critical value for a two-sided test with $\alpha = .10$ (probability of .05 in each tail) is $t^* = 1.746$, which is greater than $t = 1.625$. You estimate the P-value from the table using the fact that for $df = 16$, $t = 1.625$ falls between $t^* = 1.337$ and $t^* = 1.746$. Thus, the (two-sided) P-value must be between $2(.10) = .20$ and $2(.05) = .10$. This is consistent with the value from the calculator.

df	P	
	.10	.05
16	1.337	1.746

Conclusion in context. Whether you are using the calculator or the table, you will conclude that because the P-value for a two-sided test is greater than $\alpha = .10$, you do not reject the null hypothesis. There is insufficient evidence to claim that the mean aldrin concentration at the bottom of the Wolf River is different from the mid-depth concentration. In other words, although it appears from the stem-and-leaf plot that aldrin concentrations are greater near the bottom, with these sample sizes, the difference is not large enough to rule out chance variation as a possible explanation.

Display 9.31 shows computer printouts from three commonly used software packages for the analysis of this example.

Data Desk

```
2-Sample t-Test of μ1-μ2
No Selector
Individual Alpha Level 0.10
H0: μ1-μ2 = 0  Ha: μ1-μ2 ≠ 0

bottom - middepth
Test H0: μ(bottom)-μ(middepth) = 0 vs Ha: μ(bottom)-μ(middepth) ≠ 0
Difference Between Means = 0.99000000 t-Statistic = 1.625 w/16 df
Fail to reject H0 at Alpha = 0.10
p = 0.1236
```

Fathom

Test of AldrinTest	Compare Means ▼

First attribute (continuous): Bottom

Second attribute (continuous or categorical): Mid_depth

Ho: Population mean of **Bottom** equals that of **Mid_depth**

Ha: Population mean of **Bottom** is not equal to that of **Mid_depth**

	Bottom	**Mid_depth**
Count:	**10**	**10**
Mean:	**6.04**	**5.05**
Std dev:	**1.57917**	**1.10378**
Std error:	**0.499377**	**0.349046**

Using unpooled variances

Student's t:	**1.625**
DF:	**16.0994**
P-value:	**0.12**

Minitab

```
TWOSAMPLE T FOR Bottom VS Middepth
            N   Mean   StDev   SEMean
Bottom     10   6.04   1.58     0.50
Middepth   10   5.05   1.10     0.35

TTEST MU Bottom = MU Middepth (VS NE): T= 1.62 P=0.12 DF=16
```

Display 9.31 Printouts of significance tests from three statistical software packages. ■

For the aldrin data, it makes sense to use a two-sided alternative. (Researchers wanted to know whether samples taken at mid-depth would give essentially the same results as samples taken near the bottom.) For the walking babies experiment, a one-sided alternative makes more sense because the researchers wanted to test whether the special exercises helped kids walk *sooner*. They weren't interested in the possibility that the special exercises would slow down the kids.

Example: Walking Babies 2

Test the null hypothesis that the mean age at first unaided steps are equal for the special exercise and exercise control groups against the alternate hypothesis that the mean age for the special exercises group is shorter. Use $\alpha = .05$.

Solution

Check conditions. The conditions were checked in the *Walking Babies 1* example.

State your hypotheses. The null hypothesis is

$$H_0: \mu_{special\ ex} = \mu_{control\ ex} \quad \text{or, equivalently,} \quad H_0: \mu_{special\ ex} - \mu_{control\ ex} = 0$$

where $\mu_{special\ ex}$ is the mean age that the babies in the experiment learn to walk if they could all have been given the special exercises and $\mu_{control\ ex}$ is the mean age that the babies learn to walk if all babies in the experiment could have received the exercise control treatment. (You may also use the symbols μ_1 and μ_2 as long as you define the symbols.)

For this one-sided test, the alternate hypothesis is

$$H_a: \mu_{special\ ex} < \mu_{control\ ex} \quad \text{or, equivalently,} \quad H_a: \mu_{special\ ex} - \mu_{control\ ex} < 0$$

Compute the test statistic and draw a sketch.

Special exercises:	$\bar{x}_1 = 10.125$	$n_1 = 6$	$s_1 = 1.447$
Exercise control:	$\bar{x}_2 = 11.375$	$n_2 = 6$	$s_2 = 1.896$

The test statistic is

$$t = \frac{\bar{x}_1 - \bar{x}_2}{\sqrt{\dfrac{s_1^2}{n_1} + \dfrac{s_2^2}{n_2}}}$$

$$= \frac{10.125 - 11.375}{\sqrt{\dfrac{1.447^2}{6} + \dfrac{1.896^2}{6}}}$$

$$t \approx -1.284$$

$P = .115$, -1.284, 0

From the calculator, $df = 9.35$ and the P-value is .115.

To justify, use Table B with $df = 9$ to get a critical value of $|t^*| = 1.833$, which is larger than $|t| = 1.284$. To get a P-value, note that $|t| = 1.284$ is between the t-values cutting off tail probabilities of .15 and .10; you can observe only that the P-value is between .10 and .15.

	Tail Probability P	
df	.15	.10
9	1.100	1.383

Conclusion linked to computations and in context. Because the P-value of .115 is greater than $\alpha = .05$, you do not reject the null hypothesis. The evidence isn't convincing that babies who are given the special exercises walk at an earlier age than babies who are given the control exercise instructions. In other words, if the researchers had been able to give both the special exercises and the control exercises to all of the babies in the experiment, it would have been reasonably likely to get a sample difference of $10.125 - 11.375$, or -1.25, even if the special exercises made no difference.

Discussion: Significance Test for $\mu_1 - \mu_2$

D36. Refer to the walking babies data in Display 9.27 on page 533.

a. Test to see if the special exercises group produces a significant gain in mean walking time over the weekly report group. Use $\alpha = .05$.

b. Test to see if the exercise control group produces a significant gain in mean walking time over the weekly report group. Use $\alpha = .05$.

c. State a conclusion about comparisons between the three groups.

■ Practice

P27. Refer to P26. Is there evidence of a significant difference in mean weekly study hours for males and females at the 5% level of significance?

A Special Case: Pooling When $\sigma_1 = \sigma_2$

Suppose you are taking two independent samples from the same population for the purpose of comparing means. This happens, for example, when you randomly divide available experimental units into two groups for the purpose of comparing two treatments. If the true treatment means really do not differ (the usual null hypothesis), then the true variances of the sample measurements should not differ either. You then have two sample variances (that probably will differ) to estimate the single population variance. One way to combine these two estimates is simply to average the sample variances and use this average to estimate the population variance. (This should be a weighted average if the sample sizes are not equal.) This process is called *pooling,* and it gives you another way to calculate a confidence interval or test statistic based on *t*.

Your calculator and computer software give you the choice of "pooled" or "unpooled" when doing two-sample *t*-procedures. "Pooled" should be used only when you know that the population standard deviations are equal. Even if you know this to be true, the two-sample (unpooled) procedure discussed in this chapter works almost as well as the pooled, especially if the sample sizes are equal. The only situation in which the pooled procedure has a definite advantage is one that has equal population standard deviations but unequal sample sizes. That being the case, we won't cover the pooled procedure in this text. So unless you encounter a problem that specifically tells you to assume that $\sigma_1 = \sigma_2$, choose the unpooled procedure.

But you should know that there are extensions of the two-sample procedures that allow comparisons among more than two means. The most common of these procedures (analysis of variance, or ANOVA) is a generalization of the pooled procedure for two samples. So it is good to know when pooling works and when it doesn't.

Discussion: The Pooled *t*-Procedure

D37. ***To pool or not to pool?*** Here are three different situations for two independent samples.

I. $n_1 = 5, n_2 = 25, \sigma_1 = 10, \sigma_2 = 10$ II. $n_1 = 10, n_2 = 10, \sigma_1 = 10, \sigma_2 = 10$

III. $n_1 = 5, n_2 = 25, \sigma_1 = 10, \sigma_2 = 1$

For one of these situations, pooling is wrong. For a second situation, pooling, although not wrong, is not likely to offer an advantage over the unpooled approach. And in the third situation, pooling is not only appropriate but will also likely give narrower intervals than the unpooled approach. Which is which?

Summary 9.5: Comparing Two Means

In this section, you learned to construct a confidence interval for $\mu_1 - \mu_2$ and to perform a significance test that $\mu_1 = \mu_2$. Whether doing a confidence interval or a significance test, the conditions to check are the same.

- Samples have been randomly and independently selected from two different populations (or two treatments were randomly assigned).

- The two populations are approximately normally distributed or the sample sizes are large enough. (The 15/40 rule can be applied to each sample, although it is a bit conservative.)

- The standard deviations σ_1 and σ_2 are unknown. (In the unusual case when you know σ_1 and σ_2, use z-procedures.)

A confidence interval for the difference $\mu_1 - \mu_2$ between the means of two populations has the form

$$(\bar{x}_1 - \bar{x}_2) \pm t^* \cdot \sqrt{\frac{s_1^2}{n_1} + \frac{s_2^2}{n_2}}$$

The test statistic for a test of significance is

$$t = \frac{(\bar{x}_1 - \bar{x}_2) - 0}{\sqrt{\frac{s_1^2}{n_1} + \frac{s_2^2}{n_2}}}$$

This statistic is called t, but it doesn't have exactly a t-distribution. Fortunately, it is reasonably accurate to proceed using the t-distribution with df approximated by a rather complicated rule. So your calculator or software is the most accurate method of computing a confidence interval and of finding a P-value. Always select the unpooled t-procedure unless you are asked to do otherwise.

■ Exercises

E26. Is the average body temperature under normal conditions the same for both men and women? Medical researchers interested in this question collected data on samples of men and women. Part of the data are recorded in Display 9.7 on page 507.

a. From these data, estimate the difference in mean body temperature between

men and women in a 90% confidence interval.

b. From these data, test the hypothesis that women have higher mean body temperatures than men do. Be sure to state your conclusion in context.

E27. How about heart rates? Is there sufficient evidence from the random samples of heart rates for men and women under normal conditions shown in Display 9.32 to say that the mean heart rates differ for the two groups? Analyze the given data in two different ways (confidence interval and test of significance) before coming to your conclusion, and then state your conclusion carefully.

Heart Rate (beats per minute)

Men	Women
74	75
80	66
75	57
69	87
58	89
76	65
78	69
78	79
86	85
84	59
71	65
80	80
75	74

Display 9.32 Heart rates.

Source: *Journal of Statistics Education Data Archive,* www.amstat.org/publications/jse/datasets/normtext.txt, April 15, 2002.

E28. Refer to the data about different car models in E16, Display 9.12.

a. Estimate the difference in mean model price between family sedans and SUVs in a 95% confidence interval.

b. Estimate the difference in mean model weight between family sedans and SUVs in a 95% confidence interval.

c. Estimate the difference in mean highway mpg between models of family sedans and models of SUVs in a 95% confidence interval.

d. Which of the confidence intervals in parts a–c is the least informative? Explain.

e. In which of the confidence intervals in parts a–c do you have the least "confidence"? Explain.

E29. The data on mercury content of fish in the lakes of Maine in E21 are augmented in Display 9.33 by two other variables: whether or not the lake is formed behind a dam and the oxygen content of the water.

a. Some environmentalists claim that dams are a cause of high mercury levels. Is there sufficient evidence to conclude that the mean mercury level of the lakes behind dams is larger than the mean mercury level of the lakes without dams?

b. Type 1 lakes are *oligotrophic* (balanced between decaying vegetation and living organisms), type 2 lakes are *eutrophic* (high decay rate and little oxygen), and type 3 lakes are *mesotrophic* (between the other two states). Comparing lake types 1 and 2, what is your best estimate of the difference in mean mercury levels for these two types of lakes?

c. Is there sufficient evidence to conclude that the mean mercury level differs for lakes of type 2 as compared to lakes of type 3?

Display 9.33 Mercury content of fish in Maine lakes.

Mercury (Hg) (ppm)	Lake Type	Dam (1 = yes; 0 = no)
1.050	2	1
0.230	2	1
0.100	3	0
0.770	2	1
0.910	2	1
0.250	2	1
0.130	1	1
0.290	2	0
0.410	3	1
0.210	3	0
0.940	2	0
0.360	1	1
1.220	2	0

(continued)

Display 9.33 **(continued)**

Mercury (Hg) (ppm)	Lake Type	Dam (1 = yes; 0 = no)
0.240	1	1
0.900	3	0
2.500	2	1
0.340	3	0
0.400	3	1
0.450	2	1
1.120	3	1
0.320	2	0
0.370	3	0
0.540	3	0
0.860	3	0
0.770	2	0
0.670	3	0
0.600	3	1
0.680	2	1
0.220	3	1
0.470	3	1
0.370	3	1
0.290	2	0
0.430	2	1
0.160	1	0
0.490	3	0

Source: R. Peck, L. Haugh, and A. Goodman, *Statistical Case Studies*, 1998, ASA-SIAM, 1–14.

E30. Refer to the data summaries on length of stay in hospitals found in E22 on page 525. An independent random sample of 396 cases from Insurer B gave the results on length of stay summarized in Display 9.34.

Summary of Insurer B: Length of Stay	
Count	396
Mean	2.91
SD	1.58

Display 9.34 Data plot and summaries for lengths of stay in hospitals for Insurer B.

a. Estimate the difference between the mean lengths of stay for the two insurance companies in a 95% confidence interval. Is there evidence of a difference between the population means for the two companies?

b. Many other variables could contribute to the difference in mean length of stay for the two insurers, and one of them is the number of full-time staff per bed. The respective sample means for this variable are 4.63 for Insurer A and 6.13 for Insurer B. The respective sample standard deviations are 1.70 and 2.40. Is this difference in means statistically significant? If so, provide a practical explanation of how this might be related to the difference in mean length of stay.

c. Another contributing variable is the percentage of private hospitals versus public hospitals among the patients in each insurer's sample. For Insurer A, 93.6% of the sampled patients are in private hospitals, whereas for Insurer B, that percentage is 73.0%. Is this a statistically significant difference? If so, how might it be related to the difference in mean length of stay?

E31. An important factor in the performance of a pharmaceutical product is how fast the product dissolves in vivo (in the body). This is measured by a dissolution test, which yields the percentage of the label strength (%LS) released after certain elapsed times. Laboratory tests of this type are conducted in vessels that simulate the action of the stomach. Display 9.35 shows %LS at certain time intervals for an analgesic (pain killer) tested in laboratories in New Jersey and Puerto Rico. Time is measured in minutes.

New Jersey Elapsed Time (min)	Vessel No. (%LS)					
	V1	**V2**	**V3**	**V4**	**V5**	**V6**
0	0	0	0	0	0	0
20	5	10	2	7	6	0
40	72	79	81	70	72	73
60	96	99	93	95	96	99
120	99	99	96	100	98	100

Puerto Rico Elapsed Time (min)	Vessel No. (%LS)					
	V1	**V2**	**V3**	**V4**	**V5**	**V6**
0	0	0	0	0	0	0
20	10	12	7	3	5	14
40	65	66	71	70	74	69
60	95	99	98	94	90	92
120	100	102	98	99	97	100

Display 9.35 Percent label strength of analgesic for dissolution tests in two laboratories.

Source: R. Peck, L. Haugh, and A. Goodman, *Statistical Case Studies*, 1998, ASA-SIAM, 37–44.

a. Use a 90% confidence interval to estimate the difference in mean %LS at 40 minutes for New Jersey as compared to Puerto Rico.

b. Use a 90% confidence interval to estimate the difference in mean %LS at 60 minutes for New Jersey as compared to Puerto Rico.

c. Good manufacturing practices call for "equivalence limits" of 15 percentage points for dissolution percentages below 90% and 7 percentage points for dissolution percentages above 90%. That is, if the 90% confidence interval for the mean difference is within the equivalence limits (within an interval from −15% to +15% for the lower percentages), then the two sets of results are accepted as being equivalent.

Will the results in parts a and b be accepted as equivalent at 40 minutes? At 60 minutes?

E32. There is much controversy about whether or not coaching programs improve the scores on SAT exams by more than a minimal amount. Although this exercise is not going to settle the argument, it does present a practical use of a test of significance.

The College Board reports that coaching programs improve SAT mathematics scores by about 25 points on the average. A sample of 50 students in a "control" group who took the SAT math test as juniors and than again as seniors had a 13-point average gain (which is reported to be about the national average gain without coaching), with an approximate standard deviation of about 30 points. A sample of 9 students who were coached between their two exams had an average gain of 60 points, with a standard deviation of 42.

Source: Jack Kaplan, "An SAT Coaching Program That Works," *Chance* 15 (1) (2002): 12–17.

a. Is there evidence that the difference in mean point gain between the coached group and the control group exceeds the 25 points that is to be expected (according to the College Board)?

b. Another sample of 12 students from a coaching program had an average gain of 73 points, with a standard deviation of 42. Is there evidence that the difference in mean point gain here exceeds the 25 points that is to be expected?

c. What further questions would you like to ask about this study?

9.6   Paired Comparisons

Now that you've seen the methods for comparing means, it's time to put them to work. In this section, you'll see confidence intervals and significance tests in action. Keep in mind, though, that a *t*-test is no smarter than a chainsaw. Neither one has any brains of its own. A chainsaw can't tell whether it's cutting an old dead tree into firewood or a valuable antique table into scrapwood. A *t*-test is every bit as oblivious. The difference between thoughtful and careless use is up to you, the operator. This final section looks at four issues you need to know about in order to use your statistical tools with care.

- Do you really have two samples or only one sample of paired data?
- What if shapes aren't normal?
- Is it meaningful to compare means?
- Does your inference have the chance it needs?

One Sample or Two?

One of the recurring themes in statistics is how important it is to pay attention to data *production*. This theme was developed in Chapter 4, where the entire emphasis was on designing studies. Now the theme returns in the context of inference for means, where one of the key questions is, "Paired data or independent samples?" The next activity will help you see how pairing changes the standard errors of the differences between means.

Activity 9.5   Handspans

What you'll need: a ruler marked in millimeters

The detective Sherlock Holmes amazed a man by relating "obvious facts" about him, such as that he had at some time done manual labor: "'How did you know, for example, that I did manual labour? It is as true as gospel, for I began as a ship's carpenter.' Sherlock replied, 'Your hands, my dear sir. Your right hand is quite a size larger than your left. You have worked with it, and the muscles are more developed.'" [Source: Sir Arthur Conan Doyle, *The Adventures of Sherlock Holmes,* ed., Richard Lancelyn Green, Oxford World Classics, (Oxford & New York, 1988).] In fact, people's right hands tend to be bigger than their left, even if they are left-handed and even if they haven't done manual labor. But the difference is small, so to detect it, you will have to design your study carefully.

1. Measure your left and right handspans, in millimeters. (An easy way to do this is to spread your hand as wide as possible, place it directly on a ruler, and get the distance between the end of your little finger and the end of your thumb.) Record the data for each student in

(continued)

your class in a table with a column for the left hand and a column for the right. There should be one row for each person.

2. For each row in the table, calculate the differences, *right – left*. Find the mean of the differences and the standard error of the differences using the formula $s_d = s/\sqrt{n}$.

3. Now make a new table, but this time randomize the order of the right handspans so that people's left handspans are no longer matched with their right. Then repeat step 2 with the new table. But before you do, do you think the standard error will be larger or smaller than in step 2?

4. Finally, treat the left handspans and right handspans as independent samples. Calculate the difference between the two sample means and the standard error of that difference using the formula

$$s_{\bar{x}_R - \bar{x}_L} = \sqrt{\frac{s_R^2}{n_R} + \frac{s_L^2}{n_L}}$$

5. Compare the standard errors from steps 2, 3, and 4. Which is the smallest? Which two are closest to the same size?

6. Suppose you make scatterplots of the data in your two tables from steps 1 and 3, with the data for the left handspans on the horizontal axis. Which would you expect to have the higher correlation? Why? Make the scatterplots, and calculate the correlations to check your answer.

In Activity 4.5, you compared sitting and standing pulse rates using three different designs. In the completely randomized design, you randomly selected half of the class to sit and the other half to stand. These data may be analyzed using techniques for two independent samples, as in Section 9.5. In the matched pairs design, you matched pairs of students on a preliminary measure of pulse rate. Then you randomly assigned sitting to one student in each pair and standing to the other. In the repeated measures design, you had each person sit and stand, with the order randomly assigned. The data from the last two designs should not be analyzed using the techniques for two independent samples. In this section, you will learn how to analyze them using the differences in pulse rates for each pair. Display 9.36 on the following page shows the data for pulse rates, in beats per minute, from a class that worked through this activity.

Completely Randomized Design

Sit	Stand
70	66
88	82
82	86
88	102
66	62
70	70
72	50
86	62
74	56
86	104
80	86
46	86
54	80
86	96

Matched Pairs Design

Sit	Stand	Difference (*stand − sit*)
62	68	6
74	78	4
82	80	−2
88	92	4
82	58	−24
66	96	30
64	72	8
84	100	16
72	82	10
82	76	−6
80	92	12
72	74	2
64	60	−4
62	58	−4

Repeated Measures Design

Sit	Stand	Difference (*stand − sit*)	Sit	Stand	Difference (*stand − sit*)
60	64	4	64	66	2
70	72	2	70	76	6
72	76	4	76	86	10
78	82	4	70	88	18
80	92	12	88	96	8
84	98	14	80	86	6
60	68	8	54	56	2
62	64	2	68	82	14
66	70	4	86	96	10
72	86	14	68	74	6
82	100	18	68	80	12
74	80	6	48	58	10
50	58	8	64	72	8
52	54	2	74	94	20

Summary Statistics

	CRD		MPD			RMD		
	Sit	Stand	Sit	Stand	Difference	Sit	Stand	Difference
Mean	74.86	77.71	73.86	77.57	3.71	69.29	77.64	8.36
SD	13.00	17.04	9.13	13.86	12.38	10.67	13.57	5.28

Display 9.36 Pulse measurements from a class experiment.

Completely Randomized Design (Two Independent Samples)

We will now work through the analysis of each of the three sets of experimental results, beginning with the completely randomized design. The scatterplot in Display 9.37 shows the relationship between the sitting and standing measurements for the arbitrary pairing of values from the two independent samples in the *Completely Randomized Design* table in Display 9.36. The correlation is near 0, as it should be, because these are independent measurements in arbitrary order.

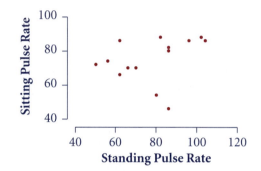

Display 9.37 Scatterplot of sitting versus standing pulse rates in a completely randomized design.

To analyze the data from this completely randomized design, use the methods of Section 9.5. A 95% confidence interval ($df \approx 24.79$) for the difference between the mean pulse rates for sitting and standing is

$$(\bar{x}_1 - \bar{x}_2) \pm t^* \cdot \sqrt{\frac{s_1^2}{n_1} + \frac{s_2^2}{n_2}} = (77.71 - 74.86) \pm t^* \sqrt{\frac{17.04^2}{14} + \frac{13.00^2}{14}}$$

or

$$-8.96 < \mu(stand) - \mu(sit) < 14.67$$

This interval overlaps 0, so there is no reason to suggest that one of these treatments should produce a higher mean than the other if every subject had been given both treatments.

Matched Pairs Design

Display 9.38 shows the relationship between the sitting and standing pulse rates for the matched pairs design. This plot gives a hint of a linear trend (correlation is .48), as it should, because these are dependent measurements based on pairing people with similar resting pulse rates.

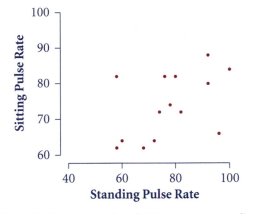

Display 9.38 Scatterplot of sitting versus standing pulse rates in a matched pairs design.

The matched pairs design has dependent observations within a pair, so the two-sample *t*-procedure is not a valid option. You can, however, look at the differences between the standing and sitting pulse rates for each pair and estimate the mean difference with a one-sample procedure. Observe in Display 9.36 that the difference between the sample means, or 77.57 – 73.86, is the same as the mean of the differences, or 3.71, so the latter is a legitimate estimator of the true difference between the population means. Thus, the two samples are reduced to one sample of differences. The summary statistics for the observed differences are

Mean $\bar{d} = 3.714$
Standard deviation $s_d = 12.375$

With 13 degrees of freedom, the t^*-value for a 95% confidence interval is $t^* = 2.160$. The confidence interval estimate of the true mean difference is

$$\bar{d} \pm t^* \cdot \frac{s_d}{\sqrt{n}} \qquad \text{or} \qquad 3.714 \pm 2.160 \cdot \frac{12.375}{\sqrt{14}}$$

which yields an interval of (–3.43, 10.86). Any value of the true mean difference between standing and sitting pulse rates in this interval could have produced the observed mean difference as a reasonably likely outcome. This interval includes zero, so there is not sufficient evidence to say that the standing mean would differ from the sitting mean if all subjects were measured under both conditions. Although this interval overlaps zero, zero is proportionally closer to the endpoint than in the one for the completely randomized design.

Repeated Measures Design

The scatterplot for the data from the repeated measures design, Display 9.39, shows a strong linear tend (correlation .93). These are paired measurements from the same person and should be highly correlated.

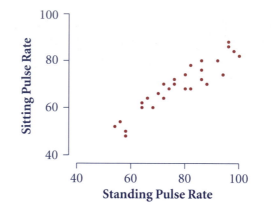

Display 9.39 Scatterplot of sitting versus standing pulse rates in a repeated measures design.

A similar analysis of the mean differences for the repeated measures design yields a 95% confidence interval (27 degrees of freedom), of

$$\bar{d} \pm t^\star \cdot \frac{s_d}{\sqrt{n}}$$

or

$$8.357 \pm 2.052 \cdot \frac{5.279}{\sqrt{28}}$$

resulting in an interval of (6.31, 10.40). Now you see an estimate of the mean difference in pulse rates that does not overlap zero, so you can say that the evidence supports the conclusion that the mean standing pulse rate is higher than the mean sitting pulse rate. That is, the mean difference is larger than we would expect to see from just rerandomizing the order in which the treatments were administered to each subject.

You may be wondering why we did not check assumptions by looking at a plot of the data. Well, we will do that now. For the analyses of differences, it is the differences, not the original two samples, that must come from an approximately normal distribution. Display 9.40 provides boxplots of these differences for both the matched pairs design and repeated measures design. They show that there is no obvious reason to be concerned about non-normality in either case.

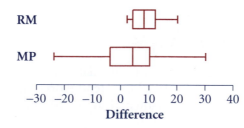

Display 9.40 Boxplots of differences in pulse rates.

Confidence Interval for the Difference Between Means from Paired Comparisons, σ_1 and σ_2 Unknown

A confidence interval for the difference between two means, or $\mu_d = \mu_1 - \mu_2$, has the form

$$\bar{d} \pm t^* \cdot \frac{s_d}{\sqrt{n}}$$

where $\bar{d}$ is the mean of the differences from pairs of measurements in a random sample of n differences and s_d is the estimated standard deviation of the differences. The value of t^* depends on the confidence level and $n - 1$ degrees of freedom.

These are the conditions that must be met:

- A random sample is from one population, with two measurements on each unit, where a "unit" might consist, say, of a pair of twins (or, in an experiment, randomly assigning two treatments to paired subjects or two treatments in random order to the same subject).

- The differences are approximately normally distributed or the sample size is large enough. (The 15/40 rule can be applied to the differences.)

- The standard deviation of the differences is unknown but is estimated by the sample standard deviation, s_d.

You could also make a decision about whether standing increases the mean pulse rate using a test of significance for the differences from the repeated measures design. The research hypothesis that standing increases the mean pulse rate implies that the mean difference is positive (if the sitting measurement is subtracted from the standing). So here are the components of a test of significance.

Check conditions. Because the order in which each subject receives the two treatments is randomized, we may treat this as a random assignment of treatments to subjects. Although the boxplot in Display 9.40 shows that the distribution of differences is not quite symmetric, there is no reason to rule out the normal distribution as a possible model for producing these differences.

State your hypotheses.

$$\text{H}_0\colon \mu_d = 0 \quad \text{versus} \quad \text{H}_a\colon \mu_d > 0$$

where μ_d is the theoretical mean difference between standing and sitting pulse rates for this group of subjects.

Compute the test statistic.

$$t = \frac{\bar{d} - \mu_d}{s_d/\sqrt{n}} = \frac{8.36 - 0}{5.28/\sqrt{28}} = 8.38$$

With 27 degrees of freedom, the *P*-value for this large *t*-statistic is essentially 0.

Conclusion in context. The very small *P*-value indicates that there is sufficient evidence to conclude that the mean pulse rate for persons standing is higher than the mean pulse rate for those same persons sitting. The result applies only to the persons (subjects) in this experiment and cannot be generalized to other persons based on these data alone.

Components of a Test for the Difference Between Two Means Based on Paired Differences

1. *Name the test and check conditions.* For a test of significance of a difference between two means based on paired differences, the method requires the same conditions as for a confidence interval.

   • A random sample is taken from one population with two measurements on each unit, where a "unit" might consist, say, of a pair of twins (or, in an experiment, randomly assigning two treatments to paired subjects or two treatments in random order to the same subject).

   • The differences are approximately normally distributed or the sample size is large enough. (The 15/40 rule can be applied to the differences.)

   • The standard deviation of the differences is unknown but is estimated by the sample standard deviation, s_d.

2. *State the hypotheses.* The null hypothesis is, ordinarily, that in the entire population, the two means are equal, which is equivalent to the mean of the differences being zero. In symbols, H_0: $\mu_d = 0$.

   There are three forms of the alternate or research hypothesis, given by

   $$H_a\text{: } \mu_d \neq 0 \quad H_a\text{: } \mu_d < 0 \quad H_a\text{: } \mu_d > 0$$

3. *Compute the test statistic and draw a sketch.* Compute the difference between the mean of the sampled differences and μ_d, then divide by the estimated standard error:

   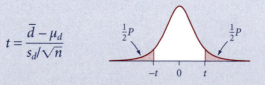

   $$t = \frac{\bar{d} - \mu_d}{s_d/\sqrt{n}}$$

   The *P*-value is based on $n - 1$ degrees of freedom.

4. *Conclusion linked to computations and in context.* Report the *P*-value. Reject H_0 if the *P*-value is less than the level of significance, α. Alternatively, if the alternate hypothesis is two-sided, you reject H_0 if $t > t^\star$ or $t < -t^\star$. If it is one-sided, reject H_0 if t falls in the appropriate critical region.

Discussion: Two Samples or Paired Data?

D38. Based on what you recall from Chapter 4 and what you saw in the analysis of the pulse rate data presented earlier, discuss the relative merits of completely randomized, matched pairs, and repeated measures designs. Under what conditions will the analysis of differences between pairs pay big dividends over the analysis of differences of independent means?

D39. *Guessing distances.* Imagine guessing the width of your classroom. Do you think you'd be as accurate guessing distances in meters as in feet? Not long after the metric system was officially introduced into Australia, a college professor asked one group of his students to guess the width of their lecture hall to the nearest meter. He asked a different group to guess the width to the nearest foot. Display 9.41 presents summary data for their guesses.

	Feet Estimate	**Meters Estimate**
Sample Size	69	44
Sample Mean	43.68	16.02
Sample *SD*	12.51	7.14

Display 9.41 Summary data for guessing widths of a classroom.

Source: M. Hills and the M345 course team, *M345 Statistical Methods, Unit 1: Data, Distributions, and Uncertainty*, in Milton Keynes, *The Open University* (1986).

a. Do the numbers come from a single sample of paired values or from two independent samples?

b. Tell how to gather data in the other format.

c. Why do you think the study was done in the way it was instead of in the way you described it in part b?

d. A test of the null hypothesis that the two population means are equal rejects that hypothesis with a *P*-value less than .0001. Is this a meaningful comparison? Why or why not?

D40. *Hospital carpets.* If you had to spend time in a hospital, would you want your room to have carpeting or a bare tile floor? Carpeting would keep down the noise level, which would certainly make the atmosphere more restful, but, on the other hand, bare floors might be easier to keep free of germs. One way to measure the bacteria level in a room would be to pump air from the room over a growth medium, incubate the medium, and count the number of colonies of bacteria you get per cubic foot of air. This method was in fact used to compare bacteria levels for 16 rooms at a Montana hospital. Eight randomly chosen rooms had carpet installed; the floors in the other 8 rooms were left bare. The data are shown in Display 9.42.

Carpeted Floors		Bare Floors	
Room	Colonies/ft^3	Room	Colonies/ft^3
212	11.8	210	12.1
216	8.2	214	8.3
220	7.1	215	3.8
223	13.0	217	7.2
225	10.8	221	12.0
226	10.1	222	11.2
227	14.6	224	10.1
228	14.0	229	13.7

Display 9.42 Bacterial levels in hospital rooms.

Source: W. G. Walter and A. Stober, "Microbial Air Sampling in a Carpeted Hospital," *Journal of Environmental Health* 30 (1968): 405.

a. For this study, half of the 16 rooms were randomly assigned to the treatment (carpet) group; the other half were left bare. There is no random sampling in the sense of choosing a small set of individuals from a large population. Which of the two designs (two independent samples or paired data) gives the more appropriate model, and why? Give the reasons for your choice.

b. Tell how to run the experiment using the other design.

■ Practice

P28. Display 9.43 shows a set of data on pulse rates, in beats per minute, from another class, with the same three designs employed as discussed for Display 9.36. Analyze these data and point out the similarities to and differences from the results found on the first data set of this type.

Display 9.43 Pulse rate data from another class experiment.

Summary Statistics

	CRD		MPD		RMD	
	Sit	Stand	Sit	Stand	Sit	Stand
Mean	64.57	75.71	67.29	73.29	68.29	72.93
SD	9.33	11.68	9.37	12.71	10.24	10.38

(continued)

Display 9.43 **(continued)**

Completely Randomized Design

Sit	Stand
78	76
64	68
50	82
58	80
50	68
70	64
70	58
64	90
66	72
72	78
80	60
56	100
58	80
68	84

Matched Pairs Design

Sit	Stand
74	78
74	76
58	60
80	96
78	90
62	64
74	74
62	70
68	66
64	74
60	80
56	58
52	52
80	88

Repeated Measures Design

Sit	Stand	Sit	Stand
62	72	60	64
76	76	58	66
76	82	78	88
50	50	62	70
66	74	78	78
58	62	74	74
68	76	88	92
52	62	66	68
68	74	76	84
80	86	82	86
78	84	54	58
60	64	72	78
82	82	58	60
60	66	70	66

What If the Populations Are Not Normal?

You discovered in Section 9.4 that the true capture rate of a nominal 95% confidence interval is less than 95% if the population is skewed and the sample size is not large enough for the sampling distribution of $\bar{x}$ to be approximately normal. The same is true for the confidence intervals for the differences between two means, except that the differencing operation sometimes pulls in the extremes and adds symmetry to the sampling distribution so that the situation is not quite as bad as in the single-sample case. Nevertheless, if the sample differences appear to come from a skewed distribution, it is usually better to transform to a new scale that reduces the skewness.

Example: Transforming Paired Data

The data in Display 9.44 are counts of leprosy bacilli colonies at specified sites on human subjects. The columns labeled *B* and *A* show the counts on the same subjects before and after an antiseptic was applied. The order of the measurements in each pair cannot be randomized, but these subjects were randomly selected from a larger group. The research question: Is the antiseptic effective in reducing bacteria counts?

B	A	$B - A$	$\sqrt{B} - \sqrt{A}$
6	0	6	2.45
6	2	4	1.04
7	3	4	0.91
8	1	7	1.83
18	18	0	0.00
8	4	4	0.83
19	14	5	0.62
8	9	−1	−0.17
5	1	4	1.24
15	9	6	0.87

Display 9.44   Counts of leprosy bacilli colonies.

Source: G. Snedecor and W. Cochran, *Statistical Methods,* 6th ed. (Ames, IA: Iowa State University Press, 1967), p. 422.

Check conditions. This particular group of subjects is a random sample from a larger group, so the randomness condition is satisfied. But the distributions of the original counts and the distribution of their differences don't appear normal, as you can see in Display 9.45. Counts of this type are notorious for having skewed distributions because they must be non-negative but could get very large. (Think of counting the number of insects on each plant in a garden.) Display 9.45 also shows that these particular counts have positively skewed distributions, whereas their differences are skewed in the negative direction. A transformation is needed. Although the condition is that the distribution of differences is normal, statisticians typically transform the original values rather than the differences.

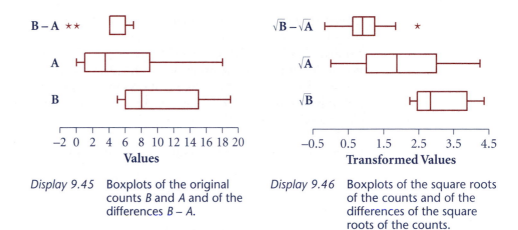

Display 9.45 Boxplots of the original counts *B* and *A* and of the differences *B* − *A*.

Display 9.46 Boxplots of the square roots of the counts and of the differences of the square roots of the counts.

If you take the square root of each of the counts in *B* and *A*, you get the more symmetric boxplots in Display 9.46. The square root transformation typically does a good job of making the distributions of counts more symmetric.

State your hypotheses. The purpose of applying the antiseptic was, of course, to reduce the bacteria counts. Evidence that the treatment was effective can be produced through a *t*-test on the differences of the square roots. The data in the table have the "after" measurements subtracted from the "before," so the hypotheses are

$$H_0: \mu_d = 0 \quad \text{and} \quad H_a: \mu_d > 0$$

where μ_d is the mean of the population of differences from which this sample came.

Compute the test statistic and draw a sketch. The statistical summary for the differences of the square roots is

Mean of differences of square roots $\qquad \bar{d} = 0.96$
Standard deviation of differences of square roots $\quad s_d = 0.77$

The test statistic has the same pattern as in the *t*-procedure for a single mean:

$$t = \frac{\bar{d} - \mu_d}{s_d/\sqrt{n}}$$

$$= \frac{0.96 - 0}{0.77/\sqrt{10}}$$

$$= 3.943$$

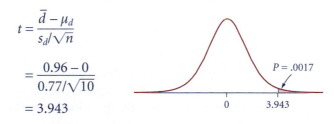

With *df* = 9, the *P*-value is .0017.

Conclusion in context. The small *P*-value of .0017 gives ample evidence to support the conclusion that the bacteria counts were lower after the antiseptic was applied; that is, the true mean difference is positive. However, because there was no randomization in the order of treatments, perhaps you can't attribute the decrease to the antiseptic. There is always the possibility that the bacteria count might have gone down over time anyway. (See D42 to find out how they eliminated this possibility in the actual experiment.)

Discussion: What If Your Population Is Not Normal?

D41. When analyzing differences taken on skewed data, the data could be transformed before the differences were calculated or the differences could be calculated first and then transformed. Does the order of doing these tasks matter? Why or why not?

D42. In the experiment described in the preceding example, a control group received a treatment that had no medical value. Which subjects were in the antiseptic group and which were in the control group was determined by random selection. The results for the control group are given in this table. Is the control "treatment" effective in reducing bacteria counts?

Before	After
16	13
13	10
11	18
9	5
21	23
16	12
12	5
12	16
7	1
12	20

■ Practice

P29. Display 9.47 extends the data used in the *Gallons per Mile* example in Display 9.19. The *H* denotes highway driving, and the *C* denotes city driving. Is there evidence of a significant difference between mean fuel economy for city driving as compared to highway driving for small car models? Answer the question by working through these steps.

a. Plot the differences using *miles per gallon*, and describe the shape of the distribution.

b. Plot the differences using *gallons per mile*, and describe the shape of the distribution.

c. Choose the measure of fuel economy that better meets the conditions of a *t*-procedure, and answer the question about mean fuel economy.

Manufacturer	Model	mpg: *H*	gpm: *H*	mpg: *C*	gpm: *C*
Acura	Integra	31	0.032	25	0.040
Dodge	Colt	33	0.030	29	0.034
Dodge	Shadow	29	0.034	23	0.043
Ford	Festiva	33	0.030	31	0.032
Ford	Escort	30	0.033	23	0.043
Geo	Metro	50	0.020	46	0.022
Honda	Civic	46	0.022	42	0.024
Hyundai	Excel	33	0.030	29	0.034
Hyundai	Elantra	29	0.034	22	0.045
Mazda	323	37	0.027	29	0.034
Mazda	Protege	36	0.028	28	0.036
Mitsubishi	Mirage	33	0.030	29	0.034
Nissan	Sentra	33	0.030	29	0.034
Pontiac	LeMans	41	0.024	31	0.032
Saturn	SL	38	0.026	28	0.036
Subaru	Justy	37	0.027	33	0.030
Subaru	Loyale	30	0.033	25	0.040
Suzuki	Swift	43	0.023	39	0.026
Toyota	Tercel	37	0.027	32	0.031
Volkswagen	Fox	33	0.030	25	0.040

Display 9.47   Gas mileage in city and highway driving for small-car models.

P30. ***Fish or fowl, which are smarter?*** Here are samples of brain and body weights for birds and fish selected from the data in Display 9.15.

Birds	Brain Weight (g)	Body Weight (kg)	Fish	Brain Weight (g)	Body Weight (kg)
Canary	0.85	0.017	Barracuda	3.83	5.978
Crow	9.30	0.337	Brown trout	0.57	0.292
Flamingo	8.05	1.598	Catfish	1.84	2.894
Loon	6.12	1.530	Mackerel	0.64	0.765
Pheasant	3.29	0.625	Northern trout	1.23	2.500
Pigeon	2.69	0.282	Salmon	1.26	3.930
Vulture	19.60	5.270	Tuna	3.09	5.210

Display 9.48   Brain and body weights for a sample of birds and fish.

a. Assuming these are random samples, is there evidence to say that there is a difference in mean brain weight for the two groups?

b. Assuming these are random samples, is there evidence to say that there is a difference in mean body weight for the two groups?

Is It Meaningful to Compare Means?

"When your only tool is a hammer, everything looks like a nail." By now you actually have quite a variety of statistical tools, but it's still worth reminding yourself to be thoughtful about when and how you use them. Before you start any computations, always ask yourself, "What is the question of interest?" Then ask whether it makes sense to try to answer that question by comparing means.

For example, business firms often give aptitude tests to job applicants to see if they have the necessary skills and interests to adapt well to the training they may have to complete. Suppose a company desiring to use such a test has two versions that they want to compare, Test A and Test B. To see how well the exams predict aptitude for a job, the company decides to try both exams on a sample of recently hired employees who have completed the training (and for whom they now know the aptitude).

Is the company interested in comparing the mean scores for the two groups? Probably not. The two tests might be graded on different scales, so a comparison of means would be meaningless anyway. More importantly, the company is really interested in how well the tests select the applicants with greatest aptitude, so they might want to compare the top 10%, say, on each exam to see if they really are the employees with greatest aptitude. In general, the validity of the test scores is more important than the mean. Those with strong aptitude should score high; those with weak aptitude should score low.

Discussion: Is It Meaningful to Compare Means?

D43. ***Old Faithful.*** The geyser Old Faithful, in Yellowstone National Park, got its name from the predictability of its eruptions, which used to occur about every 66 minutes. For decades, visitors throughout the warm months of the year have crowded the large circle of benches surrounding the geyser as each eruption time approaches, waiting to see a 150-foot-high spout of superheated steam and water shoot into the air. In recent years, however, scientists have noticed that the times between eruptions have become somewhat more variable, and the distribution of times in fact looks bimodal, as seen in Display 9.49.

Additional study suggests that the time until the next eruption may depend on the duration of the previous eruption. To test this hypothesis, a science class decides to classify eruptions as long or short. They then will compare the mean time until the next eruption for the long eruptions and the short eruptions.

a. Is this a meaningful comparison? Why or why not?

b. Is the data on times between eruptions taken as just described paired or unpaired?

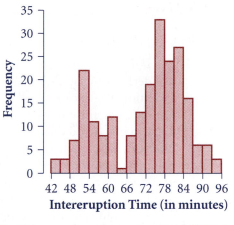

Display 9.49 Histogram based on a sample of 222 intereruption times taken during August 1978 and August 1979.

Source: S. Weisberg, *Applied Linear Regression,* 2nd. ed. (New York: John Wiley and Sons, 1985), pp. 231 and 234.

D44. ***Hens' eggs.*** In the late 1960s, Harvard statistician Arthur Dempster went into his kitchen and measured the length, width, and volume of a dozen hens' eggs. Display 9.50 gives the lengths and widths in inches. It would be possible to use these numbers to test the null hypothesis that the lengths and widths are equal. Do you think that's a sensible comparison? Why or why not?

Length	Width	Length	Width
2.15	1.89	2.17	1.85
2.09	1.86	2.16	1.87
2.10	1.87	2.20	1.84
2.14	1.87	2.15	1.84
2.19	1.87	2.16	1.85
2.13	1.85	2.11	1.87

Display 9.50 Length and width measurements of a dozen eggs.

Source: A. P. Dempster, *Elements of Continuous Multivariate Analysis* (Reading, MA: Addison-Wesley, 1969), p. 151.

D45. Review D39–D44 and summarize what you have learned about comparing means. What are some of the kinds of questions you can't answer by comparing means? What are some of the things to be careful of when thinking about designing a study to compare means?

■ Practice

P31. One of the important measures of quality for a product that consists of a mixture of ingredients, such as a cake mix, lawn fertilizer, or powdered medications, is how well the components are mixed. This is tested by taking small samples of the product from various places in the production process and measuring the proportions of various components in the samples. The sample data, then, will be a set of proportions for Component A, a set of proportions for Component B, and so on. Does a comparison of the means of these sets of data help measure the degree of mixing? If not, suggest another way to assess the degree of mixing.

P32. **Too tight?** You may have lived your entire life (until now) without once wondering what statistics has to do with the screw cap on a bottle of hair conditioner. (Brace yourself!) The machine that puts on the cap must apply just the right amount of turning force: too little and conditioner can leak out; too much and the cap may be damaged and will be hard to get off. Imagine that you are in charge of quality control for a hair conditioner manufacturer. An engineer suggests that new settings on the capping machine will reduce the variability in the force applied to the bottle caps so that fewer will be either too tight or too loose, with more in the acceptable range. He offers to set up a comparison: 10 batches of bottles will be capped by each process, and the force needed to unscrew the cap will be measured. To make the comparison scientific, statistical methods will be used to test

$$H_0: \ \mu_{new} = \mu_{old} \quad \text{versus} \quad H_a: \ \mu_{new} > \mu_{old}$$

where the μ's are the underlying true means for the amount of force. What do you say?

Does Your Inference Have the Chance It Needs?

In statistics, exploratory methods look for patterns but make no assumptions about the process that created the data; with exploratory methods, what you see is all you get. Inference can deliver more than exploration, going beyond just saying, "Here are some interesting patterns." For inference to be justified, however, you need the right kind of data. Provided your numbers come from random samples or randomized experiments, you can use probability theory and the predictable regularities of chancelike behavior to draw conclusions not only about the data you see but also about the unseen population it came from or about the treatments that were assigned to experimental units.

Discussion: Does Your Inference Have a Chance?

These scenarios are presented as examples of inference. As you read each one, try to decide whether it makes good sense to assume the data come from random samples or a randomized experiment. The final discussion question in this set will ask you to summarize what you have found.

D46. **Hospital carpets.** Does the data production process described in D40 justify using these data to test $H_0: \ \mu_{carpet} = \mu_{bare}$?

D47. **Old Faithful revisited.** Suppose you have the data in Display 9.49 on times between eruptions but you don't have the data on the corresponding durations. "Aha!" you think. "We want to know whether the difference between the two modes is a 'real' pattern, that is, too big to be due just to chance variation. To test this, I'll just divide the histogram in half at the low point between the two modes and compare the means of the upper and lower halves of the data using a statistical test." Is this a reasonable use of hypothesis testing?

D48. Make a list of ways your data can fall short of the requirements for statistical inference on means.

■ Practice

P33. An educational researcher is interested in determining whether calculus students who study with music on perform differently from those who study with no music on.

  a. She asks for volunteers to participate in the study and finds a large number of students who study with music on and a large number who study with no music on. She then compares the mean scores on the next calculus exam with a *t*-test. Is that an appropriate design and analysis for the problem at hand? Explain your reasoning.

  b. Realizing that the abilities and backgrounds of the students may be important in this study, the researcher pairs students (one for each treatment group) from the volunteer groups on the basis of their current grade in the class. She then compares the mean scores on the next calculus exam with a *t*-test. Is that an appropriate design and analysis for this problem? Is it any better than the one in part a?

  c. Explain how you would design a study to compare the effects of the two treatments on performance in calculus.

Summary 9.6: One Sample or Two?

Textbook illustrations of statistical procedures are always neater than the real-world applications of those same procedures. When confronted with a real problem involving two samples that has not been "sanitized" for textbook use, there are some key questions you should ask.

  • "Are these paired data or two independent samples?" If the data are paired, they are likely to be correlated and the two-sample procedures are not the correct methods. The better method is to analyze the differences between pairs as a single random sample using the methods of Section 9.3.

  • "How skewed are the populations from which these samples came?" If you have two independent samples, the true capture rate of a nominal 95% confidence interval for the difference between two means is less than 95% if the populations are skewed and the sample sizes are not large enough for the normality of the sample means to take over. The differencing operation itself adds some symmetry to the sampling distribution, however, so the situation is not quite as bad as in the single-sample case. If you have paired differences, check to see if it's reasonable to assume that the differences are a random sample from a normal population. If not, try a transformation on the original values.

  • "Is this a meaningful comparison?" Armed with statistical software, any two sets of data look like candidates for a comparison of means. But some may result in comparing the proverbial apples and oranges, and some may involve comparing samples that were constructed to have different means in the first place.

- "Is there any chance mechanism underlying the selection of these data?" Statistical inference is based on probability theory, and the procedures work only under the condition that the data come from random samples or randomized experiments. You can then draw conclusions not only about the data you see but also about the unseen population they came from or about the reason why treatment groups differ.

■ Exercises

E33. An undesirable side effect of some antihistamines is drowsiness, which can be measured by the *flicker frequency* of patients (number of flicks of the eyelids per minute). Low flicker frequency is related to drowsiness because the eyes are staying shut too long. One study reported data for nine subjects (Display 9.51), each subjected to meclastine (A), a placebo (B), and promethazine (C), in random order. At the time of the study, A was a new drug and C was a standard drug known to cause drowsiness. (In the actual experiment, the drugs were administered on three different days, but any possible day effect will be ignored here.)

Patient	Drug A	Drug B	Drug C
1	31.25	33.12	31.25
2	26.63	26.00	25.87
3	24.87	26.13	23.75
4	28.75	29.63	29.87
5	28.63	28.37	24.50
6	30.63	31.25	29.37
7	24.00	25.50	23.87
8	30.12	28.50	27.87
9	25.13	27.00	24.63

Display 9.51 Flicker frequency of patients.

Source: D. J. Hand et al., *A Handbook of Small Data Sets* (London: Chapman and Hall, 1994), p. 8.

a. Is there sufficient evidence to say that drug A causes more drowsiness than drug B?

b. Estimate the difference between the mean rates for treatments B and C.

c. Estimate the difference between the mean rates for treatments C and A.

d. Write a summary of what you learned about the three treatments in this analysis.

E34. **Buying jeans.** Would you guess that people tend to buy more jeans when the weather is warm or when it is cold? Display 9.52 shows the total sales, in thousands of pairs, for jeans sold in the United Kingdom in January and in June, for the years 1980–85. A test of H_0: $\mu_{Jan} = \mu_{Jun}$ gives a *P*-value of .06. Is this a reasonable use of hypothesis testing?

			Year			
Month	**1980**	**1981**	**1982**	**1983**	**1984**	**1985**
Jan	1998	1924	1969	2149	2319	2137
Jun	2286	1979	2375	2369	3126	2117

Display 9.52 Number of jeans sold in different seasons.

Source: S. Conrad, *Assignments in Applied Statistics* (Chichester, U.K.: John Wiley and Sons, 1989), p. 78.

E35. **Radioactive twins.** (Further details of this study are in Chapter 4, page 255.) Most people believe that country air is better to breathe than city air, but how would you prove it? You might start by choosing a response that narrows down what you mean by "better." One feature of healthy lungs is that they are quick to get rid of nasty stuff they breathe in, such as particles of dust and smoke. This study used as its response variable the rate of tracheobronchial clearance, that is, how quickly the lungs got rid of nasty stuff.

Investigators managed to find seven pairs of identical twins that satisfied two requirements: (1) one twin from each pair lived in the country and the other lived in a city; (2) both twins in the pair were willing to inhale an aerosol of radioactive Teflon particles! That was how the investigators measured tracheobronchial clearance. The level of radioactivity was measured twice for each person: right after inhaling and then again an hour later. The response was the percent of original radioactivity still remaining one hour after inhaling. The data are shown in Display 9.53.

| | Environment | |
	Rural	Urban
Twin Pair	**Rural**	**Urban**
1	10.1	28.1
2	51.8	36.2
3	33.5	40.7
4	32.8	38.8
5	69.0	71.0
6	38.8	47.0
7	54.6	57.0

Display 9.53 Twin data on radioactivity.

Source: Per Camner and Klas Phillipson, "Urban Factor and Tracheobronchial Clearance," *Archives of Environmental Health* 27 (1973): 82.

a. Construct a scatterplot of the data. Then decide whether a statistical test is appropriate here. If not, tell why not. If so, carry out the test and state your conclusion in ordinary English.

b. Tell how to design a study that uses two independent samples (as an alternative to paired samples) to compare clearance rates for people living in urban and rural environments.

c. What are the advantages of using paired data from a single sample?

d. What are the advantages of using two independent samples?

E36. *Bee stings.* Beekeepers sometimes use smoke from burning cardboard to reduce their risk of getting stung. It seems to work, and you might suppose the smoke acts like an insect repellent, but that's not the case. When J. B. Free, at Rothamsted Experimental Station, jerked a set of muslin-wrapped cotton balls in front of a hive of angry bees, the numbers of stingers left by the bees made it quite clear they were just as ready to sting smoke-treated balls as they were to sting the untreated controls. Yet in tests using control balls and others treated with the repellent citronellol, the bees avoided the repellent: On 70 trials out of 80, the control balls gathered more stings. What, then, is the effect of the smoke? One hypothesis is that smoke masks some other odor that induces bees to sting; in particular, smoke might mask some odor a bee leaves behind along with his stinger when he drills his target, an odor that tells other bees, "Sting here." To test this hypothesis, Free suspended 16 cotton balls on threads from a square wooden board, in a four-by-four arrangement. Eight had been freshly stung, the other 8 were pristine and served as controls. Free jerked the square up and down over a hive that was open at the top and then counted the number of new stingers left in the treated and the control balls. He repeated this whole procedure eight more times, each time starting with 8 fresh and 8 previously stung cotton balls and each time counting the numbers of new stingers left. For each of the nine occasions, he lumped together the 8 balls of each kind and took as his response variable the total number of new stingers. His data are shown in Display 9.54.

| | Cotton Balls | |
Occasion	Previously Stung	Fresh
1	27	33
2	9	9
3	33	21
4	33	15
5	4	6
6	22	16
7	21	19
8	33	15
9	70	10

Display 9.54 Number of new bee stingers left behind in cotton balls.

Source: J. B. Free, "The Stinging Response of Honeybees," *Animal Behavior* 9 (1961): 193–196.

Do an exploratory analysis and use it, together with the description of the data, to decide whether a formal hypothesis test is appropriate. If not, give your reasons. If so, state your null and alternate hypotheses in words and in symbols, carry out your test, and then state your conclusions in language a newspaper reader would understand.

E37. Auditors are often required to use sampling to check various aspects of accounts they are auditing. For example, suppose the accounts receivable held by a firm are being audited. Auditors will typically sample some of these accounts and compare the value the firm has on its books (*book value*) with what they agree is the correct amount (*audit value*), usually corroborated by talking with the customer. (Obtaining the audit values is a fair amount of work, hence the need for sampling.) Data of this sort are privileged information in most firms, but Display 9.55 shows what a typical data set might look like. (The data are in dollars.) Note that many of the book values and audit values agree, as they should. There is no question about the fact

Book Value (B) ($)	Audit Value (A) ($)	B – A ($)
389	389	0
419	419	0
336	336	0
427	427	0
418	418	0
355	355	0
293	293	0
406	406	0
392	392	0
333	333	0
394	394	0
374	374	0
446	446	0
464	464	0
338	338	0
390	390	0
372	372	0
406	406	0
364	364	0
433	433	0
426	426	0
417	417	0
415	415	0
461	461	0
431	431	0
337	367	–30
429	392	37
433	403	30
412	379	33
383	341	42
484	462	22
471	428	43
400	430	–30
409	435	–26
386	361	25
370	375	–5
486	452	34
393	339	54
382	404	–22
480	432	48

Display 9.55 Book versus audit values.

that these are paired data, so an estimate of the mean difference between the book and audit values should proceed from the differences. The question is what to do with the zeros.

a. Should the estimate of the mean difference between book and audit values be based on the entire sample of 40 or on just the 15 nonzero values? Answer this question by working through these steps.

   i. Construct a 95% confidence interval estimate using all 40 differences.

   ii. Construct a 95% confidence interval estimate using only the 15 nonzero differences.

   iii. How do the intervals compare? Which one do you think provides the better answer? Why?

b. If the estimate of the mean difference is based on the nonzero values in the sample, explain how you would use the result to estimate the total amount by which the book values exceed the audit values for all accounts in the population. The firm knows how many accounts receivable it has; assume this number is 1000. (Keep in mind that the total could be negative if audit values tend to exceed book values.)

Chapter Summary

To use the confidence intervals and significance tests of this chapter, you need either

- a random sample from a population
- two independent random samples from two distinct populations (If the study is an experiment, you should have two treatments randomly assigned to the available subjects.)

(When there is no randomness involved, you proceed with the test only if you state loudly and clearly the limitations of what you have done. If you reject the null hypothesis in such a case, all you can conclude is that something happened that can't reasonably be attributed to chance.)

You use a confidence interval if you want to find a range of plausible values for

- μ, the mean of your single population
- $\mu_1 - \mu_2$, the difference between the means of your two populations

Both of the confidence intervals you studied have the same form:

estimate from the sample ± critical value · estimated SE

The degrees of freedom for the two-sample t-procedure must be approximated from a rather cumbersome formula. For that reason, the two-sample t-procedures should be done with the aid of technology.

All of the significance tests you have studied include the same steps.

1. Justify your reasons for choosing this particular test. Discuss whether the conditions are met and decide whether it is okay to proceed if they are not strictly met.

2. State the null hypothesis and the alternate hypothesis.

3. Compute the test statistic and find the *P*-value. Draw a sketch of the situation.

4. Use the computations to decide whether to reject or not to reject the null hypothesis by comparing the test statistic to the critical value or by comparing the *P*-value to the level of significance, α. Then state your conclusion in terms of the context of the situation. ("Reject H_0" is not sufficient.) Mention any doubts you have about the validity of that conclusion.

■ Review Exercises

E38. The data in Display 9.56 show the life expectancies (in years) for males and females for a random sample of African countries. The statistical summaries are given below the data table. Is there evidence of a significant difference between the life expectancies of males and females in the countries of Africa? Give statistical justification for your answer and a careful explanation of your analysis.

| | Life Expectancy (in years) | |
	Male	Female
Algeria	65.7	67.7
Gambia	43.4	46.6
Ghana	54.2	57.8
Kenya	57.1	60.8
Liberia	54.0	57.0
Libya	61.6	65.0
Mauritius	65.0	73.0
Morocco	65.1	68.6
W. Sahara	43.3	45.3
Togo	54.0	58.0

	N	Mean	Median	StDev	SEMean
Male	10	56.34	55.65	8.30	2.63
Female	10	59.98	59.40	9.08	2.87
Female-Male	10	3.64	3.45	1.67	0.53

Display 9.56 Life expectancies (in years) for a sample of African countries.

Source: *Statistical Abstract of the United States,* 1996.

E39. ***Pesticides.*** Reread the description of the Wolf River data in Activity 9.4 on page 528.

a. Explain why a one-sided alternate hypothesis might be appropriate here. Then define suitable notation and state H_0 and H_a.

b. Tell whether the design of the study justifies the use of a probability model, giving reasons for your answers.

c. Do an informal diagnostic analysis. Display the distributions, describe the shape in words, and tell whether the shapes and sample sizes raise doubts about a *t*-test. If so, make an appropriate transformation of the data to conduct the analysis.

d. Carry out the test, find the *P*-value, and tell what your conclusion is if you take the results at face value.

e. What do you conclude from your analysis? Is the evidence strong enough to be convincing that samples should be taken from the bottom rather than from mid-depth?

E40. ***Old Faithful.*** (Refer to D43 for a description of the data.) Display 9.57 presents summary statistics and side-by-side boxplots comparing the distributions of times until the next eruption following short eruptions (duration < 3 min) and long eruptions (duration > 3 min).

	Short Eruptions (in min) ($n = 67$)	Long Eruptions (in min) ($n = 155$)
Mean	54.46	78.16
SD	6.30	6.89

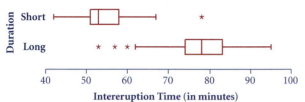

Display 9.57 Summary statistics and boxplots of times until next eruption following short (< 3 min) or long (> 3 min) eruptions.

a. State appropriate null and alternate hypotheses in words. Then define the notation you need and restate H_0 and H_a in symbols.

b. Tell whether the design of the study justifies the use of a probability model, giving reasons for your answers.

c. Based on the summary statistics and plot, tell whether the shapes of the distributions raise doubts about a *t*-test.

d. Carry out the computations, find the *P*-value, and tell what your conclusion is if you take the results at face value.

e. Now tell what you think the test really means for this particular data set.

E41. When considering all regular players (not counting pitchers) on major league baseball teams over recent years, the season batting averages are approximately normally distributed, with a mean around .250 and a standard deviation around .050. Carefully explain each of the components of a fixed-level test in the context of the following questions.

a. The New York Yankees had a team batting average of .267 in 2001, with 15 regular players contributing. Is this a higher average than would be expected for a random sample of 15 players?

b. The Seattle Mariners had a team batting average of .288 in 2001, again with about 15 regular players. Can this team be considered above average in hitting?

Source: www.mlb.com, 10/01.

E42–E44. *Passive smoking.* As the harmful effects of inhaling smoke from other people's cigarettes becomes well established and generally known, more and more places are becoming smoke free. Some of the scientific evidence is surprising. For example, did you know that sidestream smoke (what the smoker doesn't inhale) can have higher concentrations of nicotine and carbon monoxide than mainstream smoke (what goes directly into the smoker's lungs)?

E42. *Carbon monoxide.* The data in Display 9.58 give the "yield" of carbon monoxide for sidestream and mainstream smoke from eight brands of Canadian cigarettes.

Cigarette Brand	CO from Smoke	
	Sidestream (mg)	Mainstream (mg)
A	40.5	18.6
B	59.8	20.5
C	42.9	16.8
D	42.0	17.8
E	60.8	19.8
F	45.1	16.4
G	43.9	13.1
H	67.3	12.4

Display 9.58 Carbon monoxide yields from eight brands of Canadian cigarettes.

Source: Jay Devore and Roxy Peck, *Statistics: The Exploration and Analysis of Data* (Minneapolis, MN: West Publishing Co., 1986). Their source: *American Journal of Public Health* (1984): 228–231.

a. Do you have paired data or independent samples? How can you tell?

b. What is your impression from looking at the data: Do you need formal inference to decide whether carbon monoxide levels are higher in side-stream smoke? Guess whether the *P*-value for a two-sided test will be closest to .25, .10, .05, .01, or .001.

c. To what extent is a probability model for the data justified? What is the basis for the justification? What kinds of inferences are justified: From a sample to a population? About cause and effect? Something else?

d. Make a scatterplot of the data and describe the pattern in words. Which brands, if any, stand out? Does the shape of the plot suggest that your decision in part a will have a big influence on the *P*-value?

e. Compute the difference $(CO_{side} - CO_{main})$ for the eight brands and show them in a stemplot. Describe the shape of the distribution, and tell whether the shape raises concerns about the accuracy of *P*-values for inferences based on the normal distribution.

f. Carry out both the paired and two-sample *t*-tests to check your guess in part d. Are the results here consistent with your prediction based on the shape of the scatterplot? How close was your guess about the *P*-value in part b?

g. Write two or three sentences telling your conclusion about carbon monoxide in mainstream and sidestream smoke.

E43. **Nicotine.** Display 9.59 gives the nicotine yields for the same eight brands of cigarettes.

a. Carry out an exploratory analysis. Compare the two distributions for sidestream and mainstream smoke, and use a scatterplot to display the relationship. Describe the patterns in words.

b. Now compute the differences $(N_{side} - N_{main})$, show their distribution in a stemplot, and tell what the shape suggests about the trustworthiness of *P*-values based on the normal distribution.

c. State null and alternate hypotheses, and carry out the arithmetic to get a *P*-value.

d. Write a short summary telling what you conclude about nicotine.

E44. **Tar.** Display 9.60 gives the tar yields for the same eight brands of cigarettes. Notice that for three brands, the readings are higher for mainstream smoke, so perhaps for tar the evidence is less clear-cut than it is for carbon monoxide and nicotine.

	Nicotine from Smoke	
Cigarette Brand	Sidestream (mg)	Mainstream (mg)
A	2.8	1.2
B	2.7	1.1
C	3.7	1.2
D	2.8	1.0
E	4.3	1.1
F	3.9	1.2
G	3.3	1.0
H	4.6	1.0

Display 9.59 Nicotine yields from eight brands of Canadian cigarettes.

	Tar in Smoke	
Cigarette Brand	Sidestream (mg)	Mainstream (mg)
A	15.8	18.5
B	16.9	17.0
C	21.6	17.2
D	18.8	19.4
E	29.3	15.6
F	20.7	16.4
G	18.9	13.3
H	25.0	10.2

Display 9.60 Tar yields from eight brands of Canadian cigarettes.

Following the steps in E42, analyze the tar yields. Your analysis should include both exploration and formal inference and should end with a summary of what you conclude based on the data.

E45. ***Altitude and alcohol.*** At the beginning of every commercial passenger flight, there is an announcement that tells what to do if the cabin loses pressure. At high altitudes, the air is thin, with little oxygen, and you can lose consciousness in a short time. In 1965, the *Journal of the American Medical Association* published a paper that reports the effects of alcohol on the length of time subjects stayed conscious at high altitudes.

There were 10 subjects. Each was put in an environment equivalent to an altitude of 25,000 feet and then monitored to see how long the subject could continue to perform a set of assigned tasks. As soon as performance deteriorated (the end of "useful consciousness"), the time was recorded and the environment was returned to normal.

Three days later, each subject drank a dose of whiskey based on body weight—1 cc of 100-proof alcohol for every two pounds— and then, after waiting one hour for the whiskey to take effect, was returned to the simulated altitude of 25,000 feet for another test. Display 9.61 gives the times (in seconds) until the end of useful consciousness.

a. How can you tell, just from the description of the data, even before seeing the numbers, that this study gives you paired data from one sample rather than two independent samples?

b. Carry out an exploratory analysis. Include stem or boxplots of the readings under each of the two conditions and a stem or boxplot of the differences, as well as a scatterplot. Describe the patterns in words, and tell what questions, if any, these patterns raise about the validity of formal inference based on the normal distribution.

Time Until End of Useful Consciousness (sec)

Subject	Control	Alcohol
1	261	185
2	565	375
3	900	310
4	630	240
5	280	215
6	365	420
7	400	405
8	735	205
9	430	255
10	900	900

Display 9.61 The effect of alcohol consumption on useful consciousness at high altitudes.

Source: Jay Devore and Roxy Peck, *Statistics: The Exploration and Analysis of Data* (Minneapolis, MN: West Publishing Co., 1986). Their source: "Effects of Alcohol on Hypoxia," *Journal of American Medical Association* (December 13, 1965): 135.

c. Find a 95% confidence interval for the "true" difference $\mu_C - \mu_A$. What does the "true" difference refer to in this context? What do you conclude about the effect of alcohol on the length of useful consciousness at high altitude?

E46. Heights of men in the United States have a standard deviation of about 2.5 inches. Suppose you wanted to estimate the mean height of all men in your city from a random sample of size n at the 95% confidence level. What sample size would you need in order to estimate the mean height to within

a. two inches?

b. one inch?

c. one-half inch?

E47. ***Boosting your SATs?*** Several commercial companies offer special courses designed to help you get a higher score on the SAT test. Suppose market research shows that students are willing to take such a course only if, on average, it raises SAT scores by at least 30 points. Here's a way to test a claim that a course is able to do that. You'll need a sample of volunteers, the more representative the better. Randomly divide them into two groups. Those in the first group take the special course; those in the second group serve as controls. Once the course is over, both groups take the SATs.

Let $\bar{x}_1$ and $\bar{x}_2$ be the sample means for the two groups, let s_1 and s_2 be the sample standard deviations, and let n_1 and n_2 be the sample sizes.

a. Define suitable notation, and state the null and alternate hypotheses twice, first in words and then in symbols.

b. If you construct a 95% confidence interval for $\mu_1 - \mu_2$, under what circumstances do you reject H_0?

c. Carry out the test using the t-statistic if $\bar{x}_1 = 1100$, $\bar{x}_2 = 1060$, $s_1 = 100$, $s_2 = 80$, $n_1 = n_2 = 16$. Check your conclusion by constructing a confidence interval.

E48. **Nicotine.** Use the data from E43 to test the hypothesis that the nicotine yield in sidestream smoke is at least 2 mg higher than in mainstream smoke. Based on the outcome of your test, together with your work in E43, what do you conclude?

E49. An economist was stranded on a desert island when a case of canned goods washed ashore. "If only we had a way to open the cans," his statistician companion lamented. "No problem," said the economist. "I have a standard method for dealing with situations like this. Step One: Assume we have a can opener . . ."

Explain how this story is related to performing a test of significance on data from a voluntary-response sample.

E50. Arthritis is painful, and those who suffer from this disease often take pain relief and anti-inflammatory medication for long periods of time. One of the side effects of such medications is that they often cause stomach damage, such as lesions and ulcers. The goal of research, then, is to find an anti-inflammatory drug that causes minimal damage to the stomach. The data in Display 9.62 come from an experiment that tested two treatments (old and new) against a placebo by measuring their effect on lesions in the stomachs of laboratory rats. Rats with similar stomach conditions

Placebo	Old Treatment	New Treatment	Placebo	Old Treatment	New Treatment
0	0	3.16	0.08	0.361	0.227
0.364	7.997	0.044	0.572	0.036	0.756
0.768	8.21	8.233	1.194	2.541	1.029
0.851	9.897	0	0	0	0
1.1	12.873	1.329	0	4.587	0.416
6.04	0	3.447	0	1.41	1.418
1.785	9.177	0	3.112	0.272	6.027
0.103	0.784	0.643	0	1.789	2.053
0	4.323	0.748	1.777	20.273	2.107
0.087	6.16	4.637	0.907	0	0.912
0	2.51	0.384	14.447	2.443	2.552
0	3.272	2.487	0	3.83	11.563
2.932	1.94	1.536	0.171	3.241	0.215
4.603	0.637	0	1.023	0.712	4.82
0.449	0.345	0.773	0	5.97	0.901
1.793	0.211	0.53	0		2.861
6.847	3.157	1.288			1.645

Display 9.62 Lesion lengths (mm) for three arthritis treatment groups.

Source: R. Peck, L. Haugh, and A. Goodman, *Statistical Case Studies*, 1998, ASA-SIAM, 65–76.

at the start of the experimental period were randomly assigned to one of three treatments. Total length of stomach lesions (in millimeters) was measured in each rat after a two-week treatment period.

a. Do you see any problems with making comparisons of means with these data by use of standard methods?

b. With 90% confidence, estimate the difference between mean lesion length for the placebo and the old treatment.

c. With 90% confidence, estimate the difference between mean lesion length for the placebo and the new treatment.

d. With 90% confidence, estimate the difference between mean lesion length for the new and the old treatments.

e. Combine your results from parts a–d to write a brief report on the results of this experiment that could be published in a newspaper.

CHAPTER 10
CHI-SQUARE TESTS

Is it necessary to have a child at some point in your life in order to feel fulfilled? A poll asked this question of samples of adults in various countries. You can use a chi-square test to decide if the distribution of answers (yes, no, and undecided) reasonably could be the same for the adults in each country.

You have now completed two chapters on statistical inference. Chapter 8 considered inference for proportions constructed from categorical variables, and Chapter 9 considered inference for means constructed from measurement (quantitative) variables. Chapter 10 returns to the theme of categorical variables. You can think of it as a generalization of the results of Chapter 8.

In Chapter 8, there were only two categories—success or failure, heads or tails, and so on—and you used significance tests to answer questions such as:

Fair coin? You flip a coin 100 times and get 64 heads. Do you have evidence that the coin is unfair?

Who watches the Super Bowl? In a random sample of 100 college graduates and 100 high school graduates, 40% of the college graduates and 49% of the high school graduates watched the last Super Bowl. Is this convincing evidence that the percentage of college graduates and the percentage of high school graduates who watch the Super Bowl are not the same?

In this chapter, you will use chi-square tests to answer similar questions but that usually involve more than two categories:

Fair die? You roll a die 60 times. You get 12 ones, 9 twos, 10 threes, 6 fours, 11 fives, and 12 sixes. Do you have evidence that the die is unfair?

Super Bowl again. In addition to the random samples of college and high school graduates, you have a random sample of 100 people who didn't graduate from high school. Of this sample, 37% watched the last Super Bowl. Do the percentages of people watching the Super Bowl change with educational level?

You will use the chi-square technique to answer a new type of problem as well:

Better grades in the morning? If, for each student taking statistics at your school, the letter grade on the first statistics exam and class hour are recorded, does there appear to be an association between grade and hour, or do these two variables appear to be independent?

In this chapter, you will learn about three chi-square tests:

- *Goodness of fit.* Are the proportions of the different outcomes in this population equal to these hypothesized proportions?

- *Homogeneity of proportions.* Are the proportions of the different outcomes in one population equal to those in another population?

- *Independence.* Are two or more outcomes associated in this population?

The procedure in each case is almost the same. The difference is in the type of question asked and the kinds of conclusions that you can draw.

10.1 Testing a Probability Model: The Chi-Square Goodness-of-Fit Test

A **goodness-of-fit test** determines if it is reasonable to assume that your sample came from a population where, for each category, the actual proportion of outcomes is equal to some hypothesized proportions. If the result from the sample is very different from the expected results, then you have to conclude that the hypothesized proportions are wrong. To determine how different is "very different," you will need a test statistic and its probability distribution.

A Test Statistic

Example: Fair Die?

Suppose you roll a die 60 times and get 12 ones, 9 twos, 10 threes, 6 fours, 11 fives, and 12 sixes. Do you have evidence that the die is unfair?

If a fair die is rolled 60 times, you "expect" to get each face of the die on $\frac{1}{6}$ of the 60 rolls, or 10 times each. You can display the situation in a table.

Outcome	Observed Frequency, O	Expected Frequency, E
1	12	10
2	9	10
3	10	10
4	6	10
5	11	10
6	12	10
Total	**60**	**60**

A test statistic is a measure of the distance between observation and model.

You will conclude that the die is unfair if the observed frequencies are far from the expected frequencies. Are they? Here's where you need a test statistic. That is, you need to condense the information in the table into a single number that acts as an index of how far away the observed frequencies are from the expected frequencies.

Discussion: Toward a Test Statistic

D1. What is the null hypothesis for the *Fair Die* example? What is the alternate hypothesis?

D2. One test statistic that might be constructed for the data from the *Fair Die* example is

$$\sum (O - E)$$

where O represents an observed frequency and E represents the corresponding expected frequency.

    a. Compute the value of this test statistic for the *Fair Die* example.

    b. Will your result in part a always happen? Prove your answer.

    c. Is this a good test statistic?

D3. You have seen two other situations where the sum of the differences always turned out to be zero. What were those situations? What did you do in those situations?

D4. Another test statistic that might be constructed is

$$\sum (O - E)^2$$

    a. Compute the value of this test statistic for the *Fair Die* example.

    b. The two tables below show the results from the rolls of two different dice. Which table gives stronger evidence that the die is unfair?

    c. Compute and compare the values of $\sum (O - E)^2$ for Die A and Die B. Does this appear to be a reasonable test statistic? Explain.

	Die A		Die B	
Outcome	Observed Frequency	Expected Frequency	Observed Frequency	Expected Frequency
1	5	10	995	1000
2	16	10	1006	1000
3	18	10	1008	1000
4	4	10	994	1000
5	5	10	995	1000
6	12	10	1002	1000

Both the squared difference $(O - E)^2$ and the relative difference (or proportional difference) $(O - E)/E$ seem to be important in determining how "far" the observed frequency is from the expected frequency. And the test statistic involves both.

The symbol χ^2 is read "chi-square." The χ^2 statistic was first proposed by the English statistician Karl Pearson in 1900.

> **The test statistic used for chi-square tests is**
>
> $$\chi^2 = \sum \frac{(O - E)^2}{E}$$
>
> Here, O is the observed frequency in each category and E is the corresponding expected frequency.

You can compute the value of χ^2 for the *Fair Die* problem using a table.

Outcome	Observed Frequency, O	Expected Frequency, E	$O - E$	$\dfrac{(O - E)^2}{E}$
1	12	10	2	0.4
2	9	10	−1	0.1
3	10	10	0	0
4	6	10	−4	1.6
5	11	10	1	0.1
6	12	10	2	0.4
Total	**60**	**60**	**0**	**2.6**

The value of χ^2 is the sum of the last column, $(O - E)^2/E$, or $\chi^2 = 2.6$.

Discussion: A Test Statistic

D5. Will you reject the hypothesis that the die is fair if χ^2 is relatively large or if it is relatively small or both? How might you determine whether a χ^2 value of 2.6 is relatively large?

D6. What is χ^2 for Die A and for Die B in D4? Did the larger value of χ^2 correspond to the die that you thought seemed more unfair? What purpose does dividing by E serve in the formula?

D7. If you are given the observed frequencies and hypothesized proportions, how do you find the expected frequencies?

D8. The χ^2 statistic involves a sum of squared differences. What other statistics have you seen that involve a sum of squared differences?

■ Practice

P1. Suppose you want to test whether a tetrahedral (four-sided) die is fair. You roll it 50 times and observe a one 14 times, a two 17 times, a three 9 times, and a four 10 times. How many do you "expect" in each category? Compute the value of χ^2.

P2. If each observed frequency equals the expected frequency, what is the value of χ^2?

The Distribution of Chi-Square

In the *Fair Die* example, you saw that χ^2 turned out to be 2.6. If the observed and expected frequencies had been exactly equal, χ^2 would have been 0 and you would have had no reason to doubt that the die is fair. If the observed and expected frequencies had been much farther apart than they were, χ^2 would have been much larger than 2.6 and you would have had evidence that the die is not fair.

How can you assess whether a value of $\chi^2 = 2.6$ is large enough to reject the hypothesis that the die is fair? You need to see how much variation there is in the value of χ^2 when a die is fair. In Activity 10.1, you will do this.

Activity 10.1   Generating the Chi-Square Distribution

What you'll need: dice

1. Form five groups in your class. Each group should roll fair six-sided dice until your group has a total of 60 rolls. Compute χ^2 for your group's results. Get the values of χ^2 from the other four groups. From just these five results, does it appear that $\chi^2 = 2.6$ is a reasonably likely outcome when a fair die is rolled?

2. Describe how to use your calculator to simulate 60 rolls of a fair die.

3. Use your calculator to simulate 60 rolls of a fair die. Compute χ^2 for your results.

4. With your class, compute 200 values of χ^2, where each χ^2 is computed from 60 rolls of a fair die. Display them in a histogram and describe the shape of the histogram.

5. Should a one-tailed or a two-tailed test be used to test that a die is fair? Using the results from your simulation, estimate the P-value for $\chi^2 = 2.6$. Is a value of $\chi^2 = 2.6$ a reasonably likely outcome if the die is fair? Explain your decision.

The histogram in Display 10.1 shows a distribution of 2000 values of χ^2. Each value was computed from the results of 1000 rolls of a fair die. This histogram should look very much like the one your class generated in Activity 10.1 for only 60 rolls of a fair die.

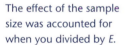
The effect of the sample size was accounted for when you divided by E.

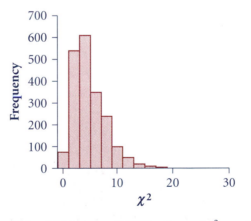

Display 10.1   A histogram of 2000 values of χ^2, $n = 1000$.

One of the properties of the χ^2 distribution is that it does not change much with a change in sample size. The distribution does depend on the number of categories, however, as you can see in Display 10.2.

Each histogram in Display 10.2 shows a distribution of 5000 values of χ^2. Each of the 5000 values was computed from the results of 60 rolls of a fair die. However, each histogram was made using a die with a different number of sides. Notice how the distribution changes as the number of categories changes.

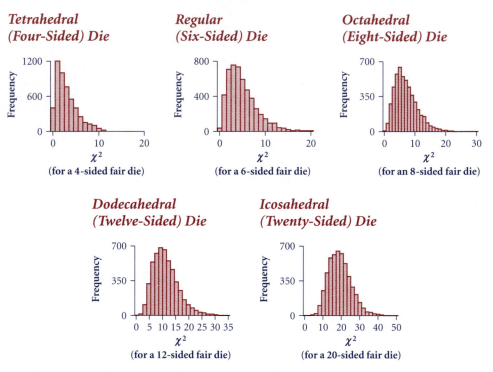

Display 10.2 Five histograms of 5000 values of χ^2, each value computed from 60 rolls of a fair die.

Discussion: The Chi-Square Distribution

D9. Describe how the distribution of χ^2 changes as the number of categories increases. Discuss these changes in terms of shape, center, and spread.

D10. For which die is it most likely to get a value of χ^2 of 15 or more? For which die is it least likely?

D11. For each die, make an approximation of the smallest value of χ^2 that would be a rare event (that falls in the upper 5% of the distribution).

Using a Table of Chi-Square Values

To find the critical values of χ^2 that cut off the upper 5% of the distribution (or another percentage), you can use the chi-square table (Table C) in the appendix. A partial table is shown in Display 10.3. In this table, the *df*, or **degrees of freedom**, is equal to the number of categories minus 1.

	Upper Tail Probability p						
df	.10	.05	...	.01	.005	...	.001
1	2.71	3.84	...	6.63	7.88	...	10.83
2	4.61	5.99	...	9.21	10.60	...	13.82
3	6.25	7.81	...	11.34	12.84	...	16.27
4	7.78	9.49	...	13.23	14.86	...	18.47
5	9.24	11.07	...	15.09	16.75	...	20.51
6	10.64	12.53	...	16.81	18.55	...	22.46
7	12.02	14.07	...	18.48	20.28	...	24.32
8	13.36	15.51	...	20.09	21.95	...	26.12
9	14.68	16.92	...	21.67	23.59	...	27.83
10	15.99	18.31	...	23.21	25.19	...	29.59
11	17.29	19.68	...	24.72	26.76	...	31.26
12	18.55	21.03	...	26.22	28.30	...	32.91

Display 10.3   A partial table of critical values of χ^2 from Table C.

To find, for example, the value of χ^2 that cuts off the upper 5% of the distribution when χ^2 is computed for rolling a six-sided die, go to the row $df = 6 - 1$, or 5. Then go over to the column heading of .05. The critical value is 11.07. The computed value of χ^2 from the *Fair Die* example is only 2.6, so you have no evidence at the $\alpha = .05$ level that the die is unfair. As illustrated in Display 10.4, it is quite likely to get a value of 2.6 or greater with a fair die.

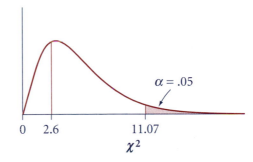

Display 10.4   Chi-square distribution with 5 degrees of freedom and a critical value of 11.07.

■ Practice: Using the Chi-Square Table and Your Calculator

P3. Suppose you roll a 12-sided die 300 times to see if it is fair. You compute χ^2 and get 21.3. Approximate a *P*-value for this test. What is your conclusion if you are using $\alpha = .05$? $\alpha = .01$?

P4. Suppose you roll an eight-sided die 100 times to see if it is fair. You compute χ^2 and get 21.3. Approximate a *P*-value for this test. What is your conclusion if you are using $\alpha = .05$? $\alpha = .01$?

P5. Learn to use your calculator to find *P*-values and then find a *P*-value for the case of

a. rolling a regular die, where $df = 5$ and $\chi^2 = 2.6$

b. rolling a 12-sided die, where $df = 11$ and $\chi^2 = 21.3$

The Chi-Square Goodness-of-Fit Test

So far in this section, you have used the *Fair Die* example to develop the ideas of a chi-square goodness-of-fit test. However, you can apply this test in many situations where you wish to assess how well a given probability model fits your data. To perform a chi-square goodness-of-fit test, you should go through the same steps as for any test of significance.

Chi-Square Goodness-of-Fit Test

1. ***Name the test and check conditions.***

   - Each outcome in your population falls into exactly one of a fixed number of categories.

   - You have a model that gives the hypothesized proportion of outcomes in the population that fall into each category.

   - You have a random sample from your population.

   - The expected frequency in each category is 5 or greater.

2. ***State the hypotheses.***

   H_0: The actual proportions in the population are equal to those in your model.

   H_a: At least one actual proportion in the population is not equal to that in your model.

3. ***Compute the value of the test statistic, approximate the P-value, and draw a sketch.*** The test statistic is

   $$\chi^2 = \sum \frac{(O - E)^2}{E}$$

   where O is the observed frequency in each category and E is the corresponding expected frequency.

   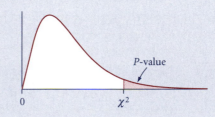

   The *P*-value is the probability of getting a value of χ^2 as extreme as or even more extreme than the one in the sample, assuming the null hypothesis is true. Approximate the *P*-value by comparing the value of χ^2 to the appropriate value of χ^2 in Table C, where *df* = *number of categories* − 1.

4. ***Write your conclusion linked to your computations and in the context of the problem.*** If the *P*-value is smaller than α (or, equivalently, if χ^2 is larger than the critical value for the given α), then reject the null hypothesis. If not, then there is no evidence that the null hypothesis is false and so you do not reject it. (Remember that you don't say that you "accept" it. This is because you don't know that it is true; you just don't have any evidence that it is false.)

Example: Plain Versus Peanut

Suppose, in a random sample of 75 peanut M&M's, you get the distribution of colors shown here. Is the distribution of peanut M&M's the same as the distribution of plain M&M's?

Color	Observed Number of M&M's
Red	11
Yellow	16
Green	8
Orange	5
Brown	17
Blue	18
Total	**75**

Solution

If the distribution of colors in peanut M&M's is the same as for plain M&M's, you would expect the numbers of each color to be as shown in Display 10.5. Note that it is not necessary that all of the expected frequencies be the same.

Color	Percentage in Plain M&M's	Expected Number of M&M's
Red	20%	$0.2(75) = 15$
Yellow	20%	$0.2(75) = 15$
Green	10%	$0.1(75) = 7.5$
Orange	10%	$0.1(75) = 7.5$
Brown	30%	$0.3(75) = 22.5$
Blue	10%	$0.1(75) = 7.5$
Total	**100%**	**75**

Display 10.5   Expected number of M&M's in the *Plain Versus Peanut* example.

Source: global.mms.com/us/about/products/milkchocolate.jsp (June 2002).

Check conditions. The conditions are met in this situation for a chi-square goodness-of-fit test. There were 75 outcomes in a random sample. Each outcome was only one color. You can compute the expected number of each color because you know the distribution of colors in a bag of plain M&M's. All of the expected counts were at least 5.

State the hypotheses. The null hypothesis is

H_0: The distribution of colors in peanut M&M's is the same as in plain M&M's. That is, there are 20% red, 20% yellow, 10% green, 10% orange, 30% brown, and 10% blue.

The alternate hypothesis is

> H_a: The distribution of colors in peanut M&M's is not the same as in plain M&M's. That is, at least one proportion is different.

Compute the test statistic, find the P-value, and draw a sketch. The test statistic is

$$\chi^2 = \sum \frac{(O-E)^2}{E}$$

$$= \frac{(11-15)^2}{15} + \frac{(16-15)^2}{15} + \frac{(8-7.5)^2}{7.5}$$

$$+ \frac{(5-7.5)^2}{7.5} + \frac{(17-22.5)^2}{22.5} + \frac{(18-7.5)^2}{7.5}$$

$$\approx 18.04$$

The value of χ^2 from the sample, 18.04, is very far out in the tail of the chi-square distribution with $6 - 1$, or 5, degrees of freedom. In fact, it is in the upper 0.003 of the tail.

Conclusion written in the context of the situation. Reject the null hypothesis. You cannot attribute the difference in the expected and observed counts to variation in sampling alone. A value of χ^2 this large is very unlikely to occur in random samples of this size if peanut M&M's have the same distribution of colors as plain M&M's. Conclude that the distribution of colors in peanut M&M's is different from that for plain M&M's. Display 10.6 shows a Fathom printout for this test.

Test of M&M's		Goodness of Fit ▼	
Attribute: (categorical): Color			
		Count	Probability
Color	Red	11	0.200
	Yellow	16	0.200
	Green	8	0.100
	Orange	5	0.100
	Brown	17	0.300
	Blue	18	0.100
Column Summary		75	1.000

Ho: Categories of Color have probabilities given above

Number of categories:	6
Chi-square:	18.04
DF:	5
P-value:	0.0029

Display 10.6 Fathom printout of the *Plain Versus Peanut* example. ∎

Discussion: Goodness-of-Fit Tests

D12. Sometimes data can be *too* close to what is expected. A student decides at the last minute to do a project about dice rolling. He reports rolling a die 600 times and getting these results.

Outcome	Frequency
1	97
2	99
3	102
4	101
5	100
6	101

Do you suspect that the student may have "dry-labbed" (that is, fabricated) his results?

D13. Which of these statements are properties of the chi-square goodness-of-fit test?

A. If you switch around the order of the categories, the value of the χ^2 statistic does not change.

B. The "observed" numbers are always whole numbers.

C. The "expected" numbers are always whole numbers.

D. The number of degrees of freedom is 1 less than the sample size.

E. A high value of χ^2 indicates a high agreement between the observed frequencies and the expected frequencies.

D14. Why is a chi-square test typically one-tailed?

■ Practice

P6. It is sometimes said that older people are overrepresented on juries. Display 10.7 gives information about people on grand juries in Alameda County, California. Does it appear that these jurors were selected at random from the adult population of Alameda County?

Age	Countywide Percentage	Number of Grand Jurors
21–40	42	5
41–50	23	9
51–60	16	19
61 or older	19	33
Total	**100**	**66**

Display 10.7   Distribution of ages for jurors in Alameda County, California.

Source: David Freedman et al., *Statistics,* 2nd ed. (New York: Norton, 1991), p. 484. Their source: *UCLA Law Review.*

P7. The 2002 *World Almanac and Book of Facts* gives a list of 104 "Major Rivers in North America" with their lengths in miles (pages 459–460). These data come from the U.S. Geological Survey. For example, the length of the Hudson River is given as 306 miles, the Columbia as 1243, and the Missouri as 2315. The final digit of the length of each of these three rivers is 6, 3, and 5, respectively.

a. What distribution of final digits would you expect if the U.S. Geological Survey measures to the nearest mile?

b. The distribution of final digits appears here. Test the hypothesis that the digits are equally likely. If they do not appear to be equally likely, what possible explanation can you offer?

Final Digit	Frequency	Final Digit	Frequency
0	44	5	15
1	6	6	5
2	9	7	3
3	5	8	3
4	5	9	9

Why Must Each Expected Value Be 5 or More?

The theoretical χ^2 distribution used in the table is a continuous distribution. For example, Display 10.8 shows a (continuous) χ^2 distribution for 10 categories ($df = 9$) and one for 6 categories ($df = 5$).

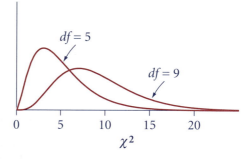

Display 10.8 Chi-square distributions for $df = 5$ and $df = 9$.

The distribution of χ^2 computed from repeated sampling is discrete, however, because there is a limited number of distinct values of χ^2 that can be calculated for a given number of categories and a given sample size. Like the normal approximation to the binomial distribution, the χ^2 distribution is a continuous distribution that can be used to approximate a discrete distribution. The larger the expected frequencies, the closer the distribution of possible values of χ^2 comes to a continuous distribution. In order to have a reasonable approximation, the expected frequency in each category should be 5 or greater. (This is a conservative rule, but it works well in most cases.)

Chi-Square Test Versus the z-Test

The chi-square test can be thought of as an extension of the z-test to more than two categories. Suppose you were checking a coin, rather than a die, to see if it was fair. If you flipped the coin n times, you would expect to see $\frac{n}{2}$ heads, under the hypothesis that the coin was fair. Letting x_1 denote the number of heads in the sample of n tosses and x_2 the number of tails, the table for constructing a χ^2 statistic for the data would look like this.

Category	Observed Frequency, O	Expected Frequency, E
Heads	x_1	$\dfrac{n}{2}$
Tails	x_2	$\dfrac{n}{2}$
Totals	n	n

The test statistic for testing the null hypothesis that the coin is fair becomes

$$\chi^2 = \frac{\left(x_1 - \frac{n}{2}\right)^2}{\frac{n}{2}} + \frac{\left(x_2 - \frac{n}{2}\right)^2}{\frac{n}{2}}$$

The sum of the number of heads and tails must be n, so $x_2 = n - x_1$ and you can write the test statistic as

$$\chi^2 = \frac{\left(x_1 - \frac{n}{2}\right)^2}{\frac{n}{2}} + \frac{\left[(n - x_1) - \frac{n}{2}\right]^2}{\frac{n}{2}} = 2\frac{\left(x_1 - \frac{n}{2}\right)^2}{\frac{n}{2}} = \frac{\left(x_1 - \frac{n}{2}\right)^2}{n\left(\frac{1}{2}\right)\left(\frac{1}{2}\right)}$$

But you already know another way of testing the hypothesis that a coin is fair. For large n, the test statistic for this hypothesis could be the familiar z-statistic from Chapter 8, given by

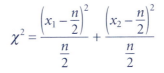

$$z = \frac{\hat{p} - p}{\sqrt{\dfrac{pq}{n}}} = \frac{\left(\dfrac{x_1}{n} - \dfrac{1}{2}\right)}{\sqrt{\dfrac{\left(\frac{1}{2}\right)\left(\frac{1}{2}\right)}{n}}} = \frac{\left(x_1 - \dfrac{n}{2}\right)}{\sqrt{n\left(\frac{1}{2}\right)\left(\frac{1}{2}\right)}}$$

You can now see that $\chi^2 = z^2$. The square of a z-statistic has a χ^2 distribution with 1 degree of freedom. This equality holds in general, even if p isn't $\frac{1}{2}$.

In summary, if there are only two types of outcomes (success and failure), you are back in the binomial situation and the z-test from Chapter 8 is equivalent to the chi-square test of this chapter. This implies, among other things, that the assumptions for the chi-square test are the same as those for the z-test: the sample must be random and large enough so that the sample proportion has approximately a normal distribution.

Discussion: Chi-Square Test Versus z-Test

D15. Discuss the similarities and differences between a *z*-test using proportions and a chi-square test using counts.

D16. It is hypothesized that among the students at your school who will be seniors next year, 20% will buy their lunch on campus, 30% will carry their own lunch, and 50% will eat off campus. A random sample of these students will be surveyed to check the claim that only half will eat off campus. How would you construct this test? Be ready to defend your choice.

Summary 10.1: The Chi-Square Goodness-of-Fit Test

The chi-square goodness-of-fit test is used when

- each outcome in your population falls into exactly one of a fixed number of categories
- you have a random sample from your population
- you want to know if it's reasonable to assume that the actual proportion of outcomes that fall into each category is equal to those in some hypothesized model

To perform a chi-square goodness-of-fit test, you go through the same steps as for any test of significance (see page 584). The expected frequencies are calculated from the probabilities specified in the null hypothesis and *df* = *number of categories* − 1.

If the result from your sample is very different from the results expected from the model, then χ^2 will be large and you reject the hypothesis that the model is the correct one for your population.

This section has concentrated on the model for a fair die because it is easy to understand. In the exercises, you will see examples of how this test is used in the real world.

■ Exercises

E1. Answer the question in the introduction to this chapter: You flip a coin 100 times and get 64 heads. Do you have evidence that the coin is unfair?

a. Use inference for a proportion, as in Chapter 8.

b. Use a chi-square goodness-of-fit test.

c. Compare the two results.

E2. In 1882, R. Wolf tossed a die 20,000 times. The results are recorded in Display 10.9. Is this evidence that the die was unfair? Or is this just about what you would expect from a fair die?

Outcome	Frequency
1	3407
2	3631
3	3176
4	2916
5	3448
6	3422

Display 10.9 Results from 20,000 rolls of a die.
Source: D. J. Hand et al., *Small Data Sets* (London: Chapman & Hall, 1994), p. 29.

E3. A study attempted to find a relationship between people's birthdays and dates of admission for treatment of alcoholism. In a sample of 200 admissions, 11 were within 7 days of the person's birthday, 24 were between 8 and 30 days inclusive, 69 were between 31 and 90 days inclusive, and 96 were more than 90 days from the person's birthday. Find appropriate expected numbers of admissions, and perform a chi-square goodness-of-fit test to test the hypothesis that admission is unrelated to birthday.

Source: Jay Devore and Roxy Peck, *Statistics: The Exploration and Analysis of Data*, 3rd ed. (Belmont, CA: Duxbury Press, 1997), pp. 565–566. Their source: *Psychological Reports*, 1992, pp. 944–946.

E4. You read in Chapter 6 about Gregor Mendel's work in genetics. Mendel performed experiments to try to validate the predictions of his theory. Mendel's experimental results match his theoretical results quite closely. In fact, statisticians think his results match *too* closely. In one experiment, Mendel predicted that he would get a 9:3:3:1 ratio between smooth yellow peas, wrinkled yellow peas, smooth green peas, and wrinkled green peas. His experiment resulted in 315 smooth yellow, 101 wrinkled yellow, 108 smooth green, and 32 wrinkled green peas. Show that these results are closer than you would reasonably expect just from variation in sampling.

E5. Display 10.10 gives the number of births for each month in a hospital in Switzerland. Only the 700 women who were having their first baby were included. Is there evidence that first births do not occur evenly throughout the year?

Month	Number of Births
January	66
February	63
March	64
April	48
May	64
June	74
July	70
August	59
September	54
October	51
November	45
December	42

Display 10.10 The number of births per month in a hospital in Switzerland.

Source: D. J. Hand et al., *A Handbook of Small Data Sets* (London: Chapman & Hall, 1994), p. 77. Their source: P. Walser, "Untersuchung über die Verteilung der Geburtstermine bei der mehrgebärenden Frau," *Helvetica Paediatrica Acta,* Suppl. XX ad 24: 3, 1–30.

E6. A sign on a barrel of nuts in a supermarket says that it contains 30% cashews, 30% hazelnuts, and 40% peanuts by weight. You mix up the nuts and scoop out 20 pounds. When you weigh the nuts, you find that you have 6 pounds of cashews, 5 pounds of hazelnuts, and 9 pounds of peanuts.

a. Do you have evidence to doubt the supermarket's claim?

b. If you have worked this problem using the chi-square test, convert everything to ounces and recalculate.

c. What is your conclusion now?

d. Do you see any problem in using a chi-square test on data like these?

E7. One of the basic principles you have learned in this textbook is that, all else being equal, it is better to have a larger sample size than a smaller one.

a. Explain why a larger sample is better.

b. As long as each expected frequency is at least 5, is there any advantage to having a larger sample size in a chi-square goodness-of-fit test?

E8. You can use statistical software to simulate the distribution of χ^2 values for any given number of degrees of freedom. This dot plot shows values of a χ^2 statistic for 2 degrees of freedom. Each dot represents 9 points.

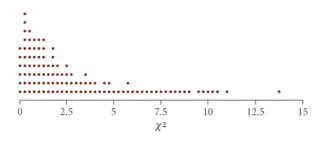

Use software to construct a distribution of χ^2 values for various numbers of degrees of freedom. Describe how these distributions change as the number of degrees of freedom increases.

E9. Many of the simulations in this textbook have been based on the assumption that the random-digit function of your calculator is equally likely to produce each of the 10 digits. Design and carry out an experiment to check whether this is indeed the case.

E10. Write a research hypothesis about data that can be analyzed using a chi-square goodness-of-fit test. For example, "People are less likely to be born on some days of the week than on others." Design and carry out a survey about this hypothesis using a random sample of a specified population.

E11. If you look at the leading digits in a table of data, you might expect each digit to occur about $\frac{1}{9}$ of the time. (There are only 9 possibilities because the digit 0 is never a leading digit.) However, smaller digits tend to occur more frequently than larger digits. Benford's Law says that the digit k occurs with relative frequency

$$\log_{10}\left(\frac{k+1}{k}\right)$$

a. Make a chart that shows the relative frequencies expected for the digits 1 through 9.

b. Prove algebraically that the sum of the relative frequencies is 1.

c. Examine tables of real data, such as populations of towns in your state, and determine if the distribution of the leading digits follows Benford's Law.

10.2 The Chi-Square Test of Homogeneity

Are proportions equal across several populations?

A chi-square test of homogeneity tests whether it is reasonable to believe that when several different populations are broken down into the same categories, they have the same proportion of members in each category. Such populations are called **homogeneous.** Before learning the formal procedure, you will work through an example that is based on a common student project.

Categorical Data with Two Variables

Example: Paper Towels

For her project, Justine decides to test the dry strength of three types of paper towels. Getting a random sample of towels from each brand is pretty much impossible on a student's budget, but she believes it is reasonable for her to assume that all towels of the same brand are pretty much identical. So Justine buys one roll of each of the three brands and uses the first 25 towels from each.

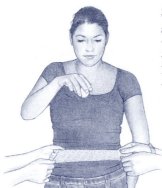

She stretches each towel tightly, drops a golf ball on it from a height of 12 inches, and records whether the golf ball goes through the towel. Justine tests the 75 towels in random order because she realizes that it will be impossible to hold her testing procedure completely constant. Display 10.11 shows the results from Justine's tests.

		Brand			
		Wipe-Ups	Wipe-Its	Wipe-Outs	Total
Towel Breaks?	Yes	18	7	5	30
	No	7	18	20	45
	Total	25	25	25	75

Display 10.11   Two-way table for the *Paper Towels* example.

The table in Display 10.11 is called a **two-way table** because each outcome is classified in two ways, according to the type of paper towel (population it comes from) and whether or not it breaks. Disregarding the labels and totals, it has two rows, three columns, and six cells. The data are called "categorical" because the only information recorded about each response is which category (or class) it falls into. A cell of the table consists of a count (or frequency). For example, the 20 in the second row and third column is the number of paper towels in Justine's sample of the 25 Wipe-Outs that did not break.

Are the results from Justine's three samples consistent with the hypothesis that the percentage of towels that break is the same for each brand?

As in any analysis, you should look at a graphical display of the data first. One possible plot is given in Display 10.12. This plot is called a **stacked** or **segmented bar graph.** For each population, the frequencies in each category are stacked on top of each other.

A segmented bar graph stacks the categorial frequencies on top of each other.

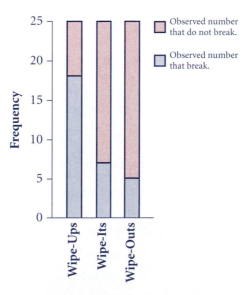

Display 10.12   A stacked bar graph for the *Paper Towels* example.   ■

Discussion

D17. From the plot only, do you believe that the three types of towels are equally likely to break?

D18. Suppose that the probability a paper towel breaks is the same for all three brands.

a. Using the data in Display 10.11, what is your best estimate of the probability that a paper towel will break?

b. Use your answer to part a to construct a two-way table for the expected results from this experiment under the hypothesis that all three brands have the same probability of breaking.

c. What would the bar graph look like if it were true that the probability a paper towel will break is the same for all three brands and if the results in the sample happened to come out exactly as expected?

Computing Expected Counts

Suppose that Justine tested a different number of towels for each brand and her results are as shown in Display 10.13.

Brand

		Wipe-Ups	Wipe-Its	Wipe-Outs	Total
Towel Breaks?	Yes	18	7	5	30
	No	14	11	20	45
	Total	32	18	25	75

Display 10.13   Two-way table for modified Paper Towels example.

If the null hypothesis is true that the proportion that break is the same for all three types of paper towels, what is the expected frequency in each cell?

Your best estimate of the overall proportion that will break is

$$\frac{\text{total number that break}}{\text{total number of towels tested}} = \frac{\text{row total for "yes"}}{\text{grand total}} = \frac{30}{75} = .4$$

The expected frequency, E, of the 32 Wipe-Ups that break is

(overall proportion that break) $\cdot$ (total number of Wipe-Ups) $= (.4)(32) = 12.8$

Following this same pattern, the general formula for E is

Formula for the Expected Count, *E*

$$E = \frac{(\text{row total})}{\text{grand total}} \cdot (\text{column total}) = \frac{(\text{row total})(\text{column total})}{\text{grand total}}$$

The completed table of expected counts appears in Display 10.14.

Brand

		Wipe-Ups	Wipe-Its	Wipe-Outs	Total
Towel Breaks?	Yes	12.8	7.2	10	30
	No	19.2	10.8	15	45
	Total	32	18	25	75

Display 10.14   Expected counts for the modified Paper Towels *example.*

After you finish computing a table of expected counts, always check to be sure the row and column totals are the same as in your table of observed counts.

Discussion: Computing Expected Counts

D19. Assuming gender would not affect the probability of survival, construct a table of expected counts for the observed *Titanic* data given in Display 10.15.

Gender

		Male	Female	Total
Survived?	Yes	367	344	711
	No	1364	126	1490
	Total	1731	470	2201

Display 10.15   Observed counts for the survivors on the Titanic.

Source: Robert J. MacG. Dawson, "The 'Unusual Episode' Data Revisited," *Journal of Statistics Education* 3, no. 3 (1995).

■ Practice

P8. This table gives the number of observations that fall into each of five categories in random samples from three populations.

Sample (observed frequencies)

		I	II	III	Total
Category	A	20	5	15	40
	B	21	50	29	100
	C	18	38	4	60
	D	20	30	0	50
	E	23	75	12	110
	Total	102	198	60	360

a. Make a segmented bar graph of the information given in the table.

b. Suppose the populations are homogeneous. What is your best estimate of the proportion of each population that falls into Category A? Category B? Category C? Category D? Category E?

   c. Compute the expected number that would fall into each cell if the populations are homogeneous.

   d. Does it appear that the three populations have the same proportion of members that fall into any given category?

Computing the Chi-Square Statistic

Homogeneous populations have the same proportion of members in any given category.

A chi-square test of homogeneity can be used to test whether the three brands of towels are equally likely to break. This test is quite similar to a chi-square test for goodness-of-fit. The value of χ^2 is computed in the same way using the observed and expected values in the six cells of the table. But in this test, the expected frequencies are estimated from the sample data. This is different from the goodness-of-fit test, where the probabilities were specified in the null hypothesis.

Example: Testing Paper Towels

You can now complete the computation of χ^2 for Justine's data in Display 10.11. The work is easily organized in a table, as shown in Display 10.16.

Outcome	Observed Frequency, O	Estimated Expected Frequency, E	$O - E$	$\dfrac{(O - E)^2}{E}$
Wipe-Ups break	18	$.4(25) = 10$	8	6.4
Wipe-Ups don't break	7	$.6(25) = 15$	−8	4.27
Wipe-Its break	7	$.4(25) = 10$	−3	0.9
Wipe-Its don't break	18	$.6(25) = 15$	3	0.6
Wipe-Outs break	5	$.4(25) = 10$	−5	2.5
Wipe-Outs don't break	20	$.6(25) = 15$	5	1.67
Total	**75**	**75**	**0**	**16.34**

Display 10.16   Calculating χ^2 for the Paper Towels example.

As before,

$$\chi^2 = \sum \frac{(O - E)^2}{E}$$

which is the sum of the last column, so $\chi^2 \approx 16.34$. ■

Before you can find the *P*-value for this test, you need to determine the number of degrees of freedom, *df*.

For a chi-square test of homogeneity, the number of degrees of freedom is

$$df = (r - 1)(c - 1)$$

where r is the number of rows in the table of observed values and c is the number of columns (not counting the headings in either case).

Example: Paper Towels, Degrees of Freedom

There are two rows (whether the towel breaks or not) and three columns (the three types of paper towels). The number of degrees of freedom is

$$df = (r - 1)(c - 1)$$
$$= (2 - 1)(3 - 1)$$
$$= 2$$

Using the calculator to determine the *P*-value for this test, where $\chi^2 \approx 16.34$, you get approximately .00028. Alternatively, if you look at χ^2 values in Table C of the appendix, you will find that this value of χ^2 is significant at the .001 level, as shown in Display 10.17. Either way, you would reject the hypothesis that the probability of breaking is the same for all three types of paper towels.

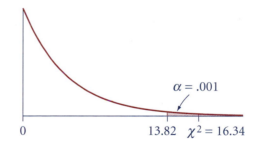

Display 10.17   Chi-square distribution with *df* = 2, α = .001. ■

Practice: Computing the Chi-Square Statistic

P9. Compute χ^2 for the table in P8. What is the number of degrees of freedom for that table? Is χ^2 significant at the $\alpha = .05$ level?

Procedure for a Chi-Square Test of Homogeneity

The steps in a chi-square test of homogeneity are much the same as the steps in the chi-square goodness-of-fit test.

Chi-Square Test of Homogeneity

1. *Name the test and check conditions.*

   - Independent simple random samples of fixed sizes (but not necessarily equal sizes) are taken from two or more large populations (or two or more treatments are randomly assigned to two or more types of available subjects).

   - Each outcome falls into exactly one of several categories, with the categories being the same in all populations.

   - The expected frequency is 5 or greater in each cell.

2. *State the hypotheses.*

   H_0: The proportion that falls into each category is the same for every population.

   H_a: For at least one category, it is not the case that each population has the same proportion in that category. That is, in some category, the proportion for at least one population is different from that for the others.

3. *Compute the value of the test statistic, approximate the P-value, and draw a sketch.* The test statistic is

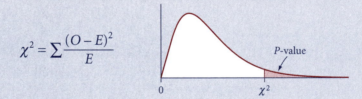

$$\chi^2 = \sum \frac{(O-E)^2}{E}$$

For each cell in a two-way frequency table, O is the observed count and E is the expected count:

$$E = \frac{(\text{row total})(\text{column total})}{\text{grand total}}$$

The *P*-value is the probability of getting a value of χ^2 as extreme as or more extreme than the one in the sample, assuming the null hypothesis is true. Approximate the *P*-value by comparing this value of χ^2 to the appropriate value of χ^2 in Table C with

$$df = (r-1)(c-1)$$

where r is the number of categories and c is the number of populations.

4. *Write your conclusion linked to your computations and in the context of the problem.* If the P-value is smaller than α (or, equivalently, if χ^2 is larger than the critical value for the given α), then reject the null hypothesis. If not, there is no evidence that the null hypothesis is false and so you do not reject it.

Example: Family Values

In November 1997, the Gallup Organization released the results of a poll on family values for different nations of the world. Results from each country were based on samples of "typically 1000 or more" from the adult population. (Assume that the sample size in each country was 1000.) One of the questions asked was this: "For you personally, do you think it is necessary or not necessary to have a child at some point in your life in order to feel fulfilled?"

The number of adults in some large countries who gave various answers are given in a two-way table and displayed in the bar graph in Display 10.18.

Table of Observed Counts
Country

		U.S.	India	Mexico	Canada	Germany	Total
Response	**Yes**	460	930	610	590	490	3080
	No	510	60	380	370	450	1770
	Undecided	30	10	10	40	60	150
	Total	1000	1000	1000	1000	1000	5000

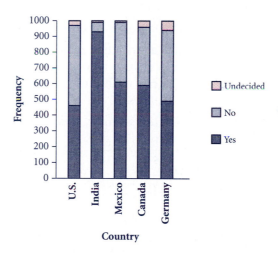

Display 10.18 Two-way table and segmented bar graph for the Gallup poll results.

Source: www.gallup.com/poll/specialreports/pollsummaries/family.asp (May 2002).

From this plot, you can see that the percentages do not appear to be the same for each country. These do not look like random samples from five populations with the same percentages in each category. The difference between India and the United States is too great. Test the hypothesis that the proportion of adults who would give each answer is the same for each country.

Solution

Under this hypothesis, the expected counts are given in this table.

Table of Expected Counts
Country

		U.S.	India	Mexico	Canada	Germany	Total
	Yes	616	616	616	616	616	3080
Response	No	354	354	354	354	354	1770
	Undecided	30	30	30	30	30	150
	Total	1000	1000	1000	1000	1000	5000

Check conditions. The conditions are met in this situation for a chi-square test of homogeneity. There are five large populations, and a random sample of size 1000 was taken from each population. Each answer falls into exactly one of three categories. All of the expected counts are at least five.

State the hypotheses. The null hypothesis is

H_0: If you could ask all adults, the distribution of answers would be the same for each country.

The alternate hypothesis is

H_a: The distribution of answers is not the same in each of the five countries. That is, in at least one country, the proportion of all adults who would give one of the answers is different from the proportion in other countries.

Compute the test statistic, compare it to a critical value, and draw a sketch. The test statistic is

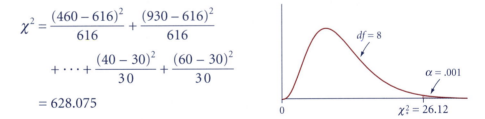

$$\chi^2 = \frac{(460-616)^2}{616} + \frac{(930-616)^2}{616}$$

$$+ \cdots + \frac{(40-30)^2}{30} + \frac{(60-30)^2}{30}$$

$$= 628.075$$

Comparing the test statistic to the χ^2 distribution with $df = (3-1)(5-1)$ or 8, you can see that the value of χ^2 from the sample, 628.075, is extremely far out in the tail and certainly greater than 26.12 ($\alpha = .001$). The *P*-value is approximately 0. Display 10.19 shows a Minitab printout of this test.

```
Chi-Square Test
Expected counts are printed below observed counts
          U.S.    India    Mexico    Canada    Germany    Total
    1      460      930       610       590        490      3080
        616.00   616.00    616.00    616.00     616.00

    2      510       60       380       370        450      1770
        354.00   354.00    354.00    354.00     354.00

    3       30       10        10        40         60       150
         30.00    30.00     30.00     30.00      30.00

Total     1000     1000      1000      1000       1000      5000
```

ChiSq = 39.506 + 160.058 + 0.058 + 1.097 + 25.773 + 68.746 + 244.169
      + 1.910 + 0.723 + 26.034 + 0.000 + 13.333 + 13.333 + 3.333 + 30.000
      = 628.075
df = 8, p = 0.000

Display 10.19 Minitab printout for the *Family Values* example.

Conclusion written in the context of the situation. Reject the null hypothesis. You cannot attribute the differences to the fact that we only have a sample of adults from each country and not the entire adult population. A value of χ^2 this large is extremely unlikely to occur in five samples of this size if the distribution of answers is the same in each country. Conclude that if you asked all people in these countries, the distribution of answers to this question would be different. ■

Discussion: Procedures for a Test of Homogeneity

D20. Verify the table of expected values in the *Family Values* example.

■ Practice

P10. Suppose that the value of χ^2 for the *Family Values* example had turned out to be 11.35. Find the *P*-value for such a test. Rewrite the conclusion.

Multiple z-Tests Versus One Chi-Square Test

The decision that the brands of paper towels were not of equal strength in the *Paper Towels* example was based on counts and a chi-square test. Suppose, however, that the data were rearranged to display the sample proportions of breaking towels, as in the next table. It appears that a Wipe-Ups paper towel has a much higher probability of breaking compared to the other populations in this study.

| | | **Brand** | | |
		Wipe-Ups Population 1	Wipe-Its Population 2	Wipe-Outs Population 3
Proportion of Breaks	**Sample**	.72	.28	.20
	Population	p_1	p_2	p_3

Based on the results of Chapter 8, you might be inclined to test the equality of these proportions using a *z*-test, and this would be the natural thing to do if there were only two populations. Based on the given sample data, testing the null hypothesis that $p_1 = p_2$ with a z-test will lead to the conclusion that these two proportions differ. This two-sample z-test is equivalent to a chi-square test of homogeneity using only the first two columns of data (a test with one degree of freedom). As in goodness-of-fit tests, $\chi^2 = z^2$. But there are still two more comparisons to be made: p_2 versus p_3 and p_1 versus p_3.

So why not use multiple z-tests rather than one chi-square test to determine whether or not the population proportions are homogeneous? This is a deep question, with the answer all tied up in a bunch of statistical theory, but the simple answer is that you have more of a chance of coming to an erroneous conclusion with multiple tests than you do with one. A guiding statistical principle for hypothesis testing is never use more tests than you absolutely need. The chi-square test is a clever way of combining many z-tests into one overall test.

Never use more tests than you need.

Degrees of Freedom: Information About Error

Jodain: I've just been letting this "degrees of freedom" thing slide, but in this section they've gone too far.

Dr. C: What's the trouble?

Jodain: Well, just for a starter, back when we did the t-test, *df* was *sample size* – 1. I was clueless why we subtracted 1, but never mind that now. Then in Section 10.1, we had almost the same formula, *number of categories* – 1. But why was it number of categories, not sample size?

Rest of class: Now we get this hugely different formula $df = (r-1)(c-1)$! (*grumble, grumble*)

Dr. C: Come on, it's a good question. And there's going to be yet another formula for *df* in the next chapter, on regression.

Jodain: You keep saying I'm not supposed to just memorize stuff, but . . .

Dr. C: Okay, the entire statistical story is way out there, but I'll try to explain. It has to do with the amount of information in the error term, or *SE*. Do you remember the *SE* from the t-test?

Jodain: Sure, I remember, it's

$$\frac{s}{\sqrt{n}} \qquad \text{where } s = \sqrt{\frac{\sum(x - \bar{x})^2}{n - 1}}$$

Rest of class: Figures! Jodain remembers everything!

Dr. C: Think about the deviations: $(x - \bar{x})$. There are n of them. Because the sum of the deviations from the mean is 0, if you know all deviations but one, you can figure out the last one. So you need only $n-1$ of the deviations to get all of the information about the size of a typical deviation from the mean.

Rest of class:	It's kind of like if the probability it'll rain tomorrow is .4, then you don't learn anything new if they also tell you that the probability it won't rain is .6. You could have already figured that out. So you really have only one piece of information.
Dr. C:	Well, that's sort of close.
Jodain:	Then the reason we didn't have to bother with *df* when we knew σ in Section 9.2 was that we didn't have to do any estimating of an error term using squared deviations?
Dr. C:	Right; you didn't compute a sum of squared deviations from the "center." When you can use a *z*-test, you don't need to worry about *df*.
Rest of class:	Like with the tests involving proportions! We like *z*-tests!
Dr. C:	What does this have to do with *df = number of categories* – 1 for a chi-square goodness-of-fit test?
Jodain:	Hmmm. The χ^2 statistic itself looks like one big error term. If you are testing whether a six-sided die is fair, there are six categories and six deviations $(O - E)$. I suppose that if you know all but one of them, you can figure out the other.
Rest of class:	Huh?
Dr. C:	Jodain's right. The last deviation is determined by the others because the sum of the deviations from the "center" is always equal to 0:

$$\sum(O - E) = \sum O - \sum E = n - n = 0$$

	Because the last deviation doesn't give you any new information about the size of the deviations from the center,

$$df = number\ of\ categories - 1$$

Jodain:	But what about the formula in this section, $df = (r - 1)(c - 1)$?
Rest of class:	(*groan*) Jodain, why do you do this to us? (*mumble, mumble*)
Dr. C:	Well, how many deviations, $O - E$, are you using in the formula?
Rest of class:	We know! The number of rows times the number of columns!
Dr. C:	Great answer! Now all you have to do is figure out how many of these give you new information about the size of $(O - E)$ and how many are redundant.
Jodain:	Well, with the *Paper Towels* example, there were $2 \cdot 3$, or 6, values of $(O - E)$. They have to sum to zero in each row and each column.

So, for example, if you know the deviations I've put in this table, you can figure out the rest, meaning that they don't tell you anything new.

Deviations by Brand

		Wipe-Ups $O - E$	Wipe-Its $O - E$	Wipe-Outs $O - E$	Total
Towel Breaks?	Yes	8	−3	–?–	0
	No	–?–	–?–	–?–	0
	Total	0	0	0	0

Rest of class:	We can be missing an entire row and an entire column's worth of deviations!
Dr. C:	That means you have only $(r - 1)(c - 1)$ independent estimates of the deviation.
Jodain:	Here's my rule: The concept of *df* applies when I need to use a sum of squared deviations from a parameter or parameters in my test statistic and I must estimate the parameter or parameters from the data. I count the number of deviations that are free to vary. This is the number of degrees of freedom.
Dr. C:	That covers it for everything you'll see in this class.
Rest of class:	We have a different rule: Just give us the formula!

Summary 10.2: The Chi-Square Test of Homogeneity

Use the chi-square test of homogeneity when

- you have independent samples from two or more populations
- you can classify the response from each member of the sample into exactly one of several categories
- you want to know if it's reasonable to assume that the proportion that falls into each category is the same for each population

This may sound familiar. In Section 8.4 you learned how to test whether the difference between two proportions is statistically significant. This is the same as a chi-square test of homogeneity when there are two populations and two categories. When there are more than two populations or more than two categories, the chi-square test of homogeneity can test the differences among the populations all in one test.

In a chi-square test of homogeneity, the expected counts are calculated from the sample data and

$$E = \frac{(\text{row total})(\text{column total})}{\text{grand total}}$$

The degrees of freedom for a two-way table of *r* rows and *c* columns is

$$df = (r - 1)(c - 1).$$

■ Exercises

E12. A recent Gallup poll asked the same question that has been asked every year for many years: "What do you think is the most important problem facing this country today?" Display 10.20 shows the percentage responses for three major concerns over the years from 1995 through 1998.

Most Important Problem
(in percent by year)

		Jobs	Crime	Health Care	Other
	1998	11	20	6	63
	1997	18	20	6	56
Year	1996	20	25	8	47
	1995	21	25	7	47

Display 10.20 Results from a Gallup poll: Percentage responses for three major concerns, 1995–1998.

Source: www.gallup.com/poll/.

a. Assume that there were 1000 people surveyed each year and convert the table to one displaying counts. Make and interpret a segmented bar graph of these counts.

b. What are the populations in this case?

c. Perform a chi-square test of homogeneity. Use $\alpha = .05$.

E13. Joseph Lister (1827–1912), surgeon at the Glasgow Royal Infirmary, was one of the first to believe in Pasteur's germ theory of infection. He experimented with using carbolic acid to disinfect operating rooms during amputations. Of 40 patients operated on using carbolic acid, 34 lived. Of 35 patients operated on not using carbolic acid, 19 lived.

Source: Richard Larson and Donna Stroup, *Statistics in the Real World: A Book of Examples* (New York: Macmillan, 1976), pp. 205–207. Their reference: Charles Winslow, *The Conquest of Epidemic Diseases* (Princeton, NJ: Princeton University Press, 1943), p. 303.

a. Display these data in a two-way table and in a segmented bar graph.

b. Compute the value of the test statistic, z, for a test of the difference of two proportions. Find the P-value for this test.

c. Compute the value of χ^2 for a test of homogeneity. Find the P-value for this test.

d. Compare the two P-values. Compare the values of z^2 and χ^2.

E14. This two-way table shows the general form of the data for a test for the difference of two proportions.

Population

Results	1	2	Total
Successes	x_1	x_2	$x_1 + x_2$
Failures	$n_1 - x_1$	$n_2 - x_2$	$(n_1 - x_1) + (n_2 - x_2)$
Total	n_1	n_2	$n_1 + n_2$

a. Using the symbols given in the table, compute the expected values.

b. Using the symbols given in the table, compute the value of the test statistic, z, for a significance test of the difference of two proportions.

E15. Suppose an experiment is designed to determine the effects of a bandage on the amount of pain a child perceives in a skinned knee. Of the first 60 children who came to a school nurse for a skinned knee, 20 were selected at random to receive no bandage, 20 to receive a flesh-colored bandage (matched to their skin color), and 20 to receive a brightly colored bandage. After 15 minutes, each child was asked, "Is the pain gone, almost gone, or still there?" Of the 20 children who got no bandage, none said the pain was gone, 5 said it was almost gone, and 15 said it was still there. Of the 20 children who got a flesh-colored bandage, 9 said the pain was gone, 9 said it was almost gone, and 2 said it was still there. Of the 20 children who got a brightly colored bandage, 15 said the pain was gone, 2 said it was almost gone, and 3 said it was still there. Organize these data into a table, display them in a plot, and perform a chi-square test of homogeneity, showing all steps.

E16. *The Gallup Poll Monthly* reported on a survey built around the question "What is your opinion regarding smoking in public places?" On the workplace part of the survey, respondents were asked to make one of three choices as to the policy they favored on smoking in the workplace. The percentages of the sample responding to these three choices across four geographic regions are shown in Display 10.21 below. Display these data in a plot. Perform a chi-square test of homogeneity to test if the proportion of people who choose each response is the same in the four regions of the country.

E17. Gallup polls of 1997 and 2001 each asked a randomly selected national sample of approximately 1000 adults, 18 years and older, this question: "Which of the following statements reflects your view of when the effects of global warming will begin to happen? (1) They have already begun to happen, (2) they will start happening within a few years, (3) they will start happening within your lifetime, (4) they will not happen within your lifetime, but they will affect future generations, or (5) they will never happen." The responses appear in Display 10.22. Organize the data, display them in a segmented bar graph, write an appropriate null hypothesis, and perform a chi-square test of homogeneity, showing all steps.

Response	Year 1997	Year 2001
Already begun	41%	54%
Within a few years	5%	4%
In my lifetime	11%	13%
Future generations	21%	18%
Never	14%	7%
No opinion	8%	4%
Total	**100%**	**100%**

Display 10.22   Results from a Gallup poll on global warming.

Source: www.americans-world.org/digest/global_issues/ global_warming/gw1.cfm (June 2002).

E18. Look up in a dictionary the definition of *homogeneous*. Explain what two homogeneous populations would be.

		Designated Area (percent)	Ban Altogether (percent)	No Restrictions (percent)	Sample Size
Region	East	65	31	4	246
	Midwest	61	36	3	256
	South	68	27	5	306
	West	60	36	4	199

Opinion

Display 10.21   Results from the Gallup poll on smoking in public places.

Source: *The Gallup Poll Monthly*, March 1994.

E19. Select one of these two ideas for a project, or make up your own.

Count the number of M&M's of each color in a bag of peanut M&M's and in a bag of plain M&M's. Write and test an appropriate hypothesis.

On your campus, observe 50 males and 50 females. Count the number who are wearing a belt, who aren't wearing a belt, and who have their shirt over their waist so you can't tell. Write and test an appropriate hypothesis.

E20. Describe how you can tell, on a given set of data, whether to use a chi-square goodness-of-fit test or a chi-square test of homogeneity.

10.3 The Chi-Square Test of Independence

Suppose you take a random sample of people from your community. From this sample alone, you could probably tell that being over 40 and having some gray hair are not independent characteristics in the population of all people. If your community is typical, people who are over 40 are more likely to have some gray hair than people 40 or younger.

Suppose that all you can observe is a sample from a population, like the sample in your community. You can use a chi-square test of independence to decide whether it is reasonable to believe that two different variables are independent in that population. The categorical variables for this possible sample are age (with categories *over 40* and *40 or younger*) and hair (with categories *has some gray hair* or *doesn't have gray hair*). For each of these two variables, each member of the population falls into exactly one category.

If categorical variables aren't independent, we call them *dependent* or *associated*.

Discussion

D21. Which of the following pairs of variables do you believe are independent in the population of U.S. students?

A. hair color and eye color

B. type of music preferred and ethnicity

C. gender and color of shirt

D. type of movie preferred and gender

E. eye color and class year

F. class year and whether taking statistics

D22. Why do you need a "test" of independence when you already have one from Chapter 6? In Chapter 6, you showed that two events A and B are independent by verifying that one of these statements is true. (And if one statement is true, all three are true.)

$$P(A) = P(A|B) \qquad P(B) = P(B|A) \qquad P(A \text{ and } B) = P(A) \cdot P(B)$$

In Activity 10.2, you will collect some data that will be used to illustrate the ideas in this section.

Activity 10.2:   Independent or Not?

1. In the next step, you will count the number of females and males in your class. You will also ask if the last digit of their phone number is even or odd. Do you think that the categorical variables of *gender* and *even/odd phone number* are independent? Explain.

2. Collect the data described in step 1 and organize the data in a two-way table.

3. Using the definition of independence in Chapter 6 (given in D22 on the previous page), determine if *gender* and *even/odd phone number* are independent events in the selection of a student at random from your class.

4. Based on your answer to step 3, what are you willing to conclude about the independence of the two variables in the population of all students? Explain.

5. Even if *gender* and *even/odd phone number* are independent events in the population of all students, why are you likely to find that this is not the case in your class?

6. In Activity 6.4, you determined whether you are right-eye dominant or left-eye dominant. Retrieve the table you made then or make another two-way table for your class where one categorical variable is *eye dominance* (*left or right*) and one categorical variable is *hand dominance* (*left or right*).

7. For a randomly selected student from your class, are these two traits independent according to the definition of independence in Chapter 6?

Discussion

D23. Suppose you have data from a sample and want to determine if two categorical variables are independent. If you use the definition of independence from Chapter 6, will you almost always find that they are independent or that they are not independent? Explain.

D24. Why do you think a chi-square test of independence is sometimes called a test of "association"?

Graphical Display of the Data

In this section, you will see what two-way tables and graphical displays look like when two variables are independent.

The segmented bar chart was a natural graphic to use for a chi-square test of homogeneity. In that situation, you had different populations that could be compared with side-by-side bars.

In a chi-square test of independence, there is only one population. It is broken down according to two different categorical variables. In Chapter 6, you saw the data on the *Titanic* disaster given in Display 10.23.

Class of Travel

Survived?		First	Second	Third	Total
	Yes	203	118	178	**499**
	No	122	167	528	**817**
	Total	**325**	**285**	**706**	**1316**

Display 10.23   Titanic survival data.

Source: *Journal of Statistics Education* 3(3) (1995).

In this case, the population under study is passengers of the *Titanic*, and they are classified according to the categorical variables of *class of travel* and *survival status*. Class of travel and whether a person survived obviously aren't independent. First-class passengers were more likely to have survived than third-class passengers.

You can construct a segmented bar graph based on either variable. These graphs are shown in Display 10.24. You can tell that the two variables are not independent by observing that the bars aren't divided into segments according to the same proportions. For example, in the left-hand graph, the third bar is more than two-thirds "Didn't survive," whereas the first bar is only about one-third "Didn't survive."

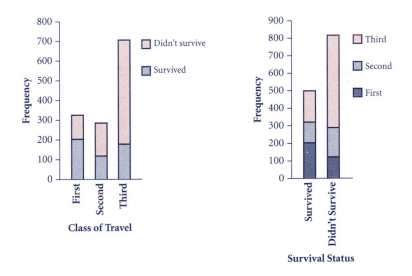

Display 10.24   Segmented bar graphs for the *Titanic* survival data.

The **column chart** shown in Display 10.25 treats the two variables symmetrically.

Display 10.25 A column chart for the *Titanic* survival data.

Discussion: Graphical Displays

D25. For each plot in Displays 10.24 and 10.25, give one fact that the plot shows more clearly than the other two plots.

D26. How can you tell from the right-hand segmented bar graph that *class of travel* and *survival status* aren't independent?

D27. How can you tell from the column chart that *class of travel* and *survival status* aren't independent?

D28. Which of the column charts in Display 10.26 display variables that are independent?

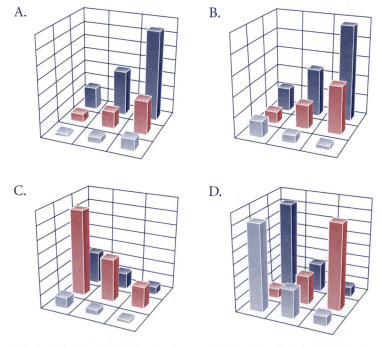

Display 10.26 Four column charts. Which ones display independent variables?

■ **Practice**

P11. What does a column chart look like if two variables are independent?

P12. Make two different segmented bars graphs and one column chart to display the data your class collected in Activity 10.2 on *gender* and *even/odd phone number*. How do your plots show whether or not the variables are independent?

Expected Values in a Chi-Square Test of Independence

In a chi-square test of independence, the null hypothesis is that the variables are independent. As you learned in Chapter 6, events A and B are independent if and only if $P(A \text{ and } B) = P(A) \cdot P(B)$. With this definition, you can fill in the two-way table of expected values in Display 10.27 if you assume that *handedness* and *eye color* are independent.

Handedness

		Right	Left	Total
	Blue	–?–	–?–	10
Eye Color	**Brown**	–?–	–?–	20
	Total	24	6	30

Display 10.27 A possible two-way table for handedness and eye color, $n = 30$.

For this example, the probability that a randomly selected person is right-handed and blue-eyed is

$$P(A \text{ and } B) = P(A) \cdot P(B)$$

$$P(\text{right-handed and blue-eyed}) = P(\text{right-handed}) \cdot P(\text{blue-eyed})$$

$$= \frac{24}{30} \cdot \frac{10}{30}$$

$$= \frac{4}{15}$$

The expected number of people who are right-handed and blue-eyed is then

$$P(\text{right-handed and blue-eyed}) \cdot n = \frac{4}{15}(30) = 8$$

In general, then, if two variables are independent, the probability that a randomly selected observation falls into the cell in column C and row R is

$$P(R \text{ and } C) = P(R) \cdot P(C) = \frac{\text{row } R \text{ total}}{\text{grand total}} \cdot \frac{\text{column } C \text{ total}}{\text{grand total}}$$

Thus, the expected number of observations that fall into this cell is

$$P(R \text{ and } C) \cdot n = \frac{\text{row } R \text{ total}}{\text{grand total}} \cdot \frac{\text{column } C \text{ total}}{\text{grand total}} \cdot (\text{grand total})$$

$$= \frac{(\text{row } R \text{ total}) \cdot (\text{column } C \text{ total})}{\text{grand total}}$$

Discussion: Computing Expected Values

D29. Complete an expected values table for your class assuming *gender* and *even/odd phone number* are independent.

D30. Complete an expected values table for your class assuming *handedness* and *eyedness* are independent.

D31. For which of the two tables in D29 and D30 are the expected and observed values farther apart?

D32. Is the formula for a cell's expected value in a chi-square test of independence the same as the expected value for the test of homogeneity?

■ Practice

P13. Here is a possible two-way table for a sample of size 160.

a. Assuming that the two variables are independent, find the expected values for each of the cells.

b. Enter a set of possible observed values into a copy of the table where the variables clearly aren't independent.

c. Enter a set of possible observed values into a copy of the table where the variables clearly are independent.

Variable 1

		I	II	Total
	A	–?–	–?–	36
	B	–?–	–?–	18
Variable 2	C	–?–	–?–	95
	D	–?–	–?–	11
	Total	100	60	160

Procedure for a Chi-Square Test of Independence

Now that you have reviewed the idea of independence, you will return to the problem of examining a sample from a population in order to decide if two variables are independent in that population.

The steps in a chi-square test of independence are much the same as the steps in the chi-square test of homogeneity.

Chi-Square Test of Independence

1. *Name the test and check conditions.*

   - A simple random sample is taken from one large population.

   - Each outcome can be classified into one of several categories on one variable and into one of several categories on a second variable.

   - The expected number in each cell is 5 or more.

2. *State the hypotheses.*

   H$_0$: The two variables are independent. That is, suppose one member is selected at random from the population. Then the probability that the member falls into both category A and category B, where A is a category from the first variable and B is a category from the second variable, is equal to $P(A) \cdot P(B)$.

   H$_a$: The two variables are not independent.

3. *Compute the value of the test statistic, approximate the P-value, and draw a sketch.* The test statistic is

$$\chi^2 = \sum \frac{(O - E)^2}{E}$$

For each cell in the table, O is the observed count and E is the expected count:

$$E = \frac{(\text{row total})(\text{column total})}{\text{grand total}}$$

The *P*-value is the probability of getting a value of χ^2 as extreme as or more extreme than the one in the sample, assuming the null hypothesis is true. Approximate the *P*-value by comparing the value of χ^2 to the appropriate value of χ^2 in Table C with $df = (r - 1)(c - 1)$, where r is the number of categories for one variable and c is the number of categories for the other variable.

4. *Write your conclusion linked to your computations and in the context of the problem.* If the *P*-value is smaller than α (or, equivalently, if χ^2 is larger than the critical value for the given α), then reject the null hypothesis. If not, there is no evidence that the null hypothesis is false and so you do not reject it.

Example: Scottish Children

The information given in Display 10.28 shows the eye color and hair color of 5387 Scottish children. You might suspect that these two categorical variables are not independent because people with darker hair colors tend to have darker eye colors, whereas people with lighter hair colors tend to have lighter eye colors. Test to see if there is an association between hair color and eye color.

		Hair Color				
		Fair	Red	Medium	Dark	Black
Eye Color	**Blue**	326	38	241	110	3
	Light	688	116	584	188	4
	Medium	343	84	909	412	26
	Dark	98	48	403	681	85

Display 10.28 Eye color and hair color data from a sample of Scottish children.

Source: D. J. Hand et al., *A Handbook of Small Data Sets* (London: Chapman & Hall, 1994), p. 146. Their source: L. A. Goodman, "Association Models and Canonical Correlation in the Analysis of Cross-Classifications Having Ordered Categories," *Journal of the American Statistical Association* 76 (1981): 320–334.

Solution

From the column chart in Display 10.29, you can see that it does indeed appear to be the case that children with darker hair colors tend to have darker eye colors, whereas children with lighter hair colors tend to have lighter eye colors.

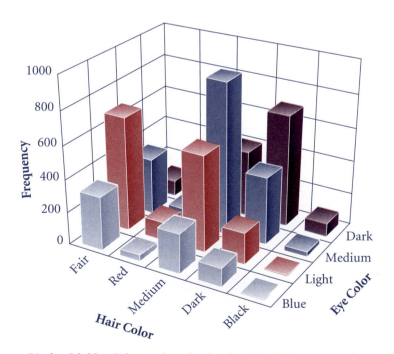

Display 10.29 Column chart for the *Scottish Children* example.

Check conditions. This situation satisfies the conditions for a chi-square test of independence if you can reasonably assume that the children can be considered a simple random sample taken from one large population. Each child in the sample falls into one hair color category and one eye color category. The Data Desk printout in Display 10.30 shows the expected values under the assumption of independence. The expected number in each cell is 5 or more.

State the hypotheses. The null hypothesis is

H_0: Eye color and hair color are independent.

The alternate hypothesis is

H_a: Eye color and hair color are not independent.

```
Rows are levels of      Eye Color: Count
Columns are levels of   Hair Color: Count
No Selector

           Fair       Red       Medium     Dark      Black      Total

Blue       326        38        241        110       3          718
           22.4       13.3      11.3       7.91      2.54       13.3
           193.928    38.1192   284.828    185.398   15.7275    718

Light      688        116       584        188       4          1580
           47.3       40.6      27.3       13.5      3.39       29.3
           426.750    83.8834   626.779    407.978   34.6092    1580

Medium     343        84        909        412       26         1774
           23.6       29.4      42.5       29.6      22.0       32.9
           479.148    94.1830   703.738    458.072   38.8587    1774

Dark       98         48        403        681       85         1315
           6.74       16.8      18.9       49.0      72.0       24.4
           355.174    69.8144   521.655    339.552   28.8045    1315

total      1455       286       2137       1391      118        5387
           100        100       100        100       100        100
           1455       286       2137       1391      118        5387

Table contents:
Count
Percent of Column Total
Expected Values

Chi-square = 1240     with    12     df
p ≤ 0.0001
```

Display 10.30   Data Desk printout of a chi-square test of independence for the *Scottish Children* example.

Compute the test statistic, calculate the P-value, and draw a sketch. Display 10.30 also gives the computation of the test statistic χ^2 and the *P*-value. Comparing the test statistic to the χ^2 distribution with 12 degrees of freedom, you can see from the printout and the sketch in Display 10.31 that the value of χ^2 from the sample, 1240, is extremely far out in the tail. The *P*-value is close to 0.

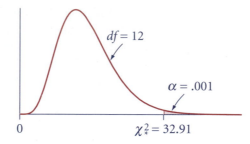

Display 10.31 Chi-square distribution with 12 degrees of freedom.

Conclusion written in the context of the situation. Reject the null hypothesis. These are not the results you would expect for a sample from a population where there is no association between eye color and hair color. A value of χ^2 this large is extremely unlikely to occur in a sample of this size if hair color and eye color are independent. Examining the table and column chart, it appears that darker eye colors tend to go with darker hair colors and lighter eye colors tend to go with lighter hair colors.

Discussion: Procedures for a Test

D33. Assuming your class is a random sample of students, test the hypothesis that *gender* and *even/odd phone number* are independent characteristics in students.

■ Practice

P14. Assuming your class is a random sample of students, test the hypothesis that handedness and eyedness are independent characteristics in students.

Homogeneity Versus Independence

Dr. C: Did you notice that a chi-square test of homogeneity and a chi-square test of independence are performed in exactly the same way, except for the wording of the hypotheses and the conclusion? The two null hypotheses say exactly the same thing about the table: the columns are proportional.

Class: So how do we know which test to use?

Dr. C: You can't tell from the table itself. You have to find out how the data were collected. Did they take one sample of a fixed size from one population and then classify each person according to two categorical variables? If so, it's a test of independence. Or did they sample separately from two or more populations and then classify each person according to one categorical variable? If so, it's a test of homogeneity.

Class: Huh?

Dr. C: It's simple. Suppose you want to find out whether there is an association between age and whether people wear blue jeans. You take a random sample of 50 people under 40. Then you take a random sample of 50 people 40 or older. Your table might look like this:

		Age	
		Under 40	**40 or Older**
Wear Blue Jeans?	**Now**	21	20
	Not now, but sometimes	23	25
	Never	6	5
	Total	**50**	**50**

Class: We get it! We have to do a test of homogeneity because we sampled separately from two populations: people under 40 and old people. There are two populations and one variable about jeans.

Jodain: Now I see what you mean by the columns being proportional. For people under 40, 42% are wearing jeans now, 46% are not wearing them now, and 12% never wear them. That's about the same distribution as for people 40 or older, so we won't be able to reject the hypothesis that these populations are homogeneous with respect to wearing jeans. Jeans-wearing behavior is about the same for both age groups.

Dr. C: Right. In the homogeneity case, you have one variable but several populations. The column (or sometimes the row) proportions represent the separate distributions for those populations.

Now suppose you go out and take a random sample of 100 people, classify them, and get this table. This time you sampled from just one population.

		Age		
		Under 40	**40 or Older**	**Total**
Wear Blue Jeans?	**Now**	23	17	**40**
	Not now, but sometimes	28	22	**50**
	Never	6	4	**10**
	Total	**57**	**43**	**100**

Class: We use a test of independence because that's the only other possibility.

Jodain: We use a test of independence because there is only one sample, from the population of all people. We categorize each person in the sample on the two variables of *age* and *jeans-wearing behavior*. The columns are roughly proportional, so we'll conclude that we can't reject the hypothesis that the two variables are independent.

Jeans-wearing behavior and age aren't associated. But this time, we didn't predetermine how many people would be in each age group.

Dr. C: Right again, Jodain.

Class: Jodain is always right!

Dr. C: Also, now it makes sense to talk about conditional distributions for the rows. You can estimate that around 58% of people wearing jeans are under 40 and around 42% are 40 or older. The percentages are about the same for people not wearing jeans now and for people who never wear jeans. That statement wouldn't have made sense in the homogeneity case.

Jodain: Then why wouldn't we always design our study as a test of independence? With that we can look at the table both ways, and as a bonus we get an estimate of the percentage of people in each age group!

Dr. C: It depends on what you want to find out. Suppose that your research hypothesis is that the proportion who wear jeans differs for the two age groups. You would design the study as a test of homogeneity, taking an equal sample size from each age group. That gives you the best chance of rejecting a false null hypothesis.

Class: How old are you, Dr. C?

Discussion: Homogeneity Versus Independence

D34. Suppose you want to decide if people with straight hair and people with curly hair are equally likely to use blow-dryers. You are considering two designs for your survey:

   I. You take a random sample of 50 people with straight hair and ask them whether or not they use a blow-dryer. You take a random sample of 50 people with curly hair and ask them whether or not they use a blow-dryer.

   II. You take a random sample of 100 people, note if they have curly hair or straight hair, and ask them whether or not they use a blow-dryer.

   a. Which of the designs would result in a test of homogeneity, and which would result in a test of independence?

   b. How are the data likely to be different?

   c. Is there any reason one design would be better than the other?

   d. Which of the two designs could also be correctly analyzed by a z-test for the difference between two proportions?

■ Practice

P15. A *Time* magazine poll asked a random sample of people from across America, "How much effort are you making to eat a healthy and nutritionally balanced diet?" The data are given in Display 10.32.

	Gender	
	Men	**Women**
Very serious	31%	43%
Somewhat serious	47%	43%
Not very serious	12%	9%
Don't really try	10%	5%

Effort labels the left side of the rows.

Display 10.32 Results of a poll on eating habits.

Source: *Time,* January 30, 1995.

a. Explain what the percentages measure. Given the way the data were collected, would you do a test of homogeneity or a test of independence?

b. Suppose that the sample size was 1000 and it just happened to be equally split between men and women. Is there evidence of dependence between gender and an effort toward a healthy diet?

c. Suppose the sample size was only 500 and it just happened to be equally split between men and women. Is there evidence of dependence between gender and an effort toward a healthy diet?

Strength of Association and Sample Size

Evidence of association is not the same as a measure of strength.

A chi-square test of independence tells you if there is evidence of an association between two categorical variables. As is typical of a test of significance, it does not tell you about the size of any difference in the proportions from column to column (the **strength of the association**) or about whether these differences are of practical importance. As an illustration, Display 10.33 shows three separate samples of responses on the same two variables. Samples 1, 2, and 3 are of size 200, 400, and 4000, respectively.

	Sample 1		Sample 2		Sample 3	
	A	**B**	**A**	**B**	**A**	**B**
Yes	48	52	96	104	960	1040
No	52	48	104	96	1040	960
Totals	100	100	200	200	2000	2000

Response labels the left side of the Yes/No rows.

	Sample 1	Sample 2	Sample 3
Difference between proportions of A and B saying "yes"	.04	.04	.04
χ^2	0.32	0.64	6.40
P-value	.57	.42	.011

Display 10.33 Three two-way tables showing how sample size can affect the test statistic.

In each sample, the columns are almost proportional—the difference between the proportions in the first and second columns is only .04. This is not a very strong association; in most applications, a difference of .04 between two proportions would be of little practical significance. However, if you do a chi-square test of independence on each of the three samples, you will see that as the sample size increases, the test statistic gets larger and the *P*-value smaller. At a sample size of 4000, the association between the variables becomes highly significant statistically, but the strength of this association is no larger than for the smaller samples. For fixed conditional distributions, the numerical value of the χ^2 statistic is directly proportional to the sample size.

> Statistical significance is not the same as practical significance.

Two bits of advice are in order: Be wary when using this test (or any test) with very large data sets, and always compute and compare proportions to see if the statistical significance has any practical value.

■ Practice: Strength of Association

P16. In the state of Maine, 71,552 passengers in cars and light trucks were involved in accidents during 1996. Display 10.34 is a two-way table of the two variables of *seat belt use* and *injury*.

Injury?

		Yes	No
Seat Belt?	**Yes**	3,984	57,107
	No	2,323	8,138
	Total	**6,307**	**65,245**

Display 10.34 Two-way table for seat belt use versus injury.

Source: Maine Department of Public Safety, *Do Seat Belts Reduce Injuries? Maine Crash Facts, 1996.* www.state.me.us/dps/Bhs/seatbelts1996.htm (Augusta, Maine).

a. Is there evidence of an association between *seat belt use* and *injury*? If so, what is the nature of the association?

b. Make a case for the use of seat belts based on these data.

c. What lurking variables might account for this association?

Summary 10.3: The Chi-Square Test of Independence

Suppose there is no reason to suspect anything other than complete independence between two variables in a population. Yet it would be a near-miracle to get a random sample in which the variables satisfy the definition of independence in Chapter 6. Requiring $P(A \text{ and } B)$ to be exactly equal to $P(A) \cdot P(B)$ for every cell of the table is too rigid a position to take when dealing with random samples. A chi-square test of independence is much more useful: Is it reasonable to assume that the random sample came from a population in which there is no association between two variables? Use the chi-square test of independence when

- a simple random sample of a fixed size is taken from one large population
- each outcome can be classified into one of several categories on one variable and into one of several categories on a second variable
- you want to know if it's reasonable to assume that this sample came from a population in which these two categorical variables are independent

In a chi-square test of independence, the expected counts are calculated from the sample data:

$$E = \frac{(\text{row total}) \cdot (\text{column total})}{\text{grand total}}$$

Each value of E should be 5 or greater.

The degrees of freedom for a two-way table of r rows and c columns is $df = (r-1)(c-1)$.

If the result from your sample is very different from the results expected if there is no association, then you reject the hypothesis that the variables are independent. However, even if the test tells you there is evidence of an association, it does not tell you anything about the strength of the association or whether it is of practical importance.

■ Exercises

E21. According to a National Center for Education Statistics report, 47,213,000 children were enrolled in the nation's K–12 public schools in 1999–2000. About 75.8% were enrolled in grades K–8 and 24.2% in grades 9–12. About 24.7% of the students were in the Northeast, 26.3% in the Midwest, 30.6% in the South, and 18.4% in the West.

Source: *The Condition of Education 2002*, Tables 2–1, 2–3, at nces.ed.gov/pubs2002/2002025_App1.pdf (June 2002).

a. Do you believe it is reasonable to assume that *region of the country* and *grade level* are independent?

b. Construct a two-way table showing the *percentage* of students who fall into each cell under the assumption of independence.

c. Construct a two-way table showing the *number* of students who fall into each cell under the assumption of independence.

E22. A student surveyed a random sample of 300 students in her large college and collected the data below on the variables of *class year* and *favorite team sport*.

Favorite Team Sport

		Basketball	Soccer	Baseball/Softball	Football	Total
Class Year	**Freshman**	12	40	10	1	63
	Sophomore	12	44	16	8	80
	Junior	9	43	11	11	74
	Senior	10	49	18	6	83
	Total	43	176	55	26	300

a. Perform a chi-square test to determine if *class year* and *favorite team sport* are independent.

b. Is it possible that you have made a Type I error?

c. Describe how the survey should have been designed if she had wanted to do a chi-square test of homogeneity.

E23. Display 10.35 (repeated from Chapter 6) is an example of a single population, *Titanic* passengers, that has been sorted by two variables: *gender* and *survival status*. These data cannot reasonably be considered a random sample from any well-defined population.

Gender

Survived?	Male	Female	Total
Yes	367	344	711
No	1364	126	1490
Total	1731	470	2201

Display 10.35 *Titanic* passengers sorted by gender and survival status.

a. Make a plot that displays these data so you can see whether *gender* and *survival status* appear to be independent.

b. Test to see if the association between the variables *gender* and *survival status* can reasonably be attributed to chance or if you should look for some other explanation.

E24. The fate of the members of the Donner party, who were trapped in the Sierra Nevada Mountains over the winter of 1846–1847, is shown in Display 10.36. Again, these data cannot reasonably be considered a random sample from any well-defined population.

Gender

Survived?	Male	Female
Yes	23	25
No	32	9

Display 10.36 The Donner party survival data.

Source: www.utahcrossroads.org/DonnerParty/Statistics.htm (June 2002).

a. Make a plot that displays these data so you can see whether *gender* and *survival status* appear to be independent.

b. Test to see if the association between the variables *gender* and *survival status* can reasonably be attributed to chance or if you should look for some other explanation.

c. Is it possible that you have made a Type I error?

E25. Display 10.37 gives the smoking behavior by occupation type for a sample of white males back in 1976. Assuming this can be considered a random sample of all white males in the United States in 1976, perform a chi-square test of independence to see if it is reasonable to assume that *occupation type* and *smoking behavior* are independent.

Occupation (percent)

Smoker?	Blue Collar	Professional	Other	Total
Yes	43.2	6.3	50.5	13,112
Former	30.9	11.0	58.1	8,509
Never	30.5	11.8	58.7	9,694

Display 10.37 Smoking behavior by type of occupation in percent.

Source: D. J. Hand et al., *A Handbook of Small Data Sets* (London: Chapman & Hall, 1994), p. 146. Their source: T. D. Sterling and J. J. Weinkam, "Smoking Characteristics by Type of Employment," *Journal of Occupational Medicine* 18 (1976): 743–753.

Chapter Summary

In this chapter, you have learned about three chi-square tests: a test of goodness-of-fit, a test of homogeneity, and a test of independence. Although each test is conducted in exactly the same way, the questions they answer are different.

In a chi-square goodness-of-fit test, you ask, "Does this look like a random sample from a population in which the proportions that fall into these categories are the same as those hypothesized?" This test is an extension of the test for a single proportion developed for the binomial case.

In a chi-square test of homogeneity, you ask, "Do these samples from different populations look like samples from populations in which the proportions that fall into these categories are equal?" This test is an extension of the test for the equality of two binomial proportions.

In a chi-square test of independence, you ask, "Does this sample look like it came from a population in which these two categorical variables are independent (not associated)?" This test is not equivalent to any test developed earlier in this book. It is, however, a forerunner of a similar idea that will be developed in Chapter 11 for quantitative variables—regression analysis.

■ Review Exercises

E26. When testing a random-digit generator, one criterion is to be sure that each digit occurs $\frac{1}{10}$ of the time.

a. Generate 100 random digits on your calculator, and use a chi-square test to see if it's reasonable to assume that this criterion is met.

b. This criterion isn't sufficient. Give an example of a sequence of digits that is clearly not random but in which each digit occurs $\frac{1}{10}$ of the time. How might you test against this possibility?

E27. Many years ago, Smith College, a residential college, switched to an unusual academic schedule that made it fairly easy for students to take most or all of their classes on the first three days of the week. (Smith has long since abandoned this experiment.) At the time, the infirmary staff wanted to know about after-hours use of the infirmary under this schedule. They gathered data for an entire academic year, recording the time and day of the week of each after-hours visit, along with the nature of the problem, which they later classified as belonging to one of four categories: a visit to the infirmary was *unnecessary* for the problem, the problem required a *nurse's* attention, the problem required a *doctor's* attention, the problem required *admission* to the infirmary or to the local hospital. The results appear in Display 10.38.

Nature of Problem	Day of Week				
	Mon	Tues	Wed	Thurs	Fri
Visit unnecessary	77	67	77	78	70
Nurse needed	80	66	53	73	62
Doctor needed	90	71	76	95	75
Admitted	61	49	42	52	28

Display 10.38 After-hours use of the infirmary.

Source: George W. Cobb, *Design and Analysis of Experiments* (New York: Springer, 1998), pp. 239–240.

a. Are there any interesting trends in the table? What might explain them?

b. Could you use a test of independence on these data? A test of homogeneity? Does the design of the study fit the assumptions of these tests?

c. Perform the test you think is best, showing all steps and giving any necessary caveats (cautions) about your conclusion.

d. How might you group the data to see whether the weekdays students attend class differs from the weekdays when students don't attend class?

e. Perform a chi-square test on your regrouped data, showing all steps.

E28. These questions were asked on a 2002 Gallup poll to a randomly selected national sample of approximately 1000 adults, 18 years and older: "Do you think that global warming will pose a serious threat to you or your way of life in your lifetime? Do you think that global warming will pose a serious threat to your children or the next generation of Americans in their lifetime?" The results are shown in Display 10.39.

Response	Your Lifetime	Your Children's Lifetime
Yes	31%	35%
No	66%	57%
No opinion	3%	8%
Total	100%	100%

Display 10.39   Results of Gallup poll on the threat of global warning.

Source: www.americans-world.org/digest/global_issues/global_warming/gw1.cfm (June 2002).

a. Does this situation call for a chi-square test of homogeneity, a chi-square test of independence, or neither?

b. Perform the test you selected, or explain why neither test is appropriate.

E29. In a random sample of 100 college graduates, 100 high school graduates, and 100 people who didn't graduate from high school, 40% of the college graduates watched the last Super Bowl, as did 49% of the high school graduates and 37% of the people who didn't graduate from high school. Is there a relationship between watching the Super Bowl and a person's educational level?

a. Does the design of this study suggest a test of goodness-of-fit, homogeneity, or independence?

b. Perform the test you selected, showing all steps.

E30. A common Gallup poll question that is often asked of a sample of residents of the United States is, "In general, are you satisfied or dissatisfied with the way things are going in the United States at this time?" Display 10.40 shows the results (in percent) for five polls taken in 1998.

a. Why don't the row percentages add to 1?

b. Suppose each poll samples about 1000 residents (which is about the size of most Gallup polls). Is there evidence that the level of satisfaction changed significantly across 1998? If so, describe the pattern of change.

c. Suppose each poll samples about 500 residents. Is there evidence that the level of satisfaction changed significantly across 1998? If so, describe the pattern of change.

Date of Poll	Percent Satisfied	Percent Dissatisfied
August 1998	63	34
May 1998	59	36
April 1998	58	38
February 1998	64	32
January 1998	63	35

Display 10.40   The results (in percent) for five polls taken in 1998.

Source: www.gallup.com/poll/news/971202.html.

Political Ideology

		Liberal	Moderate	Conservative
Party Affiliation	Democrat	313	387	174
	Independent	287	469	292
	Republican	79	185	407

Display 10.41 Results from the General Social Survey of the United States for the year 2000.

Source: www.icpsr.umich.edu/GSS/ (June 2002).

E31. The data in Display 10.41, from the General Social Survey of the United States, shows how people describe their political ideology and whether they usually think of themselves as Republican, Independent, or Democrat. Does political ideology appear to be independent of party affiliation? If not, explain something of the nature of the dependency.

E32. Vehicles can turn right, turn left, or continue straight ahead at a given intersection. It is hypothesized that vehicles entering the intersection from the south will continue straight ahead 50% of the time and that a vehicle not continuing straight ahead is just as likely to turn left as to turn right. A sample of 50 vehicles gave these counts. Do the data support the hypothesis?

	Straight	Left Turn	Right Turn
Number of vehicles	28	12	10

E33. Foresters are often interested in the patterns of trees in a forest. One common problem is to gauge the integration or segregation of two species of trees (say, I and II) by looking at a sample of one species and then observing the species of its nearest neighbor. Data from such a study can be arrayed on a two-way table such as this one.

Tree Type of Its Nearest Neighbor

		I	II
Tree Type	I	A	B
	II	C	D

a. Which cell counts, among A, B, C, and D, would be large if segregation of species is high? Would this lead to a large or small value of a χ^2 statistic in a test of independence?

b. Which cell counts, among A, B, C, and D, would be large if segregation of species is low? Would this lead to a large or small value of a χ^2 statistic in a test of independence?

c. What situation among the species would lead to a small value of the χ^2 statistic?

| | 1996 Exam | | 2001 Exam | |
	Number of Students	Percentage	Number of Students	Percentage
5 (extremely well qualified)	13,491	12.8	30	15.0
4 (well qualified)	21,924	20.8	51	25.5
3 (qualified)	28,459	27.0	44	22.0
2 (possibly qualified)	21,186	20.1	35	17.5
1 (no recommendation)	20,343	19.3	40	20.0

Examination Grade

	1996 Exam	2001 Exam
Total number of students	105,403	200
Mean grade	2.88	2.98
Standard deviation	1.30	1.36

Display 10.42 Summary statistics for the AP Calculus exams for 1996 and 2001.

Source: For the 1996 data: *AP Calculus Course Description,* May 1998, May 1999, p. 76. For the 2001 data: apcentral.collegeboard.com/repository/01_national_8128.pdf (June 2002).

E34. Display 10.42 gives information about the scores on the 1996 and 2001 Calculus AB Examinations. The data for 1996 were obtained from all students who took the exam. The data for 2001 were obtained from a random sample of 200 students.

a. Suppose you wish to investigate whether these data provide evidence of a change in the distribution of grades from 1996 to 2001. Is a chi-square test appropriate? If so, perform the test you selected, showing all steps. If not, explain why not.

b. Suppose you wish to investigate whether these data provide evidence of an increase in the mean grade from 1996 to 2001. What test is appropriate? Perform this test, showing all steps.

E35. In this chapter, you have seen three chi-square tests. Match each test here with its description and its design. (One description and one design will be left over.)

Test:

a. chi-square test of goodness-of-fit

b. chi-square test of homogeneity

c. chi-square test of independence

Description:

1. You test to see if two populations have equal proportions of members that fall into each of a given set of categories.

2. You test to see if a population has the same proportion that falls into each of a given set of categories as some hypothesized distribution.

3. You test to see if you can predict the result for one categorical variable better if you know the result for the other.

4. You test to see if two populations are independent.

Design:

I. one sample from one population sorted according to one categorical variable

II. one sample from one population sorted according to two categorical variables

III. two samples from two populations sorted according to one categorical variable

IV. two samples from two populations sorted according to two categorical variables

E36. The data in Display 10.43 show the *Titanic* passengers sorted by two variables: *class of travel* and *survival status*. These data cannot reasonably be considered a random sample. They are the population itself, so you must be careful stating the hypotheses and conclusion to any test of significance.

Class of Travel

Survived?	First	Second	Third	Total
Yes	203	118	178	**499**
No	122	167	528	**817**
Total	**325**	**285**	**706**	**1316**

Display 10.43   *Titanic* passengers sorted by class of travel and survival status.

Test to see if the lack of independence between the variables *class of travel* and *survival status* can reasonably be attributed to chance or whether you should look for some other explanation. If the latter, what is that explanation?

E37. In 1988, there were 540 spouse-murder cases in the 75 largest counties in the United States. Display 10.44 gives the outcomes.

Results	Defendant	
	Husband	Wife
Not prosecuted	35	35
Pleaded guilty	146	87
Convicted at trial	130	69
Acquitted at trial	7	31

Display 10.44   Table of spouse-murder defendants in large urban counties.

Source: Larry J. Kitchens, *Exploring Statistics*, 2nd ed. (Pacific Grove, CA: Duxbury Press, 1998). Their source: Bureau of Justice Statistics, *Spouse Murder Defendants in Large Urban Counties, Executive Summary*, NCJ-156831 (September 1995).

a. What are the populations?

b. Make a suitable plot to display the data. Describe what the plot shows.

c. Write suitable hypotheses, and perform a chi-square test showing all steps.

CHAPTER 11

INFERENCE FOR REGRESSION

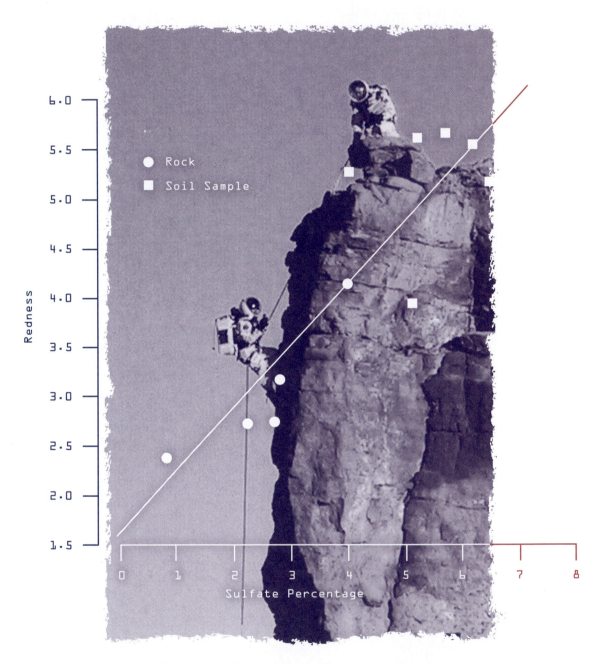

Is there evidence of a linear relationship between *percentage of sulfate* and *redness* for rock and soil samples from Mars? Scientists use statistics to explore questions like this from the Pathfinder Mission data.

Trivia Question 1: Who are Shark, Barnacle Bill, Half Dome, Wedge, and Yogi?

Answer: Mars rocks, found at the landing site of the Pathfinder Mars probe in July 1997.

Trivia Question 2: Mars is often called the red planet. What makes it red?

Answer: Sulfur?

The Sojourner robot rover rolls out of the Mars Pathfinder to meet Barnacle Bill (immediate left) and Yogi (upper right).

The table and scatterplot in Display 11.1 show how the redness of the five Mars rocks is related to their sulfur content. The response variable, *redness,* is the ratio of red to blue in a spectral analysis; the higher the value, the redder the rock. The explanatory variable is the percentage, by weight, of sulfate in the rock.

Name of Rock	Sulfate %	Redness
Shark	0.80	2.39
Barnacle Bill	2.25	2.73
Half Dome	2.72	2.75
Wedge	2.82	3.18
Yogi	4.01	4.14

Display 11.1 Scatterplot and data for the relationship between redness and sulfate content of Mars rocks.

Source: Harry Y. McSween Jr. and Scott L. Murchie, "Rocks at the Mars Pathfinder Landing Site," *The American Scientist* 87 (Jan.–Feb. 1999): 36–45. Data taken from a graph.

Shark: Have you figured out how change in the sulfate content is reflected in change in redness?

Yogi: According to my calculations, the slope of the regression line is 0.525. Rocks that differ by one percentage point in their sulfate content will differ by about 0.5, on average, in their measure of redness. I learned how to do this in Chapter 3.

Shark: That's the story if you take the numbers at face value.

Yogi: Why shouldn't I?

Shark: Well, after all, we aren't the only rocks on the planet. The slope of 0.525 is an estimate, based on a sample—us! If that cute little rolling Sojourner robot had nuzzled up to different rocks, you would have a different estimate. So the goal of this chapter is to show you how to construct confidence intervals and test hypotheses for the slope of a regression line.

Yogi: You need a whole chapter just for that? I know from Chapter 3 how to estimate the slope. So if you tell me how to find the standard error for the slope, I can do the rest. To get a confidence interval, I just do the usual: $0.525 \pm t^* \cdot SE$. To test a null hypothesis, I simply compute

$$t = \frac{0.525 - parameter}{SE}$$

and use the *t*-table to find the *P*-value.

Shark: Sounds familiar, all right. Maybe this can be a short chapter!

In this chapter, you will learn how to make inferences about the unknown true relationship between two quantitative variables. The methods and logic you will learn in this chapter apply to a broad range of such questions.

Specifically, you will learn

■ that the slope of a regression line fitted from sample data will vary from sample to sample and what things affect this variability

■ how to compute the standard error of the estimated slope

■ how to construct and interpret a confidence interval for the slope

■ how to do a significance test that the slope is equal to a given value

■ how to know when to trust confidence intervals and tests

■ how to transform variables to make inferences more trustworthy

11.1 Variation in the Estimated Slope

As you know by now, statistical thinking, although powerful, is unfortunately never as easy or automatic as just plugging numbers into formulas. In order to use the methods appropriately, you need to understand their logic, not just the computing rules. The logic in this chapter is designed for bivariate populations that you can think of as modeled by "linear fit plus random deviation." That is, all points don't lie exactly on the line of best fit but cluster around it, forming an elliptical cloud.

Bivariate: two variables, usually called x and y.

Back in Chapter 3, you learned to summarize a linear relationship with a least squares regression line,

$$\hat{y} = b_0 + b_1 x$$

And that's a complete description if you have the entire population. But if you have only a random sample, the values of b_0 and b_1 are estimates of the true population parameters. That is, there is some underlying "true" linear relationship that you are trying to estimate, just as you use $\bar{x}$ as an estimate of μ. The notation for a linear relationship is

response = true regression line + random variation

$$y = (\beta_0 + \beta_1 x) + \varepsilon$$

where the Greek letters β_0 and β_1 refer, respectively, to the intercept and slope for a line that you don't ordinarily get to see—the true regression line that you would get if you had the whole population for your data instead of just a sample. The letter ε indicates the size of the random variation—how far a point actually falls above or below the true regression line. The true regression line, sometimes called the **line of means** or the **line of averages,** is written

ε is the observed value minus the value predicted by the true regression line.

$$\mu_y = \beta_0 + \beta_1 x$$

Activity 11.1 is designed to show you how these two parallel equations operate.

Activity 11.1 How Fast Do Kids Grow?

What you'll need: a copy of Display 11.2

On the average, kids from the ages of 8 to 13 grow taller at the rate of 2 inches per year. Heights of 8-year-olds average about 51 inches. At each age, the heights are approximately normal, with standard deviation roughly 2 inches.

Source: National Health and Nutrition Examination Survey, May 30, 2000, www.cdc.gov/nchs/about/major/nhanes/growthcharts/charts.htm.

1. Use the information just given to fill in the second column of your copy of the table in Display 11.2.

2. Find the equation of the true regression line, $\mu_y = \beta_0 + \beta_1 x$, that relates average height, y, to age, x. Interpret β_0 and β_1.

(continued)

3. Now suppose you have a randomly selected child of each age. To find how much the height of your "child" of each age deviates from the average height, use your calculator to randomly select a deviation ε from a normal distribution with mean 0 and *SD* 2. Record ε and then add ε to the average height. Record the sum in the column "Observed height." Do this for each child.

4. Fit a regression equation to the (x, y) pairs where y is the observed height. Record your estimated slope, b_1. Is it close to β_1?

5. Collect the values of b_1 found by the members of your class and make a plot of these values. Discuss the shape, center, and spread of this plot of estimated slopes.

6. The population of deviations given in step 3 had mean 0. Why?

7. Suppose that instead of creating your data by simulation, you used actual data by choosing one child from each age group at random and measuring their heights. In what ways is the model in step 2 a reasonable model for this situation? In what ways is it not so reasonable?

Age, x	Average Height from Model, μ_y	Random Deviation, ε	Observed Height, y, of Your Child
8	51	–?–	–?–
.	.	.	.
.	.	.	.
.	.	.	.
13	–?–	–?–	–?–

Display 11.2 Partial data table for Activity 11.1.

Yogi: You don't expect me to believe that real data get created the way we did it in the last activity, do you?

Shark: No, at best, this model is a good approximation, and it's up to you to judge how well it describes the process that created your data. But precisely because the model is a *simplified* version of reality, when it's reasonable, it's quite useful.

The **conditional distribution of *y* given *x*** refers to all the values of y for a fixed value of x. For example, think of the distribution of children's heights described in Activity 11.1. If you were to make a scatterplot of (*age, height*) for a sample of hundreds of children, the vertical column of points for all of the 8-year-old children would be the conditional distribution of the height, y, given that the age, x, is 8, as shown in Display 11.3. Each conditional distribution of height for a given age has a mean, called μ_y for a population and $\bar{y}$ for a sample, and a measure of variability, called σ for a population and s for a sample. A linear model is appropriate for a set of data if the conditional means fall near a line and the variability is about the same for each conditional distribution.

This is one reason why you can think of the regression line as the graph of the average *y*-values.

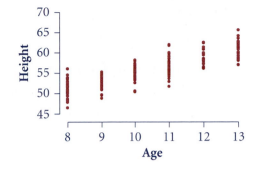

Display 11.3   Scatterplot of children's height versus age. Conditional distributions of height given age are the vertical columns of dots.

Discussion: Theoretical Linear Models

D1. Refer to the information about children's heights in Activity 11.1. What is the mean and standard deviation of the conditional distribution of the heights given the age is 10? Given the age is 12?

D2. Every spring, visitors eagerly await the opening of the spectacular Going-to-the-Sun Road in Glacier National Park, Montana. A typical range of yearly snowfall for the area is from 30 to 70 inches. The amount of snow is measured at Flattop Mountain near the top of the road on the first Monday in April and is given in *swe* (snow water equivalent: the water content obtained from melting). From analysis of past data, if there was 30 inches of *swe*, the road opened, on average, on the 150th day of the year. Every additional 0.57 inches of *swe* measured at Flattop Mountain meant another day on average until the road opened.

Source: "Spring Opening of the Going-to-the-Sun Road and Flattop Mountain SNOTEL Data," Northern Rocky Mountain Science Center, U.S. Geological Survey, December 2001, www.nrmsc.usgs.gov/research/gtsr.htm.

a. If you write an equation summarizing the given information, should you use the form $y = \beta_0 + \beta_1 x$ or $\hat{y} = b_0 + b_1 x$? Explain.

b. Write an equation that you can use to predict when the road will be open given the *swe*. In this situation, what is the response y? The predictor x?

c. In 2002, the Flattop Mountain station recorded 51.7 inches of *swe*. What date would you predict the road will open?

d. Do you think that the random variation, ε, in this situation would be relatively large or small? Make a guess as to what it might be.

D3. Explain the difference between the linear model $y = \beta_0 + \beta_1 x + \varepsilon$ and the fitted equation $\hat{y} = b_0 + b_1 x$. In particular, what is the difference between β_1 and b_1? What is the difference between the random deviation ε from the model and the observed residual, $y - \hat{y}$?

■ **Practice**

P1. The scatterplot of data on pizzas in Display 3.31 on page 130 shows the number of calories versus the number of grams of fat in one serving, for several kinds of pizza. Fat contains 9 calories per gram.

　　a. What would be the theoretical slope of a line for such data?

　　b. What does the intercept tell you?

　　c. What are some of the reasons that not all of the points fall exactly on a line with that slope?

P2. According to Leonardo da Vinci, a person's arm span and height are about equal. The data in Display 11.4 give height and arm span measurements for a sample of 15 high school students.

Arm Span (cm)	Height (cm)	Arm Span (cm)	Height (cm)
168.0	170.5	129.0	132.5
172.0	170.0	169.0	165.0
101.0	107.0	175.0	179.0
161.0	159.0	154.0	149.0
166.0	166.0	142.0	143.0
174.0	175.0	156.5	158.0
153.5	158.0	164.0	161.0
95.0	95.5		

Display 11.4　Height versus arm span for 15 high school students.

　　a. What is the theoretical regression line, $\mu_y = \beta_0 + \beta_1 x$, that Leonardo is proposing for this situation? Use *arm span* as the explanatory variable.

　　b. Find the least squares regression line, $\hat{y} = b_0 + b_1 x$, for these data. Interpret the slope. Compare the estimated slope and intercept to the theoretical slope and intercept. Are they close?

　　c. Calculate the random deviation, ε, for each student, using the theoretical regression line. Plot these random deviations against the arm span and comment on the pattern.

　　d. Calculate the residual from the estimated regression line for each student. Plot the residuals against the arm span and comment on the pattern. Is it a pattern similar to the pattern for the random deviations?

What Affects the Variability of b_1?

In Activity 11.1, you actually knew the value of the theoretical slope β_1, so when you found an estimate b_1 based on the data from your sample, you could tell how far that estimate was from the true value. In real situations, of course, you don't know the true slope. You must use your data to estimate not only β_1 but also the standard error of b_1. Then, if a linear model is appropriate, you can rely on the fact that 95% of the time, the estimated slope will be within approximately two standard errors of the true slope.

To move toward finding a formula for the standard error of b_1, think about the situations when b_1 would tend to be close to β_1 and when it would tend to be farther away. As you may have guessed, with a larger sample size, b_1 tends to be closer to β_1. Activity 11.2 is designed to help you see what other factors affect the variation in b_1 when the standard model ("true regression line plus random variation") applies.

Activity 11.2   What Affects the Variation in b_1?

In this activity, you will create four scatterplots. For each plot, there will be only two values of the predictor, x, and you will generate four values of the response, y, for each x. These response values will come from normal distributions with the specified means and standard deviations.

1. For each value of x given in the cases below, generate four values of y from a normal distribution with the given mean and standard deviation. Plot the resulting eight ordered pairs for each case on a scatterplot. Fit a regression line and record the slope.

   Case 1
      $x = 0$; conditional distribution of y has mean 10 and *SD* 3
      $x = 1$; conditional distribution of y has mean 12 and *SD* 3

   Case 2
      $x = 0$; conditional distribution of y has mean 10 and *SD* 3
      $x = 4$; conditional distribution of y has mean 18 and *SD* 3

   Case 3
      $x = 0$; conditional distribution of y has mean 10 and *SD* 5
      $x = 1$; conditional distribution of y has mean 12 and *SD* 5

   Case 4
      $x = 0$; conditional distribution of y has mean 10 and *SD* 5
      $x = 4$; conditional distribution of y has mean 18 and *SD* 5

2. Collect the estimated slopes from your class. Construct four dot plots of these estimated slopes, one plot for each case. What is the theoretical slope β_1 in each case? For each case, estimate the mean of the distribution of estimated slopes and compare it to β_1.

3. How does the variation in the estimated slopes change with an increase in the variation of the responses, y? How does the variation in the estimated slopes change with an increase in the spread of the explanatory variable, x?

You learned in Activity 11.2 that the variation in the estimated slopes depends not only on how much the values of y vary for each fixed x but also on the spread of the x-values.

In the methods of inference in this chapter, the variability in y is assumed to be the same for each conditional distribution. That is, if you pick a value of x and compute the standard deviation of all of the associated values of y in the population, you'd get the same number, σ, as you would if you picked any other x, as shown in Display 11.5. This implies that σ is also equal to the variability of *all* values of y *about the true regression line*. You can use this fact to estimate σ from your data.

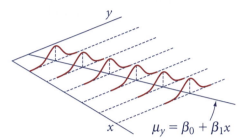

Display 11.5 The variability of all values of y about the true regression line is the same as the variability of the y-values at each fixed x.

> ### Variability in *x* and *y*
>
> Note that *s* is equal to the variability of the residuals.
>
> The common variability of y at each x is called σ. It is estimated by s, which you can compute using all n values of y in the sample:
>
> $$\sigma \approx s = \sqrt{\frac{\sum(y_i - \hat{y}_i)^2}{n-2}} = \sqrt{\frac{SSE}{n-2}}$$
>
> This would be the standard deviation, if we just divided by $n - 1$.
>
> The spread in the values of x is measured by $\sqrt{\sum(x_i - \bar{x})^2}$, where $\bar{x}$ is computed using all n values of x in the sample.

The variability in the slope depends on these two quantities. You can see why by looking at Display 11.6, which shows some combined results from Activity 11.2. Each plot shows five regression lines for one of the four cases—so, for example, Plot I shows five regression lines for Case 1. Each line was constructed using the instructions in step 1 of the activity. Plots I and II (or III and IV) show that a wider spread in x results in regression lines with less variability in their slope—even though the values of y vary equally for each x. By comparing plot I with plot III (or plot II with plot IV), you can see something more expected: Larger variability in each conditional distribution of y means more variability in the slope b_1.

Display 11.6 Regression lines relating variability in b_1 to variation in y and to spread in x.

I. $x = 0$, $\bar{y} = 10$; $x = 1$, $\bar{y} = 12$; II. $x = 0$, $\bar{y} = 10$; $x = 4$, $\bar{y} = 18$;
$\sigma = 3$ $\sigma = 3$

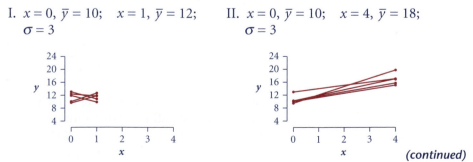

(continued)

Display 11.6 (continued)

III. $x = 0$, $\bar{y} = 10$; $x = 1$, $\bar{y} = 12$;   IV. $x = 0$, $\bar{y} = 10$; $x = 4$, $\bar{y} = 18$;

  $\sigma = 5$ $\sigma = 5$

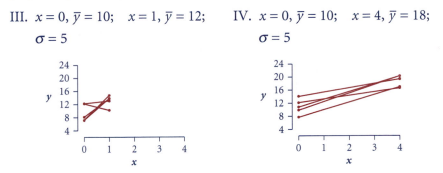

Discussion: Variability in b_1

D4. In the "scatterplots" in Display 11.7, imagine taking a sample of responses within each one of the rectangles. That is, the rectangles define the regions where the responses lie for each value of x. Suppose you take a sample and calculate a regression line. Then you repeat the process many times. Which of the three "plots" should produce regression lines with the smallest variation in slopes? Which should produce regression lines with the largest variation in slopes?

I.   II.   III.

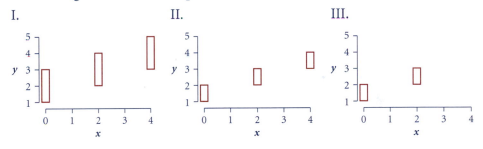

Display 11.7   Three "scatterplots" for D4.

D5. If samples were to be selected from the rectangles in Display 11.8, would you have any concerns about using a straight line to model the relationship between x and y? Why or why not?

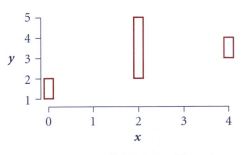

Display 11.8   A "scatterplot" for D5.

■ Practice

P3. Refer to the table in Display 11.1, which gives the sulfate percentage, x, and redness, y, of the Mars rocks. Compute the measure of spread in the x-values, and compute s, the variability in y.

P4. Five quantities are listed here, along with two possible values of each. For each quantity, decide which value will give you the larger variability for b_1 (assuming all other things stay the same), and give a reason why.

a. the standard deviation, σ, of the individual response values of y at each value of x: 3 or 5

b. the spread of the x-values: 3 or 10

c. the number of observations, n: 10 or 20

d. the true slope, β_1: 1 or 3

e. the true intercept, β_0: 1 or 7

The Standard Error of the Estimated Slope

Now you know what causes variation in the estimated slope b_1: the variability, s, of y from the fitted line and the spread in the x-values.

If you had enough time (and lots of patience!), you could use a large, fancy version of the simulations in Activity 11.2 to generate data and regression methods to analyze the results, and in that way find the quantitative relationships between the standard error of b_1 and the spreads in x and y. As an alternative, you could use a line of reasoning that relies on what you already know about the standard error for the difference between two means. Either way, you would eventually end up at the same place.

Formula for Estimating the Standard Error of the Slope

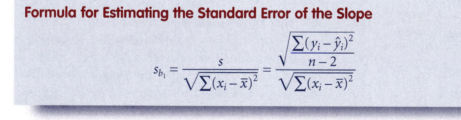

$$s_{b_1} = \frac{s}{\sqrt{\sum(x_i - \bar{x})^2}} = \frac{\sqrt{\dfrac{\sum(y_i - \hat{y}_i)^2}{n-2}}}{\sqrt{\sum(x_i - \bar{x})^2}}$$

This formula does what you would expect: The more y tends to vary from the regression line, the more the estimated slope varies; the bigger the spread in the values of x, the smaller the variation in the estimated slope.

Discussion: Estimating the Standard Error

D6. You plan to collect data in order to predict the actual temperature of your oven from the temperature given on its thermostat. Suppose that this relationship is linear. How will you design this study? What have you learned in this section that will help you?

D7. Use Display 11.6 to explain why it makes sense for the expression with the x's to be in the denominator of the formula for s_{b_1}.

■ **Practice**

P5. Look again at the Mars rocks data in Display 11.1.

    a. Compute the estimate of the standard error of the slope, s_{b_1}.

    b. On the computer output for the regression in Display 11.9, find this value. Then find your estimate from P3 of the variation in y about the line. What is the equation of the regression line?

    c. If the values in the "Sulfate" row were missing in Display 11.9, explain how you could compute each one.

```
Dependent variable is:    Redness
No Selector
R squared = 80.9%  R squared (adjusted) = 74.6%
s = 0.3414 with  5 - 2 = 3 degrees of freedom
```

Source	Sum of Squares	df	Mean Square	F-ratio
Regression	1.48269	1	1.48269	12.7
Residual	0.349593	3	0.116531	

Variable	Coefficient	s.e. of Coeff	t-ratio	prob
Constant	1.71525	0.4010	4.28	0.0235
Sulfate	0.524901	0.1472	3.57	0.0376

Display 11.9 Regression analysis for the Mars rocks data.

P6. Using the same underlying distribution as in Plot II of Display 11.6, 10 observations of y were generated at each value of x. The results are given in the table and scatterplot of Display 11.10.

    a. Find both the true regression line, $\mu_y = \beta_0 + \beta_1 x$, and the least squares regression line, $\hat{y} = b_0 + b_1 x$, and compare them.

    b. Compute s and compare it to $\sigma = 3$.

The true value of *SE* is
$$\frac{\sigma}{\sqrt{\Sigma(x - \bar{x})^2}}.$$

    c. Compute s_{b_1}, the estimate of the standard error of b_1, and compare it to the true value of the standard error.

    d. Does the least squares regression line go through the mean of the responses at both $x = 0$ and $x = 4$?

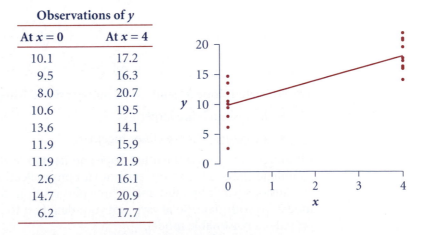

Observations of y

At $x = 0$	At $x = 4$
10.1	17.2
9.5	16.3
8.0	20.7
10.6	19.5
13.6	14.1
11.9	15.9
11.9	21.9
2.6	16.1
14.7	20.9
6.2	17.7

Display 11.10 Data and scatterplot for P6.

P7. Each of these lists gives the values of x used to compute a regression line. Assuming $\sigma = 1$ in all cases, order the lists from the one with the largest estimated standard error of the slope to the one with the smallest. (You should be able to rank these without doing much, if any, computation.)

    I. $x = 1, 2, 3, 4, 5, 6, 7, 8, 9, 10, 11, 12$

    II. $x = 1, 2, 3, 4, 5, 6, 7, 8, 9, 10$

    III. $x = 1, 1, 1, 1, 1, 1, 12, 12, 12, 12, 12, 12$

    IV. $x = 1, 1, 1, 4, 4, 4, 9, 9, 9, 12, 12, 12$

    V. $x = 1, 2, 3, 4, 5, 6, 7$

P8. As you did in P7, order these lists from largest to smallest value of s_{b_1}. (For these, you may need to use the formula and do some computation.)

    I. $\sigma = 2; x = 1, 1, 1, 1, 3, 3, 3, 3$

    II. $\sigma = 1; x = 1, 1, 1, 1, 2, 2, 2, 2, 3, 3, 3, 3$

    III. $\sigma = 3; x = 1, 1, 2, 2, 3, 3, 4, 4, 5, 5, 6, 6$

    IV. $\sigma = 1; x = 1, 1, 2, 2, 3, 3, 4, 4, 5, 5, 6, 6$

Summary 11.1: Variation in the Estimated Slope

Suppose you want to estimate some underlying linear relationship,

$$y = \beta_0 + \beta_1 x + \varepsilon$$

where $\mu_y = \beta_0 + \beta_1 x$ is the true regression line. The intercept, b_0, and slope, b_1, in the least squares regression equation

$$\hat{y} = b_0 + b_1 x$$

serve as your estimates of the true population parameters β_0 and β_1. For reasons that will become clearer later, you typically are most interested in how closely b_1 estimates β_1.

Use the values in your sample and this formula to estimate the standard error of the slope:

$$s_{b_1} = \frac{s}{\sqrt{\sum(x_i - \bar{x})^2}} = \frac{\sqrt{\dfrac{\sum(y_i - \hat{y}_i)^2}{n - 2}}}{\sqrt{\sum(x_i - \bar{x})^2}}$$

The estimated slope b_1 tends to be more variable when either

- the residuals are larger
- the values of x are closer together

Although you can use the fitted slope and its standard error to compute confidence intervals and test statistics for any collection of (x, y) pairs, your conclusions won't be valid unless "line plus random deviations" is the appropriate model for your data. So always plot your data first to judge whether a line provides a reasonable model.

■ Exercises

E1. The table and scatterplots in Display 11.11 give some information about a random sample of motor vehicle models commonly sold in the United States. The variables are

price typical selling price, in thousands of dollars
mpg typical highway mileage, in miles per gallon
liter size of the engine, in liters
hp horsepower rating of the vehicle
rpm maximum revolutions per minute the engine is designed to produce

Consider these pairs of variables:

$x = hp, y = price$ $x = hp, y = mpg$
$x = liter, y = mpg$ $x = rpm, y = mpg$

a. For each pair of variables, tell whether you think a line gives a suitable summary of their relationship.

b. By looking at the scatterplots, estimate which of the four pairs of variables has the largest standard error for the slope and which has the smallest.

c. Compute the estimated slope for $x = hp$, $y = mpg$. Interpret this slope in context. Compute the estimated standard error of the slope.

d. Find your values from part c on the computer printout in Display 11.12. Also, find s, the estimate of σ.

Manufacturer Model	price	mpg	liter	hp	rpm
Chevrolet Cavalier	13.4	36	2.2	110	5200
Chevrolet Lumina APV	16.3	23	3.8	170	4800
Chevrolet Astro	16.6	20	4.3	165	4000
Dodge Shadow	11.3	29	2.2	93	4800
Dodge Caravan	19.0	21	3.0	142	5000
Eagle Vision	19.3	28	3.5	214	5800
Ford Probe	14.0	30	2.0	115	5500
Hyundai Elantra	10.0	29	1.8	124	6000
Lexus SC300	35.2	23	3.0	225	6000
Mazda RX-7	32.5	25	1.3	255	6500
Oldsmobile Achieva	13.5	31	2.3	155	6000
Pontiac Grand Prix	18.5	27	3.4	200	5000
Suzuki Swift	8.6	43	1.3	70	6000
Volkswagen Fox	9.1	33	1.8	81	5500
Volvo 850	26.7	28	2.4	168	6200

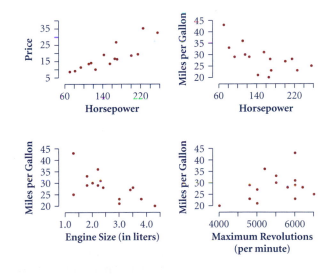

Display 11.11 Table and scatterplots of data about car models.

Source: *Journal of Statistics Education Data Archives,* June 2002.

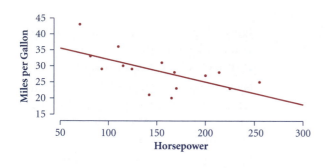

Linear Fit
MPG = 38.980466 - 0.0693953 HP

Summary of Fit
RSquare	0.405843
RSquare Adj	0.360138
Root Mean Square Error	4.778485
Mean of Response	28.4
Observations (or Sum Wgts)	15

Analysis of Variance
Source	DF	Sum of Squares	Mean Square	F Ratio
Model	1	202.75910	202.759	8.8797
Error	13	296.84090	22.834	Prob > F
C. Total	14	499.60000		0.0106

Parameter Estimates
Term	Estimate	Std Error	t Ratio	Prob>|t|
Intercept	38.980466	3.758883	10.37	<.0001
HP	-0.069395	0.023288	-2.98	0.0106

Display 11.12 Regression analysis for the car model problem.

E2. Suppose you collect a dozen cups, glasses, round bowls, and plates of various sizes and then plot the distance around the rim versus the distance across, both measured in centimeters.

    a. What is the theoretical slope of a line fitted to such data? Interpret this slope in context.

    b. What are some of the reasons why not all of the points fall exactly on a line with that slope?

E3. The data in Display 11.13 give the sulfate content and redness for six soil samples from Mars. (Sorry, Yogi and Shark.)

Soil Sample	Sulfate %	Redness
1	4.04	5.27
2	5.17	3.94
3	5.24	5.61
4	5.73	5.66
5	6.20	5.53
6	6.50	5.17

Display 11.13 Sulfate percentage and redness in six Mars soil samples.

    a. Compare this plot to the one for the rocks in Display 11.1. Which s_{b_1} do you expect to be larger—the one for the five rocks or that for the six soil samples? Give your reason.

    b. Compute s_{b_1} for the soil samples, and check your conjecture from part a. (You computed s_{b_1} for the rocks in P5.)

E4. Suppose you were to create a model for the Mars rocks data using the same approach as in Activity 11.1. Write out directions telling how to use the model to find the redness for a rock that has the same sulfur content as Half Dome. (To do this, you will need to make some assumptions about things you don't know.) In what ways do you consider your model a reasonable one for the Mars rocks data? In what ways is it not so reasonable?

E5. Going back to the heights of children ages 8 to 13, the response is *height*, the predictor is *age*, and the average height for 8-year-olds is 51 in. Begin by assuming that on average children grow 2 in. per year and that at any one age, the distribution of heights is approximately normal. Sketch a scatterplot that shows the sort of data you would expect to get in each of these situations.

a. $\sigma = 3$ b. $\sigma = 1$

c. This time, use different σ for different ages: Age 8, $\sigma = 1$ in.; age 9, $\sigma = 2$ in.; age 10, $\sigma = 3$ in.; ...; age 13, $\sigma = 6$ in.

d. Keep everything the same as in part a, except this time assume that children grow faster and faster as they get older so that the average height at age x is given by $50 + (x-8)^2$.

e. Keep everything the same as in part d, except this time use the values of σ from part c.

E6. As you did in P7, order the following lists according to the size of the standard error of the slope. (No computation should be necessary.)

I. $\sigma = 5$; $x = 1, 1, 1, 1, 3, 3, 3, 3$

II. $\sigma = 2$; $x = 1, 1, 1, 1, 2, 2, 2, 2, 3, 3, 3, 3$

III. $\sigma = 3$; $x = 1, 1, 2, 2, 3, 3, 4, 4, 5, 5, 6, 6$

IV. $\sigma = 2$; $x = 1, 1, 2, 2, 3, 3, 4, 4, 5, 5, 6, 6$

V. $\sigma = 7$; $x = 1, 2, 2, 3$

11.2 Making Inferences About Slopes

This section has two parts. The first part presents and illustrates a hypothesis test for a slope, and the second part gives a confidence interval. The logic here is important, but you've seen it before, and the computations follow a familiar pattern. Pay special attention to checking whether and how well the model fits. This is the part of inference that separates human beings from computers. Any old box of microchips can compute a *P*-value, but it takes experience and judgment to figure out whether the *P*-value really tells you anything.

The Test Statistic for a Slope

Often there is no positive or negative linear relationship between two variables. This is true, for example, of the sum of the last four digits of a person's phone number and the number of letters in his or her full name.

> ### Activity 11.3 Phone Numbers and Names
>
> Let x be the sum of the last four digits of a person's phone number and y be the number of letters in their full name. Record x and y for each student in your class. Then make a scatterplot and compute the equation of the regression line.

No association between x and y.

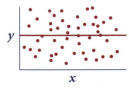

Although the scatterplot from the activity probably shows little or no obvious association, chances are excellent that b_1 isn't exactly equal to 0. Even when the true slope β_1 is 0, the estimate b_1 will usually turn out to be different from 0. In such cases, the estimated slope is not "significant" and differs from 0 simply because cases were picked at random. If another class did the activity, the value of b_1 would probably not be 0 either and would probably be different from the b_1 from your class as well. A significance test for a regression slope asks, "Is that trend real, or could the numbers come out the way they did by chance?"

The key question is "How far is b_1 from 0 (or some other hypothesized value of β_1) in terms of the standard error?"

The test statistic for the slope is the difference between the estimated slope b_1 and the hypothesized slope β_1, measured in standard errors:

$$t = \frac{b_1 - \beta_1}{s_{b_1}}$$

If a linear model is correct and the null hypothesis is true, then the test statistic has a t-distribution with $n - 2$ degrees of freedom.

Yogi: What a relief! I was waiting for them to introduce yet another new distribution: the z, the t, the χ^2, the *blah-blah-cube*.

Shark: But aren't you worried about one little thing? Why does this statistic have a t-distribution and not something else?

Yogi: Worried? I said I was relieved! I know where the t-distribution is on my calculator. The *df* rule is easy. I am happy, and now you are trying to raise problems. You have not had enough sulfur in your diet.

Shark: Okay, we'll let it go until your next course in statistics. For now, just notice that the sample slope behaves something like a mean.

Alberto Behar works on Rocky, an early prototype for the Mars Rover.

Discussion: The Test Statistic

D8. Using the printout for the Mars rocks in Display 11.9 on page 639, verify the value of the t-statistic. Then verify *df*, and use Table B to check that the P-value in the computer printout is consistent with the table.

D9. Display 11.14 shows the combined data for the five Mars rocks and six soil samples. In relation to the t-statistics for the rocks and for the soil samples, where do you expect the value of the t-statistic for the combined data to fall? Why?

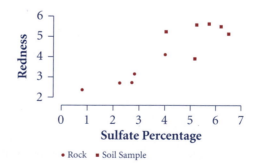

Display 11.14 Redness versus sulfate content for the five Mars rocks and six soil samples together.

■ **Practice**

P9. The regression analysis in Display 11.15 is for the six samples of soil from Mars. The value of the test statistic and the corresponding *P*-value for "Sulfate" are missing. Use the information in the rest of the printout to find these values. What is your conclusion?

```
Dependent variable is:    Redness
No Selector
R squared = 3.3%  R squared (adjusted) = -20.9%
s = 0.7095  with  6 - 2 = 4 degrees of freedom

Source        Sum of Squares   df   Mean Square   F-ratio
Regression    0.068487          1     0.068487     0.136
Residual      2.01345           4     0.503362

Variable   Coefficient   s.e. of Coeff   t-ratio    prob
Constant   4.46564       2.003                      2.23    0.0897
Sulfate    0.133399      0.3617                      -?-     -?-
```

Display 11.15 Regression analysis for the Mars soil samples.

Significance Test for a Slope

Here are the steps for testing the significance of a slope. Generally, you will use this test when you have sample data that show two variables that appear to have a positive (or negative) linear association and you want to establish that this association is "real." That is, you want to determine that the nonzero correlation you see didn't happen just by chance—that there actually is a true linear relationship with a nonzero slope and so knowing the value of *x* is helpful in predicting *y*.

Significance Test for a Slope

1. ***Check conditions.*** For the test to work, the conditional distributions of *y* for fixed values of *x* must be approximately normal, with means that lie on a line and standard deviations that are constant across all values of *x*. Of course, you can't check the population for this; you'll have to use the sample. You'll need to do several things to check conditions:

   • Verify that you have one of these situations:

     i. a single random sample from a bivariate population

     ii. a set of independent random samples, one for each fixed value of the explanatory variable *x*

     iii. an experiment with a random assignment of treatments to units

   • Make a residual plot and check to see if the relationship looks linear.

   • Check the residual plot to be sure the residuals stay about the same size across all values of *x*.

   • Make a univariate plot (dot, stem, or box) of the residuals to see if it's reasonable to assume that they came from a normal distribution.

(continued)

2. **State the hypotheses.** The null and alternate hypotheses usually will be H_0: $\beta_1 = 0$ and H_a: $\beta_1 \neq 0$, where β_1 is the slope of the linear relationship in the population.

   However, the test may be one-sided and the hypothesized value may be some other constant. (Section 11.3 shows you an example.)

3. **Compute the value of the test statistic, find the P-value, and draw a sketch.** The test statistic is

$$t = \frac{b_1 - \beta_1}{s_{b_1}}$$

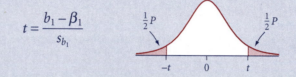

   Here, b_1 and s_{b_1} are computed from your sample. To find the P-value, use your calculator's t-distribution with $n - 2$ degrees of freedom, where n is the number of ordered pairs in your sample.

4. **Write your conclusion linked to your computations and in context.** The smaller the P-value, the stronger the evidence against the null hypothesis. Reject H_0 if the P-value is less than the given value of α, typically .05. Alternatively, compare the value of t to the critical value t^*. Reject H_0 if $|t| \geq t^*$, for a two-tailed test.

Yogi: *Four* conditions! What happened to good old "line plus random variation"?

Shark: It's still there. But "variation" takes in a lot of territory. The variation about the line has to be both random and regular. "Regular" here means (1) the vertical spread is the same as you go from left to right across your scatterplot, and (2) the distribution of points in each vertical slice is roughly normal.

Yogi: This is starting to sound complicated.

Shark: Not really. Sometimes a violation of the conditions will be obvious from the scatterplot. But to be safe—or if you happen to be taking some important test—you should look at a residual plot as well as a dot plot or boxplot of the residuals.

Yogi: That still leaves one more condition. Surely you're not going to tell me I can check randomness by looking at a plot?

Shark: No. For that condition, you need to check how the data got collected. The observations should have been selected randomly, which means, partly, that they should have been selected independently.

Yogi: (*Loud sigh.*)

Shark: Just read the next example, and you'll see how easy it is.

Example: Price Versus Horsepower

Back in Section 11.1, E1, you were asked about *price* versus *horsepower* for a random sample of car models. Display 11.16 shows the data and a scatterplot. On the face of it, the relationship looks like a strong one, strong enough to conclude that the pattern is not simply the result of random sampling: In the population as a whole, there really must be a relationship between *price* and *horsepower*. You would expect a formal hypothesis test to lead to the same conclusion, and it does.

Horsepower	Price (thousands of dollars)
110	13.4
170	16.3
165	16.6
93	11.3
142	19.0
214	19.3
115	14.0
124	10.0
225	35.2
255	32.5
155	13.5
200	18.5
70	8.6
81	9.1
168	26.7

$price = 0.126 \cdot hp - 1.5; r^2 = .72$

Display 11.16   Price versus horsepower for a random sample of 15 car models.

Solution

Check conditions. You have a random sample. The equation of the least squares regression line through these points is

$$\hat{y} = -1.544 + 0.126x$$

If you examine the residual plot in Display 11.17, you can see that while the relationship appears generally linear, the variation from the regression line tends to grow with *x*. That is, the values of *y* tend to fan out and get farther from the regression line as *x* increases. In the next section, you will see how a transformation helps fix this violation of the conditions for inference.

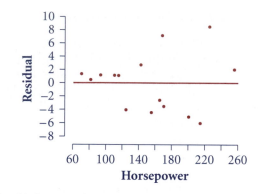

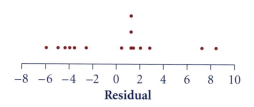

Display 11.17   Residual plot from the regression analysis of automobile price versus horsepower.

If you make a plot of the residuals themselves, as in Display 11.18, they look like they could reasonably have come from a normal distribution, but the gaps are a little worrisome.

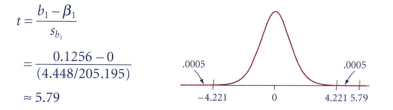

Display 11.18   Dot plot of residuals.

All in all, you should be suspicious about whether this population meets the conditions for a significance test for a slope.

State the hypotheses. The hypotheses are H_0: $\beta_1 = 0$ versus H_a: $\beta_1 \neq 0$ where β_1 is the slope of the true linear relationship between the price and the horsepower for all car models.

Computations and sketch. The *t*-statistic is

$$t = \frac{b_1 - \beta_1}{s_{b_1}}$$

$$= \frac{0.1256 - 0}{(4.448/205.195)}$$

$$\approx 5.79$$

With 15 − 2, or 13, degrees of freedom, the *P*-value corresponding to the observed value of *t* is less than .001.

Conclusion in context. There is strong evidence to reject the null hypothesis and say that the slope of the regression line of *price* versus *horsepower* is different from zero. A linear model having zero for the true slope probably would not have produced these data. ■

These results are summarized in the Data Desk printout of a regression analysis shown in Display 11.19. Note that *s* is the standard deviation of the residuals and is equal to the square root of the mean square for residuals. The regression coefficients are given in the "Coefficient" column. The standard error of b_1 is shown in the "Horsepower" line under the column "s.e. of Coeff." The *t*-ratio and its corresponding *P*-value finish out this line.

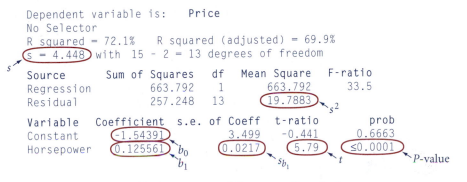

```
Dependent variable is:    Price
No Selector
R squared = 72.1%    R squared (adjusted) = 69.9%
s = 4.448 with 15 - 2 = 13 degrees of freedom

Source        Sum of Squares   df   Mean Square   F-ratio
Regression          663.792     1       663.792      33.5
Residual            257.248    13        19.7883

Variable    Coefficient   s.e. of Coeff   t-ratio      prob
Constant       -1.54391            3.499    -0.441    0.6663
Horsepower      0.125561           0.0217     5.79   ≤0.0001
```

Display 11.19 Data Desk summary of the regression of price versus horsepower for a random sample of cars.

Yogi: *F*-ratio! Aaargh! See, there is another distribution!

Shark: True, but you can ignore the *F*-ratio; here it is equivalent to the *t*-test.

Here, as for all statistical tests, there are three ways to get such a tiny *P*-value: The model could be unsuitable, the sample could be truly unusual, or the null hypothesis could be untrue.

- *Unsuitable model?* The relationship looks linear, and the sample is random. But we are a bit worried about the fact that the variability of the *y*'s seems to grow with the value of *x*.

- *Unusual sample?* The *P*-value tells just how unusual the sample would be if the null hypothesis were true: here, less than one sample in 10,000 would give such a large value of the test statistic. Even the Tooth Fairy has a larger *P*-value than that!

- *False H_0?* By a process of elimination, this is the most reasonable explanation.

What's the bottom line? In the population as a whole, there's a positive linear relationship between *price* and *horsepower*.

Discussion: Significance Test for a Slope

D10. Consider the data on phone numbers and names that you collected in Activity 11.3 as coming from a random sample of students. Test to see if there is a linear relationship between the sum of the last four digits of a student's phone number and the number of letters in his or her full name. Learn to use the *t*-test on your calculator to do the computations. Enter the data into statistical software, if you have it available, and check the computations there as well.

D11. Suppose you have a standard computer output for a regression. What is being tested in the line that includes a *t*-statistic for the coefficient of *x*? What does the *P*-value for this test represent? What is the conclusion if the *P*-value is very small, say, .001?

■ Practice

P10. For the Mars rocks and soil samples taken together, the fitted slope is 0.6268, with an estimated standard error of 0.11. Use your calculator to compute the *t*-statistic and find the *P*-value. State your conclusion.

P11. *Chirping.* The data set in Display 11.20 shows the number of cricket chirps per second and the air temperature at the time of the chirping. Some people claim that they can tell the temperature just by counting cricket chirps.

Chirps per Second	Temperature (°F)
15.4	69.4
14.7	69.7
16.0	71.6
15.5	75.2
14.4	76.3
15.0	79.6
17.1	80.6
16.0	80.6
17.1	82.0
17.2	82.6
16.2	83.3
17.0	83.5
18.4	84.3
20.0	88.6
19.8	93.3

Display 11.20 Temperature and cricket chirps per second.

Source: Richard J. Larsen and Donna Fox Stroup, *Statistics in the Real World: A Book of Examples* (New York: Macmillan, 1976), pp. 187–189. Their source: George W. Pierce, *The Songs of Insects* (Cambridge, MA: Harvard University Press, 1949), pp. 12–21.

a. Fit a straight line that could be used to predict the temperature from the chirp rate. Interpret the slope.

b. Make a residual plot of (*chirp rate, residual*) and a dot plot of the residuals to check conditions.

c. Use the Minitab printout in Display 11.21 to test to see whether there is a linear relationship between the chirp rate and the temperature.

```
The regression equation is

Temp = 25.2 + 3.29 Chirps

Predictor     Coef    Stdev   t-ratio      p
Constant     25.23    10.06      2.51  0.026
Chirps      3.2911   0.6012      5.47  0.000

s = 3.829  R-sq = 69.7%  R-sq(adj) = 67.4%

Analysis of Variance

SOURCE      DF       SS       MS       F       p
Regression   1   439.29   439.29   29.97   0.000
Error       13   190.55    14.66
Total       14   629.84
```

Display 11.21   Regression analysis for the *Chirping* problem.

Confidence Interval Estimation

Now that you know that price increases with horsepower, the next question might be "How much?" Whenever you reject a null hypothesis that a slope is zero, it is good practice to construct a confidence interval. If the interval is extremely wide, due to large variation and small sample size, that tells you that the estimate b_1 is practically useless. For example, an estimated increase in annual income of $50 to $10,000 for every additional year of experience doesn't tell you much about what an individual employee's increase might be. It also may happen that the slope is so small as to be practically meaningless, even though a large sample size makes it "statistically significant." For example, a special exercise that lowers your blood pressure by 1 to 1.5 points for every additional 5 hours per week you spend exercising will not attract many takers, even if a huge study proves that the decrease is real. Always temper statistical significance with practical significance.

As in past chapters, a confidence interval takes the form

> Be careful! Statistical significance doesn't always mean the results are useful.

$$estimate \pm t^* \cdot (estimated\ standard\ error\ of\ the\ estimate)$$

Confidence Interval for a Slope

1. *Check conditions.*   To get a capture rate equal to the advertised rate, the conditional distributions of y for fixed values of x must be approximately normal, with means that lie on a line and standard deviations that are constant across all values of x. Of course, you can't check the population for these; you'll have to use the sample.

   - Verify that you have a random sample from a bivariate population (or a set of independent random samples, one for each fixed value of x). Or verify that treatments were assigned randomly to subjects in an experiment.

   - Make a residual plot and check to see if the relationship looks linear.

   - Check the residual plot to be sure the residuals stay about the same size across all values of x.

   - Make a univariate plot of the residuals to see if it's reasonable to assume that they came from a normal distribution.

(continued)

2. **Computations.** The confidence interval is

$$b_1 \pm t^* \cdot s_{b_1}$$

The value of t^* depends upon the confidence level and the degrees of freedom, which is $df = n - 2$.

3. **Interpretation in context.** For a 95% confidence interval, you would say that you are 95% confident that the slope of the underlying linear relationship lies in the interval. By 95% confidence, you mean that out of every 100 such confidence intervals you construct from random samples, you expect the true value of β_1 to be in 95 of them.

Some calculators do not give you the value of s_{b_1}. However, when testing for the significance of the slope, they will give you t and b_1, so you can compute s_{b_1} from

$$t = \frac{b_1}{s_{b_1}} \qquad \text{or} \qquad s_{b_1} = \frac{b_1}{t}$$

Example: Price Versus Horsepower

Find a 95% confidence level for the slope of the line that relates *price* to *horsepower*.

Solution

Check conditions. The conditions are the same as for a test of significance and were checked in the previous example.

Computations. You have $15 - 2$, or 13, degrees of freedom, so the value of t^* from Table B is 2.160. Thus, a 95% confidence interval estimate of the true slope is given by

$$b_1 \pm t^* \cdot s_{b_1} \approx 0.126 \pm 2.160(0.0217) \approx 0.126 \pm 0.047$$

or $(0.079, 0.173)$.

Interpretation in context. You are 95% confident that the slope of the true linear relationship between *price* and *horsepower* is between 0.079 and 0.173. Converting from thousands of dollars to dollars, the increase in the cost of a car per unit increase in horsepower is somewhere between \$79 and \$173. In other words, if one model has 1 horsepower more than another model, its price tends to be between \$79 and \$173 more. This result means that any true slope β_1 between 0.079 and 0.173 could have produced such data as a reasonably likely outcome. A value of β_1 outside the confidence interval could not have produced numbers like the actual data as a reasonably likely outcome. ■

Discussion: Confidence Interval for β_1

D12. Find 95% confidence intervals for the slopes of the regression lines

    I. for predicting *redness* from *sulfur content* for the Mars rocks (use the information in the printout of P5 on page 639)

    II. for the Mars soil samples (use the information in the printout of P9 on page 645)

    III. for the rocks and soil samples taken together

Which interval is narrowest? Which is widest? How do you explain the differing interval widths?

D13. Suppose a test of significance of the null hypothesis that the true slope is 0 produces a *P*-value of .02. What do you know about the 95% confidence interval estimate of the slope?

■ Practice

P12. Mars rocks contain a high proportion of silicon dioxide (the predominant compound in sand and glass) and much smaller amounts of titanium dioxide (similar to silicon dioxide but with titanium replacing the silicon). Display 11.22 shows a computer printout for the regression of *titanium dioxide percentage* versus *silicon dioxide percentage* for the five Mars rocks.

    a. Find a 90% confidence interval for the slope.

    b. What does the value of $s = 0.0257$ tell you?

```
Dependent variable is:    %TiO2
No Selector
5 total cases
R squared = 96.1%    R squared (adjusted) = 94.8%
s = 0.0257  with  5 - 2 = 3 degrees of freedom
```

Source	Sum of Squares	df	Mean Square	F-ratio
Regression	0.049095	1	0.049095	74.2
Residual	0.001985	3	0.000662	

Variable	Coefficient	s.e. of Coeff	t-ratio	prob
Constant	2.76962	0.2217	12.5	0.0011
%SiO2	-0.033718	0.0039	-8.61	0.0033

Display 11.22 Regression of titanium dioxide percentage versus silicon dioxide percentage for the five Mars rocks.

P13. **Soil samples.** Here is information for the six soil samples that is similar to that in P12 for the Mars rocks. Find a 90% confidence interval for the slope. Based on your confidence intervals for the rocks and the soil samples, would you say the relationship between silicon and titanium content is the same or different for rocks and soil?

Variable	Coefficient	s.e. of Coeff	t-ratio	prob
Constant	-0.177459	2.289	-0.078	0.9419
%SiO2	0.026856	0.0461	0.582	0.5916

More About the Degrees of Freedom

Yogi: I was wondering just a teeny little bit about why there are $n - 2$ degrees of freedom.

Shark: Did you read the discussion about *df* in Section 10.3?

Yogi: That Jodain is pretty sharp! How much sulfur does Jodain have?

Shark: Never mind that. Can you use Jodain's idea to see why there are $n - 2$ degrees of freedom in the test of significance for a slope?

Yogi: The concept of *df* should apply here because I am using a sum of squared deviations from the regression line to estimate σ:

$$\sigma \approx s = \sqrt{\frac{\sum(y_i - \hat{y}_i)^2}{n - 2}}$$

Shark: Right, the "center" is your regression line. How many deviations (residuals) are there?

Yogi: There are n values of $(y_i - \hat{y}_i)$, where n is the number of (x, y) pairs in the sample. I suppose that two of them must be redundant. If I know $n - 2$ of the residuals, I can figure out the other two?

Shark: Right. Try it with this example, where I won't tell you either the values of y or the regression equation. The two missing residuals are R and S. Can you figure out what they are?

x	y	$y - \hat{y}$
1	–?–	1
2	–?–	–2.5
3	–?–	R
4	–?–	S

Yogi: Well, I know that $\sum(y - \hat{y}) = 0$ because that's always the case. So $1 - 2.5 + R + S = 0$. But that's not enough to get R and S.

Shark: Correct. You need a second condition on the residuals. That condition is that x and the residuals are uncorrelated—the residuals don't grow or shrink as x increases. If you check the formula for the correlation, you will see that for it to be 0 means that

$$\sum(x - \bar{x})(y - \hat{y} - 0) = \sum(x - \bar{x})(y - \hat{y}) = 0$$

Yogi: Okay. That gives me two linear equations in two unknowns:

$$\sum(y - \hat{y}) = 1 - 2.5 + R + S = 0$$

$$\sum(x - \bar{x})(y - \hat{y}) = (1 - 2.5)1 + (2 - 2.5)(-2.5) + (3 - 2.5)R + (4 - 2.5)S = 0$$

The missing residuals are $R = 2$ and $S = -0.5$.

Shark: Right! So do you see why there are $n - 2$ degrees of freedom?

> *Yogi:*   Anytime you give me all but two of the residuals, I can find the two missing ones—so they don't give me any new information about the spread of the points from the regression line. I'll bet even Jodain doesn't know that!

Summary 11.2: Inference for Slope

For both a significance test for a slope and a confidence interval for a slope, there are four conditions you should check, which you can remember as brief statements:

- You have a random sample.

- The relationship between the variables is linear.

- Residuals have equal standard deviations across all values of x.

- Residuals are normal at each fixed x.

You check the first condition by finding out how the observations were collected. Check the second and third conditions by making a residual plot. And check the last condition by making a dot plot or boxplot of the residuals.

As you'll see in the next section, all is not lost if the last three conditions are not met. But if the observations weren't selected at random from the population (or, in an experiment, treatments weren't randomly assigned), then if you proceed with inference, you must state your conclusions very, very cautiously.

To test H_0: $\beta_1 = 0$, compute the test statistic

$$t = \frac{b_1 - 0}{s_{b_1}}$$

To find the *P*-value, compare the value of the test statistic with a *t*-distribution with $n - 2$ degrees of freedom.

For a confidence interval, compute $b_1 \pm t^* \cdot s_{b_1}$. Again, use $n - 2$ degrees of freedom.

As a rule, first do a test to answer the question "Is there an effect?" Then, if you reject H_0, construct a confidence interval to answer the question "How big is the effect?"

Don't confuse statistical significance with practical importance. *Significant* means "big enough to be detected with the data available"; *important* means "big enough to care about."

You should not use the techniques of this section for time-series data, that is, for cases that correspond to consecutive points in time. In these situations, the individual observations typically aren't selected at random and so are highly dependent. Today's temperature depends on yesterday's. The unemployment rate next quarter is unlikely to be very far from the rate this quarter. If your cases have a natural order in time, chances are good you should use special inference methods for analysis of *time series* rather than use the methods of this chapter for inferences about a fitted slope.

■ Exercises

E7. Display 11.23 shows the gas mileage, *mpg*, and horsepower ratings, *hp*, for the random sample of car models in E1.

hp	mpg
110	36
170	23
165	20
93	29
142	21
214	28
115	30
124	29
225	23
255	25
155	31
200	27
70	43
81	33
168	28

Display 11.23 Gas mileage and horsepower ratings.

The scatterplot and printout for the regression of *mpg* versus *hp* are shown in Display 11.24.

a. Does it appear that a straight line is a good model for this relationship?

b. Find the equation for this regression line.

c. Find the standard deviation of the residuals.

d. Find the estimated standard error of b_1.

e. Is there evidence to say that the true slope for this model is different from zero? Use $\alpha = .05$.

f. Use a 90% confidence interval to estimate the true slope. Interpret the result in the context of these variables.

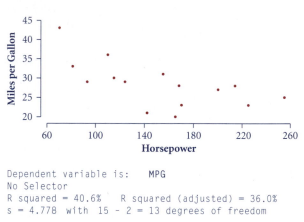

```
Dependent variable is:   MPG
No Selector
R squared = 40.6%   R squared (adjusted) = 36.0%
s = 4.778  with  15 - 2 = 13 degrees of freedom

Source       Sum of Squares  df  Mean Square  F-ratio
Regression   202.759          1    202.759      8.88
Residual     296.841         13     22.8339

Variable    Coefficient  s.e. of Coeff  t-ratio    prob
Constant     38.9805       3.759                 10.4  ≤0.0001
Horsepower  -0.069395      0.0233              -2.98   0.0106
```

Display 11.24 Regression analysis for mileage versus horsepower rating.

E8. Display 11.25 shows the gas mileage, *mpg*, and maximum revolutions per minute the engine is designed to produce, *rpm*, for the random sample of car models in E1. Is there evidence that the mean *mpg* is a linear function of the maximum *rpm*'s? Do all four steps of the test of significance.

rpm	mpg	rpm	mpg
5200	36	6000	23
4800	23	6500	25
4000	20	6000	31
4800	29	5000	27
5000	21	6000	43
5800	28	5500	33
5500	30	6200	28
6000	29		

Display 11.25 Mileage per gallon of gas and maximum revolutions per minute for the car models problem.

E9. Refer to the data on *chirp rate* in P11 on page 650.

    a. Find (and interpret, as always) a 95% confidence interval for the slope of the regression line

        i. for predicting the temperature from the chirp rate

        ii. for predicting the chirp rate from the temperature

    b. Explain why the interval widths in part a are not the same.

E10. Return to the *height* versus *arm span* data of Display 11.4 in P2. Is there evidence in the data to refute Leonardo's claim that, on the average, *height* is equal to *arm span*?

E11. To test the potential effectiveness of laetisaric acid in controlling fungal diseases in crop plants, various concentrations of laetisaric acid were applied to 12 petri dishes containing the fungus *Pythium ultimum*. After 24 hours, an average radius of each fungus colony was calculated. The data are in Display 11.26. Find the linear regression equation and interpret the slope. Then determine if the slope is statistically significant.

E12–E14. The data set in Display 11.27 on the next page shows the selling price of houses, the area of the houses, the number of bedrooms, and the number of bathrooms for a sample of used houses resold in Gainesville, FL.

E12. Construct a model that you think would be good for predicting *price* from *area*. Interpret the slope of the regression line relating these two variables.

E13. Fit a least squares regression model to *price* as a function of *number of bedrooms*. Interpret the slope of the estimated line of fit. Do you see any weaknesses in this analysis?

E14. Is it appropriate to model *price* as a linear function of the *number of bathrooms*? If so, fit a regression line, interpret the slope, and determine its significance. Now analyze the relationship between these two variables using the techniques of Chapter 9 and compare your two analyses.

Acid Concentration (μg/ml)	Colony Radius (mm)
0	33.3
0	31.0
3	29.8
3	27.8
6	28.0
6	29.0
10	25.5
10	23.8
20	18.3
20	15.5
30	11.7
30	10.0

Display 11.26   Acid concentration and average radius of each of 12 fungus colonies.

Source: Myra L. Samuels and Jeffrey A. Witmer, *Statistics for the Life Sciences*, 2nd ed. (Upper Saddle River, NJ: Prentice Hall, 1999), pp. 512–513. Their source: W. S. Bowers, H. C. Hoch, P. H. Evans, and M. Katayama, "Thallophytic Allelopathy: Isolation and Identification of Laetisaric Acid," *Science* 232 (1986): 105–106.

Price (thousands of dollars)	Area (thousands of square feet)	Number of Bedrooms	Number of Bathrooms
115.0	1.67	3	2
115.0	2.07	3	2
69.9	1.52	3	2
76.0	1.15	2	2
88.0	1.55	3	2
100.1	1.85	3	2
87.9	1.68	3	2
150.0	2.04	4	2
107.5	1.85	3	2
34.8	0.78	3	1
72.0	1.36	3	1
98.5	1.51	3	2
142.5	2.40	4	2
83.5	1.40	3	2
68.0	1.45	3	2
28.0	0.84	3	1
113.4	1.98	4	2
65.9	1.22	3	2
101.9	1.92	3	2
81.8	1.33	3	2

Display 11.27   House prices and square footage.

Source: Board of Realtors, Gainesville, FL, January 1996.

11.3   Transforming for a Better Fit

Yogi: What are we doing in another section? I already know everything about inference for regression.

Shark: Remember how the last section gave four conditions that are part of the standard regression model?

Yogi: I learned how to check those—verify that it was a random sample plus look at a residual plot and a dot plot of the residuals. Let's move on.

Shark: But what do you do if your data don't meet the conditions?

Yogi: That's easy. I say the data don't meet the conditions of inference for regression and I go on to the next exercise.

Shark: Not so fast. What if you were trying to find a cure for tree blight and those were the only data you had?

Yogi: Rocks don't get tree blight.

Shark: (*Silence*)

Yogi: (*Mumble/grumble*) Okay, what do I do?

> *Shark:* Sometimes, if one or more of the conditions fail to fit your data, you can rescue the situation using the logarithm button on your calculator.
>
> *Yogi:* Great. I just love that button!

Checking the Fit of the Model

More than any other part of inference, model checking is what makes the difference between statistical *thinking* and mindless number-crunching. Model checking is also the hardest part of inference because often the answers are not clear-cut: a plot shows a hint of curvature at one end, a point is only somewhat apart from its neighbors; a sample is not strictly random, but there are other reasons to think it is representative. If you are the sort of person who likes definite answers, you may have to push yourself to do justice to model checking.

For inference about a regression slope, there are four conditions to check.

1. You have a *random sample* or *a random assignment of treatments to subjects.* (Find out how the data were collected.)

2. The relationship is *linear.* (Check the residual plot. Sometimes nonlinearity is obvious from the original scatterplot and so it is unnecessary to make a residual plot.)

3. Residuals have *equal standard deviations* across values of x. (Check the residual plot. Again, sometimes it is unnecessary to make a residual plot because it's obvious from the original scatterplot that the residuals tend to grow or shrink as x increases.)

4. Residuals are approximately *normal* at each fixed x. (Make a dot plot or boxplot of just the residuals themselves.)

In some instances, a condition will be clearly violated or clearly satisfied. In others, there may be patterns in the data that merely raise questions without providing clear answers. In still others, there may be so little evidence that there's not much else you can say. In general, you can be more confident about your judgment when you have a larger number of points than when you have fewer points, but there is no simple rule.

Discussion: Checking the Model

D14. Four scatterplots with regression line, residual plot, and dot plot and boxplot of the residuals are shown in Display 11.28.

   a. Which plots raise questions about whether the relationship is linear?

   b. What features of a plot suggest that the variation in the response is not constant but depends on x? Which plots show this?

   c. What features of a plot suggest that the conditional distribution of y given x isn't normal? Which plots show this?

   d. Which plots contain an influential point? Imagine fitting a least squares line to this plot using all of the data points and then removing the influential point and refitting the line. How will the two lines differ?

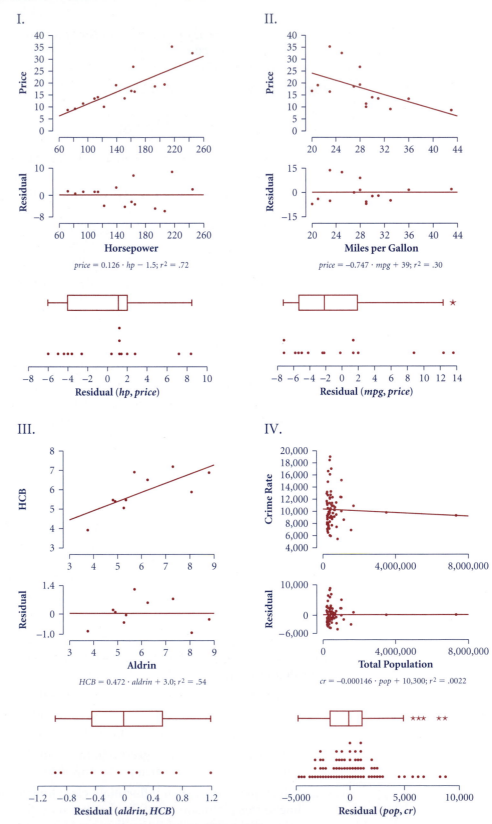

Display 11.28 Four scatterplots with their residual plots and plots of residuals.

Transformations to Improve Linearity

In Display 11.28 of D14, you saw both curvature and heteroscedasticity—the points tended to fan out at one end of the scatterplot. You may remember that Chapter 3 gave two basic transformations to straighten curved data. If you have reason to believe the underlying relationship is

- a power function, $y = ax^b$, try replacing (x, y) with $(\log x, \log y)$
- an exponential function (either growth or decay), $y = ab^x$, try replacing (x, y) with $(x, \log y)$

But what do you do about heteroscedasticity, which is a very common situation with bivariate data? For example, think about the weights of humans plotted against their heights. Almost all babies who are around 20 inches tall will be within a few pounds of each other. However, adults who are 6 feet tall can be hundreds of pounds apart in weight. If you wanted to make an inference for the slope of the regression line in a situation like this, the data would not meet the conditions. As you will see in *The World's Women* example, a transformation using logs often helps with this problem as well as straightening curvature.

Example: Largemouth Bass

The length of a captured fish is easy for a wildlife biologist to measure with a minimum of handling, but weighing the fish is a bit more difficult. A model to help the biologist predict weight from length would save both time and the health of the fish. Also, such a model would allow the biologist to predict how weight tends to change as a fish grows in length.

Solution

Display 11.29 shows the length and weight measurements for a sample of 11 largemouth bass, along with a scatterplot and a residual plot of these data. You can see from the plot that the relationship between mean weight and length should not be modeled by the straight line. You can also figure this out by thinking about the physical situation. Length is a linear measure, and weight is more closely connected to volume, a cubic measure. As you learned in Chapter 3, you can linearize power functions by taking the log of each value of x and of each value of y. (You can use either logarithms to base 10 or natural logarithms for your change of scale.) As you can see from the scatterplot, the residual plot, and the plot of the residuals in Display 11.30, ln(*weight*) versus ln(*length*) is linear and the residuals are small and scattered randomly about the line.

Replacing (x, y) with $(\log x, \log y)$ or $(\ln x, \ln y)$ is called a log-log transformation.

Length (inches)	Weight (pounds)
9.4	0.4
11.1	0.7
12.1	0.9
13	1.2
14.3	1.6
15.8	2.3
16.6	2.6
17.1	3.1
17.8	3.5
18.8	4
19.7	4.9

$weight = 0.429 \cdot length - 4.2; r^2 = .95$

Display 11.29 Length and weight of largemouth bass.

Source: *The Mathematics Teacher* 90 (November 1997): 666.

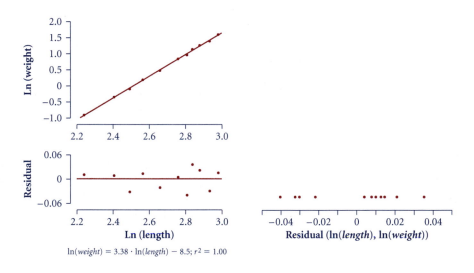

$\ln(weight) = 3.38 \cdot \ln(length) - 8.5; r^2 = 1.00$

```
Dependent variable is:    lnWeight

R squared = 99.9%
s = 0.0264 with 11 - 2 = 9 degrees of freedom

Source       Sum of Squares   df   Mean Square   F-ratio
Regression   6.38006          1    6.38006       9156
Residual     0.006271         9    0.000697

Variable   Coeff      s.e. of Coeff   t-ratio      prob
Constant   -8.50305   0.0953          -89.2        ≤0.0001
lnLength   3.38051    0.0353          95.7         ≤0.0001
```

Display 11.30 Scatterplot, residual plot, and dot plot of residuals, and printout for the least squares regression of ln(*weight*) versus ln(*length*) for largemouth bass.

Display 11.30 also shows a printout summarizing the regression analysis. The estimated slope is $b_1 = 3.38051$, with an estimated standard error of 0.0353. This is the slope of the linear relationship of ln(*weight*) versus ln(*length*):

$$\ln(weight) \approx -8.50305 + 3.38051\ \ln(length)$$

> The slope of the regression line for (ln(*length*), ln(*weight*)) is the exponent in the original power function.

That is,

$$weight \approx \text{constant} \cdot length^{3.38051}$$

It looks like the mean weight of largemouth bass is proportional to the length raised to a power that is just a little higher than 3.

Assuming the sample was randomly selected, can you reject the hypothesis that the relationship between length and weight is cubic? With 9 degrees of freedom for the estimation of error, the 95% confidence interval for the true slope is given by

$$b_1 \pm t^* \cdot s_{b_1} = 3.38051 \pm (2.262)(0.0353)$$

or (3.301, 3.460). So a slope of 3 is not a plausible value. The relationship is a little bit stronger than cubic. ■

Example: The World's Women

This example is based on the data set in Display 11.31, which provides information on population and economic variables for women in a random sample of 30 countries from around the world. What would be the appropriate model for predicting *infant mortality rate for girls* from *fertility rate*?

ps	percentage of parliamentary seats in single or lower chamber occupied by women
fr	total fertility rate (average number of births per woman)
se	girl's share of secondary school (middle school and high school) enrollment
lew	life expectancy at birth for women
lem	life expectancy at birth for men
img	infant mortality rate for girls (per 1000 live births)
imb	infant mortality rate for boys (per 1000 live births)

Solution

The plots in Display 11.32 show the regression of the infant mortality for girls, *img*, versus the fertility rate, *fr*. The association is positive and looks like it follows a linear trend, but the plot is heteroscedastic—countries with larger fertility rates tend to have larger variation in the mortality rate for infant girls. Further, the boxplot of the residuals is skewed left and shows four outliers.

Country	ps	fr	se	lew	lem	img	imb
Afghanistan	—	6.8	25	44	43	157	166
Armenia	3	1.1	—	76	70	14	17
Austria	27	1.2	47	81	75	4	5
Bahamas	15	2.3	52	74	65	15	20
Bahrain	—	2.3	51	76	72	13	16
Bangladesh	—	3.6	—	61	61	68	66
Bolivia	12	3.9	—	65	62	51	60
Denmark	37	1.7	49	79	74	5	5
Dominican Republic	16	2.7	57	70	64	31	42
Egypt	2	2.9	45	70	67	38	43
Estonia	18	1.2	52	76	66	8	11
Equatorial Guinea	5	5.9	35	54	50	91	106
French Guiana	—	3.9	50	—	—	—	—
Gabon	9	5.4	47	54	52	74	86
Jordan	1	4.3	50	73	70	21	24
Mali	12	7.0	34	53	51	116	125
Mexico	16	2.5	49	76	70	26	30
Moldova	13	1.4	50	70	63	18	23
Namibia	25	4.9	54	44	44	61	69
Netherlands Antilles	—	2.1	—	79	73	10	15
New Zealand	31	2.0	50	81	75	6	6
Oman	—	5.5	48	73	70	21	25
Qatar	—	3.3	49	72	69	9	13
Slovakia	14	1.3	49	78	70	8	8
Solomon Islands	2	5.3	38	71	68	21	21
Somalia	—	7.3	—	51	47	104	121
Tunisia	12	2.1	48	72	70	24	27
United Arab Emirates	0	2.9	50	78	74	11	11
United Kingdom	18	1.6	52	81	76	5	6
Uruguay	12	2.3	—	79	72	11	15

Display 11.31   Data on population and economic variables for women in a random sample of countries from around the world.

Source: *The World's Women 2000: Trends and Statistics,* United Nations Department of Economic and Social Affairs, unstats.un.org/unsd/demographic/ww2000/index.htm.

Compare Display 11.32 with Display 11.33, which shows the same information for the regressions of log(*img*) versus *fr* and log(*img*) versus log(*fr*). In each case, the transformation has eliminated the heteroscedasticity and the outliers in the residuals. By bringing in the larger values and spreading out the smaller ones, these transformations unsquish the cluster of points at the lower left of the scatterplot in Display 11.32. The log transformation gives a beautifully symmetrical boxplot of

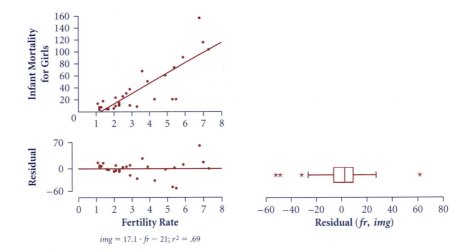

Display 11.32 Scatterplot, residual plot, and boxplot of residuals for *img* versus *fr*.

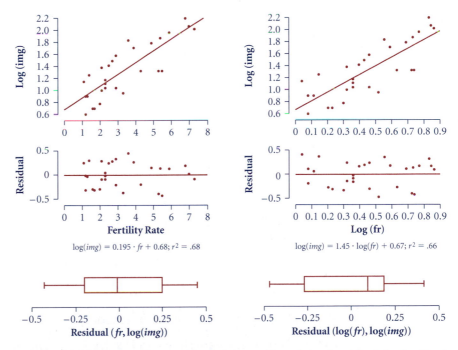

Display 11.33 Scatterplots, residual plots, and boxplots of residuals for log(*img*) versus *fr* and log(*img*) versus log(*fr*).

the residuals, but the log-log transformation gives a more randomly scattered residual plot. Either transformation would be acceptable from a mathematical point of view, so the decision would have to be made on the basis of whether a power model or an exponential model makes more sense in the context. ∎

Logistic functions model the spread of an epidemic, for example. They have the form

$$y = \frac{c}{1 + ae^{-bx}}$$

Yogi: I've got a great idea! Forget all of this transformation stuff. I just go to the Stat Calc menu on my calculator, fit every function in the list—linear, quadratic, cubic, quartic, log, exponential, power, logistic (whatever that is), and sine—and see which gives me the largest value of r. That's a whole lot easier than "log this" and "log that."

Shark: Chapter 3 was a long time ago, but . . .

Yogi: Oh, right. They wouldn't let me do that there either. Remind me again why not. After all, r does tell me how closely the points cluster about my function.

Shark: You can get a very high value of r even though the equation you used isn't a good fit to the data. Look at Display 11.29 on largemouth bass, which gives a value for r^2 of .95. That tells you that the points cluster closely to the line. But as you saw, a line isn't a good model for these data. They clearly follow a curve, not a line. You can see that best from the residual plot—not from r. Also, you can get a very low value of r even though the points form a (fat) elliptical cloud and a line is a perfectly appropriate model.

Yogi: Well, okay. I see why I shouldn't pay much attention to r. But if I think the points follow an exponential curve, why can't I just use my calculator to fit an exponential equation instead of converting all of the y's to log y's and fitting a straight line? I promise to check the residual plot.

Shark: Good statisticians transform for linearity because linear functions are simpler to deal with than curves—and simpler to understand. You know pretty well what the slope and y-intercept mean for a line, but it's much harder to interpret the parameters for other types of equations.

Yogi: Hmmm. You must be right because my calculator only tests for the significance of a slope for a line! And the statistical software we are using only fits lines, not other types of functions.

Shark: Good point.

Discussion: Transformations for Linearity

D15. Two analyses are shown in Display 11.34. The first shows *percentage of renters* versus *total population* of the 77 largest U.S. cities. The five largest cities were then removed from the analysis, and the results are shown in the second analysis.

  a. Compare the two sets of results. Did eliminating the five largest cities improve the conditions for inference? Just how influential did the five cities turn out to be?

Analysis with All Cities

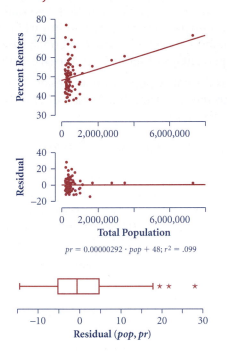

$pr = 0.00000292 \cdot pop + 48; r^2 = .099$

Test of USCities   **Test Slope** ▼

Independent attribute (continuous): Total_Population

Dependent attribute (continuous): Percent_Renters

Independent attribute: **Total_Population**
Dependent attribute: **Percent_Renters**
Sample Count: **77**

Equation of least-squares regression line:
Percent_Renters = 2.91679e-06 Total_Population + 48.332

Alternative hypothesis: The slope of the least squares regression line is **not equal to** 0.

The test statistic, Student's t, is **2.864**. There are **75** degrees of freedom (two less than the sample size).

If it were true that the slope of the regression line were equal to 0 (the null hypothesis), and the sampling process were performed repeatedly, the probability of getting a value for Student's t **with an absolute value this great or greater** would be **0.0054**.

Analysis with the Five Largest Cities Removed

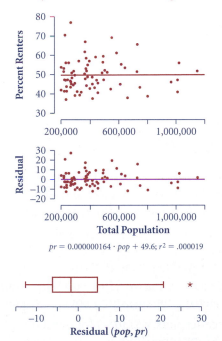

$pr = 0.000000164 \cdot pop + 49.6; r^2 = .000019$

Test of USCities-5   **Test Slope** ▼

Independent attribute (continuous): TotalPopulation

Dependent attribute (continuous): PercentRenters

Independent attribute: **TotalPopulation**
Dependent attribute: **PercentRenters**
Sample Count: **72**

Equation of least-squares regression line:
PercentRenters = 1.64209e-07 TotalPopulation + 49.618

Alternative hypothesis: The slope of the least squares regression line is **not equal to** 0.

The test statistic, Student's t, is **0.03653**. There are **70** degrees of freedom (two less than the sample size).

If it were true that the slope of the regression line were equal to 0 (the null hypothesis), and the sampling process were performed repeatedly, the probability of getting a value for Student's t **with an absolute value this great or greater** would be **0.97**.

Display 11.34   Analyses for percentage of renters versus total population.

Source: T. Erikson, *Data in Depth*, CD-ROM (Emeryville, CA: Key Curriculum Press), 2001.

b. The analysis in Display 11.35 shows a log-log transformation of the data with the five outlying cities removed. Did making a log-log transformation further improve the conditions for inference? Did it change the analysis much? What is your conclusion on the relationship between the percentage of renters and the size of the city?

c. Because this is the entire population of largest U.S. cities, is there any sense in which the regression line is meaningful? In which inference for the slope is meaningful?

Log-Log Analysis with the Five Largest Cities Removed

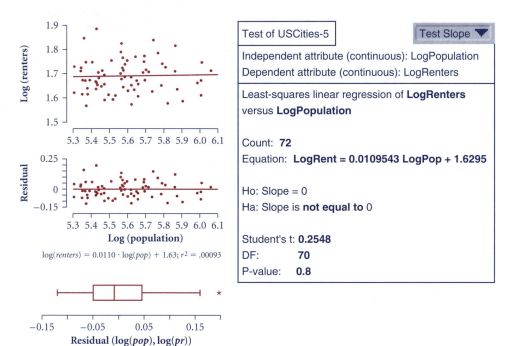

Display 11.35 Log-log analysis for D15, part b.

D16. Using the data from Display 11.29 on largemouth bass and the regression analysis in Display 11.30, conduct a test of the null hypothesis (using a *t*-statistic) that the true power is 3 versus the alternate hypothesis that the true power is not 3.

■ Practice

P14. Suppose a wildlife biologist is working with black crappies instead of largemouth bass, getting the measurements in Display 11.36.

a. Examine the scatterplots, residual plots, and dot plots of the residuals in Display 11.37 for three different models: (*length*, *weight*), (*length*, ln(*weight*)), and (ln(*length*), ln(*weight*)). Give any strengths and weaknesses of each model. Is there any reason to prefer one model over the others?

Length (in)	Weight (lb)
4.4	0.1
6.6	0.2
8.4	0.4
10.0	0.7
10.5	0.7
11.1	0.9
11.4	1.0
12.0	1.1
12.5	1.3

Display 11.36 Length and weight of black crappies.

Source: *The Mathematics Teacher* 90 (November 1997): 671.

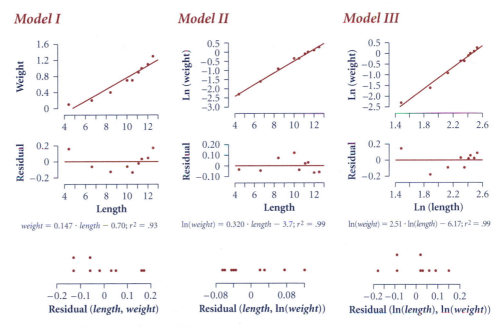

Display 11.37 Scatterplots, residual plots, and dot plots of the residuals for three models for black crappies.

b. This printout is for the (ln(*length*), ln(*weight*)) model and is for the usual test that the slope is significantly different from 0. Use the information to perform a test that the slope is significantly different from 3.

```
The regression equation is
LnWeight = -6.17 + 2.51 LnLength

Predictor     Coef    Stdev   t-ratio      p
Constant   -6.1693   0.2538   -24.31   0.000
LnLength    2.5115   0.1130    22.22   0.000
```

c. Use the printout in part b to find a 95% confidence interval for the slope. How does your model differ from the one for largemouth bass?

P15. ***World's Women (continued).*** The regression analysis for the log of infant mortality for girls, log(*img*), versus fertility rate, *fr*, appears in Display 11.38.

a. What would you predict for log(*img*) for a country that had a fertility rate of 5 children per woman? Based on these data, describe the distribution of log(*img*) for countries with *fr* = 5. What would you predict for the *img* for this country?

b. Find a 95% confidence interval for the slope of the regression line. What does this slope indicate about the relationship between *fr* and *img*?

```
The regression equation is
Log IMG = 0.679 + 0.195 FR

29 cases used 1 cases contain missing values

Predictor     Coef     Stdev   t-ratio       p
Constant    0.67929   0.09830     6.91   0.000
FR          0.19502   0.02569     7.59   0.000

s = 0.2588 R-sq = 68.1% R-sq(adj) = 66.9%

Analysis of Variance

SOURCE       DF      SS       MS      F       p
Regression    1   3.8612   3.8612  57.63   0.000
Error        27   1.8090   0.0670
Total        28   5.6702
```

Display 11.38 Regression analysis for log(*img*) versus *fr*.

Summary 11.3: Transforming for a Better Fit

If a scatterplot or residual plot indicates that your data do not fit the conditions for inference very well—if the relationship is not linear or the residuals look far from normal or standard deviations at each fixed *x* are unequal—changing to a new scale can often put your data in a form that gives a better fit to these conditions.

- If a power model $y = ax^b$ fits your data, then plotting log *y* versus log *x* will show a linear relationship. Often if transforming to logs improves the fit of the linearity assumption, the transformation will also give residuals that are more nearly normal and standard deviations that are more nearly constant.

- If an exponential model $y = ab^x$ fits your data, then plotting log *y* versus *x* will show a linear relationship and there is a good chance that the other two assumptions—that residuals are roughly normal and standard deviations are equal—will also fit better.

■ Exercises

E15. A test of the significance of the slope of the regression line for *crime rate* versus *total population* for the U.S. cities from Display 11.28, part IV, has a *P*-value of .69 with $df = 74$ and $t = -0.4053$. The crime rate for Chicago was missing, so that's why $df = 74$ rather than $77 - 2$, or 75. The remaining four of the largest cities were removed, and the resulting analysis appears in Display 11.39. Next to that is an analysis on $(\log(population), \log(crime\ rate))$.

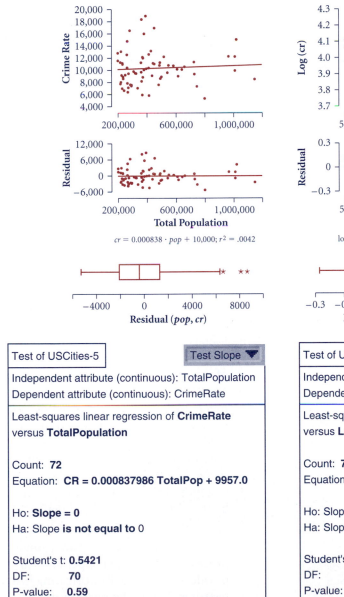

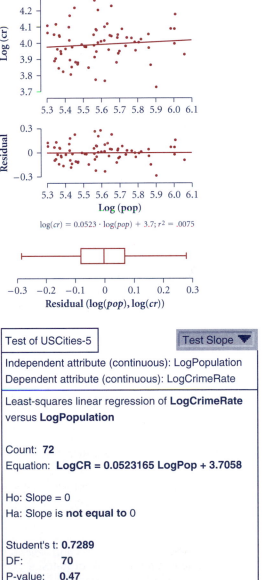

$cr = 0.000838 \cdot pop + 10{,}000; r^2 = .0042$

$\log(cr) = 0.0523 \cdot \log(pop) + 3.7; r^2 = .0075$

Test of USCities-5	Test Slope ▼
Independent attribute (continuous): TotalPopulation
Dependent attribute (continuous): CrimeRate
Least-squares linear regression of **CrimeRate** versus **TotalPopulation**
Count: **72**
Equation: **CR = 0.000837986 TotalPop + 9957.0**
Ho: **Slope = 0**
Ha: Slope **is not equal to** 0
Student's t: **0.5421**
DF: **70**
P-value: **0.59**

Test of USCities-5	Test Slope ▼
Independent attribute (continuous): LogPopulation
Dependent attribute (continuous): LogCrimeRate
Least-squares linear regression of **LogCrimeRate** versus **LogPopulation**
Count: **72**
Equation: **LogCR = 0.0523165 LogPop + 3.7058**
Ho: Slope = 0
Ha: Slope is **not equal to** 0
Student's t: **0.7289**
DF: **70**
P-value: **0.47**

Display 11.39 Scatterplot with regression line, residual plot, and boxplot of residuals for U.S. cities with the five largest cities removed for *(population, crime rate)* and *(log(population), log(crime rate))*.

a. Compare these with the results from the complete set of cities. Did eliminating the five largest cities improve the conditions for inference? Did it change the analysis much? Just how influential did the five cities turn out to be? Did making a log-log transformation further improve the conditions for inference? Did it change the analysis much? What is your conclusion on the relationship between *crime rate* and *total population*?

b. Because this is the entire population of largest U.S. cities, in what sense is the regression line meaningful? Are there any situations in which inference for the slope is meaningful?

E16. How does the number of police officers in a state relate to the violent-crime rate?

a. For the sample of states shown in Display 11.40, find a good-fitting model relating the number of police officers to the violent-crime rate.

b. Find a 95% confidence interval for the slope and interpret it.

State	Police (thousands)	Violent Crimes (per 100,000 population)
CA	86.2	1090
CO	9.2	559
FL	45.0	1184
IA	6.0	303
IL	39.9	1039
LA	11.8	951
ME	2.9	132
MO	14.6	763
NJ	30.5	635
TN	12.3	726
TX	46.2	840
VA	15.2	373
WA	10.9	523

Display 11.40 Number of police officers and violent-crime rate for a sample of states.

Source: *Statistical Abstract of the United States*, 1994.

E17. To measure the effect of certain toxicants found in water, concentrations that kill 50% of the fish in a tank over a fixed period of time (LC50's) are determined in laboratories. There are two methods for conducting these experiments. One method has water continuously flowing through the tanks, and the other has static water conditions. The Environmental Protection Agency (EPA) wants to adjust all results to the flow-through conditions. Given the data in Display 11.41 on a sample of 10 toxicants, establish a model that will allow adjustment of the static values to corresponding flow-through values.

Flow	Static
23.00	39.00
22.30	37.50
9.40	22.20
9.70	17.50
0.15	0.64
0.28	0.45
0.75	2.62
0.51	2.36
28.00	32.00
0.39	0.77

Display 11.41 Flow-through and static LC50's.

Source: R. Sheaffer and James McClave, *Probability and Statistics for Engineers* (Boston: Duxbury Press, 1995), p. 486.

E18. Metals, especially copper and its alloys, are often made by heating (sintering) powders of essentially pure ore. As the generally round particles of powder soften and merge together, the spaces (voids) between them become smaller. The volume of the voids per unit volume of the metal is a measure of the porosity of the sintered metal, and this porosity is related to the strength of the metal. Porosity is measured by the weight of liquid wax that is taken up by the sintered metal.

The data in Display 11.42 are from a process for sintering copper and tin powders to make self-lubricating bronze bearings. Model the relationship between *sintering time* and *weight of wax* taken up by the bearings. Find a confidence interval for the slope of this line and interpret it.

Sintering Time (min)	Wax Weight (g)
7	0.615
7	0.606
7	0.611
9	0.586
11	0.511
11	0.454
11	0.440
13	0.393
15	0.322
15	0.343
15	0.341

Display 11.42   Sintering time and weight of wax from a sintering process.

Source: R. Sheaffer and James McClave, *Probability and Statistics for Engineers* (Boston: Duxbury Press, 1995), p. 536.

E19. Display 11.43 shows a scatterplot of the percentage of parliamentary seats in a single or lower chamber occupied by women, *ps*, versus the girls' share of secondary school (middle school and high school) enrollment, *se*, from the *World's Women* example. Perform a signficance test for the slope. Make the necessary dot plot of the residuals by estimating them from the plots. Be sure to interpret your result.

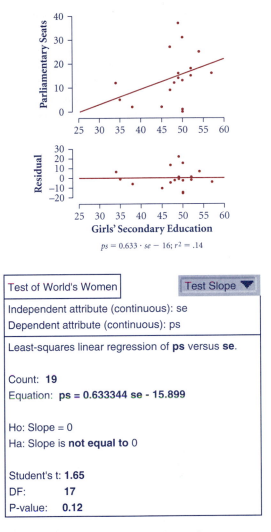

$$ps = 0.633 \cdot se - 16;\ r^2 = .14$$

Test of World's Women   Test Slope ▼

Independent attribute (continuous): se
Dependent attribute (continuous): ps

Least-squares linear regression of **ps** versus **se**.

Count: **19**
Equation: **ps = 0.633344 se - 15.899**

Ho: Slope = 0
Ha: Slope is **not equal to** 0

Student's t: **1.65**
DF: **17**
P-value: **0.12**

Display 11.43   Regression analysis of *ps* versus *se*.

Chapter Summary

When two variables are both quantitative, you can display their relationship in a scatterplot and you may be able to summarize that relationship by fitting a line to the original data or to the data after a transformation. Although many relationships are more complicated than this, simple relationships do occur often enough that many questions can be recast as questions about the slope of a fitted line.

That is old news. What is new in this chapter is that you can use the *t*-statistic to test whether the slope of the apparent linear relationship is zero. If you can reject that possibility, your regression line will be useful in predicting *y* given *x*.

If you can't, then it's plausible that the positive or negative trend that you see is due solely to the chance variation that always results when you have only a sample.

You also have learned to find a confidence interval for the slope, which is necessary when you want to estimate the true linear relationship between two variables and to decide whether your results have practical significance.

One way to put the methods of this chapter into a larger picture is to remind yourself of the inference methods you've learned so far: for one proportion and for the difference of two proportions (Chapter 8), for one mean and for the difference beween two means (Chapter 9), for the relationship between two categorical variables (Chapter 10), and, now, for the relationship between two quantitative variables. In the next chapter, you will work on three case studies that put all of these ideas together.

■ Review Exercises

E20. Study the scatterplots shown in Display 11.44.

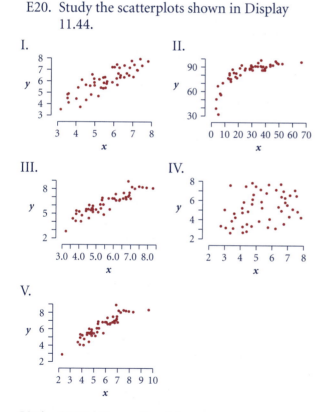

Display 11.44   Five scatterplots.

a. In which scatterplots could the relationship between y and x be modeled by a straight line?

b. If you fit a line through each scatterplot by the method of least squares, which plot will give a line with slope closest to zero?

c. If you fit a least squares line through each scatterplot, which plot will give a line with correlation coefficient closest to 1?

d. For each of the scatterplots that do not look like they should be modeled by a straight line, suggest a way to modify the data to make the shape of the plot more nearly linear.

E21. *More on pesticides in the Wolf River.* Display 11.45 presents four scatterplots of *HCB concentration* versus *aldrin concentration.* The four scatterplots are for the measurements taken at the surface, at mid-depth, on the bottom, and all three locations together. Based on the plots, which depth do you expect to give the narrowest confidence interval for the slope? The widest? Give your reasoning.

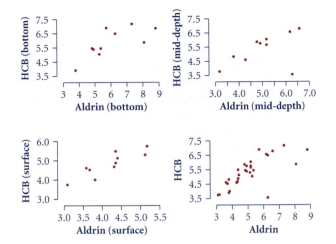

Display 11.45 Four scatterplots of HCB concentration versus aldrin concentration.

Source: R. V. Hogg and J. Ledolter, *Engineering Statistics* (New York: Macmillan, 1987). Their source: P. R. Jaffe, F. L. Parker, and D. J. Wilson, "Distribution of Toxic Substances in Rivers," *J. Env. Eng. Div.* 108 (1982): 639–649.

E22. "If I change to a brand of pizza with lower fat, will I also reduce calories?" Display 11.46 shows a printout for a significance test of *calories* versus *fat* per 5-ounce serving for the popular brands of pizza that you saw in E21, Section 3.2, page 134.

  a. Use Display 11.46 to estimate, in a 95% confidence interval, the reduction in calories you can expect per 1-gram decrease in the fat content of the pizza.

```
Dependent variable is:     Calories

No Selector
R squared = 67.8%    R squared (adjusted) = 66.3%
s = 16.92  with  24 - 2 = 22 degrees of freedom

Source      Sum of Squares  df  Mean Square  F-ratio
Regression  13240.6          1    13240.6      46.3
Residual    6297.21         22    286.237

Variable  Coefficient  s.e. of Coeff  t-ratio    prob
Constant  253.078      12.10            20.9   ≤0.0001
Fat       6.47992       0.9527           6.80   ≤0.0001
```

Display 11.46 Regression analysis for the pizza data in Display 3.37 in Section 3.2 on page 134.

  b. Use your answer to part a to estimate, in a 95% confidence interval, the reduction in calories you can expect per 5-gram decrease in the fat content of the pizza.

  c. Use a significance test to decide if reducing the fat content will change the *cost* in any significant way.

E23. ***Cerebral blood flow and anesthesia.*** Adequate and consistent cerebral blood flow (CBF) is essential to anyone's health but is an especially critical measurement for persons under anesthesia. A study of the possible effects on CBF of three different anesthetics (pentobarbitol, chloralose, and halothane) was conducted by treating laboratory rats with one of the three anesthetics (assigned randomly) and measuring two aspects of CBF. A CBF plot over time looks like the kind of wavy plot common to EKG's, EEG's, and other physiological measurements. These plots are characterized by the amplitude (height of the peaks) and the frequency of oscillations. The CBF data were categorized according to these two measures, which were analyzed separately, with the amplitude measured in percent of the mean CBF value and the frequency measured in Hertz (oscillations per second). These measurements are also affected by diastolic blood pressure, which can be controlled somewhat by the experimenter. The question investigated here is, how do the CBF measures of amplitude and frequency relate to blood pressure for the three anesthetics being examined? The data are given in Display 11.47.

  a. Analyze the relationships between *amplitude* and *blood pressure* for the three anesthetics.

  b. Analyze the relationships between *frequency* and *blood pressure* for the three anesthetics.

Treatment	Blood Pressure	Amplitude	Frequency
C	65	12.9	.089
C	55	10.3	.081
H	68	5.1	.102
H	76	3.9	.078
H	65	9.8	.063
H	65	5.8	.078
P	55	11.8	.070
P	60	7.3	.094
P	60	3.4	.094
P	70	8.2	.094
P	72	3.7	.125
C	50	7.5	.086
C	52	7.0	.125
H	75	4.5	.094
C	60	4.5	.125
C	60	14.6	.093
C	75	6.8	.110
H	58	6.0	.096
H	67	9.3	.078
P	57	5.9	.078
P	52	4.7	.102
P	75	5.4	.109
P	75	2.1	.125
C	70	6.4	.133
C	70	4.1	.109
H	67	3.7	.063
C	72	6.8	.070
H	52	6.1	.094
P	57	7.1	.094
P	70	2.0	.094
C	65	10.7	.102
C	70	3.7	.109
C	82	6.1	.078
C	58	11.4	.086
H	60	4.7	.078
C	85	5.3	.078

Display 11.47 Data for a study of the possible effects of three different anesthetics on cerebral blood flow.

Source: R. Peck, L. D. Haugh, and A. Goodman, *Statistical Case Studies*, ASA and SIAM, 1998, pp. 203–216.

E24. The data in Display 11.48 extend the data on number of police officers and violent-crime rate in Display 11.40 and come from a random sample of the 50 United States. *Expenditures* shows the amount (in millions of dollars) it takes to keep the given number of police officers on the job for one year. *Population* is the population of the state in millions of people.

State	Police (thousands)	Expenditures (millions of dollars)	Population (millions)
CA	86.2	5402	30.4
CO	9.2	435	3.4
FL	45.0	2162	13.2
IA	6.0	251	2.8
IL	39.9	1605	11.5
LA	11.8	515	4.2
ME	2.9	100	1.2
MO	14.6	515	5.2
NJ	30.5	1241	7.7
TN	12.3	437	5.0
TX	46.2	1840	17.3
VA	15.2	714	6.3
WA	10.9	564	5.0

Display 11.48 Number of police officers, expenditures, and population.

Source: *Statistical Abstract of the United States*, 1994.

a. If one state spends a million dollars more on police than does another state, how many more police officers would the first state be expected to have?

b. Does removal of the most influential data point in the set have much effect on your answer to part a?

c. If one state has a million more people than another state, how many more police officers would the first state be expected to have?

d. Does removing the most influential data point have much effect on your answer to part c?

E25. In the population, what is the relationship between the standard deviation of the responses and the standard deviation of the errors?

Chapter Summary **677**

Country	Live Births (per 1000 population)	GNP (thousands of dollars per capita)
Algeria	29.0	1.6
Argentina	19.5	4.0
Australia	14.1	16.6
Brazil	21.2	2.6
Canada	13.7	20.8
China	17.8	1.3
Cuba	14.5	1.6
Denmark	12.4	24.2
Egypt	28.7	0.5
France	13.0	24.1
Germany	11.0	19.8
India	27.8	0.3
Iraq	43.6	0.7
Israel	20.4	13.6
Japan	10.7	27.3
Malaysia	28.0	2.5
Mexico	26.6	3.1
Nigeria	43.3	0.2
Pakistan	41.8	0.4
Philippines	30.4	0.7
Russia	12.6	8.6
South Africa	33.4	2.6
Spain	11.2	13.4
United Kingdom	13.2	17.4
United States	15.2	22.6

E26. Display 11.49 shows a table and a scatterplot of *birth rate* versus *gross national product* for a sample of countries. The overall shape looks roughly like exponential decay (exponential with a negative exponent), except for one problem: The curve should snuggle down to the x-axis as it moves out into the right tail, a pattern that would correspond to birth rates dropping to zero as richer and richer countries are considered. That doesn't happen for the countries in the data set, not even approximately.

a. How might you first transform the data to give an exponential model a chance of fitting?

b. Make this transformation first and then transform again to see if an exponential model is a reasonable one for these data.

c. What birth rate would you predict for a country with a gross national product of 15?

d. Does the same principle apply with the scatterplot in Display 11.50 of *life expectancy at birth for men* and *infant mortality rate for boys* for the sample of countries in Display 11.31?

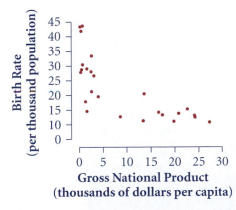

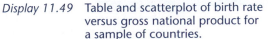

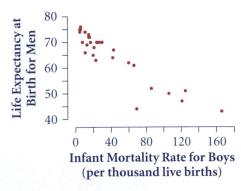

Display 11.49 Table and scatterplot of birth rate versus gross national product for a sample of countries.

Source: *Statistical Abstract of the United States*, 1995, and *Human Development Report* (New York: Oxford University Press, 1995).

Display 11.50 Scatterplot of men's life expectancy at birth versus infant mortality rate for boys for a random sample of countries.

CHAPTER 12
STATISTICS IN ACTION: CASE STUDIES

Can you grow bigger flowers by reducing the length of their stems? Plant scientists designed an experiment to test different growth inhibitors to see which ones were most effective in reducing the length of stems, anticipating that reduced stem growth will enhance the quality of the flower.

Statistics today is big business. Nearly every large commercial enterprise and governmental agency in the United States needs employees who understand how to collect data, analyze it, and report conclusions. The language and techniques of survey and experimental design are part of politics, medicine, industry, advertising and marketing, and even, as you will see, flower growing. Consequently, a statistics course typically is required of college students who major in mathematics or in fields that use data, such as business, sociology, psychology, biology, and health science.

In this chapter, you will examine three examples of statistical practice in today's world, case studies about how to

- produce bigger and better flowers
- evaluate salaries in major league baseball
- establish discrimination in employment

In each case, you decide the appropriate statistical techniques to use and practice many of the techniques you have learned. This chapter will show you more extensive examples of how statistics is used and will provide a review for your course.

12.1 Mum's the Word!

The commercial flower industry is very large and growing in the United States, particularly so in Florida. The chrysanthemum, or mum, is a popular flower for bouquets and potted plants because it is hardy, long-lasting, and colorful. If you visit a flower shop to choose a bouquet of mums for someone special, you would probably look for the largest, most fully developed, most colorful blooms you could find. Well, flowers with these characteristics are in demand, and they don't just happen to show up in your flower shop by accident. Considerable research goes into producing fine blooms.

Plants can produce and use only so much energy. So within the same species, a plant with a long stem is likely to have a smaller flower than a plant with a short stem. The Environmental Horticulture Department of the University of Florida experimented with growth inhibitors to see which ones were most effective in reducing the length of stems, anticipating that reduced stem growth would enhance the quality of the flower. There were many treatments in the complete study, but this investigation will consider only seven of them. The patterns you will see here remain much the same when all treatments are considered.

Individual plants of the same age were grown under nearly identical conditions, except for the growth-inhibitor treatment. Each treatment was randomly assigned to 10 plants, and their heights (in centimeters) were measured at the outset of the experiment (Ht_i) and after a period of 10 weeks (Ht_f). (Actually, heights were measured at intervening times as well, but those measurements are not part of this analysis.)

You can find the raw data for this experiment—the treatment each of the 70 plants received and its height before and after treatment—at the end of this section in Display 12.3.

Discussion: Analyzing the Experiment

D1. Can you suggest possible improvements to this experimental design?

D2. What response variable would be best to use for comparing treatment groups in this experiment?

D3. What statistical method(s) would you suggest to compare the effectiveness of, say, Treatments 1 and 2? What conditions (assumptions) must you check before doing this test?

Display 12.1 shows the statistical summaries of the growth, $Ht_f - Ht_i$, during the 10 weeks for each of the seven treatment groups. Notice that the means fluctuate quite a lot from group to group, and so do the variances.

```
Summary of              Htf - Hti
For categories in       treatment
No Selector

Group   Count   Mean       Median    Variance   StdDev
01      10      30.8500    29.5000   58.4472    7.64508
02      10      35.9500    38        40.6917    6.37900
03      10      43.0500    44        25.4139    5.04122
04      10      52.7500    53.7500   18.2917    4.27688
05      10      28.4500    29.5000   58.3028    7.63563
06      10      35.9000    38.7500   58.7111    7.66232
07      10      39.7500    40.7500   20.9028    4.57196
```

Display 12.1 Summary statistics for plant growth, $Ht_f - Ht_i$, for seven treatment groups.

Source: University of Florida Institute for Food and Agricultural Sciences, 1997.

In Display 12.2, the boxplots of the difference measurements, $Ht_f - Ht_i$, by treatment give a better view of the key features of the data.

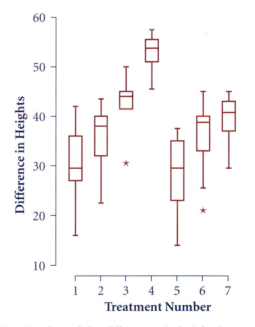

Display 12.2 Boxplots of the differences in height (in centimeters) of mums for the seven treatment groups.

■ Practice

P1. Which two treatments appear to be the most effective? Least effective? Explain your choices.

P2. Describe how the distributions compare in terms of center and spread. Do you see any pattern relating the spreads to the centers? Make a statistical graphic that shows how the mean and standard deviation are related. Describe this relationship. Is this the kind of relationship you would have expected? Explain.

P3. Use an appropriate statistical technique to answer this question: Approximately what increase or decrease in standard deviation would you expect for a 1-centimeter increase in mean growth?

P4. Is the difference in mean growth between the two most effective treatments statistically significant? Be sure to check the conditions for the test that you select. Also, write the hypotheses in words.

P5. The two treatments that differ most appear to be Treatments 4 and 5. Why is it clear from the plot that the difference in mean growth between these two treatments is statistically significant? Construct a 95% confidence interval to support your reasoning.

P6. By hand, make boxplots of the initial heights of the mums assigned to Treatments 4 and 5. Show any outliers determined by the $1.5 \cdot IQR$ rule. Describe any differences in the initial heights of the plants assigned to these two treatments.

Display 12.3 The treatment that each of the 70 plants received and its height before and after treatment.

Treatment	Initial Height (cm) Ht_i	Final Height (cm) Ht_f	Treatment	Initial Height (cm) Ht_i	Final Height (cm) Ht_f
1	6	48	3	3	46
1	3	32.5	3	3.5	47.5
1	4	20	3	2	32.5
1	3	32	4	4	55
1	1	37	4	2.5	57.5
1	2	29	4	1.5	58.5
1	2.5	34	4	2.5	60
1	4.5	46	4	3	56
1	3.5	33	4	2.5	48
1	3.5	30	4	1	54
2	2	33.5	4	2	56.5
2	3	35.5	4	2	57.5
2	3.5	47	4	2.5	48
2	2.5	41	5	2	32
2	4.5	42.5	5	2	37
2	2	24.5	5	1.5	15.5
2	4.5	47.5	5	4	41
2	2	42	5	2	28
2	5	37	5	2	31
2	2	40	5	2	25
3	1	47	5	3	23.5
3	2.5	44	5	3	35.5
3	3	48	5	5	42.5
3	2.5	46.5	6	3	48
3	3	44.5	6	3.5	43.5
3	4	54	6	3	40.5
3	3.5	48.5	6	4	38

(continued)

Display 12.3 (continued)

Treatment	Initial Height (cm) Ht_i	Final Height (cm) Ht_f	Treatment	Initial Height (cm) Ht_i	Final Height (cm) Ht_f
6	3	43	7	2.5	45.5
6	4	44	7	3	42.5
6	1	26.5	7	1	43
6	2	23	7	2	45
6	4	47	7	1.5	38
6	4	37	7	1	46
7	1.5	44.5	7	1.5	38.5
7	0	39	7	2.5	32

Source: University of Florida Institute for Food and Agricultural Sciences, 1997.

12.2 Baseball: Does Money Buy Success?

Salaries of professional athletes are astronomical, but huge payrolls are not necessarily bad for team owners if the salaries are associated with great team performance or with other variables that improve income, like attendance. What are the relationships between some of these key variables for major league baseball?

Display 12.4 gives the total payroll, the average salary per player, the total attendance, the team batting average (percentage of hits, without the decimal), and the percentage of games won in regular season play (without the decimal) of teams for the 2001 season. Teams are divided into two major leagues (the American and the National), and each league is divided into three divisions (East, Central, and West).

Display 12.4 Data on major league baseball for 2001.

Team	Payroll (millions of dollars)	Average Salary (dollars)	Attendance (thousands)	Batting Average	% Wins (in tenths of a percentage)	League	Division
New York—AL	109.792	3,541,674	3167	267	594	A	E
Boston	109.559	3,423,716	2592	266	509	A	E
Los Angeles	108.981	3,757,964	3017	255	531	N	W
New York—NL	93.174	3,327,658	2618	249	506	N	E
Cleveland	91.975	3,065,833	3182	278	562	A	C
Atlanta	91.852	2,962,958	2779	260	543	N	E
Texas	88.504	2,854,981	2831	275	451	A	W
Arizona	81.207	2,900,233	2736	267	568	N	W
St. Louis	77.271	2,664,512	3110	270	574	N	C
Toronto	75.798	2,707,089	1915	263	494	A	E
Seattle	75.652	2,701,875	3512	288	716	A	W
Baltimore	72.426	2,497,460	3064	248	391	A	E
Colorado	71.068	2,632,148	3168	292	451	N	W
Chicago—NL	64.016	2,462,147	2779	261	543	N	C
San Francisco	63.333	2,345,654	3269	266	556	N	W
Chicago—AL	62.363	2,309,741	1766	268	512	A	C

(continued)

Display 12.4 (continued)

Team	Payroll (millions of dollars)	Average Salary (dollars)	Attendance (thousands)	Batting Average	% Wins (in tenths of a percentage)	League	Division
Houston	60.383	2,236,395	2906	271	574	N	C
Tampa Bay	54.952	2,035,245	1298	258	383	A	E
Pittsburgh	52.698	1,699,946	2436	247	383	N	C
Detroit	49.831	1,779,685	1921	260	407	A	C
Anaheim	46.568	1,502,199	1998	261	463	A	W
Cincinnati	45.228	1,739,534	1880	262	407	N	C
Milwaukee	43.089	1,595,901	2811	251	420	N	C
Philadelphia	41.664	1,602,468	1792	260	531	N	E
San Diego	38.333	1,419,745	2378	252	488	N	W
Kansas City	35.643	1,229,069	1536	266	401	A	C
Florida	35.504	1,183,472	1261	264	469	N	E
Montreal	34.775	1,159,150	643	253	420	N	E
Oakland	33.811	1,252,250	2133	264	630	A	W
Minnesota	24.350	901,852	1783	272	525	A	C

Sources: www.CNNSI.com; www.slam.ca/Baseball/attendance.html; www.mlb.com, 6/20/02.

Discussion: Exploring the Table of Data

D4. Looking only at Display 12.4, do you see any interesting features of the data? Do you see any possible patterns or associations between variables?

D5. Using statistical software, explore these data and comment on any interesting patterns or possible relationships you find. Any surprises?

D6. Which variable was most strongly associated with payroll? Does it matter much whether you use total team payroll or average salary per player in the analysis of these data? Explain.

D7. Discuss whether you think it is appropriate to do inferential analyses on these data.

Is Inference Appropriate?

This is an observational study because there was no random sampling and no random assignment of treatments. (Owners would probably object to a random assignment of a payroll to each team.) The analysis should start, as you have done, with the exploration of possible relationships among the variables, but some inference may be possible if the nature of the inference is carefully phrased. Payroll is certainly not random; for the most part, these values are fixed at the beginning of the season. Even though attendance, batting averages, and wins are somewhat random events that occur during the season, they were neither randomly sampled nor randomly assigned. You can, however, still ask, "How likely is it to observe by mere chance a pattern like the one seen in the data?" In studying bivariate relationships, the null hypothesis in the significance test is that the data look like a random sample of y's for fixed x's from a population with no linear trend. You can use a test to see whether this is a reasonable model or whether a linear trend appears to be a better explanation. In comparing two means, the null hypothesis is that the data look like independent random samples from two populations with the same mean.

This is similar in construction to common investigations in industrial processes in which yields are measured after setting temperature and pressure gauges at fixed levels determined by an engineer. Often these temperature settings cannot be randomized because of other considerations in the process, like the time it takes to get a process up to the required temperature and pressure levels. Even with no randomness in the design, the statistical analyses of yields does help decide whether or not an observed association can be attributed to chance alone. But it cannot answer questions of cause and effect.

Winning from League to League

Of interest in this study is whether possible associations among the variables are different for the two leagues. Display 12.5 shows a scatterplot, residual plot, and regression analysis for *percent wins* versus *payroll* for all 30 teams.

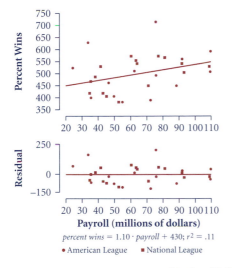

```
Dependent variable is:    %Wins
No Selector
R squared = 11.4%     R squared (adjusted) = 8.3%
s = 77.09  with  30 - 2 = 28 degrees of freedom

Source        Sum of Squares   df   Mean Square   F-ratio
Regression    21516.7           1    21516.5       3.62
Residual      166407           28    5943.12

Variable   Coefficient  s.e. of Coeff  t-ratio     prob
Constant   428.994      39.92            10.7    ≤0.0001
Payroll    1.10258      0.5795            1.90    0.0674
```

Display 12.5 Percent wins versus payroll for all teams.

The observed trend is increasing, but a closer scrutiny of a test of significance for the slope shows that the null hypothesis of a zero slope has a two-sided *P*-value of .0674, as shown in Display 12.5. This is small, but not small enough to make a bold declaration in favor of a linear trend with nonzero slope. In other words, the trend here is no more pronounced than you would expect if you randomly assigned the winning percentages to the different teams while keeping their payroll fixed.

When separating the leagues, however, we get the plots and analyses shown in Display 12.6. The American League shows no significant linear trend, whereas the National League shows a significant trend. That is, if payroll were kept constant for each team in the American League but percentage of wins were reassigned at random to the teams, then it is quite likely that you would get a slope as far from 0 as the American League's 0.782. If you did the same thing for the National League, it isn't likely that you would get a slope as far from 0 as 1.49, indicating that the positive trend in the National League cannot reasonably be explained by chance.

The *P*-value for the American League is .4545, whereas for the National League it's .0350.

American League

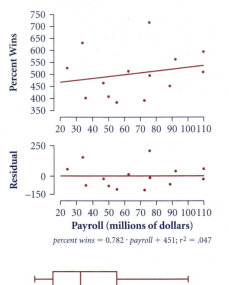

```
Dependent variable is:    %Wins
cases selected according to    Selected  MLB  payroll-tab
30 total cases of which 16 are missing
R squared = 4.7%    R squared (adjusted) = -3.2%
s = 99.54  with  14 - 2 = 12 degrees of freedom
```

Source	Sum of Squares	df	Mean Square	F-ratio
Regression	5919.36	1	5919.36	0.597
Residual	118890	12	9907.46	

Variable	Coefficient	s.e. of Coeff	t-ratio	prob
Constant	450.726	72.33	6.23	≤0.0001
pay	0.781595	1.011	0.773	0.4545

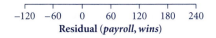

National League

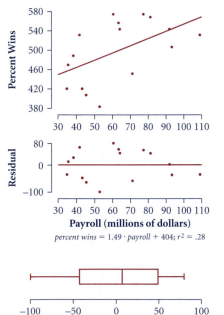

```
Dependent variable is:    %Wins
cases selected according to    Selected  MLB  payroll-tab
30 total cases of which 14 are missing
R squared = 28.0%    R squared (adjusted) = 22.9%
s = 56.89  with  16 - 2 = 14 degrees of freedom
```

Source	Sum of Squares	df	Mean Square	F-ratio
Regression	17623.7	1	17623.7	5.45
Residual	45307.3	14	3236.23	

Variable	Coefficient	s.e. of Coeff	t-ratio	prob
Constant	404.149	42.56	9.50	≤0.0001
pay	1.49376	0.6401	2.33	0.0350

Display 12.6 Percent wins versus payroll for the American and National Leagues.

Discussion: Differences Between the Leagues

D8. Compare the plots in Display 12.6 for the two leagues. What is there about these data that produces the drastic difference in results for the two leagues?

D9. Besides having a random sample, what other conditions need to be met in a regression analysis? Using the plots in Display 12.6, does it seem reasonable to assume these other conditions are met?

■ Practice

In P7 through P9, analyze the given relationship (a) for all teams, (b) for the American League, and (c) for the National League.

P7. *attendance* versus *payroll*

P8. *batting average* versus *payroll*

P9. *percentage of wins* versus *batting average*

P10. Use an appropriate test to answer these questions about means.

a. Is the difference between mean attendance for the two leagues statistically significant?

b. Is the difference between mean batting average for the two leagues statistically significant?

P11. In the 2001 World Series, the New York Yankees lost to the Arizona Diamondbacks. Is the difference in their percentage of wins during the regular 162-game season of 2001 statistically significant?

P12. For the 2001 regular season of 162 games, the three National League division winners were Atlanta, St. Louis, and Arizona. For the American League, the division winners were New York, Cleveland, and Seattle.

a. Is the difference between the proportions of games won by the National League division winners statistically significant?

b. Is the difference between the proportions of games won by the American League division winners statistically different?

Simulating a *P*-Value

The null hypothesis for the *percent wins* versus *payroll* example just given is:

The observed slope is no farther from 0 than you would be reasonably likely to get if you randomly reassigned the values of *percent wins* to different teams while keeping each team's *payroll* fixed.

The *P*-value for the test of significant slope measures how unusual the observed slope would be under those conditions. You can estimate the *P*-value by repeatedly rearranging the *y*'s and observing what happens to the slopes. Using this idea, Display 12.7 shows a set of 100 slopes found by 100 rerandomizations of the values for *percent wins*, done for the American League.

```
Stem-and-leaf of slopes for American League
Leaf Unit = 0.10
     1    -2  2
     7    -1  976655
    18    -1  44433211000
    31    -0  9988877665555
   (22)   -0  4333333333322211110000
    47     0  000001112233333444
    29     0  55566677888899
    15     1  0011122444
     5     1  779
     2     2  01
```

Display 12.7 Stem-and-leaf plot of 100 simulated slopes for percent wins versus payroll for the American League.

In the 100 trials, the observed slope of 0.782 for the American League was equaled or exceeded 21 times, giving a simulated one-sided *P*-value of .21. The *t*-test in Display 12.6 gives a two-sided *P*-value of .4545, which is almost equal to the *P*-value from the simulation. So with either method, the question of whether or not the observed pattern could have been generated by mere chance is answered in the affirmative for the American League.

Discussion: Simulating *P*-Values

D10. The stem-and-leaf plot in Display 12.8 shows a simulation for the National League that is parallel to the simulation for the American League.

a. Describe how the simulation was conducted.

b. Conduct one more trial and show where it would go on the stemplot.

c. What is the simulated one-sided *P*-value for these 101 trials? How does it compare to the two-sided *P*-value given in Display 12.6? What is your conclusion about the observed pattern for the National League?

```
Stem-and-leaf of slopes for National League
Leaf Unit = 0.10
     2    -1  44
     2    -1
     3    -1  1
     8    -0  99988
    14    -0  766666
    18    -0  4444
    31    -0  3333333332222
    45    -0  11111111100000
   (13)    0  0000000000011
    42     0  2222222233
    32     0  44445555
    24     0  677777
    18     0  888999
    12     1  00000111
     4     1  3
     3     1  4
     2     1
     2     1  89
```

Display 12.8 One hundred simulated slopes for percent wins versus payroll for the National League.

Ecological Correlations

Each of the two major baseball leagues is divided into three divisions, so another way of looking at the baseball data is to explore what happens at the division level. Display 12.9 shows the averages by division. (All teams play a 162-game schedule, with similar numbers of turns at bat for the season, so it is fair to take simple averages of team batting averages and winning percentages.)

Division	Payroll (millions of dollars)	Attendance (thousands)	Batting Average	% Wins (in tenths of a percentage)
AC	52.8	2037.6	269	481
AE	84.5	2407.2	260	474
AW	61.1	2618.5	272	565
NC	57.1	2653.7	260	483.5
NE	59.4	1818.6	257	494
NW	72.6	2913.6	266	519

Display 12.9 Data on major league baseball, by division, in 2001.

In analyzing how *percent wins* is related to *batting average*, you could use the teams as cases or the divisions as cases. The scatterplot for each kind of analysis is shown in Display 12.10.

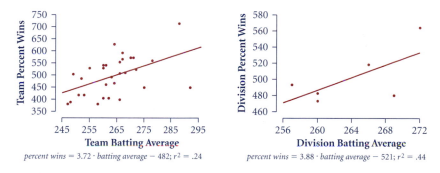

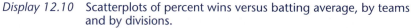
percent wins = 3.72 · batting average − 482; r² = .24   *percent wins = 3.88 · batting average − 521; r² = .44*

Display 12.10 Scatterplots of percent wins versus batting average, by teams and by divisions.

Each plot shows an increasing trend, but the correlation in the first is $\sqrt{.241}$, or about .491, while the correlation in the second is $\sqrt{.445}$, or about .667. In this situation, using the group averages instead of the individual team data inflates the correlation and makes the linear trend appear to be stronger.

The correlation computed using the divisions as cases and the division averages as variables is an example of an **ecological correlation,** or a correlation between group averages. They are often used in subjects like sociology and political science, where it is easier to get information about groups of people rather than individuals. (It's easy to find the percentage of people in your state who voted Republican in the last presidential election, but it's almost impossible to get the same information for each individual voter.) As you have now seen, correlations based on groups and averages can be quite different from correlations based on individuals. The mistake of using group-level correlations to support conclusions about individuals is called the **ecological fallacy.**

Ecological Fallacy

Ecological correlations use groups as cases and averages as variables. For many situations, you would get quite different values than if you used individuals as cases. The mistake of using ecological correlations for conclusions about individuals is called the ecological fallacy.

Of course, the team statistics are themselves averages (or totals), and we could have analyzed the relationship between salaries and batting averages by using individual players as cases. For the purpose of studying teams as business entities—taking into account variables like *attendance* and *team winning percentage*—using teams as cases makes sense. These team statistics are of little use, however, in studying player performance. So the choice of what to use as cases depends critically on the objectives of the study.

Discussion: Ecological Correlations

D11. Do both of the linear trends for the scatterplots in Display 12.10 have slopes that are significantly greater than zero? Does this suggest another problem with using ecological correlations?

D12. Calculate the correlation between *batting average* and *payroll* using teams as cases and then using divisions as cases. Comment on possible reasons for any differences you see. Which do you think is the better way to present the analysis?

D13. Calculate the correlation between *percent wins* and *payroll* using teams as cases and then using divisions as cases. Does the latter correlation appear to have a meaningful interpretation?

12.3 *Martin v. Westvaco* Revisited: Testing for Employment Discrimination

In Chapter 1 you first read about Robert Martin, who was laid off from his job at the Westvaco Corporation when he was 54. The statistical analysis presented during the lawsuit was quite a bit more involved than the simplified version you saw back then. With what you have learned, you can now carry out a much more thorough analysis while reviewing some inferential techniques and some ideas from probability theory as well.

In the Westvaco case, Martin claimed that he had been terminated because of his age, which, if true, would be against the law. His case was eventually settled out of court, before going to trial, but not until after a lot of statistical analysis and some arguments about the statistics.

Display 1.1 in Section 1.1 on page 5 shows the data used in the lawsuit, arranged in a standard "cases by variables" format. Each row of the table represents a case, one of the 50 people who worked in the engineering department of the envelope division of Westvaco on January 1, 1991.

Comparing Termination Rates for Two Age Groups

By law, all employees age 40 or older belong to what is called a "protected class": to discriminate against them on the basis of their age is against the law. At the time of the layoffs at Westvaco, 36 of the 50 people working in the engineering department were age 40 or older. A total of 28 workers were terminated; 21 of them were age 40 or older.

■ Practice

P13. Construct a table to satisfy each of the descriptions given here. In each table, keep the marginal totals the same as in the Westvaco case, presented in Display 12.11.

a. Make a copy of Display 12.11 and fill in the cells in order to present the strongest possible case for discrimination against those age 40 or over.

Terminated?

		No	Yes	Total	Percentage Terminated
Age	Under 40	–?–	–?–	14	–?–
	40 or Older	–?–	–?–	36	–?–
	Total	22	28	50	–?–

Display 12.11   Two-way table of job status and age, with marginal totals from the Westvaco case.

b. Make another table, but this time fill in the cells in order to make the variables *terminated* and *age* as close to independent as you can. If you select an employee at random from those represented in your table, are *terminated* and *age* independent events according to the definition in Chapter 6?

c. Make a final table using the actual data from Westvaco. Does it look more like the table in part a or the table in part b? Do you have stong evidence that older workers were more likely to be chosen for layoff?

Discussion: Which Test of Significance?

D14. Consider which statistical test would justify your opinion in part c of P13 about whether it looks like an older worker was more likely to be chosen for layoff.

a. Which tests of significance are possibilities in this situation? Give the strengths and weaknesses of each test in this scenario. (Don't worry about conditions for now. Should you use a one-sided or a two-sided test?)

b. Using the data from your table in part c of P13, find the *P*-value for each of these tests and compare them.

c. Using the test you think is most appropriate, what is your conclusion if you take the test at face value (that is, if you don't worry about the conditions being met)?

D15. Consider two companies, Seniors, Inc., with almost all of its 50 employees over age 40, and Youth Enterprises, with roughly half of its 50 employees under age 40. Suppose both companies discriminate against older workers in a layoff. If you use a significance test for the difference between two proportions, will you be more likely to detect the discrimination at Seniors, Inc., or at Youth Enterprises? Explain your reasoning.

Conditions Rarely Match Reality

The major differences between reality and a model are often the basis for heated arguments between opposing lawyers in discrimination cases. One of the tests you might have chosen in D14 is the *z*-test for the difference between two proportions. This test is based on several assumptions.

Independent Random Samples

A simple random sample of size n_1 is taken from a large population with proportion of successes p_1. The proportion of successes in this sample is $\hat{p}_1$. A second and independent random sample of size n_2 is taken from a large population with proportion of successes p_2. The proportion of successes in this sample is $\hat{p}_2$.

Large Samples from Even Larger Populations

In this test, you use the normal curve to approximate the sampling distribution for $\hat{p}_1 - \hat{p}_2$. This approximation is reasonable provided that all of

$$n_1\hat{p}_1, \qquad n_1(1-\hat{p}_1), \qquad n_2\hat{p}_2, \qquad \text{and} \qquad n_2(1-\hat{p}_2)$$

are at least 5 and that each population is at least 10 times the sample size. On the surface, the mismatch between the two sets of conditions is striking:

z-Test for a Difference Between Two Proportions	The Westvaco Case
Two populations that are • both very large	One population that is • small
Two samples that are	Two samples (over and under age 40) that are
• large enough • randomly selected • from the two different populations • independent of one another	• large enough • not randomly selected • from the same population • as dependent as can be, because if you know one sample, you automatically know the other

This does not make the test invalid, however. You can still use the test to answer this question: "If the process had been random, how likely would it be to get a difference in proportions as big as Westvaco got just by chance?" As long as it is made clear that this is the question being answered, the test is quite valid and can be very informative.

So you may proceed with a significance test, but you must make the limitations of what you have done very clear. If you reject the null hypothesis, all you can conclude is that something happened that can't reasonably be attributed to chance alone.

Discussion: Checking Conditions

D16. Evaluate the conditions necessary for your significance test in part c of D14. Give a careful statement of the conclusion you can make.

D17. Agree or disagree, and tell why: "A hypothesis test is based on a probability model. Like all probability models, it assumes certain outcomes are random. But in the Westvaco case, the decisions about which people to lay off weren't random. There's no probability model, and so a statistical test is invalid."

Defining "Older Worker"

The test you did in D14 was based on dividing workers into two age groups, "under 40" and "40 or older." In effect, this replaces a quantitative variable (*age*) with a categorical variable (*age group*). The age you use to define your categories can have quite an effect on your analysis and conclusions. In this section, you will explore how dividing workers into the age groups "under 50" and "50 or older" can change the analysis.

■ Practice: Cutoff Age

P14. A stem-and-leaf plot of the ages of the 50 Westvaco employees is shown in Display 12.12. Use it to construct a boxplot of the ages. Then use both plots to describe the shape of the distribution in words.

```
2 | 2 3
· | 5 9
3 | 0 1 1 2 2 3 4
· | 5 7 8
4 | 2 2
· | 7 8 8 8 8 9
5 | 0 2 3 3 3 4 4 4
· | 5 5 5 5 5 6 6 6 7 9 9 9
6 | 0 1 1 3 4 4
· | 6 9
```

Display 12.12 A stem-and-leaf plot of the ages of the 50 Westvaco employees.

P15. If you use 40 as your cutoff for defining your age groups, what percentage of the workers are in the older age group? If you use 50 instead of 40, what percentage are in the older age group?

P16. Of the 28 workers who were 50 or older, 19 were laid off. Use this information to construct a table similar to that in part c of P13, but this time use age 50 as your cutoff age.

P17. Using your table in P16, state appropriate null and alternate hypotheses, and carry out a test of the null hypothesis. What do you conclude if you take the results at face value?

Discussion: Cutoff Age

D18. Which cutoff value, 40 or 50, leads to stronger evidence of discrimination? Which test do you think is more informative about what actually happened —using age 40 or age 50 as your cutoff age? Explain your reasoning.

D19. According to the U.S. Supreme Court, if you do a statistical test of discrimination, you should reject the null hypothesis if your P-value from a one-sided test is less than or equal to .025 or if your P-value from a two-sided test is less than or equal to .05.

 a. According to this standard, should the null hypothesis be rejected in either of the two tests so far?

 b. Which of these statements correctly completes the phrase: If you use a P-value of .025 to determine "guilt," then

 I. 2.5% of "not guilty" companies would be declared "guilty" by the test

 II. 2.5% of "guilty" companies would be declared "not guilty" by the test

 c. Is 2.5% the probability of a Type I or a Type II error?

 d. Describe a Type II error in this situation.

Looking for a Better Approach

The cutoff age that defines who is classified as an "older" worker can be arbitrary and, as you have seen, can change the results of the analysis. Looking at the mean ages of those laid off and those retained can avoid such an arbitrary decision. You took that approach in Chapter 1 when you used the average age of the workers laid off by Westvaco as the test statistic.

■ Practice

P18. Construct a back-to-back stemplot of the ages of those laid off and those retained. Compare the two distributions. (Use split stems: 2│ for 20 – 24, and ·│ for 25 – 29, and so on.)

P19. Use your stemplot to construct side-by-side boxplots.

P20. Here are the means and standard deviations for the two groups. Use the summary data to carry out a two-sample *t*-test. (Again, don't worry about conditions for now. Just take the results at face value. Should you use a one-sided or a two-sided test?)

	n	**Mean Age**	*SD*
Laid off	28	49.86	13.40
Retained	22	46.18	11.00
All employees	50	48.24	12.42

Conditions for a *t*-Test

The *t*-test for the difference between two means that you used in P20 is based on random samples from large populations, or, more specifically in this case, on the condition that those terminated and those retained are independent random samples from two very large, approximately normal populations. In actual fact, there is only one population, which is finite rather than infinite, and this population gets sorted into two groups. Again, the mismatch between the two sets of conditions is striking:

t-Test for a Difference Between Means	The Westvaco Case
Two populations that are • both very large • both normal	One population that is • small • not normal
Two samples that are • randomly selected • from two different populations • independent of one another	Two samples that are • not randomly selected • from the same population • as dependent as can be, because if you know one sample—those laid off—you automatically know the other

On the surface, there's no apparent reason to think the *t*-test is appropriate. Remarkably, though, extensive simulations have shown that despite the mismatch between the two sets of conditions, the *t*-test nevertheless tends to give a good approximation of the *P*-value you would get using an approach that satisfies the actual set of conditions for situations like the Westvaco case.

You used that approach in Chapter 1, Section 2. Using simulation, you selected three workers for layoff using a completely random process and computed their average age. After you did this step many times, you compared the average age of workers actually laid off at Westvaco to your distribution of the average ages of workers selected at random for layoff. This approach is called the *randomization test* or *permutation test*. If the sampling distribution is constructed theoretically rather than by simulation, it is called *Fisher's exact test*.

For these tests, the hypothesis of no discrimination says that the process used by Westvaco was equivalent to randomly choosing the employees for layoff from the total population of 50 employees.

About all we can ever show by using statistical methods in discrimination cases is that the process doesn't look like random selection. Statistical analysis alerts us to questionable situations, but it cannot reconstruct the intent of the people who did the layoffs. Knowing the intent is crucial because it may be perfectly legal: Perhaps employees in obsolete jobs were the ones picked for termination and it just happens that the obsolete jobs were held by older employees.

Fisher's exact test requires few assumptions and uses no approximations. In contrast, the *z*-test for the difference between two proportions is an approximation and requires strong assumptions. The amazing thing is that the *z*-test works so well! So why don't we always use the randomization test or Fisher's exact test? Because both require a great deal of computing power. As technology becomes more available and more powerful, statisticians increasingly are turning to these methods rather than using approximations based on the normal distribution.

Discussion

D20. Describe how to use the randomization test to determine if the average age of the 28 workers laid off at Westvaco, 49.86 years, can reasonably be attributed to chance variation. Each student in your class should perform one trial of your simulation.

D21. In light of the earlier discussion and the design of your simulation in D20, restate your hypotheses and the conclusion that you gave in P20.

D22. Refer to the stemplot for the ages of the 50 employees given in P14 in Display 12.12. One condition for a *t*-test is that your data come from a normal distribution. How appropriate is that assumption here? (In what ways is the shape of the distribution different from normal? How important do you consider these departures from normal?)

The End of the Story

The planning that led up to the layoffs at Westvaco took place in several stages. In the first stage, the head of the engineering department drew up a list of 11 employees to lay off. His boss reviewed the list and decided it was too short: They needed to reduce the size of their workforce even further. The department head added a second group of people to the list and checked again with his boss: still too few. He added a third group of names, then a fourth, and, finally, one more person was added in the fifth round of planning. Display 1.1 on page 5 shows this information in the column headed RIF (for "reduction in force"):

An entry of a 1 means "chosen for layoff in Round 1 of the planning," and similarly for 2, 3, 4, and 5. An entry of a 0 means "Retained."

As it turned out, older employees fared much worse in the earlier rounds of the planning than in the later rounds, as you can see in Display 12.13.

Terminated in Round 1 or 2?

		No	Yes	Total	Percentage Terminated
Age	**Under 50**	18	4	22	18.18%
	50 or Older	12	16	28	57.14%
	Total	30	20	50	

Terminated in Round 3, 4, or 5?

		No	Yes	Total	Percentage Terminated
Age	**Under 50**	13	5	18	27.78%
	50 or Older	9	3	12	25.00%
	Total	22	8	30	

Display 12.13 Breakdown of reduction in force by round and age group.

Time plots give another view of the same phenomenon. Each plot in Display 12.14 shows the numbers of employees remaining at Westvaco after each round of the planning.

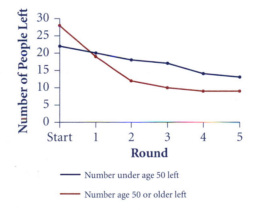

Display 12.14 A time plot of the reduction in force by round and age group.

Discussion: Reduction over Time

D23. Describe the patterns in Display 12.14. What do they suggest to you?

The Westvaco case never got as far as a jury. Just before it was about to go to trial, the two sides agreed on a settlement. Details of such settlements are not public information, so, as with many problems based on statistics, this case has no "final answer."

Appendix: Statistical Tables

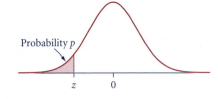

Table entry for *z* is the probability lying below *z*.

Probability *p*

TABLE A   Standard Normal Probabilities

z	.00	.01	.02	.03	.04	.05	.06	.07	.08	.09
−3.8	.0001	.0001	.0001	.0001	.0001	.0001	.0001	.0001	.0001	.0001
−3.7	.0001	.0001	.0001	.0001	.0001	.0001	.0001	.0001	.0001	.0001
−3.6	.0002	.0002	.0001	.0001	.0001	.0001	.0001	.0001	.0001	.0001
−3.5	.0002	.0002	.0002	.0002	.0002	.0002	.0002	.0002	.0002	.0002
−3.4	.0003	.0003	.0003	.0003	.0003	.0003	.0003	.0003	.0003	.0002
−3.3	.0005	.0005	.0005	.0004	.0004	.0004	.0004	.0004	.0004	.0003
−3.2	.0007	.0007	.0006	.0006	.0006	.0006	.0006	.0005	.0005	.0005
−3.1	.0010	.0009	.0009	.0009	.0008	.0008	.0008	.0008	.0007	.0007
−3.0	.0013	.0013	.0013	.0012	.0012	.0011	.0011	.0011	.0010	.0010
−2.9	.0019	.0018	.0018	.0017	.0016	.0016	.0015	.0015	.0014	.0014
−2.8	.0026	.0025	.0024	.0023	.0023	.0022	.0021	.0021	.0020	.0019
−2.7	.0035	.0034	.0033	.0032	.0031	.0030	.0029	.0028	.0027	.0026
−2.6	.0047	.0045	.0044	.0043	.0041	.0040	.0039	.0038	.0037	.0036
−2.5	.0062	.0060	.0059	.0057	.0055	.0054	.0052	.0051	.0049	.0048
−2.4	.0082	.0080	.0078	.0075	.0073	.0071	.0069	.0068	.0066	.0064
−2.3	.0107	.0104	.0102	.0099	.0096	.0094	.0091	.0089	.0087	.0084
−2.2	.0139	.0136	.0132	.0129	.0125	.0122	.0119	.0116	.0113	.0110
−2.1	.0179	.0174	.0170	.0166	.0162	.0158	.0154	.0150	.0146	.0143
−2.0	.0228	.0222	.0217	.0212	.0207	.0202	.0197	.0192	.0188	.0183
−1.9	.0287	.0281	.0274	.0268	.0262	.0256	.0250	.0244	.0239	.0233
−1.8	.0359	.0351	.0344	.0336	.0329	.0322	.0314	.0307	.0301	.0294
−1.7	.0446	.0436	.0427	.0418	.0409	.0401	.0392	.0384	.0375	.0367
−1.6	.0548	.0537	.0526	.0516	.0505	.0495	.0485	.0475	.0465	.0455
−1.5	.0668	.0655	.0643	.0630	.0618	.0606	.0594	.0582	.0571	.0559
−1.4	.0808	.0793	.0778	.0764	.0749	.0735	.0721	.0708	.0694	.0681
−1.3	.0968	.0951	.0934	.0918	.0901	.0885	.0869	.0853	.0838	.0823
−1.2	.1151	.1131	.1112	.1093	.1075	.1056	.1038	.1020	.1003	.0985
−1.1	.1357	.1335	.1314	.1292	.1271	.1251	.1230	.1210	.1190	.1170
−1.0	.1587	.1562	.1539	.1515	.1492	.1469	.1446	.1423	.1401	.1379
−0.9	.1841	.1814	.1788	.1762	.1736	.1711	.1685	.1660	.1635	.1611
−0.8	.2119	.2090	.2061	.2033	.2005	.1977	.1949	.1922	.1894	.1867
−0.7	.2420	.2389	.2358	.2327	.2296	.2266	.2236	.2206	.2177	.2148
−0.6	.2743	.2709	.2676	.2643	.2611	.2578	.2546	.2514	.2483	.2451
−0.5	.3085	.3050	.3015	.2981	.2946	.2912	.2877	.2843	.2810	.2776
−0.4	.3446	.3409	.3372	.3336	.3300	.3264	.3228	.3192	.3156	.3121
−0.3	.3821	.3783	.3745	.3707	.3669	.3632	.3594	.3557	.3520	.3483
−0.2	.4207	.4168	.4129	.4090	.4052	.4013	.3974	.3936	.3897	.3859
−0.1	.4602	.4562	.4522	.4483	.4443	.4404	.4364	.4325	.4286	.4247
0.0	.5000	.4960	.4920	.4880	.4840	.4801	.4761	.4721	.4681	.4641

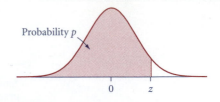

Table entry for *z* is the
probability lying below *z*.

Probability *p*

TABLE A   Standard Normal Probabilities (continued)

z	.00	.01	.02	.03	.04	.05	.06	.07	.08	.09
0.0	.5000	.5040	.5080	.5120	.5160	.5199	.5239	.5279	.5319	.5359
0.1	.5398	.5438	.5478	.5517	.5557	.5596	.5636	.5675	.5714	.5753
0.2	.5793	.5832	.5871	.5910	.5948	.5987	.6026	.6064	.6103	.6141
0.3	.6179	.6217	.6255	.6293	.6331	.6368	.6406	.6443	.6480	.6517
0.4	.6554	.6591	.6628	.6664	.6700	.6736	.6772	.6808	.6844	.6879
0.5	.6915	.6950	.6985	.7019	.7054	.7088	.7123	.7157	.7190	.7224
0.6	.7257	.7291	.7324	.7357	.7389	.7422	.7454	.7486	.7517	.7549
0.7	.7580	.7611	.7642	.7673	.7704	.7734	.7764	.7794	.7823	.7852
0.8	.7881	.7910	.7939	.7967	.7995	.8023	.8051	.8078	.8106	.8133
0.9	.8159	.8186	.8212	.8238	.8264	.8289	.8315	.8340	.8365	.8389
1.0	.8413	.8438	.8461	.8485	.8508	.8531	.8554	.8577	.8599	.8621
1.1	.8643	.8665	.8686	.8708	.8729	.8749	.8770	.8790	.8810	.8830
1.2	.8849	.8869	.8888	.8907	.8925	.8944	.8962	.8980	.8997	.9015
1.3	.9032	.9049	.9066	.9082	.9099	.9115	.9131	.9147	.9162	.9177
1.4	.9192	.9207	.9222	.9236	.9251	.9265	.9279	.9292	.9306	.9319
1.5	.9332	.9345	.9357	.9370	.9382	.9394	.9406	.9418	.9429	.9441
1.6	.9452	.9463	.9474	.9484	.9495	.9505	.9515	.9525	.9535	.9545
1.7	.9554	.9564	.9573	.9582	.9591	.9599	.9608	.9616	.9625	.9633
1.8	.9641	.9649	.9656	.9664	.9671	.9678	.9686	.9693	.9699	.9706
1.9	.9713	.9719	.9726	.9732	.9738	.9744	.9750	.9756	.9761	.9767
2.0	.9772	.9778	.9783	.9788	.9793	.9798	.9803	.9808	.9812	.9817
2.1	.9821	.9826	.9830	.9834	.9838	.9842	.9846	.9850	.9854	.9857
2.2	.9861	.9864	.9868	.9871	.9875	.9878	.9881	.9884	.9887	.9890
2.3	.9893	.9896	.9898	.9901	.9904	.9906	.9909	.9911	.9913	.9916
2.4	.9918	.9920	.9922	.9925	.9927	.9929	.9931	.9932	.9934	.9936
2.5	.9938	.9940	.9941	.9943	.9945	.9946	.9948	.9949	.9951	.9952
2.6	.9953	.9955	.9956	.9957	.9959	.9960	.9961	.9962	.9963	.9964
2.7	.9965	.9966	.9967	.9968	.9969	.9970	.9971	.9972	.9973	.9974
2.8	.9974	.9975	.9976	.9977	.9977	.9978	.9979	.9979	.9980	.9981
2.9	.9981	.9982	.9982	.9983	.9984	.9984	.9985	.9985	.9986	.9986
3.0	.9987	.9987	.9987	.9988	.9988	.9989	.9989	.9989	.9990	.9990
3.1	.9990	.9991	.9991	.9991	.9992	.9992	.9992	.9992	.9993	.9993
3.2	.9993	.9993	.9994	.9994	.9994	.9994	.9994	.9995	.9995	.9995
3.3	.9995	.9995	.9995	.9996	.9996	.9996	.9996	.9996	.9996	.9997
3.4	.9997	.9997	.9997	.9997	.9997	.9997	.9997	.9997	.9997	.9998
3.5	.9998	.9998	.9998	.9998	.9998	.9998	.9998	.9998	.9998	.9998
3.6	.9998	.9998	.9999	.9999	.9999	.9999	.9999	.9999	.9999	.9999
3.7	.9999	.9999	.9999	.9999	.9999	.9999	.9999	.9999	.9999	.9999
3.8	.9999	.9999	.9999	.9999	.9999	.9999	.9999	.9999	.9999	.9999

Table entry for *p* and *C* is the point *t** with probability *p* lying above it and confidence level *C* given by the probability of lying between –*t** and *t**.

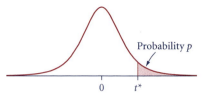

Probability *p*

0   *t**

TABLE B *t*-Distribution Critical Values

						Tail Probability *p*						
df	.25	.20	.15	.10	.05	.025	.02	.01	.005	.0025	.001	.0005
1	1.000	1.376	1.963	3.078	6.314	12.71	15.89	31.82	63.66	127.3	318.3	636.6
2	.816	1.061	1.386	1.886	2.920	4.303	4.849	6.965	9.925	14.09	22.33	31.60
3	.765	.978	1.250	1.638	2.353	3.182	3.482	4.541	5.841	7.453	10.21	12.92
4	.741	.941	1.190	1.533	2.132	2.776	2.999	3.747	4.604	5.598	7.173	8.610
5	.727	.920	1.156	1.476	2.015	2.571	2.757	3.365	4.032	4.773	5.893	6.869
6	.718	.906	1.134	1.440	1.943	2.447	2.612	3.143	3.707	4.317	5.208	5.959
7	.711	.896	1.119	1.415	1.895	2.365	2.517	2.998	3.499	4.029	4.785	5.408
8	.706	.889	1.108	1.397	1.860	2.306	2.449	2.896	3.355	3.833	4.501	5.041
9	.703	.883	1.100	1.383	1.833	2.262	2.398	2.821	3.250	3.690	4.297	4.781
10	.700	.879	1.093	1.372	1.812	2.228	2.359	2.764	3.169	3.581	4.144	4.587
11	.697	.876	1.088	1.363	1.796	2.201	2.328	2.718	3.106	3.497	4.025	4.437
12	.695	.873	1.083	1.356	1.782	2.179	2.303	2.681	3.055	3.428	3.930	4.318
13	.694	.870	1.079	1.350	1.771	2.160	2.282	2.650	3.012	3.372	3.852	4.221
14	.692	.868	1.076	1.345	1.761	2.145	2.264	2.624	2.977	3.326	3.787	4.140
15	.691	.866	1.074	1.341	1.753	2.131	2.249	2.602	2.947	3.286	3.733	4.073
16	.690	.865	1.071	1.337	1.746	2.120	2.235	2.583	2.921	3.252	3.686	4.015
17	.689	.863	1.069	1.333	1.740	2.110	2.224	2.567	2.898	3.222	3.646	3.965
18	.688	.862	1.067	1.330	1.734	2.101	2.214	2.552	2.878	3.197	3.611	3.922
19	.688	.861	1.066	1.328	1.729	2.093	2.205	2.539	2.861	3.174	3.579	3.883
20	.687	.860	1.064	1.325	1.725	2.086	2.197	2.528	2.845	3.153	3.552	3.850
21	.686	.859	1.063	1.323	1.721	2.080	2.189	2.518	2.831	3.135	3.527	3.819
22	.686	.858	1.061	1.321	1.717	2.074	2.183	2.508	2.819	3.119	3.505	3.792
23	.685	.858	1.060	1.319	1.714	2.069	2.177	2.500	2.807	3.104	3.485	3.768
24	.685	.857	1.059	1.318	1.711	2.064	2.172	2.492	2.797	3.091	3.467	3.745
25	.684	.856	1.058	1.316	1.708	2.060	2.167	2.485	2.787	3.078	3.450	3.725
26	.684	.856	1.058	1.315	1.706	2.056	2.162	2.479	2.779	3.067	3.435	3.707
27	.684	.855	1.057	1.314	1.703	2.052	2.150	2.473	2.771	3.057	3.421	3.690
28	.683	.855	1.056	1.313	1.701	2.048	2.154	2.467	2.763	3.047	3.408	3.674
29	.683	.854	1.055	1.311	1.699	2.045	2.150	2.462	2.756	3.038	3.396	3.659
30	.683	.854	1.055	1.310	1.697	2.042	2.147	2.457	2.750	3.030	3.385	3.646
40	.681	.851	1.050	1.303	1.684	2.021	2.123	2.423	2.704	2.971	3.307	3.551
50	.679	.849	1.047	1.295	1.676	2.009	2.109	2.403	2.678	2.937	3.261	3.496
60	.679	.848	1.045	1.296	1.671	2.000	2.099	2.390	2.660	2.915	3.232	3.460
80	.678	.846	1.043	1.292	1.664	1.990	2.088	2.374	2.639	2.887	3.195	3.416
100	.677	.845	1.042	1.290	1.660	1.984	2.081	2.364	2.626	2.871	3.174	3.390
1000	.675	.842	1.037	1.282	1.646	1.962	2.056	2.330	2.581	2.813	3.098	3.300
∞	.674	.841	1.036	1.282	1.645	1.960	2.054	2.326	2.576	2.807	3.091	3.291
	50%	60%	70%	80%	90%	95%	96%	98%	99%	99.5%	99.8%	99.9%

Confidence level *C*

Table entry for p is the point χ^2 with probability p lying above it.

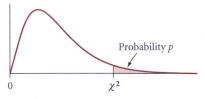

Probability p

TABLE C χ^2 Critical Values

					Tail Probability p						
df	.25	.20	.15	.10	.05	.025	.02	.01	.005	.0025	.001
1	1.32	1.64	2.07	2.71	3.84	5.02	5.41	6.63	7.88	9.14	10.83
2	2.77	3.22	3.79	4.61	5.99	7.38	7.82	9.21	10.60	11.98	13.82
3	4.11	4.64	5.32	6.25	7.81	9.35	9.84	11.34	12.84	14.32	16.27
4	5.39	5.59	6.74	7.78	9.49	11.14	11.67	13.23	14.86	16.42	18.47
5	6.63	7.29	8.12	9.24	11.07	12.83	13.33	15.09	16.75	18.39	20.51
6	7.84	8.56	9.45	10.64	12.53	14.45	15.03	16.81	18.55	20.25	22.46
7	9.04	9.80	10.75	12.02	14.07	16.01	16.62	18.48	20.28	22.04	24.32
8	10.22	11.03	12.03	13.36	15.51	17.53	18.17	20.09	21.95	23.77	26.12
9	11.39	12.24	13.29	14.68	16.92	19.02	19.63	21.67	23.59	25.46	27.83
10	12.55	13.44	14.53	15.99	18.31	20.48	21.16	23.21	25.19	27.11	29.59
11	13.70	14.63	15.77	17.29	19.68	21.92	22.62	24.72	26.76	28.73	31.26
12	14.85	15.81	16.99	18.55	21.03	23.34	24.05	26.22	28.30	30.32	32.91
13	15.93	15.58	18.90	19.81	22.36	24.74	25.47	27.69	29.82	31.88	34.53
14	17.12	18.15	19.4	21.06	23.68	26.12	26.87	29.14	31.32	33.43	36.12
15	18.25	19.31	20.60	22.31	25.00	27.49	28.26	30.58	32.80	34.95	37.70
16	19.37	20.47	21.79	23.54	26.30	28.85	29.63	32.00	34.27	36.46	39.25
17	20.49	21.61	22.98	24.77	27.59	30.19	31.00	33.41	35.72	37.95	40.79
18	21.60	22.76	24.16	25.99	28.87	31.53	32.35	34.81	37.16	39.42	42.31
19	22.72	23.90	25.33	27.20	30.14	32.85	33.69	36.19	38.58	40.88	43.82
20	23.83	25.04	26.50	28.41	31.41	34.17	35.02	37.57	40.00	42.34	45.31
21	24.93	26.17	27.66	29.62	32.67	35.48	36.34	38.93	41.40	43.78	46.80
22	26.04	27.30	28.82	30.81	33.92	36.78	37.66	40.29	42.80	45.20	48.27
23	27.14	28.43	29.98	32.01	35.17	38.08	38.97	41.64	44.18	46.62	49.73
24	28.24	29.55	31.13	33.20	36.42	39.36	40.27	42.98	45.56	48.03	51.18
25	29.34	30.68	32.28	34.38	37.65	40.65	41.57	44.31	46.93	49.44	52.62
26	30.43	31.79	33.43	35.56	38.89	41.92	42.86	45.64	48.29	50.83	54.05
27	31.53	32.91	34.57	36.74	40.11	43.19	44.14	46.96	49.64	52.22	55.48
28	32.62	34.03	35.71	37.92	41.34	44.46	45.42	48.28	50.99	53.59	56.89
29	33.71	35.14	36.85	39.09	42.56	45.72	46.69	49.59	52.34	54.97	58.30
30	34.80	36.25	37.99	40.26	43.77	46.98	47.96	50.89	53.67	56.33	59.70
40	45.62	47.27	49.24	51.81	55.76	59.34	60.44	63.69	66.77	69.70	73.40
50	56.33	53.16	60.35	63.17	67.50	71.42	72.61	76.15	79.49	82.66	86.66
60	66.98	68.97	71.34	74.40	79.08	83.30	84.58	88.38	91.95	95.34	99.61
80	88.13	90.41	93.11	96.58	101.9	106.6	108.1	112.3	116.3	120.1	124.8
100	109.1	111.7	114.7	118.5	124.3	129.6	131.1	135.8	140.2	144.3	149.4

TABLE D Random Digits

Row									
1	10097	32533	76520	13586	34673	54876	80959	09117	39292	74945
2	37542	04805	64894	74296	24805	24037	20636	10402	00822	91665
3	08422	68953	19645	09303	23209	02560	15953	34764	35080	33606
4	99019	02529	09376	70715	38311	31165	88676	74397	04436	27659
5	12807	99970	80157	36147	64032	36653	98951	16877	12171	76833
6	66065	74717	34072	76850	36697	36170	65813	39885	11199	29170
7	31060	10805	45571	82406	35303	42614	86799	07439	23403	09732
8	85269	77602	02051	65692	68665	74818	73053	85247	18623	88579
9	63573	32135	05325	47048	90553	57548	28468	28709	83491	25624
10	73796	45753	03529	64778	35808	34282	60935	20344	35273	88435
11	98520	17767	14905	68607	22109	40558	60970	93433	50500	73998
12	11805	05431	39808	27732	50725	68248	29405	24201	52775	67851
13	83452	99634	06288	98083	13746	70078	18475	40610	68711	77817
14	88685	40200	86507	58401	36766	67951	90364	76493	29609	11062
15	99594	67348	87517	64969	91826	08928	93785	61368	23478	34113
16	65481	17674	17468	50950	58047	76974	73039	57186	40218	16544
17	80124	35635	17727	08015	45318	22374	21115	78253	14385	53763
18	74350	99817	77402	77214	43236	00210	45521	64237	96286	02655
19	69916	26803	66252	29148	36936	87203	76621	13990	94400	56418
20	09893	20505	14225	68514	46427	56788	96297	78822	54382	14598
21	91499	14523	68479	27686	46162	83554	94750	89923	37089	20048
22	80336	94598	26940	36858	70297	34135	53140	33340	42050	82341
23	44104	81949	85157	47954	32979	26575	57600	40881	22222	06413
24	12550	73742	11100	02040	12860	74697	96644	89439	28707	25815
25	63606	49329	16505	34484	40219	52563	43651	77082	07207	31790
26	61196	90446	26457	47774	51924	33729	65394	59593	42582	60527
27	15474	45266	95270	79953	59367	83848	82396	10118	33211	59466
28	94557	28573	67897	54387	54622	44431	91190	42592	92927	45973
29	42481	16213	97344	08721	16868	48767	03071	12059	25701	46670
30	23523	78317	73208	89837	68935	91416	26252	29663	05522	82562
31	04493	52494	75246	33824	45862	51025	61962	79335	65337	12472
32	00549	97654	64051	88159	96119	63896	54692	82391	23287	29529
33	35963	15307	26898	09354	33351	35462	77974	50024	90103	39333
34	59808	08391	45427	26842	83609	49700	13021	24892	78565	20106
35	46058	85236	01390	92286	77281	44077	93910	83647	70617	42941
36	32179	00597	87379	25241	05567	07007	86743	17157	85394	11838
37	69234	61406	20117	45204	15956	60000	18743	92423	97118	96338
38	19565	41430	01758	75379	40419	21585	66674	36806	84962	85207
39	45155	14938	19476	07246	43667	94543	59047	90033	20826	69541
40	94864	31994	36168	10851	34888	81553	01540	35456	05014	51176
41	98086	24826	45240	28404	44999	08896	39094	73407	35441	31880
42	33185	16232	41941	50949	89435	48581	88695	41994	37548	73043
43	80951	00406	96382	70774	20151	23387	25016	25298	94624	61171
44	79752	49140	71961	28296	69861	02591	74852	20539	00387	59579
45	18633	32537	98145	06571	31010	24674	05455	61427	77938	91936
46	74029	43902	77557	32270	97790	17119	52527	58021	80814	51748
47	54178	45611	80993	37143	05335	12969	56127	19255	36040	90324
48	11664	49883	52079	84827	59381	71539	09973	33440	88461	23356
49	48324	77928	31249	64710	02295	36870	32307	57546	15020	09994
50	69074	94138	87637	91976	35584	04401	10518	21615	01848	76938

Glossary

Symbols

b_0	the y-intercept of the sample regression line
b_1	the slope of the sample regression line
E	margin of error
$E(X)$	the expected value (mean) of the random variable X
H_0	the null hypothesis
H_a	the alternative hypothesis (sometimes H_1)
H_1	the alternative hypothesis
IQR	the interquartile range
n	the sample size or the number of trials
N	the population size
p	the population proportion or the probability of a success on any one trial
$\hat{p}$	sample proportion
$P(A)$	the probability that event A happens
$P(A \text{ or } B)$	the probability that event A happens or event B happens or both
$P(A \text{ and } B)$	the probability that event A and event B both happen
$P(B \mid A)$	the conditional probability that B happens given that A happens
$P(x)$	the probability that the random variable X takes on the value x
p_0	the hypothesized value of the population proportion
Q_1	the first or lower quartile
Q_3	the third or upper quartile
r	the sample correlation
s	the sample standard deviation
s^2	the sample variance
s_{b_1}	the estimated standard error of the slope of the regression line
SD	standard deviation
SE	standard error
SRS	simple random sample
SSE	the sum of squared errors
t	a test statistic using a t-distribution

$\text{Var}(X)$	the variance of the random variable X
$\bar{x}$	the sample mean
x	the observed value of a variable
X	a random variable
y	the observed value of a variable
$\hat{y}$	the predicted value of a variable
z	the standardized value (z-score)
z^*	the critical value

Greek Letters

α	the significance level
β_0	the intercept of the population regression line
β_1	the slope of the population regression line
ε	the difference of the value of y for a point and the value predicted by the population regression line
μ	the population mean
$\mu_{\bar{x}}$	the mean of the sampling distribution of the sample mean
μ_X	the expected value (mean) of the random variable X
$\mu_{y\mid x}$	the mean of the conditional distribution of y given x
σ	the population standard deviation
$\sigma_{\bar{x}}$	the standard error of the mean
σ^2	the population variance
σ_X^2	the variance of the random variable X
χ^2	the chi-squared statistic

---

addition rule　For any two events A and B, $P(A \text{ or } B) = P(A) + P(B) - P(A \text{ and } B)$. If A and B are disjoint, then $P(A \text{ or } B) = P(A) + P(B)$.

adjacent values　On a modified boxplot, the largest and smallest non-outliers.

balanced design　A design in which each treatment has the same number of units assigned to it.

bar graph (or bar chart)　A plot that shows frequencies for categorical data as heights or lengths of bars, with one bar for each category.

bias The difference between the actual value (or parameter) being estimated and the average value of an estimator of that parameter.

bias due to sampling See **sample selection bias.**

bimodal A distribution with two peaks.

binomial distribution The random variable X has a binomial distribution if X represents the number of successes in n independent trials, where the probability of a success is the same on each trial.

bivariate data Data that involve two variables per case. For quantitative variables, often displayed on a scatterplot. For categorical variables, often displayed in a two-way table.

blind A type of experiment in which the subjects do not know which treatment they received.

blocks In an experiment, groups of similar units.

boxplot (or box-and-whisker plot) A graphical display of the five-number summary. The "box" extends from the lower quartile to the upper quartile, with a line across it at the median. The "whiskers" run from the quartiles to the minimum and maximum.

capture rate The proportion of random samples for which the resulting confidence interval captures the true (population) value. See **confidence level.**

case The subject (or object) on which a measurement is made in statistical examination.

categorical variable A variable that can be grouped into categories, such as "yes" and "no." Categories sometimes can be ordered as with "small," "medium," and "large."

census An examination of all individuals in the entire population.

Central Limit Theorem The shape of the sampling distribution of the sample mean becomes more normal-like as n increases.

chance model See **probability model.**

chi-square test of homogeneity A chi-square test used to determine whether it is reasonable to believe that when several different populations are broken down into the same categories, they have the same proportion of members in each category.

chi-square test of independence A chi-square test of the hypothesis that two categorical variables on the same population are independent of each other.

clinical trial A randomized experiment comparing medical treatments.

cluster On a plot, a group of data "clustering" close to the same value, away from other groups.

cluster sampling When the units in the population are assigned to clusters (such as classrooms) and you take a simple random sample of the clusters.

coefficient of determination The square of the correlation r. Tells the proportion of the total variation in y that can be explained by the relationship with x.

comparison group In an experiment, a group that receives one of the treatments, often the standard treatment.

completely randomized design An experimental design in which treatments are randomly assigned to subjects without restriction.

conditional distribution of y given x With bivariate data, the distribution of the values of y for a fixed value of x.

conditional probability The notion that a probability can change if you are given additional information. The conditional probability that A happens given that B happens is $P(A \mid B) = \frac{P(A \text{ and } B)}{P(B)}$ as long as $P(B) > 0$.

confidence interval A set of plausible values for a population parameter, any one of which could be used to define a population for which the observed sample statistic would be a reasonably likely outcome.

confidence level The probability that the method used will give a confidence interval that captures the parameter.

confounding The phenomenon that occurs when an experimental design has mixed together two possible influences on a response in such a way that it is impossible to separate their effects.

continuous variable A quantitative variable that can take on any value on an interval of real numbers.

control group In an experiment, a group that provides a standard for comparison to evaluate the effectiveness of a treatment; often given the placebo.

convenience sample A sample in which the units chosen from the population are the units that are easy to include.

correlation A numerical value between -1 and 1 inclusive that measures the strength and direction of a linear relationship between two variables.

critical value The value to which a test statistic is compared in order to decide whether to reject the null hypothesis.

cumulative percentage plot   A plot of ordered pairs where each value x in the distribution is plotted with the cumulative percentage of values that are less than or equal to x.

cumulative relative frequency plot   A plot of ordered pairs where each value x in the distribution is plotted with its cumulative relative frequency: the proportion of all values that are less than or equal to x.

data   A set of numbers or observations with a context and drawn from a real-life situation.

data analysis   See **statistics**.

dependent events   Two events that are not independent.

deviation   The difference from the mean, $x - \bar{x}$, or other measure of center.

disjoint events (or **mutually exclusive events**)   Events that cannot occur on the same trial. Event A and event B are disjoint if and only if $P(A \text{ and } B) = 0$.

distribution, data   The set of values that a variable takes on in a sample or population, together with how frequently each value occurs.

distribution, probability   The set of values that a random variable takes on, together with a means of determining the probability of each value (or interval of values in the case of a continuous distribution).

dot plot   A graphical display that shows the values of a variable along a number line.

double-blind   A type of experiment in which neither the subjects nor the researcher making the measurements knows which treatment the subjects received.

ecological fallacy   The mistake of using ecological (group level) correlations to support conclusions about individuals.

event   Any subset of a sample space.

expected value, μ_X, or $E(X)$ The mean of the probability distribution for the random variable X.

experimental protocol   A set of rules and steps to follow in an experiment in order to keep conditions as nearly constant as possible.

experimental units   In an experiment, the subjects to which treatments are assigned.

explanatory variable (or **predictor**)   A variable used to predict (or explain) the value of the response variable. Placed on the x-axis in a regression problem.

exploratory analysis (or **data exploration**)   An investigation to find patterns in data, using such tools as tables, statistical graphics, and summary statistics to display and summarize distributions.

extrapolation   Making a prediction when the value of the explanatory variable x falls outside the range of the observed data.

factor   An explanatory variable, usually categorical, in a randomized experiment or an observational study.

first quartile, Q_1   See **lower quartile**.

fitted value   See **predicted value**.

five-number summary   A summary that lists the minimum and maximum values, the median, and the lower and upper quartiles.

fixed-level test   A test in which the null hypothesis is rejected or not rejected based on comparison of the test statistic with the critical value for some predetermined level of significance.

frequency (or **count**)   The number of times a value occurs in the distribution.

frequency table   A table that gives the values and their frequencies.

Fundamental Principle of Counting   If there are k stages in a process, with n_i possible outcomes for stage i, then the number of possible outcomes for all k stages taken together is $n_1 n_2 n_3 \cdots n_k$.

gap   On a plot, the space that separates clusters of data.

geometric (waiting time) distribution   The random variable X has a geometric distribution if X represents the number of trials needed to get the first success in a series of independent trials, where the probability of a success is the same on each trial.

goodness-of-fit test   A chi-square test used to determine whether it is reasonable to assume that a sample came from a population in which, for each category, the actual proportion of outcomes is equal to some hypothesized proportions.

heteroscedasticity   The tendency of points on a scatterplot to fan out at one end, indicating that the relationship varies in strength.

histogram   A plot for quantitative variables that groups cases into rectangles or bars. The height of the bar shows the frequency.

homogeneous population   Two or more populations that have nearly equal proportions of members in all categories of study.

hypothesis test   See **test of significance.**

incorrect response bias   A bias resulting from responses that are systematically wrong, such as from intentional lying, inaccurate measurement devices, or faulty memories.

independent events   Events A and B are independent if the probability of event A happening doesn't depend on whether event B happens. Events A and B are independent if and only if $P(A \mid B) = P(A)$ or, equivalently, $P(B \mid A) = P(B)$ or, equivalently, $P(A \text{ and } B) = P(A)P(B)$.

inference (or inferential statistics)   Using results from a random sample to draw conclusions about a population or using results from a randomized experiment to compare treatments.

influential point   On a scatterplot, a point that strongly influences the regression equation and correlation. To judge a point's influence, fit a line and compute a correlation first with, and then without, the point in question.

interpolation   Making a prediction when the value of the explanatory variable x falls inside the range of the observed data.

interquartile range (*IQR*)   A measure of spread equal to the distance between the upper and lower quartiles: $IQR = Q_3 - Q_1$.

judgment sample   A sample selected using the judgment of an expert to choose units that he or she considers representative of a population.

Law of Large Numbers   A law that guarantees that the proportion of successes in a random sample will converge to the population proportion of successes as the sample size increases. In other words, the difference between a sample proportion and a population proportion must get smaller (except in rare instances) as the sample size gets larger.

least squares line   See **regression line.**

level   One of the values or categories making up a factor.

level of significance, α, or **alpha**   The maximum P-value for which the null hypothesis will be rejected.

line of means   Another term for the regression line, if points form an elliptical cloud. Contains the means of the points in each vertical strip (the conditional distribution of y at each value of x).

linear shape (or **linear trend**)   The characteristic of a cloud of points where the mean of each vertical strip

(the conditional distribution of y given x) falls on a single line.

lower quartile (or **first quartile, Q_1**)   In a distribution, the value that separates the lower quarter of the values from the upper three-quarters of the values. The median of the lower half of the values.

lurking variable   A variable other than the ones being plotted that can possibly cause or help explain the behavior of the pattern on a scatterplot. A variable not included in the analysis but, once identified, could explain the relationship between the other variables.

margin of error, E   Half the length of a confidence interval: $E = $ (critical value)(standard error).

matched pairs design   See **randomized paired comparison design (matched pairs)**

maximum   The largest value in a set of data.

mean, $\bar{x}$ A measure of center often called the "average" and computed by adding all the values of x and dividing by the number of values, n. On a plot, it is the place where you would put a finger below the horizontal axis in order to balance the distribution.

measure of center   A single-number summary that measures the "center" of a distribution; usually the mean (or average) is used. Median, midrange, mode, and trimmed mean are also measures of center.

median   A measure of center that is the value that divides an ordered set of data values into two equal halves. To find it, list all of the values in order, and select the middle one or the average of the two middle ones. If there are n values, the median is at position $\frac{n+1}{2}$. On a plot, it is the value that divides the area between the distribution curve and the x-axis in half.

method of least squares   A general approach to fitting functions to data that minimizes the sum of the squared residuals (or errors).

midrange   The midpoint between the minimum and maximum values in a data set; $\frac{min + max}{2}$.

minimum   The smallest value in a set of data.

mode   A measure of center that is the value in a distribution with the highest frequency. On a plot, it occurs at the highest (maximum) peak.

modified boxplot   A graphical display like the basic boxplot, except that the whiskers extend only as far as the largest and smallest non-outliers (sometimes called adjacent values) and any outliers appear as individual dots or other symbols.

Multiplication Rule For any two events A and B, $P(A \text{ and } B) = P(A) \cdot P(B \mid A) = P(B) \cdot P(A \mid B)$. If A and B are independent, then $P(A \text{ and } B) = P(A) \cdot P(B)$.

mutually exclusive events See **disjoint events.**

negative trend The tendency of a cloud of points to slope down as you go from left to right, or the tendency of y to get smaller as x gets larger.

nonresponse bias A bias that occurs when people selected for the sample do not respond to the survey.

normal distribution A useful probability distribution which has a symmetric bell or mound shape and tails extending infinitely far in both directions.

observational study A study in which the conditions of interest are already built into the units being studied and are not randomly assigned.

one-tailed (one-sided) test of significance A test in which the P-value is computed from one tail of the sampling distribution. Used when the investigator has an indication of which way any change from the standard should go.

outlier A value that stands apart from the bulk of the data.

parameter A summary number describing a population (usually unknown) or a probability distribution.

placebo A nontreatment that mimics the treatment(s) being studied in all essential ways except it does not involve the crucial component.

placebo effect The phenomenon that when people believe they are receiving special treatment, they tend to do better even if they are receiving the placebo.

plot of distribution (or graphical display or statistical graphic) A graphical display of the distribution of a variable that provides a sense of its shape, center, and spread.

point of averages The point $(\bar{x}, \bar{y})$, where $\bar{x}$ is the mean of the explanatory variable and $\bar{y}$ is the mean of the response variable. This point falls on the regression line.

population The set of people or things (units) that you want to know about.

population size The number of units in the population.

population standard deviation, σ See **standard deviation of a population.**

positive trend The tendency of a cloud of points to slope up as you go from left to right, or the tendency of y to get larger as x gets larger.

power of a test The probability of rejecting the null hypothesis.

predicted fitted value The value ($\hat{y}$) of the response variable that is being predicted from the known value of the explanatory variable x, often by using a regression line.

prediction error The difference between the actual value of y and the value of y predicted from a regression line. Usually unknown, except for the points used to construct the regression line, whose prediction errors are called residuals.

predictor See **explanatory variable.**

probability A number between 0 and 1, inclusive (or between 0% and 100%) that measures how likely it is for a chance event to happen. At one extreme, events that can't happen have probability 0. At the other extreme, events that are certain to happen have probability 1.

probability density A probability distribution, such as the normal or χ^2, where x is a continuous variable and probabilities are identified as areas under a curve.

probability model (or **chance model**) A description that approximates—or simulates—the random behavior of a real situation, often by giving a description of all possible outcomes with an assignment of probabilities.

probability sample A sample in which each unit in the population has a fixed probability of ending up in the sample.

P-value (or observed significance value) For a test, the probability of seeing a result from a random sample that is as extreme as or more extreme than the one you computed from your random sample if the null hypothesis is true.

quantitative variable (or numerical variable) A variable that results in numerical measurements.

quartiles Three numbers that divide an ordered set of data values into four groups of equal size.

questionnaire bias Bias that arises from *how* the interviewer asks and words the survey questions.

random sample A sample in which individuals are selected by using some chance process. Sometimes used synonymously with **simple random sample.**

random variable A variable that takes on numerical values governed by a chance process.

randomization (or **random assignment**) Assigning subjects to different treatment groups using a random procedure.

randomized block design   An experimental design in which treatments are randomly assigned to units within blocks.

randomized comparative experiment   An experiment in which two or more treatments are randomly assigned to experimental units for the purpose of making comparisons among treatments.

randomized paired comparison design (matched pairs)   An experimental design in which two different treatments are randomly assigned within pairs of similar units.

randomized paired comparison design (repeated measures)   An experimental design in which each treatment is assigned (in random order) to each unit.

range   A measure of spread equal to the difference between the maximum and minimum values in a set of data.

rare events   Values that lie in the outer 5% of a distribution or the outcomes in the upper 2.5% and the lower 2.5% of the distribution of all possible outcomes. Compare **reasonably likely.**

reasonably likely   Values or outcomes in the middle 95% of a distribution or the distribution of all possible outcomes. Compare **rare events.**

recentering   Adding the same number c to all the values in a distribution. This procedure doesn't change the shape or spread but slides the entire distribution by the amount c, adding c to measures of center.

regression   The statistical study of the relationship between two (or more) quantitative variables, such as fitting a line to bivariate data.

regression effect (or regression toward the mean)   On a scatterplot, the difference between the regression line and the major axis of the elliptical cloud.

regression line (or least squares line or least squares regression line)   The line for which the sum of squared errors (residuals), or SSE, is as small as possible.

regression toward the mean   See **regression effect.**

relative frequency histogram   A histogram in which the length of each bar shows proportions (or relative frequencies) instead of frequencies.

repeated measures   See **randomized paired comparison design (repeated measures)**

replication   Repetition of the same treatment on different units.

rescaling   Multiplying all the values in a distribution by the same nonzero number d. This process doesn't change the basic shape, but stretches or shrinks the distribution, multiplying the *IQR* and standard deviation by $|d|$ and multiplying the measures of center by d.

residual (or error)   For points used to construct the regression line, the difference between the observed value of y and the predicted value of y, that is, $y - \hat{y}$.

residual plot   A scatterplot of residuals, $y - \hat{y}$, versus predictor values, x, or versus predicted values, $\hat{y}$. A diagnostic plot used to uncover nonlinear trends in a relationship between two variables.

resistant to outliers   The characteristic of a summary statistic that it does not change very much when an outlier is removed from the set of data.

response variable   The outcome variable used to compare conditions during experiments or the outcome variable that is predicted by the explanatory variable or variables. Placed on the y-axis in a regression problem.

robustness   The comparative insensitivity of a statistical procedure from the assumptions on which the procedure is based.

sample   The set of units (from the population) selected for study.

sample selection bias (or sampling bias or **bias due to sampling)**   The extent to which a sampling procedure produces samples that tend to result in numerical summaries that are systematically too high or too low.

sample space   A complete list or description of disjoint (mutually exclusive) outcomes of chance process.

sampling bias   See **sample selection bias.**

sampling distribution   The distribution of a sample statistic.

sampling distribution of a sample proportion, $\hat{p}$
The theoretical distribution of the sample proportion in random sampling.

sampling distribution of the sample mean, $\bar{x}$ The theoretical distribution of the sample mean in random sampling.

sampling frame (or frame)   The units from which the sample is selected.

sampling with replacement   In sequentially sampling units from a population, a procedure in which each sampled unit is placed back into the population before the next is selected.

sampling without replacement   In sequentially sampling units from a population, a procedure in which each sampled unit is not placed back into the population before the next is selected.

scatterplot   A plot that shows the relationship between two quantitative variables, usually with each case represented by a dot.

segmented bar graph (or **stacked bar graph**)   A plot in which categorical frequencies are stacked on top of each other.

sensitive to outliers   The characteristic of a summary statistic that it changes considerably when an outlier is removed from the set of data.

shape   One of the characteristics, along with center and spread, that is used to describe distributions. Univariate distributions sometimes have a standard shape such as normal, uniform, or skewed. Bivariate distributions may form an elliptical cloud. When describing shape include outliers, clusters, and gaps.

simple random sample, SRS   A type of probability sample where all possible samples of a given fixed size are equally likely.

simulation   A procedure that uses a chance model to imitate a real situation. Often used to compare an actual result with the results that are reasonable to expect from random behavior.

size bias   A type of sample selection bias that gives units with a larger value of the variable a higher chance of being selected.

skewed   Distributions that show bunching at one end and a long tail stretching out in the other direction. Often happens because the values "bump up against a wall" and hit either a minimum that values can't go below or a maximum that values can't go above.

skewed left   A skewed distribution with a tail that stretches left toward the smaller values.

skewed right   A skewed distribution with a tail that stretches right toward the larger values.

slope   For linear relationships, the change in y (rise) per unit change in x (run).

split stem   A stem-and-leaf plot in which the leaves for each stem are split onto two or more lines. For example, if the second digit is 0, 1, 2, 3, or 4, it is placed on the first line for that stem. If the second digit is 5, 6, 7, 8, or 9, it is placed on the second line for that stem.

spread   See **variability**.

stacked bar graph   See **segmented bar graph**.

standard deviation, *SD*, or σ of a population   A measure of spread equal to the square root of the sum of the squared deviations divided by n.

standard deviation, *SD*, or *s* of a sample   A measure of spread equal to the square root of the sum of the squared deviations divided by $n - 1$, that is, $s = \sqrt{\sum(x - \bar{x})^2/(n - 1)}$.

standard error   The standard deviation of a sampling distribution.

standard error of the mean, $\sigma_{\bar{x}}$ The standard deviation of the sampling distribution of $\bar{x}$, which equals the standard deviation σ of the population divided by the square root of the sample size n, that is, $\sigma/\sqrt{n}$.

standard error of the mean (estimated)   The estimated standard deviation of the sampling distribution of $\bar{x}$, or $s/\sqrt{n}$.

standard normal distribution   A normal distribution with mean 0 and standard deviation 1. The variable along the horizontal axis is called a z-score.

standard units, *z*   The number of standard deviations that a given value lies above or below the mean:

$$z = \frac{value - mean}{standard\ deviation}$$

standardizing   Converting to standard units; the two-step process of recentering and rescaling that turns any normal distribution into a standard normal.

statistic   A summary number describing a sample taken from a population.

statistically significant   When the difference between the estimate and the hypothesized parameter is too big to reasonably attribute to chance variation.

statistics (or **data analysis**)   The study of the production, summarization, and analysis of data.

stem-and-leaf plot (or **stemplot**)   A graphical display with "stems" showing the leftmost digit of the values separated from "leaves" showing the next digit or set of digits.

stratification   A classification of the units in the population into homogeneous subgroups, known as strata, prior to the sampling.

stratified random sampling   Stratifying the population and then taking a simple random sample from within each strata.

strength Two variables have a strong relationship if there is little variation within each vertical strip (conditional distribution of y given x). If there is a lot of variation, the relationship is weak.

sum of squared errors (or **SSE**) The sum of the squared residuals: $\sum(y - \hat{y})^2$.

summary statistic A single number used to "condense" data, often to measure center or spread of a distribution.

systematic sampling with random start A sample selected by taking every nth member of the population, starting at a random spot. For example, by having people count off and then pick one of the numbers at random.

t-distribution The distribution, for example, of the statistic

$$t = \frac{\bar{x} - \mu}{s/\sqrt{n}}$$

The shape becomes more normal as n increases.

test of significance (or **hypothesis test**) A procedure that compares the results from a sample to some predetermined standard in order to decide whether the standard should be rejected.

test statistic Typically, in significance testing, the distance between the estimate and the hypothesized parameter, measured in standard errors.

third quartile, Q_3 See **upper quartile.**

treatment Condition(s) assigned during experiments to determine whether subjects respond differently to them. Compare with **control group.**

treatment group In an experiment, a group that receives an actual treatment being studied.

tree diagram A diagram used to calculate probabilities for sequential events.

trend On a scatterplot, the path of the means of the vertical strips (conditional distributions of y given x) as you move from left to right.

true regression line The regression line that would be computed if you had the entire population.

two-stage sampling A sampling procedure that involves two steps. For example, taking a random sample of clusters and then taking a random sample from each of those clusters.

two-tailed (two-sided) test of significance A test in which the P-value is computed from both tails of the sampling distribution. Used if the investigator is interested in detecting a change from the standard in either direction.

Type I error The error made when the null hypothesis is true and you reject it.

Type II error The error made when the null hypothesis is false and you fail to reject it.

uniform distribution A distribution whose frequency changes little across the possible values. Its plot is rectangular-shaped.

unimodal A distribution with only one peak.

units (or **individuals**) Individual elements of the population, from which cases are selected.

univariate data Data that involve a single variable per case. For quantitative variables, often displayed on a histogram. For categorical variables, often displayed on a bar chart.

upper quartile (or **third quartile, Q_3**) In a distribution, the value that separates the lower three-quarters of the values from the upper quarter of the values. The median of the upper half of the values.

variability (or **spread**) The degree to which values in a distribution differ. Measures of variability for quantitative variables include the standard deviation, interquartile range, and range.

variability due to sampling (or variation in sampling) A description of how the estimate varies from sample to sample.

variable A characteristic that differs from case to case and defines what is to be measured or classified.

variance A measure of spread equal to the square of the standard deviation.

voluntary response sample A sample made up of people who volunteer to be in it.

z-score See **standard normal distribution** and **standard units.**

Brief Answers to Selected Problems

Chapter 1

Section 1.1

P1. Older workers were far more likely to be laid off.

P2. a. Using 50 to define "older" shows stronger evidence of possible age bias.

Laid Off?

Under 40?	Yes	No	Total	% Yes
Yes	3	2	5	60.0
No	7	2	9	77.8
Total	10	4	14	71.4

Laid Off?

Under 50?	Yes	No	Total	% Yes
Yes	3	3	6	50.0
No	7	1	8	87.5
Total	10	4	14	71.4

b. Older workers were more likely to be laid off; the difference is more pronounced using age 50.

E1. Hourly workers were more likely to lose their jobs.

E3. a.

Round	Laid Off	40 or Older	Percent
1	11	9	82%
2	9	8	89%
3	3	2	67%
4	4	2	50%
5	1	0	0%

b. Most layoffs came early, and early rounds hit older workers harder.

E5. • Cases are the teams. Variables include games won, games lost, winning percentage, and games behind the leader.

• Cases are the companies listed. Variables include the opening price for the day, the low, the high, the close, and the net change from the previous day's close.

• Cases are the nutritional components—like fat, carbohydrates, protein, cholesterol, and various vitamins and minerals. Variables include amount per serving and percentage of the recommended daily allowance.

Section 1.2

P3. a. .155

b. There is no evidence of age discrimination because an average age of 52.67 or larger is relatively easy to get just by chance.

P4. a. 34

b. Repeatedly draw two at random from 25, 33, 35, 38, 48, 56 and compute the average age each time. Find the proportion of times you get an average age of 34 or larger.

c. about 80%; no

E7. b. If selections are done randomly, the probability of getting an average age this large is only .019. This is strong evidence in Martin's favor.

E9. the second choice

Review Exercises

E11. a. B looks better, but there is a lot of variability and this is a small sample.

c. 0.746; 1.002

E13. c. The actual difference of 0.256 will be exceeded a high proportion of the time. A proportion over about .05 suggests that the actual difference could be the result of chance alone.

E15. Evidence of age bias is much weaker as there are 10 ways out of 120 to get an average age as large as or larger than 55.

Chapter 2

Section 2.1

P1. Number of deaths is fairly constant at 190,000 per month. The exception is January, which shows the highest number of deaths.

P2. a. 1.0
b. 0.5, 1.0, and 1.5
c. 0.5 and 1.5
d. 0.15
e. 0.05 and 1.95

P4. a. 500, give or take 100
b. 20, give or take 5
c. 65, give or take 2.5
d. .260 or .270, give or take .040

P5. a. IV   b. II   c. V
d. III   e. I

P6. Plot should be skewed right.

P7. The lower quartile is about 2.9, the median about 3.35, and the upper quartile about 3.7. The middle 50% of students had GPAs between 2.9 and 3.7, with half above 3.35 and half below.

E1. a. distribution is strongly skewed left
   b. strongly skewed right
   c. distribution is approximately normal
   d. skewed right

E5. a. case: one officer who attained the rank of colonel; only one variable: the age at which the officer became a colonel
   b. skewed left with no outliers, gaps, or clusters The median is 52, and quartiles are 50 and 53. The middle half of the ages are between 50 and 53, with half above 52 and half below.
   c. mandatory retirement age requirement or age discrimination against people younger than 47 and older than 54

E7. Distribution is approximately normal, except that it has too many outliers and is a bit too "peaked" to be traced by a normal curve.

E9. The up and down nature of the plot is due to the fact that some months have more days than others. The number of deaths seems to be generally declining over the months until October, when it goes back up. There are more deaths in the colder months.

E11. a. ABC: symmetrical and mound-shaped. CBS: slightly skewed right. FOX: strongly skewed right. NBC: rectangular except for three outliers. UPN: very few shows (and very few viewers), and all but one are stacked on one point. WB: slightly skewed right. There are no outliers with the exception of NBC. There are no clusters or gaps.
   b. Median: FOX is 7 or 8; NBC is 12 or 13. Quartiles: FOX is 6 and 14; NBC is 9 and 16. The middle half of the ratings have about the same spread for the two networks. The ratings are centered higher for NBC and lowest for UPN.
   c. NBC has the most variability, and UPN has the least.

Section 2.2

P8. quantitative variables: year of birth, year of hire, RIF stage, and age; categorical variables: row number, job title, and pay category (hourly or salaried)

P9. Skewed right with no obvious gaps or clusters. There is a wall at 0 days because no mammal can have a smaller gestation period. The elephant is the only outlier. Large mammals have longer gestation periods.

P10. Average longevity is skewed right with two possible outliers at 35–40 and 40–45. The distribution of maximum longevity is more uniform but with a peak at 20–30 years and an outlier at 70–80 years. The center of the distribution of maximum longevity is much higher, and its spread is much larger.

P11. The shapes do not change, only the scale on the *y*-axis.

P12. The mean will be smaller than the median.

P14. There is an outlier on the high side for predators and two outliers on the low side for the nonpredators. The median of the predator distribution is larger at 40.5. The spreads are the same. There are no slow predators.

P15. cases: individual male members of the labor force; variable: educational attainment The distributions for males and females have much the same shape. The female labor force is overall a bit better educated.

E13. a. ages of a set of pennies collected by a statistics class; case: a penny; variable: age of the penny
   b. strongly skewed right; median: 8 years; spread: large with the middle half of the ages of pennies falling between 3 and 15 years
   c. If the same number of pennies is produced each year, and each year a certain percentage of them goes out of circulation, you get a shape like this.

E15. These are bar graphs.

E19. a. mean: 514; standard deviation: 113
   b. about 67%; about 95%; 100%

E21. U.S.: There is a population bulge around the ages of 30 to 50, with decreasing percentages in the age groups above 50.
   Mexico: Largest segments of the population are young children, with a regularly decreasing pattern in percentage of the population as the age increases.

Section 2.3

P17. a. mean: 2.5; median: 2.5
   b. mean: 3; median: 3
   c. mean: 3.5; median: 3.5

d. mean: 49.5; median: 49.5

e. mean: 50; median: 50

P18. The mean height will increase by about four inches. The median should not change much because it will still be one of the 3rd graders, who all are about 4 feet tall.

P19. a. 51; 73

b. Africa: The mean is larger than the median because of the skewness toward the larger values.

Europe: The mean is about the same as the median; there is no obvious skewness.

P20. a. quartiles: 2 and 5; *IQR:* 3

b. quartiles: 2 and 6; *IQR:* 4

c. quartiles: 2.5 and 6.5; *IQR:* 4

d. quartiles: 2.5 and 7.5; *IQR:* 5

P21. a. predators: 7, 12, 15; nonpredators: 8, 12, 15

b. predators: mound-shaped, centered at about 12 years, with 50% of the values falling between 7 and 15 years; nonpredators: centered at exactly the same place and has about the same spread but has two outliers on the high side

P22. a. *Seinfeld, Seinfeld Clips,* and *ER*

b. *Touched by an Angel*

P24. a. minimum: 1

lower quartile (Q_1): 8

median: 12

upper quartile (Q_3): 15

maximum: 41

b. $IQR = 15 - 8 = 7$

c. $Q_1 - 1.5 \cdot IQR = 8 - 1.5(7) = -2.5$
There are no outliers on the lower end.

d. $Q_3 + 1.5 \cdot IQR = 25.5$; The life spans of 35 years for the elephant and 41 years for the hippopotamus are outliers. The smallest value that isn't an outlier is 25.

P26. yes, when the minimum and first quartile are equal.

P27. quartiles: 425 and 590

median: 505

IQR: $590 - 425 = 165$

P28. standard deviation: 3.21

P29. a. 0   b. 0.577

c. 1.581   d. 5.774

e. 0.058   f. 3.162

g. 3.606

P30. The mean is larger than the median because house values tend to be skewed right. Total amount of taxes collected will be ≈ $43,964,124. This is an average of $4,508.68 per house.

P31. a. The distribution of car ages is strongly skewed right.

P32. a. mean: 4 feet

median: 3.75 feet

standard deviation: 0.2 feet

interquartile range: 0.25 feet

b. mean: 50 inches

median: 47 inches

The standard deviation and interquartile range do not change.

c. mean: 4.33 feet

median: 4.083 feet

standard deviation: 0.2 feet

interquartile range: 0.25 feet

P33. a. mean: 2; *SD:* 1

b. mean: 12; *SD:* 1

c. mean: 20; *SD:* 10

d. mean: 110; *SD:* 5

e. mean: −900; *SD:* 100

P34. a. Outliers occur above $-30 + 1.5(21) = 1.5$.
Hawaii, at 12, is an outlier.

b.

Count	49
Mean	-41.5
Median	-40
Std.Dev	16.2
Min	-80
Max	-2
Range	78
Lower ith %tile	-51
Upper ith %tile	-32

P35. a. For 1997, the mean is 2.3, standard deviation is 1.84.

b. 2

c. position of quartiles: 25.5 and 75.5
For 1967, the quartiles are 2 and 4. For 1997, the quartiles are 1 and 3. In both cases, the *IQR* is 2.

d. Both the median and mean number went down by about 1 child per family. The distributions kept the same shape and spread.

P36. a. about 10

b. $\bar{x} = \dfrac{1 \cdot 5 + 5 \cdot 6 + 10 \cdot 4 + 25 \cdot 5}{20} = \dfrac{200}{20} = 10$

c. closest to 10

d. 9.4

E23. In each part, it depends on the reason for using a summary statistic. For example, in part b, the mean will help estimate the total for the state, while the median would tell an individual farmer if his or her yield was fairly typical.

E25. Back-to-back stemplot is better because there are only a few values.

E27. 10

E29. about 30

E31. a. the mean length of a generation
b. the average speed
c. yes, if you know the number of trees

E33. b.
Variable	N	Mean	Median	TrMean	StDev
Highest	50	45.61	45.56	45.53	3.72

Variable	Min	Max	Q1	Q3
Highest	37.78	56.67	43.33	47.78

c. Yes. There is an outlier on the high side.

E35. a. 3.11 grams
b. 0.04 gram
c. yes

E37. Workers laid off: median age of 53.5; quartiles 42 and 61
Workers who were kept had a median age of 48 with quartiles 37 and 55.

E39. two; three; the spread for domestic mammals is smaller than for all the mammals

E41. Subtract 5478 and then multiply by 10 before computing.

Section 2.4

P37. a. Unknown percentage problem. You need to find the percentage of scores over 1300.
b. Unknown value problem. You need to find the value that cuts off the bottom 75% of the distribution.

P38. a. −0.47
b. −0.23
c. 1.13
d. 1.555

P39. a. 0.0129
b. 0.0475
c. 0.3446
d. 0.7881

P40. a. $.9279 − .0721 = .8558$, or 85.58%
b. $.9987 − .0013 = .9974$, or 99.74%

P41. a. −1.645 to 1.645
b. −1.96 to 1.96

P42. the death rate for cancer with $z = −1.10$

P43. a. the death rate for cancer with $z = −1.84$
b. the death rate for heart disease in Colorado with $z = −1.94$

P44. a. 2   b. 1   c. 1.5
d. 3   e. −1   f. −2.5

P45. a. 30   b. 22
c. 85   d. −9.5

P46. a. about 24%
b. 63.84 inches

P47. $1100 \pm (1.96)180$ or roughly 747 and 1453

P48. a. z-score: −0.91, not outside either interval
b. z-score: −1.10, not outside either interval
c. z-score: −3.69, outside both intervals
d. z-score: −3.74, outside both intervals

Review Exercises

E47. a. i. .6319   ii. .0395   iii. .3101
b. 287 to 723

E49. 76% of men and 87% of women

E51. You can't use normal distribution because the distribution of ages of cars is not approximately normal.

E53. a. about 145 points
b. about 25 points
c. between about 90 and about 200
d. .4207, .0359

E55. $Q_1 = 6.65$; $Q_3 = 13.35$
mean = 150; $SD \approx 44.78$
mean = 106.7; $Q_3 = 113.4$
$SD \approx 1.5$; $Q_1 = 9$

E57. b. minimum: 0
Q_1: 2
median: 10
Q_3: 35
maximum: 232
c. Outliers fall below −47.5 or above 84.5. There are two outliers, Florida and Texas.
e. The stemplot is most informative.
f. Strongly skewed right with two outliers, Florida at 95 and Texas at 232. Median is 10 with the middle half of the states having between 2 and 35.

E59. a. 23 or 24 cents

b. 10, 65, and about 55 cents

c. skewed right

d. no

E61. a. 1.16; 90th percentile

b. 1.33; 80th percentile

E63. stems: first three digits of the year

leaves: final digit of the year

distribution: Highs have a spike in 1936. Lows are relatively uniform over the years since 1890. The center and spread of the two distributions are about the same.

E65. a. Region 1: Africa

Region 2: South and Central America

Region 3: Asia

Region 4: Europe

b. Distributions 1, 2, and 4 are skewed left.

c. The dot plots are in the same order as the boxplots: A is Africa; B is South and Central America; C is Asia; and D is Europe.

d. The outlier shown in the boxplot in Region 1 doesn't appear to be an outlier in the dot plot. Region 4 does not look skewed in the dot plot even though it appears so in the boxplot. The number of countries plotted is small, and the values vary a lot, so the locations of the quartiles might change quite a bit with small changes in the data. Dot plots give the better picture here.

E67. a. median: 47

b. Outliers are above 178.5: Chicago, Los Angeles, and New York. These are the three largest cities.

c. A stemplot is a very good choice. It reveals the three outliers and shows that the distribution is skewed right.

d. Note what happened in part b. If the data were presented in rates per 100,000 population, these three cities might not be outliers. In fact, their rates of pedestrian death might be relatively small.

E69. a. yes

b. .270 and .030

c. .200 to .320

E73. An example is {1, 1, 1, 1, 1, 2, 2, 10}, which has mean 2.375 and standard deviation of about 3.11.

E75. a. 39.29 and 30.425

b. 10.54

Chapter 3

Section 3.1

P1. b. shape linear; trend positive; strength very strong

c. yes; no

d. An increase in age is associated with growth in variables such as height.

P2. a. worst for baggage: America West; worst for on-time: United

b. upper left; best for baggage: Delta; best for on-time: Northwest

c. false

d. negative and moderate

e. no; no; probably

E1. a. positive, moderately strong, linear, varying strength

b. negative, moderately strong, linear, varying strength

c. positive, moderate, linear, one possible outlier

d. negative, moderate, linear, one outlier

e. positive, strong, linear, one outlier

f. negative, strong, curved, one or two outliers

g. negative, strong, curved, one outlier

h. positive, strong, curved, one outlier

E3. a. ii   b. iv

c. iii   d. i

E5. a. cases: employees at time of layoffs

variables: age of hire, year of hire

shape: triangular

trend: positive

strength: weak

b. no; as older people hired in early years may have retired

c. Everyone hired before the early 1960s was laid off, but not all of the older people were laid off. The people with more seniority may earn more or have obsolete skills.

E7. c. A300-600 (Airbus) is unusually slow for both cargo carried and number of seats.

E9. a. i. First 3 plots are linear; 4th isn't.

ii. First 3 have one cluster; last has two.

iii. First and third have outliers.

b. A has a positive linear trend; B has a negative linear trend; C has a negative linear trend; D has little or no trend.

c. Alumni giving is best; SAT scores are good; ranking in high school is almost useless.

d. Probably not, given that these are the top 50.

Section 3.2

P3. 1.9; approximate price per gallon of fuel

P4. radiologist's pay goes up about 2.75 times faster than a family doctor's; $y = -58 + 2.7x$

P5. a. response: giving percentage
explanatory: student/faculty ratio

b. A run of 4 in the direction of the x-axis corresponds to a drop of 8 percentage points in the direction of the y-axis.

c. A university with no students has a giving rate of 55%, which makes no sense.

d. 23%; expect a large error

e. 1; 16

f. −11

g. positive

P6. a. fewest calories: Pizza Hut Hand Tossed, Little Caesars Pan! Pan!; least fat: Pizza Hut Hand Tossed, Domino's Hand Tossed
Pizzas with the most fat occur to the right.

b. I. C   II. E   III. A
IV. B   V. D

c. i. A   ii. E   iii. B
iv. D   v. C

P7. b. $y = 279.75 + 2.75x$

c. slope: each additional gram of fat adds about 2.75 calories
y-intercept: 5 ounces of pizza with no fat has 279.75 calories

d. The point of averages is (11, 310), which satisfies $279.75 + 2.75(11) = 310$.

e. $0.5 + (-1.0) + (0.5) = 0$

P8. *% on-time* $= 97.0 - 5.08$ *mishandled baggage*
Residuals are 6.6099, −2.9260, 7.7093, −0.2896, 0.6501, −1.1108, 1.1770, −5.6494, 3.8536, −10.0241.

P9. yes; SSE = 1.5 is Sum of Squares for Error

E11. a. 320, 360 calories

b. roughly 8.9; roughly $\hat{y} = 226.5 + 8.9x$

c. *calories* $= 227 + 8.97$ *fat*

d. *calories* $= 227 + 8.97$ *fat*

e. Slope is close to 9.

E13. No, there is no linear trend in either.

E15. a. points all lie on a line

b. 0

c. 1.1. Each extra 10 mph requires about 11 feet more to react.

d. $\hat{y} = 1.1x$

e. 60.5 ft, 82.5 ft

f. $\hat{y} = 1.47x$

E17. a. true

b. second graph linear, first graph is curved, opening up

E19. b.

Midpoint	Median *cost*
500	1878.5
1500	3195
2500	5081
3500	7075

c. *cost* $= 812 + 1.75$ *fuel*

d. yes

E21. a. linear relationship; $\hat{y} = 253.1 + 6.48x$; for every additional gram of fat, pizzas tend to have 6.48 more calories

b. weak linear relationship; $\hat{y} = 15.68 - 3.27x$; pizza costing \$1 more tends to have 3.27 g less fat

c. no association

Section 3.3

P10. a. −.5   b. .5   c. .95

d. 0   e. −.95

P11. $r = .885$

P12. $r \approx .884$; all positive

P13. a. positive

b. top right point, ($\approx 4.7, \approx 4.9$)

c. Quadrants I and III; 20 points (3 pts have 0)

d. Quadrants II and IV; 7 points

P14. a. $r = .650$

b. *Exam 2* $= 48.94 + 0.368$ *Exam 1*; predicted score is 78.38

c. *Exam 1* $= -14.10 + 1.149$ *Exam 2*

P15. a. size of city's population

b. divide each number by population

P16. age of child is lurking variable

P17. A lurking variable is affluence.

P18. a. $r^2 = .5534$
$r = -.744$

b. Slope of −0.61 tells us that if one state has a graduation rate 1 percentage point higher than another, we expect its poverty rate to be lower by 0.61 percentage points.

c. no

d. x is percentage of people; y is percentage of families; b_1 is percentage of families per percentage of people; r has no units.

P19. flatter; shows regression effect well

E23. a. .66 b. .25 c. −.06
 d. .40 e. .85 f. .52
 g. .90 h. .74

E25. a. $r = .707$
 b. $r = .707$

E27. No.

E29. a. no
 b. yes

E31. a. true
 b. $s_x = 25$ and $s_y = 50$
 c. $r = .081$
 d. $b_1 = 0.0183$

E33. a. false conclusion: large brain means smarter at keeping alive; lurking variable: size
 b. false conclusion: cheeseburger price affects college fees; lurking variable: inflation
 c. false conclusion: Internet helps business; lurking variable: time

E37. a. $r^2 = .8163$; $r = .903$
 b. −0.0523
 c. yes, but r alone does not tell us that
 d. degrees Fahrenheit; pints per person; pints per person per degree Fahrenheit; no units
 e. $MS = SS/DF$. MS stands for "Mean Square."

E39. a. graduation from high school; poverty rate
 b.

	Metro	White	HS Grads	Poverty
White	−.337			
HS Grads	−.004	.338		
Poverty	−.061	−.389	−.744	
S Parent	.260	−.657	−.220	.549

Section 3.4

P21. a. Data are flat except for the *Titanic* point.
 b. $\hat{y} = -409.80 + 2.28x$; $r = .758$
 c. $\hat{y} = 359.7 + 0.03x$; $r = 0$; has large influence

P22. a. A-IV; B-II; C-I; D-III
 b. increasing slope: residual plot curves upward
 decreasing slope: residual plot curves downward
 unequal variation: residual plot fans out or in
 two linear: V-shape in residual plot
 c. Plot D; residual plot III

P23. b. predicted: 0.5, 0.5, 1, 2
 residual: −0.5, 0.5, 0, 0
 d. The residual plot straightens out the tilt in the scatterplot so that the residuals can be seen as deviations above and below zero.

E41. a. $r = .79$, $\hat{y} = -3.27 + 1.07x$: complete set
 $r = .92$, $\hat{y} = -12.51 + 1.24x$: without radiology, slope increases slightly and correlation increases significantly
 b. $r = -.09$, $\hat{y} = 68.6 - 0.157x$: slope becomes negative, correlation drops significantly
 c. Not much change, except for decrease in radiology.

E43. a.

Airplane	Seats	Cost/hr	Fitted	Residual
DC9-20	100	1750	1737.5	12.5
DC9-40	110	1800	1825.0	−25.0
DC9-50	120	1925	1912.5	12.5

 b. the scale on x-axis

E45. a. stays roughly constant
 b. spread increases
 c. yes; no
 d. spread increases with larger x
 e. Larger numbers tend to have larger variation.

E47. no

E51. b. largest positive: Domino's Hand Tossed
 largest negative: Pizza Hut's Pan
 None are exceptional.
 c. Removing Domino's Deep Dish lowers the correlation and slope considerably.

Section 3.5

P24. a. ii. 3
 b. ii. −1
 c. ii. 0.5

P25. 1/3; −1; 2; they are reciprocals

P26. a. $y = \pi x^2$; 0.5
 b. $y = x^3$; 1/3
 c. $y = 2\pi x^2$; 0.5

P27. Diameter squared versus age shows more of a fan shape. Use the other plot for more of an oval.

P28. a. Exponent is less than 1.
 b. nearly linear
 c. $\log(brain) = 0.76 \log(body) + 0.91$, so $y = 8.09(x)^{0.76}$; yes

P29. $\log y = 1 + x$

P30. a. $\log y = .5x$
   b. $\log y = 2x - 14$

P31. for P29, $y = 10(10)^x$
   for P30a, $y = 10^{(0.5x)}$
   for P30b, $y = 10^{-14} \cdot 10^{2x}$

P32. *flight length* $= 1.96(101)^{speed}$

P33. a. log transformation
   b. $\ln(pop) = -55.525 + 0.361$ *year*;
      rate of growth $= 3.7\%$

P34. c. $\ln(pop) = 5.22 - 0.435$ *roll number*;
      dying rate $= 0.35$
   d. The residual plot shows some curvature,
      indicating a death rate of a little more than
      0.35 in the early stages and a little less than
      0.35 in the later stages.

E53. no as it's still curved

E55. Split data into two groups, ages 1–6 and 7–14.

E57. Airplanes separate into two groups on all
   variables relating to size.

E59. a. decreasing; nonlinear
   b. $\ln(birth\ rate) = 3.32 - 0.25 \ln(GNP)$

E61. a. exponential decay; rate of decreasing of
      smoking is about 2.4% per year.
   b. exponential decay; spread is larger in later
      years
   c. linear, so no

E63. decreasing with curvature; log-log transformation straightens.

Review Exercises

E65. a. *arm span* $= -5.81 + 1.03$ *height*
      kneeling $= 2.19 + 0.73$ *height*
      hand $= -2.97 + 0.12$ *height*
      Leonardo's rules seem good.
   b. For every 1 cm increase in height, arm span
      tends to increase by 1.03 cm, kneeling height
      by 0.73 cm, and hand length by 0.12 cm.
   c. *arm span:* $r = .992$ (strongest)
      kneeling: $r = .989$
      hand: $r = .961$ (weakest)

E67. a. $r = .308$
   b. *Exam 2* $= 45.83 + 0.51$ *Exam 1*

E69. a. zero correlation: A-B, B-E, B-F, C-D, E-F
      .02 correlation: D-E
   b. A-E, D-F, A-C
   c. A-D, A-F, B-C, C-F

   d. A-B has a pattern even though $r = 0$; A-D and
      A-F have same correlation but different
      shapes.

E71. a. about $-.86$
   b. Lose location and knowledge that there are
      2 clusters.
   c. V-shaped

E73. a. I. $+, +, +$ II. $-, +, -$
        III. $-, -, +$ IV. $+, -, -$
   b. points near the lines $y = \bar{y}$ and $x = \bar{x}$

E75. a. no
   b. Cause-and-effect is different from correlation.
      Both types of store pay about the same
      wholesale price for a book.

E77. a. $r = .761$
   b. $r = .577$; maximum longevity
   c. yes positively if you remove the 3 outliers
   d. outliers, no

E79. a. linear model: *price* $= -25.2 + 75.6$ *area* and
      $r = .899$
      new: *price* $= -48.4 + 96.0$ *area*
      old: *price* $= -16.6 + 66.6$ *area*
      These are very different, so you should not use
      the same equation for both.
   b. The two largest houses clearly are influential
      points.
      without points: $r = .867$ and *price* $=$
      $-10.3 + 65.7$ *area*
      This is quite a change in the model, the price
      increasing $10,000 less per increase of 1,000
      square feet.
   c. $50,000; $116,600. The first prediction,
      because the spread is less for smaller houses
      than for larger ones.
   d. The number of bathrooms is strongly related
      to the selling price. A lurking variable is
      square footage, which is very strongly related
      to both price and the number of bathrooms.
      Not appropriate because of skewness.

E81. a. no relationship: no clear best buy
   b. The bottom-left point is influential.
      with point: *price* $= 220 - 0.81$ *rating*
      without point: *price* $= 424 - 3.66$ *rating*
      They are dramatically different, the first one
      reflecting a weak relationship and the second
      reflecting a strong negative relationship.

E83. A transformation of $\frac{1}{price}$ does a fairly good job.
$\frac{1}{price} = 0.00541 + 0.000039$ *days*, or

$price = \frac{1}{(0.00541 + 0.000039 \; days)}$

That is, the number of days affects the price very little according to this model. No transformation will give a good model because of the two clusters of outliers in opposite corners of the plot. Thus, the best model is to say there is no relationship between the day these passengers bought their tickets and the price they paid, with the exception of five passengers who bought their tickets within nine days of the flight and who paid more than double any other passenger.

Chapter 4

Section 4.1

P1. a. size bias
b. judgment sample
c. voluntary response bias
d. size bias
e. voluntary response bias

P2. too high

P3. Question I

P4. bias from incorrect response

E1. a. too high
b. At suppertime, adults are more likely to be at home.

E3. sampling bias; too high

E5. too high; sampling bias with resulting estimate too high

E7. Probably impossible to do.

E9. Overestimate due to convenience sampling

Section 4.2

P5. a. no   b. no
c. no   d. yes
e. no   f. no

P6. For 5%, choose a random start from 1 to 20 and take every 20th person. For 20%, choose a random start between 1 and 5 and take every 5th person.

P7. Choose a random start between 1 and 17 and take every 17th member.

P8. Should be based on sampling from something like DMV records and not on observing cars.

P9. a. This is closest to three-stage cluster sampling.
b. Use more lots and containers.

P10. 0.82

E13. Frame misses undiagnosed heart disease, heart disease diagnosed more than five years ago, and heart disease cases that aren't hospitalized. The estimate may be too large due to smokers' worse health. Sampling bias. Overestimate due to convenience sampling.

E15. An SRS with male and female strata is probably best.

E17. Choose farms as the five strata and take a random sample of acres from each strata, adjusted for size of farm.

E19. Age and gender stratification may be appropriate.

Section 4.3

P11. a. population: students taking the SAT
response: SAT score
treatments: taking course or not
b. no, no random assignment
c. no; selection bias

P12. a. Lurking variable is age.
b. I causes II.
c. II causes I.

P13.

		Motivation	
		High	**Low**
Course?	**Yes**	Higher SAT	No evidence
	No	No evidence	Lower SAT

P14. a. No. Pipe and cigar smokers are older.
b. observational study
c. factor: smoking behavior
levels: nonsmoking, cigarette smoking and pipe or cigar smoking
response: deaths per 1000 men per year

P15. Pipe and cigar smokers are older and so more likely to die. The new factor is age.

P16. a. observational
b. factor: legal drinking age
levels: age groups
response: highway death rate
c. possibly population density of a state

P17. factors: popcorn brand and popper type
levels: brand/generic and air/oil
treatments: the four brand-type combinations tested in random order
confounding variable: many possible answers

P18. Control group was adults who died of nonrespiratory causes.
Placebo effect could have been a factor.
Impossible to be blind or double-blind.

P19. a. No. People could have felt how much turning had been done.
b. group with magnets
c. if the assignments were random
d. It is better than before.
e. not blind or double-blind

P20. treatments: the two textbooks
units: the 10 classes
sample size: 10

P21. experimental units: plants
Randomly assign the new product to about half of a number of plants and leave the others growing under standard conditions.
control: plants receiving the standard

E21. observational

E23. a. dormitories
b. 20
c. experimental

E25. Population sizes are different. Confounding factor is average age.

Section 4.4

P22. a. Some dorms may, for example, have healthier students than others given the same type of soap.
b. to equalize variation between the treatments

P23. a. randomized paired comparison with repeated measures
b. randomized paired comparison with matched pairs
c. matched pairs; the two treatments may have residual effects

P24. randomized paired comparison with repeated measures

P25. a. Blocking based on students' ability to memorize seems appropriate.
b. Blocking on size of customer, for example, might be good.

P26. In a completely randomized design, treatments are assigned entirely at random. In a block design, subjects are divided into blocks that are alike and then treatments are assigned randomly *within each block*.

P27. Blocking helps reduce variability due to individual differences in subjects.

P28. a. Too much variability to answer question.
b. The students are improving as most points are above the line. The scatterplot is easier to read.
c. There is considerable variation between students.

E27. completely randomized
experimental units: students
levels of factor: course or no course
response: SAT score
blocks: none

E29. design: randomized block with repeated measures on the same subject
response: rate of finger tapping
treatment: caffeine, theobromine, or placebo
unit: one day of a subject's time
block: subject (more precisely, three days of a subject's time)

E31. completely at random
response: quantity eaten relative to body weight
unit: hornworm
treatments: regular food or cellulose
blocks: none

Review Exercises

E33. a. reasonably representative
b. not reasonably representative; too low
c. possibly reasonably representative
d. not reasonably representative as all students are in a statistics class.
e. not reasonably representative; too low

E35. Advise against this frame; it will give too high an estimate.

E37. Frame misses owners who have resold their 1998 models to someone other than the dealer or have traded their model to a dealer who is not a dealer for that brand of car.

E39. The method samples only large surface-swimming fish.

E41. *Times* readers may have higher incomes and be more educated than the average New Yorker.

E43. Stage 1: Choose an SRS of states. Stage 2: Choose an SRS of districts from these states. Stage 3: Choose an SRS of precincts from these districts. Stage 4: Choose an SRS of voters from these precincts.

E45. You might block on whether the person uses corrective lenses, for example.

E47. One possible answer is to have a block as a student with repeated treatments.

Chapter 5

Section 5.1

P1. Triples 001–503 represent 0 children; 506–708 1 child, 709–899 2 children, 900–972 3 children, 973–000 4 or more; 0, 0, 3.

P2. The mean is 0.916, and the standard deviation is 1.11.

E1. a. $\mu = \Sigma x \cdot P(x) = 2.16$
   $\sigma = \sqrt{\Sigma(x - \mu)^2 \cdot P(x)} \approx 2.221$

   b. Divide the random digits into groups of two because there are two decimal places in each proportion. Then 01–45 represents a score of 1, 46–54 of 2, 55–58 of 3, 59–73 of 4, and 74–00 of 5.

E3. Divide the random digits into groups of one. Let digits 1, 2, and 3 represent a parent who occasionally calls. The other digits represent parents who do not call. Checking the first five digits, we find two "parents" who occasionally call.

E5. sample with replacement

Section 5.2

P3. a. I. B   II. C   III. A
   b. no

P4. b. The midrange has a mound-shaped, nearly symmetric sampling distribution; the center is well above the population mean.
   c. No. It is greatly influenced by outliers.

P5. a. A is the lower quartile, B is the maximum, C is the midrange, and D is the minimum.
   b. D has the largest standard error; A has the second largest.

E7. I. B   II. C
   III. A   IV. D

E9. b. Possible samples: (62, 63); (62, 64); (62, 64); (62, 65); (63, 64); (63, 64); (63, 65); (64, 64); (64, 65); (64, 65)

c. Possible means are 62.5, 63, 63, 63.5, 63.5, 63.5, 64, 64, 64.5, and 64.5. Mean = 63.6, $SE = 0.62$, Pop mean = 63.6, $SD = 1.02$.

d. The sample maximums for the 10 possible samples are 63, 64, 64, 65, 64, 64, 65, 64, 65, 65.

e. The sample ranges are 1, 2, 2, 3, 1, 1, 2, 0, 1, and 1.

f. The plots for the mean and the range are closer to normal shape.

E11. a. There are only three possible choices for the number of pounds of spoiled fish. The sample will contain 0, 24, or 48 pounds of spoiled fish, but we don't know the probability of each.

   b. This is not a good sampling plan. A better plan would sample more cartons.

E13. Larger if divided by $n - 1$. Center of the sampling distribution is much closer to the population standard deviation with $n - 1$.

E15. Yes, sampling distributions of medians will have less variation. For simulation, answers will vary.

Section 5.3

P6. a. Take a random sample of size 10 using methods discussed in Section 5.1. Compute the mean of the sample. Repeat and plot these sample means.

   b. Same process as above, but $n = 1$; distribution looks like population.

P7. approximately normal, with mean 266 and $SE = 9.553$

P8. a. approximately normal with mean 0.916 and $SE = 0.035$
   b. .324

P9. a. $z = 2.393$. The probability is about .0084. Almost no chance the network will get 1000 children.
   b. yes; probability is now .995; $z = -2.580$; $P(x \geq 1000) = .995$

P10. a. (0.481, 1.351)
   b. (0.698, 1.134)
   c. (0.847, 0.985)
   d. (0.882, 0.950)

P11. a. Assign each household a three-digit number. Select five three-digit numbers at random. Find the five households corresponding to these five numbers. If one or more of the numbers is a repeat, use that household again anyway.

b. If a number repeats, discard it and go to the next triple.

P12. almost no change

E17. To get a sample of size 5, divide the random digits into groups of two and use the assignments given in the third column of this table.

Value	Percentage	Assignment of Random Numbers
0	26	1–26
1	18	27–44
2	16	45–60
3	10	61–70
4	8	71–78
5	8	79–86
6	6	87–92
7	4	93–96
8	2	97–98
9	2	99–00

E19. Divide random digits into groups of two and use the assignments given in the third column of this table.

Value	Percentage	Assignment of Random Numbers
0	18	1–18
1	14	19–32
2	10	33–42
3	6	43–48
4	2	49–50
5	2	51–52
6	6	53–58
7	10	59–68
8	14	69–82
9	18	83–00

E21. a. no (about 15% chance)
 b. the third plot; the second plot
 c. yes, if days are selected randomly and independently.
 d. yes, if days are selected randomly and independently.
 e. A particular period of four or eight days may not look like a random sample; no, weather on consecutive days is not independent.

E23. a. 400
 b. 1600
 c. 40,000

E25. a. The sample mean remains at the population mean for all n. The SE decreases by a factor of 1 divided by the square root of n.
 b. The mean increases by a factor of n. The standard error increases by a factor of the square root of n.

E27. a. (11.8, 34.0)
 b. (69.4, 113.8)
 c. (845.8, 986.2)
 d. (3523.6, 3804.4)

Section 5.4

P13. a. shape is approximately normal; mean is 0.53; $SE = 0.05$
 b. No; the probability is close to 0.

P14. a. The probability is close to 0.
 b. Yes. Possible reasons include the fact that people do not always vote for the candidate of their party.

P15. a. (.445, .755)
 b. (.502, .698)
 c. (.551, .649)

P16. By both methods, the probability is about .645.

E29. a. $\mu_{\hat{p}} = 0.92$; $\sigma_{\hat{p}} = .0086$
 b. $\mu_{sum} = 920$; $\sigma_{sum} = 8.58$
 c. .0099
 d. .280
 e. Rare events would be numbers less than 903 or more than 937 or proportions less than .903 or more than .937.

E31. a. .5
 b. .000011
 c. No. The group is special in that they are all old enough to have a job.

E33. The rule that np and $n(1 - p)$ must both be at least 10 is reasonable.

Section 5.5

P17. a. triangular
 b. mean is 0; variance is 5.83
 c. mean is 0; variance is 5.83

P18. a.

Difference	Probability
−3	1/24
−2	2/24
−1	3/24
0	4/24
1	4/24
2	4/24
3	3/24
4	2/24
5	1/24

 b. $\mu_{difference} = 1$; $\sigma^2_{difference} = 4.167$

P19. .474

E35. a. mean is 1017; $SE = 157.69$
 b. about .084
 c. between approximately 702 and 1332
 d. .249
 e. mean is 1017; SE and shape are unpredictable because scores are not independent

E37. a. 12.8 and 4.56; 8 and 4
 b. 20.8 and 6.96
 c. Mean agrees, variance does not. Variance should not agree because times were not independent.

E39. a. Section 5.3: The formula for the sum.
 b. mean is 10.5; variance is 8.751
 c. mean is 24.5; variance is 20.419
 d. mean is 10,100; variance is 246,420; probability is about .420

E41. a. approximately normal with mean 150.8 kg and SE 10.44 kg
 b. .289
 c. .568
 d. .316

E43. roughly triangular with mean 0 and SE 1.71

Review Exercises

E45. a. Pick three cities at random, count the number with API < 100. API. Repeat process many times. Plot the results.
 b. mean is 1.6875; $SE = 0.8592$
 c. no, $np < 10$.

E47. a. about .632
 b. No. Sample is not random.

E49. a. 3/144
 b. mean is 6.5; $SD = 3.452$
 c. mean is 13; $SE = 4.882$

E51. a. 37,833
 b. 37,649 to 38,016
 c. Vehicles are not equally likely to go in each of the three directions.

E53. a. Select two integers at random from 0 through 9 and compute their mean. Repeat this. Subtract the second mean from the first.
 b. The shape will still be basically triangular but rounding out slightly, with mean 0 and SE 2.87.

E55. a. $n = 10$: .55 to 1.05, proportion = 30/33 ≈ .909
 $n = 20$: .62 to .98, proportion = 41/45 ≈ .911
 $n = 40$: .67 to .93, proportion = 67/72 ≈ .931
 $n = 80$: .71 to .89, proportion = 68/72 ≈ .944
 b. about 95%, the above values are close

E57. a. shape: more approximately normal
 mean: stays fixed at the population mean
 standard error: proportional to the reciprocal of the square root of n.
 b. shape: more approximately normal
 mean: proportional to n
 standard error: proportional to the square root of n

Chapter 6

Section 6.1

P1.

Number Who Chose T	Probability
0	1/16
1	4/16
2	6/16
3	4/16
4	1/16

Probability that all choose correctly is $\frac{1}{16}$ or .0625; Sample is not large enough to ease Jack's concern.

P2. Set is complete but not disjoint. First two are more likely than the third. No; probability is 3/4, not 2/3.

P3. the 10 disjoint outcomes: (28, 35), (28, 41), (28, 47), (28, 55), (35, 41), (35, 47), (35, 55), (41, 47), (41, 55), (47, 55); probability: 1/10

P4. $P(2H) = 6/16$

Number of Heads	Probability
0	1/16
1	4/16
2	6/16
3	4/16
4	1/16

P5. $P(sum = 7) = 1/6$

Sum of Two Dice	Probability
2	1/36
3	2/36
4	3/36
5	4/36
6	5/36
7	6/36
8	5/36
9	4/36
10	3/36
11	2/36
12	1/36

P6. probably not

P7. number of pairs: 21; $P(favorite) = 1/21$

E1.

Number of Heads	Probability
0	1/32
1	5/32
2	10/32
3	10/32
4	5/32
5	1/32

E3. a. Yes. If both foods taste the same, then preference decisions are made at random and the chance of preferring A to B is 0.5.
b. No. You need the data to know what percentage of math majors are women.

E5. a. 24
c. 48

E7. a. 24
c. 2/6; 1/4; 2/24; 12/24

E9. a. disjoint and complete; 9/16; 6/16; 1/16
b. Not disjoint but complete; "the second roll is a four" might include the outcome "the first roll is a four."
c. Disjoint but not complete; only outcomes with fours are included.
d. Disjoint but not complete; the sum equal to 2 is not included.
e. Disjoint but not complete; the second die being a 4 is not included.

Section 6.2

P8. a. complete but not disjoint
b. The complete disjoint categories are: received full amount, received partial amount, did not receive payments, and not supposed to receive payments in 1991.
c. .38

P9. b and c

P10.

		Fish in Great Lakes?		
		Yes	No	Total
Fish in Other Freshwater?	Yes	1,697	28,489	30,186
	No	855	—	855
	Total	2,552	28,489	31,041

$P(Great\ Lakes$ and *other freshwater*$) = .055$

P11. a. 8/36   b. 12/36   c. 11/36

P12. 3/4

E11. a and c

E13.

	Yes	No	Total
Yes	26	46	72
No	112	32	144
Total	138	78	216

$P(no\ dog$ or *no beeper*$) = .88$

E15. $P(masters$ or *doctorate*$) = .88$

	Bachelors	Masters	Doctorate	Total
Full Time	76	6,214	1,288	7,578
Part Time	2,568	10,699	999	14,266
Total	2,644	16,913	2,287	21,844

E17. Jill has not computed correctly because these two events are *not* mutually exclusive.

E19. Each of the following has to be true:
$P(A \text{ and } B) = 0$ $P(B \text{ and } C) = 0$
$P(A \text{ and } C) = 0$ $P(A \text{ and } B \text{ and } C) = 0$

Section 6.3

P13. a. 1/2   b. 3/4   c. 1/3   d. 1/3

P14. a. 1/2   b. 1/13   c. 1/4

P15. 1/17; 1/16

P16. 2/4, 1/3, 1/6

P17. 1/36

P18. $P(right\ handed \mid brown\ eyes) = 16/20$

	Right-Handed	Left-Handed	Total
Blue Eyes	8	2	10
Brown Eyes	16	4	20
Total	24	6	30

P19. 57%

P20. .25; .80; .60; .33. Not a good screening test; it is neither very sensitive nor very specific.

P21. false positive: $P(actually\ not\ guilty \mid found\ guilty)$
false negative: $P(actually\ guilty \mid found\ not\ guilty)$

E21. a. .34   b. .16   c. .14   d. .44

E23. 3/5, 2/5, 1/2, 1/2, 3/4, 1/4

E25. If you select an adult at random from the United States, what is the probability that the person wants to go to Mars and is under 25 years old?

E27. a.

	Male	Female	Total
Right	.44	.48	**.92**
Left	.08	0.00	**.08**
Total	**.52**	**.48**	**1.00**

b. .15   c. 0

E29. a.

Test Result

Disease	Positive	Negative	Total
Yes	2.985	$.005 \cdot 3 = 0.015$	**3**
No	$.06 \cdot 97 = 5.82$	91.180	**97**
Total	**8.805**	**91.195**	**100**

b. .088

c. .0582/.088 = .66

E31. Answers will vary. Consider the following table for a screening test that is 90% accurate (both sensitivity and specificity are .9).

Test Result

Disease Status	Positive	Negative	Total
Present	72	8	80
Absent	2	18	20
Total	74	26	100

The false positive rate is 2/74 = .03; the false negative rate is 8/26 = .31.

E33. a. 1/2   b. 1   c. 1/3
d. 2/3   e. 1   f. 0

Section 6.4

P22. no

P23. a. independent
b. not independent
c. independent

P24. a. .64
b. .32

Second Person

First Person		Grad	Not Grad	Total
	Grad	.64	.16	**.80**
	Not Grad	.16	.04	**.20**
	Total	**.80**	**.20**	**1.00**

c. The tree diagram will have four paths. The two paths where one student graduates are *GN* and *NG*, each of which has probability $(.8)(.2) = .16$. Thus, $P(GN \text{ or } NG) = .32$.

P25.

One Parent

Other Parent		T	T
	t	tT	tT
	t	tT	tT

All of the offspring will be tall.

E35. a. not independent
b. independent
c. not independent

E37. a. *BB, BG, GB,* and *GG*
  b. two boys; two girls
  c. .2601, assuming births are independent
  d. They may mean that their probability of getting a girl is higher than the population percentage for girls (or equivalently for boys). Under these conditions, *GG* would be the event with the highest probability.

E39. She forgot to account for the event *not 4 on first* and *4 on second*.

E41. 9/16; 3/16; independence of shape and color.

E43. not independent

	B	Not B	Total
A	1/8	1/8	1/4
Not A	3/16	9/16	3/4
Total	5/16	11/16	1.000

E45. The probability that they both fail is .015 only if the surgery results are assumed to be independent (a very improbable assumption).

E47. $P(A$ and $not\ B) + P(A$ and $B) = P(A)$, so $P(A$ and $not\ B) = P(A) - P(A$ and $B) = P(A) - P(A) \cdot P(B) = P(A) \cdot [1 - P(B)] = P(A) \cdot P(not\ B)$.

Review Exercises

E49. Randomly select a digit from the set {1, 2, 3, 4, 5, 6, 7, 8, 9}. Repeat this process seven times (with replacement). See if the number 3 is among the seven digits selected. If so, the repetition was a "success." Repeat this process many times.

E51. 85%

Merry-Go-Round?

		% Yes	% No	Total %
Roller Coaster?	% Yes	40	5	45
	% No	40	15	55
	Total %	80	20	100

80%

Merry-Go-Round?

		% Yes	% No	Total %
Roller Coaster?	% Yes	5	45	50
	% No	30	20	50
	Total %	35	65	100

E55. a. No. Disjoint events are always dependent.
  b. Yes. Independent events cannot be disjoint.
  c. Yes. Disjoint events are always dependent.
  d. Possible: Nondisjoint events can be dependent.

E57. a. 1/6 or about .17
  b. .249999975
  c. no; no; part b

E59. a.

Results	Test I	Test II	Total %
False Positive	6	2	8
Other Results	54	38	92
Total %	60	40	100

  b. .75

E63. a. 2700
  b. .27

Chapter 7

Section 7.1

P1. The first is triangular with expected value 7 and *SD* 2.415. The second is increasing at a constant rate with expected value 4.472 and *SD* 1.404.

P2. a. $1.20
  b. $3.60

P3. $136

P4. Expect to save $1060.80 with *SD* $106.58.

E1. a. largest profit: $204.50; largest loss: $4795.50
  b. $209.50
  c. Answers will vary but could include: whether or not there is a burglar alarm; whether there are good locks on the doors; whether the house contains expensive items; what proportion of the time someone is at home.

E3. a. $0.185
  b. $315,000
  c. 37%

E5. a. $148,861,425
  b. $29,194,142.50
  c. about $0.149

E7. AT&T: $54.49; Verizon: $73.75; Voicestream: $53.74; Sprint PCS: $66.49

Section 7.2

P5. $P(X = 3) \approx .21875$; $P(x = 2) \approx .109375$; $P(X \geq 7) \approx .03516$

P6. a. $P(0) \approx .5522$

   b. $P(X \geq 1) \approx .4478$

   c.
Number of Dropouts	Probability
0	.552160
1	.348209
2	.087836
3	.011078
4	.000699
5	.000018

P7. For example, assign 001 to 112 to represent dropouts.

P8. a. .1875

   b. i. 8 households

      ii. $SD = 2$

      iii. .227

   c. This result is very rare; suspect that the sample was not selected at random.

P9. a five-question quiz

E9. a. .0202

   b. .01883. The answers are quite close.

E11. a. .2614

   b. .2087. The answers are fairly close.

E13. a. $ 500,000   b. $948,683

   c. .3487   d. .2639

E15. a. $n = 7$ b. $n = 14$

E17. It is the average number of college graduates that would appear in many random samples of size 3.

Section 7.3

P10. a. first roll: 1/6; second roll: 5/36

   c. less than 1; there are more possible values of X to the right of the last bar, but these values have small probabilities

P11. a. .00737   b. .999

P12. a. .4

   b. .24

   c. .0864

   d. independence

P13. a. 10

   b. about 9.487

   c. 20 donations; 30 donations

E19. a. 1/4

   b. 1/64

   c. 4/3

   d. On average, only 0.5 people will be left after the third round.

E21. a. 10

   b. $E(C) = \$500,000$; $SD \approx \$474,342$

   c. $P(X \geq 5) = .6561$; $P(X \geq 15) = .2288$

E23. The probabilities for a geometric distribution follow exactly this same pattern: $p, qp, q^2p, q^3p, q^4p, q^5p, \ldots$

E25. a. $E(x) = 1p + 2pq + 3pq^2 + 4pq^3 + \cdots$

   c. the sum of the rows are, respectively, $1/p, q/p, q^2/p, q^3/p \ldots$

   d. $1/p$

Review Exercises

E27. a. $1100

   b. $310

   c. $500,000

E29. Mean is 12,000; $SD \approx 104$. Students' choices are not independent (they apply with their friends).

E31. a. $P(X > 430) \approx .480$

   b. $E(C) = \$18,600$

   c. about 14.286

E33. a. p

   b. $1 - p$

   c. $(1 - p)^k$

E35. b. $(n = 25, a = 5)$

   c. $(n = 25, a = 5)$

E37. a. 2 flips

   b. 20 individual flips or about 4.73 sets of flips

   c. 1.001

Chapter 8

Section 8.1

P1. between .264 and .536; 25 out of 50 is reasonably likely

P3. from 14 through 26 heads

P4. No. Getting 32 out of 40 still in school would be a rare event.

P5. about 26 to 38 high school graduates; from .65 to .95

P6. about 15% to 40%

P7. 30% to 60% or 45% ± 15%

P8. a. Using the chart, about 50% to 75%.

b. Without the chart, about (.475, .775).

P9. a. probably random; $n\hat{p} = 610.08$;
$n(1 - \hat{p}) = 134$; $10n <$ number of teens

b. (.792, .848); $E = 2.8\%$

c. (.797, .843); $E = 2.3\%$

d. The first; we need a larger interval to increase our confidence that it contains p.

P10. The horizontal line segments will be shorter; the confidence intervals will be shorter.

P11. Use $z = 1.28$. Intervals will be shorter.

P12. Use a sample size 9 times larger.

P13. a. The method of sampling is not given, so you can't verify that it was a simple random sample. $n\hat{p} = 535$; $n(1 - \hat{p}) \approx 219$; $10n <$ number of secondary students.

b. The 95% confidence interval is $.71 \pm .032$.

c. the proportion of *all* students age 12–17 with Internet access who rely on Internet access the most when completing a project

d. If we could ask all students age 12–17 with Internet access if they relied on the Internet the most in completing a project, we are 95% confident that the proportion who would say "yes" would be somewhere in the interval 68% to 74%.

e. Suppose we could take 100 random samples from this population and construct the 100 resulting confidence intervals. We expect that the true proportion of all students age 12–17 with Internet access who would say they relied on the Internet the most in completing a project will be in 95 of these intervals.

P14. Nonresponse, response bias, errors in recording responses, for example.

P15. 2,401; 16,590; 27,061

P16. 897

E1. a. no   b. .65; .35
c. about 30% to 55%

E3. It is likely some randomization was involved. $n\hat{p} = 810$ and $n(1 - \hat{p}) = 190$; $10n <$ number of adults. The confidence interval would be between 78.6% and 83.4%, which means that we are 95% confident that if we were to ask *all* adults from the general public if they thought TV contributed to a decline in family values, the percentage who agree would be in this interval.

E7. $.6 \pm .096$; Display 8.2 cannot be used in this case because it applies only to $n = 40$.

E9. The symbol p is the population proportion and is the unknown parameter; $\hat{p}$ is the sample proportion and is always in the center of the confidence interval.

E13. a. because we want to maximize $p(1 - p)$; p is restricted to [0,1]

b. parabola opening down

c. at the vertex, where $x = 1/2$, $y = 1/4$

Section 8.2

P18. a. $p_0 = .5$

b. $p > .5$

c. $\hat{p} = .75$

d. It is statistically significant. It strongly suggests (but doesn't prove) the student wasn't guessing.

P19. b. $z \approx -5.01$; reject H_0

P20. b. $z \approx 1.34$; don't reject H_0

P21. a. ± 1.55

b. .0836

P22. H_0: $p = .5$ (p is the proportion of all U.S. bookstores selling CDs)
H_a: $p \ne .5$
$z \approx 1.34$; $\alpha = .05$; $z^* = \pm 1.96$; do not reject H_0: do not reject the hypothesis that half of all U.S. bookstores sell CDs.

P23. a. H_0: $p = .2$ (p is the chance of getting the correct answer on each question)

b. H_0: $p = .5$ (p is the proportion of male, or female, newscasters)

c. H_0: $p = 1/7$ (p is the proportion of people who wash their car once a week and who wash them on Saturday)

P24. B

P25. a. Hila does not reject H_0; $z = 1.43$

b. no error

P26. a. Yes. They each have a .05 chance of making this error.

b. No, because H_0 is true.

P27. a. No, because H_0 isn't true.

b. Yes. Both could but Jeffrey is more likely to make a Type II error.

P28. a. The sample is simple random.
$np_0 = n(1 - p_0) = 20 > 10$. If there are more than $10 \cdot 40 = 400$ students, we should proceed.

b. H_0: $p = .5$; H_a: $p \neq .5$ where p is the proportion of all students in the school who carry a backpack.

c. $z = 2.53$; the P-value is .011

d. There is evidence to reject the hypothesis that half the students carry a backpack to class.

P29. P-value is .0016. If spinning a penny is fair, there is only a .0016 chance of getting fewer than 10 heads or more than 30 heads.

P30. $z = 19.17$ and P-value ≈ 0, so reject H_0. It isn't reasonable to assume that the observed difference in the percentage of swallows with mutations is due to chance alone.

P31. Conditions are met if there are more than 1690 teens in your community. H_0: $p = .51$; H_a: $p > .51$; $z = 1.048$; P-value $\approx .147$, so do not reject H_0

P32. a. false

b. false

c. false

d. true

E15. All conditions are met, $z = -1.75$, P-value $\approx .08$, so do not reject H_0 that $p = 2/3$ at .05 level of significance. Reject at the .10 level of significance.

E17. All conditions are met. H_0: $p = .25$; H_a: $p > .25$; $z = 3.66$; P-value $= .00128$; reject H_0

E19. Your null hypothesis is very unlikely to be *exactly* true.

E21. a. $z = .449$; P-value is .65; do not reject H_0

b. $z = -1.35$; P-value is .18; do not reject H_0

c. We don't have enough evidence to reject either H_0, but we know that we have made a Type II error in at least one of these tests.

d. $z = -1.42$; P-value is .155; do not reject H_0

e. $z = -4.26$; P-value $= .00002$; reject H_0

f. Larger samples give better evidence.

Section 8.3

P33. Simulations will vary. The shape is approximately normal with mean $-.10$ and standard error .134. Reasonably likely values are between $-.36$ and .16 (middle 95%).

P34. All conditions are met. 1.7% to 9.1%. You are 99% confident that the difference between the percentage of all Americans who used the Internet in 2001 and the percentage in 2000 is in the interval 1.7% to 9.1%.

P35. Increasing sample sizes will shrink the length of the confidence interval.

P36. Conditions are met. You are 95% confident that if all subjects could have been given both treatments, the difference in the proportion who would get an ulcer is in the interval $-.0003$ to .0059.

E23. Conditions are met. $(.117, .147)$. Because 0 isn't included in the confidence interval, it is acceptable to use the term "significantly more likely."

E25. You are 99% confident that the difference in the percentage of all men and the percentage of all women who prefer to be addressed by their last name is in the interval $-.189$ to $-.031$.

E27. a. $(.0114, .1868)$. You are 95% confident that if all the skiers in the experiment were given the placebo and all the skiers in the experiment were given vitamin C, then the difference in the proportions who get colds is in this interval.

b. $-.0162$ to .2144. Because 0 is in this confidence interval, it is plausible that the same proportion of these skiers would get a cold whether or not they took vitamin C.

E29. Provided the usual conditions are satisfied, the sampling distribution of the estimate $\hat{p}_1 - \hat{p}_2$ is approximately normal.

Section 8.4

P37. c. mean is 0, $SE \approx .135$

P38. P-value $\approx .174$; cannot conclude that there has been a drop in support.

P39. Type I

E31. a. $z = 3.195$; P-value ≈ 0; you are confident that more girls than boys would say yes.

b. Yes; some people this age have more than one phone and busy people aren't available to answer the phone as often.

c. 65.4% to 73.4%; you need to assume that there are equal numbers of males and females in the teen population.

E33. Yes; $z = -5.2$; the P-value is almost 0. If all the men could have been given both treatments, there is almost no chance of this large a difference in the proportions having heart attacks unless aspirin makes a difference.

E35. The one-sided P-value is about .017; the difference can't be reasonably attributed to chance.

E37. a. A random sample is selected from each of the two populations.

   b. The subjects are randomly assigned to one of two treatments.

E39. .0645

E41. Bus B, as this isn't a statistically significant difference in favor of Bus A by 2%.

Review Exercises

E43. a. .925 to .955

   b. You are 92% confident that if you were to ask *all* students age 12–17 with Internet access, between 92.5% and 95.5% would say that the Internet helped them the most with their homework.

   c. If you repeated the survey 100 times with sample size 754 and constructed a confidence interval for each sample, then you expect the (unknown) population proportion to lie in 92 of the 100 intervals.

E45. a. The *P*-value is close to 0; $z = 5.81$; statistically significant. But note these aren't random samples; they are the entire populations at these three ski areas.

   b. No. You still need to know the total number of skiers and snowboarders.

E47. a. 5

   b. .00592

E49. 90% of the time

Chapter 9

Note that these brief answers do not include checking conditions, but be sure you do not skip that step.

Section 9.1

P1. a. (1028, 1192); (1012, 1208); (981, 1239)

   b. no; yes; yes

P2. To shorten a confidence interval, increase sample size or decrease confidence. For vital decisions, increase sample size.

P3. B

P4. 1600 athletes

P5. 576 students

E1. a. about 18; (34, 70)

   b. The variable is the acceptance rate per college. The cases are the colleges (not the individual students within those colleges).

   c. You are 95% confident that the mean acceptance rate of the top 50 colleges is between 34% and 70%.

E3. B

E5. a. 4

   b. 16

   c. 64

E7. (15.93, 16.01). No, they are 95% confident that the current mean is in the interval (15.93, 16.01), which includes 16.

Section 9.2

P6. $z = 0.22$, for a *P*-value of .8421 for a two-sided test. There is insufficient evidence to reject the H_0 that the mean acceptance rate is 50%.

P7. $z = 1.0 < z^* = 1.96$. There is insufficient evidence to reject H_0 that this sample is from a university with a mean SAT score of 1300.

P8. $z = -6.0 < z^* = -1.645$. Reject H_0. It's not plausible that this sample comes from a university with a mean SAT score of 1300.

P9. $z = -1.70$, so reject at 10%; insufficient evidence at 5%; insufficient evidence at 1%

P10. for P6, .8259; for P7, .3173; for P8, about 0

P11. a. Both $\bar{x} = 1150$ and $\bar{x} = 1200$ would give $|z| = 1.645$ and *P*-value = .1.

   b. Both $\bar{x} = 1146$ and $\bar{x} = 1204$ would give $|z| = 1.96$ and *P*-value = .05.

   c. Both $\bar{x} = 1136$ and $\bar{x} = 1214$ would give $|z| = 2.576$ and *P*-value = .01.

P12. Use a one-sided, left-tailed hypothesis test, H_0: $\mu = 212°F$ versus H_a: $\mu < 212°F$, where μ is the true boiling point of water in Denver.

P13. H_a: $\mu > 35$: a manufacturer claims that its car gets greater than 35 mpg.
   H_a: $\mu < 35$: an environmental group claims that a car gets less than 35 mpg.

E9. D

E11. a. yes; $z \approx -2.27$; one-sided *P*-value $\approx$ .0115

   b. yes; $z \approx 2.27$; two-sided *P*-value $\approx$ .023

   c. Yes. The hypothesized value of 500 does not lie inside the 95% confidence interval, (510, 640).

E13. $z = -1.58$ and the two-sided *P*-value is .1138. The machine should not be readjusted.

Section 9.3

P14. a. (4.32, 5.68)

    b. The centers of the confidence intervals will not change, but it turns out that s will be smaller than σ more often than it is larger, so the capture rate will be smaller than the advertised value.

P15. a. 2.262

    b. 2.447

    c. 3.106

    d. 2.920

    e. 63.66

    f. 5.841

P16. a. (7.908, 46.092); calculator: (7.905, 46.095)

    b. (3.694, 8.306)

    c. (−16.552, 34.572); calculator (−16.58, 34.577)

P17. (4.21, 5.79). This is slightly longer than the interval found in P14.

P18. $t = 2.875 > t^* = 2.262$. There is sufficient evidence to reject the H_0 that the mean aldrin level is 4.

P19. a. P-value = .0183

    b. P-value = .9908

    c. P-value = .0092

P20. For each test, the chance of a Type I error is .05.

P21. A

E15. a. (9.39, 12.47); We are 90% confident that if we asked all females taking this course, the mean number of hours studied per week is in the interval (9.39, 12.47); no; possible outliers.

    b. $t = -1.826$; one-sided P-value is .045. There is sufficient evidence to reject the H_0 that males study the same or more hours per week as females, on average.

    c. We are 95% confident that the mean number of hours studied per week for all students in this course is in the interval (8.70, 11.82).

E17. a. yes; $t = 0.84$; P-value = .4893 for a two-sided test; insufficient evidence to reject the H_0 that the mean price is $3.04.

    b. with H_0: $\mu(hot\ dog\ price) = 3.04$ and H_a: $\mu(hot\ dog\ price) > 3.04$, $P = .2447$; insufficient evidence to reject H_0. Conclusion: the sample data are consistent with a mean hot dog price of $3.04.

c. Sampling 10 teams would violate the rule that the sample size must be at most 10% of N or 10% of (31) = 3.1; thus, procedures would not work properly.

E19. a. n, 0

    b. no

    c. The second deviation is −3.

    d. The third deviation is −2.

    e. nth deviation = −(sum of the other $n - 1$ deviations)

    f. For a sample size of n, $n - 1$ deviations give independent information; one deviation is redundant or can be determined from the other $n - 1$ deviations.

Section 9.4

P22. a. III   b. I   c. IV   d. III

E21. a. Because the data are highly skewed toward the larger values, perform the analysis using the logarithms of the mercury levels.

    b. For a one-tailed t-test with H_0: $\mu \le \ln(.5)$, $df = 34$, $t = -.8533$, and the P-value is .8003. There is insufficient evidence to suggest that the mean mercury level exceeds the 0.5 ppm level. If the mean mercury level must exceed 0.5 ppm before a health advisory warning is issued, then the headline is not appropriate. However, if the warning is issued when the 0.5 ppm boundary is reached, then the headline may be justified.

E23. a.

Report

		Verbal	Visual
Task	Verbal	12.85	13.69
	Visual	9.01	18.17

The data are consistent with the theory.

    b. Boxplots show skewness in *response times*, with a few outliers, suggesting a reciprocal transformation from time (how long?) to 1/time (how frequently?).

    c. For untransformed data, Vis/Vis: (15.805, 20.541); Vis/Ver: (7.708, 10.316) Ver/Vis: (12.15, 15.23); Ver/Ver: (10.85, 14.85). If assignment of treatments was properly randomized, then the fact that the confidence intervals don't overlap supports the theory that for both kinds of reporting methods, average times were slower when the task was of the same kind.

E25. b. (1.46, 3.67)

　　c. (0.355, 0.559)

　　d. (1.79, 2.82); the interval is shorter than the one found in part b.

　　e. housing units per person

Section 9.5

P23. a. (−15.4, 95.4)

　　b. The observed difference can be reasonably attributed to chance (that is, variability in sampling).

P24. 50 students

P25. 150

P26. We are 90% confident that the difference in the mean number of hours all females in this course study per week and the mean number all males study is in the interval (−0.32, 5.78). However, if the two outliers are eliminated, the interval is (1.26, 5.73), which no longer overlaps 0. (A log transformation would be better.)

P27. $t = 1.53$ with $df = 24.8$, *P*-value = .14. There is insufficient evidence to reject H_0.

E27. (−4.71, 9.94) or $t = 0.742$ with $df = 21$, *P*-value = .466. As 0 is in the confidence interval and the *P*-value is large, there is insufficient evidence to say that the mean heart rate for men differs from the mean heart rate for women.

E29. a. No. Due to the skewness of the data, the analysis is performed on the transformed data $\ln(HG)$; $t = 0.396$; *P*-value = .3474

　　b. (0.37, 1.73)

　　c. no, insufficient evidence; $t = 1.292$; *P*-value = .207

E31. a. (1.22, 9.45)

　　b. (−1.46, 4.79)

　　c. yes for both

Section 9.6

P28. CRD: A scatterplot shows no clear association. The boxplots show no reason for concern. $t = 2.788$, $df = 24.79$, *P*-value = .005. Reject H_0.

MPD: A scatterplot shows a fairly strong positive association. The boxplot shows a little skewness toward the higher values. $t = 3.421$, $df = 13$, *P*-value = .0023. Reject H_0.

RMD: Scatterplot shows a strong positive trend; boxplot shows little skewness. $t = 6.800$, $df = 27$, *P*-value < .0001. Reject H_0.

P29. a. A boxplot of the difference in highway and city mpg shows that the differences are only slightly skewed and that there are no outliers.

　　b. A boxplot of the difference in highway and city gpm shows that the differences are very slightly skewed and that there are no outliers.

　　c. The 95% confidence interval for the mean difference in highway and city gallons per mile is (−.0074, −.0046). $t = -8.670$ with $df = 19$ and *P*-value < .0001. There is sufficient evidence to reject H_0 and conclude that highway driving saves between 0.005 and 0.007 gallons per mile, on the average.

P30. a. Yes. Due to the skewness of the data, the two-sample *t*-test is performed on the transformed data $\ln(brain\ weight)$. $t = 2.604$; *P*-value = .025. Reject H_0 and conclude that there is a difference in mean brain weight in the two groups.

　　b. The data and transformed data give $t = 1.61$ and $t = 1.70$, respectively. Dropping the vulture gives $t = 2.765$—these differences reveal a problem that has no easy solution.

P31. No, to measure the degree of mixing, compute the standard deviations of the proportions for each component and compare to a degree of variation that is acceptable.

P33. a. No, treatments were not randomly assigned to the volunteers so inference should not be done on such data.

　　b. This design is better than the design in part a because it partially accounts for the abilities of the students and reduces the student-to-student variability in the results. However, because this design does not randomize treatments, meaningful inference will not be possible.

　　c. Take a group of volunteer students, pair them by past performance in math, and then randomly divide each pair into two treatment groups—*radio on* or *radio off*. Follow the groups over time to ensure that they are staying on their treatment. Compare mean scores of the two groups to assess the treatment effect.

E33. a. no, using a one-sample *t*-test, $t = 1.511$; *P*-value = .0846

　　b. (0.63, 2.59)

　　c. (−0.15, 2.16)

　　d. C appears to cause more drowsiness than the placebo.

E35. a. This is not an experiment with treatments (rural/urban) randomly assigned to subjects. Further, an outlier changes the results of a one-sample *t*-test: including the outlier, the *P*-value is .33, and without the outlier, the *P*-value is .028.

b. Independent samples would require the independent selection of random samples from clearly defined rural and urban populations of interest. Each subject would need to agree to breathe Teflon particles. Clearance rates could then be measured and the means compared by a two-sample procedure.

c. Paired data reduces the person-to-person variations that may occur in the independent sample design.

d. There is no need for a criterion on which to pair the subjects. Also, it may be difficult to find twins for the study.

E37. a. i. (−0.06, 12.82)

ii. (0.06, 33.94)

iii. These intervals are estimating different mean differences: the first for all accounts, and the second only for accounts with nonzero differences. The interval in part ii is the more reliable estimate because the conditions of the *t* procedure are better met as there are no outliers.

b. The proportion of the accounts with nonzero difference is (15/40) = .375. If there are 1000 accounts, then you would expect about 375 of those accounts to have nonzero differences. A confidence interval estimate for the total can be found by multiplying both ends of the interval for the mean of the nonzero differences by 375 to obtain (22, 12,727). This interval is so huge that it is practically meaningless. You need a much bigger sample to estimate the total precisely if a large portion of the sample is going to yield zero differences.

Review Exercises

E39. a. To test the theory that the mean concentration of pesticides is higher near the river bottom than at mid-depth, use the one-sided hypothesis, H_0: $\mu_M = \mu_B$ versus H_a: $\mu_M < \mu_B$, where μ_M and μ_B are the mean concentrations at mid-depth and bottom.

b. No explicit randomization was used, so caution is in order.

c. The bottom sample shows a center that tends to be somewhat higher and shows a greater spread and a slight skew toward the larger values. No outliers.

d. $t = -1.625$; *P*-value = .06. There is insufficient evidence to reject the null hypothesis of equal mean concentrations.

e. The borderline results cast suspicion on the null hypothesis. However, the analysis of just one pollutant on one river limits the ability to generalize the results.

E41. a. no, because $z = 1.32 < z^* = 1.645$

b. yes, because $z = 2.94 > z^* = 1.645$

E43. a. A scatterplot shows little correlation. Boxplots show a much larger mean and spread for the sidestream data: the yields are large, and the spread is also large. For mainstream smoke, the yields tend to be less than one-third as large and show hardly any variation.

b.

Stem	Leaf
1	6 6 8
2	3 5 7
3	2 6

where 1 | 6 represents 1.6

The stemplot shows none of the skewness seen in the boxplots; using a paired *t*-test is appropriate.

c. H_0: $\mu_{S-M} = 0$; H_a: $\mu_{S-M} \neq 0$; *P*-value < .0001 ($t = 9.211$ and the *P*-value = .00004)

d. Sidestream smoke contains higher levels of nicotine than mainstream smoke.

E45. a. There is one sample of subjects with two measurements on each subject—control and alcohol.

b. scatterplot: little association and one outlier; boxplots: one outlier, slight skewness toward higher values, and unequal spread, with the larger spread being associated with the group with larger values (that is, the control); stemplot of differences: skewed to the right but no outlier.

c. (30.7, 360.5); "True" can't be defined here as there was no random selection of subjects or randomization of the order of treatments. Thus the results can't be generalized.

E47. a. H_0: $\mu_{course} - \mu_{control} = 30$ and H_a: $\mu_{course} - \mu_{control} > 30$.

b. You reject the null hypothesis if the confidence interval falls to the right of 30.

c. $(-25.52, 105.52)$; $t = 0.312$; insufficient evidence to reject H_0 because 30 is in this interval

E49. The techniques of inference require randomness, and imagining you have randomness will not work. It is wiser to think of randomness as a required condition rather than assuming it.

Chapter 10

Section 10.1

P1. $E = 12.5$; $\chi^2 = 3.28$

P2. 0

P3. From the table, P-value $\approx .025$. Reject H_0 at $\alpha = .05$, and do not reject H_0 at $\alpha = .01$.

P4. P-value $= .003$. Reject H_0 at $\alpha = .05$ and $\alpha = .01$.

P5. a. P-value $= .761$

b. P-value $= .030$

P6. $\chi^2 = 61.27$, P-value $= 3.15 \times 10^{-13}$. Reject H_0; it appears that the jurors were not selected at random from the adult population of Alameda County.

P7. a. $1/10$

b. $\chi^2 = 131.77$, P-value $= 5.13 \times 10^{-24}$. Reject H_0; it appears that measurements were approximated in some cases and not measured to the nearest mile.

E1. a. $z = 2.8$, P-value $= .005$. Reject the H_0 that the coin is fair.

b. $\chi^2 = 7.84$ with $1 = df$, P-value $= .005$. Reject the H_0 that the coin is fair.

c. $z^2 = (2.8)^2 = \chi^2 = 7.84$, and the P-values are the same.

E3.

	Observed	Expected
Within 7 days	11	8.219
8–30 days	24	25.205
31–90 days	69	65.753
More than 90 days	96	100.822

$\chi^2 = 1.39$ with $3 = df$, P-value $= .708$. There is insufficient evidence to reject H_0. The observed differences can be attributed to random variation.

E5. $\chi^2 = 20.468$, $df = 11$, P-value $= .039$. Reject H_0. Given that the P-value is very close to .05, conclude that there is marginal evidence to support that first births do not occur evenly throughout the year.

E7. a. As sample size increases, the value of sample statistics such as the mean of the sample or the proportion of successes in the sample tend to get closer to the mean or proportion of the population. This property gives a test of significance more power to reject a false H_0. If the difference between the estimate and parameter stays the same as the sample size n increases, the standard deviation of the estimate will decrease and, therefore, the test statistic will increase, resulting in a lower P-value.

b. Yes, suppose H_0 is false.

E9. Results will vary. If $\chi^2 > 16.919$ with 9 df, then reject H_0 at $\alpha = .05$ and conclude that there is evidence that the random digit function of your calculator is not random.

E11. a.

Digit k	$\log((k + 1)/k)$
1	0.301
2	0.176
3	0.125
4	0.097
5	0.079
6	0.067
7	0.058
8	0.051
9	0.046
Total	**1.000**

b. $\log_{10}((1 + 1)/1) + \log_{10}((2 + 1)/2) + \cdots + \log_{10}((9 + 1)/9)$
$= \log_{10}(2/1 \cdot 3/2 \cdot 4/3 \cdots 10/9)$
$= \log_{10}10$
$= 1$

Section 10.2

P8. b.

Category	Expected Proportion
A	.111
B	.278
C	.167
D	.139
E	.306

c.

	Sample (expected frequencies)		
Category	I	II	III
A	11.33	22.00	6.67
B	28.33	55.00	16.67
C	17.00	33.00	10.00
D	14.17	27.50	8.33
E	31.17	60.50	18.33

d. No; population 3 appears to have too many that fall into category B.

P9. $\chi^2 = 64.84$, $df = 8$, P-value $< .0001$; yes. Conclude that at least one population does not have the same proportion in each category as the other populations.

P10. $\chi^2 = 11.35$; $df = 8$; P-value $= .183$. Insufficient evidence to reject the H_0. Conclude that the data do not suggest that the distribution of answers is different among the countries.

E13. a.

	Carbolic Acid Used	Carbolic Acid Not Used	Total
Patient Lived	34	19	53
Patient Died	6	16	22
Total	40	35	75

b. $z = 2.914$, P-value $= .0036$

c. $\chi^2 = 8.495$ with $df = 1$, P-value $= .0036$

d. The two P-values are equal, and $z^2 = \chi^2$.

E15. $\chi^2 = 34.58$ with 4 df, P-value $< .0001$. Reject H_0. Conclude that for at least one response, the proportion of children who give that response is not the same for all three types of bandages.

E17. H_0: The proportion of responses that fall into each category is the same for both years. H_a: For at least one response, the proportion of the population that gives that response is not the same for both years. $\chi^2 = 59.54$ with 5 df, P-value $< .0001$. Reject H_0. Conclude that the distribution of responses is not the same for both years.

Section 10.3

P11. All column heights in any row are the same multiple of the height in another row.

P13. a. Expected values:

		Variable 1		
		I	II	Total
Variable 2	A	22.5	13.5	36.0
	B	11.2	6.8	18.0
	C	59.4	35.6	95.0
	D	6.9	4.1	11.0
	Total	100.0	60.0	160.0

b. Table of possible observed values:

		Variable 1		
		I	II	Total
Variable 2	A	36	0	36
	B	18	0	18
	C	35	60	95
	D	11	0	11
	Total	100	60	160

P15. a. This was a single sample, so use a test for independence.

b. yes; $\chi^2 = 21.09$, P-value $= .0001$

c. yes; to about $\chi^2 = 11$ depending on rounding; P-value $\approx .012$

P16. a. Yes; $\chi^2 = 2734$, P-value $< .0001$. Thus, seat belt use and injuries are not independent.

b. Among passengers wearing seat belts, 6.5% sustained injuries, whereas among passengers not wearing seat belts, 22% sustained injuries. The use of seat belts appears to reduce injuries by about 70%.

E21. a. Yes, it is a reasonable assumption.

b. For the percentage of students who fall into each cell under the assumption of independence, see table on page 737.

c. For the number of students who fall into each cell under the assumption of independence, see table on page 737.

E23. b. $\chi^2 = 456.9$ with 1 df, P-value $< .0001$. Reject the H_0 that gender and survival are independent and that this difference can reasonably be attributed to chance variation.

E25. $\chi^2 = 636.6$ with 4 *df*, *P*-value < .0001. Reject H_0 and conclude that smoking behavior and occupation are not independent. By examining the table, it appears that professionals are less likely than other groups to smoke.

Review Exercises

E27. a. There were relatively few admissions on Friday and relatively many admissions and doctor-needed visits on Monday. One explanation is that students had a problem on the weekend when the infirmary wasn't open and waited until Monday to be checked.

   b. A test of homogeneity requires that independent random samples be drawn from two populations. A test of independence requires a simple random sample from one population. The test of independence is the more appropriate test.

   c. $\chi^2 = 11.02$ with 12 *df*, *P*-value = .53; insufficient evidence to reject H_0. Conclude that the observed values are typical of what you would expect to see if there were no association between severity of problem and day of the week.

   d. Place Monday, Tuesday, and Wednesday in one group and place Thursday and Friday in another group.

   e. $\chi^2 = 3.463$ with 3 *df*, *P*-value = .33; insufficient evidence to reject H_0. Conclude that the severity of the problem and whether students attend class that day appear to be independent.

E29. a. a test of homogeneity

   b. $\chi^2 = 3.202$ with 2 *df*, *P*-value = .20; insufficient evidence to reject H_0. Conclude that the observed differences are random and don't appear to represent a true difference in the proportions of all people at each educational level who watch the Super Bowl.

E31. Yes. $\chi^2 = 325.22$, *df* = 4, *P*-value = 0. There is almost no chance of getting a χ^2 this large unless there is an association between political ideology and political party.

E33. a. If segregation is high and an adequate sample size is taken, cell counts for A and D would be large relative to B and C and the value of χ^2 would tend to be large.

   b. If integration is high and an adequate sample size is taken, cell counts for B and C would be large relative to A and D and the value of χ^2 would tend to be large.

   c. χ^2 would be small if integration is moderate (or random) and the sample sizes for both species are the same.

E35. a. Description 2, design I
   b. Description 1, design III
   c. Description 3, design II

Section 10.3, E21b

		Region				
		Northeast	**Midwest**	**South**	**West**	**Total**
Grade	K–8	.187	.199	.232	.139	**.758**
	9–12	.060	.064	.074	.045	**.242**
	Total	**.247**	**.263**	**.306**	**.184**	**1.000**

Section 10.3, E21c

		Region				
		Northeast	**Midwest**	**South**	**West**	**Total**
Grade	K–8	8,839,501	9,412,100	10,950,961	6,584,892	**35,787,454**
	9–12	2,822,110	3,004,919	3,496,217	2,102,300	**11,425,546**
	Total	**11,661,611**	**12,417,019**	**14,447,178**	**8,687,192**	**47,213,000**

E37. a. The two populations are husbands who have murdered their wives and wives who have murdered their husbands.

b. The plot appears to show that the proportion in each category is about the same for husbands as for wives.

c. $\chi^2 = 32.77$ with 3 *df*, *P*-value $< .0001$. Reject H_0 and conclude that for at least one type of case outcome the proportion of husbands and wives falling into that category are different. It appears that wives are more likely to be acquitted at trial.

Chapter 11

Section 11.1

P1. a. $\beta_1 = 9$ calories per gram of fat

b. the number of calories associated with no grams of fat

c. There are calories in pizza from carbohydrates and protein as well as from fat.

P2. a. $y = x + error$ or $height = armspan + error$

b. $height = 7.915 + 0.952 \ arm \ span$; height increases 0.952 inch for each 1-inch increase in arm span. The estimated slope is very close to Leonardo's theory of 1.

c. The pattern should reveal that all of the negative deviations are for larger arm spans and the plot shows a downward linear trend.

d. The shape of the plot is similar to the shape of the plot in part c, except in this plot there is no downward trend and it looks more random.

P3. The spread in *x*-values is 2.3198; $s_{b_1} \approx 0.3414$.

P4. a. 5, because the greater variability in the conditional distribution of *y* results in greater variability in the slope.

b. 3, because the smaller variability in the value of *x* results in a greater variability in the slope.

c. 10, because a larger (random) sample size always results in a smaller amount of variability.

d. The theoretical slope does not matter, everything else being equal.

e. The theoretical intercept does not matter, all else being equal.

P5. a. 0.1472

b. The *SE* for the slope appears in the row "Variable Sulfate" and column "s.e. of Coeff." The estimate *s* of the variability in *y* about the line is found in the fourth row as $s = 0.3414$. The equation of the regression line is $\hat{y} = 1.71525 + 0.524901x$.

c. The first value is the slope of the regression line, b_1. The second value is the estimated *SE* of that estimated slope. The third value is the first divided by the second, $t = \frac{0.524901}{0.1472} \approx 3.57$. The final value, .0376, is a two-sided *P*-value for $t = 3.57$, computed with $df = 3$.

P6. a. $y = 10 + 2x$, $\hat{y} = 9.91 + 2.03x$; The estimated slope is very close to 2, and the estimated intercept is close to 10.

b. $s = 3.127$, which is very close to the theoretical value of 3.

c. The standard deviation of b_1 is 0.3496. In theory, the slope should have $SE = 0.335$.

d. yes; $\bar{y}_{x=0} = 9.91$ and $\bar{y}_{x=4} = 18.03$

P7. V, II, I, IV, III

P8. I, III, II, IV

E1. a. The plot of (*hp*, *price*) has a fairly strong, positive linear trend, although there is a tendency for the spread in *y* to increase as *x* increases. The plot of (*hp*, *mpg*) has a fairly strong negative trend, although there is a hint of curvature. Try a transformation. The plot of (*l*, *mpg*) has a fairly strong negative trend, but there are two potentially influential data points on the left side. Fit with and without these points to check their influence. The plot of (*rpm*, *mpg*) shows that this is not a good data set for supporting a linear regression model. Although there is a slight positive trend in the data, the variation in responses is too great at the larger values of *x* (*rpm*) for inferential techniques to be correct or useful.

b. The largest *SE* should be for (*rpm*, *mpg*) and the smallest for (*hp*, *price*).

c. The estimated slope is $b_1 = -0.0693953$. For every increase of 1 unit in horsepower, the gas mileage tends to drop by 0.0693953 miles per gallon. $s_{b_1} \approx 0.023288$.

d. Under "Parameter Estimates," the estimated slope appears under "Estimate" in row "HP" and the *SE* is under "Std Error" in row "HP"; $s = $ Root Mean Square Error $= 4.778485$.

E3. a. The soil samples should have the larger variability in the slope because the distance of y from the regression line tends to be larger proportional to the spread in x.

b. The s_{b_1} for the soil samples is 0.36165, which is much larger than the corresponding SE for the rock data.

Section 11.2

P9. $t \approx 0.3688$ with $df = 4$, P-value = .731. There is insufficient evidence to reject H_0. There is no evidence here of a linear relationship.

P10. $t \approx 5.6982$, $df = 9$, P-value = .00029. Reject the hypothesis that there is no linear relationship between percentage of sulfur and redness.

P11. a. The equation is $\hat{y} = 25.2 + 3.29x$. You can expect the temperature to have risen 3.29 degrees if the number of chirps per second increases by 1.

b. A residual plot shows no obvious pattern, so a linear model fits the data well.

c. The P-value is close to 0. Reject the hypothesis that there is no linear relationship between rate of chirping and temperature.

P12. a. $(-0.0429, -0.0245)$

b. The value $s = 0.0257$ is an estimate of the SD of the residuals from the regression line.

P13. $(-0.0714, 0.1252)$. Because the second confidence interval entirely overlaps the first, you can't conclude the slopes are different.

E7. a. Yes, although there may be one or two influential points.

b. $\hat{y} = 38.9805 - 0.0694x$

c. $s = 4.778$

d. $s_{b_1} = 0.0233$

e. yes; $t = -2.98$, P-value = 0.0106; the slope is significantly different from zero. There is a negative association between *horsepower* and *miles per gallon* that can't simply be attributed to chance variation in b_1.

f. $(-0.1107, -0.0281)$. An increase of 100 horsepower is expected to decrease gas mileage by somewhere between 2.8 and 11 mpg.

E9. a. i. $(1.9925, 4.5897)$

ii. $(0.12831, 0.29553)$

b. When you reverse the roles of chirp rate and temperature, the entire regression line changes. The unit of the slope changes from degrees per chirp to chirps per degree. The

sizes of the residuals change too because they are measured in different units (difference in temperature versus difference in chirp rate).

E11. $size = 31.83 - 0.712\ acid$. The slope of -0.712 means that for every increase of 1 μg/ml in acid concentration, the radius of the fungus colony tends to decrease by 0.712 mm. For $t = -19.84$, $df = 10$, P-value < .0001. Reject H_0. There is a negative linear relationship between colony radius and acid concentration.

E13. $price = 39.4553\ bedrooms - 32.326$. There is a positive trend with a statistically significant slope. The expected increase in price is about \$39,455 per bedroom. A weakness of this analysis is that there are relatively fewer data on two-bedroom and four-bedroom houses than on three-bedroom houses.

Section 11.3

P14. a. The untranslated data clearly have curvature. Both the log transformation and the log-log transformation straighten the pattern considerably. Because there is reason from the physical situation and experience with largemouth bass to believe the relationship is close to cubic, use the log-log transformation.

b. $t = -4.3230$; $df = 7$; P-value = .0035. Reject H_0; conclude that the slope of the true regression line is not 3.

c. $(2.24, 2.78)$. For widemouth bass: $(3.301, 3.460)$. The power function that relates length to weight of a black crappie has a smaller exponent than that for widemouth bass.

P15. a. 1.654. The distribution of log(*img*) should be approximately normal with mean 1.654 and $SD = 0.2588$; 45.082

b. (0.1950 ± -0.0525). The power function that best fits the relationship between *fr* and *img* has an exponent in the interval (0.1950 ± -0.0525).

E15. a. Yes. The residuals are more evenly scattered rather than being lumped up to the left. The slope of the regression line has gone from negative to positive, which seems like a change. However, in both cases the slope is not significantly different from 0. A log-log transformation improved the conditions for inference. The residual plot shows random scatter about the line 0, and the boxplot of the residuals is symmetric. There appears to be no linear relationship with nonzero slope between *crime rate* and *total population* of the city.

b. The regression line is meaningful whether you have a sample or the entire population. Possible wording of conclusion: The slope of the relationship between *crime rate* and *total population* of largest cities is no different than you would expect if the crime rate had been assigned at random to the cities.

E17. Plotting *flow* versus *static* yields a plot with some curvature and with increasing variation in *y* for increasing *x*'s. The ln-ln transformation: $ln(flow) = -1.0 = 1.12\ ln(static)$, which may be used to change static readings to flow readings. Result can then be exponentiated back to the original scale, if necessary.

E19. $t = 1.65$, $df = 17$, P-value = .12; insufficient evidence to reject H_0. Conclude that there does not appear to be a linear relationship between the percentage of parliamentary seats occupied by women and girls' share of secondary enrollment. Three points may be influential, so the analysis should be redone without them to check their influence.

Review Exercises

E21. narrowest: all depths combined; widest: mid-depth measurements

E23. a. There are no significant linear trends for *amplitude* versus *blood pressure* for any of the three anesthetics individually. Combined, however, the slope is significant.

b. Pentobarbital shows a significant increase in *frequency* with increasing *blood pressure,* while the others show no significant trend.

E25. They are equal.

Chapter 12

Section 12.1

P1. most effective: Treatments 1 and 5, because they are centered at the lowest values; least effective: Treatment 4

P2. The centers vary widely, as do the spreads. As the mean increases, the spreads tend to decrease. That is, spreads for the smaller means appear to be larger than the spreads for the larger means. Answers will vary.

P3. an average of a 0.16-centimeter decrease in standard deviation for each 1-centimeter increase in growth

P4. H_0: If all plants could have been given both treatments, the mean growth in height for Treatment 1 would equal that for Treatment 5 or $\mu_1 - \mu_5 = 0$.
H_a: If all plants could have been given both treatments, the mean growth for these two treatments would have been different or $\mu_1 - \mu_5 \neq 0$. $t = 0.702$ with 18 df and P-value = .4914. There is insufficient evidence to reject H_0. There is no evidence to suggest that the treatment means differ.

P5. The boxplots of Treatments 4 and 5 do not overlap; they appear to have widely differing means; (18.4, 30.2). This interval does not include 0, so reject the hypothesis that the mean growth would have been the same if all plants could have been given both treatments.

P6.

	Treatment 4	Treatment 5
Q_1	2.0	2.0
Median	2.5	2.0
Q_3	2.5	3.0
IQR	0.5	1.0

Outliers for Treatment 4: 1 and 4; outliers for Treatment 5: 5; Treatment 5 is skewed right, whereas Treatment 4 is fairly symmetric.

Section 12.2

P7. a. A statistically significant linear positive relationship ($t = 4.56$; $df = 28$; P-value < .0001). On the average, there is an 18.8-thousand increase in attendance for each million-dollar increase in payroll.

b. A statistically significant increasing trend of about 17.9 thousand in increased attendance for each million-dollar increase in payroll, on the average. ($t = 3.313$; $df = 12$; P-value = .0062)

c. The scatterplot indicates serious curvature, and the slope should not be interpreted. Data indicate a quadratic effect.

P8. a.–c. For all teams, there appears to be little association between payroll and batting average. *P*-values are:

  a. .3378   b. .4760   c. .6208

P9. a. A significant positive linear relationship. If the two teams with unusually high batting averages (Colorado and Seattle) are removed, the estimate of the slope increases from 3.72 to 4.5 percentage points per 1-point increase in team batting average and the *P*-value remains about the same.

b. The average increase in *percent wins* is 7.2 per 1-point increase in *batting average*. However, if Seattle is dropped from the analysis, the estimate of the slope, 5.3, is only marginally significant at *P*-value = .071.

c. With Colorado, there is no significant linear relationship. Without Colorado, the slope jumps to an estimated 5.8 increase in *percent wins* for each 1-point increase in *batting average* and *P* drops to .006.

P10. a. No; *P*-value = .6. Conclude that the difference in mean attendance is not statistically significant.

b. No; *P*-value = .16. Conclude that the difference in the mean batting average is not statistically significant.

P11. There is no evidence of a real difference between the winning percentages of these two teams; a difference this small can reasonably be attributed to chance. $z \approx 0.4743$, *P*-value $\approx$ .6353.

P12. Use a test of homogeneity. The outcomes are not strictly independent; however, of the 162 games played, only a small number are played among each other, so the dependence is not too serious. Refrain from placing too much weight on the actual *P*-value that arises.

a. No; $\chi^2 \approx 0.351$ with *df* = 2, *P*-value $\approx$ .84. The observed differences could very well be produced by chance alone.

b. $\chi^2 \approx 9.203$ with *df* = 2, *P*-value $\approx$ .01. There is strong evidence to support that the differences in the percentages of wins could not have been attributed to chance alone.

Section 12.3

P13. a.

Age Group	Terminated? No	Yes	Total	Percentage Terminated
Under 40	14	0	14	0
40 or Older	8	28	36	77.78
Total	22	28	50	

b. Not independent by Chapter 6's definition. For example,
$P(under\ 40\ and\ no) \neq P(under\ 40) \cdot P(no)$
or .12 $\neq$ .1232.

Age Group	Terminated? No	Yes	Total	Percentage Terminated
Under 40	6	8	14	57.14
40 or Older	16	20	36	55.56
Total	22	28	50	

c. The table below looks more like the one in part b, illustrating near independence or little association.

Age Group	Terminated? No	Yes	Total	Percentage Terminated
Under 40	7	7	14	50.00
40 or Older	15	21	36	58.33
Total	22	28	50	56.00

P14. Min = 22, Q_1 = 37, Median = 53, Q_3 = 56, Max = 69. The distribution is bimodal, with one peak in the low 30s and a second, higher peak in the upper 50s. The distribution is skewed toward the low values. There are no outliers.

P15. With age 40 as the cutoff, 36 of 50, or 72%, are older. With age 50 as the cutoff, 28 of 50, or 56%, are older.

P16.

Age Group	Terminated? No	Yes	Total	Percentage Terminated
Under 50	13	9	22	40.91
50 or Older	9	19	28	67.86
Total	22	28	50	56.00

P17. z = 1.9055, *P*-value = .02836; reject H_0. If Westvaco were picking people at random to lay off, a difference of .270 between the proportion of people 50 or older who were laid off and the proportion of people under 50 who were laid off is not reasonably likely to occur.

P18. The shapes of the distributions of ages are similar: somewhat skewed toward the smaller ages. The spread of the ages of those terminated is slightly greater than for those retained. The average age of those terminated is higher than the average age of those retained.

P19. The boxplots indicate some skewness but no outliers. The boxplot for terminated workers shows a slightly larger spread and a higher center.

	Min	Q_1	Median	Q_2	Max
Terminated	22	38.5	54.5	59	69
Retained	25	37.0	48.0	55	61

P20. Using a one-sided test, $t = 1066$ and P-value = .146. There is insufficient evidence to reject H_0. If you can assume that the 28 and 22 employees are like random samples from larger pools of employees laid off and retained and there is difference in the mean ages of these two larger pools, then it is reasonably likely to get a difference of 3.68 years in the average age of employees in a random sample from the pool of employees laid off minus the average age of the employees in a random sample from the pool of employees retained.

Index

Photo Credits

Chapter 1
2: John Olson/Corbis. **4:** Zigy Kaluzny/Stone/Getty Images. **10:** Jeff Greenberg/The Image Works. **20:** Larry Mulvehill/Photo Researchers Inc.

Chapter 2
22: Rick Stewart/Getty Images. **31:** Kennan Ward/Corbis. **39:** Rich Kirchner/Photo Researchers Inc.

Chapter 3
100: Peter Hvizdak/The Image Works. **108:** Andy Lyons/Getty Images. **123:** James A. Sugar/Corbis. **147:** Corbis Images/Picture Quest RF. **166:** Galen Rowell/Corbis.

Chapter 4
208: Fredrik D. Bodin/Stock Boston/PictureQuest. **212:** Tom McCarthy/Photo Network/PictureQuest. **219:** © 2002 The New Yorker Collection from cartoonbank.com. All Rights Reserved. **220:** CALVIN AND HOBBES © 1995 Watterson. Reprinted with permission of UNIVERSAL PRESS SYNDICATE. All rights reserved. **223:** © Larry Gonick. **261:** David Young Wolff/Stone.

Chapter 5
266: The Image Works. **290:** Bettmann/Corbis. **301:** Martyn Goddard/Corbis. **322:** Keren Su/Corbis.

Chapter 6
326: Hulton Deutsch Collection/Corbis. **343:** Richard Cummins/Corbis. **349:** Ralph White/Corbis. **366:** Steve Chenn/Corbis.

Chapter 7
378: Paul A. Souders/Corbis. **381:** Jose Luis Pelaez Inc./Corbis. **394:** Archive Photos/PictureQuest. **400:** AFP/Corbis.

Chapter 8
412: Getty Images. **427:** Vic Bider/eStock Photography/PictureQuest. **441:** Paul Barton/Corbis. **461:** Souder/Corbis.

Chapter 9
478: John Zoiner/Corbis. **492:** Flash Light/Stock Boston. **533:** Jose Luis Pelaez Inc./Corbis. **562:** Corbis.

Chapter 10
576: Steven Starr/Stock Boston. **588:** Bob Daemmrich/Stock Boston Inc./PictureQuest. **599:** Reuters NewMedia Inc./Corbis. **614:** Corbis.

Chapter 11
628: AP/Wide World Photos. **629:** NASA. **633:** Lowell Georgia/Corbis. **644:** Peter Menzel/Stock Boston Inc./PictureQuest. **663:** Mark Downey/Lucid Images/PictureQuest.

Chapter 12
678: Roger Ball/Corbis. **684:** Joseph Sohm/ChromoSohm Inc./Corbis.